Parasitism

The Diversity and Ecology of Animal Para

SECOND EDITION

Reflecting the enormous advances made in the field over the past 10 years, this text synthesizes the latest developments in the ecology and evolution of animal parasites against a backdrop of parallel advances in parasite systematics, biodiversity, and life cycles. It has been thoroughly revised to meet the needs of a new generation of parasitology students, whether their interest is in ecology, conservation biology, evolution, immunology, or health sciences.

Balancing traditional approaches in parasitology with modern studies in parasite ecology and evolution, the authors present basic ecological principles as a unifying framework to help students understand the complex phenomenon of parasitism. Richly illustrated with over 300 figures, the text is accompanied by case study boxes designed to help students appreciate the complexity and diversity of parasites and the scientists who study them. This unique approach, which is presented clearly and with a minimum of jargon and mathematical detail, encourages students to think generally and conceptually about parasites and parasitism.

Timothy M. Goater is Professor and former Chair in the Biology Department at Vancouver Island University, British Columbia, Canada. During the past 20 years he has taught courses in introductory biology, parasitology, ecological parasitology, invertebrate zoology, and ento-mology. His research interests focus on the population and community ecology of parasites.

Cameron P. Goater is Associate Professor and former Chair in the Department of Biological Sciences at the University of Lethbridge, Alberta, Canada. His parasitological research roots are in the community ecology of helminths of waterfowl on the Canadian prairies, and over the past 15 years he has taught courses in introductory biology, invertebrate biology, field biology, and symbiotic interactions. His current research interests are in the experimental ecology of helminth–host interactions.

Gerald W. Esch is Charles M. Allen Professor of Biology at Wake Forest University, North Carolina, USA, where he has taught for 47 years. He is widely regarded to be one of the world's leading ecological parasitologists, and served as Editor of the *Journal of Parasitology* for 19 years.

Advance praise for *Parasitism: The Diversity and Ecology of Animal Parasites, Second Edition*

"Their approach is synthetic, refreshingly original and effectively blends coverage of long-standing fundamentals of parasitology with modern advances in the field."

Janine N. Caira, University of Connecticut, USA

"This is an extremely well written book that does an excellent job of integrating conceptual and organismal aspects of parasitology."

Dale H. Clayton, University of Utah, USA

"There is a wealth of detail for well-selected examples, building on the rich experience of the authors as top-notch researchers and educators."

Mark R. Forbes, Carleton University, Canada

"*Parasitism* gives the student both the systematic and zoological background to understand parasitology and the ecological and evolutionary context to understand why it is important... As a team, their approach is clear and scholarly, with many important updates since the first edition."

Kevin D. Lafferty, US Geological Survey, University of California, Santa Barbara, USA

"This new edition will be a wonderful resource for teachers of undergraduate parasitology courses. The well-illustrated and easy-to-read text is unrivalled at the moment and will be a great tool to turn on a new generation of young minds to the wonders of parasitic organisms. A true parasitological tour de force!"

Robert Poulin, University of Otago, New Zealand

"A well-organized integration of the diversity of ideas and methods that characterize this new field of parasite ecology. The style is easily readable, the details extraordinary, and the story is told from the perspective of evolutionary thought... even the pros will learn from this book."

Michael V. K Sukhdeo, Rutgers University, USA

Parasitism

The Diversity and Ecology of Animal Parasites

SECOND EDITION

TIMOTHY M. GOATER
Vancouver Island University, British Columbia, Canada

CAMERON P. GOATER
University of Lethbridge, Alberta, Canada

GERALD W. ESCH
Wake Forest University, North Carolina, USA

CAMBRIDGE
UNIVERSITY PRESS

CAMBRIDGE
UNIVERSITY PRESS

University Printing House, Cambridge CB2 8BS, United Kingdom

Cambridge University Press is part of the University of Cambridge.

It furthers the University's mission by disseminating knowledge in the pursuit of education, learning, and research at the highest international levels of excellence.

www.cambridge.org
Information on this title: www.cambridge.org/9780521190282

© Cambridge University Press 2001, 2014

First published 2001
Second edition 2014
Reprinted 2015

Printed in the United Kingdom by TJ International Ltd, Padstow, Cornwall

A catalog record for this publication is available from the British Library

Library of Congress Cataloging in Publication data
Goater, Timothy M., 1959–
Parasitism : the diversity and ecology of animal parasites / Timothy M. Goater, Cameron
P. Goater, Gerald W. Esch. – Second edition.
    pages   cm.
First published as: Parasitism / Albert O. Bush . . . [and others], 2001.
Includes bibliographical references and index.
ISBN 978-0-521-19028-2 (hardback) – ISBN 978-0-521-12205-4 (paperback)
1. Parasites – Textbooks.   2. Parasitism – Textbooks.   3. Parasites – Ecology – Textbooks.
4. Biodiversity – Textbooks.   5. Parasitology – Textbooks.   I. Goater, Cameron P.
II. Esch, Gerald W.   III. Parasitism.   IV. Title.
QL757.P287   2013
578.6′5–dc23   2013016194

ISBN 978-0-521-19028-2 Hardback
ISBN 978-0-521-12205-4 Paperback

Additional resources for this publication at www.cambridge.org/parasitism

We dedicate this book to our students,
past, present, and future

CONTENTS

The color plates are located between pages 248 and 249.

BOXES

FOREWORD

The ability of parasites to cause disease has always been an important reason to study them, and the teaching of parasitology has almost always been stimulated by conditions conducive to disease, such as war or climate change. Currently, zoonotic diseases emerging from altered ecosystems, or carried by arthropod vectors spreading their ranges due to climate changes, supply that stimulation. However, most of us who teach, or have taught, parasitology have chosen that topic because of the fascinating life cycles of many parasites and their complex interactions with their hosts. Much of that fascination stemmed from learning how parasites can affect the population dynamics of their hosts, or the behavior of the hosts, or even the evolution of their hosts. In addition, that fascination was based on how much parasites could tell us about the life of their hosts, such as their diet, travels, or evolution. Or even of the earth itself – some of the earliest evidence for continental drift was the similarity in parasites of amphibians in Africa and South America. Examples of all of these influences are provided in this book.

Many of the systems that parasitologists have used to show these fascinating features have become relatively easy to study due to new techniques, such as those in genomics and proteomics, which have provided new and more powerful ways to study systematics, evolution, and host–parasite relationships. This has attracted the attention of biologists with a wide variety of backgrounds, so that much of the very interesting work done on host–parasite systems recently has been done by those trained in other specialties, such as ecology, behavior, neurophysiology, and evolutionary biology. Very few of the students in senior-level parasitology courses will go on for further study in parasitology, but many more will go on for further study in other biological specialties. Our courses, books, readings, and other materials used in our classes should be chosen to expose those students to the usefulness of parasites in investigations in their chosen fields.

This book is the best I have seen for that purpose. The authors have provided a wide-ranging review of the diversity of parasites, emphasizing those which provide examples of the insights provided by the use of the new techniques or examples of how parasites can provide new and exciting insights into other aspects of biology. One of the best features of this book is that it emphasizes the complexity of host–parasite systems, with full recognition that most of the outcomes are markedly dependent on the conditions in which that system is embedded. This emphasis on complexity starts with a chapter on immunity, which is the best and most succinct coverage I have ever seen of those aspects of immunity that are important in host–parasite interactions. This emphasis is most apparent in the most integrative chapters – those on the influence of parasites on their hosts, and parasite evolutionary ecology.

This is the book I would have loved to have been available when I was teaching. But, of course, it could not have been written then. Most of the more provocative insights, and especially the evidence for complexity and conditional outcomes of host–parasite encounters, have come in the past two decades since I retired. The field of parasitology has become increasingly fascinating, and its implications for other fields of biology more significant, in those two decades. Enjoy this book, as I have, and see where it leads you.

John C. Holmes

PREFACE AND ACKNOWLEDGMENTS

In this second edition, we stay true to the philosophical approach that was adopted in the first. Thus, we continue to see a need for a single text with dual focus on the diversity *and* ecology/evolution of parasites. At the core, we feel that an ideal strategy for senior undergraduate and beginning graduate students to understand and appreciate breakthroughs in parasite ecology is through a solid understanding of parallel advances in parasite diversity, life-cycle variation, systematics, and functional morphology. By way of example, we suggest that an understanding of the role of falciparum malaria in determining the worldwide distribution of the human sickle-cell gene, and thus the role of parasites in mediating natural selection (Chapter 16), comes from an understanding of life-cycle variation, functional morphology, and biodiversity of the apicomplexans (Chapter 3). Likewise, real understanding of the evidence in support of the parasite hypothesis for the evolution and maintenance of sexual reproduction in molluscs (Chapter 16) comes from a detailed understanding of variation in life cycles and life histories of the platyhelminths (Chapter 6). This dual focus, under one cover, is the hallmark of this text.

Our aim is to provide students with a synthetic understanding of the biodiversity, ecology, and evolution of animal parasites. Thus, throughout most of the text, we unabashedly take a parasite-centered view of the phenomenon of parasitism. Yet, we also aim to provide insights on the nature of the host–parasite interaction itself. It is for this reason that following a brief introductory chapter, we provide an overview of vertebrate and invertebrate immunity, and the new discipline of ecological immunology. We turn again and again to the importance of fundamental immunological principles throughout the text.

There are now nine biodiversity chapters (Chapters 3–11). We have added chapters on the Myxozoa, Microsporida, and Nematomorpha,

reflecting developments in their systematics, and their value as models in parasite ecology and evolution. By necessity, the 'phylogenetic relationships and classification' sections for all of the diversity chapters have been updated, adopting the most current molecular-based taxonomic schemes. The protist chapter in particular has been completely revised from the first edition, reflecting the monumental changes in protist systematics. New text boxes that highlight key areas of development, and the scientists behind them, are integrated into each of these chapters. New life cycle diagrams and dozens of new photographs and micrographs have also been incorporated. A color plate section has been added, showcasing dramatic photographs of parasites in or on their hosts.

Armed with a solid background in parasite biodiversity, systematics, and functional biology, Chapters 12–17 cover advances in the ecology and evolution of parasites. The titles and content of these chapters have been completely revised from the first edition, reflecting in part, the interests and backgrounds of the new authors. Yet the substantial revisions also reflect the pace of development in methodologies and in overall approaches that have matured the field over the past decade. While some of these developments have confirmed earlier ideas, others have revolutionized our understanding of even the most fundamental aspects of the parasitic way of life. Thus, the incorporation of new model host–parasite interactions that are amenable to manipulation in the laboratory and field have provided key insights into how parasite populations are regulated and how they are distributed among hosts in space and time (Chapter 12). Studies at the community level (Chapter 13) have also benefited from rigorous empirical approaches involving key model systems where the composition of component species can be manipulated. In Chapter 14, we see how advances in molecular biology, genomics, and remote sensing have transformed our understanding of parasite biogeography and phylogeography. Coverage in Chapter 15 is focused on the diverse manner in which parasites can affect the biology of their hosts, whether it is at the level of the host individual, or on the structure of entire host ecosystems. Again, key advances stemming from empirical, hypothesis-testing approaches involving selected model systems have markedly advanced our understanding of the magnitude of these effects, and their underlying mechanisms. The focus in Chapter 16 takes the next logical step, covering the manner in which parasites affect the evolutionary and coevolutionary trajectory of their hosts. We conclude the text by summarizing the nature of the parasite/human/ habitat interface, and how the multidisciplinary field of environmental parasitology (Chapter 17) can assist in interpreting the nature of host–parasite interactions in the face of anthropogenic change.

As with all projects of this scope, this book is a collaborative effort. We extend sincere thanks to the authors of the first edition, Al Bush, Jackie Fernández, and Dick Seed for their initial vision and dedication. Several of their line drawings and photographs, incorporating the image-editing skills of Maggie Bush, have been retained here. Numerous colleagues offered valuable suggestions on specific sections/chapters, especially Carter Atkinson, Mark Blaxter, Katharina Dittmar, Eric Hoberg, Jens Høeg, Kayla King, David Marcogliese, Jim Mertins, Beth Okamura, George Poinar, John Webster, Chris Whipps, and Stephen Yanoviak. We also appreciate the insightful comments John Holmes provided for several chapters. Several of our former students, especially Martin Anglestad, Melissa Beck, Aaron Jex, Chelsea Matisz, Phillip Morrison, Vanessa Phillips, Brad van Paridon, and Chris Whipps helped to review and edit chapters. Their perspectives helped clarify and focus our efforts.

This revision contains many new drawings, as well as new photographs and micrographs. Bill Pennell spent many hours of his retirement taking several new photographs, as well as editing countless others. Doug Bray and Brad van Paridon took several of the new scanning electron micrographs. We thank our colleagues for contributing extensive new data figures, photographs, and micrographs for the new edition. Their generous contributions are acknowledged in the figure captions. John Sullivan is especially thanked for sharing several of his photographs from his excellent parasitological resource, *A Color Atlas of Parasitology*. Several new life-cycle diagrams and line drawings are incorporated into this edition, thanks to Chelsea Matisz, Lisa Esch McCall, and Danielle Morrison. Danielle, in particular, is thanked for her patience and dedication in preparing, labeling, and editing many of the new figures and photographs.

Vancouver Island University is thanked for providing Tim Goater the sabbatical and professional development funds that enabled this revision to take shape. Special thanks also to Mike Steele, David Marcogliese, and Herman Eure for providing office space during his sabbatical, as well as Eric Demers, Larissa Nelson, Wendy Simms, and Jane Watson for their enthusiastic encouragement throughout the project. Likewise, Cam thanks Dean Chris Nicol and Chair Brent Selinger for moral support and teaching relief during the peak phases of this revision, and colleagues Doug Bray, Doug Colwell, Andy Hurly, Joe Rasmussen, and Brian Wisenden for their constant support. Cam also extends thanks to Barb Johnson and staff at Waterton Lakes National Park for access to their cabin during key writing phases. Most sincere thanks also to Lori Goater for her monumental patience and support and to Ben and Ali for frequently reminding their dad, and their uncle, that parasite ecologists come in all ages.

Our primary editor, Katrina Halliday and her assistant, Megan Waddington at Cambridge University Press are thanked for all of their help addressing our many queries and, especially, for their patience and devotion to seeing this project to its completion.

Timothy M. Goater
Cameron P. Goater

About this edition

The first edition of our book was published in 2001. Al Bush, Jackie Fernández, and Dick Seed were co-authors, along with myself. Sadly, Al died in 2010. Further, Jackie stepped aside to raise two sons and Dick retired from his faculty position at the University of North Carolina-Chapel Hill. As the only original author that was still active professionally, it became my responsibility to recruit new co-authors. Given the overall theme and target audience of the text, my choice fell upon brothers, Tim and Cam Goater. Tim was a former Ph.D. student of mine at Wake Forest University, and Cam was a former Ph.D. student with Clive Kennedy at the University of Exeter. Both brothers were mentored by Al Bush at Brandon University in Manitoba, Canada. Tim and Cam have extensive experience teaching senior undergraduate courses in parasitology and ecology, and Cam extends his teaching perspectives to the mentoring of graduate students. Both have diverse and complementary research backgrounds that, together, span most areas of modern parasitology.

Throughout the writing of both editions, I maintained my duties as Editor of the *Journal of Parasitology*, as well as my teaching. Mrs. Vickie Hennings, my Editorial Assistant for the Journal, continued her responsibilities while I was occasionally subsumed by the book. Cindy Davis and Zella Johnson, both long-term secretaries for the Department of Biology at Wake Forest University, are thanked for their help as well. I especially express my appreciation to Ann for being such a marvelous 'listener' and for her constant support from the book's inception.

Gerald W. Esch

1 Introduction

1.1 Encounters with parasites

On a fateful spring day in a small northern Canadian town in the 1970s, two of the authors (the two that are related) of this text came upon a sickly red fox. Following some foolhardy thinking, they handled the fox and carried it home. A few days later, health officials diagnosed the fox with rabies. To avoid the fatal consequences of the disease, the brothers required daily intramuscular injections of the prophylactic drug that was used at the time. We recall the episode with memories of pain, dismay from parents, and ruthless teasing from our friends. And so goes our introduction to the world of parasites. So too goes our introduction to the phenomenon of parasitism. Readers might envision two teenagers discussing how their predicament arose: How did that fox get infected? Why was the fox population, but not the racoon population, so heavily infected that year? How does the virus migrate from the site of a wound, to the brain, to saliva? How, and why, does it transform a normally secretive and nocturnal animal into one that is aggressive and diurnal? There are obvious parallels between these early queries and modern questions associated with host specificity, parasite site selection, the geographical mosaic of coevolution, and mechanisms of alterations in host behavior.

We hope that your introduction to parasites was (is) not as dramatic, or as dangerous, as it was for two of us! Indeed, for many, initial exposure to the concept of parasitism likely originated from media reports that describe human mortality and morbidity caused by diseases such as malaria, or other parasitic diseases that are so common in developing countries. Or, perhaps you have heard about certain parasites that are transmitted via ingestion of untreated water, or swimming in it, or from eating poorly cooked meat.

For the pet and livestock owner, parasite encounters may have occurred when a veterinarian requested a fecal sample for diagnosis of eggs/larval stages of intestinal worms. Perhaps, as a hunter or a fisherman, you have queried the identity of that animal wriggling in wild game meat or fish. In recent years, these common or at least dramatic parasites of humans, their livestock, or their pets have been made famous in the popular media (e.g., Zimmer, 2000a; 2000b), even including in situ video footage on YouTube™ and on prime-time television shows.

As undergraduate students, your first encounters with parasites and with the phenomenon of parasitism likely occurred in your introductory courses. At each of our universities, majors in many of the life sciences require an introductory course that describes the diversity and unity of the Tree of Life. In a course such as this, it would be impossible for instructors to sample that diversity without covering examples of parasites, although coverage is likely restricted to key human parasites – a protist, a fluke, a cestode, and so on. Likewise, our majors are required to take an introductory course that covers basic principles of ecology and evolution. One encouraging sign of the expanding reach of studies on the phenomenon of parasitism is its increased coverage in mainstream ecology and evolution texts (e.g., Begon et al., 2006; Freeman & Herron, 2007). Nonetheless, time constraints in a single-semester introductory course likely limit coverage of examples involving parasites.

In the chapters that follow, our coverage assumes that you have encountered parasites, both anecdotally and academically. Thus, we assume that senior students in the life or medical sciences have an appreciation for basic principles of classification and phylogeny and an appreciation for variation in the life cycles and general biology of a few animal parasites.

We also assume a general understanding of basic concepts of ecology and of the fundamental and unifying nature of evolutionary processes. Although we do not emphasize the mathematical underpinnings of host–parasite interactions, we do assume that senior students have a numeracy background consistent with introductory courses in calculus, linear algebra, and/or statistics. We do not assume a strong background in immunology or pathology.

1.2 Scope

Our first aim is to provide students with an appreciation for the biodiversity of animal parasites. From the perspective of understanding our planet's biodiversity, and understanding factors leading to its loss, an appreciation for the diversity of parasites is important. Parasitism is recognized as the most common strategy used by animals to obtain nutrients (Price, 1980; de Meeus *et al.*, 1998; de Meeus & Renaud, 2002), ubiquitous across the Tree of Life. Poulin & Morand (2004) consider that there have been at least 60 independent evolutionary transitions from free-living to obligately parasitic animals. Estimates of the overall biodiversity of parasites vary depending on how inclusive we define 'parasite,' but approximately 30–50% of described animal species are parasitic at some stage during their life cycle (Price, 1980; Poulin & Morand, 2004). Given that virtually all metazoan species are infected with at least one species of parasite (most species contain many more), that all viruses and many prokaryotes and fungi are parasitic, and that we underestimate the biodiversity of groups such as nematodes and mites (see Chapters 8 and 11), these rough estimates are undoubtedly low. Clearly, knowledge of parasite biodiversity equates to knowledge of key branches of the Tree of Life.

The biodiversity section of the text (Chapters 3–11) provides an overview of the main taxa of protist and metazoan parasites. Our focus is on characterization of key features that define each group, followed by coverage of how natural selection has shaped variation in their morphologies, in their life cycles and life histories, and in their strategies for nutrient acquisition. Our intent through this section is to provide insight on 'the art of being a parasite,' a phrase coined by Claude Combes (2005) to describe the manner in which parasites of all types solve the unifying problems of entering a host ('getting in, or on'), remaining in a host ('staying in'), and reproducing ('getting out'). Our taxonomic scope is broad, with emphasis on the traditional protists and 'worms' but also on lesser-known groups such as the microsporidians, myxozoans, hairworms, and pentastomes. Much of our coverage through this section distils material that is covered in parasitology texts (e.g., Noble *et al.*, 1989; Kearn, 1998; Roberts & Janovy, 2009). However, relative to these excellent texts, we restrict our taxonomic scope to key families or orders within each group, and we emphasize those groups that provide models for enquiries on the ecology and evolution of parasitism that we cover in later chapters.

Our second aim is to develop in students an appreciation for the phenomenon of parasitism. And from our perspective, we view the core of the phenomenon to be ecological in nature. Thus, whether interest is in understanding the innumerable rates that define the outcome of host–parasite relationships (e.g., rates of exposure to infective stages, rates of within-host migration, rates of parasite-induced host mortality, rates of dispersal, and so on), or in the dynamics of the molecular exchange that occurs at the host–parasite interface, or in the global distribution of parasites, basic ecological principles can be applied to help focus our thinking about host–parasite interactions. It is this perspective that lies at the roots of 'parasite ecology' as a subdiscipline within the ecological sciences. These roots were developed and formalized some 30–40 years ago following the coincident publications of seminal works by empirical field biologists (Kennedy, 1975; Price, 1980) and quantitative ecologists (Crofton, 1971; Anderson & May, 1979). The dynamic tension between their alternative perspectives continues to richly define the direction of a field that is now seeing an unprecedented level of activity.

Over the 10 years since the first edition of our text was published, key advances have been made in virtually all areas of parasite ecology and evolution. These include the epidemiology of wildlife disease (Hudson *et al.*, 2001), parasite phylogeny and phylogenetics (Brooks & McLennan, 2002), parasites and host behavior (Moore, 2002), parasite biodiversity (Poulin & Morand, 2004), evolutionary ecology (Frank, 2002; Poulin, 2007; Thomas *et al.*, 2009; Schmid-Hempel, 2011), and parasite biogeography (Morand & Krasnov, 2010). Two texts that synthesize general advances in parasite ecology and evolution, one from an empirical standpoint (Combes, 2001) and one from a conceptual standpoint (Poulin, 2007), are especially notable. Over the past 10 years or so, the new subdisciplines of ecological immunology, landscape epidemiology, emerging diseases, and environmental parasitology have blossomed as exciting 'hot topics.' This surge in interest is partly due to the explosion in the use of modern molecular methods, enabling advances in our understanding of parasite biodiversity, phylogenetics, population genetics, and host–parasite coevolution that would have been unthinkable even 10 years ago. Yet, the pace of advance is also due to the rapid increase in the use of experimental model systems to test key hypotheses regarding the ecology of host–parasite interactions. While traditional model systems involving laboratory rats and mice as hosts continue to provide important insights, major recent advances have arisen from models involving parasites of hosts such as sticklebacks, guppies, water fleas, songbirds, and wild small mammals. Indeed, we view the multidisciplinarity arising between parasitologists and ecologists, so long called for by the fathers of parasite ecology, that is perhaps most responsible for the unparalleled advances we are currently witnessing in the field (review in Poulin, 2007).

Our aim in this section of the text is to provide an overview of modern parasite ecology, evolution, and coevolution. In this edition, we update our earlier treatment by taking into account results originating from modern advances in molecular methodologies and from experimental models on a wide range of host–parasite interactions. Our overall approach through this section is empirical, rather than conceptual. Thus, we develop our arguments based primarily upon observations from field-based and laboratory-based experiments, although we incorporate key results from field surveys of particular hosts when warranted. Although we cover the mathematical and conceptual framework of certain areas of enquiry, our perspective is empirical and rests strongly on the background that we developed in the biodiversity section of the text. Readers seeking advances in more quantitative aspects of parasite ecology and epidemiology should consult Hudson *et al.* (2001), Ebert (2005), or Schmid-Hempel (2011).

Despite the enormous strides made in methods and approaches, modern studies in parasite biodiversity and ecology continue to be influenced by traditional approaches in parasitology, in which parasites of humans and their domesticated animals have played a key role. Throughout the biodiversity section of the text, we retain some of that traditional coverage. We do so because the history of discovery in parasitology provides the roots of current enquiry, and is itself a fascinating story of human endeavour (Box 1.1). For further account of key historical developments in parasitology, readers are directed to Esch (2007). We also retain some emphasis on select human parasites because the discovery in these groups has provided unmatched opportunities for increased understanding of ecological and evolutionary phenomena. For instance, results of studies on the interaction between falciparum malaria and the gene responsible for sickle-cell anemia provide one of the best examples of parasite-mediated natural selection (Chapter 16). Yet, this example stems from years of dedicated effort that enabled detection of the single amino acid substitution that alters the structure of the hemoglobin molecule. We cover similar examples throughout the text, not necessarily in the context of human disease, but in the context of central questions regarding the ecology and evolution of host–parasite interactions.

Box 1.1 | **A brief historical perspective of parasitology: pioneering scientists and their ground-breaking parasitological discoveries**

Sometime around 1500 BC, an Egyptian physician assembled a large body of medical information regarding the diagnosis and treatment of diseases known to occur at the time. Written in hieroglyphics on papyrus and sealed in a tomb, it was discovered in 1872. It was translated by Georg Ebers in 1873, becoming known as the Ebers' Papyrus among Egyptologists. This volume became an invaluable source that documented the medical profession that existed in the ancient world.

Based on these writings, we now know that early Egyptian physicians were aware of at least two parasitic helminths infecting humans. One of these was a nematode, probably *Ascaris*. The recommended treatment for infection by this apparently common worm included turpentine and goose fat! The second parasite was a tapeworm, most likely *Taenia saginata*, for which a special poultice applied to the abdomen was the recommended treatment. Whereas the digenean, *Schistosoma haematobium*, was not described per se, the bloody urine produced by this parasite was a well-known symptom. Moreover, eggs of this worm have since been identified in mummies from the thirteenth century BC (Grove, 1990). It is also possible that the hookworm nematode, *Ancylostoma duodenale*, was present based on descriptions in the Ebers' Papyrus of a 'deathly pallor' in some patients, a condition that may have been caused by hookworm-induced anemia.

Concurrently, another group of ancients was acquainted with a number of helminth parasites in the Nile Valley. Thus, for example, consider Numbers 21:6–9, which refers to 'the Fiery Serpent,' now recognized as the nematode *Dracunculus medinensis*. When the Israelites misbehaved during their trek out of Egypt, they were directed by God, through Moses, to "make a serpent of brass and put it upon a pole." And, "when he beheld the serpent of brass, he lived." This treatment is still used today, that is, to remove the large female nematode from its subcutaneous site of infection, and then to slowly twist the parasite on a stick, until it is removed intact. Many feel the Hebrew law against eating the flesh of an 'unclean' animal, e.g., a pig, can be traced to the nematode *Trichinella spiralis* or the cestode *Taenia solium*. On the other hand, the Talmud (a sacred Jewish book), written in AD 390, referenced the hydatid cysts of the tapeworm *Echinococcus granulosus*, indicating that they were not fatal.

Periodic fevers due to malaria were mentioned in Chinese writings from around 2700 BC and in every civilization since. Hippocrates (460 BC–377 BC) provided the earliest detailed description of these periodic fevers. Both Hippocrates and Aristotle (383–322 BC) were aware of 'worms' and refer to cucumber and melon seeds in the 'dung' of humans. Both references are probably to the gravid proglottids of *Taenia saginata*. Galen (AD 130–200) referred to the intestinal phases of what were probably *Ascaris lumbricoides* and *Enterobius vermicularis*, saying that the former worms preferred the upper portion of the gut whereas the latter were closer to the anus. He found that tapeworms, on the other hand, were found throughout the length of the intestine. These observations, so long ago, may be the first reference to the site specificity exhibited by parasites.

Box 1.1 | (continued)

The earliest use of the microscope, by Antony von Leeuwenhoek in the seventeenth century, provided a phenomenal breakthrough for the biological sciences and parasitology. He actually observed, and described, the unicellular protist parasite *Giardia lamblia*, apparently from his own feces! Also in the seventeenth century, several scientists prepared detailed drawings of a number of parasitic helminths. One father of parasitology was Francesco Redi (1626–1697), who not only determined that mites could make one itch, but apparently was also an inveterate collector, describing some 108 species of parasites. Perhaps Redi's greatest contribution was that he showed that parasites produce eggs, dispelling the widespread myth that parasites developed through spontaneous generation. The idea of spontaneous generation persisted for many years, however, and it took Louis Pasteur's now classic experiments in nineteenth-century Paris to quash the notion.

The late nineteenth and early twentieth centuries were times of major discoveries dealing with some of the protist and helminth scourges of humans, including *Wuchereria bancrofti* as the causative agent for elephantiasis and tsetse flies as the vectors for African trypanosomiasis. Ronald Ross, while working in India in 1897, demonstrated that mosquitoes vectored *Plasmodium*, winning the Nobel Prize for physiology in 1902. At the turn of the century, Paul Erlich described the first chemotherapeutic agents for African trypanosomiasis and syphilis. With this discovery, he hypothesized that organic molecules with selective toxicity to parasitic organisms would be found. For this, he is considered the father of modern chemotherapy. Between 1907 and 1912, Carlos Chagas determined the identity of trypanosomes that cause Chagas' disease and worked out the parasite's life cycle in the reduviid bug vector. In the early 1880s, Algernon Thomas and Rudolph Leuckart independently completed stages in the life cycle of the liver fluke, *Fasciola hepatica*, including detailed descriptions of the swimming behaviors of the 'embryos' that hatched from eggs, their penetration into snails, and their subsequent intramolluscan development. Thomas and Leuckart and others will be remembered for their many contributions (review in Esch, 2007), and paving the way for all those who resolved so many other parasitological mysteries.

1.3 Terminology

We often start our courses with a request for students to define 'parasite.' This is always an interesting and engaging exercise. Often, the discussion rapidly deteriorates into a mix of vague terminology, examples, and counter-examples. Are mosquitoes and vampire bats parasites? Are leeches parasites? Is my brother a parasite? Is a fetus a parasite? To direct the discussion, we might offer a classic dictionary definition. *Webster's Third New International Dictionary of the English Language* defines 'parasite' as follows:

An organism living in or on another living organism obtaining from it part or all of its organic nutrient, and commonly exhibiting some degree of adaptive structural modification – such an organism that causes some degree of real damage to its host.

It is here where some students may voice discomfort, especially those with an ecological background. In this characteristic definition, vague and unquantifiable terms such as 'part,' 'some,' and 'damage' are prominent. In our courses, we do not offer a solution to the fundamental vagueness that characterizes the definition of parasite. Nor do we do so in this text (for complete discussions, see Zelmer, 1998; Combes, 2001). In his influential text on coevolution, Thompson (1994) emphasizes that all definitions dealing with interspecific interactions are necessarily vague. From our perspective, we consider that a parasite has a metabolic commitment to its host, has evolved morphological and physiological adaptations to living in, or on it, and has the potential to decrease host fitness. As we have indicated previously, our focus is on the familiar parasitic protists, worms, and arthropods, although we extend our coverage to include lesser-known taxa because they provide splendid models in parasite ecology. The extent to which groups such as phytophagous insects, molecular parasites, blood-sucking leeches and flies, and brood parasites (e.g., cuckoos) apply to our coverage of animal parasitism, provides an excellent topic for discussion in our classes, but they lie outside the scope of this text.

Parasitism is one of at least four complex symbiotic relationships. Symbiosis, a term coined by de Bary in 1879, literally means 'living together of differently named organisms.' It describes the relationship in which a symbiont lives in, or on, another living host. Symbiotic interactions, or **symbioses**, include a tremendous variety of intimate partnerships in nature. In the broadest sense, there is no implication with respect to the length or outcome of the association, nor does it imply physiological dependence or benefit or harm to the symbionts involved in the partnership. Given such a broad definition of symbiosis, a functional separation can be made in relation to the feeding biology of one or both of the symbiotic partners, as well as the degree of host exploitation. Thus, categories of symbiosis relate to trophic relationships, and if and how energy is transferred between the partners. Such categories are best viewed as a continuum with overlapping boundaries (Fig. 1.1).

If there is no trophic interaction involved in the symbiotic interaction, then the relationship is called

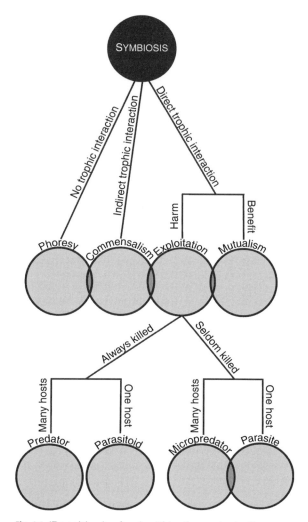

Fig. 1.1 'Parasitism's place' within the context of symbiotic relationships. This is one way of looking at parasitism and it is based, initially, on trophic relationships, followed by 'harm,' and finally, quantity of hosts involved. The final criterion, number of hosts attacked, is meaningful only if restricted to a single life history stage. For example, adult parasitoids may parasitize many host individuals but their larvae live in, and consume, only a single individual. Likewise, a typical helminth parasite may have both intermediate and definitive hosts, but each life-cycle stage will infect only a single host individual. These categories are arbitrary and, often, there is considerable overlap between many of the relationships. (Figure courtesy of Al Bush.)

phoresy (Fig. 1.1). In this case, the symbiont (=phoront) merely travels with its host; there is no metabolic commitment by either partner. Protists or fungal microbes that are mechanically carried by insects are examples of phoretic associations. Similarly, even though whale and turtle barnacles are often described as ectoparasites, there is no metabolic commitment. Functionally, they are phoronts. Phoresy grades into **commensalism**, a symbiotic interaction that implies a trophic relationship between the partners (Fig. 1.1). Commensalism means 'eating at the same table.' Here the benefit gained is unidirectional. The smaller commensal partner typically benefits via food transfer and increased dispersal opportunities, while the host is neither harmed nor benefited. When sharks feed on large prey, they scatter fragments of food that are made available to remoras. Yet, some remoras also feed on ectoparasites of their shark hosts, implying an indirect metabolic linkage. Commensalism therefore grades into **mutualism** in many cases (Fig. 1.1). Many mites are commensals, hitching a ride and sharing food with hosts as diverse as insects and molluscs to birds and mammals.

When there is a direct transfer of energy between the partners, the interaction may be either mutualistic or exploitative (Fig. 1.1). Obligate mutualists are metabolically dependent on one another. A classic example of an obligatory mutualism is the diverse microfauna of protists and prokaryotes in the intestines of wood-eating termites. A single species of flagellated protist, *Trichonympha campanula* (Fig. 1.2), may account for up to one-third of the biomass of an individual termite. These flagellates produce enzymes that digest cellulose, enabling the host to survive on a diet of wood. The mutualistic relationship between ruminant mammals and the ciliated protists and microbes in their stomach is similar. The biochemical complexity of these, and many other mutualistic associations found throughout nature, is the product of a long coevolutionary history between the partners. Such coevolved mutualisms are regarded as being creative forces in the adaptive radiations of many taxa (Thompson, 1994; Price, 1996).

In many exploitative interactions, however, benefit is unidirectional and, moreover, some form of disadvantage, or harm, is the outcome for the other partner.

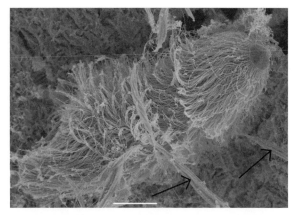

Fig. 1.2 Scanning electron micrograph of the mutualistic hypermastigote flagellate *Trichonympha campanula* from the intestine of a termite. Another, much smaller flagellate *Streblomastix* sp. (arrows) is also present. (Micrograph courtesy of Ron Hathaway.)

Several major categories of this kind of exploitation can be recognized, based primarily on the number of hosts attacked by the symbiont and the subsequent fate of the organism assaulted (Fig. 1.1). If more than one organism is attacked, but typically is not killed, then the aggressor is called a **micropredator.** Hematophagous organisms such as mosquitoes, and some leeches and biting flies, for example, are considered micropredators, taking frequent blood meals from several hosts. Some micropredators are often considered as ectoparasites, e.g., leeches. If more than one organism (considered as prey) is attacked and always killed, then the aggressor is considered a predator. If only one specific host is attacked and is almost always killed, then the aggressor is usually referred to as a **parasitoid,** most of which are wasps and flies.

If only one host is attacked, but typically is not killed outright, the aggressor is a parasite (Fig. 1.1). **Endoparasites** include those that are confined within the host's body. They include the protists, microsporidian and myxozoans, as well as the 'worm' parasites such as flukes, tapeworms, acanthocephalans, and nematodes. A variety of holdfast adaptations often serve to anchor these endoparasites to specific sites within their specific hosts. The holdfasts of elasmobranch cestodes, for example, are often exquisitely adapted to match the

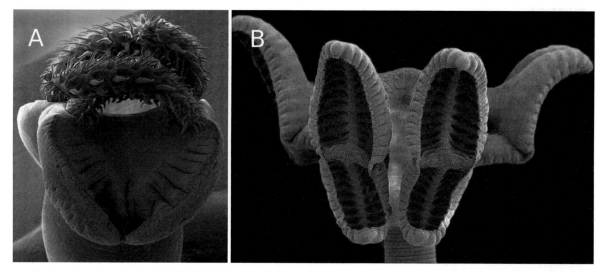

Fig. 1.3 Scanning electron micrographs illustrating the elaborate holdfasts of host-specific tapeworms of elasmobranchs. (A) Scolex of the trypanorhynch cestode *Paragrillotia similis* from the spiral intestine of the Atlantic nurse shark *Ginglymostoma cirratum*; (B) scolex of the rhinebothriidean cestode *Rhinebothrium megacanthophallus* from the spiral intestine of the freshwater whipray *Himantura polylepis*. (Micrographs courtesy of Janine Caira (A) and Claire Healy (B).)

microstructure of the intestines of their specific elasmobranch hosts (Fig. 1.3). Parasites found on the surface of the host's body are called **ectoparasites**. Most parasitic arthropods and monogeneans are ectoparasitic. There are also some parasites that are classified as **mesoparasites** (Kabata, 1979). The pennellid copepods, for example, are endoparasitic in the sense that they have elaborate holdfasts that extend deeply into their host's tissues. However, their highly modified trunk regions and egg sacs extend outside the host (Fig. 1.4; Color plate Figs. 4.2, 4.3).

Anderson and May (1979) went further, highlighting key differences within groups of parasitic organisms. **Macroparasites** are large (usually visible to the eye), have generation times approximating those of their hosts, generate a low-to-moderate immune response, and the pathology they cause to their hosts is tied to the numbers of parasites present. These are typically the classical 'worms' (trematodes, cestodes, and nematodes) and the arthropods, such as copepods, fleas, lice, and mites. They can be endoparasitic or ectoparasitic. The nematode *Heligmosomoides polygyrus* (Fig. 1.5) is an example of an endoparasitic macroparasite infecting mice. The ectoparasitic mite,

Fig. 1.4 Host and site specificity exhibited by the mesoparasitic copepod *Phrixocephalus cincinnatus* attached to the eye of an arrowtooth flounder *Atheresthes stomias*. A metamorphosed female develops an elaborate holdfast, penetrating deeply into the eye of the specific fish host, while the egg sacs and trunk region extend out of the eye. (Photograph courtesy of Dane Stabel.)

Varroa destructor, provides another example (Fig. 1.6). **Microparasites** are much smaller (typically microscopic), have generation times much shorter than their hosts, are capable of asexual replication within their hosts, and typically induce strong acquired immunity in recovered and re-exposed hosts. They can be

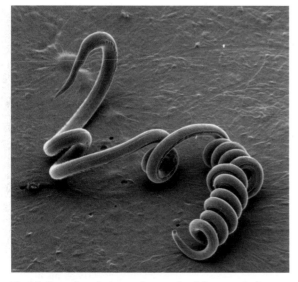

Fig. 1.5 Scanning electron micrograph of the nematode *Heligmosomoides polygyrus* from the intestine of a mouse. This macroparasite–host system is a widely used model in experimental parasitology. (Micrograph courtesy of Doug Colwell.)

Fig. 1.6 Female of the ectoparasitic mite *Varroa destructor* attached to the abdomen of a developing honey bee. (Photograph courtesy of Scott Bauer, USDA Agricultural Research Service, Bugwood.org.)

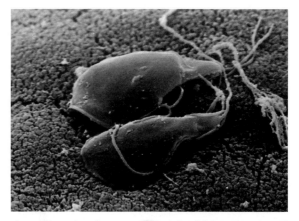

Fig. 1.7 Scanning electron micrograph of the trophozoites of the flagellated protist *Giardia muris* attached to the villi of the small intestine of an experimentally infected mouse. This microparasite reproduces asexually via binary fission. (Micrograph courtesy of Břetislav Koudela.)

ectoparasitic or endoparasitic. They are typically intracellular, i.e., adapted to recognize, penetrate, and reproduce within host cells, or they may exploit extracellular tissues, or both. Eukaryotic microparasites include protists, microsporidians, and myxozoans. In the case of the protist, *Giardia* spp., (Fig. 1.7), ingestion of a single cyst originating from untreated drinking water can lead to massive numbers of feeding stages in the intestine of a range of vertebrate hosts.

Parasites can have parasites too! The parasites living in/on other parasites are called **hyperparasites**. Parasite biodiversity will increase exponentially when we fully understand how common hyperparasitism is in nature. The sea louse, *Lepeophtheirus salmonis*, for example, is a common skin ectoparasite of salmonid species. A monogenean fluke, *Udonella caligorum*, is a hyperparasite of the egg sacs of sea lice. In addition, microsporidians such as *Desmozoon leopeophtherii*, have recently been described as intracellular hyperparasites of *L. salmonis*.

The organism in, or on, which a parasite reaches sexual maturity is the **definitive host**. Many parasites have a simple, direct life cycle, requiring only one host for transmission to occur. All monogeneans, and many nematode and arthropod parasites, have direct life cycles. Many animal parasites, however, have obligate **intermediate hosts** in which the parasites undergo developmental and morphological changes. Intermediate hosts may be the prey of the predatory definitive host in the life cycles of parasites. Thus, parasites with intermediate hosts in their complex life cycles are often transmitted trophically to definitive hosts via food web interactions. Life cycles in which more than one host are required are referred to as indirect life cycles. Many parasites have remarkably complex life cycles with several hosts and both free-living and obligate parasitic larval stages.

Some protists and filarial nematodes have **vectors** as hosts. Vectors are micropredators that transmit infective stages from one host to another. A vector may be an intermediate or a definitive host, depending on whether the sexual phase of the parasite's life cycle occurs in it or not. For example, the insect vectors for species of *Plasmodium*, the causative agents of malaria, are certain species of female mosquitoes that actively inoculate infective stages of the parasite into the next host during their blood meals. Sexual reproduction occurs within the stomach of the mosquito; consequently, mosquitoes are the definitive hosts for the parasite.

A number of parasites may use hosts in which there is no development and that are not always obligatory for the completion of a parasite's life cycle. These are called **paratenic** or **transport hosts**. Such hosts are most frequently used to bridge an ecological, or trophic, gap. For example, adults of the fluke, *Halipegus occidualis*, live under the tongue of green frogs (Color plate Fig. 1.1). Snails in the genus *Helisoma* are obligate first intermediate hosts, whereas aquatic microcrustaceans such as ostracods are obligate second intermediate hosts. But green frogs do not normally consume these small crustaceans. It turns out that various species of odonate (dragonflies and damselflies) prey upon ostracods, thus acting as paratenic hosts for this trematode. Thus, frogs are exposed to *Halipegus* larvae when they prey on nymphs and metamorphosed odonates. In this four-host life cycle, the parasite exploits two predator–prey interactions to enhance transmission (see Fig. 6.23).

A number of animals are normal hosts for parasites that may also infect humans. These are called **reservoir hosts**. These non-human hosts act as reservoirs of infection for certain parasites. Diseases of animals that are transmissible to humans are called zoonotic diseases, or **zoonoses**. Thus, giardiasis, trichinellosis, and schistosomiasis are examples of zoonoses. Similarly, rats are important reservoir hosts for the nematode,

Trichinella spiralis, and the human blood fluke, *Schistosoma japonicum*. Ecologically, reservoir hosts are similar to paratenic hosts since they may greatly increase transmission rates, and also help prevent local extinction of the parasite. The potential for controlling zoonotic diseases in humans is greatly complicated by the presence of these reservoir hosts. Furthermore, reservoir hosts greatly complicate the zoonotic parasite's **epidemiology**. Epidemiology is the study of all the many complex, inter-related ecological factors responsible for the transmission and distribution of a human disease. A related term is **epizootiology**. Epizootiology usually refers to the factors involved in the transmission and distribution of non-human parasites, often in reference to epizootics. The epidemiology of human parasites and the epizootiology of parasites of fishery, veterinary, and wildlife importance will be stressed throughout the upcoming chapters.

1.4 Overview

All eukaryotes have the capacity to recognize invading cells or organisms as non-self. It follows that a fundamental feature of the parasitic life style lies in the host's ability to defend itself against a limitless diversity of invaders. It is the strength and duration of host defenses that defines such features of the host–parasite relationship as specificity, parasite-induced pathology and phenotype alteration, site selection, parasite-mediated natural selection, and coevolution. Given the fundamental importance of host defenses to the ecology and evolution of parasitism, we provide an introduction to the massive field of immunology in Chapter 2. Our coverage is not meant to replace an introductory course in immunology. Instead, our aim is to cover the basics of innate and adaptive immunity in both invertebrates and vertebrates, and also provide an introduction to the interdisciplinary fields of immunoparasitology and ecological immunology.

The following nine chapters cover the biodiversity of animal parasites. In each of these chapters we take a parasite-centered approach to describe functional morphology, life-cycle variation, and biodiversity. Our coverage within these chapters is selective, and not meant to parallel the coverage of this material in modern parasitology texts. Our aim is to provide examples that are representative of key groups, especially those that provide models that are used to ask leading ecological or evolutionary questions. To conclude each of these chapters, we cover phylogenetic classification. Our aim is to summarize the tremendous advances that have been made over the past decade in parasite **systematics**, paralleling the increased use of molecular classification schemes across the entire Tree of Life. These new tools have, in many cases, transformed traditional views of parasite biodiversity and have led to a re-interpretation of the course of evolution in several of the main groups. These new phylogenetic schemes also have implications for how we diagnose and treat some key parasites of humans. Thus, we emphasize modern phylogenetic classifications of the main groups of animal parasites because they provide an evolutionary context for the chapters that follow, and because they have led to new insights into the evolution of parasitism. Box 1.2 provides a

---

Box 1.2 | **Parasite systematics: a phylogenetics primer**

Deciphering the evolutionary relationships of parasitic animals can be a daunting, yet exciting challenge. Consider briefly three enigmatic taxa that have long intrigued invertebrate biologists and parasitologists. The pentastomids (Chapter 10) are primarily obligate parasites of the lungs of reptiles. These parasites have been shown to be most closely related to a group of arthropods known as branchiuran crustaceans. A crustacean in a snake's lung seems bizarre to say the least! Or, consider the Myxozoa described in Chapter 5. Previously classified with single-celled organisms, it turns out that these multicellular fish endoparasites share evolutionary affinities with cnidarians. This makes the cnidarians far more diverse, both in terms of morphology and habitat exploitation, than was ever appreciated. Perhaps equally remarkable is that the endoparasitic thorny-headed worms (Acanthocephala) are evolutionarily linked to certain free-living rotifers (see Box 7.4). To fully understand how biologists have made such monumental changes to the Tree of Life you need to master some of the terminology of the modern systematist.

The focus of phylogenetics research is **homology**. Morphological features that share a common evolutionary origin are said to be homologous. Thus, any differences in homologous features among different parasite groups reflect descent from ancestors with modification (Pechenik, 2010). Homologous features can then be used in construction of phylogenetic trees and taxonomic classifications. Unfortunately, homologous features are not always easily recognized. Morphologies, for example, can independently evolve from very different ancestors, to give a close resemblance by convergence. These morphologies may appear to be homologous, but they are not. Another problematic issue concerns the direction of evolutionary change. Which of two homologous characters is the original, or ancestral, state, and which represents the more derived state? This question can be particularly difficult to decipher for parasites. We will see that much of the debate concerning the systematics of a given group of parasites centers on these issues.

Cladistics is a widely used approach to deduce evolutionary relationships. The terminology in this field of systematics is complex. We introduce here as few terms as possible in order to help you understand the basics. Cladistics use so-called **synapomorphies** (Greek, *syn*shared,

Box 1.2 (continued)

*apo*separate/derived, *morphē*form) to establish evolutionary relationships. These are shared, evolutionarily novel features derived from a common ancestor in which the characters originated. These evolutionary novelties are used to construct a **cladogram** that depicts evolutionary relationships (Fig. 1.8). A cladogram is a representation of branching sequences that are characterized by unique changes in key morphological or molecular characteristics (=synapomorphies). A cladogram shows the least complex, most parsimonius way of explaining the evolutionary history of the groups. A **clade** is a group of organisms that includes the most recent ancestor of all its members and importantly, all descendants of that ancestor. By this definition, a valid clade forms a **monophyletic** taxon (Fig. 1.8a). Synapomorphies are vital because they identify monophyletic groups. If species and groups are classified in a way that reflects evolutionary history, only monophyletic taxa should be given names. Determining monophyly, then, is a primary goal for systematists. Monophyly can be elusive, and instead, there may be inconsistencies between taxonomic schemes and estimates of phylogeny. A taxon is **paraphyletic** if it contains the most recent common ancestor of all members of the group and some, but not all, of its descendants (Fig. 1.8b). Finally, a **polyphyletic** taxon is an incorrect grouping containing species that descended from two or more ancestors (Fig. 1.8c). Thus, members of polyphyletic groups do not share the same immediate ancestor. It is important to recognize that these inconsistencies are an artefact of a naming scheme that is an inaccurate estimate of phylogeny, or vice versa. Both cladists and evolutionary taxonomists reject polyphyletic groups in taxonomic classifications because of this mixed ancestry. It is for this reason that some formerly recognized taxa are no longer considered in classification schemes or they are reclassified to generate a monophyletic group. For example, the 'old' protist phylum Sarcomastigophora, comprised of amoeboid and flagellated unicellular animals is no longer considered valid because many phylogenetic studies have shown that it is a polyphyletic taxon (see Box 3.7).

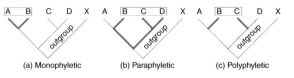

(a) Monophyletic   (b) Paraphyletic   (c) Polyphyletic

Fig. 1.8 Cladograms showing the relationships of four taxa: A, B, C, and D. (a) depicts a monophyletic group or clade, defined by a shared, derived trait or synapomorphy; note that taxa CD and ABCD are also monophyletic; (b) 1 of 4 possible paraphyletic groups is illustrated (others are ABC, ABD, ACD); (c) 1 of 4 possible polyphyletic groups is shown (others are AC, AD, BC). The outgroup (X) is phylogenetically close but not within the taxa being studied. (Figure courtesy of Danielle Morrison.)

Cladists denote the common descent of different taxa by identifying **sister taxa** or sister groups on cladograms. Sister taxa are each other's closest relatives; they arise from the same branching point (node) on a cladogram because they are derived from the same ancestor. Cladists also identify an **outgroup**, a closely related taxon that lies outside the taxa being studied (Fig 1.8). The outgroup's characteristics are assumed to represent the ancestral condition.

Box 1.2 | **(continued)**

The best phylogenetics studies consider dozens of different characters to infer evolutionary relationships. The field of phylogenetic systematics has been revolutionized by accessing the genetic code of life via the nucleotides in DNA and RNA, and comparing these gene sequences (especially 18S ribosomal RNA and mitochondrial DNA) among taxa. Indeed, throughout the upcoming chapters we will see that it is this molecular revolution that has greatly altered our taxonomic classification system of parasitic animals. Moreover, molecular data are helping us to more fully understand the origins and evolution of parasitism, and the interrelationships within the various animal parasite taxa (e.g., see Figs. 5.11, 7.12, 8.26).

While a primary focus of many phylogenetics studies is to help resolve parasite taxonomic issues, they are also widely used for several other reasons, including analyzing the geographical distributions of hosts and their parasites, and how distributions have changed over time. The research field uniting phylogeny with geography, as well as evolution and geology, is called **phylogeography** (Chapter 14). In addition, phylogenetics is widely used in studies of host–parasite **coevolution** and **cospeciation** (Chapter 16).

primer for the two approaches, classical taxonomy (or evolutionary systematics) and phylogenetic systematics (or **cladistics**), that are so often used to test hypotheses regarding the origins of parasitism and the evolutionary relationships between and among taxa.

The next six chapters summarize advances in parasite ecology and evolution, drawing upon results from modern empirical studies on a wide range of model host–parasite interactions. This section is structured hierarchically, beginning with coverage of parasite population biology and epidemiology in Chapter 12. Following a section on terminology, we briefly introduce the mathematical foundation upon which these subdisciplines rest. We then shift to a brief introduction of how parasite populations are distributed within their host populations, and how parasite population sizes are determined and regulated. We proceed along this natural hierarchy to synthesize advances in parasite community ecology in Chapter 13. Here, we first

characterize the structure of multi-species parasite assemblages and then emphasize how modern empirical approaches have uncovered some unexpected underlying mechanisms. We conclude this chapter by describing comparative patterns of parasite biodiversity among host species, ending with a discussion of the approaches that field-based parasitologists utilize to infer causation. In Chapter 14, we extend the hierarchy to include aspects of parasite biogeography and phylogeography. Our emphasis in this chapter is on how tools drawn from molecular biology, genomics, and remote sensing are revolutionizing our understanding of the geographical distribution of parasites, ranging from continent-wide scales, down to a few meters.

In the final three chapters, we shift our focus away from the parasites themselves, onto the nature of the host–parasite interaction and the complex interplay that exists between the two partners over ecological and evolutionary timescales. Given the exploitative nature of the relationship, the concept of 'harm' to hosts is central to the very definition of the parasitic way of life. In Chapter 15, we summarize the results of empirical studies that document a wide range of effects on hosts, including those on the metabolism, physiology, behavior, and reproduction of host individuals. We then cover the effects of various parasites on host populations, and on host communities and entire ecosystems. Coverage in Chapter 16 focuses on aspects of the evolution and coevolution of host–parasite interactions, including parasite-mediated natural selection and evolution, parasite population genetic structure, and host–parasite cospeciation. In our concluding chapter we introduce the multidisciplinary field of environmental parasitology, stressing the role of parasites as biological indicators of anthropogenic changes at individual parasite, and population and community scales.

References

Anderson, R. M. & May, R. M. (1979) Population biology of infectious diseases: Part I. *Nature*, **280**, 361–367.

Begon, M., Townsend, C. R. & Harper, J. L. (2006) *Ecology: From Individuals to Ecosystems*, 4th edition. Oxford: Blackwell Scientific Publishers.

Brooks, D. R. & McLennan, D. A. (2002) *The Nature of Diversity: An Evolutionary Voyage of Discovery*. Chicago: University of Chicago Press.

Combes, C. (2001) *Parasitism: The Ecology and Evolution of Intimate Interactions*. Chicago: University of Chicago Press.

Combes, C. (2005) *The Art of Being a Parasite*. Chicago: University of Chicago Press.

Crofton, H. D. (1971) A quantitative approach to parasitism. *Parasitology*, **62**, 179–193.

De Meeus, T., Michalakis, Y. & Renaud, F. (1998) Santa Rosalia revisited: or why are there so many kinds of parasites in 'The Garden of Earthly Delights'? *Parasitology Today*, **14**, 10–13.

De Meeus, T. & Renaud, F. (2002) Parasites within the new phylogeny of eukaryotes. *Trends in Parasitology*, **18**, 247–251.

Ebert, D. (2005) Ecology, epidemiology, and evolution of parasitism in *Daphnia* (Internet). Bethesda: National Center for Biotechnology Information.

Esch, G. W. (2007) *Parasites and Infectious Disease: Discovery by Serendipity, and Otherwise*. Cambridge: Cambridge University Press.

Frank, S. A. (2002) *Immunology and Evolution of Infectious Disease*. Princeton: Princeton University Press.

Freeman, S. & Herron, J. C. (2007) *Evolutionary Analysis*, 4th edition. London: Pearson Education.

Grove, D. J. (1990) *A History of Human Helminthology*. Wallingford: CAB International.

Hudson, P. J., Rizzoli, A. P., Grenfell, B. T., *et al.* (2001) *The Ecology of Wildlife Diseases*. Oxford: Oxford University Press.

Kabata, Z. (1979) *Parasitic Copepoda of British Fishes*. London: Ray Society.

Kearn, G. C. (1998) *Parasitism and the Platyhelminths*. London: Chapman and Hall.

Kennedy, C. R. (1975) *Ecological Animal Parasitology*. Oxford: Blackwell Scientific Publications.

Moore, J. (2002) *Parasites and the Behavior of Animals*. Oxford: Oxford University Press.

Morand, S. & Krasnov, B. (2010) *The Biogeography of Host–Parasite Interactions*. Oxford: Oxford University Press.

Noble, E. R., Noble, G. A., Schad, G. A. & MacInnes, A. J. (1989) *Parasitology: The Biology of Animal Parasites*, 6th edition. Philadelphia: Lea & Febiger.

Pechenik, J. A. (2010) *Biology of the Invertebrates*, 6th edition. New York: McGraw-Hill.

Poulin, R. (2007) *Evolutionary Ecology of Parasites*, 2nd edition. Princeton: Princeton University Press.

Poulin, R. & Morand, S. (2004) *Parasite Biodiversity*. Washington, D.C.: Smithsonian Books.

Price, P. W. (1980) *Evolutionary Biology of Parasites*. Princeton: Princeton University Press.

Price, P. W. (1996) *Biological Evolution*. Philadelphia: Saunders College Publishing.

Roberts, L. S. & Janovy, Jr., J. (2009) *Foundations of Parasitology*, 8th edition. New York: McGraw-Hill.

Schmid-Hempel, P. (2011) *Evolutionary Parasitology: The Integrated Study of Infections, Immunology, Ecology, and Genetics*. Oxford: Oxford University Press.

Thomas, F., Guégan, J.-F. & Renaud, F. (2009) *Ecology and Evolution of Parasitism: Hosts to Ecosystems*. Oxford: Oxford University Press.

Thompson, J. N. (1994) *The Coevolutionary Process*. Chicago: University of Chicago Press.

Zelmer, D. A. (1998) An evolutionary definition of parasitism. *International Journal for Parasitology*, 28, 531–533.

Zimmer, C. (2000a) Do parasites rule the world? *Discover Magazine*, 21, 80–85.

Zimmer, C. (2000b) *Parasite Rex: Inside the Bizarre World of Nature's Most Dangerous Creatures*. New York: Simon & Schuster.

2 Immunological aspects of parasitism

2.1 General considerations

One unifying characteristic of most organisms is their ability to distinguish entities of their own bodies from entities that are genetically different. This ability to distinguish 'self' from 'non-self' originated in the deepest roots of the Tree of Life, when natural selection favored individual prokaryotes that could phagocytize potential food and not kin or potential mating partners. Likewise, sexual reproduction, another unifying characteristic on almost all branches of the Tree of Life, requires the recognition of appropriate gametes. The processes underlying recognition of self/non-self are undoubtedly complex, as we will see, but they fundamentally require intimate, molecular-level interactions at the interface (usually on cell surfaces) between two unrelated partners. Herein lies the foundation of one of the key processes that defines biological systems: cell–cell communication, and the ability for all organisms to protect themselves from potential invaders, both abiotic and biotic, via immunity.

In the introductory chapter, we alluded to the phenomenal success of the parasitic way of life. In a sense, this success should not surprise us. Individuals that adopt a life style that avoids predators and diseases, that provides access to potentially limitless resources, that provides access to mates, and so on, should be favored by natural selection. Yet, all organisms that adopt this life style confront the constraint of avoiding (or limiting) immunological defenses (and other host defenses, see Chapter 16), many of which can drive parasite reproductive success to zero. Thus, the host immune response represents a critical selective force on individual parasites. As we will see later in this chapter, and throughout this book, the manner in which parasites evade the sophisticated host immune response has major consequences to human health and

to the development of parasite control strategies. Perhaps more fundamentally, the dynamic interplay between host defense and parasite evasion impacts almost every conceivable aspect of host–parasite interactions. Immune responses are relevant to the evolution of parasite life cycles and life histories and to the evolution of parasite specificity. Immune responses contribute to the rate at which hosts are exposed to parasites, they can determine the numbers and types of parasites in a host, and can determine parasite fitness. It follows that the effects of parasites on host individuals and thus, host populations, are also impacted by host immune responses. Finally, traits associated with host immunity provide obvious targets for parasite-mediated natural and sexual selection to act upon, and thus, can determine the trajectory of host–parasite coevolution.

Components of the immune system can be considered key host life history traits. The notion that host immune traits can be considered in the same light as traits such as body size, reproductive rate, time to reproduction, and egg size is one of the core principles of the new subdiscipline of ecological immunology (reviews in Sadd & Schmid-Hempel, 2009; Schulenburg *et al.*, 2009; Demas & Nelson, 2012). Thus, from an ecological perspective, we should expect natural selection to trade-off the expression of traits linked to host immunity with other life history traits, with important consequences. We should also expect immune traits to be costly in terms of host reproduction. By extension then, the expression of host immunity likely impacts the strength and direction of parasite-mediated host evolution. Phylogenetic analyses of the major histocompatibility complex (MHC, a key component of the parasite-recognition system in vertebrates), and its alleles, suggest that periods of host radiation were accompanied by periods of reorganization of the MHC (Klein *et al.*, 2007). In some instances, periods of host radiation were also periods of parasite radiation, leading ultimately to new adaptive responses by the host immune system. Thus, while the host immune response imposes strong selection pressure on parasites, they in turn, impose strong selection on the host immune system.

Given the diversity of ecological and evolutionary influences of host immunity on parasites, and vice versa, a basic understanding of unifying principles of immunology is necessary to understand the diversity and ecology of host–parasite interactions. This chapter is structured to provide an introduction to the enormous and expanding field of immunology. It is not meant to replace even an introductory course in immunology. Further, our emphasis is not on molecular or medical immunology, but on features that provide a framework for our subsequent coverage in the diversity, ecology, and evolution chapters. Although our focus in this chapter is on fundamental principles drawn from classical vertebrate immunology (e.g., Cox, 1993; Coico & Sunshine, 2009) and immunoparasitology (e.g., Wakelin, 1996, 1997), we extend our coverage to include advances in invertebrate immunity and in ecological immunology. Non-immunological forms of host defense (e.g., tolerance, behavioral avoidance) are covered in Chapter 16.

The basic processes involved in host immunity are best known in the jawed vertebrates. Empirical studies that utilize well-characterized model systems, often involving laboratory rats and mice as hosts, have provided enormous advances. Progress in the field over the past 20 years has been astonishing, due mostly to methodological advances associated with DNA cloning, proteomics, and genomics. Currently, entire parasite genomes are available, with enormous consequences for potential vaccine development and parasite control. Remarkably, despite this increase in activity, our understanding of the key underlying processes has remained essentially unchanged.

Immunity in the jawed vertebrates can be attributed to two distinct, but highly interconnected, sets of mechanisms (see Box 2.1 for list of terms and concepts). One is called innate immunity because the factors involved are always present, regardless of the presence or absence of parasites. One key feature of the innate immune response is that its recognition and targeting systems are non-specific. Thus, they work against a wide variety of potential parasites and pathogens, particularly those that share common structural properties (e.g., carbohydrates on bacterial cell walls). Innate mechanisms of immunity are present in all animal groups. Indeed, innate immune molecules in hosts as distantly related as humans and sponges show remarkable structural similarities (Müller & Müller, 2003). Thus, protective coverings, anti-microbial substances, and cells involved in wound repair occur throughout the animal kingdom. Hundreds of broad-spectrum anti-microbial peptides have been discovered having activity against viruses, bacteria, fungi, and other parasitic organisms in all metazoan animal groups. These innate mechanisms of immunity appear to have arisen very early in the evolutionary history of the animals (Hoffmann *et al.*, 1999).

In contrast, the second set of mechanisms, known collectively as acquired immunity or simply, the immune response, is restricted to the jawed vertebrates. These responses are highly specific to a particular parasite or pathogen (or other foreign proteins) and are inducible. Thus, in contrast to innate properties of the skin and other components of innate immunity, the cells involved tend to increase in number following invasion of the parasite, and subsequently respond only to that particular invader (antigen) following re-exposure. With this defense strategy, when a host is subsequently challenged with the same parasite, it remembers with a response that is qualitatively and quantitatively greater than during the first exposure, a phenomenon known as **immunological memory** (Fig. 2.1). The sophisticated processes that involve parasite recognition by specialized host cells, followed by their proliferation, and then memory, are the hallmarks of the vertebrate immune system.

The immune response in the higher vertebrates involves three main cell types (Box 2.1). These include antigen-presenting cells (macrophages and dendritic cells) and the B- (bone-marrow derived) and T- (thymus derived) lymphocyte cells. Following invasion by

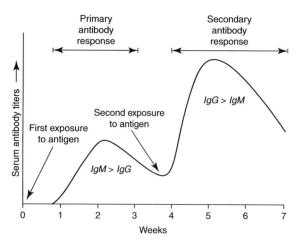

Fig. 2.1 Schematic diagram of a hypothetical primary and secondary antibody response. Note that the secondary response is more rapid and that the plateau in antibody concentration is considerably greater. In addition, the lag time for the primary response is longer than that for the secondary response. The predominant class of antibody produced during the primary response is IgM, whereas in the secondary response it is IgG. (Figure courtesy of Richard Seed.)

a parasite, it is these three cell types that, through an exchange of chemical messengers (especially cytokines), expand and respond specifically to a particular infectious agent. Macrophages are phagocytic white blood cells that, in the normal host, are always present (thus, are also part of the innate immune system) and capable of engulfing and killing infectious disease agents. In the acquired immune system, it is the macrophage that processes the engulfed parasite and presents small pieces (epitopes) of the parasite's protein to the specialized lymphocytes. This transfer of information, from macrophage to lymphocyte, induces the lymphocyte to divide and expand in number. The activated lymphocyte can then exclusively respond to the parasite that contained the information originally processed and transferred by the macrophage. In addition, the macrophage itself can be induced to become an activated, or 'angry,' macrophage. The activated macrophage is now a better 'killing machine' and is an integral part of the immune system. Finally, not only are there complex cell-to-cell interactions

---

Box 2.1 | Immunology: a terminology primer

Acquired immunity (*inducible immunity or adaptive immunity*) lymphocyte-mediated immune response characterized by inducibility, antigen specificity, and memory

Antibody (*Ab*) – an immunoglobulin molecule specific for one parasite epitope produced in response to an antigen

Antigen (*Ag*) – a molecule that can induce a specific immune response, and be a target for an antibody; antigens are composed of one or more epitopes

Antigen presentation – intracellular breakdown of antigen by specialized cells that result in antigenic fragments (epitopes) binding to a MHC receptor and then being expressed on a cell membrane

B-lymphocytes (*B-cells*) – lymphocytes that mature in the bone marrow; these recognize antigens via immunoglobulins on their cell surface, and upon antigenic stimulation, secrete antibody

Complement – a complex set of serum proteins that act in a cascade-like fashion to produce components that label target parasites, trigger inflammation responses, or assemble into pore-forming channels capable of lysing cells

Cytokines – a group of intercellular messengers that regulate immune and inflammatory responses (e.g., interferons, interleukins)

Box 2.1 | (continued)

Epitope – a small region of an antigen that is recognized by antigen-binding receptors on B- and T-lymphocytes

Granulocytes – bone marrow-derived cells such as neutrophils and eosinophils that mediate innate immunity via phagocytosis or the release of cytotoxic chemicals

Hemocytes – a class of cells found in the hemolymph of invertebrates that are involved in diverse components of cell-mediated immunity

Hypersensitivity – inflammatory reactions that are produced upon re-exposure to an antigen

Immunocompetence – generally, the ability of a host to respond immunologically to an invader

Immunoglobulin (*Ig*) – antigen-binding proteins (e.g., IgM, IgG, IgE, IgD, IgA) located on the surface of B-lymphocytes; these proteins enter the general circulation following their exposure to antigen

Inflammation – a common and complex outcome of exposure to trauma, parasites, or allergic immune responses that includes biochemical, physiological, and structural changes to tissues; typically mediated by specialized cells or their products (e.g., cytokines)

Innate immunity – non-specific host defense that is independent of lymphocytes such as skin, mucus, complement, and phagocytosis

Lymphocytes – bone marrow-derived cells that mediate vertebrate immunity; B-cells differentiate in bone barrow in mammals and the bursa fabricus in birds, T-cells differentiate in the thymus gland

Macrophages – bone marrow-derived cells that ingest non-self material by phagocytosis; key players in the presentation of antigen to T-cells via MHC receptors located on their surface

Major histocompatibility complex – a closely linked group of gene loci that code for cell surface proteins; these proteins are often expressed on macrophages, playing a key role in self recognition and presentation of antigen to T-cells

Phagocytosis – engulfment of foreign material by specialized cells such as macrophages and neutrophils

T-lymphocytes (*T-cells*) – lymphocytes that mature in the thymus; together with the MHC proteins on their surface, T-cells mediate cellular recognition of self/non-self; T-helper cells and cytotoxic T-cells are functional subsets

T-cell receptor – antigen-binding receptor found on T-lymphocytes; these interact with MHC receptors located on the surface of macrophages

(Modified from Wakelin & Apanius, 1997. Immune defense: genetic control. In *Host-Parasite Evolution: General Principles and Avian Models*, ed. D. H. Clayton & J. Moore, pp. 30–58. Oxford: Oxford University Press.)

between macrophages and the other cells of the immune system, but there is also extensive chemical communication or cross-talk via chemical messengers (e.g., hormones and cytokines) between the various organs of the body and the immune system (Fig. 2.2). Interactions between the immune system and nervous system are especially important (review in Steinman, 2004). For example, macrophages release cytokines

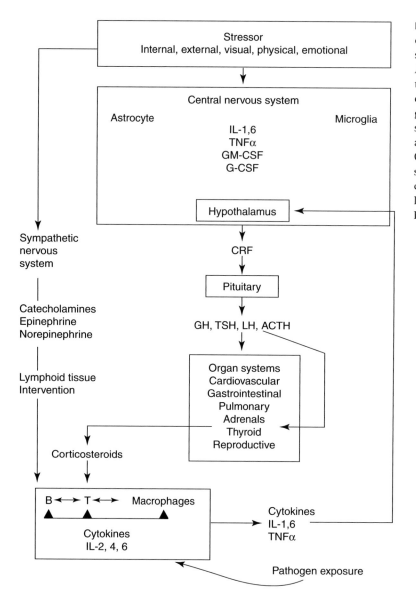

Fig. 2.2 The interaction between the central nervous system, other organ systems, and the immune system. Abbreviations: IL, interleukins; TNF, tumor necrosis factor alpha; CRF, corticotrophin releasing factor; GH, growth hormone; TSH, thyroid stimulating hormone; ACTH, adrenocorticotrophin hormone; GM-CSF, granulocyte-macrophage-colony stimulating factor; G-CSF, granulocyte-colony stimulating factor; LH, luteinizing hormone. (Figure courtesy of Richard Seed.)

called IL-1, or interleukin-1, which signal the brain to initiate a fever response (Fig. 2.2). The secretion of IL-6 by macrophages is a key component of the overall vertebrate immune response, but it also activates serotonin – a key neuromodulator. Further, through a series of hormones, the brain can stimulate the adrenal glands to produce corticosteroids. These interact with macrophages and other cell types to produce an anti-inflammatory response. Thus, as we discuss the details of various forms of host immunity, it is important to remember that natural selection has favored extensive chemical integration within the various arms of the immune system itself, and also with other organ systems in the body.

2.2 Vertebrate immunity

2.2.1 Innate immunity

Several of the most important processes involved in innate immunity occur at the site of parasite penetration or attachment such as the skin and the mucous membranes of the gastrointestinal tract and lungs or gills. Structures such as scales, feathers, and hair undoubtedly provide physical barriers against some parasites. While parasites may mediate natural selection for host traits such as hair density or scale thickness, there is little supportive evidence. If a parasite successfully navigates these external structures, they will encounter the complex vertebrate epidermis. In mammals and birds, the exteriormost component of the epidermis includes compacted layers of dead cells onto which are excreted salts, organic acids, and fatty acids from epidermal and dermal glands. These substances have known anti-parasite and anti-pathogenic properties. In aquatic vertebrates, cells within the epidermis secrete anti-microbial peptides and toxic substances such as nitric oxide that reduce the metabolic activity of prokaryotic and metazoan parasites (e.g., Magnadottir, 2006). Thus, the vertebrate epidermis acts both as a physical and a chemical barrier to many infectious agents. Humans with segments of skin removed, such as those with severe burns, or fish with high numbers of skin-damaging ectoparasites (see Chapter 11 and Box 15.2), can be highly susceptible to secondary bacterial and fungal infections.

Penetration of a parasite into deeper tissues activates a further set of innate defense mechanisms. Most parasites release chemical substances, especially certain cytokines, as part of their tissue penetration or migration activities that attract macrophages (especially neutrophils and eosinophils) to the site of invasion. As we will see, cytokines are protein hormones that are a key component of the innate and acquired immune systems, with pivotal roles in cell-to-cell communication. Parasite penetration may also cause minor breaks in lymphatic and vascular beds. This leads to an infiltration, at the site of penetration, by a variety of macromolecules involved in the complement pathways. In mammals, each of these pathways includes a series of at least 30 serum or membrane-associated proteins in which one protein or peptide in the pathway activates a second protein, and so on, in a cascading response (Walport, 2001). In the classical pathway, the end product either attacks the foreign cell membrane directly, or pokes holes in the membrane leading to cell lysis. Also, during the enzymatic cascade, products are formed that are chemotactic for macrophages. The surface macromolecules of many parasites are capable of interacting with proteins in the complement pathway, or activating the clotting mechanism, or both. In the clotting pathway, the end result is the formation of a clot consisting of proteins and cellular elements. A clot, in addition to the obvious control of bleeding, also forms a matrix of protein and cellular elements. This matrix traps some parasitic organisms, localizing them for phagocytosis by macrophages. Thus, the clotting and complement pathways are each important in innate immunity and in the general inflammation response to some parasites.

Taken together, various arms of the innate immune response ensure that early in the host–parasite interaction, there is a non-specific and fast-acting (hours to days) attempt by the host to surround the parasite and kill it by anti-parasite macromolecules, or by the phagocytic activity of macrophages, or both. These processes tend to be part of a complex inflammation response, which in combination with increased body temperature (fever), occurs whenever external barriers are breached and a parasite enters the body. Although the numbers of receptors for the recognition of parasite molecules is limited in innate immunity, it is a key component of overall immunity and a key contributor to the orchestration of the acquired immune response. Further details on vertebrate innate immunity are in Cox (1993) and Coico & Sunshine (2009).

2.2.2 Acquired immunity

If a parasite evades a host's innate defenses, it will induce a response that is much more specific. In the jawed vertebrates, there are two distinctive but highly integrated pathways, both of which are evolutionarily

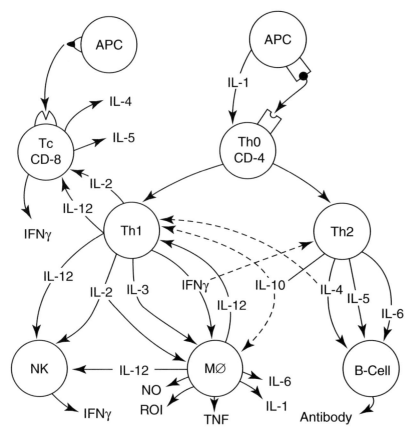

Fig. 2.3 Simplified diagram of the host cells, interleukins, and other reactive molecules involved in the host's immune response. This diagram demonstrates how dynamic the host–parasite interaction can be. For example, following the processing and presentation of parasite antigens by the antigen-presenting cells (APC) to the Th-0 T-cell, the response may be directed towards either cellular immunity (Th-1 T-cells) or a humoral response (Th-2 T-cells and, subsequently, B-cells). Note that once stimulated, the Th-1 or the Th-2 cells produce interleukins such as interferon (IFN) or interleukin-4 (IL-4) that either up-regulate (stimulate, solid lines), or down-regulate (broken lines) the other cell type. For example, if the Th1 cells are stimulated during a leishmanial infection, the Th-1 cells produce interleukins such as IL-2 and IL-12, which can activate (solid lines) natural killer (NK) cells, macrophages (MØ), or cytotoxic T-cells (Tc), but they also secrete IFN, which down-regulates the Th-2 cell population. This insures a preferential cellular (or Th1) type response. Abbreviations: NO, nitric oxide; ROI, reactive oxygen radicals. Both NO and ROI are known to have anti-microbial activity. TNF, tumor necrosis factor; IL-2, IL-3, etc., interleukins with different functions; CD4 and CD-8, T-cell populations having different biological functions and identified by having different surface markers. (Figure courtesy of Richard Seed.)

conserved. These are outlined, in simplified form, in Fig. 2.3. Initiation of the complex cascade of events begins with recognition of the parasite, followed by its presentation to thymus-derived lymphocytes, or T-cells. The initiation of this process requires the processing of parasite proteins (antigens) by special antigen-presenting cells (APCs) such as macrophages (Fig. 2.3). These cells engulf antigen, break them into smaller fragments (epitopes) within phagocytic vacuoles, and then present them on their cell surface as peptides that are recognizable by T-cells. In vertebrates, especially birds and mammals, the presentation

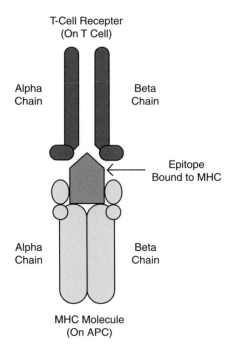

Fig. 2.4 Schematic diagram of the molecular interactions involved in antigen (epitope) presentation, showing the key integrative roles of MHC molecules (on antigen-presenting cell, APC), and T-cell receptors. (Figure courtesy of Danielle Morrison; modified from Wakelin & Apanius, 1997, Immune defense: genetic control. In *Host-Parasite Evolution: General Principles and Avian Models*, ed. D. H. Clayton & J. Moore, pp. 30–58. Oxford: Oxford University Press.)

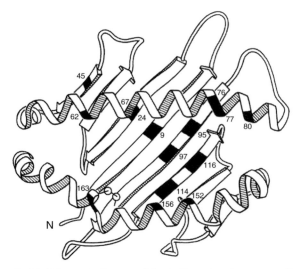

Fig. 2.5 Schematic diagram of the 3-dimensional structure of the class I MHC molecule, viewed from the top. The diagram indicates the pocket within which the binding of parasite epitopes occurs, prior to their presentation to T-cells. Numbered and darkened sections are amino acid sites that have been characterized as highly polymorphic. (Modified from Hedrick, 1994, with permission, *The American Naturalist*, 143, 945–964, University of Chicago Press.)

of parasite peptides to T-cells is determined by molecules of the MHC. These are produced in the Golgi complex of many cells, including APCs, and are transported to the cell surface in the phagocytic vacuoles. Class I MHC molecules are those that bind antigen that originate intracellularly, whereas Class II molecules bind antigen that originate extracellularly. The molecular complex (Fig. 2.4) comprising of parasite epitopes bound to MHC receptors located on the cell surface of macrophages (the APCs) is required for recognition by T-cell receptors on the T-helper cells. Following recognition of appropriate antigen, the T-cell rapidly undergoes clonal reproduction that results in cells with identical T-cell receptors. So long as the antigen is present, clonal reproduction continues so that there is

continual recognition, and continual release of regulatory cytokines that mediate the immune response (Fig. 2.3).

The presentation of parasite epitopes to T-cells is a potential bottleneck in the vertebrate immune response (Wakelin & Apanius, 1997). Given the key role that MHC genes play in this process, the structure and function of their products has received intensive interest from immunologists and more recently, from evolutionary biologists (review in Hedrick, 1994). The structure of a typical MHC molecule is presented in Fig. 2.5. The epitope recognition site on these molecules lies within a distinctive pocket that is surrounded by ridges that provide structural support. It is the sequence of amino acids located at specific sites within the pocket that is responsible for the specificity of epitope binding. Remarkably, genetic heterozygosities of amino acids at these sites are among the highest in the animal kingdom (Hedrick, 1994). In contrast, genetic polymorphism at non-functional sites that occur outside the pocket are typical of those found at sites on other molecules. The conclusion to be drawn from these structural and genetic studies is that natural selection favors extraordinarily high diversity at epitope recognition sites to correspond to the infinite number of parasite peptides that an individual may encounter over its lifetime. We return to the evolutionary significance of the MHC in our discussion of parasite-mediated natural selection and evolution in Chapter 16.

Antibody formation by bone marrow-derived lymphocytes, or B-cells, is the key outcome of one of the two arms of vertebrate immunity (Fig. 2.3). Their formation requires sophisticated and specific biochemical interactions among parasite epitopes, macrophages, T-cells, and B-cells. Once the epitope information has been passed to a B-cell, it is induced to produce antibodies to that one epitope (Fig. 2.4). In addition, the T-cells produce IL-4, IL-5, and IL-6 that signal the B-cell to divide. Thus, the B-cell, in cooperation with the T-cell, is induced to produce antibody, as well as being stimulated to divide and produce a clone of

antibody-producing cells to the one epitope. Through this ingenious evolutionary innovation, vertebrates have the immunological capability to respond to millions of different epitopes. As a consequence, although each immune response to a parasitic invader is highly specific, the immune system has enormous potential to respond to a phenomenal diversity of different parasites. Note that parasites are not composed of a single antigenic epitope, but rather, their surface is a topographical mosaic of different macromolecular antigens and, therefore, epitopes. The host will recognize a number of these different epitopes and there will be a polyclonal antibody response.

Antibodies, or immunoglobulins, are large macromolecules (150 000 daltons or greater) that are typically comprised of four peptide chains linked by disulfide bonds (Fig. 2.6). Each antibody molecule includes at least two antigen-combining sites, each of which is composed of one light and one heavy chain molecule. They are present in blood, lymph, and cerebrospinal fluid. Immunologists currently recognize five classes of antibody molecules (IgM, IgG, IgE, IgD,

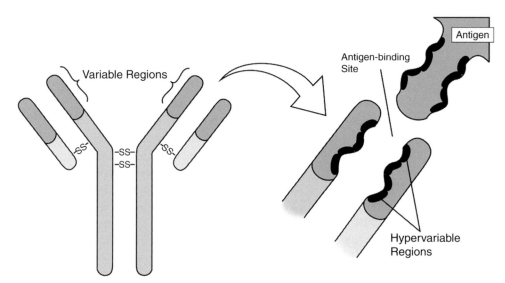

Fig. 2.6 Schematic diagram of an antibody molecule. The molecule is composed of four polypeptide chains: two shorter light chains, and two longer heavy chains, joined by disulfide bonds (SS). There are at least two combining sites, each composed of the variable region of a light and a heavy chain. The inset depicts the antigen-binding sites that are in clefts formed in the variable portions of the heavy and light chains. These hypervariable regions are responsible for the specificity of the antibody response to diverse antigens. (Figure courtesy of Danielle Morrison.)

IgA) distinguished by the characteristics of their heavy chains. Each has different biological functions based on their ability to clump or to precipitate antigens out of solution. For example, heavy chains of immuno-globulin E (IgE) bind with receptors on mast cells and platelets and, following interaction with an antigen, cause the release of small, low-molecular-weight mediators, one of which is histamine. The release of these molecules produces many of the symptoms that we recognize as an allergic response (e.g., asthma, hay fever, anaphylaxis). Yet, IgE also plays an important role in immunity to helminths. In contrast, the heavy chains of IgM do not combine with mast or platelet cell receptors. The various immunoglobulin classes are also predominantly associated with different parts of the body. For example, the large IgM is located predom-inantly in the vascular bed, whereas IgA is found in body secretions such as saliva.

The second arm of the acquired immune response is initiated by Th-1 lymphocytes, followed by the secre-tion of different subsets of regulatory cytokines (Fig. 2.3). This arm requires similar levels of coordi-nation and synchrony among T-cells, chemical mes-sengers, and macrophages, but it is independent of the production of antibodies. Traditionally, immunolo-gists refer to this defense pathway as cell-mediated immunity. The cellular response involves another set of immune cells, the CD-4, Th-1, T-cell, and the CD-8, cytotoxic T-cells (Fig. 2.3) distinguished by their sur-face markers. The CD-4, Th-1, T-cell, and the CD-8 cells contain surface receptors that interact with spe-cific epitopes on foreign antigens. The specificity of the CD-4 cellular response is similar to that induced with the antibody response. Each activated T-cell responds to a single epitope and, following combination with a specific antigen, there is clonal proliferation of these cells. There is also the release of regulatory cytokines.

As for the antibody response, cellular immunity involves antigen-presenting macrophages (APCs) and the presentation of parasite epitopes in combination with specific macrophage surface molecules (the MHC proteins) to the T-cells. Specific cytotoxic T-cells are capable of combining with epitopes on the surface of parasite cells and ultimately cause their death.

Following antigen presentation to Th-1 cells, several different cytokines are secreted, including IL-2 and gamma-interferon, IFN-γ (Fig. 2.3). IL-2 stimulates a particular subset of T-cells to grow and proliferate and become cytotoxic when they are presented with anti-gen bound to MHC molecules. When this occurs, the cytotoxic T-cells can lyse cells carrying the antigen. This is the primary defense mechanism against virus- or tumor-infected cells. In contrast, secretion of IFN-γ by Th-1 cells activates or 'angers' macrophages to become increasingly phagocytic and destructive (Fig. 2.3) through enhanced production of proteolytic enzymes, nitric oxide, tumor necrosis factor, and other highly toxic substances. These macrophages can also bind, via specific receptors, directly to immunoglobu-lins, forming a direct physical contact between a par-asite and the macrophage. In this way, activated macrophages can kill or damage large parasites.

Our brief description of the vertebrate immune response emphasizes the key role of various cytokines in the induction of B-cells, cytotoxic T-cells, and activated macrophages. Together, these three cell types are responsible for the main effector mechanisms that underlie vertebrate immunity – antibodies, cytolysis, and macrophage-mediated cell death. As we will see below, cytokines are also pivotal in the regulation of the overall immune response, and therefore, the con-trol of immunopathology. Yet cytokines, especially IL-1, also play a critical role in the determination of particular subsets of the T-helper lymphocytes, Th-1 and Th-2 (Fig. 2.3). Thus, different parasites preferen-tially stimulate Th-1 or Th-2 populations, depending in part on which cytokines they induce for secretion (Mosmann et al., 1986). Moreover, Th-subset dominance appears to be mutually suppressive. For example, mice that are resistant to Leishmania spp. induce a Th-1 dominated response with a character-istic and stable cytokine profile, whereas susceptible mice have a Th-2 dominated response. In contrast, macroparasites tend to induce Th-2 dominated responses associated with contrasting cytokine profiles.

The adaptive significance of Th-subsets and espe-cially their role in human disease is an active area of enquiry. Studies completed over the last decade or so suggest that designation of effector T-cells into the Th-1/Th-2 dichotomy (as indicated in Fig. 2.3) is too simplistic. There is wide consensus for at least two other subsets (Th-17 and regulatory T-cells (T-regs)), with others being proposed, and there is evidence for extensive plasticity among subsets (review in Locksley, 2009). One interesting outcome of this research activity is the intriguing evidence in humans and mice for causal linkages between allergy (e.g., hay fever, asthma), infection with helminths, and Th-subset bias. In mice infected with the intestinal nematode *Heligmosomoides polygyrus* (see Chapter 8), Finney *et al.* (2007) showed that infected hosts could suppress the inflammation of their airways induced by allergy. Then, when uninfected, allergic mice were injected with T-cells from infected, but allergen-naïve mice, inflammation in their airways was also suppressed (Wilson *et al.*, 2005). These results show that T-cells induced by Th-2-inducing intestinal worms can influence the outcome of host allergic responses, possibly even in humans. An extension of these causal linkages may also help explain increased incidence of Th-1-mediated autoimmune diseases (e.g., multiple sclerosis, rheumatoid arthritis, psoriasis) in 'aseptic' Western countries, where infections with Th-2-promoting worms are rare. Despite the large number of confounding factors that have to be taken into account to determine causation, there seems to be mounting immunological support for the 'hygiene hypothesis' that implies causal linkages between worms, Th subset dominance, and increasingly important human diseases. Reviews on this rapidly advancing area of immunology are in Yazdanbakhsh *et al.* (2002) and Maizels (2009).

2.2.3 Regulation of host immunity

The nature of the integrative, complex immune response requires that it be carefully regulated. Natural selection for mechanisms of regulation comes from the inherent costs associated with the maintenance and expression of host immunity (see Section 2.4.1) especially those associated with immunopathology and autoimmune disorders. In the diversity chapters that follow, we describe many examples where parasite-induced pathology is directly attributed to misdirection, or overexpression, of the immune response itself or directly to parasite immune-evasion strategies. Thus, regulation is a critical feature of the immune system, acting at all stages of innate and acquired responses.

It is clear from our simplified summary of vertebrate immunity that cytokines play a key role in the orchestration of the complex immune response – both in terms of activation and regulation (Fig. 2.3). For example, depending upon the particular cytokine released, activated cytotoxic T-cells can down-regulate the response of the helper T-cells involved in the antibody response. Similarly, helper T-cells can down-regulate the proliferation of the cytotoxic T-cells. Various cytokines also mediate the effectiveness of regulatory T-cells, known as T-regs (previously known as suppressor T-cells), which play a key role in the down-regulation of immune responses by other cells, although the molecular mechanisms are poorly understood (review in Brusco *et al.*, 2008). Research regarding the diversity and function of T-regs has expanded since approximately 2000, primarily due to their role in preventing human autoimmune diseases and in cancer therapies. Macrophages are also capable of immune suppression. The requirement to regulate and control immune responsiveness may explain the counter-intuitive observation that in some host–parasite interactions, natural selection has favored genes that down-regulate the strength of some arms of the host immune response (Mitchison & Oliveira, 1986).

2.2.4 Examples of the vertebrate immune response

We conclude our coverage of the vertebrate immune response with two examples that demonstrate its integrative nature and its linkages to immunopathology. We select two characteristic parasites of humans, a microparasite and macroparasite, to illustrate the varied and species-specific nature of the host response.

2.2.4.1 Human trypanosomes

The trypanosomes are flagellated protists that infect many organisms, including all vertebrate classes, as well as insects and plants. The human trypanosomes can cause sleeping sickness in Africa (*T. brucei*) and Chagas' disease (*T. cruzi*) in South America. Infective stages of the human 'tryps' are transmitted by insect vectors. The life cycles and biology of trypanosomes are covered in detail in Chapter 3.

Chagas' disease caused by *T. cruzi* (Fig. 2.7) advances through distinct incubation, development, and chronic phases that last approximately 2 weeks, 4 weeks, and years, respectively. In immune-compromised hosts, the acute development phase is often fatal. This suggests that there is an effective, but incomplete immunity. Studies in mice have shown that shortly after exposure, the trypanosomes enter various host cells, especially circulating macrophages, where they undergo rapid asexual reproduction. Some of these break out of their natal cell and enter the bloodstream, where almost all are attacked by antibody and complement. Those that survive have the ability to cleave immunoglobulins or produce chemicals that inhibit the action of complement. The parasites also have the ability to inhibit the production of IL-2 from

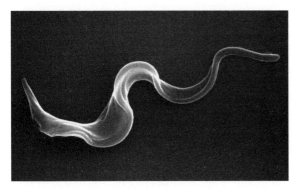

Fig. 2.7 Scanning electron micrograph of the blood-dwelling trypomastigote of *Trypanosoma cruzi*, the causative agent of Chagas' disease in humans. This parasite undergoes extensive asexual reproduction within host cells. Profound immunosuppression and immunopathology can lead to serious cardiovascular disease. (Micrograph courtesy of Cheryl Davis and John Andersland.)

Th-1 cells and remarkably, can down-regulate the expression of IL-2 receptors (Fig. 2.3). This leads to the general state of immunosuppression that characterizes *T. cruzi*-infected hosts. A further characteristic of infected host cells, including muscle and nerve cells, is the presence of 'tryp' antigens on their surface. This allows the host to recognize and kill infected host cells by Th-1-mediated pathways. In the context of immunopathology, one problem is that parasite antigen is also expressed on uninfected host cells, which are then attacked by the host. Further, these antigens often lead to the formation of antibodies to the host's own proteins, a form of autoimmunity that can damage heart tissues, leading to one of the pathological outcomes of Chagas' disease known as cardiomegaly (see Chapter 3). Overall, this example highlights a classical immunological trade-off such that many parasites are killed by effective host immunity (probably permitting host survival during infection of this rapidly reproducing microparasite), but at a cost of damage to uninfected host cells.

African trypanosomiasis is more complex in the context of host response and immunopathology. This group of trypanosomes, unlike *T. cruzi* and many other parasitic protists, have no intracellular stages, and circulate within the circulation system. Superficially, we might expect that the sophisticated vertebrate immune response could easily detect tryps in the blood and then respond via antibodies and complement. Indeed, following the inoculation of infective stages by the vector, parasitemia rapidly increases, stimulating acute responses by both arms of acquired immunity. There is rapid release of IL-1 and TNF (tumor necrosis factor) associated with Th-1 responses and also a B-cell response leading to the production of anti-tryp antibodies. This integrated onslaught is remarkably effective, leading to rapid reduction in parasitemia. But some individuals persist in the circulation system, albeit initially in very low numbers. The anti-trypanosome antibody is known to both lyse the trypanosome in the presence of complement, as well as enhance phagocytosis of the trypanosomes. The antigen–antibody complexes are then trapped in various tissue sites, including the kidneys. These complexes are known to activate complement, which attracts phagocytic cells to the site of deposition. This accumulation of phagocytes ultimately leads to a decrease in pH and oxygen tension at the site, and the release of a variety of enzymes that can cause host tissue destruction. In addition, the antigen–antibody complex leads to changes in the vascular bed that result in edema and an immediate hypersensitivity response. In African trypanosomiasis, the infection persists, and there is continued trapping of antigen–antibody complexes in the kidneys and other tissues. A persistent response can lead to severe tissue damage. Thus, the African trypanosomes provide another example of an immunological trade-off involving partial protection via immune-mediated parasite death, but at the cost of potentially severe damage to uninfected host tissues.

2.2.4.2 Human schistosomes

These 'blood flukes' are trematodes of a variety of birds and mammals. We discuss the biology and life cycles of important human schistosomes in Chapter 6. Infection is via free-swimming cercariae that are released from snails. Penetration of the skin by schistosome cercariae is followed by differentiation into migrating schistosomula, followed by metamorphosis into adults that settle into species-specific locations in the circulation system. There, a female worm, seemingly unencumbered by host immunity, releases 100s to 1000s of eggs per day for the rest her long life – up to 30 years in the case of *Schistosoma mansoni*. Although adult worms provoke strong host responses, these tend only to be effective against larvae from subsequent exposure events, not to the initial population of worms. This phenomenon is known as **concomitant immunity.**

Eggs released from female worms also provoke a strong response, and it is here where immunopathology arises (see Chapter 6, and Box 6.2 for full discussion). The host responds immunologically to the eggs that are often trapped in the liver, and to soluble egg antigens (SEA) released from the developing larva inside the egg. Within weeks of exposure to cercariae, the host produces both a B-cell and T-cell response to SEA. When an egg is trapped in the liver, there is active migration of T- and B-cells, macrophages, and eosinophils to the site of egg deposition. Although both antibody and B-cells can be observed in areas surrounding the egg, the most prominent response involves T-cells in a hypersensitivity reaction. The continued influx of T-cells, macrophages, and their associated cytokines leads to a granuloma that envelops the egg (Fig. 2.8). Since the egg often remains viable in the presence of host defenses, the granuloma continues to enlarge. If the egg is retained within the granuloma for an extended period of time, the microenvironment in the granuloma becomes fatally toxic

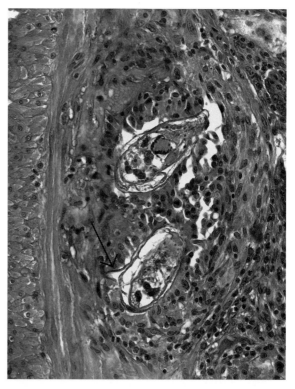

Fig. 2.8 Histological section of the intestine from a mouse infected with *Schistosoma mansoni*, the causative agent of schistosomiasis. Two eggs of the parasite are surrounded by granulomas, consisting of a complex of immune cells (e.g., lymphocytes, macrophages) due to a severe immunopathological response. Note the lateral spine (arrow), an adaptation that is thought to trap the eggs in capillaries, but also leading to blockage of blood flow and enlargement of the spleen and liver. (Photograph courtesy of John Sullivan.)

and the larvae within die. Since female worms produce eggs over a period of years, new granulomas are continually being formed. Therefore, as eggs are eventually walled off and killed, there is wound healing, with collagen deposition and egg calcification. Eventually, the granulomas, with scarring and calcified areas, join together, producing severe tissue damage and hepatic hypertension, which is a key component in the overall granulomatous phenomenon affecting liver function and morbidity (Kojima, 1998). Therefore, immunity to the schistosomes likely saves the host from complications associated with massive numbers of eggs lodging in critical host tissues such as the brain and heart. However, as with the trypanosomes, the cost of effective immunity is potentially severe immunopathology.

2.2.5 Parasite evasion mechanisms

Whereas humans and other animals have evolved complex and highly effective defense mechanisms, it is clear that parasites have evolved adaptations to avoid them. Indeed, natural selection must favor a strategy for all parasites to survive to reproduce within their natural hosts. This is a profound viewpoint when we consider that all parasites must be recognized as foreign and thus subjected to sophisticated host immune responses, especially those in vertebrates. Secondly, some parasites live as long as their hosts, indicating that their evasion mechanisms must be long-lasting. In this section, we describe some broad examples of immune-evasion strategies for protist and helminth parasites of vertebrates. More complete coverage is provided in Behnke *et al.* (1992) and Damian (1997). Coverage from an ecological and evolutionary viewpoint is provided in Schmid-Hempel (2008).

One strategy is extremely rapid asexual replication. Natural selection will favor an approach whereby replication proceeds faster than clonal reproduction of B- and T-cells, so long as the parasite has time to transmit infective stages to new hosts. Two common intestinal microparasites of humans and other vertebrates, *Giardia* spp. (Fig. 2.9) and *Cryptosporidium* spp., adopt a strategy whereby infection via a single ingested cyst

Fig. 2.9 Scanning electron micrograph of trophozoites of the flagellate *Giardia muris* in the small intestine of a mouse. This protist evades host immunity by a form of antigenic variation. Massive asexual reproduction within the intestine can cause giardiasis, characterized by nutrient malabsorption and diarrhea. (Micrograph courtesy of Břetislav Koudela.)

leads to rapid increases in clonal population size within a few hours of exposure (see Chapter 3 for detailed description). After approximately 2 weeks, hosts with normal immune systems respond vigorously with Th-1 and Th-2 responses to clear the infection. Most infected hosts are then protected upon re-exposure. This strategy would not be effective for those parasites that require a period of time to develop within their hosts.

Another strategy is to avoid immune detection by occupying sites that T-cells, B-cells, macrophages, and other components of host immunity have difficulty accessing. Many parasitic protists are highly site-specific within host cells. Some of these, such as *T. cruzi* that we discussed in the previous section, are capable of surviving within the very cells of the host that are integral to host immunity. As we describe in the next chapter, other parasitic protists invade liver cells, epithelial cells, or red blood cells (RBCs) where they are protected from some key components of the host's immune system. Malaria-causing *Plasmodium* spp. inhabit hepatocytes and RBCs, and, so, are sequestered from the immune system, and are also protected from direct attack by antibodies and cytotoxic T-cells.

Macroparasites can also reduce detection and attack when they locate in immunologically privileged sites.

Sites such as muscle and nervous tissue, and to some extent the gut, can be accessed by some components of vertebrate host immunity, but not to the same extent as those that are in direct contact with the host's vascular system. Many trematode metacercariae (see Chapter 6) of fish, amphibians, and insects encyst within regions such as the eyes, brain, and muscles that are considered immunologically privileged. Within these sites, they often develop within a double-walled cyst in which the outermost component comprises modified host cells, so are immunologically invisible. Many species of trematode that form these kinds of cysts live for the length of their intermediate hosts' life, apparently in the absence of strong host immunity. Encysted cysticerci of the pork tapeworm, *Taenia solium* (see Chapter 6), are enveloped by a collagenous cyst wall of host origin. But in this case, the wall is porous, allowing the developing worm to transport host material, apparently including host antibodies, to support its own growth. In exchange, the larvae release a cocktail of molecules that block normal immune function (White *et al.*, 1997).

The adult schistosomes that we covered in the previous section (see also Chapter 6) apparently mask themselves by mimicking key host macromolecules, or coating their own surface antigens with host-derived material. Damian (1997) reviews the fascinating, but controversial, concept of **molecular mimicry** as it applies to parasite evasion of host immunity. Following penetration of cercariae, the metamorphosed schisto-somula absorb a suite of host-derived glycoproteins and glycolipids, including those from RBCs and MHC molecules, onto their outer surface. These migrating forms, effectively disguised as host, end up in blood vessels associated with the bladder or intestine, where they mature into long-lived adults. The mechanisms they use to select specific host components and how they are bound to the outer tegument are unknown. An extension to molecular disguise comes from the observation that the outer surface of the adult worms also contains an enzyme that destroys host complement molecules. It turns out that it is the precise enzyme that the host uses to protect its own cells from attack by complement. It is not clear whether the worms absorb

the enzyme from the host, or if they have the genetic capability of producing their own.

Another parasite escape mechanism, again made famous by the African trypanosomes, is to undergo a series of temporal changes in surface antigens (review in Oladiran & Belosevic, 2012). Recall that these tryp-anosomes are intercellular within the circulation system, so should be easily detected by host immunity. Indeed, the parasites are highly immunogenic, inducing an antibody response that clumps the parasites together so that complement-mediated lyses can occur. Their strategy of evasion is to modify their entire surface coating, comprising variant surface glycoprotein (VSG), in a process called **antigenic variation** (Fig. 2.10). The VSGs provide a surface coat that envelopes the cell, protecting the invariant antigens on the surface from immune detection while also stimulating an antibody-mediated response to themselves (Oladiran & Belosevic, 2012). The VSGs also protect the cell from lysis via complement. In the case of *T. brucei*, most infective stages injected by the tsetse vector have an identical VSG, but a few are antigenically different. As we discussed previously, humans can mount a highly effective antibody-mediated response. The response kills almost all individuals that

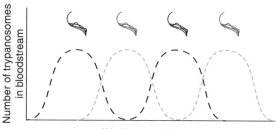

Fig. 2.10 The cyclic appearance of African trypanosomes in the human bloodstream due to antigenic variation. Each population is a clone and expresses a unique variant surface glycoprotein (VSG) coat. Host antibodies destroy most of the trypanosomes; however, a few survive and change their VSG coat and reproduce asexually until antibodies again recognize this new surface antigen. The cycle is repeated many times in chronic African trypanosomiasis. (Figure courtesy of Danielle Morrison; modified from Donelson & Turner, 1985, *Scientific American*, **252**, 44–51.)

are covered with the dominant VSG, leaving the survivors with an alternative VSG to reproduce, become dominant, and so on in a potentially limitless cycle. The wax and wane of parasitemia in blood samples of humans is the hallmark of the African trypanosomes, whereby successive peaks (7–10 days apart) are characterized by different VSGs (Fig. 2.10). In the case of *T. brucei*, over 1000 genes code for different surface antigens, only one of which is expressed by an individual at any one time. *Plasmodium* spp. and *Giardia* spp. are also known to avoid the host's immune response by antigenic variation, although the mechanisms differ (see Chapter 3).

Probably the most widespread escape mechanism involves some form of biochemical suppression of host immunity. Examples can be drawn from the literature where most of the pathways depicted in Fig. 2.3 are manipulated by parasites, either through the secretion of specific molecular blockers, or secretion of 'smoke-screen' molecules that interfere with typical cell-to-cell signaling. Two of the most common involve the induction of T-cells to produce inappropriate signals, and interference with the production of various cytokines, or their receptors. Effective immunity to *Leishmania* within macrophages is mediated by IFN-γ-activated macrophages as part of a specific Th-1 response. But during infection, the production of IFN-γ is suppressed by interleukins associated with a Th-2 response. Parasite-induced disturbance in the Th-1/Th-2 balance is seen in several other parasites, including some macroparasites such as *Schistosoma*. Similarly, *T. cruzi* and *Plasmodium* spp. inhibit the production of Th-1-mediated IL-1 and IL-2, sharply reducing the effectiveness of Th-1 responses. *Trypanosoma cruzi* extends the idea of immune evasion by producing a protein (T-lymphocyte triggering factor) that activates CD-8 cells to secrete an interferon that actually stimulates trypanosome growth (Donelson *et al.*, 1998). Finally, parasite interference with typical Th-2 responses often leads to the production of non-functional antibodies that either lack specificity or target incorrect epitopes.

In summary, parasites can survive in immunologically competent hosts by avoiding recognition,

changing their antigenic composition, or by suppressing the host's immune system in myriad ways. Many parasites likely combine evasion mechanisms. Thus, while the African trypanosomes undergo antigenic variation, they also immunosuppress their host and, in the appropriate host, can adsorb host proteins onto their surface presumably to avoid immune recognition. *Plasmodium* spp. within RBCs also undergo antigenic change, as well as cause immunosuppression. Finally, adult schistosomes avoid the immune response by masking and mimicry, or they may undergo antigenic surface changes during parasite maturation from the schistosomula stage to the adult worms, or even suppress the host's immune response.

2.3 Invertebrate immunity

Invertebrates have evolved sophisticated innate immune defense mechanisms to distinguish self from non-self and to recognize and kill foreign organisms rapidly and efficiently. Their key strategies are similar among most invertebrate groups, suggesting they have a common origin and have been conserved over millions of years (Schulenburg *et al.*, 2004). Although invertebrates lack the classic antibody–antigen system of vertebrates, there is evidence that a form of immunological memory has evolved, at least in some groups (e.g., insects; review in Schmid-Hempel, 2005). Over the last decade, the field of comparative invertebrate immunology has expanded considerably due largely to the importance of certain invertebrates as model research systems, and to advances in biochemistry, immunology, and molecular biology, as well as proteomics and genomics tools. Although many mysteries remain, the complexities of immunity in many groups, especially the cnidarians, molluscs, nematodes, annelids, and several groups of arthropods are becoming much better known (reviews in Schulenburg *et al.*, 2004; Loker *et al.*, 2004; Söderhäll, 2010). Given the importance of freshwater snails as intermediate hosts for the medically important schistosomes, and the significance of insects as vectors of human and veterinary diseases, studies involving these two taxa have been especially prominent.

2.3.1 Immunobiology of insect–parasite interactions

The insect immune system comprises both humoral and cell-mediated mechanisms (review in Beckage, 2008). Inducible humoral immune factors include a variety of glycoproteins that are important in surveying hemolymph for foreign material. These include antibacterial/ antifungal proteins and lysozymes. For example, lytic peptides called cecropins that disrupt the cell membranes of microbes have been isolated from moths. These proteins appear in the hemolymph after a primary infection. Intriguingly, insects sublethally infected with bacteria can rapidly develop increased resistance to subsequent infection, a form of immunological memory (review in Schmid-Hempel, 2005). This protection and increased immune activity can be long-lasting.

Non-inducible humoral factors include recognition molecules like lectins and, especially, phenoloxidases. The so-called pro-phenoloxidase cascade system is involved in a key defense mechanism in insects known as melanization. The phenoloxidase system, once activated by parasite antigens, causes a complex cascade of chemical events. One biochemical pathway leads to the production of quinones and then the dark brown pigment, melanin, which is deposited onto a parasite. Melanization, in cooperation with circulating hemocytes in the hemolymph can lead to encapsulation of a foreign organism, leading to its eventual destruction.

A critical aspect of the immune response of insects to parasites is via cell-mediated immunity involving an array of highly differentiated, morphologically diverse, and multifunctional hemocytes. Three primary types, plasmatocytes, granulocytes, and lamellocytes, are responsible for the three main processes in insects: phagocytosis, nodulation, and encapsulation. Following recognition of non-self, microbes in the hemolymph are engulfed by circulating hemocytes, especially the plasmatocytes and granulocytes. The activity of insect lectins 'marks' an invading foreign organism for destruction by phagocytosing

hemocytes. Phagocytosis often leads to the aggregation of entrapped materials into nodules that become attached to various host organs and tissues; these are then surrounded by plasmatocytes and destroyed.

When parasites are too large for either phagocytosis or nodulation, the complex processes leading to encapsulation by several layers of lamellocytes may occur (Romoser & Stoffolano, 1998). Upon recognition of a parasite (via specific protein recognition receptors) by a circulating granulocyte, the steps of encapsulation are initiated. The granulocyte degranulates and material sticks to the parasite, followed by additional granulocyte attachment. A hemocytic recognition factor (HRF) is released that recruits plasmatocytes and lamellocytes to attach to the parasite and HRF complex. These flatten and spread out over the parasite surface. Lamellocytes increase the number of layers around the parasite until it is no longer recognized as non-self.

A variety of genes and pathways regulate the arsenal of molecules and cells involved in insect immune defense. Insect model systems have revealed at least two important signaling pathways – the so-called Toll and Imd pathways (review in Lemaitre & Hoffmann, 2007). Both pathways can be stimulated by parasites in insects (e.g., trypanosomes) to produce potent anti-parasite peptides. Identification of immune-related genes and the complexities associated with cell-cell signaling are active research areas. For example, Schlüns et al. (2010) demonstrated that a putative pathogen-recognition gene, hemomucin, was significantly up-regulated in trypanosome-infected bumblebees. Likewise, Waterhouse et al. (2007), for example, took advantage of knowing the entire genome of the falciparum malaria-transmitting mosquito Anopheles gambiae to characterize immune-related genes and immune-signaling pathways.

Braconid and ichneumonid wasp endoparasitoid–insect systems (see Chapter 11) have taken center stage in addressing the question of parasite evasion of host immunity. One particular model system, the braconid wasp Cotesia congregata and its tobacco hornworm caterpillar host Manduca sexta, has received considerable attention. Eggs and larvae of compatible parasitoids must be able to evade the immune system of the caterpillar host in order for them to develop inside the hemocoel prior to pupation and metamorphosis into wasps. One key to their success is the induction of profound immunosuppression in their caterpillar host. It turns out that host immunosuppression is mediated by mutualistic virus particles called **polydnaviruses** (PDVs) (reviews in Lavine & Beckage, 1995; Beckage, 1998; Whitfield & Asgari, 2003). Brachoviruses and ichnoviruses are known from a variety of braconid and ichneumonid wasps. These mutualists replicate in ovarian cells within the female wasp and are injected along with the eggs and venom into the caterpillar. The combination of ovarian proteins, venom, and the PDVs act synergistically to 'knock out' virtually all aspects of the insect immune system we just discussed. As a consequence, the eggs of compatible parasitoids are not recognized and can develop in the hemocoel undetected by the immune system. For example, there are dramatic alterations in humoral responses, especially a marked reduction in phenoloxidase activity and an inhibition of melanization. There is also an inhibition of cellular-mediated immune responses, most notably a suppression of nodulation and a complete inhibition of the caterpillar's encapsulation response. The absence of encapsulation is accompanied by alterations in hemocyte structure and function (e.g., reduced adherence and spreading), and a decrease in the number of circulating hemocytes (Lavine & Beckage, 1995; Beckage, 1998). These mutualistic viruses and their immunosuppressive effects have likely contributed to the radiations of insect endoparasitoids, most notably the braconid and ichneumonid wasps. Whitfield & Asgari (2003) present a review of the possible origins of parasitoid viruses and their phylogenetics and coevolution with their hosts.

2.3.2 Immunobiology of snail–trematode interactions

Molluscs also possess an effective and complex innate immune system, comprising both humoral and cellular components. Much of our understanding of snail immunity comes from decades of extensive research involving the *Schistosoma mansoni–Biomphalaria glabrata* model system (reviews in Bayne, 2009; Loker, 2010; Yoshino & Coustau, 2011). Approximately 18 000 species of digenean trematodes use mostly specific snails as their obligatory first intermediate hosts. The details of intramolluscan development are presented in Chapter 6. Briefly, when a miracidium larva of *S. mansoni* penetrates a susceptible strain of *B. glabrata*, it transforms into a mother sporocyst, which then undergoes asexual replication, proliferating extensively throughout the snail, and eventually producing cercariae that leave the snail in water to penetrate human or rat definitive hosts. How, then, do resistant strains of *B. glabrata* recognize and then kill *S. mansoni* miracidia/sporocysts to stop further development of the parasite?

Discrimination of self from non-self in snails is mediated via a receptor-mediated process by circulating hemocytes, which are formed in an amoebocyte-producing organ. Like the insects, *Biomphalaria* hemocytes are structurally and functionally diverse (e.g., granulocytes and hyalinocytes). It is the coordinated interactions of hemocytes and complex signaling pathways that are responsible for phagocytosis and encapsulation of foreign organisms. Hemocyte-mediated encapsulation, leading to formation of a multilayered cellular capsule, and cytotoxic responses are the primary immune defenses to larval trematodes (review in Yoshino & Coustau, 2011). Hemocytic encapsulation of the mother sporocyst of *S. mansoni*-resistant *B. glabrata* results in complete destruction of the parasite's outer tegument and eventual sporocyst death (Fig. 2.11).

How do schistosome sporocysts avoid immune detection from the snail's immune system? There is evidence that the larval schistosomes (both miracidia and mother sporocysts) use molecular mimicry (Yoshino & Bayne, 1983). Subsequent studies have provided further support for molecular mimicry, specifically the potential importance of carbohydrate moieties as shared epitopes in the *B. glabrata–S. mansoni* system (review in Yoshino & Coustau, 2011). Another possible immunoevasion mechanism has

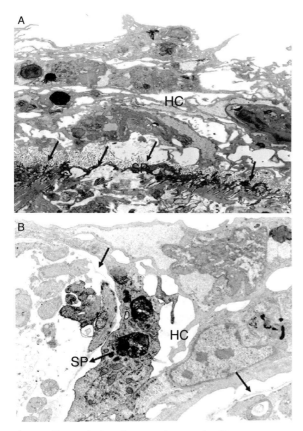

Fig. 2.11 Transmission electron micrograph of hemocyte encapsulation of *Schistosoma mansoni* sporocysts (SP) by circulating hemocytes (HC) from a schistosome-susceptible (A) and schistosome-resistant (B) strain of the snail *Biomphalaria glabrata*. After 24 hours of *in vitro* hemocyte–sporocyst cocultivation, note the presence of the intact trematode's outer surface (tegument) (arrows) at the interface with susceptible snail hemocytes (A); compared to complete destruction of the tegument and internal damage associated with sporocysts during snail encapsulation reactions (B). The arrows in (B) point to the location of the basal lamina, normally located just proximal to the sporocyst's tegument, which is missing. (Micrograph courtesy of Tim Yoshino.)

been recently described by comparing the proteomes of *S. mansoni*-compatible and -incompatible strains of *B. glabrata*. The proteomic analysis revealed a high degree of molecular polymorphism of mucin glycoproteins (Roger *et al.*, 2008). These proteins are only expressed in the larval stages that interact with the

snail host and are released during transformation of the miracidia. Thus, it has been proposed that they may bind snail-recognition receptors, serving as an immunological smoke-screen functioning to overwhelm the snail immune system (Roger *et al.*, 2008; Yoshino & Coustau, 2011).

Larval trematodes can also influence the host's cellular response by modulating or interfering with hemocyte activity, and causing immunosuppression. For example, in studies of echinostomes (e.g., *Echinostoma paraensei*) and *S. mansoni* infecting *B. glabrata* it has been shown that infected snails lose their phagocytic capabilities and their ability to adhere to surfaces. In addition, hemocytes from infected snails exhibit diminished ability to encapsulate trematode larvae (reviews in Bayne, 2009; Loker, 2010). The molecular mechanisms for interference are largely unknown; however, it is likely that digeneans produce factors that alter the activity of hemocyte signaling pathways (reviews in Loker, 2010; Yoshino & Coustau, 2011). Humphries & Yoshino (2003) review the receptors, signaling molecules, and pathways involved in hemocyte immune responses in snails.

The complete *S. mansoni* genome has now been sequenced and the sequencing of the large *B. glabrata* genome has been initiated. These genomic data will be crucial to future research efforts to fully understand the complex interactions between *B. glabrata* and *S. mansoni* at the molecular level (Yoshino & Coustau, 2011).

2.4 Ecological immunology

At first glance, a marriage between immunology and ecology seems fanciful. Traditional immunologists are concerned with characterizing host defenses and underlying mechanisms from a medical or veterinary perspective, whereas traditional ecologists and evolutionary biologists tend to have entirely different motivations. Yet, commencing in the early 1990s, distinctions between the two disciplines began to fade. An initial impetus was a seminal paper that was among the first to consider the adaptive nature of the host immune response (Behnke *et al.*, 1992). Using chronic nematode infections of laboratory mice as models, the authors developed the notion of 'optimal immunity' by evaluating the costs and benefits of host defense within an evolutionary framework. At about the same time, researchers were beginning to focus on understanding immune responses in invertebrates. This was an important development because these systems were amenable to experimental tests of linkages between 'whole organism immunity' and a hosts' lifetime reproduction. Lastly, in the mid 1990s, evolutionary biologists began to view traits associated with the immune system in the same context as other life history traits, sparking studies (especially involving birds and insects) aimed at understanding the fitness costs and benefits of particular defense strategies (Sheldon & Verhulst, 1996). In the final section of this chapter, we review the key underlying principles of the new subdiscipline of ecological immunology.

2.4.1 Costs of host immunity

It should be clear from our coverage in this chapter that one of the fundamental costs of immunological defense comes in the form of immunopathology. We emphasize this fundamental component of host–parasite interactions in subsequent chapters. Thus, much of the damage to individual hosts that is ascribed to parasites stems from misdirected immune responses, or the inability of the host to regulate immune effectors (Graham *et al.*, 2005). The autoimmune diseases of humans provide familiar examples of misdirected immunity. Likewise, the complex pathological syndromes that we described for humans infected with *T. cruzi*, and with blood flukes (*Schistosoma* spp.) are due, in part, to strong immunopathology within specific tissues. Immunopathology undoubtedly occurs in wildlife–parasite interactions as well, although it is much less studied. The destruction of uninfected RBCs in waterfowl infected with the blood parasite *Leucocytozoon* (Chapter 3) is one example (Wobeser, 1981). Larvae of a shark cestode stimulate a severe inflammation response, causing significant mortality, in introduced striped bass (Sakanari & Moser, 1990).

Activation of the phenoloxidase cascade in flour beetles is associated with host tissue damage (Sadd & Siva-Jothy, 2006). Thus, evidence supports the notion that immunopathological outcomes are a regular phenomenon of host–parasite interactions.

There also exists abundant evidence, drawn from both vertebrate and invertebrate interactions, that the maintenance and expression of the complex host immune system is associated with energetic costs. In humans, the classical fever that follows exposure to *Plasmodium*-causing malaria (see Chapter 3) is associated with a 40% increase in resting metabolism (Hall, 1985). Further evidence for metabolic costs of this magnitude come from studies involving laboratory animals and domestic stock exposed to various agents that are known to up-regulate host immunity (review in Lochmiller & Deerenberg, 2000). The administration of IL-1 to rats and typhoid vaccine to humans caused an 18% and 16% increase, respectively, in resting metabolic rates. In a review of studies involving birds and mammals injected with proteins that up-regulate various arms of immunity, increases in metabolism relative to controls varied from zero to about 30% (Schmid-Hempel, 2011). Lochmiller and Deerenberg (2000) consider that these substantial metabolic costs are due to increased protein synthesis associated with immune up-regulation and to the generation of large quantities of glucose needed to fuel an immune response.

The energetic costs of maintaining and mounting an immune response, together with the costs due to immunopathology, will constrain evolutionary responses for ever-increasing immunity (review in Goater & Holmes, 1997; see Chapter 16). Thus, given the ecological costs of immunity, evolutionary ecologists predict that traits associated with immunity will be traded off with other traits that are directly tied to host fitness. One way to test this prediction is to test for negative genetic correlations between immune traits and traits associated with features such as growth and reproduction (review in Schmid-Hempel, 2011). In an early example, Brindley & Dobson (1981) used an artificial selection experiment to selectively breed mice that could respond immunologically (high-responders) to one of their common nematodes. They compared the reproductive rates of high-responders to mice from a selected line of low-responders. Mice from the former group demonstrated enhanced immunity, as expected, but they also produced fewer litters of smaller young. Remarkably, the distinction between the two groups occurred within a single generation of artificial breeding. Negative correlations between immune effectiveness and host reproduction have been demonstrated for snails infected with larval trematodes and insects exposed to nematode and protist parasites. The consistency of these results, together with those that demonstrate negative correlations between immune traits and indirect indicators of host fitness (review in Schmid-Hempel, 2011), is striking. We re-visit the ecological and evolutionary significance of defense costs, immunological and otherwise, in our coverage of parasite-mediated natural selection and host–parasite coevolution in Chapter 16.

2.4.2 Context-dependent immunity

Parasitologists and immunologists have long-recognized that the expression of immunity can be strongly influenced by environmental factors, especially those that impact a host's overall condition. One paradigm in human epidemiology holds that the health burden of parasites in developing countries can be attributed to complex interactions among rates of exposure, malnutrition, and reduced immunity (review in Crompton, 1993). In such cases, sophisticated and effective immune responses must be under strong selection, but their full expression is constrained by limited resources. This general observation has been confirmed by numerous experimental tests on the role of various stressors, especially the quality and quantity of nutrients, on various arms of host immunity. In one clear example, Tu et al., (2008) showed that laboratory mice fed a normal diet developed a Th-2-mediated response that rapidly expelled an intestinal nematode. However, mice fed a nutritionally deficient diet up-regulated their expression of IFN-γ, leading to a reduction in Th-2 response that prolonged the life span of the worms. The impact of the nutrient deficiency was reversed when the host's nutrient status was restored. Here, the combined effect of infection and protein deficiency led to a reduction in host immune performance. These results indicate that the expression of host immunity was highly dependent upon ecological context – in this case, determined by the availability of nutrients. The context-dependent nature of host immunity is another core principle of ecological immunity.

Studies involving wild vertebrates and invertebrates have provided similar results (reviews in Schmid-Hempel, 2008; Demas & Nelson, 2012). Thus, experimental studies on insects, snails, and birds have identified an incredible range of factors that impact the expression of host immunity. Flour beetles that were temporarily deprived of food reduced their phenoloxidase activity (Siva-Jothy & Thompson, 2002) and kittiwakes reduced their concentrations of serum immunoglobulins when food was limited (Gasparini et al., 2006). Ecologists have expanded these general approaches to show that stressors associated with season (e.g., hibernation), reproductive status, host density, and habitat quality impact the expression of specific components of host immunity. In a striking example, snails that were repeatedly exposed to the risk of predation reduced their production of hemocytes (Rigby & Jokela, 2000). The reduction in immune capacity led to reduced snail survival and reproduction when the snails were exposed to water that contained parasites. Aquatic stressors such as temperature, pollution, and ultraviolet radiation affect a wide range of immune parameters, especially in fish (Tort, 2011, see Chapter 17). Taken together, these results highlight the context-dependent nature of host immunity in natural host–parasite interactions. The challenge ahead is to evaluate whether variation in specific aspects of host immunity due to ecological context is associated with variation in host fitness (review in Schmid-Hempel, 2011).

2.4.3 Optimal immunity

It is clear from our discussion of vertebrate and invertebrate immunity that hosts have evolved a sophisticated and diverse response to parasites. Further, the magnitude of these responses varies between individuals within a population, even in tightly controlled laboratory models involving congenic strains of hosts. Both laboratory and field-based studies have confirmed that this variation is under the control of specific genes, especially those associated with MHC architecture. Lastly, earlier studies on domestic animals, together with more recent studies on wild vertebrates and invertebrates, have demonstrated that immune competency can be readily altered by artificial selection. Thus, the requirements are present for strong and directed natural selection for host responses to counter parasites that are also sophisticated and diverse – and can do them harm. Perhaps then, we should expect natural selection to favor immune responses that are ever more sophisticated, ultimately leading to responses that are 100% effective (and thus, no or few, parasites!) and little parasite-induced host mortality or pathology. Yet, immunity of this sort (often known as sterilizing immunity) and outcomes of this kind are exceedingly rare in nature (Behnke et al., 1992).

Why is this so? The significant and widespread costs associated with immunopathology will constrain the evolution of sustained responses, as will the energy costs associated with the maintenance and deployment of immune effectors (review in Goater & Holmes, 1997). Further, there is clear evidence from artificial selection experiments that strong immune responders pay the ultimate fitness cost in the form of reduced productivity. It is important to emphasize that at least some of these costs must be paid, even in the absence of the parasites themselves. Thus, the high environmental complexity and variability that characterizes natural host–parasite interactions (see Chapter 12) will also constrain the evolution of sustained immunological responses.

Taken together, 'more is not better' in the context of immune responsiveness. Rather, natural selection will disfavor responses that are too strong, too specific, or too long term (Wakelin & Apanius, 1997). We should expect then that organisms will optimize levels of immunity that balance the benefits of responsiveness with its numerous and substantial costs. The picture is complicated further by the many facets of host immunity that might conceivably be optimized; the type of immune response (innate versus acquired in vertebrates, Th-1 versus Th-2 dominated, behavioral versus immunological), when, and at what strength. One of the general patterns that emerges in nature is one of incomplete control to a wide range of parasites (see also Chapter 16), often accompanied by some immunopathology. Optimization strategies of this sort are likely played-out in natural populations of hosts exposed to their countless numbers and types of parasites, contributing to the enormous between-individual heterogeneity in immune capacity observed in natural host populations (e.g., Abolins et al., 2011). As the new field of ecological immunology shifts into its second decade, understanding the functional significance of this notoriously high variation is likely to be a key area of emphasis.

References

Abolins, S. R., Pocock, M. J. O., Hafalla, J. C. R., et al. (2011) Measures of immune function in wild mice, *Mus musculus. Molecular Ecology*, 20, 881–892.

Bayne, C. J. (2009) Successful parasitism of vector snail *Biomphalaria glabrata* by the human blood fluke trematode *Schistosoma mansoni*: a 2009 assessment. *Molecular Biochemistry and Parasitology*, 165, 8–18.

Beckage, N. E. (1998) Parasitoids and polydnaviruses. *BioScience*, 48, 305–311.

Beckage, N. E. (2008) *Insect Immunology*. Amsterdam: Elsevier.

Behnke, J. M, Barnard, C. J. & Wakelin, D. (1992) Understanding chronic nematode infections: evolutionary considerations, current hypotheses, and the way forward. *International Journal for Parasitology*, 22, 861–907.

Brindley, P. J. & Dobson, C. (1981) Genetic control of liability to infection with *Nematospiroides dubius* in mice: selection of refractory and liable populations of mice. *Parasitology*, **83**, 51–65.

Brusco, T. M., Putnam, A. L. & Bluestone, J. A. (2008) Human regulatory T cells: role in autoimmune disease and therapeutic opportunities. *Immunological Reviews*, **223**, 371–390.

Cox, F. E. G. (1993) Immunology. In *Modern Parasitology*, ed. F. E. G. Cox, pp. 193–218. Oxford: Blackwell Scientific Publications.

Coico, R. & Sunshine, G. (2009) *Immunology: A Short Course*, 6th edition. New York: John Wiley & Sons.

Crompton, D. W. T. (1993) Human nutrition and parasitic infection. *Parasitology*, **107**, 1–203.

Damian, R. T. (1997) Parasite immune evasion and exploitation: reflections and projections. *Parasitology*, **115**, 169–175.

Demas, G. E. & Nelson, R. J. (2012) *Ecoimmunology*. Oxford: Oxford University Press.

Donelson, J. E., Hill, K. L. & El-Sayed, N. M. A. (1998) Multiple mechanisms of immune evasion by African trypanosomes. *Molecular and Biochemical Parasitology*, **9**, 51–66.

Finney, C. A., Taylor, M. D., Wilson, M. S. & Maizels, R. M. (2007) Expansion and activation of CD4+CD25+ regulatory T cells in *Heligmosomoides polygyrus* infection. *European Journal of Immunology*, **37**, 1874–1886.

Gasparini, J., Roulin, A., Gill, V. A., *et al.* (2006) In kittiwakes food availability partially explains the seasonal decline in humoral immunocompetence. *Functional Ecology*, **20**, 457–463.

Goater, C. P. & Holmes, J. C. (1997) Parasite-mediated natural selection. In *Host-Parasite Evolution: General Principles and Avian Models*, ed. D. H. Clayton and J. Moore, pp. 9–29. Oxford: Oxford University Press.

Graham, A. L., Allen, J. E. & Read, A. F. (2005) Evolutionary causes and consequences of immunopathology. *Annual Review of Ecology, Evolution, and Systematics*, **36**, 373–397.

Hall, A. (1985) Nutritional aspects of parasitic infection. *Progress in Food and Nutrition Science*, **9**, 227–256.

Hedrick, P. W. (1994) Evolutionary genetics of the major histocompatibility complex. *The American Naturalist*, **143**, 945–964.

Hoffmann, J. A., Kafatis, F. C., Janeway, Jr., C. A. & Ezekowitz, R. A. B. (1999) Phylogenetic perspectives in innate immunity. *Science*, **284**, 1313–1318.

Humphries, J. E. & Yoshino, T. P. (2003) Cellular receptors and signal transduction in molluscan hemocytes: connections with the innate immune system of vertebrates. *Integrative and Comparative Biology*, **43**, 305–312.

Klein, J., Sato, A. & Nikolaidis, N. (2007) MHC, TSP, and the origin of species: from immunogenetics to evolutionary genetics. *Annual Review of Genetics*, **41**, 281–304.

Kojima, S. (1998) Schistosomes.In *Topley & Wilson's Microbiology and Microbial Infections*, 9th Edition, Volume 5, *Parasitology*, ed. F. E. G. Cox, J. Kreier & D. Wakelin, pp. 479–504. London: Edward Arnold.

Lavine, M. D. & Beckage, N. E. (1995) Polydnaviruses: potent mediators of host immune dysfunction. *Parasitology Today*, **11**, 368–378.

Lemaitre, B. & Hoffmann, J. (2007) The host defense of *Drosophila melanogaster*. *Annual Review of Immunology*, **25**, 697–743.

Lochmiller, R. L. & Deerenberg, C. (2000) Trade-offs in evolutionary immunology: just what is the cost of immunity? *Oikos*, **88**, 87–98.

Locksley, R. M. (2009) Nine lives: plasticity among T helper cell subsets. *Journal of Experimental Medicine*, **206**, 1643–1646.

Loker, E. S. (2010) Gastropod immunobiology. In *Invertebrate Immunity*, ed. K. Söderhäll, pp. 17–43. Austin: Landes Bioscience.

Loker, E. S., Adema, C. M., Zhang, S. M. & Kepler, T. B. (2004) Invertebrate immune systems – not homogenous, not simple, not well understood. *Immunological Reviews*, **198**, 10–24.

Magnadottir, B. (2006) Innate immunity of fish (overview). *Fish and Shellfish Immunology*, **20**, 137–151.

Maizels, R. M. (2009) Exploring the immunology of parasitism – from surface antigens to the hygiene hypothesis. *Parasitology*, **136**, 1549–1564.

Mitchison, N. A. & Oliveira, D. B. G. (1986) Chronic infection as a major force in the evolution of the suppressor T-cell system. *Parasitology Today*, **2**, 312–313.

Mosmann, T. R., Cherwinski, H., Bond, M. W., *et al.* (1986) Two types of murine helper T cell clone. I. Definition according to profiles of lymphokine activities and secreted proteins. *Journal of Immunology*, **136**, 2348–2357.

Müller, W. E. G. & Müller, I. M. (2003) Origin of the metazoan immune system: identification of the molecules and their functions in sponges. *Integrative and Comparative Biology*, **43**, 281–292.

Oladiran, A. & Belosevic, M. (2012) Immune evasion strategies of trypanosomes: a review. *Journal of Parasitology*, **98**, 284–292.

Rigby, M. C. & Jokela, J. (2000) Predator avoidance and immune defence: costs and trade-offs in snails. *Proceedings of the Royal Society London B*, **267**, 171–176.

Roger, E., Mitta, G., Mone, Y., *et al.* (2008) Molecular determinants of compatibility polymorphism in the *Biomphalaria glabrata/Schistosoma mansoni* model: new candidates identified by a global comparative proteomics approach. *Molecular and Biochemical Parasitology*, **157**, 205–216.

Romoser, W. S. & Stoffolano, Jr., J. G. (1998) *The Science of Entomology*, 4th edition. Boston: McGraw-Hill.

Sadd, B. M. & Schmid-Hempel, P. (2009) Principles of ecological immunology. *Evolutionary Applications*, **2**, 113–121.

Sadd, B. M. & Siva-Jothy, M. T. (2006) Self-harm caused by an insect's innate immunity. *Proceedings of the Royal Society B*, **273**, 2571–2574.

Sakanari, J. A. & Moser, M. (1990) Adaptation of an introduced host to an indigenous parasite. *Journal of Parasitology*, **76**, 420–423.

Schlüns, H., Sadd, B. M., Schmid-Hempel, P. & Crozier, R. H. (2010) Infection with the trypanosome *Crithidia bombi* and expression of immune-related genes in the bumblebee *Bombus terrestris*. *Development and Comparative Immunology*, **34**, 705–709.

Schmid-Hempel, P. (2005) Evolutionary ecology of insect immune defenses. *Annual Review of Entomology*, **50**, 529–551.

Schmid-Hempel, P. (2008) Parasite immune evasion: a momentous molecular war. *Trends in Ecology and Evolution*, **23**, 318–326.

Schmid-Hempel, P. (2011) *Evolutionary Parasitology: The Integrated Study of Infections, Immunology, Ecology, and Genetics*. Oxford: Oxford University Press.

Schulenburg, H., Kurz, C. L. & Ewbank, J. J. (2004) Evolution of the innate immune response: the worm perspective. *Immunological Reviews*, **198**, 36–58.

Schulenburg, H., Kurz, J., Moret, Y. & Siva-Jothy, M. T. (2009) Introduction. Ecological Immunology. *Philosophical Transactions of the Royal Society B.*, **364**, 3–14.

Sheldon, B. C. & Verhulst, S. (1996) Ecological immunology: costly defences and trade-offs in evolutionary ecology. *Trends in Ecology and Evolution*, **11**, 317–321.

Siva-Jothy, M. T. & Thompson, J. J. W. (2002) Short-term nutrient deprivation affects immune function. *Physiological Entomology*, **27**, 206–212.

Söderhäll, K. (2010) *Invertebrate Immunity*. Austin: Landes Bioscience.

Steinman, L. (2004) Elaborate interactions between the immune and nervous systems. *Nature Immunology*, **5**, 575–81.

Tort, L. (2011) Stress and immune modulation in fish. *Developmental and Comparative Immunology*, **35**, 1366–1375.

Tu, T., Koski, K. G. & Scott, M. E. (2008). Mechanisms underlying reduced expulsion of a murine nematode infection during protein deficiency. *Parasitology*, **135**, 81–93.

Wakelin, D. (1996) *Immunity to Parasites: How Parasitic Infections are Controlled*, 2nd edition. Cambridge: Cambridge University Press.

Wakelin, D. (1997) Parasites and the immune system: conflict or compromise? *Bioscience*, **47**, 32–40.

Wakelin, D. & Apanius, V. (1997) Immune defence: genetic control. In *Host–Parasite Evolution: General Principles and Avian Models*, ed. D. H. Clayton & J. Moore, pp. 30–58. Oxford: Oxford University Press.

Walport, M. J. (2001) Advances in immunology: complement (First of two parts). *New England Journal of Medicine*, **344**, 1058–1066.

Waterhouse, R. M., Kriventseva, E. V., Meister, S., *et al.* (2007) Evolutionary dynamics of immune-related genes and pathways in disease-vector mosquitoes. *Science*, **316**, 1738–1743.

White, A. C., Robinson, P. & Kuhn, R. (1997) *Taenia solium* cysticercosis: host–parasite interactions and the immune response. *Chemical Immunology*, **66**, 209–230.

Whitfield, J. B. & Asgari, S. (2003) Virus or not? Phylogenetics of polydnaviruses and their wasp carriers. *Journal of Insect Physiology*, **49**, 397–405.

Wilson, M. S., Taylor, M., Balic, A., *et al.* (2005) Suppression of allergy airway inflammation by helminth-induced regulatory T cells. *Journal of Experimental Medicine*, **202**, 1199–1212.

Wobeser, G. A. (1981) *Diseases of Wild Waterfowl*. New York: Plenum Press.

Yazdanbakhsh, M., Kremsner, P. G. & van Ree, R. (2002) Allergy, parasites, and the hygiene hypothesis. *Science*, **296**, 490–494.

Yoshino, T. P. & Bayne, C. J. (1983) Mimicry of snail host antigens by miracidia and primary sporocysts of *Schistosoma mansoni*. *Parasite Immunology*, **5**, 317–328.

Yoshino, T. P. & Coustau, C. (2011) Immunobiology of Biomphalaria-Trematode interactions. In *Biomphalaria Snails and Larval Trematodes*, ed. R. Toledo & B. Fried, pp. 159–189. New York: Springer.

3 Protista: the unicellular eukaryotes

3.1 General considerations

The protists (Greek, *the very first*), also referred to as the protozoans (Greek, *proto*first, *zoa*animals), comprise a spectacular diversity of unicellular, eukaryotic organisms possessing organelles such as a membrane-bound nucleus, mitochondria, chloroplasts, Golgi, etc., found in the metazoan plants and animals. There is considerable evidence that eukaryotic protists evolved by a process of sequential endosymbiosis of prokaryotes (see Box 3.1, as well as Margulis (1981) for a discussion of the theory). The Kingdom Protista was erected almost 150 years ago by the famous German zoologist Ernst Haeckel in an attempt to accommodate this diversity. Today, with considerable ultrastructural, genetic, and biochemical research and the molecular phylogenetic revolution, it is now known that unicellular animals are distributed among all kingdoms. There is no longer a formal taxonomic category called the Protista. However, 'protist' is still widely used as a general term (as is 'protozoan') when referring to this diversity of unicellular eukaryotes, even though neither of these terms implies monophyletic origins.

Within the confines of a single cell membrane (= **plasmolemma**), these organisms have undergone an enormous adaptive radiation. This single cell functions as a complete organism. Protists are not simple; they feed, move, behave, and reproduce, and, thus, can be considered more complex and versatile than our own cells! Complexity arises from the specialization of organelles. Protists have evolved a bewildering array of morphologies, physiologies, behaviors, reproductive strategies, life histories, and nutritional and locomotory modes. In short, the diversity of protist form and function rivals that encountered among all other animals combined. As

stated by Pechenik (2010), "Protozoans absolutely defy tidy categorization."

Over 80 000 protist species have been described. This estimate certainly under-represents the true biodiversity of this group. Evidence from molecular sequencing data indicates that there are likely thousands of undescribed cryptic species of 'avian malaria' alone that remain to be described (Bensch *et al.*, 2004). Such 'hidden' biodiversity is likely characteristic of many other protist groups. Much of this biodiversity comprises the parasitic protists, having evolved independently in virtually all of the currently recognized protist phyla. Of course, some of the extensive biodiversity of the parasitic protists is associated with the most significant diseases of humans, wildlife, and domestic animals. Given their rich biodiversity and the enormous research effort that is devoted to pathogenic species, it is not surprising that the parasitic protists provide superb model systems in ecological parasitology, and in host–parasite coevolution and biogeography (see Chapters 12–17).

A hallmark of the parasitic protists, and a major reason for the acute diseases they may cause, is their capacity to replicate extensively within their hosts. To survive, the host must be able to control the protist's phenomenal reproductive abilities, while at the same time the protist must evolve adaptations to successfully invade, feed, and reproduce within the host, and then evade the host immune response and be transmitted to a new host. As such, the protists have evolved an exquisite array of morphological, reproductive, biochemical, and life history adaptations to parasitize a great diversity of invertebrate and vertebrate animals. Many are intracellular parasites, and have evolved elaborate mechanisms for host cell recognition and penetration. Transmission modes among the parasitic protists are as diverse as the host habitats they occupy; thus, blood-dwelling intracellular and extracellular protists may use blood-feeding vectors for transmission. Transmission in the intestinal protists may be via environmentally resistant cysts or by predator–prey interactions. Those that inhabit the host's reproductive system may be transmitted via host sexual reproduction.

From a phylogenetic perspective, no other group of organisms has changed so dramatically in recent years. The evolutionary relationships among the protist taxa remain controversial and protist taxonomy constantly changes. Older taxonomic schemes based on ultrastructure have been largely abandoned since they did not accurately reflect evolutionary relationships. More recently, the tools of molecular biology have allowed protist systematists to draw further evolutionary and phylogenetic inferences based on nucleotide-sequence homologies. Most often, these are based on the sequence homologies of small ribosomal RNA. A second procedure is to compare the sequence homology for genes coding for specific enzymes or structural proteins. The taxonomic scheme we adopt is a simplified one (see Box 3.7), based largely on Hausmann & Hülsmann (1996), Leander & Keeling (2003, 2004), and Adl *et al.* (2005).

3.2 Form and function

Given the diversity of nutritional and metabolic modes, it is impossible to generalize protist structure and function. We do so here only to introduce some key features that most protists share. Many protozoans use a gliding type of locomotion, **pseudopodia**, or external **flagella**, or **cilia** for movement. Their cell surfaces may be a naked cell membrane, may have numerous flagella, or may be covered with cilia. In addition, during their life cycles, they may form **cysts** with thick cell walls, containing stages with reduced metabolic activity. These cyst stages are usually released into the environment, and must survive outside until acquired by a new host. Protists possess the typical eukaryotic cell membrane. There is a complex infrastructure of microtubules, microfilaments, and other organelles associated with the cell membrane,

especially in flagellated and ciliated forms. The cell membrane and associated infrastructure is often referred to as the **pellicle**.

Within the cytoplasm, there is a variety of eukaryotic organelles, including mitochondria, Golgi, etc. The number, type, and even position of these organelles vary between the different taxonomic groups, and many appear to be associated with the parasite's habitat. For example, the parasitic amoeba *Entamoeba histolytica*, which lives in the anaerobic environment of the intestine, does not possess mitochondria. In contrast, the flagellated African trypanosomes, in their insect vector, have a completely functional mitochondrion. However, in the mammalian host, these trypanosomes possess a non-functional mitochondrion, one that lacks cristae and cytochromes. Metabolically, therefore, insect trypanosomes are aerobic, while those in the bloodstream of the vertebrate host are anaerobic. Although parasitic protists have a nucleus with a nuclear membrane, the presence or absence of a true mitotic apparatus and the presence of condensed chromosomes during division varies among the different phyla. Also, within some protist phyla are groups with unique cellular structures. For example, the ciliates have a subpellicle organization of microfilaments and microtubules that are associated with the cilia and are involved in the coordination of their movement; some flagellates lack mitochondria and, instead, possess **hydrogenosomes**, organelles that produce hydrogen as a metabolic by-product.

3.2.1 Nutrient uptake and metabolism

Protists obtain their nutrients by a variety of mechanisms. The most direct method is by the active transport of small molecules across the cell membrane. Some employ phagocytosis of bacteria, RBCs, other host cells, or debris that can be digested internally as nutrients. Another mechanism used in some groups is **pinocytosis**. For example, the trypanosomes transport large-molecular-weight compounds, such as transferin, from the host into the flagellar pocket where the specific receptor for transferin is located. These molecules are then ingested by pinocytosis into the cell cytoplasm where they are digested. The flagellar pocket is thus a distinct organelle with a specific function. Finally, protists such as the ciliates have a **cytostome** (or mouth) into which food particles are swept by the action of cilia.

Some protozoans have a specialized osmoregulatory organelle for maintenance of intracellular water and ion concentrations. The **contractile vacuole** of ciliates, for example, is a rhythmically pulsating vesicle that opens to the outside through a small pore in the cell membrane. The infrastructure of the contractile vacuole insures that water and wastes flow out from the vacuole into the external environment.

The metabolism among different protist groups is as varied as the organelles used for feeding and locomotion. Because of the different habitats in which protists live, this should not be surprising. We have already noted that the parasitic amoebae, which inhabit the gut, lack mitochondria. They also have an anaerobic type of metabolism in which oxidative phosphorylation does not play a role. Similarly, African trypanosomes within the mammalian host do not have functional mitochondria and obtain their energy by substrate-level phosphorylation. Although they use the oxygen within their habitat to maintain the appropriate oxidation-reduction balance, no energy is transformed and saved in ATP during this process. During the vector phase of their life cycle, however, these flagellated parasites have a functional mitochondrion and, therefore, possess an active oxidative phosphorylation pathway. Similarly, the pathways involved in carbohydrate catabolism, as well as amino acid and nucleic acid metabolism, are equally varied (Gutteridge & Coombs, 1977; Coombs & North, 1991; Bryant, 1993). However, these pathways are similar to those observed in other animal cells, e.g., glucose is catabolized to CO_2 and H_2O through glycolysis, the pentose phosphate shunt, the tricarboxylic acid cycle, and the cytochrome system. The trypanosomes and most apicomplexans do not store carbohydrates, whereas *Entamoeba* stores glycogen. Other carbohydrates such as starch and amylopectin are found in

other protozoan groups. Not long ago, apicomplexans were discovered to have a shikamate pathway for the synthesis of aromatic amino acids. This pathway occurs only in plants, fungi, and bacteria, not in animals. Evidence suggests that a line of organisms that were to become apicomplexans acquired a cell line of cyanobacteria that became endosymbiotically related to their hosts (see Box 3.1).

3.2.2 Reproduction

The process of reproduction among the parasitic protists is highly varied. Some groups employ only asexual reproduction. In its simplest form, this involves binary, or transverse, fission in which one cell simply divides equally into two daughter cells. A variant of **binary fission**, known as **multiple fission**, is associated with some parasitic protists and occurs when a second division takes place before the first division is completed. This leads to a common cytoplasm, more than one set of organelles, and long chains of connected cells; there may also be a ring, or rosette, of organisms, all joined through a common cytoplasmic bridge.

There are three types of asexual reproduction in the Apicomplexa, i.e., **merogony** (also called **schizogony**), **sporogony**, and **gametogony**. Following nuclear division in merogony (or schizogony) the individual nuclei move to the cell's periphery. When nuclear division is completed, the cytoplasmic membrane then surrounds each nucleus and the daughter cells bud from the parent. Each schizont, or **merozoite**, then contains a nucleus with the appropriate organelles and cytoplasm. In sporogony, a zygote undergoes multiple asexual divisions, with the formation of **sporozoites**. Finally, gametogony leads to the formation of gametes.

Sexual reproduction also occurs in many protist groups. It is a standard part of the life cycle of apicomplexans, ciliophorans, and some flagellates. The kinds of sexual reproduction, however, vary widely. In some, for example, **micro-** and **macrogametes** are involved and, in other groups, the gametes are identical. In the ciliates, true gametes are not formed; rather, following the temporary fusion of the two cells

Box 3.1 | A plastid in apicomplexans: a promise for new drug therapies?

There now exists overwhelming evidence for the presence of plastids, similar to the organelles responsible for photosynthesis in plants and algae, in apicomplexans. Acquisition of the **apicoplast** was probably accomplished through secondary endosymbiosis of a cyanobacterium by a heterotrophic eukaryote (see Ralph, 2005, for a full discussion). Ralph (2005) goes on to suggest that the progenitor of the apicoplast was probably a free-living red alga, and that it has long since disappeared. This is consistent with phylogenetic studies showing that the apicomplexans are nested within photosynthetic lineages having plastids derived from red algae, such as dinoflagellates (Leander & Keeling, 2003). There are, not surprisingly, competing hypotheses regarding the phylogeny of apicoplast-bearing organisms (Cavalier-Smith, 1999; Martin *et al.*, 2002). Evidence further suggests that in some apicomplexans, e.g., *Cryptosporidium* spp., the apicoplast has been lost. Although this organelle has disappeared, genes with an apicoplast origin are now part of the parasite's genome, probably having been acquired via horizontal gene transfer (Huang *et al.*, 2004).

Of course, the fairly recent finding of plastids in apicomplexans is of interest as a stand-alone discovery. There are also profound implications for possible drug therapies for malaria. One of the most significant problems facing public health workers in areas where malaria is endemic is the trouble with acquired drug resistance by *Plasmodium* spp. In recent years, the development of resistance has reached a crisis level because drug therapy is now ineffective in some parts of the world. The presence of a shikimate pathway in *Plasmodium* spp., however, opens new opportunities for the development of new drugs (Roberts *et al.*, 1998). Why? The answer is simple. This metabolic pathway is involved in the synthesis of aromatic amino acids, i.e., phenylalanine, tryptophan, and tyrosine. The only organisms known to possess such a pathway are the higher plants, fungi, and bacteria, but not vertebrate animals. According to Ralph (2005), ". . .the genome project shows that no shikimate enzymes have apicoplast-targeting leaders (Gardner *et al.*, 2002)." This means that shikimate enzymes could be 'sitting ducks' for the development of inhibitory molecules that affect the synthesis of aromatic amino acids without affecting host metabolism. While there is no direct promise this will happen, there is at least the chance.

(conjugation), there is an exchange of micronuclei. Some parasitic flagellates are believed to undergo sexual recombination in the insect vector. However, the sexual process has never been observed. Data on recombinant gene frequency suggest that sexual reproduction in these forms is a rare phenomenon and that reproduction usually takes place by binary fission.

3.3 Biodiversity and life-cycle variation

A consideration of a parasite's habitat helps to define the ecological relationship between the host and parasite. In considering the habitats of parasitic protists, it must be recognized that all of the possible different sites within, or on, a host can be occupied. These include external surfaces such as the skin and gills, as

Box 3.2 | Protists as hosts for microbial symbionts

In addition to discussing protists as parasites, it is also necessary to recognize that protists can act as hosts for a number of other microorganisms (Miles, 1988). Indeed, some of the most fascinating and intimate mutualisms involve wood-eating termites and the diverse flagellated protists and prokaryotes inhabiting their guts. Of these flagellates, *Mixotricha paradoxa* of the Australian termite *Mastotermes darwiniensis* is perhaps the most famous. This flagellate is host to at least four different bacterial ecto- and endosymbionts. Ectosymbionts include a spirochaete and a bacillus; they give the protist a most impressive, bristle-like appearance. Both bacteria are arranged on the surface of *M. paradoxa* in a highly organized pattern, actually fitting into distinct and separate surface receptors. Even more amazing is that the spirochaetes move in a synchronous, coordinated manner and are clearly involved in propelling the flagellate through its environment. This intriguing phenomenon is known as motility symbiosis; such motility is thought to be essential for protists of the gut to resist the turbulence of the gastrointestinal tract and to acquire food (Ohkuma, 2008). Remarkably, *Mixotricha paradoxa* lack mitochondria and have bacterial endosymbionts to fulfill this role.

Many other termite gut-inhabiting flagellates have been shown to harbor ecto- and endosymbiotic prokaryotes. Indeed, protist-associated prokaryotes comprise a considerable proportion of the microbial community within the termite hind gut. Consider, for example, that a single individual of the termite flagellate *Pseudotrichonympha grassii* harbors approximately 10^5 cells of the endosymbiotic bacteria, *Bacteroidales*. The protists and their associated prokaryote symbionts clearly play major roles in gut metabolism in wood-eating termites, especially in the efficient breakdown and utilization of cellulose. The fascinating functional interactions and coevolutionary relationships between termite flagellates and their mutualistic prokaryotes are reviewed by Ohkuma (2008).

An equally intimate relationship is illustrated by the ciliated protist *Paramecium bursaria* and its unicellular algal symbiont, *Chlorella* sp. Within the cytoplasm of this so-called 'green *Paramecium*' and many other ciliates are dense populations of these mutualistic green algae cells. The ciliate both protects and carries the *Chlorella* cells through its life cycle, even at cell division and also during sexual reproduction. By photosynthesis, these photobionts supplement their host's diet with excess carbohydrates. Recent genetic studies indicate four species of *Chlorella* inhabit *P. bursaria*, and that the symbionts are polyphyletic, suggesting at least four independent acquisitions and multiple origins of the symbioses (Hoshina & Imamura, 2008).

What is clear from these examples is that not only have protists evolved parasitic modes of life, but they also offer diverse habitats for other mutualistic microorganisms.

well as internal sites, e.g., the blood, lymph, brain, liver, intestine, etc. In addition, protists have successfully adapted to both extracellular and intracellular locations. It turns out that protists themselves can be hosts to diverse symbionts (Box 3.2). Their habitat may also define the environmental factors that influence the parasite's life cycle. For example, the fact that a parasite resides within the blood of a vertebrate host will largely determine the method of transmission to the next host, as well as routes of entry and exit. As we will see, habitat can also define the nature of energy resources available to the parasite.

Protists vary widely in their range of hosts and in their patterns of host specificity. *Plasmodium falciparum* parasitizes only humans and gorillas. In contrast, the African and American trypanosomes infect a wide range of mammals. The success of the African trypanosomes can be further emphasized by noting that even after large control programs to reduce the extent of human trypanosomiasis, it persists in the same geographical locations in Africa. This is also true for the *Plasmodium* species infecting humans. Despite all the malaria control programs over the decades, the prevalence of malaria is believed to be as high today as at any time in human history.

It is difficult to convey the extraordinary diversity of protists in a single chapter. Our focus here is on those protists that have been the subject of considerable research over the years, largely because of their enormous economic, veterinary, and/or medical importance. As a way of organizing and sorting their diversity, we adopt a simplified version of the taxonomic classification scheme currently recognized, without providing the phylogenetic evidence or interpretations (see Section 3.4). Regardless of their phylogenetic affinities, four 'groups' of protists can be recognized: the flagellates, the amoebae, the opalinids, and the alveolates; the latter comprises the dinoflagellates, and the Phyla Apicomplexa and Ciliophora (see Box 3.7). Rankings such as 'Super-groups' (e.g., Amebozoa) and 'Superphyla' (e.g., Superphylum Alveolata) have been erected to accommodate this new phylogenetic scheme (Adl *et al.*, 2005).

3.3.1 The flagellates

3.3.1.1 General considerations

The flagellates comprise a highly diverse group of single-celled, eukaryotic organisms that possess one or more flagella on the pellicle surface. Approximately 25% of so-called animal flagellates (=zooflagellates)

are symbiotic in invertebrates and vertebrates. Some of these cause some of the most significant parasitic diseases of humans and other vertebrates.

Most flagellates obtain nutrients by pinocytosis and by the active transport of small-molecular-weight compounds across their cell membranes. Cytostomes have been reported in some parasitic flagellates. In addition to their diversity in locomotion and methods for obtaining essential nutrients, parasitic zooflagellates occupy many diverse habitats and can differ extensively in their mode of transmission.

Zooflagellates infect almost all animal phyla, occupying a great variety of different habitats within their many hosts. In humans, for example, they are found in the intestine (*Giardia* spp.), the reproductive tract (*Trichomonas* spp.), intracellularly in deep body tissues (*Leishmania donovani*), as well as both intracellularly (*Leishmania* spp.) and extracellularly (*Trypanosoma brucei*) in the vascular system. Transmission can be direct, i.e., host to host, as occurs in the sexual transmission of *Trypanosoma equiperdum* or *Trichomonas vaginalis*. It can be indirect by the accidental ingestion of infective stages from contaminated feces (*Giardia* spp.), by scratching infective stages into the lesion produced by the bite of an insect vector (*Trypanosoma cruzi*), or by the injection of the parasite during a blood meal of an insect (*T. brucei*) or leech vector (*Cryptobia* spp.).

Reproduction occurs primarily by asexual binary fission. However, sexual reproduction has been described in a number of species, e.g., *T. brucei* (Molyneux & Ashford, 1983; Seed & Hall, 1992). There is also considerable diversity in the cell structure and the organelles observed in the different flagellate groups.

3.3.1.2 Form and function

Several zooflagellate taxa are recognized. Members of the Kinetoplastida possess one or two flagella, typically with a **paraxial rod**. They have, at some stage in their life cycle, a single mitochondrion, which contains up to 20% of the cell's DNA. Representative genera in this order include *Trypanosoma* and *Leishmania*. The

Diplomonads include those protists with one to four flagella/nucleus. Included here are *Giardia* and *Hexamita*. Trichomonads have cells with four to six flagella, sometimes with an **undulating membrane**. Examples of genera in this order include *Trichomonas* and *Histomonas*. Hypermastigidans possess numerous flagella and are mutualists of insects such as wood-eating termites, e.g., *Trichonympha*.

The prominent external cell structures present on these flagellates therefore, may include flagella and an **axostyle**, or an undulating membrane, or both (Fig. 3.1). Organisms such as *Giardia* have an adhesive disk that allows them to adhere to the epithelial lining of the gut. In kinetoplastidans, the flagellum, in addition to its function in locomotion, is also an organelle of attachment to the host. Finally, in some lower termite flagellates, there are specific surface receptors for the attachment of bacterial symbionts, such as spirochaetes (see Box 3.2).

In the Kinetoplastida, the single mitochondrion contains up to 20% of the cell's DNA that is localized in one area of the mitochondrion called the **kinetoplast** (Fig. 3.2). This so-called kDNA is arranged into densely packed strands of coiled and twisted DNA (Liu *et al*, 2005), some of which codes for a number of the mitochondrial proteins. The kinetoplast is closely associated with the flagellar pocket and the **basal body (kinetosome)** of the flagellum (Figs. 3.2A, B).

The Kinetoplastida have a second unique organelle called a **glycosome**. This unusual membrane-bound organelle contains a majority of the enzymes involved in glycolysis. Finally, the Kinetoplastida possess a ring of microtubules located just under the cell membrane (Fig. 3.2A). These microtubules extend the length of the cell and are interconnected by a series of short fibrils, creating a ladder-type appearance. This microtubule–microfibril complex is linked to the cell membrane and remains together as a discrete unit following cell lysis. It presumably gives structural support to the trypanosome's cell membrane.

The trichomonads, such as *Trichomonas vaginalis* (Fig. 3.1C), do not have a mitochondrion. They obtain all their energy by substrate-level phosphorylation and maintain their oxidation–reduction balance by reducing hydrogen in a special membrane-bound organelle called a **hydrogenosome**. In this organelle, protons are reduced to H_2. The hydrogenosome is thought to be unique to this group and appears adapted to the organism's life in an anaerobic environment (Müller, 1991).

The zooflagellates obtain most of the nutrients required for their anabolic pathways from their host by direct transport across the cell membrane. In the Kinetoplastida, however, cytostomes involved in nutrient uptake have been observed in some species. We have previously described how pinocytosis of large-molecular-weight proteins occurs in the flagellar pocket of these organisms. From this very brief description of the zooflagellates, it must be obvious that just as the external features of the individual species differ extensively, so do the organelles found in the cytoplasm and the various physiological or biochemical pathways.

3.3.1.3 Biodiversity and life-cycle variation

The life cycles of zooflagellates are also varied. *Giardia* spp. are transferred from host to host by a resistant cyst that is passed in feces. Thus, the parasite is transmitted by fecal-oral contamination, and the epidemiology of the disease giardiasis is dependent upon ingestion of the cysts in contaminated water or food (Box 3.3). Other flagellates are passed directly from one host to another during sexual intercourse, e.g., *Trichomonas vaginalis* of humans and *Trypanosoma equiperdum* of horses.

In contrast to *Giardia*, many of the Kinetoplastida have indirect (heteroxenous) life cycles. Generally, the parasite is transmitted from one vertebrate host to another by a blood-sucking invertebrate vector. In some cases, the vector acts purely in a mechanical manner, i.e., an insect feeds on an infected animal and in the process the proboscis becomes contaminated with the parasite. It then transfers the parasite mechanically to a new host at the next blood meal (e.g., *Trypanasoma evansi*). Members of the *brucei* group of African trypanosomes undergo a complex morphological and biochemical transformation in their tsetse fly vector. These trypanosomes are

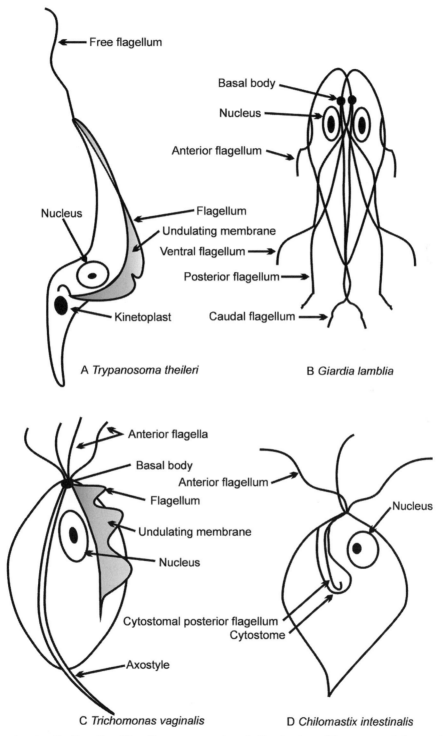

Fig. 3.1 Diagrams showing the diversity of flagellar arrangements and ultrastructure of four representative species of zooflagellates. (A) *Trypanosoma theileri*, a member of the Kinetoplastida; (B) *Giardia lamblia*, a member of the Diplomonadida; (C) *Trichomonas vaginalis*, a member of the Trichomonadida; (D) *Chilomastix intestinalis*, a member of the Retortamonadida. (Figures courtesy of Richard Seed.)

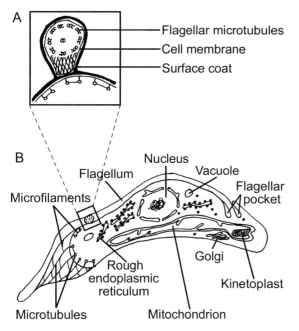

A
Flagellar microtubules
Cell membrane
Surface coat

B
Nucleus
Flagellum
Vacuole
Microfilaments
Flagellar pocket
Rough endoplasmic reticulum
Golgi
Kinetoplast
Microtubules
Mitochondrion

Fig. 3.2 Simplified diagram showing the ultrastructure of an intermediate blood stage trypomastigote of an African trypanosome. The inset box (A) is a representative cross-section depicting the relationship between the flagellum, the cell membrane, and the associated microtubule–microfilament complex of trypanosomes (B). (Figure courtesy of Richard Seed.)

acquired in a blood meal; they migrate to the hindgut where transformation begins and then to the salivary glands where it is continued until an infective stage emerges. On taking the next blood meal, the infective metacyclic trypomastigote migrates via the vector's proboscis into the bloodstream of the vertebrate host.

Two extremely important trypanosome congeners, namely *Trypanosoma cruzi*, the American trypanosome, and the African trypanosome *T. brucei*, vary considerably in several details (Molyneux & Ashford, 1983; Seed & Hall, 1992). These two species are classified as either stercorarian or salivarian, respectively, depending on the location within the insect vector in which the parasite develops to an infective stage.

The life cycle of *T. cruzi* has both intracellular and extracellular stages (Fig. 3.6). The parasite is the causative agent of the serious human disease known as American trypanosomiasis or Chagas' disease (Box 3.4). The intracellular stages lack a flagellum and are known as **amastigotes** (Fig. 3.7). These occur in a variety of mammalian cells, including macrophages and muscle. In the vertebrate host, the amastigotes reproduce by binary fission and their numbers increase forming **pseudocysts** (Fig. 3.7). Infected cells lyse and release amastigotes that can either infect other cells, or

Box 3.3 | Giardiasis: epidemiology and pathogenesis of beaver fever

This parasite has a long history of discovery, having been first described by Antoni van Leeuwenhoek from his own feces in 1681! Today, due to its medical, veterinary, and evolutionary importance, the parasite is the focus of considerable research. The species epithet of *Giardia* infecting humans has undergone a series of name changes over the years. Thus, *lamblia, intestinalis,* and *duodenalis* are now considered synonyms. Most parasitologists now recognize that *G. duodenalis* is the species infecting humans (see Thompson *et al.,* 1993; Olson *et al.,* 2004). Several strains of *G. duodenalis* are currently recognized, each varying considerably in their pathogenicity.

The pear-shaped **trophozoites** of *Giardia* spp. are striking in appearance, almost like a cartoon character. The trophozoite stage has two prominent nuclei, axonemes, four pairs of flagella, and an adhesive disk set in a convex ventral surface (Figs. 3.1B, 3.3, 3.4, 3.5). The adhesive disk attaches the trophozoite to the surface of an intestinal epithelial cell.

Box 3.3   (continued)

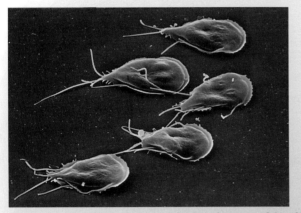

Fig. 3.3 Scanning electron micrograph of trophozoites of *Giardia muris* from a mouse. (Micrograph courtesy of Břetislav Koudela.)

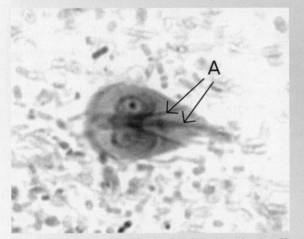

Fig. 3.4 Trophozoite of *Giardia duodenalis* in a stool smear. The two nuclei are visible, as are three of the eight axonemes (A) within the cytoplasm. (Photograph courtesy of John Sullivan.)

The life cycle is direct, with a resistant cyst as the infective stage. Transmission of cysts from human to human can be common, especially in regions of the world with poor hygiene. However, giardiasis is a zoonosis since the parasite survives in a wide range of mammalian reservoir hosts, including dogs, cats, sheep, cattle, and beaver. The latter can be a significant factor in the epidemiology of giardiasis in wilderness settings. The mature cysts must be ingested in contaminated food or water. In North America, the cysts are typically ingested by humans from drinking water directly out of streams while hiking or backpacking. It is for this reason that the disease has also come to be known as 'hiker's disease,' or the 'wilderness trots'. Serious backpackers and canoeists treat, filter, or boil their water to avoid the risk and discomfort of acquiring the disease while 'in the bush.'

Giardia duodenalis has been found in 81% of raw surface water supplies entering 66 treatment plants in 14 states in the USA and in one province in Canada. In addition, *G. duodenalis* was detected in 17% of the filtered water samples from these same plants. From 1984 through 1990, it has been estimated that there were 25 outbreaks of waterborne disease involving 3486 individuals due to *G. duodenalis*; from 1991 through 1994, nine of 36 waterborne diarrheal outbreaks, caused by a known etiological agent, were due to this parasite (Marshall *et al.*, 1997; Steiner *et al.*, 1997). According to the World Health Organization, giardiasis is now recognized as one of the 10 most common parasitic diseases of humans, infecting some 200 million people in both developed and developing countries around the world.

Box 3.3 (continued)

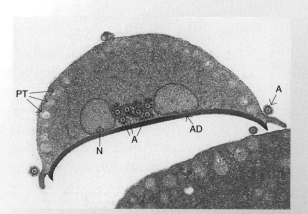

Fig. 3.5 Transmission electron micrograph of a trophozoite of *Giardia muris* from a mouse, showing the peripheral tubules (PT), two nuclei (N), adhesive disk (AD), and the axonemes (A) of the flagella. (Micrograph courtesy of Břetislav Koudela.)

Although giardiasis is not a severe pathogen, waterborne disease outbreaks can have significant direct and indirect economic consequences. There are the direct medical and pharmaceutical expenses, but also the indirect expenses due to the necessity for homes and businesses to obtain *Giardia*-free water, loss of work and leisure time while ill, loss of school time due to absenteeism, loss of restaurant business in the area affected, etc. An excellent analysis of an outbreak in Pennsylvania (USA) has been described by Harrington *et al.* (1991).

Following ingestion of cysts, excystation occurs in the small intestine, after being prompted by the acidic pH of the stomach. Following attachment via the ventral disk, the trophozoites multiply asexually by binary fission, often reaching massive numbers. Trophozoites pass into the colon and encyst. Cyst production will ensue as early as 4 days following initial infection. A single stool may contain as many as 300 million cysts that can survive for several months in water. An entire watershed can be contaminated by a single infected human host or wild mammalian reservoir host.

Giardia duodenalis is usually not a life-threatening pathogen. However, it can become severe in young children and immunocompromised individuals. A frequent clinical feature of giardiasis is a foul-smelling stool, which is affected by faulty fat absorption in the small intestine. Enormous numbers of trophozoites can also negatively impact essential nutrient and vitamin absorption. Other symptoms include malaise, abdominal cramps, weight loss, and flatulence. The disease usually runs its course in a few weeks, but the diarrhea may become chronic and last for months, with frequent relapses. Several drugs are efficacious in treating the disease, although with re-infection and retreatment in children, there is a risk of toxicity problems.

Remember, the next time you choose to drink water from that irresistible cold, clear mountain stream, you are taking a gamble!

transform into **trypomastigotes**. The trypomastigote stage (Fig. 3.8) does not divide and is found in extracellular sites, predominantly the blood. It is this stage that is acquired by the blood-sucking reduviid bug vectors (Fig. 3.6). In the vector, the trypomastigotes travel to the midgut and develop first into **promastigotes** and then into **epimastigotes** (Fig. 3.9). The epimastigotes divide by binary fission. The larger

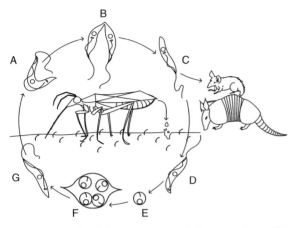

Fig. 3.6 Life cycle of *Trypanosoma cruzi*. Trypomastigotes (A) are ingested by reduviid bug vectors during a blood meal. They transform into epimastigotes (B) which multiply within the midgut of the reduviid bug vector and then develop to metacyclic trypomastigotes (C) in the hindgut. These pass with the feces and gain entry into humans if rubbed into mucous membranes. Trypomastigotes (D) in the bloodstream invade cells, e.g., cardiac muscle cells, and become amastigotes (E), which replicate extensively, forming pseudocysts (F). The life cycle is completed when reduviid bugs feed on humans containing infective trypomastigotes (G). Reservoir hosts include rats and armadillos. (Figure courtesy of Danielle Morrison.)

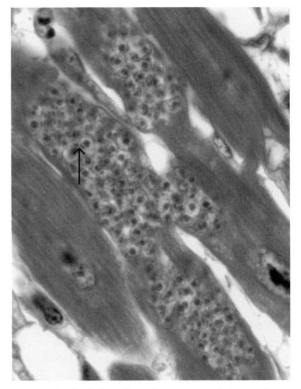

Fig. 3.7 Histological section of cardiac muscle infected with a pseudocyst of *Trypanosoma cruzi* in a muscle cell. The pseudocyst is filled with amastigotes. A nucleus and kinetoplast (arrow) are visible in some of the amastigotes. (Photograph courtesy of John Sullivan.)

epimastigote forms migrate to the rectum and adhere (via the free flagellum) to the epithelial lining of the rectal gland. Here, the epimastigotes transform into infective metacyclic ('stumpy' forms) stages that are usually found free in the lumen. The metacyclic forms do not divide. The development of the ingested trypanosomes into infective metacyclic trypomastigotes takes 1 to 2 weeks depending on the vector and the external temperature. The infective trypomastigotes are released in the vector's feces, which are deposited on the mammal's skin near the bite wound (Fig. 3.6). There is no evidence that metacyclic forms can enter through intact skin; rather, when the bite itches, and the host scratches, the metacyclic trypomastigotes in the contaminated feces are forced into the lesion left by the insect's proboscis. Infection can also occur by the ingestion of infective feces or an infected vector since metacyclic trypomastigotes can penetrate

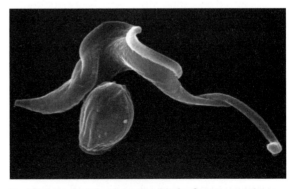

Fig. 3.8 Scanning electron micrograph of a trypomastigote and amastigote stage of *Trypanosoma cruzi* from the blood of an experimentally infected mouse. (Micrograph courtesy of Cheryl Davis and John Andersland.)

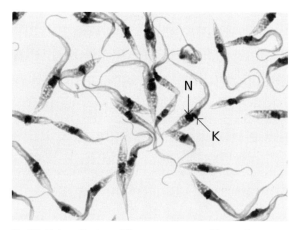

Fig. 3.9 Epimastigotes of *Trypanosoma cruzi* in a culture smear. Note the anterior location of the kinetoplast (K) relative to the nucleus (N). (Photograph courtesy of John Sullivan.)

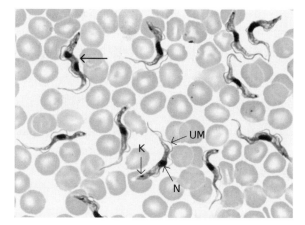

Fig. 3.10 Trypomastigotes of *Trypanosoma brucei gambiense* in a peripheral blood smear. Note the undulating membrane (UM) and posterior location of the kinetoplast (K) relative to the nucleus (N). The long, slender forms of the trypomastigotes divide by binary fission (arrow). (Photograph courtesy of John Sullivan.)

mucosal surfaces. Stercorarian trypanosomes, therefore, develop in the hindgut and rectum of the vector. Transmission is either by contamination of the bite with vector feces containing infective metacyclic trypomastigotes, or by the ingestion of the vector, or its feces.

In contrast, in salivarian trypanosomes, the development of infective metacyclic trypomastigotes occurs in the salivary glands or mouthparts of their vector. Trypanosomes of this type are transmitted during feeding by the insect vector. In the African trypanosomes, the short stumpy, non-dividing trypomastigote stage is acquired in the tsetse fly's (*Glossina* spp.) blood meal. It travels to the midgut and then to the hindgut as a procyclic form that maintains the trypomastigote morphology. After a period of development, the procyclic trypomastigotes migrate anteriorly to a region of the midgut where they divide and multiply. Between 14 and 21 days later, the parasites enter the proventriculus region of the insect's gut. From there, they migrate anteriorly into the salivary glands where they transform into epimastigotes and again multiply. The epimastigotes attach themselves by their flagella to microvilli that line the epithelium of the salivary gland. In this site, the epimastigotes transform into non-dividing, infective, metacyclic trypomastigotes

that are released from their attachment to the microvilli. They can then be found in the lumen of the salivary gland. The entire cycle from the time of the first blood meal containing short stumpy trypanosomes to infective metacyclic trypomastigotes takes 3 to 4 weeks. The metacyclic trypomastigotes are then injected in the saliva into a new host during a blood meal. Once in the mammalian host, the metacyclic forms transform to a long, slender trypomastigote, which divides rapidly by binary fission (Fig. 3.10). In the African trypanosomes, the life cycle is, therefore, totally extracellular. As the parasitemia reaches its peak, a percentage of the long slender forms transform into a short stumpy, non-dividing, trypomastigote stage that is infective for the tsetse fly vector.

Recall from Chapter 2 that the African trypanosomes are renowned for one of the most exquisite immunological evasive adaptations – antigenic variation. They avoid complete antibody-mediated destruction by continually changing their highly antigenic, variable surface glycoprotein (VSG) coat (see Fig. 2.10). The epidemiology of African trypanosomiasis is extraordinarily complex; reservoir hosts, especially native ungulates, and environmental factors influencing tsetse fly vector reproduction and human blood-feeding are the most important epidemiological aspects. Northwestern to southeastern Equatorial Africa is the tsetse-fly belt. Initial pathogenesis due to *T. brucei gambiense* (Fig. 3.10) is due to the massive numbers of trypomastigotes in the blood, leading to fever and general weakness (see also Chapter 2). The chronic stage results from invasion of the central nervous system by trypomastigotes. This causes the 'sleeping sickness' and paralysis, and may lead to coma and eventual death. It is estimated that perhaps 60–70 million people are at risk of exposure. Paradoxically, these parasites act as an 'agent of conservation' largely preventing human agriculture in vast tracts of land in Africa. In effect, they help preserve the diverse African fauna from human encroachment.

In all members of the Kinetoplastida, reproduction is primarily, if not totally, via asexual binary fission. 'Sex' has only been reported to occur in the African trypanosomes, and is assumed to occur in the vector. Even then, however, it would appear to be a rare event and has never been observed directly.

Box 3.4 | Pathogenesis, epidemiology, and diagnosis of Chagas' disease

Trypanosoma cruzi, the causative agent of American trypanosomiasis or Chagas' disease, is a zoonotic disease that can infect over 150 species of mammal, including humans (Molyneux & Ashford, 1983). It is estimated that approximately 10–12 million people in Central and South America are infected with *T. cruzi*, ranking it high in terms of the prevalence of human parasitic diseases around the world. Approximately 10% of those infected will develop a serious disease involving cardiovascular pathology, or damage to ganglia of the autonomic nervous system of the esophagus and/or colon. This can lead to malfunction and enlargement of the esophagus and colon, leading to the grave medical conditions known as megaesophagus or megacolon. In some Latin American countries, *T. cruzi* is a main cause of chronic heart disease, due to the destruction of heart muscle and conducting cells. Immunopathology is a significant component of Chagas' disease (see Chapter 2).

Reduviid bugs, commonly referred to as kissing bugs because they usually bite on the face, and usually at night, transmit the disease. The bug takes up the trypanosome during an evening blood meal and, within a relatively short time, the parasite moves to the hindgut where it multiplies and matures into an infective metacyclic form (Fig. 3.6). While the vector feeds, it defecates, depositing fecal material containing infective forms onto the skin near the bite. The insect bite itches and the individual literally scratches the infective stages into the wound. Following entry into the host cells, the parasites transform into amastigotes. The infected cells eventually burst, releasing the amastigotes, which invade surrounding cells, and the cycle is repeated. Subsequently, trypomastigote forms are released from the infected cells into the blood. These trypomastigotes do not divide, but are infective to the vector. When they reach the vector's hindgut the process of division and eventual maturation to the infective metacyclic stage is repeated (Fig. 3.6).

Chagas' disease is difficult to treat due to the lack of effective chemotherapeutic agents. A vaccine is not available. Therefore, current control measures involve methods to prevent humans

Box 3.4 | (continued)

from becoming infected. Primary emphasis is on insect control through the use of insecticides, and improved housing (screens on windows, metal roofs, etc.) in order to help eliminate the vector from homes and disrupt transmission.

Domestic dogs and cats, as well as armadillos and rats (Fig. 3.6), act as reservoir hosts and greatly complicate the epidemiology of Chagas' disease. In addition, various reduviid bug species, adapted to both domestic and natural habitats, can transmit infective stages. Infection causes serious disease in about 10% of the human population and may be chronic, lasting for a number of years, thereby ensuring a long-term carrier state that increases the opportunity for transmission. Sadly, most new infections occur in children, likely due to their sleeping habits and exposure to feeding reduviid bugs. It is no wonder that Chagas' disease is most common among peoples in the poorest of villages in Mexico and in impoverished Central and South American countries. Improved housing and education efforts will help in the future control of this serious disease.

Diagnosis of chronic Chagas' disease is not easily made using blood smears. Serological methods or in vitro culture can be used, but these are expensive and sometimes inconclusive. One of the older, but unique, methods is called xenodiagnosis. Many hospitals across Central and South America, where Chagas' disease is endemic, maintain insectaries where stocks of parasite-free reduviid bugs are maintained. In a suspected case of Chagas' disease, uninfected kissing bugs will be fed on the patient until engorged, then maintained under controlled temperature conditions. Over a period of 3 months, the feces of the reduviid will be regularly examined for metacyclic trypomastigotes, confirming the presence, or absence, of *T. cruzi* in the patient.

Another of the significant human kinetoplastids is *Leishmania* spp., causing various forms of mild to severe leishmaniasis. All *Leishmania* species are transmitted to humans by the bite of female phlebotomine sand flies. *Leishmania* spp. occur in a range of other mammals, including dogs and rodents, which, in turn, can act as reservoir hosts for human infections. There are at least 14 recognized species of *Leishmania* in South America alone, and undoubtedly many different serotypes, with most causing a unique form of clinical disease (Molyneux & Ashford, 1983).

The genus *Leishmania* includes both New and Old World species of human parasites. The Old World species include *Leishmania aethiopica, L. major*, and *L. tropica. Leishmania tropica* normally produces skin lesions (oriental sore) (Fig. 3.11) that can be self-limiting and capable of spontaneous cure. Another Old World species is *L. donovani*, which causes visceral leishmaniasis (or kala azar), a potentially fatal disease. However, it should be noted that the etiological agent involved in several recent cases of visceral leishmaniasis also has been identified as *L. tropica*.

New World species include *L. braziliensis*. The subspecies in this group cause oriental sore, as well as a devastating non-healing, mucocutaneous form of the disease. A second New World species is *L. mexicana*. Again, the subspecies in this group produce pathologies ranging from oriental sore to diffuse cutaneous leishmaniasis. *Leishmania chagasi* has been

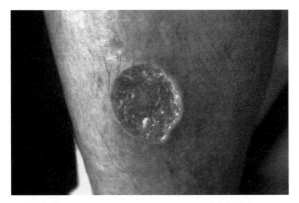

Fig. 3.11 Cutaneous lesion (oriental sore) caused by *Leishmania tropica* on an individual from French Guyana. (Photograph courtesy of Jean-Phillipe Chippaux.)

identified in the New World, where it produces a form of visceral leishmaniasis.

According to the World Health Organization, the leishmaniases are now endemic to 88 countries on five continents, with 12 million cases, and an estimated two million new infections annually. In the latter group, fully 500 000 are visceral leishmaniasis, with 90% of the victims living in Bangladesh, Brazil, India, Nepal, and Sudan. Most (90%) of the mucocutaneous infections occur in Bolivia, Brazil, and Peru. Approximately one million new cases of cutaneous leishmaniasis occur annually, with 90% of these in Afghanistan, Brazil, Iran, Peru, Saudi Arabia, and Syria. Of increasing concern is the association between *L. donovani* and HIV, especially in southern Europe, where 673 of the first 700 reported co-infections occurred. Of this number, most victims were young males and intravenous drug users.

A remarkable aspect of leishmaniasis is the propensity of the infective promastigote to invade cells of the macrophage lineage in their vertebrate hosts (Chapter 2). Thus, not only can the parasite avoid the antibody and cell-mediated immune response by exploiting an intracellular environment, but it survives in the very cells that are directly involved in the phagocytosis of microbes. Once inside the macrophage, the promastigotes transform into amastigotes that reside permanently within a **parasitophorous**

vacuole, a membrane-bound compartment inside the macrophage. This vacuole is also called the phagolysosome as it is formed by the fusion of the phagosomal vacuole containing the parasite with a lysosomal vacuole. The parasitophorous vacuole contains many amastigotes and can swell to occupy the majority of the host cell. These parasitophorous vacuoles are highly acidic and contain all the hydrolytic enzymes (proteases, nucleases, etc.) released from the lysosomal vacuole. Clearly, the environment in which the amastigotes reside is hostile. However, in the non-immune host, the parasite is able not only to survive, but to proliferate extensively (Alexander & Russell, 1992; Alexander *et al.*, 1999).

The clinical manifestations of the disease are associated with the particular species or subspecies of the parasite. In kala azar, the parasite is viscerotropic and there is an undulant fever, hepatosplenomegaly, and even dysentery. In contrast, *L. braziliensis braziliensis* is dermotropic and affects the mucocutaneous membranes of the mouth and nostrils, often causing horribly disfiguring lesions. *Leishmania tropica* is also dermotropic, producing a self-limiting cutaneous infection. Although not normally considered serious, the lesions produced by *L. tropica* can result in scarring. Since multiple lesions can be found on exposed parts of the body, scarring can be significant, and can be of cosmetic concern, particularly for children. *Leishmania mexicana*, mostly of Central America and Mexico, also produces cutaneous lesions. In Belize, the condition is called 'bay sore' and lesions can occur on the ear. Ear cartilage is poorly vascularized and the immune response is limited, making treatment and recovery from *L. mexicana* more problematic than if the lesions occur on the skin. The lesion caused by *L. mexicana* is commonly referred to as 'chiclero's ulcer.' The name of the disease comes from the so-called chicleros, forest workers who harvest gum from the chicle tree in Central America.

It is worth noting that for many centuries, the Arabs were aware the disease known as oriental sore (caused by *L. tropica*) produced permanent scarring, but not death. At some point, they learned that the disease was self-limiting, i.e., the infection would occur just once.

They also learned to scrape a pin across the periphery of a lesion and use the contaminated pin to puncture the buttocks or foot of a child, especially females. After a period of time, a lesion would appear and then disappear, but the disfiguring scar would be on part of the body well covered by clothing. This technique was essentially the same one employed by Edward Jenner in the eighteenth century when he discovered that fluid removed from cowpox pustules could prevent infection by smallpox. The treatment for the various leishmaniases around the world usually involves frequent injections with antimonials.

While our focus has been on the hemoflagellates of humans, it is important to recognize that there are also many trypanosome species that have been described from all vertebrate classes, including fish and amphibians. For many of those whose life cycles are known, blood-sucking leeches act as vectors. For example, *Cryptobia salmositica* can be a severely pathogenic kinetoplastid trypanosome infecting all species of Pacific salmonid fishes, *Onchorhynchus* spp. (Woo, 2001). The freshwater leech, *Piscicola salmositica*, acts as the vector for *C. salmositica*, and transmission to fish occurs in streams and rivers.

The ecology of blood-dwelling trypanosomes in amphibian hosts has also received considerable attention. *Trypanosoma diemyctyli* infects red-spotted newts, *Notophthalmus viridescens*, and uses a leech as vector. The leech–trypanosome–newt system has been the focus of several ecological and evolutionary studies (e.g., Mock & Gill, 1984). In the ranid frog *Rana clamitans*, from Louisiana, *Trypanosoma rotatorium* has two different cycles in its blood parasitemia. During cold winter months, when frogs are inactive, it is difficult to find trypanosomes in the peripheral circulation of the frog host (Seed *et al.*, 1968a). The frogs do not lose their infection but, rather, the trypanosomes are sequestered in the deep, vascular beds of the internal organs. When the weather warms, trypanosomes move to the peripheral circulation. During this time, the frogs spend significant time periods in the water. Since division of trypanosomes has never been observed in these frogs, this represents a true seasonal cycle of migration of the trypanosomes through the frog's vascular bed. It is important to recognize that *T. rotatorium* is leech-transmitted; therefore, it is significant that peripheral parasitemia is seasonally high when the frogs are in the water and have the greatest probability for contact with the leech vector. In addition to seasonal migration through the frog's circulation system, Seed *et al.* (1968b) also show a diurnal rhythmicity with peak parasitemia during daylight hours when contact with leech vectors is presumably highest. In short, parasite reproduction is dependent upon the coupling of activity cycles of the frogs, the leech vectors, and the trypanosome populations. Similar examples of tight synchronicity between host and parasite life cycles are provided in Chapter 12.

3.3.2 The amoebae

3.3.2.1 General considerations

Amoebae were formerly classified in the Subphylum Sarcodina. This is no longer tenable as amoeboid protists are most definitely polyphyletic and several phyla of uncertain affinities are now currently recognized. Not everyone agrees with the phylum designations and, as a result, the classification of the amoebae is in a constant state of change. Our focus here is on the amoebae parasitic in humans. Most of these parasitic species are called 'naked amoeba' because they do not produce a shell (or test).

Amoeboid movement via pseudopodia unifies the diverse amoebae. Flagella, when present, are restricted to developmental or sexual stages. For example, the opportunistic pathogen, *Naegleria* spp. has a free-living, flagellated stage, but transforms into amoebae in vertebrate hosts. Reproduction is usually asexual, by binary fission; sexual reproduction, when it occurs, usually involves flagellated gametes. However, amoeboid gametes have also been observed. Most amoebae are free-living, not parasitic. Trophozoites of free-living and parasitic amoebae feed by engulfing food particles by lobose or filamentous pseudopodia, via phagocytosis.

A majority of the parasitic amoebae live in the intestine of their hosts. They may, however, secondarily invade deeper tissues, as can occur during infection of humans by *Entamoeba histolytica*. These amoebae have been found in abscesses primarily in the liver and brain, but have also occasionally been observed in other tissues, including cutaneous sites (Parshad *et al.*, 2002). The amoebae located in the deep tissue sites are not involved in the amoeba's life cycle. They are at a dead-end since transmission always occurs by excretion of environmentally resistant cysts in the stool. Acquisition of amoebae is generally by an oral–fecal transmission route in which resistant cysts are ingested by a new host.

The most important amoeboid parasite is *Entamoeba histolytica* (Fig. 3.12). Several other species of *Entamoeba* are considered as commensals of humans. For example, *Entamoeba dispar*, a non-pathogenic parasite of humans, has been separated taxonomically from the pathogenic *E. histolytica*. This separation is based on the association of distinct biochemical and

molecular markers with pathogenic and non-pathogenic isolates (Martínez-Paloma & Espinosa-Cantellano, 1998).

Most naked amoebae (Gymnamoebae) are phago-trophic and produce resistant cyst stages for transmission. Most move only by lobose pseudopodia and never have a flagellated stage. Important genera include *Entamoeba*, *Endolimax*, and *Iodamoeba*. Species in these genera produce cysts that exit from their hosts in feces. They are directly transmitted by ingestion of feces-contaminated water or food containing cysts. An exception is *E. gingivalis*, an amoeba that inhabits the mouths of humans. This parasite is believed to be transmitted directly as a trophozoite.

So-called amoebaflagellates (Heterobosids) possess flagellated stages at some stage of their life cycles. An important heterolobosid species is *Naegleria fowleri*, which can infect humans. Species of *Naegleria* form cysts, but also have a flagellated stage during their free-living phase. Trophozoites enter the human host through the nostrils or mouth while swimming in contaminated water. Amoeboid stages migrate up the olfactory tract into the brain where they cause a devastating disease known as primary amoebic meningeoencephalitis, or PAM (Martínez & Visvesvara, 1997).

3.3.2.2 Form and function

In the amoeboid stage, there are no observable external structures outside the cell membrane. Among the heterolobisids, typical eukaryotic flagella are present during part of their life cycle. The flagellated stages of *Naegleria* have a conspicuous contractile vacuole. At the ultrastructural level, a **glycocalyx** of varying thickness can be detected. The cysts are smooth-walled, oval stages containing a variable number of nuclei. There are no obvious distinguishing external features and, except for size and internal structural details, can not be distinguished microscopically from each other at the species level.

The internal morphology of the amoebae is eukaryotic in design. Species of *Entamoeba* and *Naegleria*

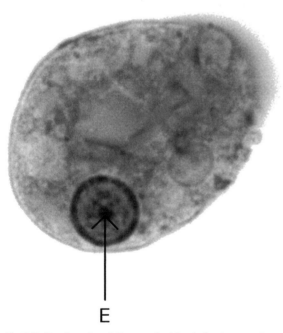

E

Fig. 3.12 Trophozoite of *Entamoeba histolytica* in a stool smear. Note the centrally located endosome (E) within the nucleus. (Photograph courtesy of John Sullivan.)

possess a nucleus that has an **endosome**, a nucleolus-like organelle (Fig. 3.12) that does not disappear during mitosis, and they have chromatin that is distributed around the inner surface of the nuclear membrane. The nuclear membrane remains intact during cell division. The cytoplasm of trophozoites contains food vacuoles, lysosomes, an endoplasmic reticulum, and ribosomes. Among species of *Entamoeba*, both the Golgi apparatus and mitochondria are absent. The absence of cytochromes and a mitochondrion limits the amoebae, metabolically, which means they must obtain energy by substrate-level phosphorylation. They are considered to be microaerophilic organisms in which glucose is catabolized to acetate, ethanol, and CO_2. They do not produce molecular hydrogen (Müller, 1991). The normally free-living amoebae, e.g., *Naegleria* spp., which are accidental parasites of humans, contain mitochondria. Chromatoid bodies (or bars), composed of ribonucleoprotein, can be observed by light microscopy in some trophozoites and in cysts when stained appropriately (Fig. 3.13). The differences in the chromatoid bodies are used diagnostically at the species level. In the pre-cyst and young cyst stages, glycogen vacuoles can also be observed. Both glycogen vacuoles and the chromatoid bodies disappear as the cysts

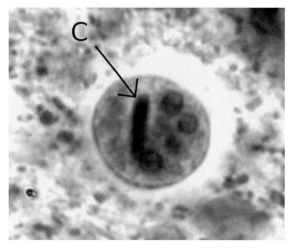

Fig. 3.13 Cyst of *Entamoeba histolytica* in a stool smear. Note the chromatoid bar (C) and three of the four nuclei. (Photograph courtesy of John Sullivan.)

mature. The free-swimming stage of *Naegleria* spp. possesses two flagella.

Species of *Entamoeba* infecting humans can be differentiated diagnostically by their cyst morphology. There are differences in cyst size, and in the number of nuclei per cyst, nuclear structure, and chromatoid bars.

3.3.2.3 Life cycle

Entamoeba spp. spend most of their life cycles as trophozoites (Fig. 3.12) that reproduce asexually in the gut of their hosts. They are extracellular, rarely penetrating beyond the lumen of the gut. When they do occupy extra-intestinal sites, they invade through the epithelial lining of the intestine. Those amoebae that invade deeper tissue can cause extensive pathology.

Most amoebae are not pathogenic but, rather, live commensally, feeding on intestinal microorganisms. They reproduce by binary fission and do not reproduce sexually. Following a period of growth and reproduction, some trophozoites begin to encyst as they are carried down the intestinal tract. During this process, they lose any undigested nutrients and condense into a sphere. This uninucleate, early stage of encystment is often referred to as the pre-cyst. Within the pre-cyst stage is a large glycogen vacuole. Chromatoid bars (Fig. 3.13) are also formed during this period. They are storage sites for ribsomes and protein synthesis. The pre-cyst secretes a thin, but tough, cyst wall. There is nuclear division, and the glycogen vacuole and chromatoid bodies usually disappear as the cyst matures. The mature cyst (or metacyst) (Fig. 3.13) is the infective stage of the amoeboid parasite; it can remain viable outside the host for several weeks. When the metacysts are ingested, they are capable of surviving the acid environment of the stomach. They only begin to excyst when they reach the small intestine and an alkaline environment. The multinucleate amoeba then begins to move in the cyst, the cyst wall weakens, and the amoeba emerges and quickly undergoes cytokinesis. The individual trophozoites pass down the intestinal tract and, upon reaching the appropriate habitat, they become established, feeding and reproducing extensively. They are site-specific, e.g., *E. histolytica* colonizes the colon of humans, primarily in the sigmoidal-rectal and ileo-caecal areas.

Entamoeba histolytica (Figs. 3.12, 3.13) is the causative agent of amoebic dysentery, a disease that affects millions of people worldwide, especially in developing countries. Trophozoites of this dangerous parasite secrete proteolytic enzymes enabling the organism to invade submucosal intestinal tissues. Often, food vacuoles within the cytoplasm of trophozoites contain RBCs. Mucosal ulceration leads to extensive tissue damage and lesions, and a grave condition known as invasive intestinal amoebiasis. Trophozoites may also invade the liver via the hepatic portal system, causing extensive damage and hepatic amoebiasis.

We have noted that *E. gingivalis* does not produce cysts, but is, instead, passed from host to host in the trophozoite stage. *Naegleria fowleri*, in contrast, will form cysts, but also has a flagellated, free-swimming stage. The flagellated stage is presumably important for dispersal of the parasite in contaminated water, thereby increasing the probability of contact with a potential host. The cyst stage, on the other hand, is assumed to increase the parasite's ability to survive under harsh environmental conditions.

Much like *Naeglaria fowleri*, *Acanthamoeba* spp. can be opportunistic parasites of the skin or central nervous system of humans. They can cause pathology, especially in immunocompromised individuals. Species of *Acanthamoeba* are ubiquitous in fresh water and soil environments. *Acanthamoeba* spp. have been shown to cause keratitis or corneal inflammation leading to corneal ulcers in humans. *Acanthamoeba* keratitis can occur in patients who sustain minor corneal trauma. The disease is typically associated with humans who wear contact lenses. Amoebae can be introduced while swimming when wearing contact lenses or by using contaminated, homemade contact lens solutions that do not kill the parasite's cysts. Contact-lens wearers who do not disinfect their lenses correctly seem to be the most at risk of acquiring *Acanthamoeba*. Rare, potentially serious, eye infections can result.

3.3.3 The opalinids

3.3.3.1 General considerations

All members of the enigmatic Order Opalinida are commensals. They are found primarily in the lower intestines of fishes and amphibians, especially frogs and toads. The opalinids have numerous longitudinal and oblique rows of cilia covering the entire body surface. Thus, superficially, opalinids resemble the ciliates. However, unlike the ciliates they do not have a cytostome and they do not undergo sexual reproduction by conjugation; instead opalinids exhibit sexual reproduction by producing micro- and macroflagellated gametes. Furthermore, opalinids only have a single type of nucleus, very unlike the dimorphic nuclear situation in ciliates. The opalinids feed using a modified form of pinocytosis, and asexual reproduction is by binary fission. Opalinid genera are distinguished from each other on the basis of their size and shape, cilia arrangement, and the number and morphology of their nuclei.

3.3.3.2 Form and function

The distinguishing external feature of the opalinids is the characteristic morphology of the cilia that are arranged in longitudinal rows over the surface of the entire cell. The cell surface structure is dominated by the cilia and associated complex infraciliature including kinetosomes, and they have the typical eukaryote organelles.

The gametes are flagellated and differ in size (micro- and macrogametes). There are also two cyst stages that can occur, a gamontocyst from which the micro- and macrogametes will develop and a smaller, thick-walled zygocyst. In *Opalina ranarum*, the gamontocysts are released into the water by adult frogs. Tadpoles then ingest the gamontocysts. Male and female gametes excyst in the intestine and fuse to form zygocysts. The zygocysts are released in the tadpole's feces. Following ingestion by the adult frogs, excystation occurs in the intestine and the trophozoites undergo binary fission.

3.3.3.3 Life cycle

The life cycle of the opalinids includes both a sexual cycle with flagellated micro- and macrogametes, and an asexual cycle in which the cells divide by longitudinal binary fission. In species of *Opalina*, the entire life cycle is remarkably synchronized with the reproductive cycle of the host (Smyth, 1994). During the non-breeding period, only trophozoites of *O. ranarum* are present in the intestine of the European frog *Rana temporaria*. During this period, the trophozoites reproduce solely by fission. However, when the frogs enter the water to mate, the opalinids increase their rate of fission. Shortly thereafter, small pre-cystic forms appear and then mature to encysted forms with an average of four nuclei per cyst. Meiosis is pre-zygotic and is believed to occur prior to cyst formation. After the frogs mate, the number of cysts (gamontocysts) increases and the cysts are passed out into the water with the feces. For the next several months, the number of cysts in the intestine slowly decreases to zero. It has been shown experimentally that adrenalin and both sex and gonadotropic hormones can induce the encystment of *Opalina* in the adult frog.

Tadpoles ingest the cysts that are voided in the feces of the adult frog. Approximately 8 hours later, excystment occurs; small, multinucleate micro- or macrogametes emerge and then migrate to the cloaca where meiotic division occurs. The haploid micro- and macrogametes ultimately fuse to form a diploid zygote. The zygote then encysts; the zygocyst is passed in the feces and is later ingested by adult frogs. In the intestine of the adult frog, the zygote excysts, a uninucleate form escapes, and divides repeatedly giving rise to trophozoites. Thus, for *O. ranarum*, the timing of cyst formation, its release into the water that coincides with the time that the adult frog enters the water, and the time at which the new tadpoles emerge in the pond are all highly synchronous.

3.3.4 Apicomplexa

3.3.4.1 General considerations

The Apicomplexa is accorded separate phylum status based on the absence of obvious external organelles involved in locomotion, although specific sexual stages may be flagellated. The presence of an **apical complex** also is diagnostic of the group, suggesting a monophyletic origin. Most notably, apicomplexans are intracellular parasites for a majority of their life cycles.

Members of the Apicomplexa reproduce both sexually and asexually, and are thought to be haploid for most of their life cycle. Many species have resistant stages that are involved in transmission. These spores or **oocysts** have a thickened protective cell wall when released into the environment. In Apicomplexa that are vector-transmitted, the oocyst cell wall that contains the infective stages is reduced, or non-existent.

In this group, there are several extremely important parasites of humans and various domesticated animals. *Plasmodium* spp., the causative agents of malaria, infect an appalling 500 million people worldwide and kill an estimated 1 to 3 million people annually, mostly children under the age of five. *Toxoplasma gondii* infects even more humans. Various estimates suggest that approximately 20% of the world's population have been exposed to *T. gondii*. In some populations, the prevalence of *T. gondii* infection, based on serological assays, can exceed 50%. Certain of the coccidian apicomplexans, especially *Eimeria* spp. are important parasites of poultry, and can produce serious losses. *Cryptosporidium parvum*, an intestinal apicomplexan of humans and other animals, caused the largest waterborne disease outbreak ever recorded, with over 400 000 people infected in Milwaukee, Wisconsin (USA) (Smith & Rose, 1998). *Babesia* and *Theileria* produce serious disease in cattle and other farm animals, again resulting in great economic losses to the agricultural industry. The three main apicomplexan genera of birds (*Plasmodium*, *Leucocytozoon*, and *Haemoproteus*) have been studied intensively by parasitologists around the world, offering excellent models for studies of the ecology and evolution of vertebrate–parasite interactions (review in Valkiunas, 2004).

3.3.4.2 Form and function

The external structure of the asexual stages of apicomplexans is unremarkable. The general form may vary from being amorphous (like an amoeba) to teardrop in shape. The sexual stages may be flagellated and the gametes (micro- and macrogametes) vary in size. Cysts, when produced, are usually thick-walled and round, but again the gross external features, except for size, are of limited diagnostic value.

Internally, the most prominent feature of apicomplexans is the apical complex located at the anterior end of the sporozoite and merozoite stages (Fig. 3.14). This feature is characteristic of the phylum. The complex has five distinct components: (1) polar rings, consisting of one or more electron-dense rings at the most anterior position of the cell; (2) the conoid, which is inside the polar ring and is composed of a number of coiled microtubules; (3) the micronemes, which are elongated tubular organelles arranged longitudinally in the anterior part of the cell; (4) rhoptries, which are tubular or saccular organelles extending from inside the conoid back longitudinally into the cell body; and (5) subpellicular microtubules, which extend away from the polar ring into the posterior part of the cell. The function of these substructures is not fully understood, although it has been suggested that the rhoptries are involved in secretion, and the entire complex may be involved in both attachment, i.e., alignment of the apical complex adjacent to the red blood cell surface membrane (in the case of *Plasmodium* spp.), and penetration of the parasite into the specific host cell (Sam-Yellowe, 1996). The subpellicular microtubules give structural support and contribute to the general shape of the cell.

The definitive hosts, the intracellular sites within the host, and the mode of transmission differ considerably between the various groups of Apicomplexa. For

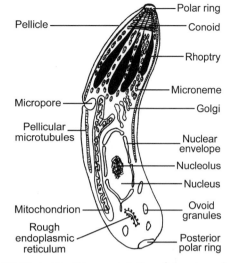

Fig. 3.14 Diagrammatic representation of an apicomplexan merozoite showing the key structures visible via electron microscopy, especially the apical complex. (Figure courtesy of Richard Seed.)

example, species within the Order Eucoccidiorida such as *Eimeria* spp. inhabit intestinal cells of vertebrates. They have no vector and transmission is direct, i.e., by the ingestion of oocysts in contaminated food or water. In contrast, invertebrate vectors transmit the blood-dwelling haemosporidian apicomplexans, e.g., *Plasmodium* spp. They are primarily parasites of vertebrates where they inhabit blood and hepatic cells. In the insect vector, the parasites occupy a number of sites, including the stomach, the body cavity, and the salivary glands.

In many species, a thick-walled oocyst is excreted into the external environment. Inside the mature oocyst, sporozoites are found within individual sporocysts. The sporozoite is the infective stage. The sporocyst wall is also thickened and the numbers of sporozoites within individual sporocysts vary. Similarly, the number of sporocysts per oocyst also varies between the different species and genera. For example, *Cyclospora* spp. have two sporocysts per oocyst, and each sporocyst contains two sporozoites. *Toxoplasma gondii* also has two sporocysts per oocyst, but each sporocyst contains four sporozoites. In contrast, *Eimeria* spp. have four sporocysts per oocyst and each sporocyst contains only two sporozoites. There may be more than 16 sporozoites per sporocyst, as well as more than 16 sporocysts per oocyst. Both the number of sporozoites per sporocyst and the number of sporocysts per oocyst are diagnostic characters at the generic level. The sporozoites have a typical nucleus, mitochondria, Golgi, endoplasmic reticulum, microtubules beneath the pellicle, and various refractory globules.

3.3.4.3 General life cycle

The basic life cycle of the apicomplexans has both sexual and asexual phases (Fig. 3.15). Sporozoites released from sporocysts, or from cells in the vector, invade a new host cell, and grow as trophozoites that, in turn, divide repeatedly by multiple fission to form merozoites. Depending on the species, many generations of merozoites may be produced. Eventually, some of the merozoites will emerge, invade new cells, and differentiate into either micro- or macrogametocytes (via gametogenesis or gametogony). The merozoite

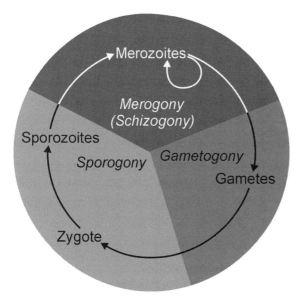

Fig. 3.15 Schematic representation of a generalized life cycle of the Apicomplexa. (Figure courtesy of Richard Seed.)

may form a single gametocyte, or it may divide repeatedly to form multiple gametocytes. Following release (or ingestion in the case of vectored types) of the micro- and macrogametocytes, and their differentiation into 'male' or 'female' gametes, the microgamete will fuse with a macrogamete forming a diploid zygote. Meiosis then occurs, followed by repeated divisions. The cycle is repeated with the formation of infective sporozoites (sporogony) and merozoites (merogony). As we will see, there are several variations on this general pattern.

3.3.4.4 Biodiversity and life-cycle variation

Currently, two classes of Apicomplexa are recognized, the Conoidasida and the Aconoidasida (see Box 3.7). This taxonomic distinction is based on the presence or absence of a truncated cone of spiral fibrils within the polar rings called the **conoid**. In reviewing diversity of conoidasidans we focus our attention on the biology of so-called gregarines within the subclass Gregarinasina and coccidian apicomplexans within the subclass Coccidiasina. The latter includes such important genera as *Cryptosporidium*, *Eimeria*, and *Toxoplasma*. *Plasmodium* spp. are by far the most important representatives of the Aconoidasida and will be the focus of our attention in this section.

Conoidasida

The subclass Gregarinasina contains numerous different species. One genus alone, *Gregarina*, contains approximately 260 species. Members are generally monoxenous, i.e., they infect only a single type of host. They are usually parasites of the digestive tract of invertebrates, particularly the insects. They are also parasites of echinoderms, molluscs, and annelids.

Most gregarines do not have a merogony phase in their life cycle. They increase in number solely through sporogony. The oocysts (spores) are released from the infected host and are transmitted to a new host by ingestion of the oocysts. Following ingestion, sporozoites emerge from the oocyst in the intestinal tract and enter an epithelial cell where they increase in size.

For most gregarines, the trophozoites eventually leave the host cell but remain temporarily attached to it by their anterior end.

The anterior end of gregarines contains a modified conoid that is involved in attachment of the trophozoite to the host cell. The anterior segment does not contain a nucleus and is separated from the posterior end by a septum. The posterior segment contains only a single nucleus and therefore the anterior and posterior segments should not be considered two cells. In addition to the modified conoid apparatus, gregarines also contain mitochondria, a Golgi apparatus, and numerous granules presumably containing stored carbohydrates, lipids, etc.

Gregarines eventually detach or break away from the host cell and wander in the gut or body cavity of the host. The wandering trophozoites will become gametocytes. Two gametocytes become attached and encyst. The gametocytes within the cyst (or gametocyst) undergo nuclear division and, eventually, individual nuclei bud off as gametes. The gametes within the cyst fuse in pairs to become zygotes. The zygotes encyst, forming an oocyst that remains within the gametocyst. Division occurs within the oocyst (now a sporocyst) forming sporozoites and, ultimately, the individual sporocysts (or oocysts) are released from the gametocyst and the host into the external environment.

Lankesteria culicis provides an example of a specific gregarine life cycle. In this case, mosquito larvae ingest oocysts, and the sporozoites are then released to invade gut cells. The sporozoite grows into a large trophozoite that eventually is released into the gut lumen. When the mosquito larva pupates, the mature trophozoites enter the Malpighian tubules where they pair and begin the process of encystment. Repeated divisions then occur and there is the formation of a large number of gametes that fuse. The zygotes then develop into oocysts with each oocyst containing eight sporozoites. The oocysts are released into the lumen of the tubules, enter the intestine, and are released into the environment with the feces. Amplification of parasite numbers occurs during the processes of gametogony and sporogony.

One order, the Eucoccidiorida, has most of the known species in the subclass Coccidiasina. Not surprisingly, it also contains a majority of species that are of medical and veterinary importance. Members of the suborder Adeleorina, or the haemogregarines, such as *Haemogregarina* spp. and *Hepatozoon* spp., are distinguished by male and female gametes that develop in association with each other in a process called syzygy. They are primarily intracellular parasites of vertebrate red and white blood cells. They have an indirect life cycle, being transmitted by leeches, ticks, mites, and biting insects. Transmission may occur when a new host consumes an infected vector, or, in some cases, the vector's feces in which sporocysts are found. Infection is thus by various mechanisms, although in most cases the sporozoites actually enter the vertebrate through the mucous membrane. Haemogregarines follow the basic apicomplexan life cycle pattern, with sexual reproduction in the vector and asexual reproduction in the vertebrate host.

There are many important families and genera in the suborder Eimeriorina, more commonly known as the coccidia. Some species complete their life cycles in a single host. Sporocysts are passed in feces or urine, and are then ingested directly by a new host. Coccidians are usually distinguished on the basis of the number of sporocysts within the oocysts, as well as on the number of sporozoites contained within each sporocyst.

Cryptosporidium (=hidden spore) spp. are enteric coccidians that have been the focus of considerable research attention (Fayer, 1997; McDonald & Bancroft, 1998). 'Crypto' is of tremendous public health and veterinary importance. Thick-walled oocysts of this waterborne parasite are transmitted by oral–fecal contamination. These oocysts are incredibly resistant, even surviving chlorinated water treatment. This has led to major outbreaks of cryptosporidiosis in humans in various regions of the world, including North America and Australia.

The life cycle of 'crypto' is characterized by having a merogony phase, micro- and macrogametes, oocysts containing four naked sporozoites, and a monoxenous life cycle (O'Donoghue, 1995; Smith & Rose, 1998). Sporocysts in the thick-walled oocysts excyst in the intestine and invade intestinal epithelial cells, producing merozoites by schizogony and forming a schizont. These rupture and invade new intestinal cells or form into gametocytes which produce micro- and macrogametes. These fuse to form a zygote and the highly protective oocyst that is shed in the feces of the mammalian host. A complicating feature of the *Cryptosporidium* life cycle may be that so-called thin-walled oocysts can also form in the host. At least under laboratory conditions, these can rupture and start the whole infection cycle over again. This process is known as autoinfection.

The species-level taxonomy of *Cryptosporidium* is complex. *Cryptosporidium* spp. are considered true coccidians and are placed in the same order as species of *Eimeria* and *Isospora*. However, recent phylogenetic studies suggest that *Cryptosporidium* spp. may be the most distantly related genus in the order. Thus, Leander and Keeling (2003) suggest a stronger phylogenetic relationship between *Cryptosporidium* and gregarines. There have been over 20 different species of *Cryptosporidium* described, many based on differences in the host species from which they were isolated. At the present time, 16 species are considered valid based on oocyst morphology (e.g., size differences), host specificity, level of virulence, and site of infection (Power & Ryan, 2008). Virulence is highly species- and strain-dependent. Host specificity is also variable, although most *Cryptosporidium* spp. are restricted to a single vertebrate class (O'Donoghue, 1995).

Pathology is often associated with *Cryptosporidium* infections in both domesticated and wild animals. The pathology appears to be dependent on the route of inoculation, the age of the host, and the particular host species infected. *Cryptosporidium parvum* can cause severe gastrointestinal distress, including massive diarrhea in humans, especially in immunocompromised individuals, in whom it can lead to death (McDonald & Bancroft, 1998). It may be one of the most common causes of both short-term and chronic diarrhea in humans. Cryptosporidiosis is a zoonosis and cattle are often implicated in the epidemiology of the disease. Given that a cow can shed several billion oocysts in a single day, the potential for contamination of water supplies by cattle manure is astounding.

Species of *Eimeria* are often restricted to a single host species. *Eimeria* spp. infect cells of the intestinal mucosa of all classes of vertebrates and may cause serious disease in a number of domesticated animals. For example, *Eimeria tenella* causes cecal coccidiosis and high mortality rates in juvenile domestic chickens. Several other species have been described from domestic animals, including *Eimeria bareillyi* in cattle (Fig. 3.16). Remarkably, there may be well over 30 000 species in the genus *Eimeria* alone, with most being highly host- and site-specific. One, or a limited number, of cycles of merogony characterize their life cycles. Merozoites (Fig. 3.16A) enter epithelial cells in the cecum and undergo gametogony, developing into micro- and macrogamonts (Figs. 3.16A, B). Oocysts are produced and excreted into the environment. The oocysts are thick-walled (Fig. 3.16D) and represent the environmentally resistant stage in the life cycle.

Species of *Isospora* have two sporocysts per oocyst and four sporozoites in each sporocyst. Molecular phylogenies show a close relationship between *Toxoplasma*, *Neospora*, and *Isospora*. A more distant relationship is seen between *Sarcocystis*, *Eimeria*, and *Cyclospora* (Carreno *et al.*, 1998). *Cryptosporidium* spp. appear to be the most taxonomically distant group based on the differences in their 18S ribosomal DNA sequences as well as in their life cycles. It has been stated that many, possibly all, of the Eimeriidae that have an oocyst containing two sporocysts, each of which contains four sporozoites, have two hosts in their life cycle. In many cases, the sexual stages occur in the definitive host, whereas the asexual stages are found in an alternate host. Carnivores often serve as the definitive host; alternate hosts are accidentally infected by ingesting oocysts excreted in the feces. Species of the *Isospora* type infect intestinal cells, as well as many other tissues in a variety of vertebrate species.

Based on molecular phylogeny, *Toxoplasma* spp. are related to, but are distinct from, the *Isospora* spp. The life cycle of *T. gondii* is complex; while cats are the definitive hosts, several species of birds and mammals

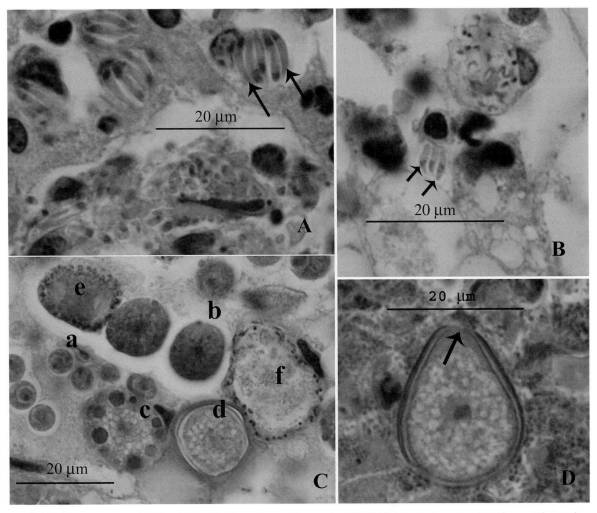

Fig. 3.16 Endogenous stages of *Eimeria bareillyi* in sections of ileum of a calf. (A) Schizonts in entrocytes. The crypt lumen is plugged with degenerating merozoites. Note elongated merozoites with terminal nucleus (arrows). (B) A small schizont with three small merozoites (arrows). Compare the sizes of merozoites in parts A and B; both photos are of comparable sizes. A microgamont is located above the schizont. (C) Gametogonic stages: a – small macrogamont with a large nucleus, b – nearly mature macrogamont with a central nucleus, c – macrogamont with eosinophilic wall-forming bodies, d – an oocyst, e – immature microgamont with peripherally located nuclei, and f – microgamont with peripheral microgametes. (D) A longitudinally cut oocyst with two oocyst walls and a micropyle (arrow). (Photographic plate courtesy of Jitender Dubey.)

can act as intermediate hosts (Cox, 1993). *Toxoplasma gondii* has even been recovered from marine mammals, and has been implicated as a source of mortality in sea otters in California (Conrad *et al.*, 2005). The parasite has a worldwide distribution and a high prevalence in humans, estimated at approximately 23% of the world's population. Fortunately, however, the infection is usually benign. On the other hand, it may cause severe pathology in a developing fetus when infected *in utero*, causing a tragic condition known as congenital toxoplasmosis. The parasite can also be a serious pathogen in severely immunosuppressed patients, e.g., those with AIDS (see Dubey, 2007).

The life cycle of *T. gondii* and other members of the Sarcocystidae have both intestinal and tissue stages. Cats become infected by either accidentally ingesting infective oocysts in contaminated soil, or by eating the tissues of infected prey in which are found tissue cysts containing **tachyzoites** (*tach*, Greek=swift). Although *T. gondii* reproduces asexually in a variety of different hosts and cells, both merogony and gametogony are restricted to the intestine of the cat. Humans become infected by accidental ingestion of oocysts from cat feces, e.g., cleaning the cat litter box, or by the ingestion of **zoitocysts** (tissue cysts, containing **bradyzoites** (*bradys*, Greek=slow)) in infected undercooked meat. Infective sporozoites released from ingested oocysts, or infective stages released from tissue cysts, penetrate the gut wall and infect macrophages. The trophozoite-infected macrophages then migrate throughout the body. The trophozoites divide repeatedly within the cell, eventually forming the tissue cyst. When the infected host cell dies and the trophozoites, now called tachyzoites, are released, they may infect cells in all tissues and organs of the body. When the proliferative stage ends, the tachyzoites enter new cells and develop slowly. The dividing stages occur in a true cyst, containing bradyzoites, with a thin, protective wall. The cysts can also be found throughout the body and the bradyzoites can remain viable in the cyst for several years. They may become active if the infected host is immunocompromised, or if the muscles containing the tissue cyst are consumed by a new host. Therefore, although the host immune response does not eliminate the infection, it would appear to control it and, as a result, the infected host may have few, or no, clinical symptoms (Alexander & Hunter, 1998).

Neospora caninum infects many cells (Fig. 3.17) of many mammalian species, especially dogs and cattle. Cell death can occur due to the rapid replication of tachyzoites (Fig. 3.17). Neosporosis is recognized as a major cause of abortion in dairy cattle. In dogs, severe infections leading to paralysis can occur in congenitally infected pups.

Species of *Sarcocystis* are parasites in a wide variety of mammals and birds. They are most

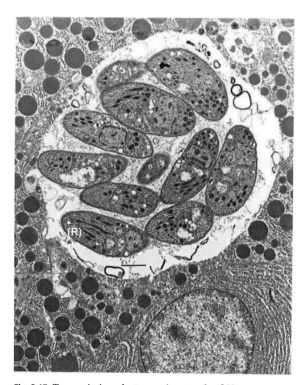

Fig. 3.17 Transmission electron micrograph of *Neospora caninum* tachyzoites in pancreatic acinar cells of a mouse. Note the cluster of tachyzoites in the cytoplasm of the host cell and the electron-dense rhoptries (R). (Scale bar = 1 μm). (Modified from Koyama *et al.* 2001, with permission, *Journal of Parasitology*, **87**, 1486–1488.)

commonly observed in cattle, horses, sheep, pigs, monkeys, ducks, chickens, and humans. Usually, the infection goes unnoticed and is only detected at necropsy. However, mild to fatal infections have been observed in mice. The stages of the life cycle in the definitive host are of the *Isospora* type in which merogony and gametogony occur in the intestinal epithelial cells of the definitive host. The oocysts contain two sporocysts and four sporozoites per sporocyst. In the intermediate host, the stage observed is the **sarcocyst**, which can be quite large and contain numerous trophozoites. The cyst wall is complex, consisting of both host and parasite material. The definitive host is always a carnivore or an omnivore.

Aconoidasida

Two orders of Aconoidasida are recognized: the Piroplasmorida and the Haemosporida (see Box 3.7 below). Our focus is on the latter. Briefly, the piroplasms include the intracellular parasite genera *Theileria* and *Babesia*. Typically, both undergo merogony in RBCs or lymphocytes and infect a variety of mammals. Some piroplasm species are of veterinary significance, e.g., *Babesia bigemina* and *Theileria parva*, and are important agents of disease (babesiosis, theilerosis) of domestic mammals, especially cattle and other ruminants (Homer *et al.*, 2000). In highly susceptible hosts death due to *Theileria parva* can occur very quickly; the pathogenicity is related to strain specificity and the host's immune response (Morrison & McKeever, 1998). Ticks acts as vectors for piroplasms and gamete maturation, fusion, and sporogony occurs in specific tick species. In some *Babesia* species, transovarial transmission occurs in the developing eggs.

Members of the order Haemosporida are also obligate intracellular parasites for most of their life cycles. They are acquired from a vertebrate host by blood-feeding insect vectors such as mosquitoes, sand flies, black flies, or midges. Merogony and gametogony occur in the vertebrate host. Merogony takes place in liver and RBCs, whereas gametogony occurs only in RBCs. Fertilization and zygote formation are restricted to the insect vector.

Parasitologists currently recognize more than 500 species (15 genera) within the order Haemosporidia, approximately 200 of which infect birds. The remainder are parasites of reptiles (especially lizards and turtles) and mammals. As for many protist groups, these estimates probably underestimate the true biodiversity of the 'malaria' parasites. Molecular phylogenetic analyses have provided extraordinary insights into the evolutionary diversification and radiation of this group, including the four main *Plasmodium* spp. of humans. Martinsen *et al.* (2008) constructed a comprehensive phylogeny of over 50 Haemosporidia from birds, mammals, and lizards based upon DNA sequences of four genes. While certain components of their phylogeny confirmed earlier classifications of genera based upon morphology, life cycle, and host taxa, other components differed strongly. Remarkably, they identified several major clades, each of which was associated with a shift into different families of dipteran vector. Thus, *Plasmodium* spp. of mammals (including *P. falciparum*) formed a single clade associated with specialization in *Anopheles* mosquitoes. The *Plasmodium* of birds and reptiles also formed a single clade with specialization in culicid mosquitoes. These results confirmed the earlier contention by Escalante & Ayala (1994) that vector host switches were central to the evolutionary history of malaria parasites based upon analysis of divergence times of various Haemosporidia and their vectors.

By far, the most well known and important haemosporidan genus is *Plasmodium*, members of which cause human malaria. Historically, it is well known that a combination of malaria and yellow fever significantly impeded progress on construction of the Panama Canal early in the twentieth century. In Ontario, Canada, the construction of the Rideau Canal which links Ottawa to Lake Ontario was associated with severe outbreaks of *P. vivax* in the early 1820s. Malaria adversely affected the human settlement of a highly malarious area of southern Europe prior to vector control. Malaria caused significantly more casualties among allied soldiers in the South Pacific during World War II than bullets, and some historians feel that the fall of the Roman Empire was in no small part due to the severe ramifications created by malaria. Malaria continues to be one of the most devastating diseases ever encountered by humankind (review by Esch, 2007).

The history of the discovery of the causative agent for malaria and then the life cycle of *Plasmodium* has been extensively described (reviews by Desowitz, 1991; Esch, 2007). Enquiry into the fascinating phenomena of relapse and recrudescence in human malaria (Box 3.5), and malaria pathophysiology, also has a long history. Indeed, the manner in which hypotheses regarding the transmission of *Plasmodium* to humans shifted from 'bacterial and enteric' to 'protist and vectored' may represent one of the most significant historical developments in parasitology (see also Box 1.1). In 1902, following a series of infection

Box 3.5 | Relapse and recrudescence in malaria

One of the peculiar phenomena unique to several species of primate *Plasmodium* (as well as species of *Haemoproteus* and *Leucocytozoon* in birds) is their ability to relapse, or, alternatively, undergo recrudescence. The relapsing phenomenon is associated with *P. vivax* and *P. ovale*, whereas recrudescence is linked to *P. falciparum* and *P. malariae*. For all four species, clinical manifestations of the disease begin after a species-specific number of days following inoculation of sporozoites by infected female anopheline mosquitoes. Four weeks to a few months after the first clinical signs of disease, the pathological manifestations of infection gradually decline and stop, and the host no longer has patent parasitemia. In the case of *P. vivax* and *P. ovale*, parasites disappear from the blood, but not from the liver. In the case of *P. falciparum* and *P. malariae*, the parasites are absent in the liver, but occur in the vascular system. For all four species, patent parasitemia is absent, but, depending on the species of parasite, it is likely to return.

Relapse has long been known to occur, but its cause was in much dispute until the early 1970s. The older line of thought held that the phenomenon was in some way related to hepatic recycling, i.e., that exoerythrocytic schizogony was cyclic and that latency in the pattern was mixed because of genetic variation among strains of the parasite. The second idea suggested that some sporozoites (also known as **hypnozoites**) within an inoculum were pre-programmed for dormancy in hepatic cells. According to Krotoski *et al.* (1982), "Hypnozoites arise from genetically pre-programmed sporozoites. Thus, each subpopulation of sporozoites would be pre-programmed to form a subpopulation of hypnozoites which then commence to grow into pre-erythrocytic schizonts in hepatocytes at a pre-determined time, some immediately, others at varying times over subsequent months." The second hypothesis seems the more widely accepted. McKenzie *et al.* (2008) state, "It is disconcerting to realise how recently the hypnozoite stage of *P. vivax* was discovered. . .and how little progress has been made since then in understanding how hypnozoite formation and reactivation is controlled. . .". They continue, "Nevertheless, the ability to genotype relapsing infections has led to confirmation that relapses show a strong tendency to be clonal and that multiple relapses in a single patient reflect reactivation of different parasite genotypes (Chen *et al.*, 2007; Imwong *et al.*, 2007)."

Recrudescence is another matter entirely. It is known that recrudescence occurs in *P. malariae* and *P. falciparum*, as well as *Leucocytozoon simondi*. The difference between recrudescence and relapse is simple. In the former, the source of re-infection is not from the liver, but from stages present in the vascular system. It is possible that non-detectable erythrocytic schizogony continues after patent parasitemia disappears, followed by resurgence of the disease, accompanied by patent parasitemia and the full range of signs and symptoms we recognize as malaria. Others suggest that recrudescence is caused by intracellular parasites that become sequestered somewhere in endothelial cells of the vascular system and, at some point in time for some

Box 3.5 | (continued)

unknown reason, are stimulated to resume development. *Leococytozoon simondi* in geese can be 90% lethal in goslings. In those that survive until the next spring, concomitant immunity will develop while the parasites will undergo 'benign' recrudescence in preparation for transmission by black flies to the next gosling cohort. McKenzie *et al.* (2008) review the remarkable capacity for relapse/recrudescence by malarial parasites.

trials involving anopholine mosquitoes allowed to feed on infected patients (typically soldiers), the British military doctor, Ronald Ross received the Nobel Prize for completing critical components of the life cycle of *P. vivax*. A few years later, the Italian parasitologist Giovanni Grassi completed the life cycle of *P. falciparum.*

The life cycles of *Plasmodium* spp. are similar (Fig. 3.18). Various species of mosquitoes act as vectors. An infected female mosquito injects sporozoites when feeding on blood from the vertebrate host. They migrate via the bloodstream to the liver, penetrate liver cells, and undergo schizogony, producing merozoites. Merozoites are released and penetrate RBCs, undergoing schizogony and producing more merozoites. These rupture from RBCs and penetrate new ones. Some transform into male microgametocytes and female macrogametocytes within RBCs.

Successful transmission requires that gametocytes be ingested by the specific mosquito vector. Following ingestion, the male gametocyte undergoes a remarkable transformation in a process called **exflagellation**, typically producing eight gametes within the mosquito midgut (Fig. 3.18). In contrast, each female gametocyte produces only one gamete. Sex chromosomes are absent in *Plasmodium* spp., yet a single clone of *P. falciparum* can produce males and females and all-male or all-female gametocytes can originate from a single sexually committed merozoite (Smith *et al.*, 2000). Parasitologists and evolutionary biologists have long been intrigued by the extensive variation that occurs in gametocyte sex ratios (which are

typically female biased) of *Plasmodium* spp., between species of vertebrate host, between geographical areas, and within an infection cycle.

The male microgametes actively locate and then fertilize female macrogametes within 30 minutes of ingestion in the blood meal, forming a motile zygote called the **ookinete**. The ookinete penetrates the mosquito's gut wall and forms an oocyst on the hemocoel side of the gut. Like other apicomplexans, sporogony occurs within oocysts. Sporozoites rupture from the oocyst and migrate to the salivary gland to begin the life cycle anew (Fig. 3.18).

There are four main species of *Plasmodium* that infect humans, although at least two others have been recognized as a causative agent (Singh *et al.*, 2004). They differ from each other in such characteristics as the morphology of the erythrocytic stages, the temporal duration of schizogony, and the nature of the clinical disease they produce. The different morphological stages of *Plasmodium* observed microscopically in stained human blood films are shown in Figures 3.19 and 3.20. Several of the diagnostic differences used to distinguish between the human *Plasmodium* spp. in stained blood films are shown in Fig. 3.21. The current tools used to diagnose human malaria are more fully discussed in Box 3.6.

The four species of *Plasmodium* that infect humans cause a serious, debilitating fever, accompanied by violent shaking. However, usually only *P. falciparum* produces fatal infections. Malaria affects some 500 million people in the world and kills about 1–3 million annually, mostly children and mostly in sub-Saharan

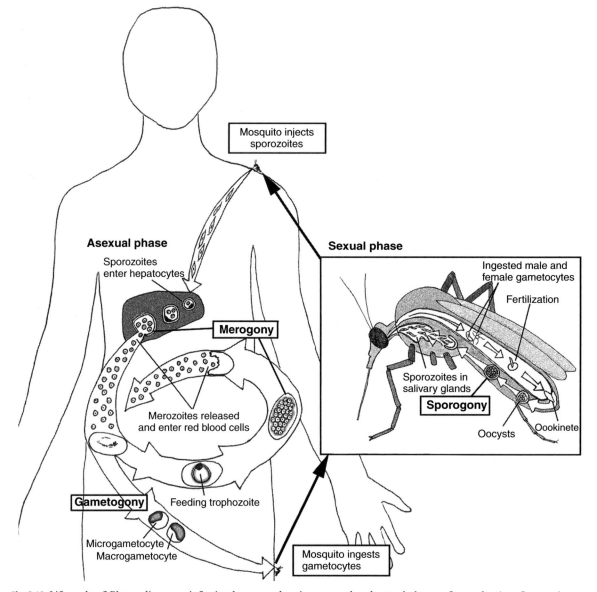

Fig. 3.18 Life cycle of *Plasmodium* spp. infecting humans, showing asexual and sexual phases of reproduction. Sporozoites are injected into humans by a mosquito vector. Merogony occurs within hepatocytes and RBCs of humans. Gametogony results in production of micro- and macrogametocytes; these are ingested during a mosquito's blood meal. Fertilization, resulting from the union of microgamete and macrogamete, occurs within the mosquito midgut. Sporogony occurs within oocysts, producing sporozoites. (Figure courtesy of Chelsea Matisz.)

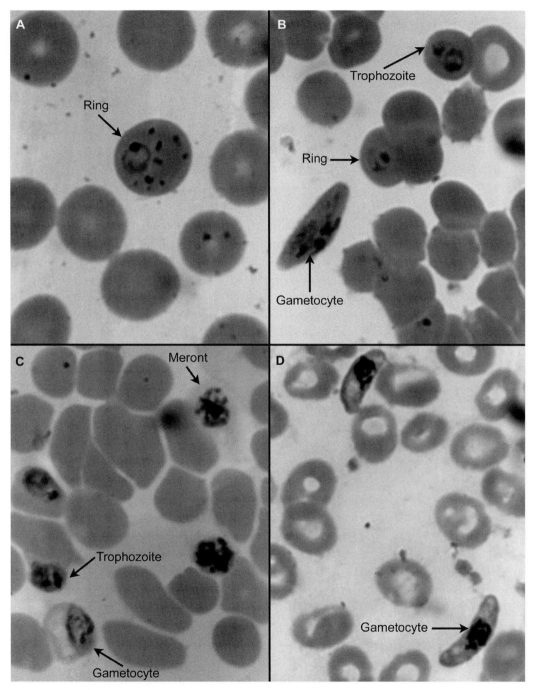

Fig. 3.19 Morphology of the various life cycle stages of *Plasmodium falciparum* in stained, thin blood smears. (A) The ring stage of *P. falciparum* in infected RBCs. Maurer's dots in the infected cells can also be seen. (B) Ring, trophozoite and the banana-shaped gametocyte stages are present. The gametocyte-infected RBC is greatly distorted in shape. (C) Trophozoite, a late-stage meront (or schizont), and an early gametocyte stage are seen. (D) Gametocyte-infected RBCs are seen. Again, note the banana shape of the infected cell. All of the different morphological stages observed in stained, thin blood smears of *P. falciparum*, except the free merozoites, can be compared. (Photographs courtesy of Purnomo Projodipuro and Michael Bangs.)

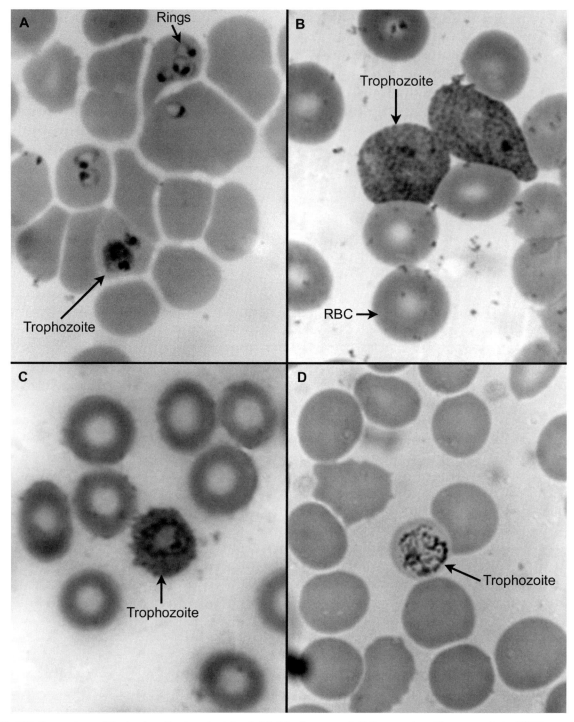

Fig. 3.20 Comparison of the trophozoite stage in stained, thin blood smears of all four species of human malaria. (A) Ring and early trophozoite stages of *P. falciparum* are present. (B) Trophozoite stage of *P. vivax*. The trophozoite-infected RBC is irregular in shape, the RBC is enlarged, and there is abundant cytoplasmic pigment (Schüffner's dots). (C) Trophozoite stage of *P. ovale*. (D) The trophozoite stage of *P. malariae*. (Photographs courtesy of Purnomo Projodipuro and Michael Bangs.)

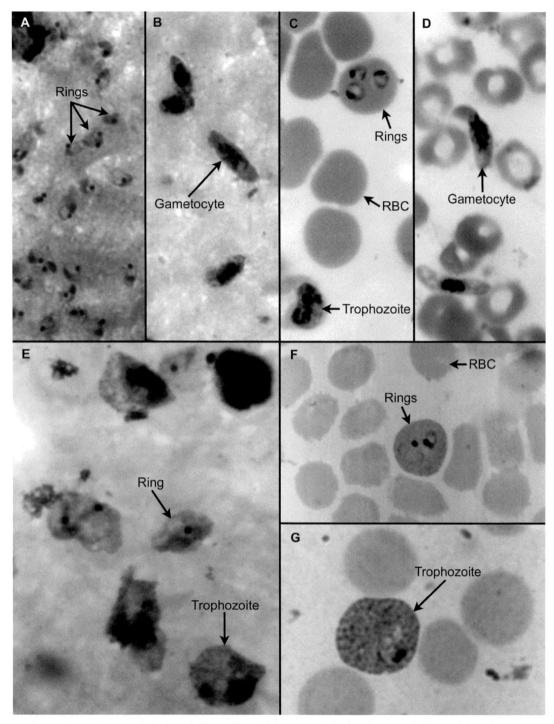

Fig. 3.21 Comparison of the morphologies of *Plasmodium falciparum* and *P. vivax* in stained thick and thin blood smear preparations. (A) Thick blood smear of *P. falciparum* in which numerous ring stages can be seen. (B) Thick blood smear of

Box 3.6 | Diagnosis of human malaria

The diagnosis of human malaria is based on three different methods, i.e., microscopic, immuno-logical, and molecular techniques (review in Garcia, 2007). The immunological protocols can be divided into those that detect antibodies to *Plasmodium* antigens or procedures that directly assay for *Plasmodium* antigens in infected blood. Antibodies to *Plasmodium* can be detected by several different serological tests. The presence of anti-*Plasmodium* antibody in infected serum indicates either a past or a present infection. It cannot distinguish between these two possibilities. It is, therefore, of limited value as a guide for treatment. In contrast, the direct detection of *Plasmodium* antigens in human serum indicates a current infection and can be used as a guide for disease management and chemotherapy. A commercial, simple-to-use antigen detection kit is now available. The test takes only 10 minutes to run, and can be easily used by health care workers.

The most sensitive assay for the detection of any parasitic organism is through the use of molecular tools. It is suggested that 20 parasites/ml of blood can be detected by real-time PCR, making this procedure relatively sensitive. This, and other molecular techniques, holds great promise. In these assays, specific parasite DNA within a blood (or tissue) sample is amplified many times and then detected by hybridization of the amplified gene(s) with known parasite-specific gene probes. However, at the present time, molecular techniques cannot be performed in the field, they require expensive equipment and reagents, and they have not achieved their potential sensitivity.

The diagnostic 'gold standard' is still the microscopic detection of *Plasmodium* in infected blood using stained thick films (Swierczynski & Gobbo, 2007). Several recent modifications permit easier detection as well as reducing the time required to examine a slide. One is by the use of acridine orange which stains the parasite. This makes it easier to detect in thick films. In addition, by centrifugation in a specifically designed and patented micro-capillary tube, the parasites are concentrated in one area of the tube (the buffy coat area). The parasites and white blood cells are then stained with acridine orange. This protocol, the quantitative buffy coat (QBC) method, is fast, easy to perform, and easy to read. However, in comparison to the thick blood film assay, it requires extra equipment and the greater

Caption for Fig. 3.21 (cont.)

P. falciparum in which gametocytes are present. (C) Thin blood smear of *P. falciparum* in which the ring stages and a trophozoite can be seen. (D) Thin blood smear in which gametocytes are present. Note that in the thick blood smears (A, B), it is more difficult to see the morphological details of the various stages; however, many more infected RBCs can be detected. An experienced microscopist more easily detects an infection in a thick blood smear, especially when the parasitemia is low. (E) Thick blood smear of *P. vivax* in which ring and trophozoite stages can be seen. (F) Thin blood smear of *P. vivax* showing rings and (G) a trophozoite stage in infected RBCs. Again, by comparing the ring stages in the thick and thin preparations, it can be observed that the morphological details are more difficult to see in thick preparations. (Photographs courtesy of Purnomo Projodipuro and Michael Bangs.)

Box 3.6   (continued)

expense of the capillary tubes. In addition, it is apparently more difficult to distinguish between the different species of *Plasmodium* than by the use of the simple thick blood film. Unfortunately, the sensitivity of the serological assays (either the antibody, or the *Plasmodium* antigen detection protocols) is, to date, no better than the microscopic examination of stained thick blood films, which do not require the expensive reagents, and/or additional equipment. The most promising new diagnostic tool would appear to be combining the colorimetric PLDH assay with the capture of PLDH from infected blood by anti-PLDH. As noted, this quick, easy-to-use assay requires no special equipment and has the sensitivity of the microscopic examination of stained thick blood films. Further work is required to determine if this assay can distinguish between the four species of *Plasmodium* and to determine if the cost of the commercial kit is within the means of those countries where infections are endemic. At the present time, it appears that the microscopic examination of stained thick blood films will continue to be used in most endemic areas for some time into the future.

Africa (Teklehaimanot & Singer, 2005). It is estimated that over 40% of the world's population is at risk of infection.

As we have noted, the morbidity associated with malaria is staggering. The debilitating nature of the disease is due to the complicated nature of the pathophysiology (reviews by Mehdis & Carter, 1995; Sherwin, 2005; Haldar *et al.*, 2007). Much is known about malaria, but a great deal remains to be learned. The pathophysiology considered here will include the paroxysm, anemia, pigmentation and organ enlargement, capillary thromboses and cerebral malaria, and renal failure.

The classic paroxysm begins with chills, convulsions, and shaking, collectively known as rigor (pronounced RYE-gor). These symptoms are abruptly followed by a burning fever, violent headache, and nausea. The body temperature may reach 39.4–40.6 °C and there will be profuse sweating. After about 10 hours, body temperature returns to normal. The attacks typically begin between midnight and noon. The fevers are recurrent and peaks appear to be synchronous with the release of the merozoites from infected RBCs. Since

the duration of the erythrocytic merogony phase differs among three of the four species of *Plasmodium*, the length of time between paroxysms and fever peaks also differs. *Plasmodium ovale* and *P. vivax* have a paroxysm and fever peak every 48 hours (tertian), whereas *P. malariae* has a longer erythrocytic merogony phase and its paroxysm occurs every 72 hours (quartan). The differences in the time interval between paroxysms and their clinical severity have led to different names for the disease produced by each *Plasmodium* spp. For example, the clinical disease produced by *P. falciparum* is referred to as malignant subtertian malaria and the disease produced by *P. vivax* as benign tertian malaria.

Evidence suggests that a malaria toxin, upon release, induces the production of a cytokine called tumor necrosis factor (TNF; see Chapter 2). The most likely toxin candidate is a lipid moiety. TNF has been suggested to play a central role in malaria pathology in that it heightens the inflammatory response, making it a proinflammatory cytokine (Mackintosh *et al.*, 2004). High serum levels of TNF have been associated with cerebral malaria. It has also been shown that the 'malaria toxin' can increase the production of interferon (IFN) *in vitro*. It has been suggested that the pre-inflammatory cytokines induce an increase in nitric oxide (NO) that, in turn, ultimately induces part of the pathology observed in human malaria (Bordmann *et al.*, 1997; Clark *et al.*, 1997; Smith *et al.*, 1998).

The increase in merozoite numbers in host blood appears to occur synchronously, and the rate of development appears to be specific to different species. Thus, in *P. falciparum*, peak parasitemia is reached in 36 to 48 hours, with the synchronous release of the merozoites and periodic bouts of fever approximately every 2 days. In contrast, the merozoite increase in *P. malariae* takes 72 hours and there is a fever peak approximately every 3 days. Peak parasitemia (merozoite and also micro- and macrogametocyte numbers), a high fever with a high metabolic rate, and increased CO_2 release have been hypothesized as important factors in parasite transmission. Peak parasitemia, including the number of gametocytes, is thought to occur during periods of maximum mosquito activity, and the increased exhalation of CO_2 may act as an attractant to vectors. Therefore, it is possible that parasite development and host pathology have evolved to maximize parasite transmission.

A serious consequence in destruction of the RBCs is anemia. There is also phagocytosis of both infected and uninfected RBCs, especially in the spleen, and this serves to exacerbate the anemia problem. Mackintosh *et al.* (2004) attribute removal of RBCs from circulation to complement-mediated lysis and phagocytosis that result "from immune complex deposition and complement activation." Some evidence suggests that iron incorporation is slowed in erythropoiesis and that RBC production may be suppressed (see also Sexton *et al.*, 2004). Multiple vascular thromboses in the brain, especially in falciparum malaria, also add to the anemia problem. Moreover, the thromboses produce plasma loss from the blood vascular system. At autopsy, both the liver and spleen will be black and the white matter in the central nervous system may be slate gray in color. The pigment accumulation in the spleen and liver is due to the phagocytosis of **haemozoin** (a by-product of hemoglobin metabolism by the trophozoite) by fixed

macrophages. In the brain, it is the result of haemozoin accumulation produced when blood vessels burst because of the multiple vascular thromboses. Splenomegaly in malaria is due to hyperplasia or an increase in the number of cells, primarily macrophages.

The surface of RBCs infected with trophozoites and merozoites of *P. falciparum* develop characteristic 'knobs' that are also associated with cerebral malaria. These structures become sticky, ultimately leading to the sequestration of RBCs to the endothelial linings of capillaries and post-capillary venules, primarily in the brain. This cytoadherence, as it is known, involves surface ligands on the infected RBCs, as well as other specialized receptors on the endothelial cells lining the capillaries. Evidence suggests that cytoadherence is mediated in some way by the spleen because it does not occur in splenectomized monkeys. The adaptive significance of this radical, parasite-induced alteration to the structure of the host cell is probably increased survival. Thus, by 'sticking' to the walls of blood vessels, individual parasites can complete their development without risking filtration by the host's spleen.

Another phenomenon, called rosetting, involves the binding of infected and uninfected RBCs. It occurs primarily in the brain of hosts in which sequestration is known to occur, e.g., in humans with *P. falciparum*. When cells become sticky, they cause the formation of multiple vascular thromboses, or clots. Following formation of the clots, pressure builds, and the vessels burst. There will then be localized hemorrhaging. When the blood supply is interrupted in these tissues, localized anoxia and cell death takes place, followed by convulsions, coma, and death. It must be emphasized that the explanation regarding mechanical blockage in cerebral malaria has a number of etiologies and that its pathophysiology is not clearly understood.

As noted, with the exception of *P. falciparum*, death rarely occurs. Eventually the fever subsides and parasites become difficult to detect in the blood. The host immune response will not eliminate the infection but, with time, will keep it under control. Host immunity is age-related, possibly strain-specific, and, to be maintained, requires continuous exposure. Immunity involves T- and B-cells, as well as both Th-1 and Th-2

T-cells and their associated cytokine responses. The immune response is directed towards both the liver and the infected erythrocytes. Adaptive transfer and depletion experiments have demonstrated that CD4 cells are required for host immunity (Smith *et al.*, 1998), yet are also probably responsible for the pathology and the release of key cytokines. It has been known for a long time from work in both human and experimental animal models that antibody can dramatically reduce parasitemia. It would appear that the IgG class of antibody molecules is the most important in immunity. In humans, there is a strong epidemiological association between the IgG1 and IgG3 subsets of antibody to the asexual blood stages and immunity. These antibodies are known to promote phagocytosis of RBCs containing the more mature stages of merogony. The opsonized infected RBCs are primarily cleared in the spleen of the host. The intact spleen appears to be critical for maintaining immunity to the *Plasmodium* spp. parasites.

If an infected adult from an endemic area who shows few, or no, signs of malaria leaves for a relatively short time (a few months to a few years), and then returns to an endemic area, they appear to be as susceptible to re-infection as a previously unexposed individual. It has, therefore, been suggested that the development of a vaccine will be most difficult (but also see Engers & Godal, 1998). Experimental animals vaccinated with sporozoite or merozoite antigens may show good protection against a homologous challenge; however, a field vaccine for use in humans has not yet been developed. There are several possible reasons for this difficulty. One is that there appear to be numerous antigenic strains of *Plasmodium* spp. and some strains occur in only a very limited geographical area (Babiker & Walliker, 1997). Another reason is that *Plasmodium* spp. are capable of undergoing antigenic variation (Chapter 2). Therefore, any vaccine for use in the field must include multiple antigenic epitopes. Moreover, the parasites, except for very short time intervals during their life cycle, reside within host cells. Although infected cells eventually possess surface antigens of parasite origin, they are expressed on the host cell surface primarily during the later stages of parasite development and the window of opportunity for the host to respond to these parasite antigens is, therefore, limited. Moreover, *Plasmodium* spp. likely invade new host RBCs before the host can respond immunologically. Finally, the host is immunosuppressed during a malarial infection. It thus appears that the parasite may manipulate the host's immune response to its own advantage, thereby increasing the opportunity to successfully complete its life cycle.

Although complete protection and long-lived immunity does not occur naturally, the host has, in addition to the short-lived and age-related immunity, also evolved several RBC genotypes that confer some resistance to *Plasmodium* spp. It appears that *Plasmodium* parasites grow poorly in RBCs from humans with a variety of RBC abnormalities. For example, human carriers of the sickle-cell anemia gene have distorted RBCs that provide suboptimal habitats for *P. falciparum* merozoites. The gene leads to severe reductions in oxygen carrying capacity of homozygous individuals, often leading to death. Not surpisingly, evidence exists for counter-selection against the sickle-cell gene in many regions of the world. The exception occurs in areas where falciparum malaria was, or is, endemic, in which case heterozygotes are protected from serious forms of pathology. Maintenance of the sickle-cell gene in malarious regions provides one of our best examples of parasite-mediated natural selection (Piel *et al.*, 2010; see Box 16.1).

There has been a marked contraction in the worldwide range of malaria over the past 100 years (Gething *et al.*, 2010). Enormous efforts at disease and vector control and economic development since 1900 have reduced malaria endemicity to zero in the USA, Canada, Europe, the former Soviet Union, and elsewhere. In contrast, the prevalence of malaria seems to have exceeded its pre-1900 level in some regions of the world, such as east Africa (Krogstad, 1996). The causes of regional resurgences are difficult to evaluate due to the difficulty in isolating effects of temperature and precipitation from concurrent changes in socio-economic factors such as access to health care and to intervention programs. Dealing with these and other confounding factors has made it especially difficult to assess the accuracy of epidemiological models that link global climate change with anticipated range extensions of human *Plasmodium* and their vectors (Gething *et al.*, 2010).

One explanation for the localized resurgence of malaria is the evolution of drug resistance by *Plasmodium* spp. Massive use of chloroquine over the years has resulted in chloroquine-resistant *P. falciparum* throughout its range and resistant *P. vivax* is known to occur. Chloroquine works by disrupting a crucial hemoglobin digestion pathway essential to the parasite's survival in RBCs (Wellems, 1992). Specifically, the drug blocks an enzyme known as haem polymerase and results in poisoning of the trophozoite's food vacuole and leading to its eventual demise. Resistant *P. falciparum* does not concentrate the drug to high enough levels to inhibit haem polymerase. Consequently, scientists are racing against rapidly evolving *Plasmodium* spp. to find new drugs, or drug cocktails, to circumvent the evolution of resistance. In addition, although millions of dollars of research money over many years have gone into attempts to develop pre-erthyrocyctic, blood stage, and transmission blocking vaccines, to date, a completely effective vaccine is not yet available. However, as of 2011, clinical trials of the RTS,S vaccine (also known as Mosquirix) in Africa have shown considerable promise in protecting approximately 50% of children from malaria.

Another note of concern is that there appears to be a possible synergy between HIV and malarial infections, and there is evidence to suggest that malaria exacerbates the severity of HIV (Abu-Raddad *et al.*, 2006). It has been shown experimentally that a *Plasmodium* antigen can increase the rate of HIV replication *in vitro*. If malaria can increase the rate of progression of an HIV-infected individual to an active HIV infection, it could have a profound impact on morbidity and mortality rates in areas such as Africa that have a high prevalence of both infections.

From the parasite's perspective, this is a remarkable success story, certainly with respect to the human *Plasmodium* spp. We note that a WHO directive in 2007 has removed DDT from the 'banned' list and now permits the controlled use of the insecticide for mosquito control. It is clear that mosquito control is critical in malaria-stricken regions. Research into genetically modified mosquitoes to control malaria transmission is a current avenue of research. Insecticide-treated bednets have been shown to be highly effective in protecting people from the ravages of malaria, but resistance to the insecticide has begun to appear.

3.3.5 Ciliophora

3.3.5.1 General considerations

The ciliates comprise the largest and most ecologically diverse group of protists. Ultrastucturally, they are also the most complex of the protists, exhibiting the highest degree of organelle specialization and cytoarchitecture. The presence of external cilia and associated complex **infraciliature** at some stage of the life cycle are unique features shared by all members of the phylum. The complex structure of the infraciliature, including cords of fibers termed **kinetodesmata**, is an important tool used in ciliate taxonomic classification. A series of membranes forms the body covering, or pellicle of ciliates. A variety of organelles called **extrusomes** (e.g., trichocysts, toxicysts) are often associated with the ciliate pellicle. Ciliates have two morphologically different nuclei, a macronucleus that is involved in regulating metabolic activities of the cell, and a micronucleus involved in sexual reproduction. The macronucleus is polyploid and morphologies vary considerably among species. Generally, both nuclei can replicate when the cell divides by binary fission. During division, however, the nuclear membrane usually remains intact, and intranuclear microtubules are involved in the separation of the chromosomes. Sexual reproduction in the ciliates never involves gamete formation; instead, it occurs by a unique and complicated process called **conjugation** in which ciliates pair and there is a temporary, partial fusion of the cells. A cytoplasmic bridge forms between the two so-called conjugants, the micronuclei divide meiotically, and genetic material is exchanged between them.

Most ciliates are free-living and comprise a significant proportion of aquatic microplankton communities, forming complex food webs in freshwater and marine ecosystems. Ciliates also display an extraordinary variety of functional morphologies and behaviors, reflecting their enormous variation in life styles from active predators to microbotrophic suspension feeders. Tremendous adaptive modifications of the oral region, especially in the ciliated cytostome and cytopharnynx, reflect the variety of diets encountered among the ciliates. Not surprisingly, patterns of ciliation of the oral region are of primary taxonomic significance.

Ciliate taxonomy is complex (see Box 3.7). Approximately one-third of all ciliates are symbiotic in, or on, other animals. Thus, there are mutualistic, commensalistic, and parasitic species, which appear to have evolved independently within the various taxonomic groups. There are ectosymbionts and endosymbionts of a variety of hosts including marine and freshwater animals, and both invertebrates and vertebrates. In contrast to the apicomplexans and flagellates, which have a diversity of parasitic forms in humans, only a single ciliophoran, *Balantidium coli* infects humans. Indeed, given their diversity it is perhaps surprising that parasitism is not a more common life style among the ciliates. Many of the freshwater ciliates and all of the parasites produce environmentally resistant cysts.

3.3.5.2 Form and function

The oral region of a ciliate is involved in food uptake and initial digestion, whereas the somatic region is involved in forming a protective coating, sensing the environment, locomotion, and/or attachment to the substrata. A large expenditure of energy is needed for these activities and mitochondria can be found located in grooves associated with the infraciliature. This close association presumably permits the most efficient transfer of energy from the mitochondria to the infraciliature and enables the metachronal coordinated beating of cilia. The cilia associated with the cytostome direct food particles towards the cytopharynx. Following ingestion, food vacuoles are formed and the

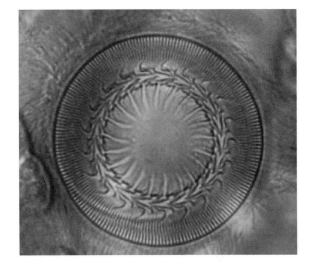

Fig. 3.22 Basal disk and ring of radially arranged proteinaceous subpellicular elements (denticles) of the ciliate *Trichodina* sp. This elaborate structure is used for attachment to its fish host's skin. (Photograph courtesy of Duane Barker.)

material within the vacuoles is digested. The surface of the ciliate cell is characterized by various morphological patterns of locomotory cilia. There is great diversity in these ciliary patterns among the Ciliophora; these patterns can give a ciliate a most exquisite morphology. The adhesive basal disk on the aboral surface of *Trichodina* spp., with its ring of radially arranged subpellicular elements (Fig. 3.22), provides an excellent example of the structural complexity encountered within the parasitic ciliates.

There is very little information available on the biochemistry and physiology of the parasitic ciliates. It is known that *Balantidium coli* in humans can use O_2 even though it normally resides in the microaerophilic environment of the large intestine. It has also been shown that *B. coli* contains hydrolytic enzymes such as hyaluronidase which may be involved in the invasion of host tissue. Numerous starch granules and ingested bacterial cells, as well as RBCs and fat droplets, have been observed in the cytoplasm. It is assumed that the stored fats and starch are reserve energy sources and that ingested bacteria, following their hydrolytic digestion, supply the cell with many of the required macromolecular building blocks.

3.3.5.3 Biodiversity and life-cycle variation

Balantidium coli is a representative of the Class Vestibuliferida. It has a cosmopolitan distribution, although it is most common in tropical countries. It is a large, colon-dwelling species that is not usually pathogenic (Zaman, 1998). *Balantidium coli* has also been isolated from monkeys, pigs, cats, and rodents. Balantidiasis is a zoonosis, with pigs suggested as the primary reservoir host (Esteban *et al.*, 1998). Normally, *B. coli* infections are asymptomatic and the organisms act as commensals. Humans appear to have a high degree of natural resistance to infection with *B. coli*. However, the ciliate can invade the mucosa and submucosa of the cecum and colon, producing ulcers. Unlike *E. histolytica*, however, it usually does not penetrate beyond the intestinal wall into deeper body tissues and other organs, although several cases involving the liver, lung, or pleural cavity have been reported. The reason for the increased susceptibility of some individuals to clinical disease is not known, but may be related to a depressed immune status.

In the colon, resistant cysts are produced and then excreted in the feces. Cysts are ingested in contaminated food or water, and excyst in the intestine. The trophozoites are phagotrophic, feeding primarily on cell debris, starch grains, etc. Trophozoites multiply by transverse fission and sexual reproduction occurs by conjugation between trophozoites in the intestinal tract.

Nyctotherus cordiformis is a representative of the order Clevelandellida. It lives in the colons of tree frogs and toads. Tree frog tadpoles ingest cysts of *N. cordiformis* which excyst, and divide repeatedly until host metamorphosis begins, and then conjugation occurs. Sexual reproduction only occurs in tadpoles undergoing metamorphosis. Following metamorphosis into froglets, the ciliate matures into the adult trophozoite stage, complete with a small micronucleus and large macronucleus. Division again occurs in the adult frog and, eventually, pre-cyst and cyst stages are excreted into the environment. Once again, this frog–ciliate system demonstrates the degree of reproductive synchrony that can occur between host and parasite.

Perhaps the best known parasitic ciliate is *Ichthyophthirius multifiliis* (a representative of the Order Hymenostomatida), and is the causative agent of a disease known as white spot disease or simply 'ich' to tropical fish enthusiasts. Mature trophozoites produce large white pustules (**trophonts**) in the skin of fish that are visible to the naked eye. It is one of the most significant pathogens of aquarium- and hatchery-raised fishes around the world. The life cycle involves a resistant cyst and free-living trophozoites called **tomites**. Infective tomites of *I. multifiliis* invade the skin and gills of fishes, become the feeding trophozoite, and form the characteristic white pustules under the epidermis. When the pustules rupture, the freed trophozoite eventually settles to the bottom, secretes a thick gelatinous coat, and divides repeatedly within the cyst forming many daughter cells. Once division is completed, free-swimming tomites are released from the cysts to contact new fish (McCartney *et al.*, 1985).

3.4 Phylogenetic relationships and classification

Given the phenomenal diversity of the unicellular eukaryotes, it should come as no surprise that attempts to classify them into evolutionarily meaningful taxonomic groups is extraordinarily difficult. Molecular phylogenetic and comparative ultrastructural studies have revolutionized our understanding of protist evolutionary relationships. As a result, the taxonomic classification scheme used for the protists has been 'turned upside down' during the last decade or so (see Box 3.7). As such, protist classification in the previous edition of this book is now completely out of date. For example, protozoologists no longer consider the 'old' phylum Sarcomastigophora, comprising the amoeboid and flagellated protists, as tenable. It is now clear that amoebas and flagellates are not monophyletic groups and, instead, belong to several taxa, with diverse evolutionary affinities not all yet determined (Hausmann & Hülsmann, 1996). Many mysteries remain, and deciphering evolutionary relationships among major protist lineages still presents a major challenge. The newer classifications divide the protists into at least 13 phyla (although some recognize over 30!). The amoebas are currently divided into two phyla: the Amoebozoa and the Heterolobosea (see Box 3.7). The clade Alveolata (superphylum Alveolata in some classifications) comprises three extremely biologically diverse groups of unicellular protists: the dinoflagellates, the ciliates, and the apicomplexans (Leander & Keeling,

2003; 2004). The unifying diagnostic feature of the alveolates is that their cell surfaces are underlain by a series of sacs called **alveoli**.

Comparative molecular studies now place the Microsporida in the Kingdom Fungi (see Chapter 4). Furthermore, the Myxozoa are no longer considered as protists; there is enough convincing morphological and molecular phylogenetic evidence that they are now regarded as metazoans (see Chapter 5).

Box 3.7 | **Classification of the Protista**

As we have stressed, the evolutionary relationships among the major protist lineages remain to be resolved. As such, there is no consensus as to how the protists are inter-related, or how they should be classified. The classification scheme we adopt for this edition represents a compromise between current evolutionary thinking and a need to present protist classification in a way that facilitates communication among scientists. It follows, for the most part, the taxonomic distinctions proposed by Hausmann & Hülsmann (1996), Patterson (1999), Lee *et al.* (2000), and Adl *et al.* (2005) (see also Cox, 1993; Baldauf, 2003; Pechenik, 2010). Herein, we include only examples of parasitic taxa used in this chapter and do not provide the lower level categories in the taxonomic hierarchy, e.g., subclasses, suborders.

The flagellates
Phylum Retortamonada
    Class Diplomonadea
        Order Diplomonadida
        Representative genus: *Giardia*
Phylum Parabasala (Axostylata in some classifications)
    Class Trichomonadida
        Order Trichomonadida
        Representative genera: *Mixotricha, Trichomonas*
        Order Trichonymphida
        Representative genus: *Trichonympha*
Phylum Euglenozoa
    Class Kinetoplasta (Subphylum in some classifications)
        Order Trypanosomatida
        Representative genera: *Cryptobia, Leishmania, Trypanosoma*

Box 3.7 | **(continued)**

The amoebas
Phylum Amoebozoa
   Representative genera: *Acanthamoeba, Entamoeba, Endolimax, Iodamoeba*
Phylum Heterolobosea
   Representative genus: *Naegleria*
The opalinids
Phylum Chromista
  Class Opalinea
    Representative genus: *Opalina*
The alveolates
Phylum Apicomplexa
  Class Conoidasida
    Order Eugregarinorida
    Representative genera: *Gregarina, Monocystis*
    Order Eucoccidiorida
    Representative genera: *Cryptosporidium, Eimeria, Isopora, Neospora, Sarcocystis, Toxoplasma*
  Class Aconoidasida
    Order Piroplasmorida
    Representative genera: *Babesia, Theileria*
    Order Haemosporida
    Representative genera: *Haemoproteus, Leucocytozoon, Plasmodium*
Phylum Ciliophora
  Class Spirotrichea
    Order Vestibuliferida
    Representative genus: *Balantidium*
    Order Clevelandellida
    Representative genus: *Nyctotherus*
  Class Oligohymenophorea
    Order Hymenostomatida
    Representative genus: *Ichthyophthirius*
    Order Mobilida
    Representative genus: *Trichodina*

References

Abu-Raddad, L. J., Patnaik, P. & Kublin, J. G. (2006) Dual infection with HIV and malaria fuels the spread of both diseases in sub-Saharan Africa. *Science*, 314, 1603–1606.

Adl, S. M., Simpson, A. G. B., Farmer, M. A., *et al.* (2005) The new higher level classification of eukaryotes with emphasis on the taxonomy of protists. *Journal of Eukaryotic Microbiology*, 52, 399–451.

Alexander, J. & Russell, D. J. (1992) The interaction of *Leishmania* species with macrophages. *Advances in Parasitology*, 31, 175–254.

Alexander, J. & Hunter, C. A. (1998) Immunoregulation during toxoplasmosis. In *Immunology of Intracellular Parasitism*, Chemical Immunological Series, vol. 70, ed. F. Y. Liew & F. E. G. Cox, pp. 81–102. Basel: Karger.

Alexander, J., Satoskare, A. R. & Russell, D. G. (1999) *Leishmania* spp: models of intracellular parasitism. *Journal of Cell Science*, 112, 2993–3002.

Babiker, H. A. & Walliker, D. (1997) Current views on the population structure of *Plasmodium falciparum*: implications for control. *Parasitology Today*, 13, 262–267.

Baldauf, S. L. (2003) The deep roots of eukaryotes. *Science*, 300, 1703–1706.

Bensch, S., Perez-Tris, J., Waldenstrom, J., & Hellgren, O. (2004) Linkage between nuclear and mitochondrial DNA sequences in avian malaria parasites: multiple cases of cryptic speciation? *Evolution*, 58, 1617–1621.

Bordmann, G., Favre, N. & Rudin, W. (1997) Malaria toxins: effects on murine spleen and bone marrow cell proliferation and cytokine production *in vitro*. *Parasitology*, 115, 475–483.

Bryant, C. (1993) Biochemistry. In *Modern Parasitology: A Textbook of Parasitology*, ed. F. E. G. Cox, pp. 117–136. Oxford: Blackwell Scientific Publications.

Carreno, R. A., Schnitzler, B. E., Jeffries, A. C., *et al.* (1998) Phylogenetic analysis of coccidia based on 18S rDNA sequence comparison indicates that *Isospora* is most closely related to *Toxoplasma* and *Neospora*. *Journal of Eukaryotic Microbiology*, 45, 184–188.

Cavalier-Smith, T. (1999) Principles of protein and lipid targeting in secondary symbiogenesis: euglenoid, dinoflagellate, and sporozoan plastid origins and the eukaryote family tree. *Journal of Eukaryotic Microbiology*, 46, 347–366.

Chen, N., Auliff, A., Rieckmann, K., *et al.* (2007) Relapses of *Plasmodium vivax* infection result from clonal hypnozoites activated at pre-determined levels. *Journal of Infectious Diseases*, 195, 934–941.

Clark, I. A., Al Yaman, F. M. & Jacobson, L. S. (1997) The biological basis of malarial disease. *International Journal for Parasitology*, 27, 1237–1249.

Coombs, G. & North, M. (1991) *Biochemical Protozoology*. London: Taylor & Francis.

Conrad, P. A., Miller, M. A., Kreuder, C., *et al.* (2005) Transmission of *Toxoplasma*: clues from the study of sea otters as sentinels of *Toxoplasma gondii* flow into the marine environment. *International Journal for Parasitology*, 35, 1155–1168.

Cox, F. E. G. (1993) Parasitic Protozoa. In *Modern Parasitology: A Textbook of Parasitology*, ed. F. E. G. Cox, pp. 1–23. Oxford: Blackwell Scientific Publications.

Desowitz, R. S. (1991) *The Malaria Capers*. New York: Norton.

Dubey, J. P. (2007) The history and life cycle of *Toxoplasma gondii*. In *Toxoplasma gondii: The Model Apicomplexan. Perspective and Methods*, ed. L. M. Weiss and K. Kani, pp. 1–17. London: Academic Press.

Engers, H. D. & Godal, T. (1998) Malaria vaccine development: current status. *Parasitology Today*, 14, 56–64.

Esch, G. W. (2007) *Parasites and Infectious Disease: Discovery by Serendipity, and Otherwise*. Cambridge: Cambridge University Press.

Escalante, A. A. & Ayala, F. J. (1994) Phylogeny of the malarial genus *Plasmodium*, derived from rRNA gene sequences. *Proceedings of the National Academy of Sciences USA*, 91, 11373–11377.

Esteban, J.-G., Aguirre, C., Angles, R., *et al.* (1998) Balantidiasis in Aymara children from the northern Bolivian Altiplano. *American Journal of Tropical Medicine and Hygiene*, 59, 922–927.

Fayer, R. (1997) *Cryptosporidium and Cryptosporidiosis*. New York: CRC Press.

Garcia, L. S. (2007) *Diagnostic Medical Parasitology*. Washington, D.C.: ASM Press.

Gardner, M. J., Hall, N., Fung, E., *et al.* (2002) Genome sequence of the human malarial parasite *Plasmodium falciparum*. *Nature*, 419, 498–511.

Gething, P. W., Smith, D. L., Patil, A. P., *et al.* (2010) Climate change and the global malaria recession. *Nature*, 465, 342–345.

Gutteridge, W. E. & Coombs, G. (1977) *Biochemistry of Parasitic Protozoa*. Baltimore: University Park Press.

Haldar, K., Murphy, S. C., Milner, D. A. & Taylor, T. E. (2007) Malaria: mechanisms of erythrocytic infection and pathological correlates of severe disease. *Annual Review of Pathology*, **2**, 217–249.

Harrington, W., Krupnick, A. J. & Spofford, Jr., W. O. (1991) *Economics and Episodic Disease: The Benefits of Preventing a Giardiasis Outbreak*. Washington, D.C.: Resources for the Future.

Hausmann, K. & Hülsmann, N. (1996) *Protozoology*. New York: Theime Medical Publishers.

Homer, M. J., Aguilar-Delfin, I., Telford III, S. R., *et al.* (2000) Babesiosis. *Clinical Microbiology Reviews*, **13**, 451–469.

Hoshina, R. & Imamura, N. (2008) Multiple origins of the symbioses in *Paramecium bursaria*. *Protist*, **159**, 53–63.

Huang, H., Mullapudi, N., Lancto, C. A., *et al.* (2004) Phylogenomic evidence supports past endosymbiosis, intracellular and horizontal gene transfer in *Cryptosporidium parvum*. *Genome Biology*, **5**, R88.

Imwong, M., Snounou, G., Pukrittayakamee, S., *et al.* (2007) Relapses of *Plasmodium vivax* infection usually result from activation of heterologous hypnozoites. *Journal of Infectious Diseases*, **195**, 927–933.

Krogstad, D. (1996) Malaria as a reemerging disease. *Epidemiological Reviews*, **18**, 77–89.

Krotoski, W., Bray, R., Garnham, P., *et al.* (1982) Observations on early and late post-sporozoite tissue stages in primate malaria. II. The hypnozoite of *Plasmodium cynomolgi bastianelli* from 3 to 105 days after infection, and detection of 36- to 40-hour pre-erythocytic forms. *American Journal of Tropical Medicine and Hygiene*, **31**, 211–235.

Leander, B. S. & Keeling, P. J. (2003) Morphostasis in alveolate evolution. *Trends in Ecology and Evolution*, **18**, 395–402.

Leander, B. S. & Keeling, P. J. (2004) Early evolution of dinoflagellates and apicomplexans inferred from HSP90 and actin phylogeny. *Journal of Phycology*, **40**, 341–350.

Lee, J. J., Leedale, G. F. & Bradbury, P. (2000) *An Illustrated Guide to the Protozoa*, 2nd edition. Lawrence: Society of Protozoologists.

Liu, B., Liu, Y., Motyka, S., *et al.* (2005) Fellowship of the rings: the replication of kinetoplast DNA. *Trends in Parasitology*, **21**, 363–369.

Mackintosh, C., Beeson, J., & Marsh, K. (2004) Clinical features and pathogenesis of severe malaria. *Trends in Parasitology*, **20**, 597–603.

Margulis, L. (1981) *Symbiosis in Cell Evolution: Life and its Environment on the Early Earth*. San Francisco: W. H. Freeman.

Marshall, M., Naumovitz, D., Ortega, Y. & Sterling, C. (1997) Waterborne protozoan pathogens. *Clinical Microbiological Reviews*, **10**, 67–85.

Martin, W., Rujan, T., Richly, E., *et al.* (2002) Evolutionary analysis of *Arabidopsis*, cyanobacterial, and chloroplast genomes reveals plastid phylogeny and thousands of cyanobacterial genes in the nucleus. *Proceedings of the National Academy of Sciences USA*, **99**, 12246–12251.

Martínez, A. J. & Visvesvara, G. S. (1997) Free-living amphizoic and opportunistic amebas. *Brain Pathology*, **7**, 583–598.

Martínez-Paloma, A. & Espinosa-Cantellano, M. (1998) Amoebiasis: new understanding and new goals. *Parasitology Today*, **14**, 1–2.

Martinsen, E. S., Perkins, S. L., & Schall, J. J. (2008) A three-genome phylogeny of malaria parasites (*Plasmodium* and closely related genera): Evolution of life-history traits and host switches. *Molecular Phylogenetics and Evolution*, **47**, 261–273.

McCartney, J. B., Fortner, G. W. & Hansen, M. F. (1985) Scanning electron microscope studies on the life cycle of *Ichthyophthirius multifiliis*. *Journal of Parasitology*, **71**, 218–226.

McDonald, V. & Bancroft, G. J. (1998) Immunological control of *Cryptosporidium* infections. In *Immunology of Intracellular Parasitism*, Chemical Immunology Series, vol. **70**, ed. F. Y. Liew & F. E. G. Cox, pp. 103–123. Basel: Karger.

McKenzie, F. E., Smith, D. L., O'Meara, W. P. & Riley, E. M. (2008) Strain theory of malaria: the first 50 years. *Advances in Parasitology*, **66**, 1–46.

Mehdis, K. N. & Carter, R. (1995) Clinical disease and pathogenesis in malaria. *Parasitology Today*, **11**, 1–16.

Miles, M. A. (1988) Viruses of parasitic protozoa. *Parasitology Today*, **4**, 289–290.

Mock, B. A. & Gill, D. E. (1984) The infrapopulation dynamics of trypanosomes in red-spotted newts. *Parasitology*, **88**, 267–282.

Molyneux, D. H. & Ashford, R. W. (1983) *The Biology of Trypanosoma and Leishmania, Parasites of Man and Domestic Animals*. London: Taylor & Francis.

Morrison, W. I. & McKeever, D. J. (1998) Immunobiology of infections with Theileria parva in cattle. In *Immunology of Intracellular Parasitism*, Chemical Immunology Series, vol. **70**, ed. F. Y. Liew & F. E. G. Cox, pp. 163–185. Basel: Karger.

Müller, M. (1991) Energy metabolism of anaerobic parasitic protists. In *Biochemical Protozoology*, ed. G. Coombs & M. North, pp. 80–91. London: Taylor & Francis.

O'Donoghue, P. J. (1995) *Cryptosporidium* and cryptosporidiosis in man and animals. *International Journal for Parasitology*, **25**, 139–195.

Ohkuma, M. (2008) Symbioses of flagellates and prokaryotes in the gut of lower termites. *Trends in Microbiology*, **16**, 345–352.

Olson, M., O'Handley, R., Ralson, B., *et al.* (2004) Update on *Cryptosporidium* and *Giardia* infections in cattle. *Trends in Parasitology*, **20**, 183–191.

Parshad, S., Grover, P., Sharma, A., Verma, D., and Sharma, A. (2002) Primary cutaneous amoebiasis: case report with review of the literature. *International Journal of Dermatology*, **41**, 676–680.

Patterson, D. J. (1999) The diversity of eukaryotes. *The American Naturalist*, **154**, S96–S124.

Pechenik, J. A. (2010) *Biology of the Invertebrates*, 6th edition. New York: McGraw-Hill.

Piel, F. B., Patil, A. P., Howes, R. E., *et al.* (2010) Global distribution of the sickle cell gene and geographical confirmation of the malaria hypothesis. *Nature Communications*, doi: 10.1038/ncomms1104. 10.1038/ncomms1104

Power, M. L., & Ryan, U. M. (2008) A new species of *Cryptosporidium* (Apicomplexa: Cryptosporidiidae) from the eastern grey kangaroos (*Macropus giganteus*). *Journal of Parasitology*, **94**, 1114–1117.

Ralph, S. A. (2005). The apicoplast. In *Molecular Approaches to Malaria*, ed. I. W. Sherman, pp. 272–289. Washington, D.C.: ASM Press.

Roberts, F., Roberts, C., Johnson, J., *et al.* (1998) Evidence for the shikimate pathway in apicomplexan parasites. *Nature*, **393**, 801–805.

Sam-Yellowe, T. Y. (1996) Rhoptry organelles of the Apicomplexa: their role in host cell invasion and intracellular survival. *Parasitology Today*, **12**, 308–316.

Seed, J. R. & Hall, J. E. (1992) Trypanosomes causing disease in man in Africa. In *Parasitic Protozoa*, 2nd edition, ed. J. P. Kreier & J. R. Baker, pp. 85–155. New York: Academic Press.

Seed, J. R., Bollinger, R. & Gam, A. A. (1968a) Studies of frog trypanosomiasis. II. Seasonal variations in the parasitemia levels of *Trypanosoma rotatorium* in *Rana clamitans* from Louisiana. *Tulane Studies in Zoology and Botany*, **15**, 54–69.

Seed, J. R., Southworth, C. & Mason, G. (1968b) Studies of frog trypanosomiasis. I. A 24 hour cycle in the parasitemia level of *Trypanosoma rotatorium* in *Rana clamitans*. *Journal of Parasitology*, **54**, 946–960.

Sexton, A., Good, R., Hansen, D., *et al.* (2004) Transcriptional profiling reveals suppressed erythropoiesis, up-regulated glycolysis, and interferon-associated responses in murine malaria. *Journal of Infectious Diseases*, **189**, 1245–1256.

Sherwin, I. W. (2005) *Molecular Approaches to Malaria*. Washington, D.C.: ASM Press.

Singh, B., Sung, L. K., Matusop, A., *et al.* 2004. A large focus of naturally acquired *Plasmodium knowlesi* infections in human beings. *Lancet*, **363**, 1017–1024.

Smith, H. V. & Rose, J. B. (1998) Waterborne cryptosporidiosis: current status. *Parasitology Today*, **14**, 14–22.

Smith, N. C., Fell, A. & Good, M. F. (1998) The immune response to asexual blood stages of malaria parasites. In *Immunology of Intracellular Parasitism*, Chemical Immunology Series, vol. **70**, ed. F. Y. Liew & F. E. G. Cox, pp. 144–162. Basel: Karger.

Smith, T. G., Lourenco, P., Carter, R., *et al.* (2000) Commitment to sexual differentiation in the human malaria parasite *Plasmodium falciparum*. *Parasitology*, **121**, 127–133.

Smyth, J. D. (1994) *Introduction to Animal Parasitology*, 3rd edition. Cambridge: Cambridge University Press.

Steiner, T. S., Thielman, N. M. & Guerrant, R. L. (1997) Protozoal agents: what are the dangers for the public water supply? *Annual Review of Medicine*, **48**, 329–340.

Swierczynski, G. & Gobbo, M. (2007) *Atlas of Human Malaria*. Sirmione, Italy: Az Color.

Teklehaimanot, A. & Singer, B. (2005) *Coming to Grips with Malaria in the New Millennium*. London: Earthscan.

Thompson, R. C. A., Reynoldson, J. A. & Mendis, A. H. W. (1993) *Giardia* and giardiasis. *Advances in Parasitology*, **32**, 71–160.

Valkiunas, G. (2004) *Avian Malarial Parasites and Other Haemosporidia*. London: Taylor & Francis.

Wellems, T. E. (1992) How chloroquine works. *Nature*, **355**, 108–109.

Woo, P. T. K. (2001) Cryptobiosis and its control in North American fishes. *International Journal for Parasitology*, **31**, 565–573.

Zaman, V. (1998) Balantidium coli. In *Topley & Wilson's Microbiology and Microbial Infections*, 9th edition, vol. 5, *Parasitology*, ed. F. E. G. Cox, J. P. Kreier & D. Wakelin, pp. 445–450. London: Edward Arnold.

4 Microsporida: the intracellular, spore-forming fungi

4.1 General considerations

Following decades of taxonomic upheaval, strong molecular phylogenetic evidence now indicates that the phylum Microspora is a monophyletic lineage within the Kingdom Fungi (review in Corradi & Keeling, 2009). Members of this clade are obligate, intracellular, spore-forming parasites. The unique and distinctive spores of these unicellular eukaryotes are minute, ranging from 2 to 20 μm in length. Although they are eukaryotic, microsporidian cells have several unusual characteristics, including a lack of organelles such as flagella, peroxisomes, mitochondria, and Golgi apparatus. Microsporidians also have 16S rather than 18S ribosomes. However, despite their relative simplicity, they can also be considered as marvels of structural and functional complexity, possessing adaptations for survival while outside their host, and also for intracellular parasitism. Within the spore is a diagnostic, exquisite extrusion apparatus, adapted for the penetration of host cells.

Louis Pasteur described the first microsporidian in the mid nineteenth century. He showed that *Nosema bombycis* caused 'pebrine disease' in silk-moth larvae, and provided recommendations to European silkworm farmers regarding control. Currently, a total of approximately 1300 species of microsporidians in 160 genera have been described. Following from modern advances in molecular diagnostics, it is likely that many more species await discovery. While most microsporidians are parasites of insects, they infect a wide range of other invertebrates, including nematodes, molluscs, annelids, and crustaceans. Microsporidians are present within all five classes of vertebrates, with 14 genera described from teleost fishes alone. Research involving microsporidians has traditionally focused on economically important species of insects (e.g., *Nosema* spp., review in Wittner & Weiss, 1999) and fish (e.g., *Loma* spp., review in Dyková, 2006). In recent years, this focus has expanded to include microsporidians that have been implicated as causative agents of opportunistic infections and emergent diseases in humans (review in Weiss, 2001).

Microsporidian–host interactions are increasingly being used as experimental models in ecology and evolutionary biology. Given that the spores of some species can be isolated and cultured under laboratory and/or field conditions, several are amenable to experimental manipulation. For example, *Pleistophora* spp. is one of several common microsporidians of the water flea, *Daphnia*. Dieter Ebert and his colleagues combine state-of-the-art experimental epidemiology and evolutionary genomics within a rigorous conceptual framework to test current ideas regarding parasite-mediated natural selection and host–parasite coevolution (e.g., Haag & Ebert, 2004; Ebert, 2008). Microsporidians of sticklebacks and amphipods are also utilized in experimental parasite ecology and evolution. Smith (2009) provides an excellent overview of the ecology and evolution of microsporidian–host interactions and their utilization as models in modern parasite ecology and evolution.

4.2 Form and function

The ovoid or elliptical spores are morphologically distinctive stages of microsporidians. Infective spores are environmentally resistant and surrounded by a thick, rigid wall (Fig. 4.1). The protective spore wall consists primarily of an outer exospore layer composed of glycoprotein and an inner endospore layer composed of a chitin–protein matrix (Fig. 4.2). Some species have spores with double nuclei (=diplokaryotic) (e.g., *Nosema* sp.), while others have a single nucleus (=monokaryotic). Within the spore are organelles specialized for the penetration of host cells, and for reproduction. Most remarkably, and the hallmark of the Microsporida, the spore houses a sophisticated extrusive apparatus consisting of a long, hollow, coiled **polar filament** (=**polar tube**). Its coils wind like a spring around the inside of the spore wall (Fig. 4.1). An amoeboid **sporoplasm**, with its nucleus and cytoplasm, lies within the polar filament coils. At the base of the filament is an anchoring disk, which attaches the filament to the spore wall. At one end of the spore is an organelle consisting of stacked membranes called the **polaroplast**. A posterior vacuole is typically found at the opposite end of the spore (Figs. 4.1, 4.2). Most of these organelles are visible only

via transmission electron microscopy although the posterior vacuole is visible via light microscopy in some species (Fig. 4.3).

The process of germination of the mature spore is spectacular. When an infective spore is ingested by its host, the spore absorbs water and germination is initiated, resulting in the explosive eversion of the polar filament, through which the sporoplasm is ejected (Fig. 4.4). The polar filament everts from the apical end of the spore with such force that it punctures the cell membrane of the host. The polaroplast

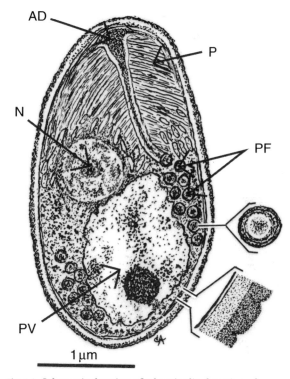

Fig. 4.2 Schematic drawing of a longitudinal section of a spore of *Kabatana rondoni* from the muscle fibers of the Amazonian teleost fish *Gymnorhamphichthys rondoni*, illustrating ultrastructural features characteristic of microsporidians. The single nucleus (N), polaroplast (P), posterior vacuole (PV), anchoring disk (AD), and sections through the coils of the polar filament (PF) are shown. Details of transverse sections of the polar filament and the two-layered spore wall are also represented. (Figure courtesy of Carlos Azevedo; from Casal *et al.*, 2010, with permission, *Journal of Parasitology*, **96**, 1155–1163.)

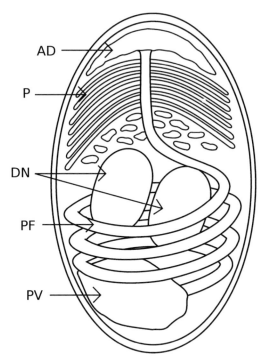

Fig. 4.1 Schematic drawing of the internal anatomy of a microsporidian spore, showing the thick wall, anchoring disk (AD), coiled polar filament (PF), wrapping around the diplokaryon nucleus (DN), and the polaroplast (P) and posterior vacuole (PV). (Figure courtesy of Danielle Morrison.)

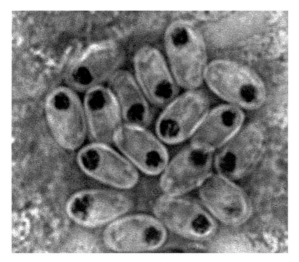

Fig. 4.3 Spores of *Pleistophora husseyi* in a smear of albumin gland tissue from the snail *Physa virgata*. The posterior vacuole is visible within spores. The spores are approximately 5 μm in length. (Photograph courtesy of John Sullivan.)

membranes swell and unfold as the filament discharges. These contribute to the extruded polar filament, greatly increasing its length relative to when it was coiled within the spore (Fig. 4.4). The posterior vacuole expands during germination of the spore, pushing the sporoplasm into the tubular polar filament for injection into the host cell, and subsequent development and reproduction.

4.3 Development and general life cycle

Development of spores can occur directly inside the host cell cytoplasm, or inside an envelope consisting of a thin membrane containing the spores called a **sporophorous vesicle**. As mature spores develop and accumulate, the infected cell expands and eventually ruptures, releasing the spores. There are three life cycle stages common to most microsporidians. First is a transmission or infective phase that includes the spore transmission stage, and, usually, the oral uptake of the mature spores by a new host. Following penetration of a specific host cell, the injected sporoplasm undergoes merogony (=schizogony), forming meronts

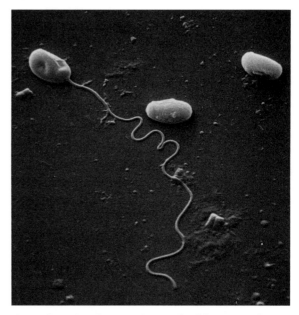

Fig. 4.4 Scanning electron micrograph of the spores of *Nosema tractabile* from the leech *Helobdella stagnalis*. The long polar filament of one of the spores has ejected, releasing the sporoplasm, consisting of two coupled nuclei – the diplokaryon. The spore and ejected polar filament measure approximately 3.5 μm and 31 μm, respectively. (Micrograph courtesy of Ronny Larsson, from Larsson, 1981, with permission, *Protistologica*, **17**, 407–422.)

(=trophozoites), reminiscent of apicomplexan protists (Chapter 3). In the second phase, the trophozoite's nuclei divide repeatedly and the parasite becomes a large multinucleate plasmodium. Cytokinesis takes place and the process is repeated. The third phase includes the sporogonic stages. Sporogony involves the development of a **sporont**, which divides to form **sporoblasts**. In each sporoblast, spores differentiate and a mass of tubules is formed. These become the polar filament and polaroplast of the mature spores. These spores exit the host, primarily in the feces or upon the host's death, to infect other hosts.

4.4 Biodiversity and life-cycle variation

The primary life-cycle stages, the meront and the spore, are found in all microsporidians. However, the complexity of the life cycle varies considerably.

Some have simple life cycles with a single sporogonic phase within host cytoplasm; others have complex life cycles with four different sporogonic sequences, some of which occur in different hosts (review in Smith, 2009).

The genus *Nosema* is perhaps the most diverse and well known of the Microsporida, including entomopathogenic species that infect pollinators such as bumblebees and honey bees. Consequently, we know a great deal about the life cycles and pathogenesis of microsporidians such as *Nosema bombi* of bumble bees and the other bee pathogens, *Nosema ceranae* and *Nosema apis* (Box 4.1). When ingested, the polar filaments of the spores of both of these latter species pierce the peritrophic membrane of the midgut and enter intestinal epithelial cells, where they undergo merogony and sporogony. Since these species are transmitted horizontally via oral–fecal transmission and ingestion of spores, *Nosema* spp. transmission can occur quickly within the social context of bee colonies. This can lead to mortality of bees and declines in colony vigor (Box 4.1). Other species may parasitize the fat body or the reproductive organs of various insects. Due to their host specificity, and their pathogenicity in insects of economic and medical importance, several *Nosema* spp. are also studied for their potential as biological control agents. *Nosema algerae*, for example, significantly reduces the number of oocysts of malaria-causing anopheline mosquitoes. *Antonospora locustae* (formerly *Nosema locustae*), develops in fat body cells of orthopterans and is used as a biological control agent for grasshopper pests in agricultural settings.

Vertical transmission, or transovarial transmission, also occurs among many microsporidians (review in Dunn *et al.*, 2001). In fact, most microsporidians, especially those infecting invertebrates, combine vertical and horizontal transmission in their life cycles. In vertical transmission, the spores are transmitted via the female adult's ovaries to the egg, and, at least in insects, eventually multiply in cells of infected larvae. *Nosema bombycis*, the widely

---

Box 4.1 | *Nosema ceranae*: an emerging pathogen of European honey bees

Honey bee keepers recognize *Nosema apis* and *Nosema ceranae* as causative agents of one of the most common diseases of honey bees. The cosmopolitan disease caused by these microsporidians is called bee dysentery, nosemosis, or *Nosema* disease. Transmission of *Nosema* in honey bee colonies occurs via oral–fecal contamination. Adult bees ingest spores with food and water. These then develop in the midgut of the bee, releasing polar filaments that transfer sporoplasms into midgut epithelial cells. The parasite multiplies astronomically within the midgut cells, with up to 50 million spores reported within a bee's midgut only 2 weeks after infection. The spores pass out of the bee in its feces and new bees are infected through the cleaning and feeding activities occurring within the honey bee colony (Chen *et al.*, 2009). In this way, an initial infection can quickly spread to queens, drones, and workers. Infection of the queen not only disseminates infection within the colony, but also aids in its spread to new colonies via swarming behavior. *Nosema* infections negatively impact honey-bee colonies in a variety of ways, including causing dysentery and shortened bee life spans, and reduced fecundity of infected queens. Weakening of heavily infected colonies and dramatic reductions in honey production can occur, leading to significant economic losses.

For many years, the only known cause of *Nosema* disease in honey bees was *N. apis*. This view changed when Fries *et al.* (1996) described a new species, *N. ceranae* from the Asian honey bee,

Box 4.1 (continued)

Apis cerana. Although most microsporidians are highly host-specific, *N. ceranae* invaded a new host, the European honey bee, *Apis mellifera*, with devastating consequences. In this economically valuable honey bee species, the parasite is now considered an emergent pathogen. Pathogen emergence of this sort is often associated with such host species 'jumps' (Klee *et al.*, 2007).

Comparative molecular phylogenetic analyses conducted by Chen *et al.* (2009) demonstrated that *N. apis* is not the closest relative of *N. ceranae*, even though they infect the same host. Rather, *N. ceranae* is more closely related to *N. vespula*, a species infecting wasps. *Nosema apis* is more closely linked to *N. bombi*, a parasite of bumblebees.

Higes *et al.* (2007) showed that *A. mellifera* experimentally infected with *N. ceranae* died within 8 days of exposure. At present, the precise mechanism for such extreme pathogenesis is not well understood. Such a high degree of virulence poses a significant threat to the beekeeping industry, prompting numerous epidemiological and pathological studies of *N. cerana* in European honey bees in laboratories around the world. Moreover, *N. ceranae* may have significant cascading ecological impacts and serious ecosystem consequences. It is clear now that *N. ceranae* adapted to and became established in *A. mellifera* as early as 1998. Throughout the USA, for example, *N. ceranae* was shown to be well established and, furthermore, was the only *Nosema* species infecting *A. mellifera* (Chen *et al.*, 2008).

Nosema ceranae has been implicated as a causative agent of colony collapse disorder (CCD), the sudden and catastrophic die-offs of honey bees which occurred in 2006, and again in 2010 in the USA. There is recent evidence indicating that co-infection of an iridovirus and *Nosema* are linked to CCD (Bromenshenk *et al.*, 2010). Laboratory trials also showed that co-infection with both pathogens was more lethal to bees than either pathogen alone.

Klee *et al.* (2007) conclude, "clearly there is a pressing need for studies on the epidemiology of *N. ceranae* in *A. mellifera* and on methods to control *N. ceranae* if its potential contribution to further honey bee losses worldwide are to be avoided". Further investigations will also focus on the roles of *N. ceranae* genetics, as well as honey-bee physiology, immunity, and behavior (e.g., Chen *et al.*, 2009; Rueppell *et al.*, 2010). Now that the entire genome of *A. mellifera* has been elucidated, many of these, and other, exciting research frontiers will be explored in greater detail in the years to come (see also Box 11.3).

studied species causing pebrine disease in the silkworm moth, has a mixed transmission strategy. In the first case, transmission is horizontal via oral–fecal contamination among feeding larvae. In the second, transmission is vertical from the adult silkworm moth to her eggs. Transovarial transmission has also been shown to occur in *Nosema pyracusta*, a species infecting the corn borer and *Nosema lymantriae*, a parasite of the gypsy moth (review in Smith, 2009). The parasites can have significant effects on the development and survival of larvae and on the reproductive success of adults, increasing their potential for use as biological control agents of these pests.

The ability to transmit via vertical transmission complicates the epidemiology of the diseases caused by microsporidians. Understanding the relative roles of vertical and horizontal transmission in relation to disease pathogenesis among the microsporidians is a key challenge (Smith, 2009). Another intriguing aspect of vertical transmission in microsporidians is associated with reproductive manipulation of the host, including feminization, sex-specific virulence, and sex ratio distortion (SRD). Although bacterial endosymbionts are well known to induce a range of these manipulations, microsporidians are the only eukaryotes in which manipulations of host reproduction occur (Dunn *et al.*, 2001). *Nosema granulosis*, a parasite of the amphipod crustacean *Gammarus duebeni*, for example, undergoes a high rate of transovarial transmission. It was the first feminizing microsporidian to be examined in detail. The parasite feminizes its host, causing genetic males to develop as phenotypic females, leading to SRD and a high female-biased sex ratio (Terry *et al.*, 1998). Since this pioneering study, evidence is accumulating showing relationships between vertical transmission and SRD, at least in amphipods (Haine *et al.*, 2004; Terry *et al.*, 2004). A bias in sex ratio towards females may enhance the growth rate of infected host populations and, thus, may also increase colonization of new habitats. For interested readers, Smith (2009) presents further details regarding the role of vertical transmission in the life cycles of the Microsporida.

A wide variety of other crustaceans are host to several microsporidians. *Thelohania contegeani*, for example, is an intramuscular cell parasite of crayfish. As the disease progresses, the musculature deteriorates and becomes whiter, leading to the common name of 'porcelain disease' in Europe. Other invertebrates, including bryozoans, nematodes, annelids, and molluscs, are host to microsporidians.

Microsporidians are also well known as parasites of fish, and are widely distributed in teleosts in freshwater and marine habitats (review in Lom & Nilsen, 2003). For those whose life cycles are known, it appears that direct horizontal transmission via oral ingestion of spores is required. Fish microsporidians infect a wide variety of host cells, ranging from connective tissue (*Glugea* spp.), intestine (*Glugea* spp.), liver (*Microgemma* spp.), muscle (*Pleistophora* spp. and *Kabatana* spp.), gills (*Loma* spp.), and gonads (*Ovipleistophora* and *Microsporidium* spp.). Several are pathogens of economically important fish species (review in Dyková, 2006). In fish microsporidian genera such as *Glugea* and *Loma*, a remarkable **xenoma** forms. In these species, the injected sporoplasm stimulates the cell to undergo enormous hypertrophy. Xenoma formation can severely impact normal functions of infected organs, and contribute to the mortality of fish. Xenomas are a combination of the intracellular parasite and the hypertrophied cell. Such cells often contain hypertrophic nuclei and have a variety of surface modifications, including micro-villi, invaginations, or thick walls (Fig. 4.5). Xenomas of some species, especially *Glugea* spp., can become so large that they form visible cyst-like structures (Fig. 4.6). Remarkably, the hypertrophic cells and their

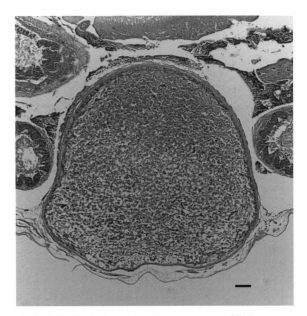

Fig. 4.5 Histological section of a mature, spore-filled xenoma of *Glugea pimephales* from a fathead minnow *Pimephales promelas*. The xenoma is located between the fish epidermis and peritoneal lining. Scale bar = 100 μm. (Photograph courtesy of Jonathon Forest.)

Fig. 4.6 Large external xenomas of *Glugea pimephales* on fathead minnows *Pimephales promelas*. (Photograph courtesy of Jonathon Forest.)

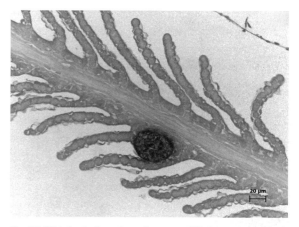

Fig. 4.7 Histological section of a spore-filled xenoma of *Loma salmonae* on gill filaments of a pink salmon *Oncorhynchus gorbuscha*. (Photograph courtesy of Catherine Thomson.)

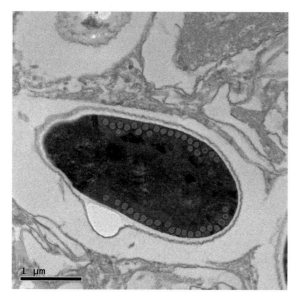

Fig. 4.8 Transmission electron micrograph of a longitudinal section of a spore of *Loma morhua* from Atlantic cod *Gadus morhua*, showing the thick spore wall, and sections through the coils of the polar filament (numbering 20). (Micrograph courtesy of Aaron Frenette.)

microsporidians integrate morphologically and physiologically, and form a separate entity with its own development (Dyková, 2006).

Loma spp. comprise another important xenoma-forming fish microsporidian genus. *Loma* spp. are mostly host- and site-specific, generally forming

xenomas in gill filaments or lamellae, although some species can also infect other organs, including the intestine, heart, and spleen. *Loma salmonae* (Fig. 4.7) of Pacific salmonids and *Loma morhua* (Figs. 4.8, 4.9) of Atlantic cod are two of the most widely studied species. Both parasites cause mortality, especially of hatchery-raised fish or fish reared in net pen aquaculture. *Loma salmonae*, for example, is the agent of microsporidial gill disease and significant mortalities in salmonids, notably chinook and coho salmon (Ramsay *et al.*, 2002). Gill filaments are distorted by formation of the xenomas (Fig. 4.7). When they rupture and release the spores, a cascade of inflammatory events occurs, leading to severe gill damage and respiratory distress. Sanchez *et al.* (2001) worked out the intricate details of *L. salmonae* development and migration. The parasite first infects the gut mucosal epithelial cells and then migrates to the heart, where merogony occurs within subendocardial macrophages. The parasite is then transferred within macrophages to the pillar cells of the gill filaments, where the xenoma forms and sporogony occurs, eventually becoming filled with spores (Fig. 4.7).

Several microsporidian genera, including *Brachiola*, *Microsporidium*, *Enterocytozoon*, and *Encephalitozoon*, have been reported to infect various tissues of humans, and some are considered as

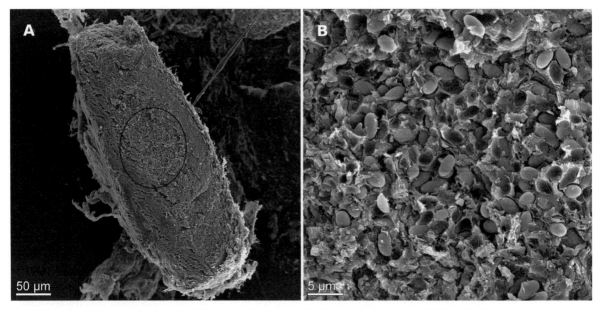

Fig 4.9 Scanning electron micrographs of *Loma morhua* from Atlantic cod *Gadus morhua*. (A) Longitudinal cross-section through a xenoma excised from spleen tissue. The circled area contains the spores of *L. morhua* localized within the center of the xenoma; (B) Higher magnification of the circled area, showing the spores of the parasite. (Micrographs courtesy of Aaron Frenette.)

emergent diseases (Weiss, 2001). Many of the important species infecting humans, such as *Encephalitozoon cuniculi*, are also found in other mammals, including dogs, cats, and rabbits. As a result, contact with infected animals may lead to transmission. As such, human microsporidiosis caused by *E. cuniculi* may be considered a zoonotic disease in some cases (Lindsay *et al.*, 2009). In humans, pathology occurs primarily in immuno-compromised individuals, including those infected with HIV and in those immunosuppressed due to organ transplantation. *Encephalitozoon cuniculi* has been reported from a wide variety of organs in immunocompromised humans, including the muscles, brain, heart, eyes, and urogenital organs. For example, chronic renal disease leading to renal failure is a clinical sign of *E. cuniculi* infections in both dogs and humans. Inflammation of the eye causing kera-toconjunctivitis is another aspect of the pathogenesis caused by microsporidians such as *Brachiola algerae* and *E. cuniculi*. For comprehensive reviews of microsporidiosis of humans, see Wittner & Weiss (1999) and Weiss (2001).

4.5 Phylogenetic relationships and classification

For many years, microsporidians were considered as primitive eukaryotes, possibly as sporozoan protists. The lack of functional mitochondria and other organelles certainly suggested an ancient origin. However, with the discovery of the **mitosome**, it is widely believed that microsporidians have relict mitochondria. As we saw with the protists, molecular data now indicate that the microsporidians are highly specialized intracellular parasites. In fact, considerable morphological and molecular phylogenetic evidence now indicates that the Microsporida are a lineage within the Kingdom Fungi (Hirt *et al.*, 1999; Keeling *et al.*, 2000; Van de Peer *et al.*, 2000; Vossbrinck & Debrunner-Vossbrinck, 2005; Fischer & Palmer, 2005; Lee *et al.*, 2008; Corradi & Keeling, 2009). The complete genomic sequencing of *E. cuniculi* added a wealth of data lending further strong support to the Microsporida–fungal connection (Katinka *et al.*, 2001; Thomarat *et al.*, 2004). In short, the Microsporida are monophyletic and are highly derived fungi, probably descended from a zygomycete ancestor (Lee *et al.*, 2008; Smith, 2009). Corradi & Keeling (2009) present a fascinating account of the history of the proposed origins, and the many taxonomic revisions the Microsporida have undergone over the years.

Box 4.2 | Classification of the Microsporida

The ultrastructure of the spore, such as the presence of a monokaryotic or diplokaryotic nucleus, the size of the spore, the nature of the spore wall, the presence of a sporophorous vacuole, the structure and number of polar filament coils, and the morphology of the polaroplast have all been used to classify microsporidians. Host species, site selection, and life cycle structure are also used in taxonomic classifications of the microsporidians. As is so often the case, however, morphological criteria do not always match those revealed by molecular phylogenetics data. *Pleistophora* and *Nosema*, for example, have been split into several genera based upon molecular studies. Thus, microsporidian taxonomy is in a constant state of change and, to date, there is no consensus hierarchical taxonomic classification system for the Microsporida.

A phylogenetic reconstruction using small unit rDNA of several microsporidian species reveals the presence of five major clades based on the hosts infected (Vossbrinck & Debrunner-Vossbrinck, 2005). Thus, one clade includes those species infecting freshwater bryozoans and oligochaete annelids, e.g., *Bacillidium*. Two clades recognize microsporidians infecting insects. One of these is the so-called dipteran clade that mainly contains parasites of flies, especially mosquitoes, e.g., *Amblyospora*. A 'lepidopteran clade' contains microsporidians of moths and butterflies, but also hymenopteran insects, e.g., *Nosema* and *Vairimorpha*. Another clade includes mainly parasites of mammals, such as *Encephalitozoon*. Finally, a clade dominated by parasites infecting fish, including *Loma*, *Kabatana*, *Glugea*, and *Pleistophora*, was recognized. These associations between host and parasite taxa are not perfect, but are very strong (Smith, 2009). The amphipod crustaceans, for example, do not appear to follow this pattern. Rather, microsporidians from amphipods are distributed through multiple branches of the phylogenetic tree, prompting Smith (2009) to suggest that perhaps the early radiation of the parasites occurred in this host group.

References

Bromenshenk, J. J., Henderson, C. B., Wick, C. H., *et al.* (2010) Iridovirus and microsporidian linked to honey bee colony decline. *PLoS One* 5(10), e13181. DOI10:1371/journal.pone.0013181.

Chen, Y. P., Evans, J. D., Smith, I. B. & Pettis, J. S. (2008) *Nosema ceranae* is a long-present and widespread microsporidian infection of the European honey bee (*Apis mellifera*) in the United States. *Journal of Invertebrate Pathology*, 97, 186–188.

Chen, Y. P., Evans, J. D., Murphy, C., *et al.* (2009) Morphological, molecular, and phylogenetic characterization of *Nosema ceranae*, a microsporidian parasite isolated from the European honey bee, *Apis mellifera*. *Journal of Eukaryotic Microbiology*, 56, 142–147.

Corradi, N. & Keeling, P. J. (2009) Microsporidia: a journey through radical taxonomical revisions. *Fungal Biology Reviews*, **23**, 1–8.

Dunn, A. M., Terry, R. S. & Smith, J. E. (2001) Transovarial transmission in the Microsporidia. *Advances in Parasitology*, **48**, 57–101.

Dyková, I. (2006) Phylum Microspora. In *Fish Diseases and Disorders*, vol. 1, *Protozoan and Metazoan Infections*, 2nd edition, ed. P. T. K. Woo, pp. 205–229. Wallingford: CAB International.

Ebert, D. (2008) Host–parasite coevolution: insights from the *Daphnia*: parasite model system. *Current Opinions in Microbiology*, **11**, 290–301.

Fischer, W. M. & Palmer, J. D. (2005) Evidence from small-subunit ribosomal RNA sequences for a fungal origin of Microsporidia. *Molecular Phylogenetics and Evolution*, **36**, 606–622.

Fries, I., Feng, F., da Silva, A., *et al.* (1996) *Nosema ceranae* n. sp. (Microspora, Nosematidae), morphological and molecular characterization of a microsporidian parasite of the Asian honey bee *Apis cerana* (Hymenoptera, Apidae). *European Journal of Protistology*, **32**, 356–365.

Haag, C. R. & Ebert, D. (2004) Parasite-mediated selection in experimental metapopulations of *Daphnia magna*. *Proceedings of the Biological Sciences*, **271**, 2149–2155.

Haine, E. R., Brondani, E., Hume, K. D., *et al.* (2004) Coexistence of three microsporidia parasites in populations of the freshwater amphipod *Gammarus roeseli*: evidence for vertical transmission and positive effect on reproduction. *International Journal for Parasitology*, **34**, 1137–1146.

Higes, M., Garcia-Palencia, P. Martin-Hernández, R., & Aránzazu, M. (2007) Experimental infection of *Apis mellifera* honeybees with *Nosema ceranae* (Microsporidia). *Journal of Invertebrate Pathology*, **94**, 211–217.

Hirt, R. P., Logsdon, Jr., J. M., Healy, B., *et al.* (1999) Microsporidia are related to fungi: evidence from the largest subunit of RNA polymerase II and other proteins. *Proceedings of the National Academy of Sciences USA*, **96**, 580–585.

Katinka, M. D., Duprat, S., Cornillot, E., *et al.* (2001) Genome sequence and gene compaction of the eukaryote parasite *Encephalatizoon cuniculi*. *Nature*, **414**, 450–453.

Keeling, P. J., Luker, M. A. & Palmer, J. D. (2000) Evidence from beta-tubulin phylogeny that microsporidia evolved from within the fungi. *Molecular Biology and Evolution*, **17**, 23–31.

Klee, J., Besana, A. M., Gensch, E., *et al.* (2007) Widespread dispersal of the microsporidian *Nosema ceranae*, an emergent pathogen of the western honey bee, *Apis mellifera*. *Journal of Invertebrate Pathology*, **96**, 1–10.

Lee, S. C., Corradi, N., Byrnes, E. J., *et al.* (2008) Microsporidia evolved from ancestral sexual fungi. *Current Biology*, **18**, 223–231.

Lindsay, D. S., Goodwin, D. G., Zajac, A. M., *et al.* (2009) Serological survey of antibodies to *Encephalitozoon cuniculi* in ownerless dogs from urban areas of Brazil and Columbia. *Journal of Parasitology*, **95**, 760–763.

Lom, J. & Nilsen, F. (2003) Fish microsporidia: fine structural diversity and phylogeny. *International Journal for Parasitology*, **33**, 107–127.

Ramsay, J. M., Speare, D. J., Dawe, S. C. & Kent, M. L. (2002) Xenoma formation during the microsporidial gill disease of salmonids caused by *Loma salmonae* is affected by host species (*Onchorhynchus tshawytscha*, *O. kisutch*, *O. mykiss*) but not by salinity. *Diseases of Aquatic Organisms*, **48**, 125–131.

Rueppell, O., Hayworth, M. K. & Ross, N. P. (2010) Altruistic self-removal of health-compromised honey bee workers from their hive. *Journal of Evolutionary Biology*, **23**, 1538–1546.

Sanchez, J. G., Speare, D. J., Markham, R. J. F., *et al.* (2001) Localization of the initial developmental stages of *Loma salmonae* in rainbow trout (*Onchorhynchus mykiss*). *Veterinary Pathology*, **38**, 540–546.

Smith, J. E. (2009) The ecology and evolution of microsporidian parasites. *Parasitology*, **136**, 1901–1914.

Terry, R. S., Smith, J. E. & Dunn, A. M. (1998) Impact of a novel feminizing microsporidian parasite on its crustacean host. *Journal of Eukaryotic Microbiology*, **45**, 497–501.

Terry, R. S., Smith, J. E., Sharpe, R. G., *et al.* (2004) Widespread vertical transmission and associated host sex-ratio distortion within the eukaryotic phylum Microspora. *Proceedings of the Royal Society Series B, Biological Sciences*, **271**, 1783–1789.

Thomarat, F., Vivarès, C. P. & Gouy, M. (2004) Phylogenetic analysis of the complete genome sequence of *Encephalitozoon cuniculi* supports the fungal origin of microsporidia and reveals a high frequency of fast-evolving genes. *Journal of Molecular Evolution*, **59**, 780–791.

Van de Peer, Y., Ben Ali, A. & Meyer, A. (2000) Microsporidia: accumulating molecular evidence that a group of amitochondriate and suspectedly primitive eukaryotes are just curious fungi. *Gene*, **246**, 1–8.

Vossbrinck, C. R. & Debrunner-Vossbrinck, B. A. (2005) Molecular phylogeny of the Microsporidia: ecological, ultrastructural and taxonomic considerations. *Folia Parasitologica*, **52**, 131–142.

Weiss, L. M. (2001) Microsporidia: Emerging pathogenic protists. *Acta Tropica*, **78**, 89–102.

Wittner, M. & Weiss, L. M. (1999) *The Microsporidia and Microsporidiosis*. Washington, D. C.: ASM Press.

Myxozoa: the spore-forming cnidarians

5.1 General considerations

The myxozoans are highly specialized, multicellular, spore-forming parasites primarily of marine and freshwater fishes. In vertebrate hosts, they are mostly site-specific and found in cavities such as the gall bladder, urinary bladder, and ureters, or in tissues such as cartilage, muscle, gills, and skin. For this reason, myxozoans are often classified as being either coelozoic (inhabiting cavities) or histozoic (within tissues). Approximately 2200 species in 62 genera have been described; many more await discovery. A tremendous amount of research has been, and is currently, devoted to these enigmatic parasites. As such, our understanding of the biodiversity, ecology, evolution, and systematics of the Myxozoa has been greatly enhanced (Okamura & Canning, 2003; Canning & Okamura, 2004; Fiala & Bartošová, 2010).

One of the reasons for the wealth of scientific interest is that several myxozoans are serious fish pathogens. *Myxobolus cerebralis*, the causative agent of whirling disease in salmonid fishes (see Box 5.1), and *Kudoa thyrsites*, the cause of post-mortem myoliquefaction in various marine fishes, are two examples of well-studied myxozoans having significant economic impacts in sports fisheries and/or the aquaculture industry (reviews in Kent *et al.*, 2001; Yokoyama, 2003; Fiest & Longshaw, 2006). Although many fish myxozoans can be pathogenic, several others are valuable as biological tags in fish stock discrimination and in determining fish migration routes.

A significant advance in myxozoan biology concerns their life cycles. Previously, the stages found in the vertebrate and the invertebrate hosts were treated as different species. So different were the two stages that they were included in different taxonomic classes within the Myxozoa when, in reality, they were different developmental stages of the same species. The landmark discovery by Wolf & Markiw (1984), who found that the life cycle of *M. cerebralis* alternates between a fish and an oligochaete annelid, *Tubifex tubifex*, resulted in increased interest in the development and life cycles of the Myxozoa. Since their pivotal finding, at least 25 further myxozoans have been shown to include annelid hosts in their life cycles (Kent *et al.*, 2001). **Myxospores** are found in various sites and tissues of vertebrate hosts, while so-called **actinospores** are found within invertebrates. Molecular studies, based on small subunit rDNA sequence comparisons, have helped to link the actinosporean stage with the myxosporean stage and to confirm several myxozoan life cycles (Andree *et al.*, 1997; Bartholomew *et al.*, 1997; Holzer *et al.*, 2004; Rangel *et al.*, 2009). Anderson *et al.* (1999) presented molecular data indicating that the actinospore of *Tetracapsuloides bryosalmonae*, the causative agent of proliferative kidney disease in salmonid fishes, was found in a freshwater bryozoan (Box 5.2). This was the first demonstration that bryozoans were involved in the life cycles of some myxozoans. This discovery was to have profound phylogenetic and taxonomic consequences (Okamura & Canning, 2003; Holland *et al.*, 2011; see Boxes 5.2, 5.3).

Historically, the myxozoans were included among the protists. Considerable molecular and morphological evidence now indicates that myxozoans are, in fact, multicellular and are true metazoans. A peculiar, worm-like animal, *Buddenbrockia plumatellae*, also found in freshwater bryozoans, was confirmed as a myxozoan (Monteiro *et al.*, 2002). Its rediscovery provided a significant clue to the origins of the Myxozoa (e.g., Zrzavý & Hypša, 2003). The molecular-based phylogeny of *B. plumatellae* indicates it is a cnidarian (Jiménez-Guri *et al.*, 2007), confirming earlier studies that suggested the Cnidaria as the most closely related taxon to the Myxozoa.

5.2 Form and function

Myxozoans are microscopic endoparasites characterized by multicellular spores. Spore morphology is complex, consisting of two or more **shell valves** (as many as 15 in some species), a **sporoplasm** that is infective to the host, and **polar filaments** coiled within one, or more, **polar capsules** (Fig. 5.1). Myxozoans, like the microsporidia, have an exquisite infection mechanism. Myxozoan shell valves are held together along a suture line. When spores come into contact with a host or an appropriate host ingests the spores, the polar filaments within the polar capsules evert rapidly and the shell valves split along the suture line to release the infective sporoplasm. This spore activation is in response to chemical cues (Kallert *et al.*, 2011), and the filaments aid in host attachment, facilitating the infection process. The outer valve cells of myxosporean spores secrete hard walls, which make them extremely resistant. Some of the spores are completely covered by a mucus envelope (Fiest & Longshaw, 2006). Several myxozoan species have characteristic spores with prominent valve extensions (Fig. 5.1B, C).

Two main types of spores are recognized, i.e., myxospores and actinospores. The myxospores occur within vertebrate hosts, while the actinospores are present within invertebrates, mostly annelid worms. In the case of myxospore development, the sporoplasm migrates to its specific site and undergoes karyokinesis, forming a multinucleate mass (**trophozoite** or **plasmodia**) that feeds on the surrounding tissues. In some species, a presporogonic developmental stage takes place in a tissue distant to the ultimate site of infection (Bjork & Bartholomew, 2010). The plasmodia of several myxosporean parasites can become quite large and cyst-like (Fig. 5.2). As development proceeds, a unique stage is formed, called the **pansporoblast**, that will produce the spores. Actinospores typically

have triradial symmetry, as in the example of the *M. cerebralis* triactinomyxon form, in reference to the three polar capsules (Fig. 5.1E, F). Often, there are also three spore valve extensions (caudal processes), giving the actinospore of *Myxobolus* species a grappling hook-like morphology (Fig. 5.1E). These may serve as floats to assist in dispersal and in maintaining the actinospore's position in the water column. These tri-actinomyxons are produced in pansporoblasts within the intestinal epithelium of annelid worms (Okamura & Canning, 2003). Actinospores are released from annelids and are infective to specific vertebrate hosts, primarily fish.

In the class Malacosporea, so-called **malacospores** in bryozoan hosts develop within hollow rounded sacs in the case of *Tetracapsuloides bryosalmonae*, or in larger worm-like structures in the case of *Buddenbrockia plumatellae*. These spores also have polar capsules and sporoplasms consistent with other Myxozoa. Outer valve cells of these malacosporeans remain soft and their sporoplasms harbor organelles known as **sporoplasmosomes**. Intriguingly, the presence of these unique organelles provided a valuable clue that parasites from bryozoans were responsible for proliferative kidney disease in salmonid fishes (Okamura & Canning, 2003; see Box. 5.2). It turned out that the proliferative stages of *T. bryosalmonae* in fish blood had similar organelles.

5.3 Development and general life cycle

Very few myxozoan life cycles have been completely resolved. The details of the developmental biology and life cycle are best known for the fish pathogen, *Myxobolus cerebralis*, the causative agent of whirling disease (Box 5.1). The life cycle of this important myxosporean was famously worked out by Markiw & Wolf (1983). They were the first to demonstrate that the parasite had a complex life cycle alternating between annelid and fish hosts. Actinospores are now known to develop in tubificid oligochaete annelids, while the myxospores develop in salmonid fishes, then reinfect the worms (Fig. 5.3).

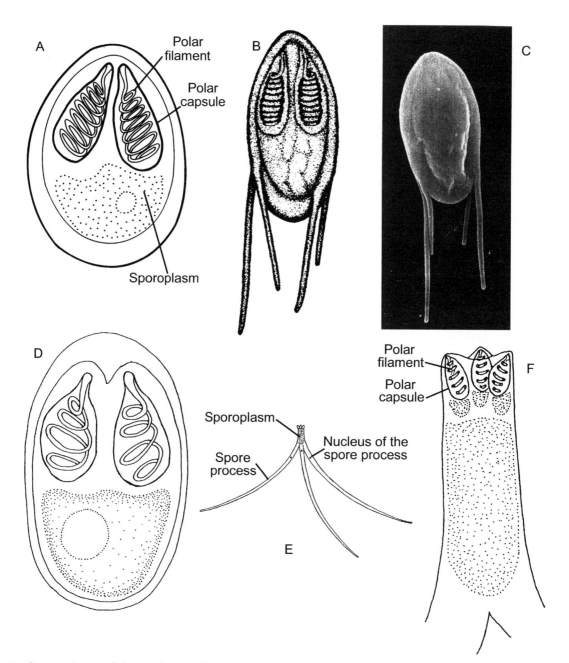

Fig. 5.1 Comparative morphologies of spores of myxozoans. (A) Myxospore of *Myxobolus rhinichthidis* from the head of a longnose dace *Rhinichthys cataractae*; (B) myxospore of *Tetrauronema desaequalis* from the ventral fin of a wolf fish *Hoplias malabaricus*; (C) scanning electron micrograph of the myxospore of *T. desaequalis*; (D) myxospore of *Myxobolus cultus* from the cartilage of a goldfish *Carassius auratus*; (E) triactinomyxon spore of *Myxobolus cultus* from the intestinal epithelium of the tubificid oligochaete *Branchiura sowerbyi*; (F) detail of the upper portion of the triactinomyxon spore *of M. cultus* showing the polar filaments within the polar capsules. ((A) Modified from Cone & Raesly, 1995, with permission, *Canadian Journal of Fisheries and Aquatic Sciences*, 52, S7–S12, NRC Research Press; (B, C) modified from Azevedo & Matos, 1996, with permission, *Journal of Parasitology*, 82, 288–291; (D, E, F) modified from Yokohama *et al.*, 1995, with permission, *Journal of Parasitology*, 81, 446–451.)

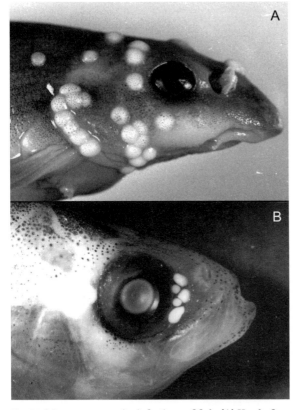

Fig. 5.2 Myxozoan parasite infections of fish. (A) Head of a longnose dace *Rhinichthys cataractae*, infected with cysts of *Myxobolus rhinichthidis*; (B) head of a threadfin shad *Dorosoma petenense*, showing cysts of *Myxobolus penetenses* in the anterior margin of the eye. ((A) Modified from Cone & Raesly, 1995, with permission, *Canadian Journal of Fisheries and Aquatic Sciences*, 52, S7–S12, NRC Research Press; from Frey *et al.*, 1998, with permission, *Journal of Parasitology*, 84, 1204–1206.)

Specifically, myxospores released by the fish host are ingested by the tubificid and the polar filaments are everted upon contact and attach to the intestinal epithelium. The attached spore valves separate and release the infective amoeboid sporoplasm, which then penetrates between the epithelial cells of the intestinal wall. Actinospores are produced within pansporoblasts that develop within the intestinal epithelium of the worms. Development in the oligochaete is complex and involves three phases, i.e.,

schizogony, gametogony, and sporogony. El-Matbouli & Hoffman (1998) determined the intricate details of intra-oligochaete development in *M. cerebralis* (see reviews in Kent *et al.*, 2001; Fiest & Longshaw, 2006). During schizogony, a myxospore sporoplasm undergoes a series of nuclear divisions to produce a multinucleate cell. It is important to note here that the process of schizogony for myxozoans has been challenged and that this cell-within-cell development begins with the engulfment of one cell by another (Morris, 2010). Internal cells then undergo repeated mitotic divisions leading to the production of capsulogenic and valvogenic primorida. These will form the polar capsules and valves of the actinospore, respectively. Further development leads to the production of the pansporoblast, which contains four actinospores. After about 3 months of development, infective triactinomyxons are released into the water with the worm's feces. Shirakashi & El-Matbouli (2009) demonstrated that *M. cerebralis* was an agent of parasite-induced castration and significantly reduced the fecundity of infected tubificid worms. Moreover, reduced feeding activity of infected tubificids was also observed.

When actinospores contact the skin of a fish, the polar capsules evert and the infective sporoplasm emerges; it invades the skin, where presporogonic replication occurs, followed by migration to the site of infection. In the case of *M. cerebralis*, this site is the spine and head cartilage. The plasmodial cells divide repeatedly, forming cavities in the cartilaginous tissue, which becomes packed with spores (Fig. 5.4). The cartilage tissue is destroyed, eventually causing the characteristic pathology of whirling disease, especially in juvenile fish (Box 5.1). Myxospores are released into the environment via a number of possible routes, depending on where the spores develop. They may simply rupture from myxosporean cysts if they are located on body surfaces or on gills of fish hosts. Alternatively, if internal, they may be released into the environment from decomposing tissues of dead and dying fish and then infect annelid hosts. The resistant myxospores of *M.*

Fig. 5.3 Life cycle of *Myxobolus cerebralis*, the causative agent of whirling disease in salmonid fish. Infected juvenile rainbow trout, showing the diagnostic blackened tail, release mature myxospores into water when the fish dies. These are ingested by benthic oligochaete annelid worms such as *Tubifex tubifex*. After a period of development in the tubificid worm, infective triactinomyxon spores are released into water to infect fish (Figure courtesy of Danielle Morrison; modified from Roberts & Janovy, 2009, *Foundations of Parasitology*, 8th edition, McGraw-Hill.)

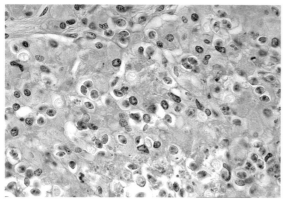

Fig. 5.4 Myxospores of *Myxobolus cerebralis* developing in tissues of juvenile rainbow trout (*Oncorhynchus mykiss*) replacing areas where cartilage had previously existed. (Photograph courtesy of Ronald Hedrick.)

cerebralis have been shown to survive passage through the intestines of piscivorous birds and fishes (Taylor & Lott, 1978; El-Matbouli & Hoffman, 1991). Thus, predation upon infected fish and subsequent release of spores through the predatory animal's feces may enhance dispersal of the myxospores. If this is a common route by which myxospores enter the environment, it would certainly complicate the parasite's epizootiology and any potential parasite-induced pathology.

5.4 Biodiversity and life-cycle variation

It is becoming clear that myxozoans infect a wider variety of host taxa than previously appreciated. Most

Box 5.1 | Pathogenesis and epizootiology of whirling disease caused by *Myxobolus cerebralis*

If you are serious about fly-fishing for wild trout, you are no doubt familiar with the tragic consequences of whirling disease. The disease affects mainly young salmonids with as yet uncalcified skeletons and where cartilage is destroyed or consumed by the trophozoites (plasmodia) of *Myxobolus cerebralis*. When the parasite attacks the cartilaginous tissue of the auditory and equilibrium organs, it interferes with coordination and equilibrium, the fish loses its sense of balance and tumbles erratically (or whirls) as it attempts to feed. As the disease progresses, the parasite invades the vertebral column, mainly that section from the anus to the tail, and prevents normal ossification, causing the tail and trunk to curve (Fig. 5.5). Simultaneously, the sympathetic nerves that control the melanocytes in that region become impaired, producing a permanent dark coloration of the caudal region known as 'black tail' (Fig. 5.3).

Fig. 5.5 Juvenile steelhead (*Oncorhynchus mykiss*) with whirling disease caused by *Myxobolus cerebralis* in left column as compared to uninfected cohorts in the right column. Infections were induced experimentally 2 months prior. (Photograph courtesy of Ronald Hedrick.)

Malformation of the skeleton is often seen among older fishes that manage to overcome and survive the infection. These malformations include retraction of the operculum, curvature of the vertebral column, permanently open or twisted jaws, and deformed heads. In reality, however, very few individuals are likely to survive because the lack of coordination, distinctive coloration, and skeletal deformities are serious handicaps that greatly increase the risk of predation in natural habitats.

As with some other pathogens and parasites, *M. cerebralis* is a good example of what can happen when organisms colonize new hosts or habitats. *Myxobolus cerebralis* is endemic to brown trout in central Europe and southeast Asia, where it causes no visible symptoms to its natural host. The disease was first reported at the turn of the century in central Europe in rainbow trout soon after they were imported from North America. In North America, the disease was first observed in 1956, probably after being introduced with imported processed fish products from Denmark. Presently, *M. cerebralis* has a near-cosmopolitan distribution, having been reported now from 26 different countries. The parasite causes high mortalities and significant losses, especially in farm- and hatchery-reared brook and rainbow trout. Other salmonids, such as sockeye, coho and chinook salmon, also are affected (Hoffman, 1990).

Box 5.1 | (continued)

The disease was not of great significance in natural bodies of water in North America until it was detected in wild rainbow trout in the Madison River of Montana in late 1994. Because of whirling disease, the wild rainbow trout population in this river plummeted from about 2100 fish per km in 1990 to about 200 fish per km in 1994. The world famous wild rainbow trout highly prized by fly fishermen were in serious danger, generating considerable amounts of research; these efforts continue to this day.

Ron Hedrick and Bill Granath are leaders in the study of whirling disease in wild trout populations (e.g., Hedrick *et al.*, 1998; Gilbert & Granath 2003; Granath & Vincent, 2010). Much of their research effort has focused on the invertebrate host in the life cycle, the aquatic oligochaete *Tubifex tubifex* (Fig. 5.3). Earlier studies found that some individuals of *T. tubifex* seemed to be resistant to infection by *M. cerebralis*. If resistant tubificids could displace susceptible ones in nature, the life cycle of *M. cerebralis* would be interrupted, and this should be enough to control the new disease. An alternative approach was directed at the temperature-dependent release of infective spores from the tubificids. At 15 °C, the tubificids release huge numbers of triactinomyxons into the water, but at 5 °C, very few spores are released. Generally, rainbow trout (the most susceptible host species) eggs hatch in May as temperatures rise, coinciding with high densities of triactinomyxon spores in the water column. If earlier spawning in rainbow trout could be induced through artificial selection, then the synchrony of trout hatching and spore production could be disrupted and parasite transmission reduced.

Long-term field studies conducted by Granath and his colleagues on trout within a Montana watershed indicate complex epizootiological patterns (Granath *et al.*, 2007; Granath & Vincent, 2010). Between 1998 and 2003, prevalence of *M. cerebralis* in rainbow trout within sentinel cages increased dramatically and its geographical range expanded within the watershed. Dramatic declines in rainbow trout densities were observed over this interval. In contrast, there is now evidence for a general decline in range and prevalence within the watershed following a peak in 2006. They did not attribute the decline to changes in the prevalence of *M. cerebralis*-infected tubificid worms, or to changes in water temperature. Rather, they conclude that the wild rainbow trout are developing resistance to the parasite, a phenomenon similar to one reported from another Montana study site (Miller & Vincent, 2008).

myxosporeans are known from a range of marine, estuarine, and freshwater teleost fishes (Figs. 5.6, 5.7). However, species have been recently reported from a diversity of other hosts including elasmobranchs, amphibians, reptiles, birds, and even terrestrial mammals. *Myxidium*, a genus comprising over 150 species, for example, are coelozoic myxosporeans infecting the gall bladder, urinary bladder, or kidneys of primarily fish hosts. A few *Myxidium* species, such as *M. melleni*, however, have been described from the gall bladder of amphibians (e.g., Jirků *et al.*, 2006). *Myxidium hardella* was reported from the kidneys of river turtles, causing renal pathology (Garner *et al.*, 2005), and *Myxidium scripta* (Fig. 5.6E) was described from red-eared slider

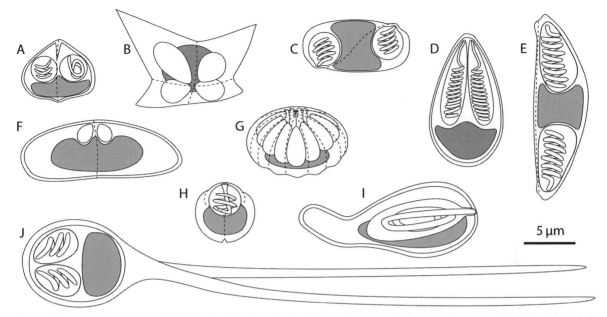

Fig. 5.6 Diverse myxospore morphologies of coelozoic and histozoic myxosporeans. All spores are composed of at least two shell valves with sutures (dashed lines) between, at least one polar capsule with filament (not shown in all examples), and a sporoplasm (gray). (A) *Wardia lucii* from the kidney of bluegill sunfish *Lepomis macrochirus*; (B) *Kudoa thyrsites* from the musculature of Pacific hake *Merluccius productus*; (C) *Ellipsomyxa* sp. from the gall bladder of naked goby *Gobiosoma bosc*; (D) *Myxobolus funduli* from the gills of killifish *Fundulus* sp.; (E) *Myxidium scripta* from the kidney of red-eared slider turtles *Trachemys scripta elegans*; (F) *Ceratomyxa kenti* from the gall bladder of scissortail sergeant *Abudefduf sexfasciatus*; (G) *Kudoa permulticapsula*, with its 13 polar capsules and valves, from the skeletal muscle of narrow-barred Spanish mackerel *Scomberomorus commerson*; (H) *Unicapsula seriolae*, with a single functional polar capsule and two degenerated capsules, from the skeletal muscle of yellowtail amberjack *Seriola lalandi*; (I) *Auerbachia caranxi* from the gall bladder of brassy trevally *Caranx papuensis*; (J) *Henneguya salminicola* from the muscle of coho salmon *Onchorhynchus kisutch*. (Drawings courtesy of Chris Whipps.)

turtles (Roberts *et al.*, 2008). Eiras (2005) provides a review of the myxosporean parasites in amphibians and reptiles.

The myxosporean, *Myxidium anatidum* was described from the bile ducts of waterfowl, expanding the host range to include endothermic hosts (Bartholomew *et al.*, 2008). The discovery of the myxozoan, *Soricimyxum fegati*, from the livers of shrews (Prunescu *et al.*, 2007) further expands the host range and environments exploited by the Myxozoa. Canning & Okamura (2004) provide a detailed review of the biodiversity of myxozoan parasites.

There is a diversity of coelozoic and histozoic myxozoans of teleost fishes (Figs. 5.6, 5.7). By far the most speciose genus is *Myxobolus* with over 800

species described from fish worldwide (Lom & Dyková, 2006). Typically, *Myxobolus* species are histozoic and tissue-specific (Fig. 5.2). Most species are found in freshwater fishes, although some have been described from coastal marine fishes. *Myxobolus* species are among the most diverse parasite lineages in freshwater fishes; as many as five species have been found on an individual fish host (Cone *et al.*, 2004).

The best-known myxosporean species are those that are pathogenic in commercially important fish, or decrease marketability due to unsightly cysts in the flesh of fish of sports fishery or aquaculture importance. Many fish myxozoans are extremely site-specific. In addition to the sites already mentioned, they infect such diverse locations as the gills, fins, skin,

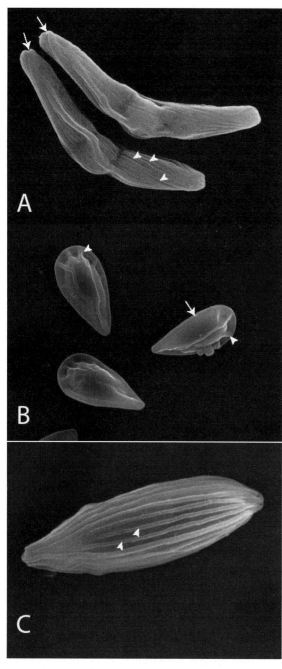

connective tissues, muscle, ovaries, central nervous system, liver, and intestine of various fish hosts. Kent *et al.* (2001), Lom & Dyková (2006), and Fiest & Longshaw (2006) provide excellent reviews of fish myxosporean diversity and the variety of conditions and pathologies they cause.

Among pathogenic species are those of gill-inhabiting *Henneguya* spp., including *Henneguya ictaluri*, the causative agent of proliferative gill disease of channel catfish, and *Henneguya exilis*, the causative agent of lamellar disease in channel catfish. *Henneguya salminicola* occurs in large cysts in the muscle of Pacific salmon species. The cysts are filled with milky fluid and millions of myxospores (Fig. 5.6J). Heavy infections render salmon fillets unmarketable.

Sphaerospora testicularis is a coelozoic myxosporean inhabiting the seminiferous tubules in the testis of European sea bass, *Dicentraarchus labrax*, a fish species of mariculture importance. Infections can lead to testicular hypertrophy and ascites in the abdominal cavity; parasitic castration of valuable broodstock males can occur (Kent *et al.*, 2001).

The multivalvulid myxosporean genus, *Kudoa* (Figs. 5.6B, G), comprises the most diverse of marine myxozoans. Over 70 species of *Kudoa* have been described from several different families of marine fishes worldwide. Most infect the somatic musculature. The most widely studied species is *Kudoa thyrsites* (Fig. 5.6B) (reviews in Moran *et al.*, 1999; Whipps & Kent, 2006). This species has been reported from at least 40 species of fish around the world. In several countries, the parasite has emerged as a serious marketing problem in Atlantic salmon reared in net pen aquaculture. Heavy infections can cause unsightly white cysts in the flesh, as well as a condition known as 'post-harvest soft flesh' or, even more specifically,

Fig. 5.7 Scanning electron micrographs of myxospores from fish. (A) *Sphaeromyxa* sp. (7000×) from the gall bladder of naked goby *Gobiosoma bosc*, showing suture line (arrows) and striated valve surface (arrowheads); (B) *Myxobolus* sp. spores (3700×) from the kidney of banded killifish *Fundulus diaphanous*, a suture line is evident on the edge of the spore (arrow) and distinctive sutural edge markings (arrowheads)

Caption for Fig. 5.7 (cont.)
are visible in this species. (C) *Acauda* sp. spores (8000×) from the kidney of bluegill sunfish *Lepomis macrochirus*, showing prominent valve ridges (arrowheads). (Micrographs courtesy of Chris Whipps.)

Fig. 5.8 Severe post-mortem myoliquefaction in an Atlantic mackerel *Scomber scombrus* heavily infected with *Kudoa thyrsites*. (Photograph courtesy of Chris Whipps.)

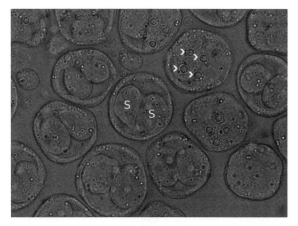

Fig. 5.9 Spores of *Tetracapsuloides bryosalmonae* from the freshwater bryozoan *Fredericella sultana*, showing the two internal sporoplasms (S) and four polar capsules (arrowheads) in some of the spores. (Photograph courtesy of Hanna Hartikainen and Beth Okamura.)

post-mortem myoliquefaction (Fig. 5.8). Following the death of the fish, enzymatic destruction of the muscle tissue occurs. The mode of transmission and the complete life cycle of this parasite are unknown.

The life cycles of myxosporeans are likely to be similar to that of *M. cerebralis*, with cycling between a vertebrate and invertebrate host. With as few as 2% of life cycles known (Atkinson & Bartholomew, 2009), there may be other life-cycle types as has been suggested for *Enteromyxum leei*, which can be transmitted directly from fish to fish (Diamant, 1997). Still, of those that are known, the most common life cycle is indirect. *Ceratomyxa shasta*, another pathogenic myxozoan of salmonids, for example, has a life cycle that involves actinospore production in freshwater polychaete annelids, *Manayunkia speciosa* (Bartholomew *et al.*, 1997). Upon infection of salmon and trout host gills, prespore stages of *C. shasta* migrate to the intestinal tissues of the fish host where plasmodia are formed and myxospore production occurs. The same polychaete species serves as the alternate host for *Parvicapsula minibicornis* and probably other marine fish myxozoans (Køie, 2002). For example, Rangel *et al.* (2009) demonstrated that *Zschokkella mugilus* undergoes early development in the intestinal epithelium of the polychaete *Nereis diversicolor*. Sporogony and gametogony occur in the coelom of the polychaete. Mature actinospores were released from the worms during their reproductive season when they die after spawning. Myxospores of this species are found in estuarine fishes of the Mediterranean Sea.

The landmark discovery that bryozoans were hosts to myxozoans led to the breakthrough that the causative agent of proliferative kidney disease (PKD) in salmonid fishes was *Tetracapsuloides bryosalmonae* (see Box 5.2). Development of *T. bryosalmonae* does not follow the same pattern observed in the myxospore–actinospore life cycle. Rather, the malacospores are developed within sac-like structures in freshwater bryozoans. The spores are much shorter-lived than are myxospores. This is because the outer valve cells of malacosporeans remain soft and, thus, unprotected (Okamura & Canning, 2003). The spores of *T. bryosalmonae* are ovoid with indistinct valves, two sporoplasms, and four polar capsules (Fig. 5.9). Sporoplasms of malacosporeans harbor sporoplasmosomes. Sporoplasmogenic cells within these organelles undergo a series of meiotic divisions leading to the production of uninucleate sporoplasms. These will differentiate to form the valve and capsulogenic cells. Further details of malacospore development are reviewed by Fiest & Longshaw (2006). Fiest *et al.* (2001) experimentally induced PKD in rainbow trout exposed to the bryozoan stages of the parasite. Entry into the fish is rapid and migration to tissues such as the kidney and spleen occurs via the blood. Sporogony occurs within the lumen of renal tubules leading to characteristic pathology (see Box 5.2). The parasite's complete life cycle has not been elucidated. In fact,

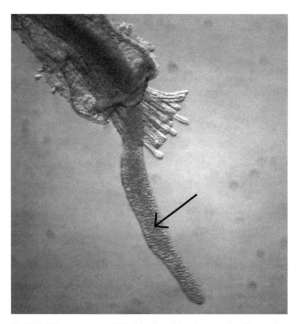

Fig 5.10 Worm stage of *Buddenbrockia plumatellae* (arrow) exiting from a zooid of the freshwater bryozoan *Plumatella fungosa*. (Figure courtesy of Beth Okamura; from Okamura & Canning, 2003, with permission, *Trends in Ecology and Evolution*, **18**, 633–639, Elsevier.)

bryozoans may be the only host for *T. bryosalmonae* and fish may be accidental hosts (Box 5.2).

Spores of the enigmatic malacosporean, *Buddenbrockia plumatellae* develop in both worm-like and sac-like stages within freshwater bryozoans (Fig. 5.10). To date, *T. bryosalmonae* and *B. plumatellae* are the only members within the Malacosporea (see Box 5.3).

5.5 Phylogenetic relationships and classification

Determining the phylogenetic relationships of the Myxozoa with other animal phyla has been the focus of considerable attention and debate (Smothers *et al.*, 1994; Siddall *et al.*, 1995; Schlegel *et al.*, 1996; Fiala, 2006). Highly divergent gene sequences, as well as shared morphological features with protists and the Bilateria and Cnidaria have obscured phylogenetic affinities. Many researchers had long remarked on the similarities between the myxozoan polar capsules and the cnidarian nematocysts. Phylogenetic analyses

---

Box 5.2 | **Proliferative kidney disease and solving the PKX organism mystery: an orphan parasite of bryozoans finds a taxonomic home in the Myxozoa**

For those who would doubt the importance of taxonomy in modern biology, several recently resolved taxonomic mysteries involving myxozoan parasites suggest otherwise. The Myxozoa have been a particularly enigmatic and controversial group to classify and to interpret evolutionary relationships. Our point of considerable taxonomic change is further made when we consider the myxozoan papers with 'demise' in them at phylum, class, and generic levels (e.g. Kent *et al.*, 1994; Siddall *et al.*, 1995; Gunter & Adlard, 2010). Clearly, change is rampant and Myxozoa taxonomy is far from stagnant.

One of the most fascinating taxonomic mysteries involves a myxozoan that infects a group of sessile, colonial invertebrate animals known as bryozoans. The phylum Bryozoa (=moss animals) comprises about 5000 species. Most are marine, although there are some that are important constituents of various freshwater habitats. They feed as suspension feeders using a tentacular crown (lophophore). Bryozoans exhibit a great variety in external morphologies and growth patterns, and a single colony can house over a million genetically identical individuals (zooids).

Box 5.2 (continued)

Freshwater species overwinter as tiny dormant, seed-like structures called statoblasts. Extensive research efforts of Elizabeth Canning and Beth Okamura on myxozoans of bryozoans have helped to determine the identity of an extremely significant fish pathogen, and their findings have revolutionized the taxonomy and phylogeny of the Myxozoa (see Box 5.3).

The mystery concerns the identity of a parasite causing a devastating disease of salmonid fishes in Europe and North America known as proliferative kidney disease. Clinical signs of the disease include swollen kidney and spleen, and accumulation of abdominal fluid (ascites). Mortality associated with PKD can be up to 100% in fish farms and hatcheries. Molecular studies confirmed that the parasite in question was a myxozoan. However, its specific identity remained unknown since mature spores were never found in fish. The causative agent of PKD became known as PKX organism. Direct fish to fish transmission could not be completed experimentally, suggesting that an alternative host was involved in the life cycle. This was not a novel idea given Wolf & Markiw's (1984) pivotal discovery that *Myxobolus cerebralis* includes annelid hosts in its life cycle.

Then, in 1996, a clue to the mystery surfaced. *Tetracapsula bryozoides*, a spore-forming, sac-like parasite was described floating in the body cavities of a freshwater bryozoan species collected from a lake in the United Kingdom. Similar myxozoan sacs were encountered in several bryozoan species from North America. Molecular analyses indicated that the small subunit rDNA sequences of these sacs within bryozoans were phylogenetically indistinguishable from those of PKX (Anderson *et al.*, 1999). Experimental transmission studies confirmed that the spores released from bryozoans infected rainbow trout and produced PKD (Fiest *et al.*, 2001). The significant discovery of bryozoans as hosts allowed PKX to be first described as *Tetracapsula bryosalmonae*, and then *Tetracapsuloides bryosalmonae* (Canning *et al.*, 2000, 2002). The class Malacosporea was erected to accommodate myxozoans of bryozoans (see Box 5.3). Outbreaks of PKD were most commonly found in association with the two bryozoan genera *Fredericella* and *Plumatella* (Okamura & Wood, 2002).

Although many aspects of its life cycle are poorly understood, it is now apparent that *T. bryosalmonae* undergoes developmental cycling within bryozoans dependent on host condition. When bryozoans undergo rapid growth, spore-producing sacs of *T. bryosalmonae* develop as overt infections. When bryozoans are in poor condition overt infections recede to covert infections, and the parasite exists as single cells associated with the body wall (Tops *et al.*, 2009). These dynamics should contribute to long-term persistence within bryozoan populations and recurrent PKD outbreaks (Tops *et al.*, 2009; Okamura *et al.*, 2011). Other features that are likely to contribute to long-term infections in bryozoan populations are vertical transmission of *T. bryosalmonae* infections by fragmentation, re-attachment of bryozoan colonies, and low virulence of infection (Okamura *et al.*, 2011). There is increasing concern that PKD may contribute to declines in wild trout populations, due to increases in freshwater bryozoan abundances from rising temperatures and/or eutrophication (Hartikainen *et al.*, 2009; Tops *et al.*, 2009) and the host condition-dependent development of overt infections (Tops *et al.*, 2009; Okamura *et al.*, 2011).

combining molecular and morphological data have provided evidence indicating affinities between the Myxozoa and the Cnidaria (Siddall *et al.*, 1995). In this analysis, the myxozoans were included within the cnidarian clade and closely related to *Polypodium hydriforme*, a cnidarian intracellular parasite of sturgeon.

The rediscovery of the bizarre bryozoan-inhabiting worm, *Buddenbrockia plumatellae* (Fig. 5.10), has led to further phylogenetic studies attempting to resolve the origins of the Myxozoa (Monteiro *et al.*, 2002). *Buddenbrockia plumatellae* is undoubtedly a myxozoan but its worm-like morphology (Okamura *et al.*, 2002) and putative myxozoan Hox genes (Anderson *et al.*, 1998) suggested the possiblity of a bilaterian affinity. Later, a phylogenomic study of *B. plumatellae* concluded that this animal was a cnidarian (Jiménez-Guri *et al.*, 2007), providing further tantalizing evidence of phylogenetic affinities of the Myxozoa with the Cnidaria. However, certain limitations of the

---

Box 5.3 | Classification of the Myxozoa

Traditionally, myxozoan taxonomy was based largely on the structure of the myxospore stage, especially the number of valves, spore shape, and the number and position of the polar capsules. The use of molecular sequence information has confirmed some of the morphological criteria for taxonomic determinations in the Myxozoa. However, it is becoming increasingly clear that there are huge discrepancies between current taxonomy based on spore morphology and phylogenetic analyses based on molecular data (Whipps *et al.*, 2004; Fiala, 2006; Lom & Dyková, 2006; Fiala & Bartošová, 2010). For example, since many long-standing genera were found to be polyphyletic, considerable taxonomic revision has taken place. *Pentacapsula* and *Hexacapsula* have now been assigned to *Kudoa* (Whipps *et al.*, 2004). The two most speciose genera, *Myxobolus* and *Henneguya*, differ in the presence (*Henneguya*; Fig 5.6J) or absence (*Myxobolus*; Figs. 5.1A, 5.6D, 5.7B) of spore tails and are completely intermixed phylogenetically, suggesting they should be collapsed into a single genus. The discrepancies are likely due to the extreme myxospore plasticity occurring during evolution of the Myxozoa (Fiala & Bartošová, 2010). Phylogenetic studies have demonstrated that at the genus level, marine taxa branch separately from genera that infect freshwater fish (Kent *et al.*, 2001) although numerous exceptions occur. Current phylogenies (Fig. 5.11) tend to support five major splits in the evolution of myxozoans, largely based on spore type, tissue specificity, and habitat. Fiala & Bartošová (2010) present a detailed analysis of myxozoan evolution; they propose that the ancestor of all myxozoans infected the renal tubules of freshwater fish.

Once it was discovered that actinospores in annelids represented an alternate stage of myxozoans in fish, the class Actinosporea was famously suppressed (Kent *et al.*, 1994). Now, two classes are recognized. The class Myxosporea, comprising the majority of myxozoans, currently consists of the two orders, Multivalvulida and the Bivalvulida. After the discovery of myxozoan parasites in bryozoans, Canning *et al.* (2000) erected the new class Malacosporea and order Malacovalvulida.

Box 5.3 (continued)

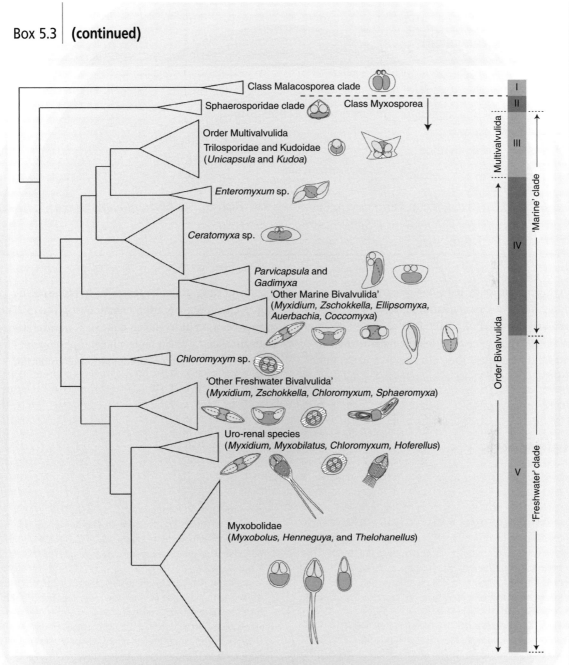

Fig. 5.11 Estimate of myxozoan phylogeny based on ribosomal DNA sequences, with representative spore diagrams. Existing species can be categorized broadly into five lineages (I–V). Note that a major split occurs between Class Malacospora and Myxosporea. Within Myxosporea, species group in the orders Multivalvulida and Bivalvulida and generally into two main groupings of species that are largely parasites of marine fishes or parasites of freshwater fishes. There are numerous exceptions to this marine and freshwater split (e.g., there are freshwater *Ceratomyxa* and *Parvicapsula* species). Challenges for reconciling morphological classifications and molecular phylogeny are apparent as species from many different genera may be intermixed (e.g., *Myxobolus, Henneguya, Thelohanellus*), and some genera appear in multiple lineages (e.g., *Myxidium* and *Chloromyxum*). (Figure courtesy of Chris Whipps.)

Box 5.3 | (continued)

Class Malacosporea
    Order Malacovalvulida
        Representative genera: *Buddenbrockia, Tetracapsuloides*
Class Myxosporea
    Order Multivalvulida
        Representative genera: *Kudoa, Unicapsula*
Order Bivalvulida
    Representative genera: *Ceratomyxa, Chloromyxum, Henneguya, Myxidium, Myxobolus, Parvicapsula, Sphaeromyxa, Sphaerospora, Zschokkella*

phylogenomic study led some researchers to conclude that myxozoan affinities have not yet been resolved (Evans *et al.*, 2008; Collins, 2009). Holland *et al.* (2011) may have settled the debate and convinced the skeptics. They discovered a novel minicollagen gene in *Tetracapsuloides bryosalmonae*. It turns out that minicollagens are phylum-specific genes encoding cnidarian nematocyst proteins. The presence of a minicollagen gene in *T. bryosalmonae* thus provides a direct molecular/morphological link between myxozoan polar capsules and cnidarian nematocysts.

References

Anderson, C. L., Canning, E. U. & Okamura, B. (1998) A triploblast origin for Myxozoa? *Nature*, 392, 346–347.

Anderson, C. L., Canning, E. U. & Okamura B. (1999) Molecular data implicate bryozoans as hosts for PKX (Phylum Myxozoa) and identify a clade of bryozoan parasites within the Myxozoa. *Parasitology*, 119, 555–561.

Andree, K. B., Gresoviac, S. J. & Hedrick, R. P. (1997) Small subunit ribosomal RNA sequences unite alternate myxosporean stages of *Myxobolus cerebralis*, the causative agent of whirling disease in salmonid fish. *Journal of Eukaryotic Microbiology*, 44, 208–215.

Atkinson, S. D. & Bartholomew, J. L. (2009) Alternate spore stages of *Myxobilatus gasterostei*, a myxosporean parasite of three-spined sticklebacks (*Gasterosteus aculeatus*) and oligochaetes (*Nais communis*). *Parasitology Research*, 104, 1173–1181.

Bartholomew, J. L., Whipple, M. J., Stevens, D. G. & Fryer, J. L. (1997) The life cycle of *Ceratomyxa shasta*, a myxosporean parasite of salmonids, requires a freshwater polychaete as an alternate host. *Journal of Parasitology*, 83, 859–868.

Bartholomew, J. L., Atkinson, S. D., Hallet, S. L., *et al.* (2008) Myxozoan parasitism in waterfowl. *International Journal for Parasitology*, 38, 1199–1207.

Bjork, S. J. & Bartholomew, J. L. (2010) Invasion of *Ceratomyxa shasta* (Myxozoa) and comparison of migration to the intestine between susceptible and resistant fish hosts. *International Journal for Parasitolology*, 40, 1087–1095.

Canning, E. U. & Okamura, B. (2004) Biodiversity and evolution of the myxozoa. *Advances in Parasitology*, 56, 43–131.

Canning, E. U., Curry, A., Feist, S. W., *et al.* (2000) A new class and order of myxozoans to accommodate parasites of bryozoans with ultrastructural observations on *Tetracapsula bryosalmonae* (PKX organism). *Journal of Eukaryotic Microbiology*, **47**, 456–468.

Canning, E. U., Tops, S., Curry, A., *et al.* (2002) Ecology, development and pathogenicity of *Buddenbrockia plumatellae* Schröder, 1910 (Myxozoa, Malacosporea) (syn. *Tetracapsula bryozoides*) and establishment of *Tetracapsuloides* n. gen. for *Tetracapsula bryosalmonae*. *Journal of Eukaryotic Microbiology*, **49**, 280–295.

Collins, A. G. (2009) Recent insights into cnidarian phylogeny. *Smithsonian Contributions to the Marine Sciences*, **38**, 139–149.

Cone, D. K., Marcogliese, D. J. & Russell. R. (2004) The myxozoan fauna of spottail shiner in the Great Lakes Basin: membership, richness, and geographical distribution. *Journal of Parasitology*, **90**, 921–932.

Diamant, A. (1997) Fish-to-fish transmission of a marine myxosporean. *Diseases of Aquatic Organisms*, **30**, 99–105.

Eiras, J. C. (2005) An overview on the myxosporean parasites in amphibians and reptiles. *Acta Parasitologica*, **50**, 267–275.

El-Matbouli, M. & Hoffman, R. W. (1991) Effects of freezing, ageing and passage through the alimentary canal of predatory animals on the viability of *Myxobolus cerebralis* spores. *Journal of Aquatic Animal Health*, **3**, 260–262.

El-Matbouli, M. & Hoffman, R. W. (1998) Light and electron microscopic study on the chronological development of *Myxobolus cerebralis* to the actinosporean stage in *Tubifex tubifex*. *International Journal for Parasitology*, **28**, 195–217.

Evans, N. M., Lindner, A., Raikova, E. V., *et al.* (2008) Phylogenetic placement of the enigmatic parasite, *Polypodium hydriforme*, within the Phylum Cnidaria. *BMC Evolutionary Biology*, **8**:139, DOI:10.1186/1471-2148-8-139.

Fiala, I. (2006) The phylogeny of Myxosporea (Myxozoa) based on small subunit ribosomal RNA gene analysis. *International Journal for Parasitology*, **36**, 1521–1534.

Fiala, I. & Bartošová, P. (2010) History of myxozoan character evolution on the basis of rDNA and EF-2 data. *BMC Evolutionary Biology*, **10**, 228.

Fiest, S. W. & Longshaw, M. (2006) Phylum Myxozoa. In *Fish Diseases and Disorders*, vol. 1, *Protozoan and Metazoan Infections*, 2nd edition. ed. P. T. K. Woo, pp. 230–296. Wallingford: CAB International.

Fiest, S. W., Longshaw, M., Canning, E. U. & Okamura, B. (2001) Induction of proliferative kidney disease (PKD) in rainbow trout (*Oncorhynchus mykiss*) via the bryozoan *Fredericella sultana*, infected with *Tetracapsula bryosalmonae*. *Diseases of Aquatic Organisms*, **45**, 61–68.

Garner, M. M., Bartholomew, J. L, Whipps, C. M., *et al.* (2005) Renal myxozoanosis in Crowned River turtles *Hardella thurjii*: description of the putative agent, *Myxidium hardella* n. sp. by histopathology, electron microscopy, and DNA sequencing. *Veterinary Pathology*, **42**, 589–595.

Gilbert, M. A. & Granath, Jr., W. O. (2003) Whirling disease of salmonid fish: life cycle, biology and disease. *Journal of Parasitology*, **89**, 658–667.

Granath, Jr., W. O., & Vincent, E. R. (2010) Epizootiology of *Myxobolus cerebralis*, the causative agent of salmonid whirling disease in the Rock Creek drainage of west-central Montana: 2004–2008. *Journal of Parasitology*, **96**, 252–257.

Granath, Jr., W. O., Gilbert, M. A., Wyatt-Pescador, E. J. & Vincent, E. R. (2007) Epizootiology of *Myxobolus cerebralis*, the causative agent of salmonid whirling disease in the Rock Creek drainage of west-central Montana. *Journal of Parasitology*, **93**, 104–119.

Gunter, N. & Adlard, R. (2010) The demise of *Leptotheca* Thelohan, 1895 (Myxozoa: Myxosporea: Ceratomyxidae) and assignment of its species to *Ceratomyxa* Thelohan, 1892 (Myxosporea: Ceratomyxidae), *Ellipsomyxa* Köie, 2003 (Myxosporea: Ceratomyxidae), *Myxobolus* Butschli, 1882 and *Sphaerospora* Thelohan, 1892 (Myxosporea: Sphaerosporidae). *Systematic Parasitology*, **75**, 81–104.

Hartikainen, H., Johnes, P., Moncrieff, C. & Okamura, B. (2009) Bryozoan populations reflect nutrient enrichment and productivity gradients in rivers. *Freshwater Biology*, **54**, 2320–2334.

Hedrick, R. P., El-Matbouli, M., Adkison, M. A. & MacConnell, E. (1998) Whirling disease: re-emergence among wild trout. *Immunological Reviews*, **166**, 365–376.

Hoffman, G. L. (1990) *Myxobolus cerebralis*, a worldwide cause of salmonid whirling disease. *Journal of Aquatic Animal Health*, **2**, 30–37.

Holland, W. H., Okamura, B., Hartikainen, H. & Secombes, C. J. (2011) A novel minicollagen gene links cnidarians and myxozoans. *Proceedings of the Royal Society B*, **278**, 546–553.

Holzer, A. S., Sommerville, C. & Wootten, R. (2004) Molecular relationships and phylogeny in a community of myxosporeans and actinosporeans based on their 18S rDNA sequences. *International Journal for Parasitology*, **34**, 1099–1111.

Jiménez-Guri, E., Philippe, H., Okamura, B. & Holland, P. W. H. (2007) *Buddenbrockia* is a cnidarian worm. *Science*, **317**, 116–118.

Jirků, M., Bolek, M. G., Whipps, C. M., *et al.* (2006) A new species of *Myxidium* (Myxosporea: Myxidiidae), from the western chorus frog, *Pseudacris triseriata triseriata*, and Blanchard's cricket frog, *Acris crepitans blanchardi* (Hylidae), from eastern Nebraska: morphology, phylogeny, and critical comments on amphibian *Myxidium* taxonomy. *Journal of Parasitology*, **92**, 611–619.

Kallert, D. M., Bauer, W., Haas, W. & El-Matbouli, M. (2011) No shot in the dark: myxozoans chemically detect fresh fish. *International Journal for Parasitology*, **41**, 271–276.

Kent, M. L., Margolis, L. & Corliss, J. O. (1994) The demise of a class of protists: taxonomic and nomenclatural revisions proposed for the protist phylum Myxozoa Grassé, 1970. *Canadian Journal of Zoology*, **72**, 932–937.

Kent, M. L., Andree, K. B., Bartholomew, J. L., *et al.* (2001) Recent advances in our knowledge of the Myxozoa. *Journal of Eukaryotic Microbiology*, **48**, 395–413.

Køie, M. (2002) Spirorbid and serpulid polychaetes are candidates as invertebrate hosts for Myxozoa. *Folia Parasitologica*, **49**, 160–162.

Lom J. & Dyková, I. (2006) Myxozoan genera: definition and notes on taxonomy, life-cycle terminology and pathogenic species. *Folia Parasitologica*, **53**, 1–36.

Markiw, M. E. & Wolf, K. J. (1983) *Myxosoma cerebralis* (Myxozoa: Myxosporea) etiologic agent of salmonid whirling disease requires tubificid worms (Annelida: Oligochaeta) in its life cycle. *Journal of Protozoology*, **30**, 561–564.

Miller, M. P. & Vincent, E. R. (2008) Rapid natural selection for resistance to an introduced parasite of rainbow trout. *Evolutionary Applications*, **1**, 336–341.

Monteiro, A. S., Okamura, B. & Holland, P. W. H. (2002) Orphan worm finds a home: *Buddenbrockia* is a myxozoan. *Molecular Biology and Evolution*, **19**, 968–971.

Moran, J. D. W., Whitaker, D. J. & Kent, M. L. (1999) A review of the myxosporean genus *Kudoa* Megalitsch, 1947, and its impact on the international aquaculture industry and commercial fisheries. *Aquaculture*, **172**, 163–196.

Morris, D. J. (2010) Cell formation by myxozoan species is not explained by dogma. *Proceedings of the Royal Society Series B, Biological Sciences*, **277**, 2565–2570.

Okamura, B. & Canning, E. U. (2003) Orphan worms and homeless parasites enhance bilaterian diversity. *Trends in Ecology and Evolution*, **18**, 633–639.

Okamura, B. & Wood, T. S. (2002) Bryozoans as hosts for *Tetracapsula bryosalmonae*, the PKX organism. *Journal of Fish Diseases*, **25**, 469–475.

Okamura, B., Curry, A., Wood, T. S. & Canning, E. U. (2002) Ultrastructure of *Buddenbrockia* identifies it as a myxozoan and verifies the bilaterian origin of the Myxozoa. *Parasitology*, **124**, 215–223.

Okamura, B., Hartikainen, H., Schmidt-Posthaus, H. & Wahli, T. (2011) Life cycle complexity, environmental change and the emerging status of salmonid proliferative kidney disease. *Freshwater Biology*, **56**, 735–753.

Prunescu, C. C., Prunescu, P., Pucek, Z. & Lom, J. (2007) The first finding of myxosporean development from plasmodia to spores in terrestrial mammals: *Soricimyxum fegati* gen. et sp. n. (Myxozoa) from *Sorex araneus* (Soricomorpha). *Folia Parasitologica*, **54**, 159–164.

Rangel, L. F., Santos, M. J. Cech, G. & Székely, C. (2009) Morphology, molecular data, and development of *Zschokkella mugilis* (Myxosporea, Bivalvulida) in a polychaete alternate host, *Nereis diversicolor. Journal of Parasitology*, **95**, 561–569.

Roberts, J. F, Whipps C. M., Bartholomew, J. L., *et al.* (2008) *Myxidium scripta* n. sp. identified in urinary and biliary tract of Louisiana-farmed red-eared slider turtles *Trachemys scripta elegans. Diseases of Aquatic Organisms*, **80**, 199–209.

Schlegel, M., Lom, L., Stechmann, A., *et al.* (1996) Phylogenetic analysis of complete small subunit ribosomal RNA coding region of *Myxidium lieberkuehni*: evidence that Myxozoa are Metazoa and related to the Bilateria. *Archiv für Protistenkunde*, **147**, 1–9.

Shirakashi, S. & El-Matbouli, M. (2009) *Myxobolus cerebralis* (Myxozoa), the causative agent of whirling disease, reduces fecundity and feeding activity of *Tubifex tubifex* (Oligochaeta). *Parasitology*, **136**, 603–613.

Siddall, M. E., Martin, D. S., Bridge, D., *et al.* (1995) The demise of a phylum of protists: Phylogeny of Myxozoa and other parasitic Cnidaria. *Journal of Parasitology*, **81**, 961–967.

Smothers, J. F., Von Dohlen, C. D., Smith, Jr., L. H. & Spall, R. D. (1994) Molecular evidence that the myxozoan protists are metazoans. *Science*, **265**, 1719–1721.

Taylor, R. L. & Lott, M. (1978) Transmission of salmonid whirling disease by birds fed trout infected with *Myxosoma cerebralis. Journal of Protozoology*, **25**, 105–106.

Tops, S., Hartikainen, H. & Okamura, B. (2009) The effects of infection by *Tetracapsuloides bryosalmonae* (Myxozoa) and temperature on *Fredericella sultana* (Bryozoa). *International Journal for Parasitology*, **39**, 1003–1010.

Whipps, C. M. & Kent, M. L. (2006) Phylogeography of the cosmopolitan marine parasite *Kudoa thrysites* (Myxozoa: Myxosporea). *Journal of Eukaryotic Microbiology*, **53**, 364–373.

Whipps, C. M., Grossel, G., Adlard, R. D., *et al.* (2004) Phylogeny of the Multivalvulidae (Myxozoa: Myxosporea) based upon comparative rDNA sequence analysis. *Journal of Parasitology*, **90**, 618–622.

Wolf, K. & Markiw, M. E. (1984) Biology contravenes taxonomy in the Myxozoa: new discoveries show alternation of invertebrate and vertebrate hosts. *Science*, **225**, 1449–1452.

Yokoyama, H. (2003) A review: gaps in our knowledge on myxozoan parasites of fish. *Fish Pathology*, **38**, 125–136.

Zrzavý, J. & Hypša, V. (2003) Myxozoa, *Polypodium*, and the origin of the Bilateria: The phylogenetic position of 'Endocnidozoa' in light of the rediscovery of *Buddenbrockia. Cladistics*, **19**, 164–169.

6 Platyhelminthes: the flatworms

6.1 General considerations

The phylum Platyhelminthes (Greek, *platy*flat, *helminthes*worm) includes at least 30 000 species. The phylum represents a large and diverse group of organisms, most of which are obligate parasites, living on, or in, most species of vertebrate and invertebrate animal. As the phylum name suggests, these worms are flattened dorsoventrally. They are without segmentation, although cestodes, or tapeworms, superficially appear otherwise. Cestodes are modular iterations, with each segment or **proglottid** being more like an individual within a colony since each is a complete sexual unit (Hughes, 1989). Moreover, there is no coelom or peritoneum as there are in truly coelomate, segmented animals such as the annelids. Platyhelminths may, or may not, possess an incomplete gut. They are without circulatory, skeletal, and respiratory systems. The functional and structural unit of their excretory/osmoregulatory system is a **protonephridium**, or flame cell (Fig. 6.1), so named for a tuft of cilia extending away from the cell body that resembles the flame of a burning candle. Most species are monoecious, but a few, such as the medically important schistosomes, are dioecious.

Parasitic platyhelminths are extraordinarily diverse in terms of their morphology, habitats, life cycles, and transmission adaptations. The ectoparasitic monogeneans of primarily fish and amphibians, for example, have direct life cycles, featuring a free-swimming **oncomiracidium** stage. On the other hand, all endoparasitic digenean trematodes (flukes) have remarkably complex life cycles with molluscs (mostly snails) as first intermediate hosts, in which free-swimming stages known as **cercariae** are produced by extensive asexual reproduction. Further, most trematodes (and many cestodes) incorporate a resting stage within a second intermediate host. This host is generally a potential prey item of the definitive vertebrate host in which the parasite matures. Thus, most trematodes and cestodes are transmitted via predator–prey interactions. To add to the complexity, and stressing the importance of trophic transmission and food web dynamics, still other trematodes and cestodes have added third intermediate, or often, paratenic hosts to their life cycles. How and why such life cycle complexity evolved in the Platyhelminthes has long been the subject of debate (reviews in Cribb *et al.*, 2003; Parker *et al.*, 2003). As we will see in this, and several subsequent chapters, many species of platyhelminth are utilized as model systems for addressing questions in ecological, evolutionary, and environmental parasitology (see Chapters 12–17).

The systematics of platyhelminths has changed significantly over the last two decades due to the application of cladistic methods that utilize molecular markers (reviews in Kuchta *et al.*, 2008; Olson *et al.*, 2008). There are at least two primary reasons why platyhelminth phylogenetics and systematics have received so much recent attention. First, as the earliest of the bilateral and triploblastic animals, they are now considered to occupy a pivotal position in metazoan evolution. Thus, understanding their evolutionary relationships to other phyla and establishing a robust phylogeny for all multicellular animals has become a 'holy grail' for zoologists (Littlewood *et al.*, 1999b). Second, in terms of biodiversity, the Platyhelminthes is a primarily parasitic phylum. If we ever hope to understand the evolution of parasitism and such an impressive adaptive radiation, we must ascertain platyhelminth evolutionary relationships. One of the keys to establishing such a phylogenetic framework for the evolution of parasitism is to identify within the free-living platyhelminths, the sister group to the parasitic taxa.

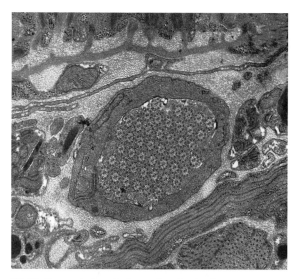

Fig. 6.1 Transmission electron micrograph of a protonephridium, showing a cross-section of the tuft of cilia extending away from the cell body. (Micrograph courtesy of Darwin Wittrock.)

Traditionally and prior to the cladistic and molecular phylogenetic revolution, the phylum Platyhelminthes was divided into four classes, namely Turbellaria, Monogenea, Trematoda, and Cestoda. Although most of these latter parasitic groups remain in newer systematic systems, several that were included within these classes now form distinct groups. The phylum Platyhelminthes is probably not monophyletic, which complicates phylogenetics and systematics. Currently, the taxon Neodermata (=new skin) is regarded as the monophyletic group to which the Monogenea, Trematoda, and Cestoda belong (see Box 6.3). The classification system used here follows Littlewood *et al.* (1999a, b), Olson & Tkach (2005), and Kuchta *et al.* (2008). The platyhelminth taxa will first be discussed with relatively little emphasis on their phylogenetic status. We initially discuss three 'minor' parasitic platyhelminth groups that feature prominently in flatworm phylogenetics (see Section 6.10), and then focus on the biology of select orders and families within the three obligate parasitic classes, the Trematoda, Monogenea, and Cestoda.

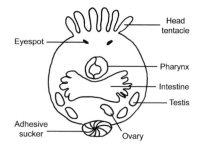

Fig. 6.2 Schematic representation of a temnocephalidean, illustrating several of the most prominent internal and external features. (Figure courtesy of Lisa Esch McCall.)

6.2 Temnocephalidea

Temnocephalids include a small group of ectocommensal organisms that live on the surface or in the branchial chamber of crayfish, prawns, isopods, other crustaceans, and certain aquatic insects, although they also have been found in turtles and molluscs. Most of their geographical distribution follows a Gondwanic pattern, as they are relatively common in South America, Australia, New Zealand, Madagascar, and India, although a few species have been reported in Europe. Temnocephalids are small and flattened, with a sucker at the posterior end and 5 to 12 finger-like tentacles at the anterior end (Fig. 6.2). They have two eyespots and are able to move about with a leech-like motion. The majority of species feed on small organisms such as bacteria, diatoms, protists, rotifers, and nematodes found in, or around, the host and, although they are mobile, they do not normally leave their hosts. If experimentally removed from their host, some individuals of some species die promptly, whereas others survive for weeks (Avenant-Oldewage, 1993). Temnocephalids are monoecious. The eggs are situated within capsules attached to the exoskeleton of the host and development is direct. A miniature of the adult hatches from the eggs and matures to adulthood.

Interestingly, *Temnocephala chilensis* was reported (Viozzi *et al.*, 2005) to serve as a second intermediate host for the digenean *Echinoparyphium* sp. in the Patagonian region of Argentina. The metacercariae recovered were fed to young chicks and ovigerous adults of *E. megacirrus* were found at necropsy.

6.3 Udonellida

The phylogenetic position of the enigmatic udonellids has been controversial for many years. Udonellids are small, nearly cylindrical, with a posterior muscular sucker without hooks or anchors, and two small anterior suckers. They have a modified excretory system that consists of many pores and canals. This feature, together with the lack of hooks and the lack of a ciliated larva, suggests separation of the udonellids from monogeneans. However, ultrastructural and molecular phylogenetic studies strongly suggest that udonellids are monogeneans (Littlewood *et al.*, 1998). We keep them separate here to draw attention to their unique biology.

Only four species of *Udonella* have been described. At least one of them, *Udonella caligorum*, has a cosmopolitan distribution. Udonellids are normally described as hyperparasites of copepods (Fig. 6.3) that parasitize marine fishes. However, it seems that the worms use copepods only for transport, feeding on the surface of the fish. Some authors argue that they should be considered parasites of fishes and not of copepods (Kabata, 1973; Byrnes, 1986; Aken'Ova & Lester, 1996). Regardless of the nature of their true host, udonellids are ubiquitous on many caligid copepods in temperate marine waters.

6.4 Aspidobothrea

The aspidobothreans are a small taxa of about 80 species that are endoparasitic in molluscs and poikilothermic aquatic vertebrates. Their sister-group status with the Digenea is well supported from their morphology, patterns of host use, and molecular-based systematics (Blair, 1993; Littlewood *et al.*, 1999a, b). They are of little

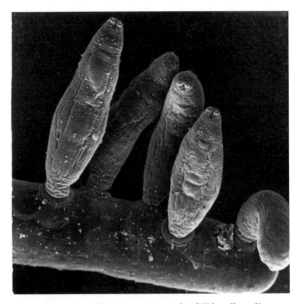

Fig. 6.3 Scanning electron micrograph of *Udonella caligorum* attached to the egg sacs of the copepod *Lepeophtheirus mugiloidis*. (Micrograph courtesy of Jackie Fernández.)

medical or economic importance, and are hence relatively poorly studied. However, those species that establish facultative parasitic relationships have provided parasitologists with intriguing insights into the transition between the free-living and parasitic platyhelminths.

There are three distinctive body forms. In some, a single large sucker divided into shallow depressions by muscular septa covers the ventral surface (Fig. 6.4). In others, there is a series of suckers distributed longitudinally, whereas in still others there is a ventral holdfast with transverse ridges. The digestive system is incomplete, like that of digeneans and monogeneans, except that it is in the form of a simple, blind sac. Digestion is extracellular. The excretory/osmoregulatory system includes numerous protonephridia and a bladder located at the posterior end. The nervous system is more complex than usual for a platyhelminth parasite, with a well-developed set of anterior nerves and a ladder-like peripheral system. Aspidobothreans also have a great diversity and number of receptors in their surface, which is a counter-trend to the generalized reduction of the nervous system of many parasitic organisms. Rohde (1989) suggested that in

Fig. 6.4 The aspidobothrean *Trigonostoma callorhynchi* from the Chilean elephant fish *Callorhynchus callorhynchus*, showing the ventral surface of the parasite modified into an adhesive structure. (Modified from Fernández *et al.*, 1986, with permission, *Biología Pesquera*, 15, 63–73.)

large aspidobothreans, such as *Lobatostoma*, there are between 20 000 and 40 000 receptors whose role is to sense the host's environment. These parasites are monoecious and their reproductive systems resemble those of digeneans. Development is direct, sharing numerous features with the monogeneans. Yet, unlike the monogeneans, several species require development within an intermediate host. Typically, a ciliated larva hatches from the eggs and develops directly into an adult. Most aspidobothreans that parasitize molluscs have simple life cycles, without intermediate hosts, but species parasitic on fishes and turtles, e.g., *Lobatostoma* spp., require a snail intermediate host. In *L. manteri*, a snail ingests the egg; a larva hatches in the stomach, and migrates into the hepatopancreas where it develops into a pre-adult. When a fish ingests an infected snail, the parasite matures.

Aspidobothreans do not exhibit a high degree of specificity for either their molluscan or vertebrate hosts. For example, *Aspidogaster conchicola* is a common parasite in the pericardium of freshwater clams in Africa, Europe, and North America. In North America alone, it has been reported from over 70 different species of freshwater hosts, mostly bivalves (Hendrix *et al.*, 1985).

6.5 Digenea

6.5.1 General considerations

Most adult digeneans (or 'flukes') are endoparasitic and are found in all classes of vertebrates, typically within the gastrointestinal tract. They usually, though not always, possess a pair of suckers, including a ventral **acetabulum** and an anterior oral sucker in the center of which is the mouth. Some have accessory suckers on either side of the oral sucker. Digeneans have an incomplete digestive system. In terms of size and general morphology, there is no such thing as a typical digenean. They occur in virtually all shapes and sizes. Some are thin and round, and others are ribbon-like. The genital openings are usually located between the two suckers, but this is variable as well. Most digeneans are hermaphrodites, with male and female organs in the same individual (Figs. 6.5, 6.6). Digeneans have complex, indirect life cycles, and almost all use molluscs as their first intermediate hosts. Most also have second intermediate hosts and utilize predator–prey interactions for transmission to their specific definitive hosts. Flukes are renowned for exhibiting remarkable life-cycle variation, ranging from those that have abbreviated two-host life cycles to those with four obligate hosts. Parker *et al.* (2003) and Cribb *et al.* (2003) explore the evolutionary causes and consequences of variation in digenean life cycles and life histories.

6.5.2 Form and function

There is an extensive literature on trematode functional morphology, starting with the classic work by Erasmus (1972). Updates are provided in Fried & Graczyk (1997). In this section, we provide an introduction to the basic digenean body plan, restricting our coverage to examples from the most important groups.

6.5.2.1 Holdfasts

With a few exceptions, there are two suckers used for attachment, including an oral sucker and a ventrally

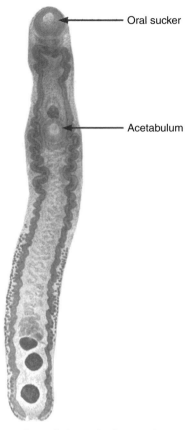

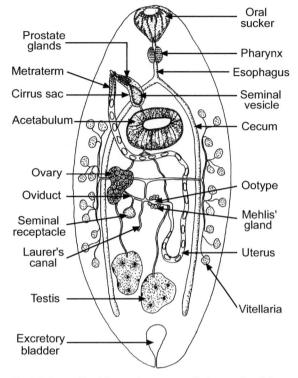

Fig. 6.6 Generalized internal anatomy of a hermaphroditic digenean. (Figure courtesy of Derek Zelmer & Michael Barger.)

Fig. 6.5 External morphology of a distome digenean, *Azygia* sp., showing the positions of the oral sucker and acetabulum. Note also the paired testes and ovary at the posterior of the fluke, as well as the intestinal cecae. (Photograph courtesy of Harvey Blankespoor.)

positioned acetabulum or ventral sucker (Figs. 6.5, 6.6). The position of the acetabulum varies among species, although it is usually located about halfway along the length of the body. The muscular suckers are covered by tegument. Many adult digeneans, especially those in the alimentary tract, are ornamented with elaborate spines along their dorsal surface that presumably assist in lodging the worms within the gut (Fig. 6.7).

6.5.2.2 Tegument

The platyhelminth's tegument was once considered a non-living cuticle. This early view provided a logical

parallel to the morphological strategy used by endoparasitic nematodes and arthropods for protection against host digestive chemicals and host immunity. We now know that the trematode tegument is a complex, living tissue that provides phenomenal functional versatility (Fig. 6.8). Indeed, it is difficult to envision a comparable type of animal tissue that provides both environmental protection and a platform to meet nutritional needs. In addition to providing key roles in protection and nutrition, the tegument also has sensory and excretory capabilities (review in Dalton *et al.*, 2004). In the case of the platyhelminths, it should be no surprise that our broad acceptance of the Neodermata as a monophyletic clade reflects support for the idea of radiation following the evolution of this complex tissue (**neodermis**) from free-living flatworms.

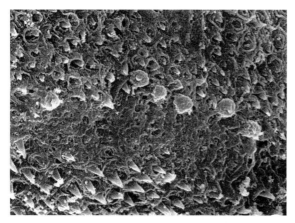

Fig. 6.7 Scanning electron micrograph of the tegument of the schistosome *Australobilharzia variglandis*, illustrating raised tubercles, papillae, and spines (Modified from Barber & Caira, 1995, with permission, *Journal of Parasitology*, **81**, 585–592.)

The distal-most layer, and the one ultimately providing the boundary between host and parasite tissue, is covered by a living plasma membrane that is interspersed with protein-rich elements that collectively form a **glycocalyx**. In *Schistosoma* spp., this layer of the tegument is continuously renewed, probably functioning to replace tissue damaged by the host's immune response. For these well-studied species, this layer also binds specific host molecules, effectively providing protection against host defenses (Gobert *et al.*, 2003). The drug praziquantel is effective against many trematodes because it affects the integrity of the tegument, especially with respect to its permeability to calcium ions.

The glycocalyx envelops a layer of distal cytoplasm that is packed with mitochondria and

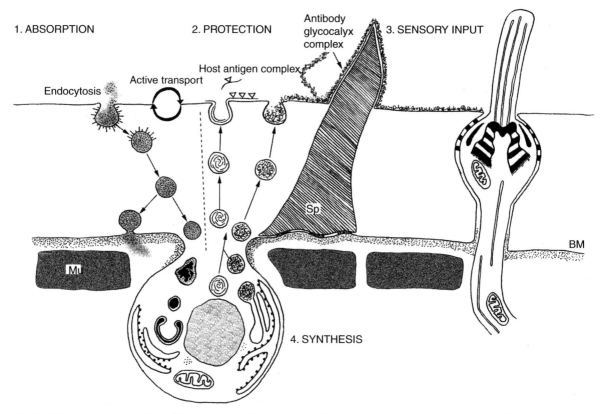

Fig. 6.8 Schematic representation of the functional ultrastructure of the platyhelminth tegument, showing the basement membrane (BM), muscle (Mu), and a spine (Sp). (Figure courtesy of Chelsea Matisz; modified from Dalton *et al.*, 2004, *Canadian Journal of Zoology*, **82**, 211–232.)

numerous secretory vesicles (Fig. 6.8). This is the highly metabolic syncytial layer (**syncytium**) that is the hallmark of the Neodermata body plan, providing opportunity for the lateral transfer of nutrients without restriction imposed by individual cell membranes. The syncytium likely also provides protection from host chemicals and defenses that would typically be targeted for features located on individual cell membranes. Importantly, the distal cytoplasm is connected via special channels to cell bodies containing nuclei and other typical cell organelles (Fig. 6.8). Between the distal cytoplasm and the proximal cell bodies lies a layer of both longitudinal and circular muscles.

6.5.2.3 Digestive system

The **parenchyma** forms a complex acellular matrix within which all the internal organs and organ systems occur. The parenchyma also functions in metabolism and circulation. The incomplete digestive system opens into a mouth within the oral sucker. In many digeneans, the gut begins with a mouth, a pre-pharynx, followed by a muscular pharynx in some species, and then an esophagus that usually splits into two ceca that may or may not extend the length of the body (Figs. 6.5, 6.6). The ceca are generally simple canals lined with columnar epithelium; they may be highly branched, and may, or may not, fuse posteriorly. The epithelium of some species is syncytial and of others is cellular. Functionally, these cells are both absorptive and secretory, releasing both proteoglycans and proteases (Fujino, 1993).

6.5.2.4 Excretory/osmoregulatory system

The removal of metabolic wastes in digeneans is effected by diffusion through the surface to the outside, via the intestinal ceca to the lumen, and by protonephridia, or flame cells located within the parenchyma. Nitrogenous wastes primarily are in the form of ammonia. The pattern of organization of the flame cells is species specific and bilateral. The complex arrangement of flame cells and their collecting ducts is illustrated in the metacercaria

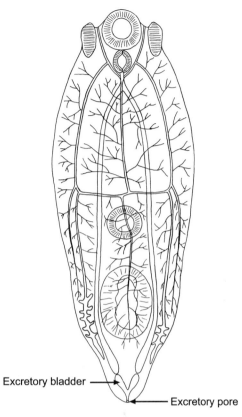

Excretory bladder

Excretory pore

Fig. 6.9 The excretory system in the metacercaria of *Tylodelphys xenopi*. Note the posterior Y-shaped excretory bladder and the excretory pore. The terminus of each line on the drawing represents the location of an individual protonephridium. (Modified from King & Van As, 1997, with permission, *Journal of Parasitology*, 83, 287–295.)

shown in Figure 6.9. The collecting ducts fuse and empty into a posterior excretory bladder, which may, or may not, be lined with epithelium. The bladder usually empties through a pore at the very posterior end of the body. There is evidence suggesting that the protonephridial system may have an osmoregulatory function as well.

6.5.2.5 Reproductive system

In hermaphroditic flukes, the body of the worm contains a complete set of male and female reproductive organs (Figs. 6.5, 6.6). In most species, sperm are produced in paired testes located in the posterior portion of the body (the number and location of the testes will vary). Generally, the testes are of similar size. Sperm leave the testes through the vas efferens, which fuse to form the vas deferens, or sperm duct. The terminal portion of the male system usually ends in a cirrus sac and protrusible **cirrus**. Associated with the cirrus sac are the prostate gland and **seminal vesicle**. The final opening to the outside may lie in a genital atrium or depression on the ventral surface. During copulation, the cirrus is inserted into the gonopore of the female and sperm pass through the female's reproductive tract to the **seminal receptacle** where they are stored. Ova are produced in a single ovary. They move into an oviduct and, as they pass by the opening of the seminal receptacle, they are fertilized. Close by, there may be a **Laurer's canal**, a vestigial structure connecting the oviduct to the surface of the digenean body. The diploid zygote continues into the **ootype** that is surrounded by the **Mehlis' gland**. Also emptying into the ootype is the vitelline duct, which is connected to **vitellaria** that are usually scattered along both lateral borders of the body. The vitellaria have a dual function in that they supply yolk as well as produce substances that will become the eggshell. Enzymes released by the Mehlis' gland harden the eggshell as the egg passes through the ootype into the uterus. Usually, enormous numbers of eggs are produced and packed into the uterus as they make their way to the outside.

6.5.2.6. Nervous system and sensory input

The central nervous system includes a primitive anterior brain from which extends at least one pair of longitudinal nerve cords connected at intervals by lateral commissures. There is also a highly organized peripheral system connected to the tegument and muscle layers, the gut, and the reproductive system. The neurotransmitters include most of those commonly associated with vertebrate animals, e.g., acetylcholine, noradrenaline, dopamine, and serotonin. Mousley *et al.* (2005) describe the rich diversity

of neuropeptides found in the platyhelminths and the role they play as targets for anti-parasite drug therapies. The sensory physiology of flukes and other parasitic flatworms has been difficult to assess because of the generally small size of the organisms. Putative functions of external sensory organs include mechanoreception, chemoreception, and osmoreception. The existence of eyespots and experimental evidence of phototactic behavior strongly suggest photoreceptive capabilities for both miracidia and cercariae. Halton *et al.* (1997) provide an authoritative review of helminth neurobiology.

6.5.3 Nutrient uptake and metabolism

The dual nature of digenean feeding is a hallmark of trematode biology (review in Dalton *et al.*, 2004). Thus, digeneans are selective absorbers via the tegument, but they also possess a functional mouth, pharynx, and cecae. Little is known regarding the relative importance of the two alternative feeding methods for particular species (Pappas, 1988). Further, it is not known whether one method dominates over the other under particular conditions, or changes ontogenetically. Radiolabeling studies have shown that certain low-molecular-weight molecules, including glucose and some amino acids, are absorbed through the tegument. Most evidence indicates that some combination of passive diffusion and active transport is involved in nutrient transfer.

Feeding and digestion via the mouth and cecae are highly variable among adult Digenea. Food may consist of blood, host tissue, mucus, host intestinal contents, or some combination of these. In most species, digestion in the gut ceca appears to be extracellular. In some species, such as the liver fluke *Fasciola hepatica*, digestion is both intra- and extracellular, whereas *Haplometra cylindracea*, a lung parasite of frogs, secretes an enzyme that predigests its food externally before ingestion, similar to some free-living flatworms. Enzymes secreted by the gut of digeneans include, among others, proteases, amino-peptidases, esterases, and phosphatases.

Whereas the energy metabolism of digeneans and cestodes is basically similar, there are several species-specific variations that are reflected in both the efficiency and nature of waste products produced. In a well-studied species such as *F. hepatica*, there is also strong evidence suggesting that intermediary carbohydrate metabolism shows some ontogenetic change as well. The primary energy resource for digeneans is glucose and the storage form of the carbohydrate is glycogen (Tielens, 1997). Indeed, the glycogen content in some species is known to be as high as 30% of the dry weight.

6.5.4 Development and general life cycle

A three-host life cycle of a typical aquatic trematode is shown in Fig. 6.10. Once fertilization has occurred and eggshell formation is under way, the zygote proceeds with embryonic development. The development of digeneans inside the molluscan host is by a specialized asexual process called **polyembryony**. This process involves a type of cloning, where more than one embryo is ultimately derived from a single zygote.

The first larval stage produced in this process is called a **miracidium**. Development of the miracidium in some species may be complete by the time the egg is shed, in which case the egg will hatch imme-diately. In others, development inside the eggshell will not proceed until the egg is released from the host. Access to the molluscan host is gained by the free-swimming, ciliated miracidium, or when an appropriate mollusc accidentally ingests an egg with an unciliated miracidium. In eggs that hatch, there is an operculated shell through which the miracidium (Fig. 6.11) emerges. For these species, the detection, recognition, and penetration of an appropriate first intermediate host are key elements of the life cycle (Box 6.1).

On shedding their cilia, miracidia migrate to species-specific sites within their molluscan hosts where they transform into **sporocysts**. These are little more than germinal sacs containing a mixed population of stem and somatic cells (Fig. 6.12). Stem cells continue to give rise to populations of both cell types, whereas somatic cells give rise to daughter sporocysts or to another developmental stage called a **redia** (Fig. 6.13). The redia resembles the sporocyst in containing populations of stem and somatic cells, yet they also possess a sucker, mouth, pharynx, and gut.

The somatic cell population may give rise to daughter rediae or to new developmental forms called cercariae. For aquatic species such as those depicted in Figure 6.10, large numbers of cercariae (typically numbering in the hundreds) emerge from their molluscan hosts each day, and then swim freely within the water column. Cercariae morphology is highly variable and species specific (Fig. 6.14). Variation in morphology is paralleled by variation in behavior, reflecting a set of remarkable adaptations that are designed to bring the short-lived larvae into contact with the second intermediate host (e.g., Fig. 6.10). Thus, some aquatic cercariae are shaped for continuous swimming, others float on the surface, others crawl along a substratum, and still others mimic the prey of their second intermediate hosts. Many species also demonstrate a bewildering diversity of generalized kinetic responses to light, gravity, temperature, and water currents rather than to specific chemical cues emitted by second intermediate hosts (see Chapter 12).

The adaptive nature of such morphological and behavioral diversity has been studied extensively, with results showing a close correspondence between the adaptation and the probability of contact with the second intermediate host. One example involves a suite of trematodes in a lagoon in the eastern Mediterranean Sea (review in Combes, 2001). Laboratory observations showed that cercariae of one species emerging from a benthic snail in the lagoon are positively phototactic and partially negatively geotactic, leading to the concentration of cercariae in the middle of the water column where they contact their fish second intermediate host. Cercariae of a second species showed the reverse taxes, crawling along the

Fig. 6.10 The three-host life cycle of the diplostomatid trematode *Uvulifer ambloplitis*. Adults are found in piscivorous birds such as belted kingfishers. Eggs pass in the feces and hatch in water, penetrating pulmonate snails, such as *Helisoma* spp. Fork-tailed cercariae are produced in daughter sporocysts within snails. These are released in water and penetrate centrarchid fishes such as bluegill sunfish, encysting as metacercariae under the skin. The life cycle is complete when kingfishers ingest infected fish. (Artwork courtesy of Lisa Esch McCall.)

substrate after release from its snail host. This species utilizes benthic marine annelids as second intermediate host. The third species was positively phototactic, remaining near the surface awaiting ingestion by surface-feeding fish that are second intermediate hosts. These adaptations appear to localize cercariae into what Combes *et al.* (1994) describe as 'host space' and they operate even in the absence of actual hosts. Equally impressive adaptations exist to concentrate cercariae in 'host time' (e.g., Théron, 1984).

Even more highly specialized cercariae are formed in some species. All hosts of *Dicroceolium* spp. are

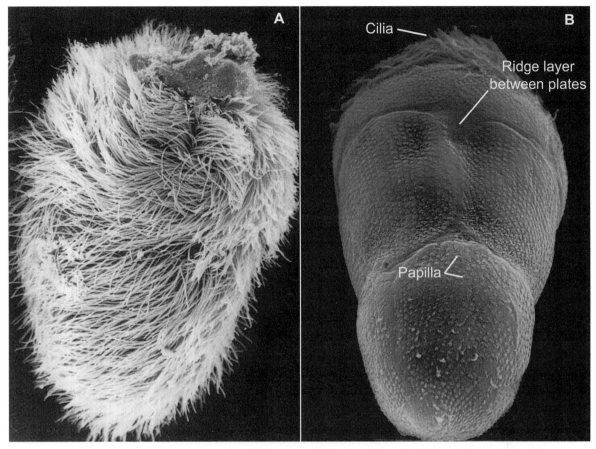

Fig. 6.11 Scanning electron micrographs of the miracidium of the digenean *Echinostoma caproni*. (A) Fully developed, newly hatched miracidium covered with cilia; (B) cilia can be removed chemically to reveal characteristic ectodermal plates and papillae. (Modified from Ataev *et al.*, 1998, with permission, *Journal of Parasitology*, **84**, 227–235.)

Box 6.1 | **'Getting in': host location, recognition, and penetration by trematode miracidia**

The life cycle of many aquatic trematodes includes a free-swimming miracidium whose function following hatching from an egg is to recognize and penetrate an appropriate molluscan host. The brief life span of these tiny, non-feeding larvae is surely a precarious component of transmission. Indeed, following hatching, most trematode miracidia have only a few hours to detect an appropriate host, a time during which they must also avoid water currents, predators, and other hazards of the aquatic environment. It is not surprising that natural selection has favored the evolution of a complex suite of adaptations, including behavioral traits that are used by miracidia to counter such constraints.

Box 6.1 | (continued)

Parasitologists recognize three general phases of miracidial host-finding behavior (review in Haas *et al.*, 1995). After emergence from the egg, behavioral responses to physical stimuli bring the larvae into the general vicinity of its snail host. Miracidia of the sheep liver fluke, *Fasciola hepatica*, are positively phototactic and negatively geotactic, responses that position the larvae along the land/water interface where its amphibious snail intermediate host, *Lymnaea truncatula*, is most common (Wright, 1971). In contrast, the miracidia of *Fasciola gigantica* are negatively phototactic and positively geotactic, behavioral responses that position the larvae in deeper water where the probability of contact with their snail intermediate hosts is enhanced. The second phase begins within the general habitat of its snail host. Movement during this phase has been described as random scanning, consisting of long, straight-line swimming with few turns. Swimming is undirected and rapid during this phase (approximately triple the speed of *Paramecium*) and is accompanied by rotation of about 90 degrees along the long axis of the body.

The third phase of miracidial host-finding behavior involves host contact, recognition, and penetration, and only occurs upon successful entry of the larvae into a habitat that has been conditioned by an appropriate snail (Haas *et al.*, 1995). When miracidia come within a few millimeters of an appropriate host's 'active space,' the straight-line movement of the second phase shifts to a marked increase in rates of change of swimming direction, including abrupt 180° changes when miracidia swim away from appropriate cues. Indeed, the new behavior adjacent to appropriate snails appears frantic and haphazard – a behavior described by Wright (1971) as the 'devil's dance.' Following physical contact with the host (typically in the head/foot region), the larva bores into host tissue, shedding its ciliated epithelium as penetration proceeds.

The nature of the chemical stimuli leading to phase three behavioral responses has interested parasitologists for many years. Meticulous experiments by Wilfried Haas and his students involving several trematode miracidia have led to important advances (review in Haas & Haberl, 1997). In these experiments, the behaviors of individual miracidia exposed to water conditioned by specific types and concentrations of potential snail cues were evaluated and compared to controls. Haas and his students have shown that a large macromolecular component of snail-condition water (SCW) elicits miricidia host-finding behaviors, rather than small organic and inorganic components (Kalbe *et al.*, 1996). Fractionation of SCW, followed by behavioral testing of miricidia, has identified highly glycosylated macromolecules (> 300 KDa) that lead to the characteristic phase three behaviors for *S. mansoni* and *S. haematobium* miracidia. These macromolecules are likely present in snail mucus.

Identification of the chemical cues that are responsible for miracidia host-finding behavior has been a 'holy grail' for parasitologists for many years. If host cues could be produced synthetically and applied safely to waterbodies containing human *Schistosoma* spp., it may be possible to interfere with natural transmission from egg to snail. In addition to the importance of

Box 6.1 | (continued)

this line of applied research, the identification of chemical cues has also provided insight into the mechanisms underlying variation in host specificity, even among different strains of parasites. The response of two strains of *Schistosoma mansoni* miracidia to water conditioned by cues from different potential snail hosts was studied by Kalbe *et al.* (1996). In this case, both strains demonstrate both behaviors (turnback, and rapid change in direction) as part of their snail-recognition strategy. The Egyptian strain chemo-orientates to its specific snail host (and secondarily to its congener). However, the Brazilian strain responds equally to all hosts, even one that it would never encounter in nature. Variation of this sort is common. Whereas *F. hepatica* miracidia can discriminate cues from different snail species, other species fail to discriminate between species of snail, and even between other aquatic invertebrates. Understanding why natural selection favors strict chemically mediated host specificity in some cases, but not in others, is an important element of ongoing studies.

terrestrial, leading to significant constraints relative to both miracidia and cercariae mobility. Cercariae of *D. dendriticum* are packaged into 'slime balls' that are released from their terrestrial snail hosts. The slime balls mimic the eggs of their snail host, a favorite food of the worker ants that are second intermediate hosts. In the hemiurid trematodes, a so-called cercaria body encysts within the tail of the cercaria and a unique delivery tube and handle apparatus are present (Fig. 6.15). Remarkably, when the handle apparatus is manipulated by the feeding appendages of an ostracod second intermediate host, it triggers the instantaneous eversion of the delivery tube (Fig 6.15B). The cercaria body passes through this structure and is 'injected' into the microcrustacean's hemocoel where it becomes an unencysted metacercaria.

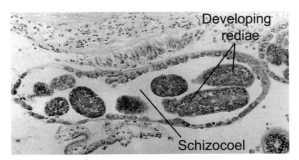

Fig. 6.12 Light micrograph of histological section of mother sporocyst of *Echinostoma caproni*, showing the schizocoel, developing rediae, and internal germinal masses. (Modified from Ataev *et al.*, 1998, with permission, *Journal of Parasitology*, **84**, 227–235.)

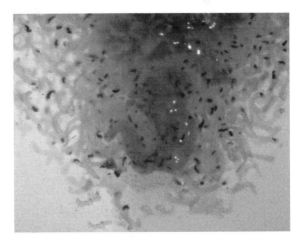

Fig. 6.13 Daughter rediae of *Haligegus occidualis* from the pulmonate snail *Helisoma anceps*. The anterior blind gut of each redia is filled with snail gonad tissue. Hundreds of cystophorous cercariae are produced asexually within the daughter rediae.

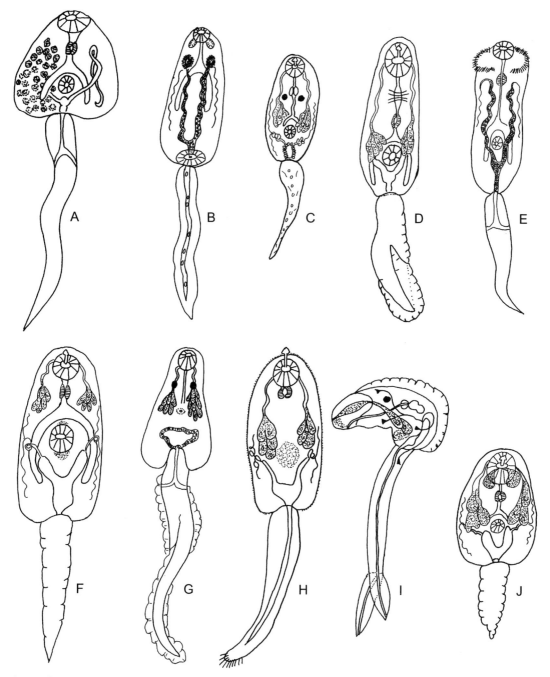

Fig. 6.14 Pronounced variation in cercariae morphology within selected taxa. Identification and naming of cercariae types are based on sucker placement, tail morphology, and a number of other morphological features such as the presence of a stylet, eyespots, pharynx, or circumoral spines. (A) Gymnophalus; (B) Amphistome; (C) Ophthalmoxiphidio; (D) Ornatae; (E) Echinostome; (F) Armatae; (G) Parapleurolophocercous; (H) Ubiquita; (I) Brevifurcate–pharyngate; (J) Virgulate. (Drawings courtesy of Lisa Esch McCall.)

A

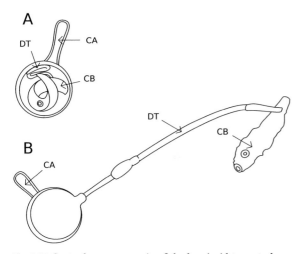

B

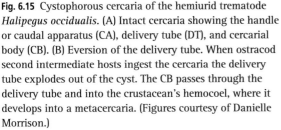

Fig. 6.15 Cystophorous cercaria of the hemiurid trematode *Halipegus occidualis*. (A) Intact cercaria showing the handle or caudal apparatus (CA), delivery tube (DT), and cercarial body (CB). (B) Eversion of the delivery tube. When ostracod second intermediate hosts ingest the cercaria the delivery tube explodes out of the cyst. The CB passes through the delivery tube and into the crustacean's hemocoel, where it develops into a metacercaria. (Figures courtesy of Danielle Morrison.)

In species such as those depicted in Figure 6.10 and those discussed in the examples above, cercariae penetrate another intermediate host where they become **metacercariae**. Whereas some metacercariae encyst immediately upon contact with appropriate aquatic substrata, e.g., rocks, vegetation, many encyst in, or on, a second intermediate host where they await ingestion by the appropriate final host. Developmentally and ecologically, this component of the trematode life cycle is typically regarded as a resting stage. The cyst wall is bi-layered in those species that encyst within intermediate hosts, consisting of a metacercariae-derived inner layer and a host-derived outer layer comprising polysaccharides, lipids, and proteins. For species such as *Fasciola* that encyst on vegetation, the cyst wall is comprised of at least four layers. The metacercaria stage is a critical component within the life cycles of most trematodes, providing an additional dispersal mechanism and

permitting survival during periods when other hosts in the life cycle are absent.

The resting nature of encysted metacercariae implies a period of quiescence during which there is little metabolic interaction with their intermediate hosts. While this is the case for species such as *Fasciola* that encyst on vegetation, other species require an extensive development period prior to encystment. Metacercariae in the Diplostomidae are examples. Penetration through the epidermis of fathead minnows by *Ornithodiplostomum* sp. cercariae is followed by complex migration of specialized metacercariae (known as **diplostomula**) through the vascular system to the liver (Matisz & Goater, 2010). Migration is followed by a 2-week period of growth within liver tissue, during which the surface of the tegument is comprised of an elaborate network of microvilli that assist in absorptive feeding. Following this period of development, microvilli on the tegument disappear and the larvae encyst within the body cavity. Similarly, metacercariae of the related trematode, *O. ptychocheilus* include a non-encysted development stage in the tissues of the optic lobes of minnows (Fig. 6.16A, B, C) followed by an encysted stage (Fig. 6.16D) in adjacent meninges (Matisz *et al.*, 2010). Results from these experimental studies indicate that for some trematodes, metacercariae development to the resting stage is much more complex than originally thought.

For those trematodes that incorporate second intermediate hosts into their life cycles, we might expect adaptations that facilitate the transmission of metacercariae to final hosts. Indeed, such adaptations are common across a range of trematodes, and often they are spectacular in their expression. The brightly colored and pulsating tentacles of certain terrestrial snails infected with the metacercariae of *Leucochloridium* spp. (Color plate Fig. 7.1), provides one well-known example. We emphasize this, and other metacercariae–host model systems, in our discussion of parasite-induced alteration of host phenotypes in Chapter 15.

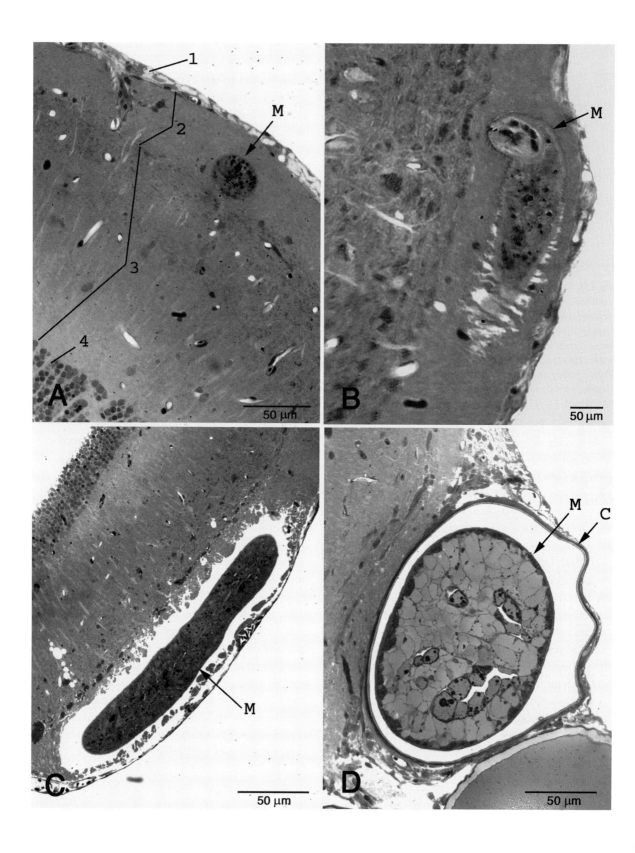

6.5.5 Biodiversity and life-cycle variation

The life cycles of most digeneans are among the most complex in nature, and are usually linked, inextricably, to the feeding behaviors of their definitive hosts. Although life cycles of the sort indicated in Figure 6.10 are the most common, there are many variations on this general theme. Some of these variations are provided in stylized form in Figure 6.17 (see Esch *et al.*, 2002). Despite the complexity, there are two features of overall consistency. First, almost all life cycles include both free-living and parasitic stages. Second, substantial complexity and diversity occurs within the molluscan intermediate host. Thus, as described above, the miracidium stage is always followed by a sporocyst, but then, depending on the species, the pattern is highly variable. Moreover, there is always an increase in parasite numbers, via asexual reproduction, within the molluscan host. From an ecological/evolutionary perspective, this increase in numbers reflects an amplification of the successful parasite genome. Many of these life-cycle traits will be detailed in the discussion of the more common or economically important families, genera, or species of digeneans. In our effort to present the diversity of digeneans within a phylogenetic framework, we adopt the three-order system proposed by several research groups and currently adopted in such seminal comprehensive taxonomic monographs as the *Keys to the Trematoda* edited by Gibson *et al.* (2002), Jones *et al.* (2005), and Bray *et al.* (2008).

6.5.5.1 Echinostomida

The Echinostomida includes a number of families that actually bear little resemblance in the adult stage. Some of the most common families are the Paramphistomidae, flukes found in all groups of vertebrates, the Fasciolidae, parasites of the liver, bile duct, and intestine of herbivorous mammals, including humans, and the Echinostomidae, intestinal parasites of reptiles, birds, and mammals.

Paramphistomidae

Many paramphistomid digeneans are robust. Their cercariae encyst on aquatic vegetation, animals, or inanimate objects, after emerging from their snail hosts. In the Paramphistomidae, the acetabulum may be either located posteriorly, e.g., amphistomes (Fig. 6.18; 6.19A), or it may be lacking (monostomes). Paramphistomids are parasitic in all groups of vertebrates. One of the most common species in cold-blooded vertebrates in North America is *Megalodiscus temperatus*, a parasite in the rectum of frogs (Fig. 6.19A). The cercariae released from certain pulmonate snails encyst on the heavily pigmented areas of a tadpole's skin. When the tadpole undergoes metamorphosis, it consumes the shed skin and infects itself with the encysted metacercariae. On occasion, a tadpole may accidentally eat cercariae before they have the chance to encyst. When this happens, the parasite encysts immediately, becoming a metacercaria, which then passes through the intestine. On reaching the rectum, it excysts, but remains undifferentiated. When the tadpole undergoes

Caption for Fig. 6.16
Light micrographs of stained sections of *Ornithodiplostomum ptychocheilus* metacercariae within the optic lobes of fathead minnows *Pimephales promelas*, demonstrating obligate changes in site selection, growth, and development. Metacercariae (M) have completed their migration through the host's central nervous system by 2 days post-infection, establishing in the outermost tissue layer (2) of the optic lobes (A). By 4 days post-infection (B), the anterior penetration spines begin to disappear and the metacercariae have approximately doubled their body volume. Metacercariae continue development within the tissues, reaching their maximum body volume at 7 days post-infection (C). By day 28 (D), metacercariae encyst within the circulatory network that envelops the optic lobes. A thick cyst wall (C) surrounds the metacercariae at this time. 1 – the endomneninx is a secretory organ that envelopes the central nervous system of fish; 2 – *O. ptychocheilus* metacercariae develop within the stratum marginale of the optic lobes; 3 – an additional layer of unmyelinated fibers within the optic lobes; 4 – a dense, granular layer composed of cell bodies (A). (Modified from Conn *et al.*, 2008, with permission, *Journal of Parasitology*, **94**, 635–642.)

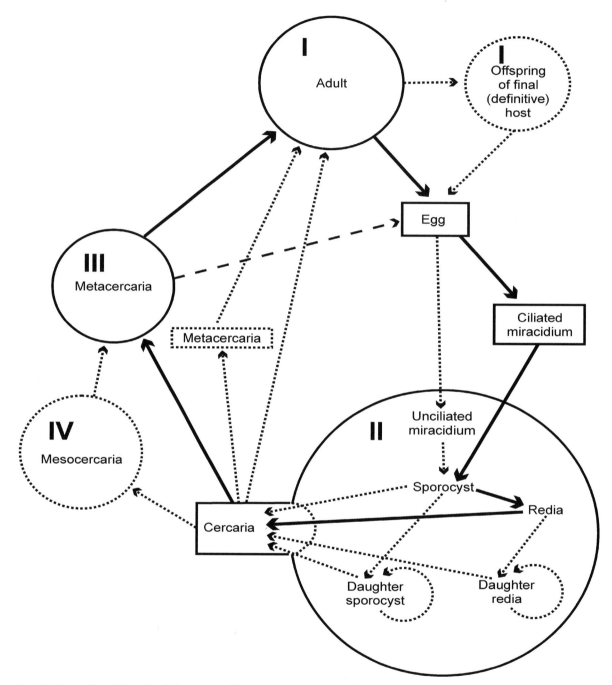

Fig. 6.17 Generalized life cycle of digeneans, with some variations on the basic pattern. Host types are represented by roman numerals (I, definitive host; II, molluscan first intermediate host; III, second intermediate host; and IV, paratenic or third intermediate host). Stages within hosts are enclosed in circles, whereas stages in the environment are enclosed within rectangles. Solid lines represent the typical three-host life cycle, whereas dotted lines represent less common variations. Note that some metacercariae are found encysted in the environment, and that some cercariae do not leave their first intermediate hosts. The dashed line represents abbreviation of the life cycle by progenesis. Some rare, alternative pathways are not included in this diagram. (Figure courtesy of Al Bush.)

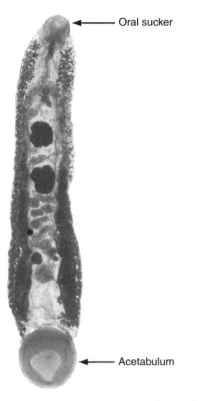

Oral sucker

Acetabulum

Fig. 6.18 The paramphistomid digenean *Wardius* sp. from the small intestine of a muskrat *Ondatra zibethicus*. Note the posterior acetabulum. (Photograph courtesy of Harvey Blankespoor.)

metamorphosis, a few may be lost, but others migrate back up into the stomach where they remain until the metamorphosed frog takes a meal. Then, they migrate back down into the rectum again and mature.

A number of species within the Paramphistomidae, e.g., *Paramphistomum cervi*, infect the stomach or liver of domesticated and wild ruminants, causing an important disease known as paramphistomiasis. When the ruminant ingests metacercariae encysted on vegetation, the worms excyst in the duodenum, penetrate the mucosa, and migrate through the tissues to the abomasum. Here, they cross the mucosa into the lumen and then migrate into the rumen where they finally attach. The migration of the juveniles causes enteritis, diarrhea, and hemorrhage. In severe cases, when large quantities of metacercariae are ingested in a short time, death of the host may occur.

Fasciolidae

With a cosmopolitan distribution, members of the Fasciolidae are almost always of large sizes, and somewhat leaf-like in appearance. A few species occur in the intestine of their definitive host, but the majority are associated with the liver, gall bladder, and bile ducts.

Fasciola hepatica (Fig. 6.20) is typically found in the major sheep-raising countries of the world (see Box 1.1). Adults occur in the bile ducts where they release eggs that are shed in the feces. The hatching of *F. hepatica* eggs is stimulated by blue–violet light (450–550 nm). When exposed to blue–violet light, the miracidium releases proteolytic enzymes that attack the cement holding the operculum in place. Simultaneously, a gel inside the shell is converted to a sol, which doubles the volume within the shell. The increased pressure 'pops' the enzyme-loosened operculum from the shell and the miracidium escapes. Interestingly, newly released eggs of *F. hepatica* can be maintained in the dark for up to a year, then hatch within minutes when exposed to light of the correct wavelength.

Aquatic snails are intermediate hosts and cercariae released from the snail encyst on vegetation or inanimate objects, becoming metacercariae. When metacercariae are ingested by herbivores, they excyst in the lumen of the small intestine under the influence of the host's digestive enzymes. The parasites migrate through the gut wall into the body cavity, then into the liver, and finally into the bile duct and gall bladder where they mature sexually. Migration through the liver can lead to severe pathology and weight loss.

There are several other fasciolid flukes infecting cattle, sheep, and wild ungulates in various areas of the world. In North America, the giant liver fluke, *Fascioloides magna*, occurs in the livers of elk and other ruminants, where the parasite reaches up to 10 cm in length (Fig. 6.21). The flukes live in pairs within host-derived fibrous cysts within the liver parenchyma. In populations of moose in northern Minnesota, USA, liver pathology associated with high numbers of *F. magna* likely restricts their

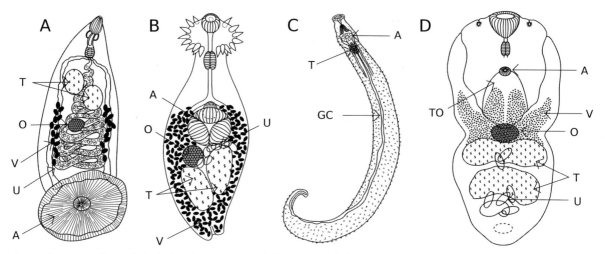

Fig. 6.19 Pronounced morphological variation among adult digeneans. (A) *Megalodiscus temperatus* (Paramphistomidae) from the rectum of amphibians; (B) *Petasiger nitidus* (Echinostomidae) from the small intestine of grebes; (C) male of *Schistosoma mansoni* (Schistosomatidae) from the mesenteric veins of mammals, showing the gynecophoral canal (GC); (D) *Alaria marcianae* (Diplostomidae) from the small intestine of canids, showing an accessory holdfast – the tribocytic organ (TO). Note the variation in relative positions of the acetabulum (A), ovary (O), vitellaria (V), and testes (T). (Drawings courtesy of Danielle Morrison.)

southern expansion into suboptimal habitats (Murray *et al.*, 2006). The introduction of *F. magna* from foci in northern US cervids (especially white-tailed deer) into wild ungulate and beef cattle populations in Canada and central Europe is a concern for wildlife managers and ranchers. Elk and white-tailed deer were imported from North America to Europe for hunting purposes during the nineteenth century. With their introduction, *F. magna* was transferred from the Nearctic zone to the Palaearctic, where it has since established local populations throughout parts of Europe (Králová-Hromadová *et al.*, 2011). In some parts of the USA, the parasites are considered a delicacy to some brave people; fried giant liver flukes (aptly called 'little livers') are actually a specialty on the menu in some restaurants! *Fasciola gigantica* and *Fasciolopsis buski* are widely distributed in cattle. The former is common in Asia, Africa, and Hawaii, whereas *F. buski* is abundant in the Far East where it infects up to 10 million humans. The usual source of infection for both cattle and humans is edible aquatic plants, on which metacercariae have encysted.

Echinostomidae

These flukes infect the intestine and bile ducts of a wide range of amphibians, reptiles, birds, and mammals. Both adults and cercariae possess either a single or a double row of spines (collar) that almost completely encircle the oral sucker and mouth (Figs. 6.14E, 6.19B, 6.22). There are at least 17 genera of echinostomatids and in excess of 100 species in *Echinostoma* alone.

At least 15 species of echinostomes infect humans, mainly in the Far East, including southeast Asia. Humans become infected by eating metacercariae-infected snails, tadpoles, or freshwater fishes. The life cycle of *E. caproni* is routinely established within laboratory rats, mice, and hamsters and also within its intermediate hosts, making it a superb model in experimental parasitology (Fried & Graczyk, 2000).

Another echinostomatid fluke that has generated much recent interest is *Ribeiroia ondatrae*, a member of the Psilostomidae. Encysted metacercariae of this species have been shown to cause grotesque limb deformities (Fig. 17.5; Color plate Fig. 7.2) and

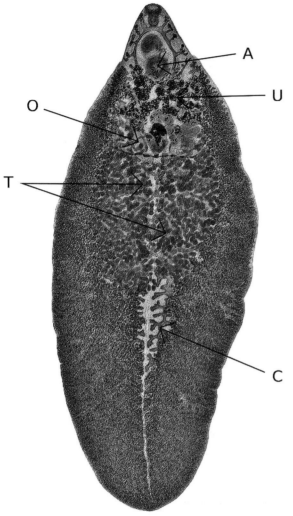

Fig. 6.21 The giant liver fluke *Fascioloides magna* from the liver of a Roosevelt elk *Cervus canadiensis roosevelti*, showing the large body size and highly dendritic intestinal cecae of this digenean. (Photograph courtesy of Bill Pennell.)

6.5.5.2 Strigeida

The Strigeida comprise an extraordinarily diverse group of digeneans. The adults parasitize all classes of vertebrates.

Spirorchiidae
Most species in this group inhabit the heart, large arteries, and other blood vessels of reptiles, primarily turtles. Their bodies are flat, lanceolate, and the ventral sucker or acetabulum may be absent. Their life cycle includes a snail intermediate host, and cercariae released from the snail penetrate the mucous membranes of the eyes, nose, mouth, and cloaca of turtles. The Sanguinicolidae are small, flat digeneans without suckers; they inhabit the blood vessels and heart of freshwater and marine fishes. *Cardicola davisi* is a common sanguinicolid of trout in North America and may cause severe mortality in hatcheries. One of the

Fig. 6.20 The fasciolid digenean *Fasciola hepatica* from the bile duct of a sheep. The dendritic testes (T) and ovary (O) are diagnostic features of fasciolid flukes. Note the positions of the acetabulum (A), uterus (U), and the intestinal cecum (C). (Photograph courtesy of John Sullivan.)

other morphological abnormalities observed in several amphibian species throughout North America (Johnson *et al.*, 1999; 2002; Johnson & Sutherland, 2003). This species and its suite of intermediate hosts provide a model for studies on host phenotypic manipulation by parasites (Chapter 15) and environmental parasitology (see Box 17.1).

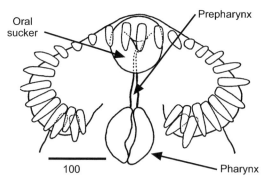

Fig. 6.22 Anterior region of the echinostomid digenean *Himasthla limnodromi*, showing the pre-pharynx and pharynx, and the collar of spines surrounding the oral sucker. Units are µm. (Modified from Didyk & Burt, 1997, with permission, *Journal of Parasitology*, 83, 1124–1127.)

problems faced by parasites in the circulatory system of vertebrates is how to get the eggs produced by the adult released into the environment. In many cases, the eggs escape from the host's blood vessels into the intestine and then into the environment with the host feces. In others, the eggs are carried by the blood into the capillaries of the gill filaments where the miracidium hatches and escapes the fish host through the very thin epithelium overlying the gill capillaries. The pathology produced by sanguinicolid flukes is caused primarily by eggs, which occlude vessels in many areas of a host's circulatory system or by miracidia exiting the gill epithelium, and not by the adult flukes themselves (Smith, 1997).

Hemiuridae

Most hemiurids are found in the gut of marine and freshwater fishes, although a few occur in the lungs or the intestines of sea snakes, and in the mouth, esophagus, and eustachian tubes of ranid frogs. Many species in this family, mainly those parasitic in marine fishes, exhibit low host specificity and have worldwide distributions. A peculiarity of many hemiurids present in the stomach of fishes is their capacity to retract the posterior end of their body. The retractable section of the body is called the tail or **ecsoma**, and it appears to function as a feeding organ that is extruded when the pH or osmolarity of the stomach content is at a tolerable level (Gibson & Bray, 1979).

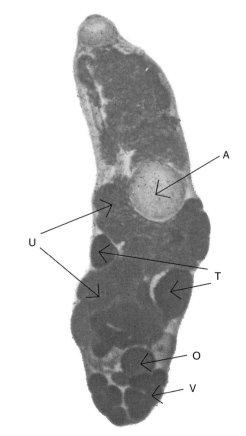

Fig. 6.23 The hemiurid digenean *Halipegus occidualis* from the buccal cavity of the green frog *Rana clamitans*, showing the acetabulum (A), uterus (U), testes (T), and ovary (O). The posterior vitelline follicles (V) are characteristic of the genus.

Halipegus spp. are common in ranid frogs of South America, North America, and Europe. *Halipegus occidualis* (Fig. 6.23; Color plate Fig. 1.1) and *H. eccentricus* occur in the buccal cavity and eustachian tubes, respectively, of ranid frogs in North America, while *H. eschi* is from a ranid frog in Costa Rica. The life cycle of these trematodes is amazingly complex, involving four hosts. In the case of *Halipegus occidualis*, pulmonate snails and ostracod crustaceans are the first and second intermediate hosts, respectively, while diverse odonates act as paratenic hosts (Fig. 6.24). Paratenic dragonfly hosts help to bridge the aquatic to terrestrial transmission barrier. *Halipegus occidualis*, in particular, provides an excellent model

Fig. 6.24 The four-host life cycle of the hemiurid trematode *Halipegus occidualis*. Adults live under the tongue in the mouth of green frogs *Rana clamitans*. Eggs are released in the feces and are ingested by pulmonate snails *Helisoma anceps*. Cystophorous cercariae are produced in daughter rediae within snails and are shed into water. Cystophorous cercariae are ingested by ostracod second intermediate hosts, becoming unencysted metacercariae in the hemocoel. Dragonfly nymphs eat infected ostracods and accumulate metacercariae within their intestine. Metamorphosing dragonflies act as paratenic hosts, bridging the aquatic–terrestrial transmission barrier. The life cycle is complete when infected odonates are eaten by frogs. (Artwork courtesy of Lisa Esch McCall.)

in parasite population ecology, in part because this trematode's populations can be fully censused without killing the hosts (see Chapters 12 and 15).

Schistosomatidae

Perhaps the best known of all the digeneans are the schistosomes. These are dioecious worms that occupy the hepatic portal and pelvic veins of birds and mammals. The size of adult schistosomes is generally in the range of 10–15 mm, although *Gigantobilharzia acotyles*, a parasite of the black-headed gull, can reach 150 mm. There are several unique features of the schistosome body plan that correspond to their dioecious life history strategy. Male schistosomes (Fig. 6.19C) are considerably larger than females and possess a **gynecophoric canal** in which the female lies

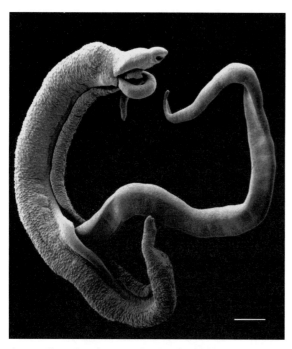

Fig. 6.25 Scanning electron micrograph of male and female of *Schistosoma mansoni in copula*. Note that the more slender female worm lies in the gynecophoral canal located in the ventral surface of the male. Scale bar = 1 mm. (Micrograph courtesy of David Halton; from Halton, 2004, with permission, *Micron*, 35, 361–390, Elsevier.)

in copula (Fig. 6.25). This canal is a ventral groove that begins immediately behind the ventral sucker. Most schistosomes have a life span of between 5 and 30 years. Female fecundity tends to be low compared to other trematodes, typically on the order of 300 eggs per day in some species. In single worm infections, males develop normally, but females remain sexually immature. A commentary on the origin of dioecy in the schistosomes is provided by Platt & Brooks (1997).

The life cycles of the various schistosome species, including those that infect humans, are characteristic (Fig. 6.26). Cercariae released from snail intermediate hosts swim for up to 48 hours, or until they come into physical contact with an appropriate host. The cercariae then actively penetrate the skin of their definitive hosts, assisted by both vigorous penetration behaviors and secretions from anterior penetration glands. There is no metacercarial stage. The tail of the

cercariae detaches during penetration, becoming a characteristic **schistosomule**. Within 24 hours, the schistosomules enter the peripheral circulation system and then enter the right side of the heart. Via pulmonary capillaries, the worms enter the left side of the heart and ultimately enter the systemic circulation. The schistosomules stop in the blood sinuses of the liver where they remain for up to 15 days before moving into small venules (as pairs) at their species-specific final site of infection. Approximately 4 weeks after infection, they begin producing eggs and the life cycle is complete.

The three primary species of human schistosome infect approximately 200 million people within 74 tropical and subtropical countries. These species, and approximately five others, can be characterized by their site preference inside the circulatory system, their geographical distribution, and the snails used as first intermediate hosts. Schistosomiasis or bilharziasis is a disease produced by the three main human schistosomes (*S. mansoni, S. haematobium*, and *S. japonicum*), leading to devastating problems for about 200 million people (85% of whom live in Africa) and to approximately 20 000 deaths per year in sub-Saharan Africa alone. According to recent estimates, reported cases of schistosomiasis are on the rise due to increasing human population densities, political and economic turmoil, and to anthropogenic modifications to aquatic habitats that contain snail intermediate hosts. The epidemiology of the disease in humans is linked to certain agriculture and cultural practices that tie humans to water, whether to drink, bathe, or irrigate crops. Poor socio-economic conditions often excacerbate rates of cercarial transmission from snails to humans, especially in children where play behavior (Fig. 6.26) and hygiene are important risk factors. Reviews of the epidemiology, diagnosis, and control of human schistosomiasis are provided in Jordan (2000) and Fenwick *et al.* (2006).

Pathology due to the schistosomes is complex and highly species- and context-dependent (review in Strickland & Ramirez, 2000). Mechanisms of pathology are unusual in that they are due almost exclusively to the eggs released by adult females, and not to the adult worms (see Chapter 2). For *S. mansoni*, pathology is directly related to the passage of eggs through

Fig. 6.26 The two-host life cycle of *Schistosoma mansoni* and the epidemiology of schistosomiasis. Adults live *in copula* within the mesenteric veins of the intestine of humans. Embryonated eggs are passed in feces and hatch into ciliated miracidia in water. These penetrate snails *Biomphalaria* spp., and two sporocyst generations develop. Furcocercous cercariae leave snails and directly penetrate human skin, migrating as schistosomules to become adults in the veins. As depicted here, children playing in water contaminated by human feces are at risk of acquiring schistosomiasis. (Artwork courtesy of Lisa Esch McCall.)

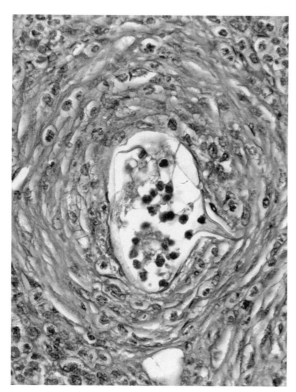

Fig. 6.27 Histological section of liver from a mouse infected with *Schistosoma mansoni*, showing a single granuloma surrounding an egg. Note the prominent lateral spine. (Photograph courtesy of John Sullivan.)

the tissues of the definitive host from the mesenteric veins to the lumen of the large intestine. One key challenge for the host is the antigenic nature of the eggshell, as well as the antigens produced and released through the shell by the miracidium developing inside. One outcome of antigenic stimulation is the production of tumor necrosis factor ([TNF]-β) by activated macrophages (see Chapter 2). Severe granulomatous reactions surrounding eggs produce pseudotubercles not unlike those seen in tuberculosis. The acute form of pathology comes with severe granulomatous inflammation and necrosis along the gut wall, followed by scarring and loss of functional integrity. Eggs may become lodged in the gut wall and produce pseudo-abscesses. The eggs also cause extensive mechanical damage to the intestine, affecting peristalsis. It should be remembered that peristalsis in the intestine involves constant contraction and relaxation of smooth muscle. This, in combination with the antigenicity of the miracidial metabolites and the eggshell, and the presence of a prominent spine (Fig. 6.27), can lead to serious problems for a host. Remarkably, it is immunopathology caused by the eggs themselves that is essential for the passage of schistosome eggs from the venous system to the lumen of the intestine (Box 6.2).

Box 6.2 | **'Getting out': the enigma of egg release in the human schistosomes**

Adult schistosomes are robust and long-lived. Although the fecundity of individual females is modest relative to some other trematodes and cestodes, an *S. mansoni* female can release approximately 3 000 000 eggs over her multi-year life span. But what is the fate of these eggs? If her eggs remain in the host's circulation system, or remain lodged within host tissue, her lifetime fitness will be zero. Thus, mechanisms must exist that encourage the passage of at least some of her eggs through host tissues, and ultimately into water that contains appropriate snail intermediate hosts. This must represent a monumental hurdle for all schistosomes. Although the manner in which they solve the problem of 'getting out' (eggs, in this case) is a parasitological marvel, their solution to this problem causes schistosomiasis, one of the most devastating human parasite diseases.

Box 6.2 | (continued)

The process of egg release is intimately tied to the development of the characteristic granuloma that envelopes individual eggs lodged within host tissues (Fig. 6.27). For *S. mansoni* and *S. haematobium*, the granulomatous response is a hypersensitive inflammatory reaction that is regulated by host immunity (review in Doenhoff, 2006; see Chapter 2). In the case of *S. mansoni* in humans, the complex granulomatous process starts at the onset of egg deposition in host tissues at 6–8 weeks post-infection. Highly antigenic excretory/secretory products originating from within the egg stimulate up-regulation of a T-helper type 2 response, a process that ultimately activates basophils to release a complex cocktail of cytokines. So begins the sophisticated immune response in tissue adjacent to individual eggs that leads to the complex of monocytes, eosinophils, macrophages, fibroblasts, and lymphocytes that make up the characteristic granuloma (Fig. 6.27). All of these cells, and the interactions among them, play important roles in the development of a functional granuloma.

The function of the immune-mediated granuloma has been a focus of debate for many years. One possibility is that a functional granuloma protects tissues of the host from toxic products released from eggs. Experimental support for this idea comes from studies involving mice infected with *S. mansoni* (review in Kusel *et al.*, 2007). Infected mice with intact immune systems form normal granulomas, show few signs of liver damage, and have high survival. In contrast, immunocompromised mice form small (or no) granulomas, suffer from a severe hepatotoxic reaction, and die earlier. Not surprisingly, the conventional explanation for the granulomatous response is the protection of the host. However, counter evidence for the 'host protection' hypothesis comes from the results of a number of experiments. First, when lymphocytes from *S. mansoni*-infected mice were transferred to immunocompromised mice, specific antibodies alone, and not the presence of a granuloma, provided the best protection against egg-derived toxins. Furthermore, immunocompromised mice infected with *S. bovis* do not suffer from hepatotoxicity and actually survive better than mice with normal immune systems. These results suggested that the conventional 'host protection' explanation for the granulomatous response may not be correct.

An alternative primary role for the granulomatous response comes from the results of experiments that manipulated granuloma structure and simultaneously monitored egg release from female worms (review in Doenhoff, 1997). Immunosuppressed mice infected with *S. mansoni* developed smaller granulomas. Yet, these hosts also released far fewer eggs in their feces compared to mice with normal immune systems. Further, when immunosuppressed mice were injected with lymphocytes from normal infected mice, they developed fully functional granulomas and released normal numbers of eggs. In humans, *S. mansoni* egg excretion is reduced in hosts with reduced T-cell counts stemming from co-infection with HIV (Karanja *et al.*, 1997). Taken together, these results demonstrate a direct linkage between worm fitness and the development of a granuloma that is formed and mediated by host immunity (see Chapter 2). This

Box 6.2 | (continued)

explanation also nicely resolves the enigma of the chronic antigenicity of schistosome eggs and miracidia. Thus, natural selection should favor sophisticated and diverse mechanisms to trigger an inflammatory immune response that enables the transport of eggs across host tissue. Damian (1987) reached a similar conclusion, hypothesizing that the exploitation of host immunity can be considered a schistosome adaptation to solve the problem of egg release.

In fact, the process of egg release probably goes further than exploitation of host immunity. Doenhoff and his colleagues provide evidence in support of the idea that *S. mansoni* eggs are known to affect gut and smooth muscle metabolism in ways that could assist in the transport of eggs across host tissue (review in Kusel *et al.*, 2007). Recent studies have also shown that schistosome eggs promote angiogenesis in tissue immediately adjacent to individual eggs. Remarkably, *S. mansoni* eggs have been shown to promote a very rapid thrombogenic response by the host, perhaps functioning to 'tether' the eggs to the endothelial surface to avoid being swept downstream by blood flow. The emerging picture is one of fascinating complexity, involving parasite integration into several of the host's physiological systems, in addition to host immunity. On the other hand, the picture is one of uniformity and simplicity, inextricably linked to the fundamental requirement for individual females to release their eggs into water.

In addition to their role as agents of human disease, a number of avian schistosomes produce cercariae that penetrate the skin of humans, provoking localized immune hypersensitivity reactions known as swimmer's itch or cercarial dermatitis. The attraction of avian schistosome cercariae to humans has been attributed to similarities in the lipid compositions of bird and human skin (Horak *et al.*, 2002). The cercariae die in mammalian skin, causing itchy papules and an irritating rash that can last several days. Each papule corresponds to the penetration site of a single avian schistosome cercaria. Often, first-time infections are asymptomatic. Upon reinfection though, a quicker and more severe inflammatory response, similar to an allergic reaction, occurs (Horak & Kolarova, 2001). In freshwater lakes of North America, swimmer's itch is attributed to species of *Trichobilharzia* or *Gigantobilharzia*, common parasites of birds, especially waterfowl such as mallards, wood ducks, and common mergansers (Leighton *et al.*, 2000). Pulmonate snail species, notably *Lymnaea* and *Physa*, are the most important intermediate hosts

for avian schistosomes. Whereas these schistosomes are not serious or dangerous pathogens in humans, the 'itch' can be intensely irritating. Swimmer's itch can limit recreational opportunities and impact lake-oriented tourism. Swimmer's itch can become so severe that in areas in which the disease is rampant, signs may actually be posted, warning bathers to be aware of the potential risk.

Diplostomidae

In this family (and a number of other strigeids) the body of the adult may be divided into an anterior flattened, spoon-shaped portion and a posterior section shaped like a cylinder. The anterior portion may have a pair of accessory suckers on each side of the oral sucker to aid in attachment, and a **tribocytic organ**, in proximity to the ventral sucker (Fig. 6.19D). Besides helping with attachment, the tribocytic organ secretes proteolytic enzymes that digest host tissues on which the parasite then may graze. The posterior end of the body contains most of the reproductive structures.

The diplostomids are cosmopolitan parasites in a number of avian hosts, primarily ducks and geese, and several mammals. *Cotylurus flabelliformis*, for example, is a common intestinal parasite of ducks in North America. Cercariae penetrate lymnaeid snails where they develop into a special kind of metacercaria called a **tetracotyle**. However, if the cercariae enter planorbid or physid snails, they actually may penetrate the sporocysts or rediae of other digenean species and then develop into tetracotyle metacercariae within them, a case of hyperparasitism. Ducks become infected when they eat snails harboring tetracotyle metacercariae.

Other diplostomatids include species of *Diplostomum*, *Alaria*, and *Uvulifer*, several of which have received substantial attention because of a distinctive characteristic associated with their life cycles, or ecology, or both. For example, *Alaria americanae* has a complicated four-host life cycle. Adults of this parasite are found in the intestines of canines in North America. Cercariae from snails penetrate the skin of tadpoles, developing into **mesocercariae**, an intermediate stage between a cercaria and a metacercaria. The mesocercaria remains within the body even after the tadpole undergoes metamorphosis to the adult stage. If a canine eats an infected tadpole or an adult frog, the life cycle can be completed. This last step is generally unlikely, however, as canines do not normally prey on tadpoles and frogs. So, how is the life cycle completed? If the tadpole or adult frog is eaten by a snake or a rodent, the mesocercaria is freed from the tissues by digestion, penetrates the gut wall, and takes up residence in the tissues. The snake or rodent thus becomes a paratenic host and, when preyed upon by canines, the parasite's life cycle is completed. This is another example of a parasite using a particular host to bridge an ecological, or trophic, gap. However, this complicated life cycle is still incomplete. Once into the canine, the mesocercaria penetrates the gut wall and moves into the coelom, passes through the diaphragm, and becomes established in the lung. There, it develops into a diplostomulum, a modified metacercariae that remains unencysted in the lungs for about 5 more weeks. It then migrates out of the lungs into the trachea and throat, where it is coughed up and swallowed, finally passing into the intestine and maturing sexually. In *Alaria marcianae* (Fig. 6.19D), mesocercariae can migrate to the mammary glands of mammalian hosts, an example of maternal or transmammary transmission (Shoop *et al.*, 1990).

Most members of the Diplostomidae use birds as their definitive hosts and fishes as their second intermediate hosts. Fish-diplostome metacercariae systems have been the focus of numerous behavioral, ecological, and evolutionary studies (see Chapters 12 and 15). Diplostomes such as *Uvulifer ambloplitis* (and several other species) often cause the familiar 'black spot disease' in their fish intermediate hosts (Fig. 6.10). Metacercariae within muscle tissue provoke a strong tissue reaction, leading to the mobilization of histiocytes and melanocytes that produce a darkly pigmented cyst wall around the parasite, sequestering it even further. Cercariae of several species of so-called 'eye fluke' penetrate the eyes of several species of fishes and develop into metacercariae on the lens of the eye, where they may induce cataractous lenses. The metacercariae of *Diplostomum spathaceum* cause cataracts within the lens of rainbow trout, leading to alterations in visual acuity and altered rates of predation by potential final hosts (Seppälä *et al.*, 2011).

Brachylaemidae

Many members of this group use terrestrial vertebrates as definitive hosts. This poses a significant problem, given that the miricidia and cercariae stages of typical trematodes require mobility through water. *Leucochloridium* spp. (Fig. 6.28) solve this problem in a spectacular fashion. Within its terrestrial snail host, a unique branched sporocyst, or brood sac, develops, first in the hepatopancreas and then secondarily in the snail's tentacles. The two portions remain connected via a tube through which embryos are passed from the sporocyst in the hepatopancreas into that part of the sporocyst located in the tentacles. Cercariae develop into metacercariae within the sporocyst in the tentacles, which become swollen, and assume a striped, brightly colored morphology

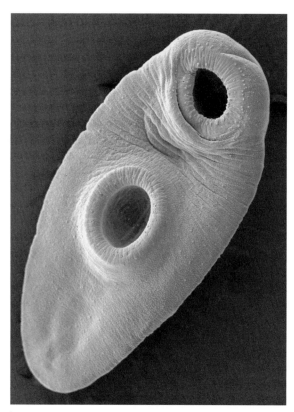

Fig. 6.28 Scanning electron micrograph of the brachylaemid digenean *Leucochloridium variae* from the cloaca of an experimental chicken, showing the large oral sucker and muscular acetabulum characteristic of the genus. (Micrograph courtesy of Doug Bray.)

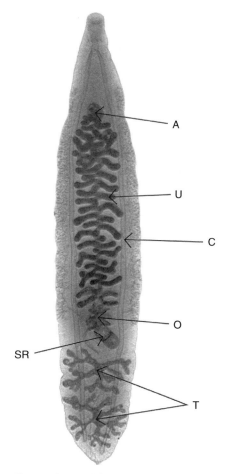

Fig. 6.29 The opisthorchiid digenean *Clonorchis sinensis* from the liver of fish-eating mammals including humans, showing the acetabulum (A), uterus (U), testes (T), ovary (O), seminal receptacle (SR), and intestinal cecum (C). (Photograph courtesy of John Sullivan.)

(Color plate Fig. 7.1). Thus, the snail is used as both first and second intermediate host. Unique among the trematodes, sporocyst-infected tentacles pulsate at particular times of the day, appearing (at least to humans) as mimics of mobile caterpillars. Further, the anterior end of *Leuchochlordium* sp.-infected sporocysts have what appear to be paired eyespots. This spectacular suite of traits that combines vivid coloration, eyespots, and altered host behavior (infected snails are photo- and geotactic at particular times of day) likely represents an extreme case of host phenotypic alteration leading to increased conspicuousness and enhanced transmission (see Chapter 15).

6.5.5.3 Plagiorchiida

The Plagiorchiida includes the largest number of families among the digeneans and is probably the most diverse in terms of morphology, with very little resemblance among those in the order.

Opisthorchiidae

Species of Opisthorchiidae infect the bile ducts, gall bladders, and livers of reptiles, birds, and mammals. Perhaps the best known is *Opisthorchis* (*Clonorchis*) *sinensis*, also called the Chinese liver fluke (Fig. 6.29).

This parasite infects nearly 30 million humans, mostly in the Far East, but also in many other parts of the world where fish are consumed raw, as sushi. These flukes occur in the liver, bile duct, and gall bladder where a single adult may produce up to 4000 eggs per day. Several species of freshwater snail serve as first intermediate hosts. Cercariae emerge from the snail, penetrate the surface of fishes, and encyst in the muscles as metacercariae. More than 80 species of cyprinid fishes are known to harbor the parasite. There are also several reservoir hosts for the fluke, including canines, felines, badgers, and mink. With the increase in aquaculture in many areas of the world, the parasite appears to be spreading. Infected humans may suffer minor to serious problems, including liver damage, vomiting, and diarrhea.

Heterophyidae

In terms of their life cycles and mammalian hosts, species in the Heterophyidae closely resemble opisthorchiids. These are diminutive intestinal parasites, however, and their numbers may reach into the thousands within a single host. In high intensities, they may cause abdominal discomfort, nausea, vomiting, and diarrhea. Adult worms may on occasion erode the intestinal mucosa and deposit eggs, which then may reach the circulatory system and spread to various parts of the body, especially the brain and heart. They may actually cause death in human hosts due to cardiac arrest if a sufficient number of eggs become entrapped in cardiac muscle. Common species in the family include *Heterophyes heterophyes* and *Metagonimus yokawagai*. The former is prevalent in Egypt, eastern Asia, and Hawaii, whereas the latter is common in the Balkan countries and the Far East. Most of these parasites are present in countries where fish is consumed raw and, like their opisthorchiid relatives, are increasing in prevalence because of the increasing consumption of raw fish, and aquaculture. A species common along the Gulf Coast of North America is *Phagicola longa*. Metacercariae of this species occur in the viscera and flesh of mullets, whereas adults are present in a number of birds, including the brown pelican.

Allocreadidae

Species of the Allocreadidae are referred to as the papillose flukes because some possess a series of muscular papillae, or head lappets, surrounding the oral sucker. These parasites are found in fishes and amphibians throughout the world. One of the better studied, and discussed in ecological and environmental context in Chapter 17, is *Crepidostomum cooperi*. It is widely distributed in North America and commonly occurs in the pyloric cecae of freshwater centrarchid fishes. Sphaeriid clams and aquatic arthropods are intermediate hosts. Several allocreadiids are neotenic, and their metacercariae develop to sexual maturity in what were once their arthropod second intermediate hosts, e.g., *Allocreadium neotenicum* in diving beetles.

Troglotrematidae

Adult troglotrematids are generally found in the lungs, nasal passages, cranial cavities, and intestines of birds and mammals. Perhaps the best known member of the family is *Paragonimus westermani*, which occurs in the lungs of a number of crab-eating mammals, including humans, in eastern Asia, parts of Africa, and Peru and Ecuador. Cercariae emerge from snail intermediate hosts usually in the late evening, which coincides with the maximum behavioral activity of crabs and crayfish, their next intermediate hosts. The cercariae crawl on the substratum much like inchworms; when they make contact with a crab or crayfish, they attach and penetrate through soft tissues in the joints, and encyst in its muscles. When a crab-eating mammal consumes the infected second intermediate host, the parasites excyst in the small intestine, penetrate the gut wall, then the diaphragm, and enter the lung where they mature sexually. Remarkably, they are usually able to locate a partner and generally encyst in pairs within the lung. Paragonimiasis in humans is a zoonotic disease because the normal hosts in nature are felines and canines, and humans are infected secondarily. Several million human infections occur in the Far East where crabmeat is eaten raw and unsalted.

Nanophyetus salmincola is a troglotrematid that occurs in northwestern North America and in Siberia where it causes salmon-poisoning disease in canines (Millemann & Knapp, 1970). Whereas the adult fluke is virtually innocuous for the canine, it is the vector for *Neorickettsia helminthoeca*, a bacterial organism that causes the disease in the definitive host. The adult parasite is embedded in the intestinal wall of canines and other mammals, including humans, although raccoons and spotted skunks are apparently the primary reservoir hosts in nature. The stream snail *Oxytrema silicula*, and a wide range of fish species, especially salmonids, are the required second intermediate hosts. The rickettsial infection in dogs is 90% lethal within 7 to 10 days, if untreated. The symptoms include a high fever, vomiting, and diarrhea. If the dog recovers, it is immune to another exposure to the rickettsial organism (Millemann & Knapp, 1970).

Another troglotrematid of interest is *Collyriclum faba*. Adults encyst in pairs subcutaneously around the cloaca of several species of birds (Blankespoor *et al.*, 1985), including English sparrows, grackles, and swifts. The cysts are about the diameter of an ordinary garden pea and appear in clusters of up to 15 in a single bird. Each cyst has a small hole that is plugged with a mucus secretion, which dissolves when the bird dips its rump in water when taking a drink. Eggs are released at that point. The life cycle is completely unknown, although winged insects are probably the second intermediate hosts since swifts are infected with the parasite and their feeding is strictly on the wing.

Dicrocoelidae

Species in the Dicrocoelidae parasitize terrestrial or semi-terrestrial vertebrates and use terrestrial snails and arthropods as first and second intermediate hosts. Several dicrocoelids are notorious for producing spectacular behavioral modifications in their arthropod hosts. The best known is *Dicrocoelium dendriticum* (Fig. 6.30) which has an apparently cosmopolitan distribution in sheep, cattle, pigs, goats, and cervids. Occurring in the bile ducts, adults produce eggs that are shed in the feces. Terrestrial snails ingest the eggs and cercariae accumulate in the snail's modified mantle. Their accumulation in the 'lung'

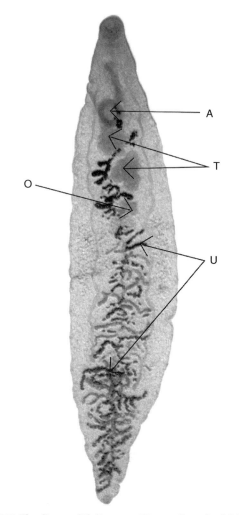

Fig. 6.30 The dicrocoelid digenean *Dicrocoelium dendriticum* from the liver of herbivorous mammals including sheep, showing the acetabulum (A), uterus (U), testes (T), and ovary (O). (Photograph courtesy of John Sullivan.)

is an irritant and causes the snail to secrete mucus in which the cercariae become entangled. The infected snail releases masses of mucus, containing up to 500 cercariae, as slime balls. Apparently, ants (*Formica* spp.) find these slime balls delectable and consume them. When this occurs, the cercariae penetrate the gut wall where they become encysted in the hemocoel as metacercariae, or at least most of the cercariae follow this route. One or two, however, migrate to the subesophageal ganglion and remain there unencysted (and uninfective). These metacercariae provoke a striking behavioral change in the ant (Color plate Fig. 8.1). Ants infected with *D. dendriticum* are unable to open their mouths when temperatures fall at dusk. If they happen to be grazing on vegetation, they become attached for the night. They remain there, fixed in position by their closed jaws, until temperatures rise the next morning. Presumably, this temperature-mediated form of 'lock jaw' increases the vulnerability to predation by herbivores. On excystment in the small intestine, the metacercariae migrate directly to the bile ducts where they take up residence and mature sexually.

Brachylecithum mosquensis is a common dicrocoelid fluke in American robins. Metacercariae of this parasite also alter the behavior of its second intermediate host, thereby increasing the probability of transmission. Metacercariae encyst in the sub-esophageal ganglia of carpenter ants, *Camponotus* spp. Normally, carpenter ants are relatively secretive, avoiding bright sunlight. When infected with this parasite, however, they prefer bright light and thus become vulnerable to predation by robins.

Plagiorchiidae

The Plagiorchiidae is a large family of digeneans infecting vertebrates of all classes. Plagiorchiids typically use snails as a first intermediate host and an insect second intermediate host. *Plagiorchis noblei* is a common parasite of yellowheaded and redwinged blackbirds in North America; lymnaeid snails and several larval insects are first and second intermediate hosts.

Haematoloechus is a widely distributed genus of plagiorchiid found in the lungs of frogs (Color plate Fig. 1.2). Several species of lung flukes have been described from North American and European frogs. Most exhibit great morphological plasticity, making their specific identification difficult. Frogs are infected upon consumption of infected odonate (damselfly and dragonfly) second intermediate hosts. Some *Haematoloechus* congeners specialize on using dragonflies as second intermediate host. Others are generalists and use both types of odonate naiad, and possibly other arthropods (Snyder & Janovy, 1996; Snyder & Tkach, 2001). These patterns of second intermediate host specificity are related to the evolution of distinct behaviors of the cercariae of the different lung fluke species. Thus, those lung flukes that are 'dragonfly specialists' have cercariae that are passively recruited into dragonfly naiads during cloacal respiration. Generalist lung fluke species have cercariae that actively penetrate the exoskeleton of both damselfly and dragonfly hosts.

Opecoelid flukes typically occur in the intestines of marine fishes, and a few are found in freshwater piscine hosts. They usually have a three-host life cycle, one that includes a snail, an arthropod, and the piscine definitive host. Barger & Esch (2000) discussed the unique *Plagioporus sinitsini* and its abbreviated life cycle in the freshwater stream snail *Elimia* sp. from a creek in the foothills of the Appalachian Mountains of North Carolina. They found snails shedding sporocysts that contained daughter sporocysts, cercariae, and sexually mature adults with eggs and viable miracidia. It seems that this abbreviated life cycle is neither neoteny nor progenesis because the adult parasites inside the sporocysts are equivalent in size to adults occurring in stream fishes in the same habitats.

6.6 Monogenea

6.6.1 General considerations

Most monogeneans are ectoparasites of marine and freshwater fishes. A few parasitize amphibians and one species, *Oculotrema hippopotami*, is found on the eyes of hippopotamases. Monogeneans typically occur on the external surfaces of their hosts. Most species exhibit remarkable host and site specificity. They possess an incomplete gut, are monoecious, and have direct life cycles. Most have a free-swimming larval stage called an oncomiracidium. They are usually small in size and equipped with a large, modified attachment organ at the posterior end, known as the **haptor (=opisthaptor)**. The morphology of this structure is highly variable, and provides insight as to the relationship between the attachment device of the parasite and the morphology of the site of infection on, or in, the host. Adult life spans range from several days to several years. Several monogeneans have been extensively studied from the standpoint of their ecology, pathogenesis, and biogeography.

6.6.2 Form and function

6.6.2.1 Holdfast and body wall adaptations

Most monogeneans are thin and flattened, ranging in size from about 0.3 mm to 20 mm. The anterior end of the body, also called the **prohaptor**, has various adhesive and feeding structures. In some species, the prohaptor may have a number of cephalic or head glands that secrete a sticky adhesive substance, and shallow muscular suckers, all used for attachment. In other species, there is an oral sucker, with various degrees of muscularization that surrounds the mouth.

The posterior haptor is responsible for host attachment (Figs. 6.31, 6.32). The variable morphology of the haptor is key to monogenean systematics and species identification. It may have suckers in various degrees of development, large hooks called anchors (or hamuli), small hooks that are remnants from the larval stage, or complex clamps that may be either muscular or sclerotized. There are normally one to three pairs of anchors, often with a connecting bar or accessory sclerites supporting them. The hooks are generally small and, like the anchors, are made of keratin. These hooks and those of eucestodes and cestodarians are similar chemically, whereas hooks of digeneans are composed of crystallized actin. Clamps work by pinching at their attachment site, clothespin-like, and their numbers vary from eight to several hundred, depending on the species.

On the anterior surface of the body there may be up to four pigmented eyespots, or photoreceptors (Fig. 6.31). There also may be an organized set of external papillae and underlying nerve endings that probably function as mechanoreceptors, especially in association with the haptor of some species,

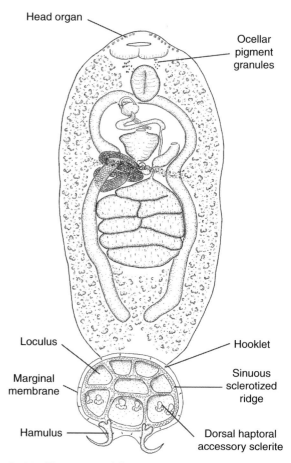

Fig. 6.31 The monocotylid monogenean *Neoheterocotyle inpristi* from the gills of smalltooth sawfish *Pristis pectinata*. Note the structural complexity of the posterior haptor and complex hermaphroditic reproductive system between the intestinal cecae (Modified from Chisolm, 1994, with permission, *Journal of Parasitology*, **80**, 960–965.)

e.g., *Entobdella* spp. The surface of monogeneans, as in digeneans and cestodes, is a syncytial tegument, with the nuclei and many other organelles located in cytons below the tegumental surface (Fig. 6.8).

6.6.2.2 Digestive system

The incomplete digestive system is simple and includes a mouth, buccal funnel, pre-pharynx, a muscular and glandular pharynx, esophagus, and paired ceca that

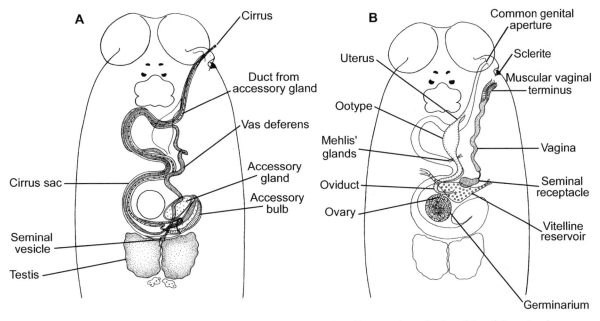

Fig. 6.32 Reproductive systems of the capsalid monogenean *Metabenedeniella parva* from the dorsal fin of the painted sweetlips *Diagramma pictum*. (A) Male; (B) female. (Modified from Horton & Whittington, 1994, with permission, *Journal of Parasitology*, **80**, 998–1007.)

may possess diverticula. The pharynx is a powerful sucking structure that brings food into the gut. The alimentary tract is in the shape of a wishbone, although sometimes the two branches may unite near the posterior end of the body (Fig. 6.31). There is no anus. Once food is internalized, digestion is both extra- and intracellular.

6.6.2.3 Excretory/osmoregulatory system

The excretory/osmoregulatory system is simple. The flame cells of the protonephridia are connected to two lateral main collecting ducts that swing down the length of the body and then back towards the anterior end, where paired contractile bladders help empty them to the surface via excretory pores. These pores are lateral and located near the anterior end of the organism. This is an important distinguishing character between monogeneans and digeneans, because in digeneans a single bladder and excretory pore are located posteriorly.

6.6.2.4 Reproductive system

Monogeneans are monoecious, but the reproductive systems are entirely separate (Fig. 6.32) and cross-fertilization is the rule in most cases. The number of testes is variable, depending on the species, with as few as one and as many as two hundred. Most species, however, have paired testes. Arising from each testis is a vas efferens that follows into an ejaculatory duct and then continues into a genital atrium. In many species, the tissues around the ejaculatory duct are thickened and form a penis-like structure, the cirrus (Fig. 6.32A), which may be armed with hooks distally. In some, there is a complex, sclerotized copulatory organ that joins with the ejaculatory duct. The morphology of the hooks of the opisthaptor and of the sclerotized copulatory organ is characteristic of each species.

The female reproductive system is complex and variable. It contains a single ovary of variable shape. Haploid ova move from the ovary into a short oviduct, which, in turn, empties into an ootype that is surrounded by the Mehlis' gland. Before the oviduct enters the ootype, the vitelline duct, the vaginal canal, and a genitointestinal canal may join it (Fig. 6.32B). Sperm stored in the seminal receptacle fertilize the ova as they pass through the oviduct on their way to the ootype. Once inside the ootype, the zygote begins development to the egg stage. Secretions from the Mehlis' gland apparently function in lubricating the ootype, facilitating movement of the developing egg. Secretions from the vitelline glands, or vitellaria, which are scattered laterally up and down the length of the body on both sides, contain a tanning protein that contributes to sclerotizing the eggshell. Egg production can be quite rapid, with several being produced within just a few seconds. Eggs leaving the ootype move into the uterus that opens into the genital atrium. The genitointestinal canal, mentioned earlier, connects the female reproductive system to the gut. Although its function is not known, there is speculation that it may be a vestigial structure through which eggs were passed into the intestine to be expelled through the mouth.

6.6.2.5 Nervous system and sensory input

The nervous system of monogeneans is similar to that of most other platyhelminths, with paired cephalic ganglia located anteriorly and usually a pair of nerve cords extending posteriorly. At intervals, the nerves are connected laterally. A variety of sensory cells and structures are located in the tegument. The most prominent of these sensory structures are eyespots on the surfaces of oncomiracidia, which usually disappear in the adult stage. The opisthaptor is well innervated internally and possesses external papillae that function in mechanoreception as well. Monogeneans are known to produce a number of neurotransmitters, particularly serotonin (Halton *et al.*, 1997).

6.6.3 Nutrient uptake and metabolism

Monogeneans feed on various host tissues, including blood, mucus, and epidermal cells. Feeding is achieved by the sucking action of the muscular pharynx. In some species, e.g., *Entobdella soleae*, the pharynx can

be everted. The pharyngeal glands of *E. soleae* secrete proteases that erode and lyse the epidermis, which is then sucked inside (Kearn, 1971). In species that feed on blood, such as *Diclidophora* spp., digestion of hemoglobin is mostly intracellular. By-products of digestion, such as hematin, or other non-digestible compounds, are evacuated through the mouth. Nutrient uptake also probably occurs through the tegument. The blood-feeding *Diclidophora merlangi*, for example, can absorb neutral amino acids through the tegument (Halton, 1978).

The metabolism of monogeneans is poorly known (Smyth, 1994). Living on the surface of fishes would suggest aerobic metabolic pathways, but this is speculative.

6.6.4 Development and general life cycle

The life cycle of most monogeneans is direct, without intermediate hosts, and includes an egg, a free-living larval stage called an oncomiracidium (Fig. 6.33), and the adult. The oncomiracidium hatches from the egg. These larvae are covered with cilia, have one or two pairs of eyespots, and possess numerous hooklets at the posterior end (Fig. 6.33), which are retained by the adult, becoming the smaller hooks on the haptor. Oncomiracidia are excellent swimmers, but their life span is brief. Mucus secretions apparently influence host location. Once attached to an appropriate host, they lose their cilia and develop directly into adults.

The exceptions to this pattern are the gyrodactylids, all of which are viviparous. Larvae are retained in the uterus until they develop into functional juveniles (Fig. 6.34). Even more intriguing, within a developing juvenile there is a second juvenile, and a third within the second, and a fourth within the third, much like nested boxes. As soon as the first juvenile is born, it begins to feed and gives birth to the juveniles remaining inside. This process can be completed very rapidly, within 24 hours. Only after the first juvenile has given birth to all the others can one of its own ova become fertilized, then to repeat the reproductive sequence.

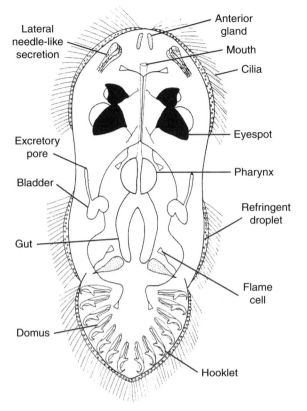

Fig. 6.33 Oncomiracidium of the monocotylid monogenean *Clemacotyle australis* from the skin of the branchial cavity of the white-spotted eagle ray, *Aetobatis narinari*. (Modified from Beverly-Burton & Whittington, 1995, with permission, *Journal of Parasitology*, **81**, 616–625.)

6.6.5 Biodiversity and life-cycle variation

Gyrodactylids are small monogeneans, measuring 1 mm or less in length, that are parasitic on the gills and skin of marine and freshwater fishes (Fig. 6.35). Their haptor is relatively simple, with two median anchors and 16 small hooks. Their viviparous life-history strategy makes them unique among the monogeneans. The lack of a swimming oncomiracidium does not appear to hinder the worm's dispersal, which occurs directly by physical contact between hosts and via water that is circulated through a host's gills, to which gyrodactylids are attached. The highest reproductive activity in gyrodactylids occurs during the breeding period of the host because the proximity

Fig. 6.34 Viviparous development in *Gyrodactylus spathulatus* from the fins of the white sucker *Catostomus commersoni*. Visible within the uterus are the large anchors of a developing juvenile. Scale bar = 60 μm. (Photograph courtesy of David Cone.)

of a potential host clearly increases the chances of successful transfer.

Many gyrodactylids are pathogenic and are known to cause fish mortalities in hatcheries and rearing ponds, mainly in Europe and the former Soviet Union, where high densities of the host facilitate parasite dispersal. The short, direct life cycle and viviparous mode of development can lead to massive numbers of 'gyro' developing rapidly on fish hosts. *Gyrodactylus salaris*, for example, was responsible for devastating

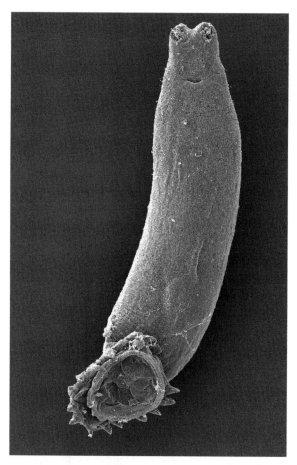

Fig. 6.35 Scanning electron micrograph of the gyrodactylid monogenean *Gyrodactyloides bychowskii* from the gills of Atlantic salmon *Salmo salar*. (Micrograph courtesy of David Cone.)

epidemics in Norway in the 1970s (Bakke *et al.*, 2007). Gyrodactylids damage the gill epithelium by the mechanical action of the anchors and hooks of the opisthaptor and via feeding activities. This causes epithelial proliferation and increases the production of mucus in the gills that may eventually lead to host death by functional failure of the respiratory epithelium, causing asphyxia (Schaperclaus, 1991). The potential fish mortality often associated with *Gyrodactylus* spp. and their viviparous mode of reproduction has led to the gyrodactylids being called the 'Russian doll killers.' Bakke *et al.* (2007)

present an extensive review of the behavior, ecology, and evolution of these fascinating parasites.

Members of the Dactylogyridae (not to be confused with the Gyrodactylidae) and, in particular, species of *Dactylogyrus* are of economic significance as pathogens of fishes in hatcheries throughout the world. They parasitize the gills and the damage inflicted is similar to that caused by gyrodactylids. Dactylogyrids are small monogeneans measuring up to two millimeters in length and, superficially, they resemble gyrodactylids. Their life cycle, however, follows the typical monogenean pattern, with an egg and a free-swimming oncomiracidium. It is fairly common to find more than one species on a given host, in which case they exhibit marked site specifity. In Europe, for example, the common carp may simultaneously harbor three different species, each with a specific location on a single gill filament. For example, *D. vastator* prefers the outer tips of the gill filaments, *D. extensus* is found halfway along the length of the gill filament, and *D. anchoratus* prefers the basal region (see Chapter 13).

The Capsalidae, including species of *Neobenedenia* (Fig. 6.36) and *Entobdella*, are relatively large monogeneans parasitic on the skin, mouth, and nostrils of marine fishes. Massive infections of some species can produce the same kind of damage as previously described for the gyrodactylids.

Members of the Loimoidae and Hexabothriidae parasitize the gills of elasmobranchs, whereas the Microbothriidae attach to the skin and nostrils. The haptor of hexabothriids, such as *Branchotenthes robinoverstreeti*, has muscular suckers with claw-like sclerites, and a muscular appendix equipped with suckers (Fig. 6.37). The Dionchidae are exclusive parasites of the gills of remoras, and the Mazocraeidae are gill parasites of herring and mackerels. The haptor of this latter group is peculiar because the clamps used for attachment are mounted directly on the sides of the body.

Although most monogeneans are parasites of marine and freshwater fishes, a few parasitize amphibians and reptiles. Species of *Sphyranura* are parasitic on the gills and skin of caudate amphibians,

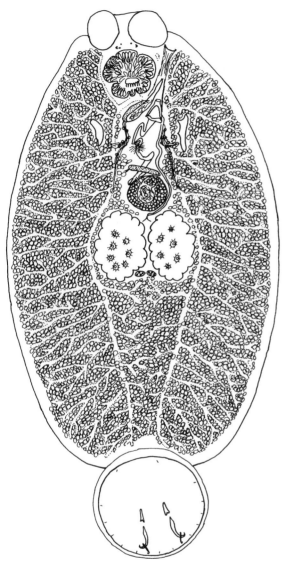

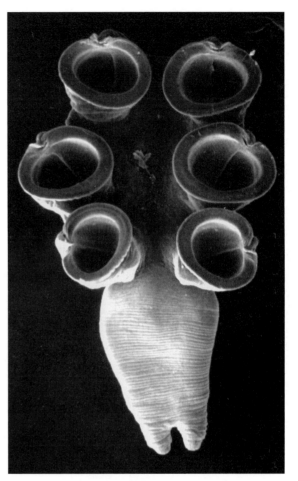

Fig. 6.37 Scanning electron micrograph of the haptor of the hexabothriid monogenean *Branchotenthes robinoverstreeti* from the gill filaments of the bowmouth guitarfish *Rhina ancylostoma*, showing the six suckers, each equipped with a sclerite, and a muscular appendix with minute suckers at the apex. (Micrograph courtesy of Ash Bullard.)

Fig. 6.36 The capsalid monogenean *Neobenedenia girellae* from the body surface, fins, and eyes of various Japanese fishes, including the Japanese flounder *Paralicthys olivaceus*. (Modified from Ogawa *et al.*, 1995, with permission, *Journal of Parasitology*, 81, 223–227.)

such as mudpuppies, whereas species of *Polystoma* parasitize the urinary bladder, nasal cavities, mouth, pharynx, or esophagus of amphibians. In those monogeneans parasitic on, or in, amphibians, the amphibious nature of their hosts imposes severe temporal constraints on the transmission of oncomiracidia (see Chapter 12). Because many amphibians only visit bodies of water during their breeding period, parasite reproduction must be closely synchronized with that of the host. In *Polystoma integerrimum*, a common species in the urinary bladders of European frogs and toads, transmission of oncomiracidia is stimulated by gonadotropins that are produced by the host during bouts of reproduction in spring. Other polystomatids of interest include *Pseudodiplorchis americanus* (Fig. 6.38), a parasite of

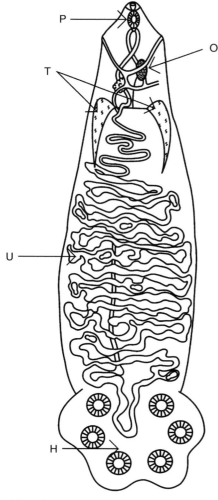

Fig. 6.38 The polystomatid monogenean *Pseudodiplorchis americanus* from the urinary bladder of the spadefoot toad *Scaphiopus couchii*, showing the pharynx (P), testes, (T), ovary (O), extensively coiled uterus (U), and six-suckered haptor (H). (Figure courtesy of Danielle Morrison.)

the spadefoot toad (*Scaphiopus couchii*) and *Neodiplorchis scaphiopodis*, a parasite of two other species of spadefoot toads (*S. mutiplicatus* and *S. bombifrons*), all of which are sympatric in the driest parts of southwestern North America. These monogeneans have a remarkable life cycle representing what might be the narrowest transmission window of any helminth (Tinsley, 1999). The transmission and population dynamics of *P. americanus* in toads are described in Chapter 12.

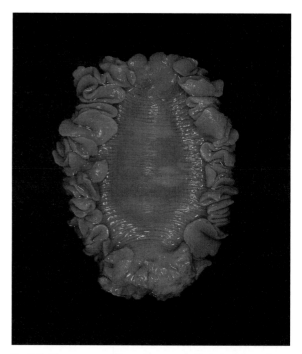

Fig. 6.39 The gyrocotylidean *Gyrocotyle fimbriata* from the spiral intestine of the ratfish *Hydrolagus colliei*. (Photograph courtesy of Bill Pennell.)

6.7 Gyrocotylidea

The gyrocotylideans are large, monozoic cestodes and are likely most closely related to the Monogenea. They are parasites in the spiral intestine of primitive cartilaginous fishes, the Holocephali, with strict specificity for their hosts. They are oblong, hermaphroditic platyhelminths, without an intestine. The anterior end has a small, muscular, cup-shaped holdfast organ. The posterior end bears an attachment organ, which may be in the form of a frilled, rosette-like structure (e.g., *Gyrocotyle*; Fig. 6.39). Nutrients are absorbed through the tegument and the protonephridium is the functional unit of the excretory system. The male reproductive system is relatively simple, with testes and a vas deferens. The female reproductive system includes an ovary located in the posterior end of the body, with an oviduct that becomes the ootype before enlarging to form the uterus. Vitelline glands, Mehlis' gland, and seminal receptacle are also present. The genital pores are near the anterior end of the body. Gyrocotylideans have a free-swimming larva, the **lycophore**, with 10 hooks in the posterior end. The life cycle is probably direct, but it has not been completed experimentally for any species. Adult gyrocotylids have at least 10 different sensory receptors and their larvae have eight. Interestingly, the receptors in the adult and the larva are different, suggesting extreme specialization for their respective habitats.

6.8 Amphilinidea

Amphilinids are flattened and unsegmented platyhelminths, with a proboscis-like holdfast at the anterior end. They live in the coelomic cavity of their hosts and some can become relatively large, measuring up to 380 mm in length. They lack a digestive system. The male reproductive system is simple, consisting of paired testes and a vas deferens. The female reproductive system consists of an ovary, oviduct, ootype, Mehlis' gland, and uterus (Fig. 6.40). A vagina is present and the vitelline follicles are poorly developed. The life cycle of amphilinids is complex, with ciliated larvae possessing 10 hooks, and a crustacean intermediate host.

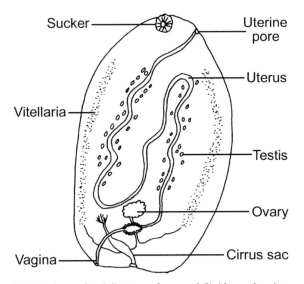

Fig. 6.40 Generalized diagram of an amphilinidean, showing the position of the reproductive organs. (Drawing courtesy of Lisa Esch McCall.)

Amphilinideans are parasites of the coelomic cavity of primitive fishes and turtles on all continents, excluding Antarctica. Species of *Amphilina* parasitize the body cavity of sturgeons (*Acipenser* spp.) and use amphipods as intermediate hosts. The parasite seems to have debilitating effects on the host, affecting the levels of hemoglobin, minerals such as zinc, and body fat. *Austramphilina elongata* parasitizes the body cavity of Australian freshwater turtles, using crayfish and freshwater shrimp as intermediate hosts. The amphilideans, as probable sister group to the Eucestoda, represent an interesting evolutionary link between the monogeneans and eucestodes.

6.9 Eucestoda

6.9.1 General considerations

With few exceptions, adult eucestodes are parasites of the intestine, or its accessory structures, of vertebrates. Except for a couple of species, all tapeworms are monoecious and are composed of unique reproductive segments known as proglottids. They range from tiny worms only a few millimeters in size, to gigantic worms exceeding 20 meters. The structure of the reproductive system within the proglottids, as well as the structure of the unique holdfast structure, the **scolex**, are extremely important in cestode systematics. All tapeworms lack an intestine and nutrients are absorbed through a syncytial, **microtriches**-covered tegument. The unit of structure and function in the excretory (osmoregulatory) system is the protonephridium. Their life cycles are complex, and transmission to vertebrate definitive hosts almost always occurs by predator–prey interactions involving one, or sometimes two, intermediate and/or paratenic hosts.

6.9.2 Form and function

6.9.2.1 Holdfast and body wall

The body plan of most cestodes includes a scolex, neck, and a **strobila**, consisting of proglottids in various stages of development (Fig. 6.41). The scolex is a

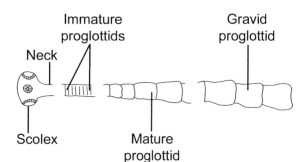

Fig. 6.41 Adult polyzoic cestode, showing the typical body plan with scolex and strobila with immature, mature, and gravid proglottids. (Figure courtesy of Lisa Esch McCall.)

specialized attachment structure located at the anterior end. The morphology is highly variable and may have suction organs, hooks, spines, tentacles, glands, or any combination of these structures. There are three main types of suction devices used for attachment, **bothria**, **bothridia**, and **acetabula**. Bothria are slit-like grooves with weak attachment power. The common number is two per scolex, as in the Diphyllobothriidae. Bothridia are highly muscular, may assume several shapes, and their margins are thin and flexible, usually matching the structure of their attachment site. The scolices of Tetraphyllidea, for example, have four bothridia. Acetabula are suckers or suction cups distributed around the scolex; there are normally four and they are characteristic of the Cyclophyllidea. Some of the Cyclophyllidea may also have a dome-shaped structure, the **rostellum**, at the tip of the scolex, which may or may not be armed with hooks for attachment. If present, hooks are arranged in one or more circles around the rostellum (Fig. 6.42).

Immediately posterior to the scolex is an unsegmented neck region. Internally, there is a large population of undifferentiated stem cells that give rise to the proglottids. The entire worm behind the scolex is collectively called the strobila. Although tapeworms with differentiated proglottids appear to be segmented, tissues such as the tegument and muscles are continuous between the proglottids, without division and, therefore, are not true segments. Immediately behind the neck are the youngest proglottids, followed by mature, and then gravid, proglottids (Fig. 6.41). As new proglottids are

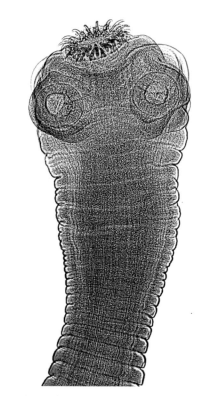

Fig. 6.42 Scolex and neck region of the cyclophyllidean cestode *Taenia solium*, showing the suckers, and the rostellum armed with hooks. (Photograph courtesy of John Sullivan.)

formed in the neck, the previous ones move posteriorly in a continuous process. In some species, eggs are shed from the strobila while the proglottid remains attached. Then, when these egg-filled, gravid proglottids are spent, they detach and rapidly disintegrate. In other species, the proglottid detaches before releasing the eggs and leaves the host intact with the feces. Almost all cestodes have many proglottids and are referred to as being polyzoic. In a few tapeworms, however, all of the reproductive organs and the scolex are within a single body, without a strobila, lacking all signs of segmentation; these tapeworms are monozoic and belong to the Caryophyllidea.

Similar to the trematodes, the cestode tegument is complex and multifunctional, with roles in nutrition, attachment, evasion of host immunity, and protection from host digestion. Its complex ultrastucture (Fig. 6.43) is similar to the trematode tegument

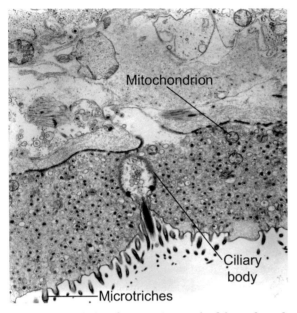

Fig. 6.43 Transmission electron micrograph of the surface of the Asian tapeworm *Bothriocephalus acheilognathi*. Externally on the tegument are microtriches and a sensory ciliary body and terminal cilium. The tegument contains numerous vesicles and mitochondria. (Modified from Granath *et al.*, 1983, with permission, *Transactions of the American Microscopical Society*, **102**, 240–250, John Wiley & Sons.)

(Fig. 6.8), with some important modifications (reviews in Jones, 1998; Dalton *et al.*, 2004). The distal-most layer is enveloped by countless microtriches. These are structurally similar to the microvilli found along the epithelial surfaces of other animals (especially the gut) and similarly, function to increase absorptive surface area. The microtriches surface is covered by a digenean-like glycocalyx that is a carbohydrate-rich layer involved with inhibition of some host digestive enzymes and absorption of cations and bile salts. Typically, ciliated sensory structures (Fig. 6.43) are scattered within the microtriches border that function as tactile receptors, or chemoreceptors. The distribution of microtriches and their associated glycocalyx shows marked regional specialization. In *Echinophallus wageneri*, found in deep-sea fish, regional specializations of microtriches between proglottids are associated with the curling of the

tapeworm along the length of the host gut (Poddubnaya *et al.*, 2007).

Immediately beneath the microtriches lies the distal cytoplasm (Figs. 6.8, 6.43), an anucleate syncytium packed with mitochondria. These mitochondria typically possess few cristae, suggestive of the anaerobic metabolism that is characteristic of cestodes. The distal cytoplasm is also characterized by large numbers of vesicles, whose function is linked to the production of microtriches and hooks. The highly metabolic syncytial layer lies above a basal matrix and a layer of subtegumental circular and longitudinal muscles (Fig. 6.43). The basal matrix is perforated by channels that connect the distal cytoplasm with the underlying **cytons** (Fig. 6.8), similar to the pattern for trematodes. Cytons contain nuclei and other typical cell organelles, e.g., ribosomes, rough endoplasmic reticulum, Golgi apparatus, etc. Apparently, the vesicles in the distal cytoplasm originate in the cytons.

6.9.2.2 Excretory/osmoregulatory system

The excretory/osmoregulatory system consists of protonephridia that connect with longitudinal collecting canals, one pair dorsal and one pair ventral, on each side of the strobila. Normally, a transverse duct at the posterior end of each proglottid connects the ventral canals. The dorsal and ventral osmoregulatory canals merge in the scolex, which means that they are actually continuous. The excretory fluid present in the canals of *Hymenolepis diminuta* contains glucose, lactic acid, soluble proteins, urea, and ammonia (Webster & Wilson, 1970). It appears that short-chain organic acids, the main end products of metabolism, may be excreted through the tegument.

6.9.2.3 Reproductive system

Most cestodes are monoecious, with each proglottid having one set of male organs and one set of female organs. In some species, e.g., *Moniezia expansa*, however, there is a duplicate set of male and female organs in each proglottid and, in a few, there are two sets of male and one set of female organs, e.g.,

Diplophallus. Rarely, others such as *Shipleya*, are dioecious. Its mechanism of sex determination may depend on the interaction between two or more strobilae. For example, if only one strobila is present in a host, it is typically female; but if two strobilae are present, one is female and the other is male.

Both self- and cross-fertilization occur. However, self-fertilization is normally avoided because either the male reproductive system matures before the female counterpart (**protandry**), or because the female reproductive organs mature before the male organs (**protogyny**). Transfer of sperm normally occurs between different proglottids of the same strobila, or between proglottids in different strobilae.

The reproductive organs of cestodes are highly variable in their structure, arrangement, and distribution among the different taxonomic groups. Figure 6.44 compares the mature proglottid morphology of four different species of cestodes, a proteocephalan, a diphyllobothriidean, and two cyclophyllideans. There are some basic differences in the plan as to ventral versus lateral openings of the reproductive systems, lateral versus posterior vitellaria, compact versus scattered vitellaria, numbers of testes, and several other more specific characters. Despite this variability, some basic structures are in common. The male reproductive system includes one or more testes, each with a vas efferens, that unite to form a common vas deferens that directs the sperm towards a cirrus located in the cirrus sac. The cirrus is a protrusible copulatory organ that evaginates through the genital atrium. The vas deferens may be a simple tube or may be modified in several ways to store sperm.

The female reproductive system is more complex than that of the male (Fig. 6.44). A single ovary is continuous with an oviduct that is, in turn, joined by the vagina carrying implanted sperm and by the vitelline duct from the vitelline glands. The oviduct then merges into an ootype surrounded by the Mehlis' gland. The ootype is connected to the uterus, which may be a single sac or a simple or a convoluted tube. In some groups, it may be replaced by other structures such as the paruterine organ of the Mesocestoididae. The vagina may be simple or may include a seminal receptacle. The vitelline glands, which contribute yolk and shell material to the embryo, may be grouped into a single mass, or may be dispersed, forming a variety of structural patterns within the proglottid. Sperm transfer normally occurs from the cirrus into the vagina. Some species, however, lack a vagina. In these cases, the cirrus is forced through the body wall and sperm are deposited into the parenchyma. It is unclear how the sperm then find their way into the female's reproductive system.

One important difference among the various groups is the structure of the egg and the process of shell formation. Both the vitelline follicles and the Mehlis' gland contribute to shell formation. In groups like the Diphyllobothriidea, a thick capsule similar to the eggs of digeneans covers the egg. In others, the egg capsule may be very thin or even absent.

6.9.2.4 Nervous system and sensory input

The nervous system in all cestodes except *Diphyllobothrium* spp. consists of a pair of cerebral ganglia located in the scolex and two main nerve cords that extend the length of the strobila (Halton & Gustafsson, 1996). Species of *Diphyllobothrium* possess a single ganglion and two nerve cords. The longitudinal nerves are connected by transversal commissures present in each proglottid. A number of smaller nerves emanating from this ladder-like system innervate the muscles, sensory structures, tegument, and various reproductive structures such as the cirrus and vagina. The nervous system also is highly developed in the scolex, with nerves associated with the various structures used for attachment, e.g., acetabula, bothria, bothridia, and rostellum. A number of sensory receptors for both chemical and physical stimuli are present in cestodes, and probably function in the same manner as those in digeneans, e.g., mechanoreception, chemoreception, and osmoreception. Eyespots or other photoreceptors apparently are lacking in cestodes.

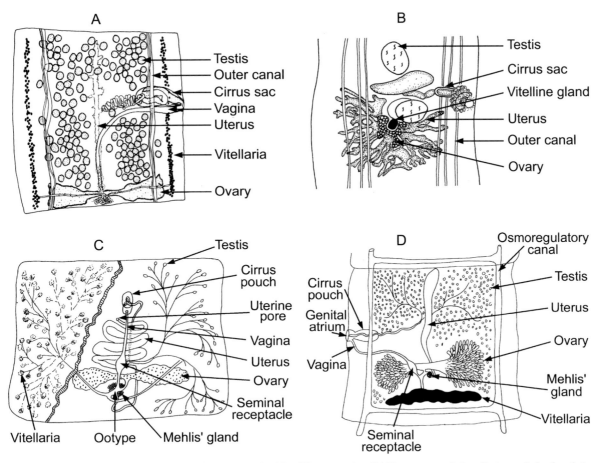

Fig. 6.44 Comparative morphologies of the mature proglottids of four cestodes. (A) The proteocephalan *Proteocephalus brooksi*; (B) the cyclophyllidean *Fimbriasacculus africanensis*; (C) generalized diphyllobothriidean proglottid; (D) generalized taeniid cyclophyllidean proglottid. ((A) Modified from García-Prieto *et al.*, 1996, with permission, *Journal of Parasitology*, 82, 992–997; (B) modified from Alexander & McLaughlin, 1996, with permission, *Journal of Parasitology*, 82, 907–909; (C, D) drawings courtesy of Lisa Esch McCall.)

6.9.3 Nutrient uptake and metabolism

Unlike the trematodes, cestodes lack a gut. Thus, all of their nutrient needs are met via absorption through their microtriches-covered, syncytial tegument. Pioneering work by Clark Read in the 1950s and 1960s (e.g., Pappas & Read, 1975) demonstrated that all required carbohydrates, fatty acids, nucleosides, and lipids are transported actively from the lumen of the host's gut into the body of the worm. Thus, separate and independent molecular carriers have been identified within the tegument of the rat tapeworm *Hymenolepis diminuta* that transport amino acids, carbohydrates, and both short- and long-chain fatty acids (Pappas & Read, 1975). The digestion of host material is also aided by endogenous hydrolytic and proteolytic enzymes that are associated with the glycocalyx. Furthermore, certain host enzymes are captured, bound to the glycocalyx, and then used to convert starch into sugars. Dalton *et al.* (2004) suggest that selection for such a diverse and efficient repertoire of transport mechanisms and digestion strategies in the tapeworms results from competition with the host (also an efficient absorber) within the nutrient-rich gut.

Respiration is primarily anaerobic. In general, intermediary carbohydrate metabolism occurs in two ways among the cestodes, making them similar to digeneans. One is of the homolactate type, wherein glucose is converted to lactate via the glycolytic pathway. The second is via malate dismutation, whereby glucose is converted to phosphoenolpyruvate (PEP), then CO_2 is fixed to PEP and malate is formed. Malate is then metabolized to either proprionate or acetate with the generation of additional ATP in the process. Some species excrete succinate. Whereas oxygen uptake can be demonstrated in cestodes, there is no evidence to indicate that oxygen is an electron acceptor as in the classical electron transfer system that operates in the mitochondria of most organisms. Since the major energy resource for tapeworms is glucose, many species store considerable quantities of glycogen – up to 50% of their dry weight. Protein and lipid metabolism are not important energy sources for tapeworms. Further information concerning the biochemistry and physiology of cestodes is presented by Smyth & McManus (1989).

6.9.4 Development and general life cycle

Most cestodes have life cycles that follow one of two major patterns, but with many variations in each (Fig. 6.45). A sample of this variation is discussed in the next section when we consider specific taxa.

In the first pattern, including those tapeworms that are mostly aquatic, are the Diphyllobothridea and Proteocephalidea (Fig. 6.45). The life cycles of species in these two orders are inextricably connected with aquatic first intermediate hosts, namely zooplanktonic microcrustaceans, mostly copepods. Within the eggs

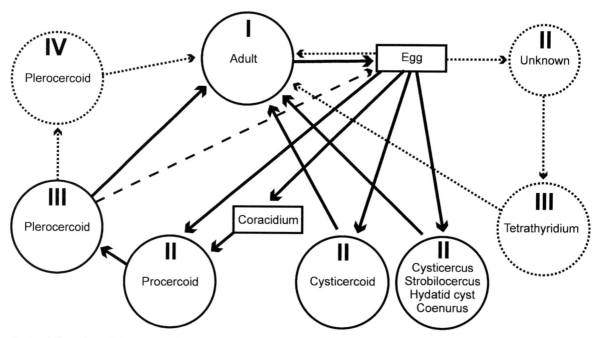

Fig. 6.45 Life-cycle variation among the cestodes. Host types are represented by roman numerals (I, final/definitive host; II, first intermediate host; III, second intermediate host; IV, paratenic or third intermediate host). Solid lines represent 'typical life cycles,' whereas dotted lines represent less common variations. Dashed line represents a form of progenesis in which the plerocercoid is thought to be neotenic, i.e., *Glaridacris* spp. Circles indicate stages found within intermediate or definitive hosts and rectangles depict the free-living stage of aquatic tapeworms such as *Diphyllobothrium* spp. (Drawing courtesy of Al Bush.)

are **oncospheres**, or **hexacanth** (six-hooked) embryos (Fig. 6.45). In some species, free-floating eggs in the water column must be consumed by the first intermediate host. In other species, the eggshell has an operculum and a larval stage hatches. This larva is called a **coracidium** and is covered by cilia that enable it to actively swim in the water column. As for miracidia and oncomiracidia, the coracidium is a short-lived stage that must infect its host within 24 to 36 hours. When the first intermediate host eats either an egg or a coracidium, the parasite penetrates the gut wall and enters the hemocoel. There, it transforms into a **procercoid**, a nondescript larval stage that resembles a cigar, possessing a knob-like structure called a **cercomer** on one end. Retained on the cercomer are the six hooks of the hexacanth embryo, a pattern similar to monogenean development where the young adult retains embryonic hooks of the oncomiracidium. Next, the infected microcrustacean must be eaten by an appropriate planktivorous host, usually a fish. The procercoid penetrates the gut wall of the fish and then migrates to species-specific locations within the new host. Once the final site of infection is reached, the procercoid grows rapidly and transforms into the next stage, known as a **plerocercoid** (Figs. 6.45, 6.46). In general, the scolex that develops on the plerocercoid closely resembles that of the adult. The overall size to which a plerocercoid will grow is also species specific. In some, they are large relative to the size of the host

Fig. 6.46 Plerocercoids of the cestode *Schistocephalus solidus* from the body cavity of a threespine stickleback *Gasterosteus aculeatus*.

and in others they remain relatively small. For such parasites, the life cycle is completed when fish intermediate hosts are eaten by avian or mammalian definitive hosts. Paratenic hosts (piscivorous fishes) are often used to bridge the ecological gap between definitive hosts and planktivorous fishes (Fig. 6.45).

The second general life cycle pattern is normally associated with a single intermediate host, and is typically terrestrial (Fig. 6.45). Among the cyclophyllideans, there are two basic body plans associated with the larval stage in the intermediate host, but with many variants. In some species, eggs are shed via apolytic (apolysis) proglottids. Within the eggs are the hexacanth embryos, or oncospheres. When these eggs are ingested, always by an invertebrate and many times an insect, the oncosphere emerges and penetrates the gut wall. On entering the hemocoel, the parasite transforms into a **cysticercoid** possessing a fully withdrawn (but not invaginated) scolex resembling that of the adult into which the parasite will eventually develop. Membranes that were present when the oncosphere was still inside the eggshell surround the scolex. To complete the life cycle, the infected arthropod intermediate host must be eaten by a definitive host.

The second type of larval body plan among cyclophyllideans is always associated with a mammalian intermediate host. In these species, the eggs are retained by the proglottid (anapolysis) when it is shed from the strobila. The proglottid then disintegrates on the outside, releasing the eggs. The eggshell consists of hundreds of tiny blocks held together with cement that will be dissolved through action of the intermediate host's digestive enzymes in the small intestine. Once freed, the oncosphere then uses its six embryonic hooks to assist in penetrating the gut wall. There, the larva is picked up in the circulatory system and transported to the appropriate site of infection. The larval stage that develops is one of several different types (Fig. 6.45), but all were probably derived evolutionarily from one, the **cysticercus**, or bladder worm. Cysticerci can be found in a number of organs and tissues in mammals, including the heart, lungs, skeletal muscles, mesenteries, and brain, depending on the species of parasite. The cysticercus may or may not be encysted by a hyaline capsule of host origin, but this is also related to the species of cestode as well as the site of infection. The cysticercus has an invaginated scolex, complete with four suckers. There is always a rostellum, generally with a crown of hooks that are in the shape of rose thorns. Large and small hooks usually alternate in two rows. Associated with the cysticercus, there is always a bladder filled with a fluid of host origin. The life cycle of the parasite is completed when flesh that contains the cysticercus is consumed by a definitive host (Fig. 6.45). Once freed from intermediate host tissue by digestion in the small intestine of the definitive host, the scolex of the cysticercus evaginates, attaches to the gut wall using the suckers and rostellar hooks, and undergoes **strobilation** to form the proglottids.

6.9.5 Biodiversity and life-cycle variation

There is tremendous diversity in cestode functional morphology and life cycles. In the following section, we select common groups to represent this diversity, following the current systematic outline as presented in Box 6.3.

6.9.5.1 Caryophyllidea

Caryophyllideans are usually parasitic in the intestine of freshwater teleost fishes, mainly cyprinids, catfishes, and catostomids, although a few parasitize the coelom of freshwater oligochaetes. The scolex is very simple, with shallow depressions or bothria. The body lacks proglottids or any sign of segmentation (monozoic). Internal anatomy resembles the pseudophyllideans. The life cycle involves two hosts, the fish definitive host and an oligochaete intermediate host infected with the procercoid stage. *Glaridacris catostomi* is a common parasite in the stomach and small intestine of suckers (*Catostomus* spp.) in North America. Species of *Caryophyllaeus*, also known as cloverworms because the scolex resembles a cloverleaf, are common parasites of carp, bream, and other cultured cyprinids in Europe. Intermediate hosts are

tubificids and other oligochaetes that are castrated when infected.

Several species of *Archigetes* mature sexually while in the oligochaete host. *Archigetes sieboldi* is a common species found in the coelom of aquatic oligochaetes in Europe and North America. To release eggs, gravid worms escape the oligochaete through the body wall and die, liberating the eggs as their body disintegrates on the substratum of a pond or lake. Species of *Archigetes*, then, appear to be neotenic.

It has been argued over the years that the non-segmented caryophyllaeids, more specifically the monozoic species, including *Archigetes*, represent the most basal eucestode order (Calentine, 1962). Alternatively, Mackiewicz (2003) views the monozoic form as secondarily derived. Olson *et al.* (2008) provide molecular evidence to support this position.

6.9.5.2 Diphyllobothriidea

Adult diphyllobothrideans are polyzoic, showing distinct segmentation. Their scolices possess paired bothria. On occasion, there may be two sets of reproductive organs within a single proglottid. Their size is variable, ranging from a few millimeters to more than 30 m in length in the case of *Hexagonoporus*, a tapeworm of sperm whales.

Diphyllobothriids infect both birds and mammals, including humans. The broad fish tapeworm, *Diphyllobothrium latum*, is a prominent member of the Diphylobothriidae, which extends in distribution from the northern USA and Canada, to parts of Scandinavia, Siberia, south into the Balkans, Japan, China, and Korea, and even South America (Revenga, 1993). Its host specificity is wide, variously infecting felids, canids, urcids, mustelids, pinnipeds, and humans. It is estimated that there are up to nine million humans infected worldwide. Humans acquire the parasite by eating raw or poorly cooked fish, including salmon, northern pike, walleye, and other fishes infected with the plerocercoid larval stage. *Diphyllobothrium latum* has a life cycle that includes a free-swimming coracidium, procercoids in copepods, plerocercoids in fishes, and adults in mammals,

Fig. 6.47 Life cycle of the cestode *Schistocephalus solidus*. Adults live in the intestine of piscivorous birds, including the great blue heron. Embryonated eggs pass in the feces and hatch in water as a ciliated coracidium larva. These are eaten by cyclopoid copepods. The six-hooked larva bores through the gut and develops into a procercoid in the hemocoel. Infected copepods are eaten by threespine stickleback second intermediate hosts. The procercoid burrows out of the intestine and becomes a large plerocercoid in the body cavity of the stickleback, often resulting in a distended abdomen. The life cycle is complete when infected fish are eaten by piscivorous birds. (Artwork courtesy of Lisa Esch McCall.)

similar to that outlined in Figures 6.45 and 6.47. Piscivorous fish at higher trophic levels act as paratenic hosts, greatly increasing opportunities for transmission. This species can reach 8 to 10 m in length and possess up to 3000 proglottids. In humans, diphyllobothriasis results in abdominal discomfort, diarrhea, and nausea. It may also cause pernicious anemia since the tapeworm is a 'sink' for vitamin B12, which, in humans, is converted into a coenzyme necessary for nucleic acid synthesis and the maturation of RBCs.

Ligula intestinalis and *Schistocephalus solidus* are common cestodes of fish-eating birds and mammals. The life cycles are similar for both species, with the procercoid stage occurring in freshwater copepods followed by the plerocercoid stage in the coelom of fishes (Fig 6.47). *Ligula* sp. plerocercoids have been reported from 49 species of freshwater host (mostly cyprinids) in the former USSR alone, and adults from 72 species of fish-eating avian definitive hosts (review in Hoole *et al.*, 2010). Adult *S. solidus* are reported from a wide range of fish-eating birds and mammals, but plerocercoids are specific to threespine stickleback, *Gasterosteus aculeatus*. Both species are perhaps best known for the enormous sizes that plerocercoids can reach relative to the sizes of their fish second intermediate hosts (Fig. 6.46; Color plate Fig. 7.3). Plerocercoids of these species may account for up to 20–50% of the biomass of their hosts. Early studies by J. D. Smyth on tapeworm growth and development utilized *Ligula* and *Schistocephalus* adults cultured *in vitro*. These classical studies provided valuable insight into platyhelminth physiology, but they also helped establish both species as experimental models for studies in phenotypic manipulation, population ecology, parasite/host–endocrine interactions, parasite-mediated natural selection, and environmental parasitology (Heins & Baker, 2008; Barber & Scharsack, 2010; Hoole *et al.*, 2010; see Chapters 12, 15–17).

6.9.5.3 Bothriocephalidea

All species of Bothriocephalideans are parasitic in fishes, amphibians, and reptiles. One well-known bothriocephalid is the Asian tapeworm *Bothriocephalus acheilognathi*, first described in Japan in the 1920s. Since that time, it has spread throughout the world, first into eastern Europe, then the United Kingdom and, by the mid 1970s, into North America, including Mexico. Adults have been reported from at least 40 different fish host species. Its procercoid larval stages have also been found in at least five different species of copepod intermediate hosts. The lack of host specificity, combined with the indiscriminate shipment of cultured fishes between countries, is probably responsible for its cosmopolitan distribution.

6.9.5.4 Haplobothriidea

The Haplobothriidea is a small group with only one genus and two species, but with some unique developmental characteristics. *Haplobothrium globuliformae* parasitizes the bowfin *Amia calva*, a primitive holostean fish in North America. The life cycle of *H. globuliformae* is similar in most respects to diphyllobothrideans. It includes a free-swimming coracidium, a procercoid within a cyclopoid copepod, a plerocercoid within bullheads (*Ameiurus* spp.), and an adult in the bowfin. Oddly, the plerocercoid contains spinose, retractable/protractile tentacles. Once the plerocercoid reaches the bowfin intestine, it strobilates. However, no sexual development occurs within proglottids of what is termed the primary strobila. Instead, these proglottids drop off from the primary strobila, form a typical bothriidean scolex at one end, and begin to develop a secondary strobila. Within each new proglottid in the secondary strobila, a complete set of male and female reproductive organs then develops. Haplobothriids thus possess two scolex types, the primary one with four tentacles resembling those of trypanorhynchs, and the secondary one that is spatulate, with paired bothria, resembling diphyllobothrideans. The reproductive system in proglottids of the secondary strobila morphologically is similar to that of a diphyllobothridean cestode (Fig. 6.44C).

6.9.5.5 Proteocephalidea

Proteocephalidean cestodes are commonly found in fishes, amphibians, and reptiles throughout the world. They resemble cyclophyllideans in having a bulbous scolex with four acetabula. Perhaps the best-studied member of the group in North America is *Proteocephalus ambloplitis*, the so-called bass tapeworm. Parenteric plerocercoids of *P. ambloplitis* can cause host castration when they penetrate the host's gonads, but also wander extensively within other organs and tissues of the visceral mass, causing substantial damage. These plerocercoids possess a terminal pore on the tip of the scolex through which they secrete histolytic enzymes that enable them to move through host tissues. The bass tapeworm has apparently spread from North America, where it was originally endemic, to other parts of the world where the largemouth bass *Micropterus salmoides* has been introduced.

6.9.5.6 Trypanorhyncha

The Trypanorhyncha is a large order of cestodes, the adults of which occur ubiquitously in the spiral intestines of elasmobranchs. In marine habitats, only the tetraphyllideans match this order in terms of numbers of species. Trypanorhynchs cause little harm to their larval or definitive hosts, although the flesh of intermediate hosts can be unsightly for commercial use. Nonetheless, members of this group have been studied extensively as models for parasite speciation and host–parasite coevolution (review in Olson *et al.*, 2010).

The scolices of these cestodes are remarkable for their elaborate four tentacles, each armed with large numbers of hooks, in addition to two or four bothridia (Fig. 6.48). Each tentacle can be retracted into an internal tentacle sheath. Two life-cycle patterns are characteristic. In some species, procercoids develop in crustaceans, plerocercoids in planktivorous fishes, and adults in elasmobranchs. *Nybelinia surmenicola*, for example, has a life cycle that includes euphausiid crustaceans as an intermediate host, squid and fish as paratenic hosts, and salmon sharks as the definitive host (Shimazu, 1999). In other species, the sequence of hosts involves a filter-feeding mollusc and an elasmobranch definitive host, or two teleost fishes and an elasmobranch. Trypanorhynch plerocercoids typically are surrounded by a fleshy capsule or vesicle called the **blastocyst**. A number of commercially important fishes serve as second intermediate or paratenic hosts. In many cases, the plerocercoids are located in the flesh and infected fishes must be discarded because of their 'wormy' condition, even though trypanorhynchs cannot infect humans. In temperate oceans of the Southern Hemisphere, species of *Hepatoxylon* and *Grillotia* are ubiquitous in teleost fishes. *Poecilancistrium caryophyllum* is common along the Gulf Coast of North America where the bull shark and related elasmobranchs are the definitive hosts, copepods are first intermediate hosts, and sea trout are the second intermediate hosts.

Fig. 6.48 Scanning electron micrograph of the scolex of the trypanorhynch cestode *Otobothrium* sp. from the spiral intestine of the sicklefin lemon shark, *Negaprion acutidens*. (Micrograph courtesy of Janine Caira.)

6.9.5.7 Tetraphyllidea

Tetraphyllideans are also exclusively parasitic in chondrichthyan fishes, primarily elasmobranchs, and have a cosmopolitan distribution. Almost all elasmobranchs examined to date are host to at least one species of tetraphyllidean. The shape of the scolex is complex, possessing highly modified bothridia (=acetabula) that may or may not include hooks, spines, and suckers (Figs. 6.49, 6.50). In tetraphyllideans, host specificity is remarkably high (Caira *et al.*, 2001). It appears that scolex morphology is responsible to some extent for this phenomenon, with the shape of the bothridia fitting perfectly among the folds of the mucosa in the spiral intestine of the host.

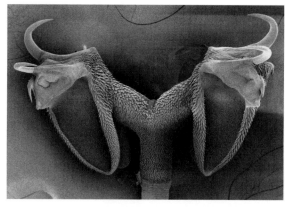

Fig. 6.49 Scanning electron micrograph of the scolex of the tetraphyllidean cestode *Yorkeria izardi* from the spiral intestine of the brownbanded bamboo shark, *Chiloscyllium punctatum*. (Micrograph courtesy of Janine Caira.)

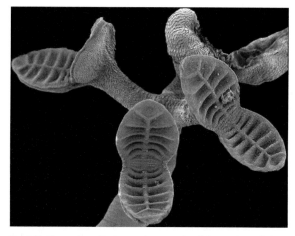

Fig. 6.51 Scanning electron micrograph of the scolex of the rhinebothriidean cestode *Rhinebothrium biorchidum* from the spiral intestine of the yellow stingray, *Urolophus jamaicensis*. (Micrograph courtesy of Claire Healy.)

Echeneibothrium maculatum, for example, occurs in the intestine of the ray *Raja montagui* but not in the closely related *R. naevus*. Similar 'lock-and-key' patterns between the morphology of the cestode's scolex and the host's mucosa have been described for other tetraphyllidean–elasmobranch associations.

6.9.5.8 Rhinebothriidea and Lecanthicepalidea

Healy *et al.* (2009) addressed, in part, the polyphyletic nature of the Tetraphyllidea, and erected a new order – the Rhinebothriidea (see also Caira *et al.*, 2001). The order includes at least 13 genera; most are strictly host specific and restricted to marine rays in the Batoidea, as well as freshwater stingrays. Intriguingly, rays have been shown to host one to four *Rhinebothrium* species and representatives of up to four rhinebothriidean genera concurrently. The rhinebothriidean scolices are also elaborate and consist of four unarmed, muscular bothridia (Fig. 6.51). The bothridia are borne on stalks, a feature that differentiates rhinebothriideans from other cestodes. Recently, a remarkable diversity of new lecanicephalidean tapeworm species has been discovered in Indo-Pacific eagle rays (Koch *et al.*, 2012).

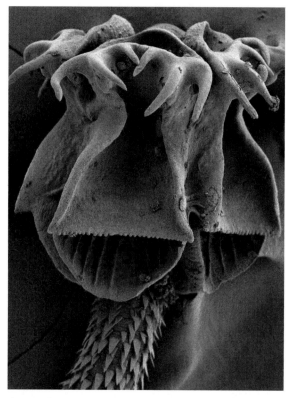

Fig. 6.50 Scanning electron micrograph of the scolex of the tetraphyllidean cestode *Phoreiobothrium* n. sp. from the spiral intestine of the bonnet head shark, *Sphyrna tiburo*. (Micrograph courtesy of Janine Caira.)

6.9.5.9 Cyclophyllidea

Species of Cyclophyllidea parasitize amphibians, reptiles, birds, and mammals throughout the world. Indeed, most of the tapeworm species found in birds and mammals belong to this order. Similarly, most of the species common in human and domesticated animals are cyclophyllideans. The scolices of cyclophyllidean cestodes are quite variable morphologically, but most are somewhat bulbous, possessing four suckers. A rostellum may or may not be present and, if present, may or may not be armed with hooks (Fig. 6.42).

Taeniidae

Taeniid cestodes are usually large in size, with the beef tapeworm of humans reaching 10 m. Most species have a rostellum that is not retractable. Some taeniids that infect humans can cause serious harm. The adults of *Taenia saginata* have about 10 000 eggs packed into each gravid proglottid when the intact proglottid is shed in the feces. On occasion, however, the free proglottid may actively crawl out through the anus. Flies and other insects can disperse eggs widely once the proglottid is shed and disintegrates (Lawson & Gemmell, 1990). On ingestion of an egg by an appropriate intermediate host, an active oncosphere emerges, penetrates the gut wall, gains access to the circulatory system, and is carried to skeletal muscle sites where it moves into the tissues and develops into a cysticercus. This bladder worm measures about 10 mm and is also called *Cysticercus bovis*, having been given both generic and species names long before the parasite's life cycle was worked out. When the larva is eaten by humans in rare or poorly cooked beef, the cycle is completed. The adult tapeworm is innocuous in humans from the standpoint of harm and little damage is inflicted by the cysticercus in infected cattle.

In contrast, *Taenia solium*, the pork tapeworm, is a dangerous human parasite. As an adult, it is about half the length of its bovine counterpart. Eggs are shed in gravid proglottids. On ingestion by hogs, the oncospheres emerge and penetrate the gut wall where they move into both cardiac and skeletal muscles before developing into cysticerci. The adult tapeworm develops in humans when they consume poorly cooked pork infected with cysticerci. The adult tapeworm in humans is benign. However, in many parts of the world where pork is a prime source of protein, and socio-economic conditions are poor, humans also are likely to ingest the eggs of *T. solium*. When this happens, the eggs hatch in the intestine and the oncospheres migrate via the circulatory system to subcutaneous sites as well as skeletal and cardiac muscle where they develop into cysticerci. If cysticerci lodge in the brain of humans the grave condition known as neurocysticerosis can result, triggering epilepsy-like symptoms.

Taenia taeniaeformis, *Taenia pisiformis*, and *Taenia crassiceps* are all cosmopolitan taeniids. *Taenia taeniaeformis* uses felines as definitive hosts, whereas the latter two occur in canines. The larval stage of *T. taeniaeformis* is a **strobilocercus**. This robust larva has a typical taeniid bladder at the posterior end and what appears to be a segmented neck, although there are no true proglottids. The intermediate hosts are rats where the larvae are confined to the livers. *Taenia pisiformis* is a common cestode in canines, with cysticerci occurring unencysted in the abdominal cavities of rabbits. *Taenia crassiceps* is unusual because the cysticerci bud exogenously in their microtine intermediate hosts. Typically, these larvae occur encysted in subcutaneous locations, but may also occur in the abdominal cavity. When inoculated by a syringe and needle into the abdominal cavity of a laboratory mouse, a single small bud will develop into an infective cysticercus that then begins budding exogenously from the region of the bladder. Budding is possible because the bladder wall possesses large populations of undifferentiated stem cells. Each daughter bud will develop into an infective cysticercus that, in turn, also produces new daughter buds. From the single bud initially inoculated, an infrapopulation may reach into the thousands before the host dies of thirst or starvation.

Tapeworms in the genus *Echinococcus* are the smallest of the taeniids, rarely exceeding 5 mm in total

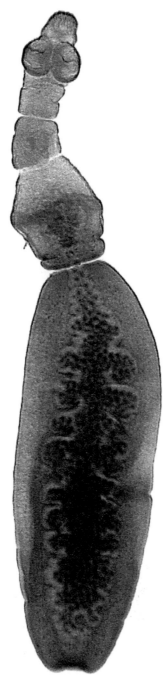

Fig. 6.52 The taeniid cestode *Echinococcus granulosus* from the intestine of a canid definitive host. Adults consist of a scolex and three proglottids. (Photograph courtesy of John Sullivan.)

length. They consist of only a scolex and an immature, mature and gravid proglottid (Fig. 6.52). The adults typically infect the gut of wild and domestic canines (secondarily felines) whereas the larvae develop within fluid-filled capsules in, or on, internal organs of a variety of herbivorous mammals (Color plate Fig. 1.3). A potentially lethal condition known as echinococcosis or hydatid disease refers to the pathological outcome induced by the capsules, or **hydatid cysts**, within intermediate hosts (Fig. 6.45). Thus, accidental ingestion of *Echinococcus* eggs passed in the feces of canine final hosts may lead to complications imposed by the developing hydatid cyst in a variety of herbivorous hosts, as well as humans.

Adults of *E. granulosus* pass gravid proglottids in the canine host feces. In the wild, or sylvatic cycle, the eggs are accidentally consumed by grazing herbivores (review in Thompson, 1986). The oncospheres emerge from the eggs, penetrate the gut wall, and are picked up in the circulation. They exit in the capillary beds of various organs of the body, primarily the liver and lung, where they develop into hydatid cysts (Color plate Fig. 1.3). A single hydatid cyst may reach 50 cm in diameter, and contain several liters of fluid, along with up to a million protoscolices. The structure of the hydatid cyst wall is laminar. The outermost layer is derived from host tissue. Immediately inside is a so-called laminated layer of parasite origin, and inside that is a very thin, germinal membrane. From the germinal membrane, the parasite produces buds that will drop off as brood capsules and float free as 'hydatid sand' in the hydatid fluid (Fig. 6.53). Within the brood capsules the infective protoscolices develop with invaginated scolices. Note that in this species, budding can occur only endogenously. The life cycle is completed when organs or tissues containing the hydatid cyst are consumed by the canine definitive host.

The tapeworm becomes dangerous for humans when the normal sylvatic cycle is interrupted by the domesticated dog in which the adult parasite can develop, or by sheep, cattle, hogs, or horses, where the hydatid cyst can develop, usually in the liver or lungs. People in sheep-herding areas of the world have a

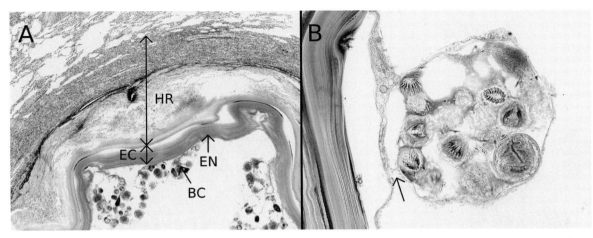

Fig. 6.53 Hydatid cyst of *Echinococcus granulosus*. (A) Histological section of the wall of a unilocular cyst. Hanging into the fluid-filled cavity are brood capsules (BC), composed of germinal membrane and containing protoscolices. The inner germinal layer of the cyst, the endocyst (EN), is surrounded by an acellular ectocyst (EC). A thick fibrous tissue host response (HR) surrounds the ectocyst. (B) A single brood capsule, attached to the germinal epithelium by a stalk (arrow). The outer margin of the brood capsule consists of germinal membrane. Protoscolices bearing hooks are clearly visible. (Photographs courtesy of John Sullivan.)

higher prevalence of hydatid disease because of close contact with their working dogs that, in turn, acquired the tapeworm from eating hydatid cysts in sheep offal (Fig. 6.54). If the eggs are accidentally ingested by humans, then hydatid cysts will develop in the liver, lungs, or brain, reaching the same size as those in hervivore intermediate hosts. In addition, the parasites are long-lived in their intermediate hosts. The long life of the parasite means that the host is exposed for extended periods to small, but persistent, quantities of parasite antigens that begin to accumulate in the hydatid fluid. If the hydatid cyst ruptures, the human host is suddenly exposed to large quantities of antigen to which it has developed a strong hypersensitivity over a long period of time. The result can be classic anaphylactic shock and rapid death. Currently, at least nine strains of *E. granulosus* are recognized (Thompson & McManus, 2002).

Also dangerous is *E. multilocularis*. The sylvatic cycle of this parasite typically involves species of microtine rodents (*Microtus* spp.) as intermediate hosts and foxes as the final hosts. The sylvatic cycle is broken when humans accidentally consume eggs passed by their domesticated dogs that have become infected by eating parasitized rodents. The hydatid cysts typically develop in the liver, lungs, and abdominal cavity. These hydatid cysts lack a laminated layer. Apparently for this reason, the cysts are able to bud exogenously, or **metastasize**, and are referred to as multilocular or alveolar cysts. Via the exogenous budding, they spread from the initial site of infection into other locations.

Mesocestoididae

The mesocestoidid cestodes are an unusual group of cyclophyllideans. They are typically found as adults in mammals such as opossums and raccoons, and birds. The scolex has four acetabula and lacks a rostellum. The larval stage is called a **tetrathyridium** and is found in the body cavities of reptiles and rodents. It is assumed that these larvae are in the second intermediate host because the hosts cannot be infected with eggs and the larva can be used to experimentally produce adults in laboratory animals. The first intermediate host is not known. Moreover, adults of some *Mesocestoides* species can undergo asexual reproduction in the definitive host when the scolex divides longitudinally (Smyth, 1987).

Fig. 6.54 The two-host life cycle of *Echinococcus granulosus* and the epidemiology of hydatid disease. Adults mature in the intestine of dogs. Embryonated eggs are passed in the feces and ingested by sheep intermediate hosts. Oncospheres penetrate the gut wall and migrate to the lungs and/or liver, forming hydatid cysts. Brood capsules containing protoscolices develop within the hydatid cysts. Dogs become infected if they have access to discarded sheep carcasses. Humans are at risk of acquiring hydatid disease if they accidentally ingest eggs from infected dogs. (Artwork courtesy of Lisa Esch McCall.)

Anoplocephalidae

As adults, cestodes of the cosmopolitan Anoplocephalidae occur in herbivorous mammals, including several of veterinary importance such as sheep and horses. They always possess two complete sets of reproductive organs in each proglottid. In some species, the adults measure 6 m in length. Commonly occurring in sheep-raising countries in the world are *Moniezia expansa* and *M. benedeni*. Eggs of both of these species infect oribatid (soil) mites where cysticercoids develop in the hemocoel. Sheep are infected when they accidentally ingest infected mites while grazing. Another anoplocephalid tapeworm of veterinary importance is *Anoplocephala perfoliata*

(Fig. 6.55). This is the most common cestode of domestic horses worldwide. The parasite is site-specific for the ileocecal junction of the intestine (Color plate Fig. 1.4). From a clinical disease perspective, the parasite has been implicated as a causative agent of colic, the most significant cause of mortality in horses. Moreover, a variety of other pathological manifestations, including cecal perforation leading to peritonitis, and intestinal obstruction have been reported (Gasser *et al.*, 2005).

Hymenolepididae

Hymenolepidid cestodes infect a wide range of birds and mammals throughout the world.

Fig. 6.55 Scanning electron micrograph of the scolex region of the anoplocephalid cestode *Anoplocephala perfoliata* from the ileocecal region of a horse. (Micrograph courtesy of Doug Colwell.)

Arthropods are obligate intermediate hosts for all species except *Vampirolepis nana* (formerly known as *Hymenolepis nana*). *Hymenolepis diminuta* is probably the most widely studied of all tapeworm species, primarily because of the ease with which it can be maintained in the laboratory and its relatively large size (up to 90 cm). Rodents, primarily rats, serve as the definitive hosts and beetles, mostly species of *Tenebrio* or *Tribolium*, are the intermediate hosts where cysticercoids develop within the hemocoel. *Vampirolepis nana* is the so-called dwarf tapeworm because of the diminutive size of its strobila, on average about 30 mm in length. It has a cosmopolitan distribution, using rodents and sometimes humans as the definitive host and grain beetles as intermediate hosts. An unusual feature of this parasite is its ability to bypass the arthropod intermediate host and complete development from an egg to the adult stage in the definitive host, a form of internal autoinfection. If eggs are ingested by a definitive host, they hatch within the intestine, develop into cysticercoids between villi of the mucosa, emerge into the lumen again, and develop rapidly into adults. Humans may become infected, with prevalences as high as 25% in endemic areas.

Dilepididae

The dilepidid cestodes infect a range of birds and mammals. One of the most widespread members of the family is *Dipylidium caninum*, a parasite of dogs, cats, and occasionally humans. It possesses a retractable rostellum, with several rows of hooks. There are two sets of reproductive organs in each proglottid. Early in its development, the uterus disappears and the eggs become enclosed within packets. The gravid proglottids are very active and can frequently be seen exiting the anus, then migrating in the fur in the perianal region of an infected canine. When the proglottids dry, the freed egg packets, which resemble tiny grains of rice, will be consumed by fleas, where cysticercoids develop in the hemocoel. When a flea bites, so does the dog, and, in the process, becomes infected when it ingests the infected flea while grooming. Humans, particularly children, can become infected via the ingestion of fleas.

6.10 Phylogenetic relationships and classification

The evolutionary history of the Platyhelminthes has not been completely deciphered. Evolutionary studies within the group, using morphological and molecular data, combined with cladistic methods, show significant concordance. These studies agree, for the most part, on a phylogenetic tree (Fig. 6.56) in which the free-living, commensal, and mutualistic groups of platyhelminths, e.g., Catenulida, Macrostomatida, Polycladida, etc., occupy a basal position, whereas the parasitic groups, i.e., Aspidobothrea, Digenea, Monogenea, Gyrocotylidea,

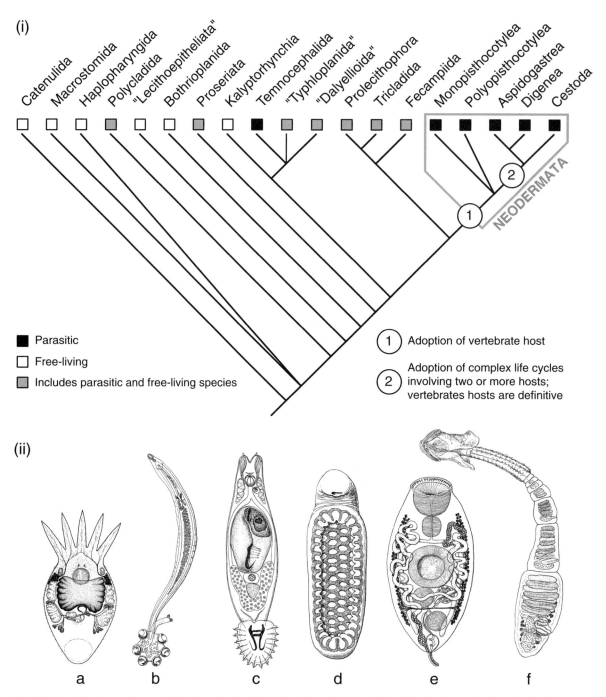

Fig. 6.56 Parasitism has arisen multiple times in the Platyhelminthes, but the origin of the Neodermata, characterized by free-living forms replacing their epidermis with a syncitial neodermis and the adoption of vertebrate hosts, represents perhaps the most successful innovation (measured by species richness) of any flatworm radiation. (i) Estimate of phylogenetic relationships among major platyhelminth groups based on various, predominantly molecular, studies. Unresolved nodes represent conflict between data sets. Temnocephalans are principally parasites of freshwater crustaceans such as crayfish.

Amphilinidea, and Eucestoda, occupy terminal positions.

All these parasitic groups, collectively now called Neodermata (see Fig. 6.56), resemble each other in structural and developmental aspects of their epidermal cilia, sensory receptors, flame cells, and sperm. Moreover, the epidermis of the larva is replaced in the adult by a syncytial **neodermis**, or 'new skin,' a characteristic feature that led to the collective name of Neodermata.

The phylogenetic relationships within the groups of parasitic Platyhelminthes have received much attention in the last couple of decades, both from the morphological and molecular points of view (Ehlers, 1985; 1986; Boeger & Kritsky, 1993; Brooks & McLennan, 1993; Rohde et al., 1993; Hoberg et al., 1997; Justine, 1998; Mariaux, 1998; Hoberg et al., 1999; Littlewood et al., 1999a, b; Hoberg et al., 2001; Baguñà & Riutort, 2004; Olson & Tkach, 2005; Waeschenbach et al., 2007; Kuchta et al., 2008; Littlewood, 2008; Olson et al., 2008). The abundance of characters generated by combining morphological and molecular data has allowed researchers to propose a more evolutionarily sound classification of the Platyhelminthes, especially of the parasitic forms (see Box 6.3). The phylogenetic relationships of the Temnocephalida and Udonellida are not yet completely resolved, although molecular evidence indicates that the udonellids belong in the Monogenea, specifically among the monopisthocotyleans (Littlewood et al., 1998). The Digenea are closely related to the Aspidogastrea and probably evolved from a common ancestor (Fig. 6.56). Although the Monogenea have been historically treated as closely related to the Digenea

and Aspidobothrea, most researchers now consider the Monogenea to be the sister group of the Cestodaria, the clade that includes the Gyrocotylidea, Amphilinidea, and Eucestoda. Still others do not consider the Monogenea as monophyletic.

Current phylogenetic and evolutionary studies of Platyhelminthes are not restricted to the major groups discussed above. A plethora of studies has been devoted to phylogenetic analyses of families and genera, mainly within the Digenea, Monogenea, and Eucestoda. Their detail is beyond the scope of this book. Interested readers should consult the primary literature, most of which has been summarized up to 1993 by Brooks & McLennan (1993). Several of the more recent works include citations in the preceding paragraph.

A basic principle of phylogenetics is that the approach hinges on hypotheses that are open to continuous testing, modification, and refinement. Recent phylogenetic studies of Platyhelminthes and other metazoans using 18S ribosomal DNA indicate that the Acoela, one of the most basal and primitive groups previously included in the Platyhelminthes are not platyhelminths (Ruiz-Trillo et al., 1999). They seem to be positioned even earlier in the evolutionary tree, representing one of the earliest branches within the bilateral animals with living descendants. The Acoela, then, are a living relic of the transition between radially symmetrical animals such as jellyfish, and the more complex bilateral organisms. These findings indicate that the Acoela should be placed in their own phylum, probably evolving just prior to the Platyhelminthes.

Caption for Fig. 6.56 (cont.)
Neodermatans use vertebrates as definitive hosts with trematodes (Aspidogastrea + Digenea) and tapeworms (Cestoda) involving various other phyla as intermediate hosts to complete their life cycles. (ii) Representative obligate parasitic flatworm genera: (a) temnocephalan *Temnocephala*; (b) polyopisthocotylean monogenean *Rajaonchotyle*; (c) monopisthocotylean monogenean *Gyrodactylus*; (d) aspidogastrean *Lophataspis*; (e) digenean *Urogonimus*; (f) cestode *Echinobothrium*. (Figure courtesy of Janine Caira & Tim Littlewood; modified from Caira & Littlewood, 2012, *Encyclopedia of Biodiversity*, 2nd edition, ed. S.A. Levin. New York: Academic Press.)

Box 6.3 | **Classification of the Platyhelminthes**

The classification of the Platyhelminthes presented here focuses on the ectocommensal and obligate parasitic taxa discussed in this chapter. This classification scheme is, for the most part, the one adopted in most current invertebrate zoology and parasitology texts, as well as the reviews by Littlewood *et al.* (1999a, b), Olson *et al.* (2003), Baguñà & Riutort (2004), Olson & Tkach (2005), Kuchta *et al.* (2008), and Littlewood (2008). We have tried to make the scheme as simple as possible. For those seeking more details of these interrelationships we recommend the above reviews and, especially, the edited volume by Littlewood & Bray (2001).

We should emphasize that since the publication of the first edition of this book, significant changes in classification have been made for the parasitic platyhelminth taxa, with most of these modifications based on extensive molecular phylogenetic studies. For example, Brooks & McLennan (1993) proposed the existence of 10 orders of digenetic trematodes. Currently, three are recognized: the Strigeida, Echinostomida and the Plagiorchiida (see Gibson *et al.*, 2002; Jones *et al.*, 2005; Bray *et al.*, 2008).

The proposed taxonomic system for tapeworms in the first edition was that of Hoberg *et al.* (1997). Several molecular studies have been undertaken for the Eucestoda since their contribution. The number of orders proposed by Hoberg *et al.* (1997) was 12. The systematic classification scheme of Waeschenbach *et al.* (2007) includes 14. Kuchta *et al.* (2008) have revised the order Pseudophyllidea, dividing it into the Diphyllobothriidea and Bothriocephalidea, based in part on molecular evidence provided by Brabec *et al.* (2006) and Waeschenbach *et al.* (2007). Additional morphological and life cycle differences are persuasive in the separation of the old order Pseudophyllidea. Kuchta *et al.*'s scheme places the new order Diphyllobothriidea closest to its sister group, the Haplobothriidea, which possesses a single genus and two species that infect the relict bowfin, *Amia calva*. The polyphyletic nature of the Tetraphyllidea has been partially resolved with the establishment of the new order Rhinebothriidea (Healy *et al.*, 2009), for a total of 15 eucestode orders.

For the Monogenea, we adopt the nine-order classification system proposed by Boeger & Kritsky (1993) and Olson & Littlewood (2002).

Please note that due to space constraints we have only included here some of the genera within the parasitic platyhelminth orders mentioned within this chapter.

Class Rhabdocoela
  Order Neorhabdocoela
    Suborder Temnocephalida
      Representative genus: *Temnocephala*
Neodermata
Class Trematoda
  Subclass Aspidobothrea (=Aspidogastrea)

Box 6.3 | (continued)

Order Aspidobothrida
  Representative genera: *Aspidogaster, Lobatostoma*
Subclass Digenea
  Order Echinostomida
    Representative genera: *Echinostoma, Fasciola, Fascioloides, Megalodiscus,*
      Paramphistomum, Ribeiroia
  Order Strigeida
    Representative genera: *Alaria, Azygia, Brachylaima, Cardicola, Clinostoma,*
      Diplostomum, Halipegus, Gigantobilharzia, Leucochloridium, Schistosoma,
      Transversotrema, Trichobilharzia
  Order Plagiorchiida
    Representative genera: *Brachylecithum, Crepidostomum, Clonorchis, Collyriclum,*
      Dicrocoelium, Haematoloechus, Heterophyes, Metagonimus, Nanophyetus,
      Paragonimus, Plagiorchoris, Prosthogonimus
Class Monogenea (Monogenoidea, of some authors)
  Order Dactylogyridea
    Representative genera: *Dactylogyrus*
  Order Gyrodactylidea
    Representative genera: *Gryodactylus,*
      Gyrodactyloides
  Order Capsalidea
    Representative genera: *Neobenedenia, Entobdella*
  Order Lagarocotylidea
  Order Monocotylidea
  Order Montchadskyellidea
  Order Polystomatidea
    Representative genera: *Polystoma, Pseudodiplorchis, Sphyranura*
  Order Mazocraeidea
    Representative genus: *Microcotyle*
  Order Diclybothriidea
  Order Chimaericolidea
Class Cestoda (Cestoidea, of some authors)

Box 6.3 | (continued)

Subclass Cestodaria
  Gyrocotylidea
    Representative genus: *Gyrocotyle*
  Amphilinidea
    Representative genera: *Amphilina,*
      Austramphilina
Subclass Eucestoda
  Order Caryophyllidea
    Representative genera: *Archigetes,*
      Caryophyllaeus
  Order Diphyllidea
  Order Lecanicephalidea
  Order Nippotaenidea
  Order Onchobothriidea
  Order Phyllobothriidea
  Order Proteocephalidea
    Representative genus: *Proteocephalus*
  Order Diphyllobothriidea
    Representative genera: *Diphyllobothrium, Ligula, Schistocephalus*
  Order Bothriocephalidea
    Representative genus: *Bothriocephalus*
  Order Haplobothriidea
    Representative genus: *Haplobothrium*
  Order Spathebothriidea
  Order Trypanorhyncha
    Representative genera: *Hepatoxylon, Grillotia, Nybelinia, Poecilancistrium*
  Order Tetraphyllidea
    Representative genera: *Acanthobothrium, Echeneibothrium*
  Order Rhinebothriidea
    Representative genus: *Rhinebothrium*
  Order Cyclophyllidea
    Representative genera: *Anoplocephala, Dipylidium, Echinococcus, Hymenolepis,*
    Moniezia, Mesocestoides, Taenia, Taeniarhynchus, Vampirolepis

References

Aken'Ova T. O. & Lester, R. J. G. (1996) *Udonella myliobati* n. comb. (Platyhelminthes: Udonellidae) and its occurrence in Australia. *Journal of Parasitology*, 82, 1017–1023.

Avenant-Oldewage, A. (1993) Occurrence of *Temnocephala chaeropsis* on *Cherax tenuimanus* imported into South Africa and notes on its infestation of an indigenous crab. *South African Journal of Science*, 89, 427–428.

Baguñà, J. & Riutort, M. (2004) Molecular phylogeny of the Platyhelminthes. *Canadian Journal of Zoology*, 82, 168–193.

Bakke, T. A., Cable, J. & Harris, P. D. (2007) The biology of gyrodactylid monogeneans: the "Russian-doll killers." *Advances in Parasitology*, 64, 161–376.

Barber, I. & Scharsack, J. P. (2010) The three-spined stickleback–*Schistocephalus solidus* system: an experimental model for investigating host–parasite interactions in fish. *Parasitology*, 137, 411–424.

Barger, M. & Esch, G. W. (2000) *Plagioporus sinitsini* (Digenea: Opecoelidae): a one-host life cycle. *Journal of Parasitology*, 86, 150–153.

Blair, D. (1993) The phylogenetic position of the Aspidobothrea within the parasitic flatworms inferred from ribosomal RNA sequence data. *International Journal for Parasitology*, 23, 169–178.

Blankespoor, H. D., Esch, G. W. & Johnson, W. C. (1985) Some observations on the biology of *Collyriclum faba* (Bremser in Schmalz, 1831). *Journal of Parasitology*, 71, 469–471.

Boeger, W. A. & Kritsky, D. C. (1993) Phylogeny and a revised classification of the Monogenoidea Bychowsky, 1937 (Platyhelminthes). *Systematic Parasitology*, 26, 1–32.

Brabec, J., Kuchta, R. & Scholz, T. (2006) Paraphyly of the Pseudophyllidea (Platyhelminthes: Cestoda): circumscription of monophyletic clades based on phylogenetic analysis of ribosomal RNA. *International Journal for Parasitology*, 36, 1535–1541.

Bray, R. A., Gibson, D. I. & Jones, A. (2008) *Keys to the Trematoda*, vol. 3. Wallingford: CAB International & The Natural History Museum, London, UK.

Brooks, D. R. & McLennan, D. A. (1993) *Parascript, Parasites and the Language of Evolution*. Washington, D. C.: Smithsonian Institution Press.

Byrnes, T. (1986) Five species of Monogenea from Australian bream, *Acanthopagrus* spp. *Australian Journal of Zoology*, 34, 65–86.

Caira, J. N., Jensen, K. & Healy, C. J. (2001) Interrelationships among tetraphyllidean and lecanicephalidean cestodes.

In *Interrelationships of the Platyhelminthes*, ed. D. T. J. Littlewood & R. Bray, pp. 135–158. London: Taylor & Francis.

Calentine, R. L. (1962) *Archigetes iowensis* n. sp. (Cestoda: Caryophyllaeidae) from *Cyprinus carpio* L. and *Limnodrilus hoffmeisteri* Claparede. *Journal of Parasitology*, 48, 513–524.

Combes, C. (2001) *Parasitism: The Ecology and Evolution of Intimate Interactions*. Chicago: University of Chicago Press.

Combes, C., Fournier, A., Moné, H. & Théron, A. (1994) Behaviors in trematode cercariae that enhance parasite transmission: patterns and processes. *Parasitology*, 109, S3–S13.

Cribb, T. H., Bray, R. A., Olson, P. D. & Littlewood D. T. J. (2003) Life cycle evolution in the Digenea: a new perspective from phylogeny. *Advances in Parasitology*, 54, 197–254.

Dalton, J. P., Skelly, P. & Halton, D. W. (2004) Role of the tegument and gut in nutrient uptake by parasitic platyhelminths. *Canadian Journal of Zoology*, 82, 211–232.

Damian, R. T. (1987) The exploitation of immune responses by parasites: presidential address. *Journal of Parasitology*, 73, 1–13.

Doenhoff, M. J. (1997) A role for granulomatous inflammation in the transmission of infectious disease: schistosomiasis and tuberculosis. *Parasitology*, 115, 113–125.

Doenhoff, M. J. (2006) Subversion of the immune response by helminths. In *Microbial Subversion of Immunity: Current Topics*, ed. P. J. Lachmann & M. B. A Oldstone, pp. 107–176. Norwich: Caister Academic Press.

Ehlers, U. (1985) Phylogenetic relationships among the Platyhelminthes. In *The Origins and Relationships of Lower Invertebrates*, ed. C. Morris, J. D. George, R. Gibson & H. M. Platt, pp. 143–158. Oxford: Oxford University Press.

Ehlers, U. (1986) Comments on a phylogenetic system of Platyhelminthes. *Hydrobiologia*, 132, 1–12.

Erasmus, D. (1972) *The Biology of Trematodes*. London: Edward Arnold.

Esch, G. W., Barger, M. A. & Fellis, K. J. (2002) The transmission of digenetic trematodes: style, elegance, complexity. *Integrative and Comparative Biology*, 42, 304–312.

Fenwick, A., Rollison, D. & Southgate, V. (2006) Implementation of human schistosomiasis control: challenges and prospects. *Advances in Parasitology*, 61, 567–622.

Fried, B. & Graczyk, T. K. (1997) *Advances in Trematode Biology*. New York: CRC Press.

Fried, B. & Graczyk, T. K. (2000) *Echinostomes as Experimental Models for Biological Research.* Norwell: Kluwer Academic Publishers.

Fujino, T. (1993) Ultrastructure and function of alimentary systems in parasitic helminths. *Japanese Journal of Parasitology*, **42**, 277–294.

Gasser, R. B., Williamson, R. M. C. & Beveridge, I. (2005) *Anoplocephala perfoliata* of horses – significant scope for further research, improved diagnosis and control. *Parasitology*, **131**, 1–13.

Gibson, D. I. & Bray, R. A. (1979) The Hemiuroidea: terminology, systematics and evolution. *Bulletin of the British Museum (Natural History), Zoology Series*, **36**, 35–146.

Gibson, D. I., Jones, A. & Bray, R. A. (2002) *Keys to the Trematoda*, vol. 1. Wallingford: CAB International & The Natural History Museum, London, UK.

Gobert, G. N., Stenzel, D. J., McManus, D. P. & Jones, M. K. (2003) The ultrastructural architecture of the adult *Schistosoma japonicum* tegument. *International Journal for Parasitology*, **33**, 1561–1575.

Haas, W. & Haberl, B. (1997) Host recognition by trematode miracidia and cercaria. In *Advances in Trematode Biology*, ed. B. Fried & T. K. Graczyk, pp. 197–228. New York: CRC Press.

Haas, W., Haberl, B., Kalbe, M. & Korner, M. (1995) Snail-host finding by miracidia and cercariae: chemical host cues. *Parasitology Today*, **11**, 468–472.

Halton, D. W. (1978) Trans-tegumental absorption of L-alanine and L-leucine by a monogenean, *Diclidophora merlangi. Parasitology*, **76**, 29–37.

Halton, D. W. & Gustafsson, M. K. S. (1996) Functional morphology of the platyhelminth nervous system. *Parasitology*, **113**, S47–S72.

Halton, D. W., Maule, A. G. & Shaw, C. (1997) Trematode neurobiology. In *Advances in Trematode Biology*, ed. B. Fried & T. K. Graczyk, pp. 345–382. New York: CRC Press.

Healy, C. J., Caira, J. N., Jensen, K., *et al.* (2009) Proposal for a new tapeworm order, Rhinebothriidea. *International Journal for Parasitology*, **39**, 497–511.

Heins, D. C. & Baker, J. A. (2008) The stickleback–*Schistocephalus* host–parasite system as a model for understanding the effect of a macroparasite on host reproduction. *Behaviour*, **145**, 625–645.

Hendrix, S. S., Vidrine, M. F. & Hantenstine, R. H. (1985) A list of records of freshwater aspidogastrids (Trematoda) and their hosts in North America. *Proceedings of the Helminthological Society of Washington*, **52**, 289–296.

Hoberg, E. P., Mariaux, J., Justine, J.-L., *et al.* (1997) Phylogeny of the orders of the Eucestoda (Cercomeromorphae) based on comparative morphology:

historical perspectives and a new working hypothesis. *Journal of Parasitology*, **83**, 1128–1147.

Hoberg, E. P., Gardner, S. L. & Campbell, R. A. (1999) Systematics of the Eucestoda: advances towards a new phylogenetic paradigm, and observations on the early diversification of tapeworms and vertebrates. *Systematic Parasitology*, **42**, 1–12.

Hoberg, E. P., Mariaux, J. & Brooks, D. R. (2001) Phylogeny among orders of the Eucestoda (Cercomeromorphae): integrating morphology, molecules and total evidence. In *Interrelationships of the Platyhelminthes*, ed. D. T. J. Littlewood & R. A. Bray, pp. 112–126. London: Taylor & Francis.

Hoole, D., Carter V. & Dufour, S. (2010) *Ligula intestinalis* (Cestoda: Pseudophyllidea): an ideal fish–metazoan parasite model? *Parasitology*, **137**, 425–438.

Horak, P. & Kolarova, L. (2001) Bird schistosomes: do they die in mammalian skin? *Trends in Parasitology*, **17**, 66–69.

Horak, P., Kolarova, L. & Adema, C. M. (2002) Biology of the schistosome genus *Trichobilharzia. Advances in Parasitology*, **52**, 155–233.

Hughes, R. N. (1989) *A Functional Biology of Clonal Animals.* London: Chapman & Hall.

Johnson, P. T. J. & Sutherland, D. R. (2003) Amphibian deformities and *Ribeiroia* infection: an emerging helminthiasis. *Trends in Parasitology*, **19**, 332–335.

Johnson, P. T. J., Lunde, K. B., Ritchie, E. G. & Launer, A. E. (1999) The effect of trematode infection on amphibian limb development and survivorship. *Science*, **284**, 802–804.

Johnson, P. T. J., Lunde, K. B, Thurman, E. M., *et al.* (2002) Parasite (*Ribeiroia ondatrae*) infection linked to amphibian malformations in the western United States. *Ecological Monographs*, **72**, 151–168.

Jones, M. K. (1998) Structure and diversity of cestode epithelia. *International Journal for Parasitology*, **28**, 913–923.

Jones, A., Bray, R. A. & Gibson, D. I. (2005) *Keys to the Trematoda*, vol. 2. Wallingford: CAB International & The Natural History Museum, London, UK.

Jordan, P. (2000) From Katayama to the Dakhla Oasis: the beginning of epidemiology and control of bilharzia. *Acta Tropica*, **77**, 9–40.

Justine, J.-L. (1998) Spermatozoa as phylogenetic characters for the Eucestoda. *Journal of Parasitology*, **84**, 385–408.

Kabata, Z. (1973) Distribution of *Udonella caligorum* Johnston, 1835 (Monogenea: Udonellidae) on *Caligus elongatus* Nordman, 1932 (Copepoda: Caligidae). *Journal of the Fisheries Research Board of Canada*, **30**, 1793–1798.

Kalbe, M., Haberl, B. & Haas, W. (1996) *Schistosoma mansoni* miracidial host-finding: species-specificity of an Egyptian strain. *Parasitology Research*, **82**, 8–13.

Karanja, D. M. S., Colley, D. G., Nahlen, B. L., *et al.* (1997) Studies on schistosomiasis in western Kenya.

1. Evidence for immune-facilitated excretion of schistosome eggs from patients with *Schistosoma mansoni* and human immunodeficiency virus coinfections. *American Journal of Tropical Medicine and Hygiene*, 56, 515–521.

Kearn, G. C. (1971) The physiology and behavior of the monogenean skin parasite Entobdella soleae in relation to its host (Solea solea). In *Ecology and Physiology of Parasites*, ed. A. M. Fallis, pp. 161–187. Toronto: University of Toronto Press.

Koch, K. R., Jensen, K. & Caira, J. N. (2012) Three new genera and six new species of lecanicephalideans (Cestoda) from eagle rays of the genus *Aetomylaeus* (Myliobatiformes: Mylobatidae) from northern Australia and Borneo. *Journal of Parasitology*, 98, 175–198.

Králová-Hromadová, I., Bazsalovicsová, E., Štefka, J., *et al.* (2011) Multiple origins of European populations of the giant liver fluke *Fascioloides magna* (Trematoda: Fasciolidae), a liver parasite of ruminants. *International Journal for Parasitology*, 41, 373–383.

Kuchta, R., Scholz, T., Brabec, J. & Bray, R. A. (2008) Suppression of the tapeworm order Pseudophyllidea (Platyhelminthes: Eucestoda) and the proposal of two new orders, Bothriocephalidea and Diphyllobothriidea. *International Journal for Parasitology*, 38, 49–55.

Kusel, J. R., Al-Adhami, B. H. & Doenhoff, M. J. (2007) The schistosome in the mammalian host: understanding the mechanisms of adaptation. *Parasitology*, 134, 1477–1526.

Lawson, J. R. & Gemmell, M. A. (1990) Transmission of taeniid eggs via blowflies to intermediate hosts. *Parasitology*, 100, 143–146.

Leighton, B. J., Zervos, S. & Webster, J. M. (2000) Ecological factors in schistosome transmission, and an environmentally benign method for controlling snails in a recreational lake with a record of schistosome dermatitis. *Parasitology International*, 49, 9–17.

Littlewood, D. T. J. (2008) Platyhelminth systematics and the emergence of new characters. *Parasite*, 15, 333–341.

Littlewood, D. T. J. & Bray, R. A. (2001) *Interrelationships of the Platyhelminthes*. London: Taylor & Francis.

Littlewood, D. T. J., Rhode, K. & Clough, K. A. (1998) The phylogenetic position of *Udonella* (Platyhelminthes). *International Journal for Parasitology*, 28, 1241–1250.

Littlewood, D. T. J., Rohde, K., Bray, R. A., & Herniou, E. (1999a) Platyhelminthes and the evolution of parasitism. *Biological Journal of the Linnaean Society*, 68, 257–287.

Littlewood, D. T. J., Rohde, K & Clough, K. A. (1999b) The interrelationships of all major groups of Platyhelminthes: phylogenetic evidence from morphology and molecules. *Biological Journal of the Linnean Society*, 66, 75–114.

Mackiewicz, J. S. (2003). Caryophyllaeidae: molecules, morphology, and evolution. *Acta Parasitologica*, 48, 143–154.

Mariaux, J. (1998) A molecular phylogeny of the Eucestoda. *Journal of Parasitology*, 84, 114–124.

Matisz, C. E. & Goater, C. P. (2010) Migration, site selection, and development of *Ornithodiplostomum* sp. (Trematoda: Digenea) metacercariae in fathead minnows (*Pimephales promelas*). *International Journal for Parasitology*, 40, 1489–1496.

Matisz, C. E., Goater, C. P. & Bray, D. (2010) Migration and site selection of *Ornithodiplostomum ptychocheilus* (Trematoda: Digenea) metacercariae in the brain of fathead minnows (*Pimephales promelas*). *Parasitology*, 137, 719–731.

Millemann, R. E. & Knapp, S. E. (1970) Biology of *Nanophyetus salmincola* and salmon poisoning disease. *Advances in Parasitology*, 8, 1–41.

Mousley, A., Maule, A. G., Halton, D. W. & Marks, N. J. (2005) Inter-phyla studies on neuropeptides: the potential for broad-spectrum anthelmintic and/or endectocide discovery. *Parasitology*, 131, 143–167.

Murray, D. L., Cox, E. W., Ballard, W. B., *et al.* (2006) Pathogens, nutritional deficiency, and climate change influences on a declining moose population. *Wildlife Monographs*, 166, 1–30.

Olson, P. D. & Littlewood, D. T. J. (2002) Phylogenetics of the Monogenea: evidence from a medley of molecules. *International Journal for Parasitology*, 32, 233–244.

Olson, P. D. & Tkach, V. V. (2005) Advances and trends in the molecular systematics of the parasitic Platyhelminthes. *Advances in Parasitology*, 60, 165–243.

Olson, P. D., Cribb, T. H., Tkach, V. V., *et al.* (2003) Phylogeny and classification of the Digenea (Platyhelminthes: Trematoda). *International Journal for Parasitology*, 33, 733–755.

Olson, P. D., Poddubnaya, L. G., Littlewood, D. T. J. & Scholz, T. (2008) On the position of *Archigetes* and its bearing on the early evolution of tapeworms. *Journal of Parasitology*, 94, 898–904.

Olson, P. D., Caira, J. N., Jensen, K., *et al.* (2010) Evolution of the trypanorhynch tapeworms: Parasite phylogeny supports independent lineages of sharks and rays. *International Journal for Parasitology*, 40, 223–242.

Pappas, P. W. (1988) The relative roles of the intestines and external surfaces in the nutrition of monogeneans, digeneans and nematodes. *Parasitology*, 96, S105–S121.

Pappas, P. W. & Read, C. P. (1975) Membrane transport in helminth parasites. *Experimental Parasitology*, 37, 469–530.

Parker, G. A., Chubb, J. C., Ball, M. A. & Roberts, G. N. (2003) Evolution of complex life cycles in helminth parasites. *Nature*, **425**, 480–484.

Platt, T. R., & Brooks, D. R. (1997) Evolution of the schistosomes (Digenea: Schistosomatoidea): the origin of dioecy and colonization of the venous system. *Journal of Parasitology*, **83**, 1035–1044.

Poddubnaya L. G., Scholz T., Kuchta R., *et al.* (2007) Ultrastructure of the proglottid tegument (neodermis) of the cestode *Echinophallus wageneri* (Pseudophyllidea: Echinophallidae), a parasite of the bathypelagic fish *Centrolophus niger*. *Parasitology Research*, **101**, 373–383.

Revenga, J. E. (1993) *Diphyllobothrium dendriticum* and *Diphyllobothrium latum* in fishes from southern Argentina: association, abundance, distribution, pathological effects, and risk of human infection. *Journal of Parasitology*, **79**, 379–383.

Rohde, K. (1989) At least eight types of sense receptors in an endoparasitic flatworm: a counter-trend to sacculinization. *Naturwissenschaften*, **76**, 383–385.

Rohde, K., Hefford, K., Ellis, J. T., *et al.* (1993) Contributions to the phylogeny of Platyhelminthes based on partial sequencing of 18S ribosomal DNA. *International Journal for Parasitology*, **23**, 705–724.

Ruiz-Trillo, I., Riutort, M., Littlewood, D. T. J., *et al.* (1999) Acoel flatworms: earliest extant bilaterian metazoans, not members of Platyhelminthes. *Science*, **283**, 1919–1923.

Schaperclaus, W. (1991) *Fish Diseases*, 5th edition. New Delhi: Oxonian Press.

Seppälä, O., Karvonen, A. & Valtonen, E. T. (2011) Eye fluke-induced cataracts in natural fish populations: is there potential for host manipulation? *Parasitology*, **138**, 209–214.

Shimazu. T. (1999) Plerocercoids with blastocysts of the trypanorhynch cestode *Nybelinia surmenicola* found in the euphausiid crustacean *Euphausia pacifica*. *Otsuchi Marine Science*, **24**, 1–4.

Shoop, W. L., Font, W. F. & Malatesta, P. F. (1990) Transmammary transmission of mesocercariae of *Alaria marcianae* (Trematoda) in experimentally infected primates. *Journal of Parasitology*, **76**, 869–873.

Smith, J. W. (1997) The blood flukes (Digenea: Sanguinicolidae and Spirorchidae) of cold-blooded vertebrates: Part 1. A review of the literature published since 1971, and bibliography. *Helminthological Abstracts*, **66**, 255–294.

Smyth, J. D. (1987) Asexual and sexual differentiation in cestodes: especially Mesocestoides and

Echinococcus. In *Molecular Paradigms for Eradicating Parasites*, ed. A. J. MacInnis, pp. 19–34. New York: Alan R. Liss.

Smyth, J. D. (1994) *Introduction to Animal Parasitology*, 3rd edition. Cambridge: Cambridge University Press.

Smyth, J. D. & McManus, D. P. (1989) *Physiology and Biochemistry of Cestodes*. Cambridge: Cambridge University Press.

Snyder, S. D. & Janovy, Jr., J. (1996) Behavioral basis of second intermediate host specificity among four species of *Haematoloechus* (Digenea: Haematoloechidae). *Journal of Parasitology*, **82**, 94–99.

Snyder, S. D. & Tkach, V. V. (2001) Phylogenetic and biogeographical relationships among some holarctic frog lung flukes (Digenea: Haematoloechidae). *Journal of Parasitology*, **87**, 1422–1440.

Strickland, G. T. & Ramirez, B. L. (2000) Schistosomiasis. In *Hunter's Tropical Medicine*, 8th edition. ed. G. T. Strickland, pp. 804–832. Philadelphia: W. B. Saunders.

Théron, A. (1984) Early and late shedding patterns of *Schistosoma mansoni* cercariae: ecological significance in transmission to human and murine hosts. *Journal of Parasitology*, **70**, 652–655.

Thompson, R. C. A. (1986) *The Biology of* Echinococcus *and Hydatid Disease*. London: Allen & Unwin.

Thompson, R. C. A. & McManus, D. P. (2002) Towards a taxonomic revision of the genus *Echinococcus*. *Trends in Parasitology*, **18**, 452–457.

Tielens, A. G. M. (1997) Biochemistry of trematodes. In *Advances in Trematode Biology*, ed. B. Fried & T. K. Graczyk, pp. 309–344. Boca Raton: CRC Press.

Tinsley, R. C. (1999) Parasite adaptations to extreme conditions in a desert environment. *Parasitology*, **119**, S31–S56.

Viozzi, G., Flores, V. & Rauque, C. (2005) An ectosymbiotic flatworm, *Temnocephala chilensis*, as second intermediate host for *Echinoparyphium megacirrus* (Digenea: Echinostomatidae) in Patagonia (Argentina). *Journal of Parasitology*, **91**, 229–231.

Waeschenbach, A., Webster, B. L., Bray, R. A. & Littlewood, D. T. J. (2007) Added resolution among ordinal level relationships of tapeworms (Platyhelminthes: Cestoda) with complete small and large subunit nuclear ribosomal RNA genes. *Molecular Phylogenetics and Evolution*, **45**, 311–325.

Webster, L. A. & Wilson, R. A. (1970) The chemical composition of protonephridial canal fluid from the cestode *Hymenolepis diminuta*. *Comparative Biochemistry and Physiology*, **35**, 201–209.

Wright, C. A. (1971) *Flukes and Snails*. London: Allen & Unwin.

7 Acanthocephala: the thorny-headed worms

7.1 General considerations

The acanthocephalans (Greek, *acantho*thorn, *cephala*head), or so-called 'thorny-headed worms,' are a relatively small group of obligatory intestinal endoparasites comprising approximately 1100 described species. Adult acanthocephalan body lengths are variable, ranging from 1 mm in size to greater than 60 cm. Acanthocephalans were first described in 1684 by Francesco Redi, an Italian physician (Box 1.1), who reported finding white worms (probably *Acanthocephalus anguillae*) with hooked anterior ends in the intestines of European eels. Since then, adult acanthocephalans have been reported from all classes of vertebrate animals, in marine, freshwater, and terrestrial habitats.

As the phylum name suggests, acanthocephalans have an anterior hooked proboscis that acts as a retractable holdfast, which anchors adults into the intestines of their vertebrate hosts. They are similar in some respects to other intestinal worms; like the tapeworms they have a tegument and lack a mouth or an intestine and are transmitted in food webs by trophic interactions. They also share features with the nematodes and possess a fluid-filled pseudoceolom, and are dioecious and sexually dimorphic, with females generally larger than males. However, acanthocephalans have a number of unique features that demonstrate their independent evolutionary history.

Acanthocephalans exhibit wide variation in morphology, body size, host distribution, host habitat, and life cycles. Such diversity offers a wealth of material for rigorous comparative studies (e.g., Poulin *et al.*, 2003; Poulin, 2007). Thus, the acanthocephalans have contributed significantly to many diverse biological disciplines – from physiology and biochemistry, to ecology and biogeography, to animal behavior

and evolutionary biology (reviews in Crompton & Nickol, 1985; Kennedy, 2006). For example, since the pioneering work of Bethel and Holmes (1973), no other group of parasites has contributed so significantly to the intriguing phenomenon of manipulation of intermediate host behavior in order to facilitate trophic transmission (Box 7.1; see Chapter 15). Moreover, their phylogenetic association with the free-living rotifers provides tantalizing clues to the evolution of parasitism in metazoan animals (Near, 2002; Herlyn *et al.*, 2003; see Box 7.4). They may be a small phylum in terms of biodiversity, but these worms pack a parasitological punch!

7.2 Form and function

7.2.1 Holdfast and body wall adaptations

The body of most acanthocephalans resembles an elongated tube, tapered at both ends, with the spined proboscis at the anterior. Externally, the body can be divided into the hook-bearing, retractile **proboscis**, a smooth neck region, and the trunk, which comprises most of the body and houses a fluid-filled body cavity containing the reproductive organs (Fig. 7.1).

The shape of the proboscis is highly variable, from spherical to cylindrical, depending on the species (Figs. 7.2, 7.3). The roots of the sclerotized hooks are embedded in a thin, muscular wall under the tegument, and their size, shape, number, and distribution may be important characters used in classification. The proboscis is a hollow structure filled with fluid that can be totally invaginated by the proboscis inverter muscles into a muscular sac called the proboscis receptacle. Evagination of the proboscis is achieved by hydraulic pressure created from within the lacunar system when the proboscis receptacle contracts. The cerebral ganglion

Box 7.1 | **Acanthocephalans: masters of phenotypic manipulation**

John Holmes and his graduate student, William Bethel, completed a series of studies involving larval acanthocephalans that would become classics in ecological parasitology. Bethel and Holmes (1973) were the first to demonstrate that the normal behavior, specifically the phototaxic response, of amphipods (*Gammarus lacustris*) was altered when infected with cystacanths of the acantho-cephalan *Polymorphus paradoxus*. Uninfected amphipods are photophobic and positively geo-tactic, diving and burrowing into the benthic substrate when disturbed. However, when infected with the infective cystacanth stage of *P. paradoxus*, the amphipods are positively phototactic and negatively geotactic. Thus, infected hosts typically skim along the water's surface and often cling to vegetation. In a laboratory experiment with captive mallards and muskrats, Bethel and Holmes confirmed that these definitive hosts ate significantly more amphipods containing *P. paradoxus*, compared to uninfected controls. Since this pioneering study, a variety of acanthocephalans have been shown to manipulate the phenotypes of diverse arthropod intermediate hosts to increase transmission to their definitive hosts. Indeed, behavioral modification of acanthocephalan-infected arthropods, often in combination with other altered phenotypes such as color or conspicuousness, has been demonstrated in all cases where it has been assessed (review in Moore, 2002).

Janice Moore is highly regarded as a leader in the field of parasite-induced behavioral manipulation. Her studies of the acanthocephalan *Plagiorhynchus cylindraceus*, which uses terrestrial isopods (pill bugs) as intermediate hosts and mature in starlings, provide a convincing case of parasite-induced phenotypic manipulation that acts to increase transmission success. Moore (1983) demonstrated that several behaviors of *P. cylindraceus* cystacanth-infected iso-pods were altered relative to uninfected controls. She showed that infected isopods were more likely to be attracted to areas of low relative humidity, to be out in the open and away from shelter, and to be found more often on light-colored substrata. The key question was whether these behavioral changes increased the parasite's chances of being transmitted to a starling. To answer this question, she took advantage of the fact that parent birds provide nestlings with their food. She then collected the prey from the nestlings to estimate the rate at which parents were delivering pill bugs to their young, and then determined the number of *P. cylindraceus* in the nestlings, as well as the prevalence of infected isopods in proximity to the starling nests. Sure enough, the starlings were not foraging for pill bugs at random. Rather, they preferentially preyed upon those that were infected with the larval acanthocephalan. The behavioral changes Moore noted in her pill bug lab experiments carried over to the field, confirming that parasite-induced behavioral changes in the intermediate host do, indeed, increase transmission success.

These pioneering studies involving larval acanthocephalans have inspired the development of what is now one of the most active and integrative subdisciplines in parasite ecology and evolution (see Chapter 15). Thus, the phenomenon of parasite-induced alteration in host behavior now includes a much wider range of parasite taxa, a much wider range of host phenotypes, and a much wider range of hosts, including humans! Yet despite this new and expanded focus onto other types of host–parasite interactions, studies involving larval

Box 7.1 | (continued)

acanthocephalans continue to play a key role in the development of this area. As we will see in Chapter 15, for example, our understanding of how parasites interfere with host biochemical, neuroendocrine, and immunological pathways to influence host behaviors rely heavily on acanthocephalan–arthropod interactions.

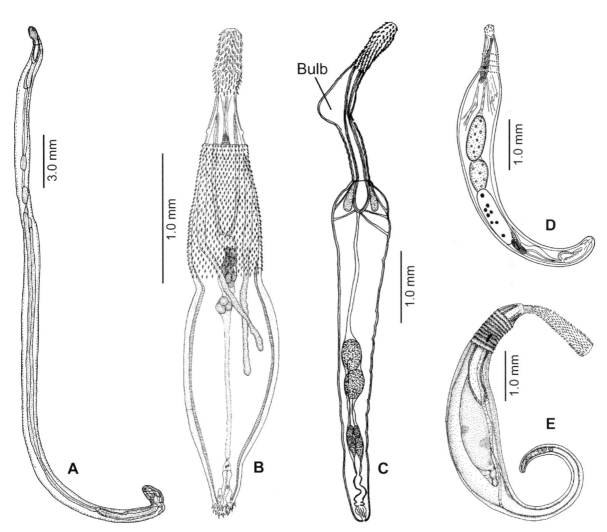

Fig. 7.1 Schematic representations of variation in male acanthocephalan body plans. (A) *Centrorhynchus conspectus* from various owls; (B) *Corynosoma bipapillum* from Bonaparte's gull *Larus philadelphia*; (C) *Pomphorhynchus patagonicus* from various freshwater fishes; (D) *Quadrigyrus nickoli* from a freshwater fish, *Hoplerythrinus unitaeniatus*; (E) *Polyacanthorhynchus kenyensis* from a paratenic fish host *Tilapia* sp. ((A) Modified from Richardson & Nickol, 1995, with permission, *Journal of Parasitology*, **81**, 767–772; (B) modified from Schmidt, 1965, with permission, *Journal of Parasitology*, **51**, 814–816; (C) modified from Ortubay *et al.*, 1991, with permission, *Journal of Parasitology*, **77**, 353–356; (D) modified from Schmidt & Hugghins, 1973, with permission, *Journal of Parasitology*, **59**, 829–835; (E) from Schmidt & Canaris, 1967, with permission, *Journal of Parasitology*, **53**, 634–637.)

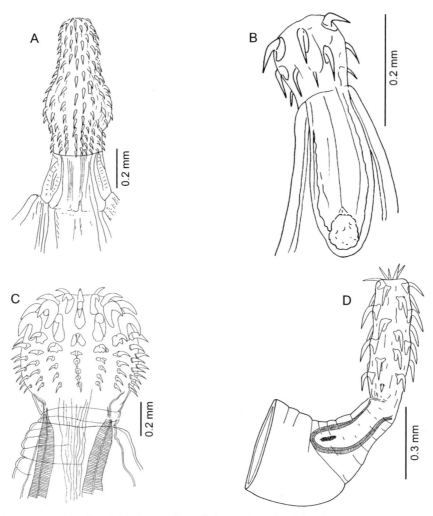

Fig. 7.2 Schematic representation of variation in acanthocephalan proboscides. (A) *Polymorphus spindlatus* from a black-crowned night heron *Nycticorax nycticorax*; (B) *Neoechinorhynchus venustus* from a western sucker *Catostomus macrocheilus*; (C) *Sphaerechinorhynchus serpenticola* from an Asian cobra *Naja naja*; (D) *Paracanthocephalus rauschi* from a sea otter *Enhydra lutris*. ((A) Modified from Amin & Heckmann, 1991, with permission, *Journal of Parasitology*, **77**, 201–205; (B) modified from Amin & Heckmann, 1992, with permission, *Journal of Parasitology*, **78**, 34–39; (C) modified from Schmidt & Kuntz, 1966, with permission, *Journal of Parasitology*, **52**, 913–916; (D) modified from Schmidt, 1969, with permission, *Canadian Journal of Zoology*, **47**, 383–385, NRC Research Press.)

is located inside the proboscis receptacle and an apical sensory organ may be present at the tip of the proboscis.

The neck is the smooth area located between the most posterior hooks of the proboscis and an infolding of the body wall that marks the beginning of the trunk (Figs. 7.1, 7.2). The neck may contain lateral sense organs and can be retracted by a pair of neck retractor muscles that attach to an infolding of the body wall. The trunk, like the proboscis and neck, is covered by a tegument with internal layers of circular and longitudinal muscles. The body wall of the trunk encloses the body cavity that contains not only the reproductive organs, but also the **ligament sacs** and the genital ganglia in the male.

In most acanthocephalans, attachment is accomplished by the insertion of the proboscis into the intestinal wall of the vertebrate host. In some species, sclerotized spines on the trunk's surface also act as accessory holdfasts and aid in attachment to the host intestine (Fig. 7.1B). Attachment is not permanent in most species; repositioning is possible by introversion of the proboscis, followed by reinflation using the hydrostatic pressure produced by the lacunar canal system. Thus, most acanthocephalans can relocate within the host's intestine in response to food availability, competition with other parasites, or to enhance mating opportunities.

The body wall of acanthocephalans is a syncytium organized into three distinct sections: an outer tegument, a middle group of circular muscles, and an inner group of longitudinal muscles (Fig. 7.4A). The tegument of acanthocephalans is ultrastructurally complex and fulfills many functions, including protection, inactivation of the host's digestive enzymes, osmoregulation, and acquisition of nutrients. The outermost layer of the tegument is a plasma membrane covered by a carbohydrate-rich glycocalyx and with numerous infoldings or crypts that open to the surface by pores. The glycocalyx and pores act as a sieve that allows particles of certain size to reach the crypts, where they undergo pinocytosis. Underlying the plasma membrane is a fibrous stratum without cellular boundaries and pierced by numerous canals. This layer is followed by a highly invaginated basal plasma membrane that may be involved in water and ion transport, and by a fine basal lamina. Mitochondria, Golgi complexes, lysosomes, ribosomes, glycogen deposits, lipid droplets, and various vesicles are found in the tegument. Rough endoplasmic reticulum is also present, but it is closely associated with the nuclei. There are few nuclei in the syncytial tegument. These are more or less fixed in position, and their numbers are approximately constant for each species, at least in early developmental stages.

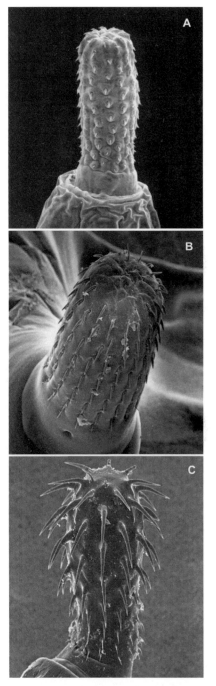

Fig. 7.3 Scanning electron micrographs of acanthocephalan proboscides. (A) *Moniliformis moniliformis* from rats; (B) *Acanthocephalus caspanensis* from toads *Bufo* sp. in Chile; (C) unidentified acanthocephalan from a flounder in Chile. ((A) Micrograph courtesy of Tom Dunagan. (B, C) Micrographs courtesy of Jackie Fernández.)

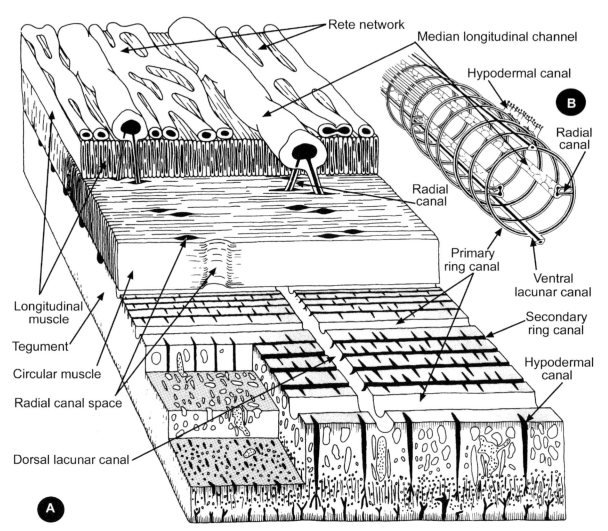

Fig. 7.4 The body wall and lacunar system of the acanthocephalan *Macracanthorhynchus hirudinaceus*. (A) Diagrammatic model of the body wall, sectioned in various ways, to show the relationship between the lacunar system, rete system, muscles, and tegument; (B) model of the lacunar system. (Modified from Dunagan & Miller, 1991, Acanthocephala. In *Microscopic Anatomy of Invertebrates*, vol. 4, *Aschelminthes*, ed. F. W. Harrison & E. E. Ruppert, pp. 299–332, with permission, John Wiley & Sons.)

The unique **lacunar system** is a complex interconnecting network of fluid-filled canals distributed throughout the tegument and muscle layers (Fig. 7.4B). The lacunar fluid is moved by muscle contraction. In the proboscis and neck, the lacunar system has canals that connect with two sac-like structures called **lemnisci**. Each lemniscus extends from the base of the neck into the body cavity. The precise function of the lemnisci is unknown, but they may be involved in the hydraulics of proboscis eversion and may serve as reservoirs for lacunar fluid. The lacunar system in the trunk of acanthocephalans is separate from that of the proboscis and neck, and the extensive network of interconnecting canals includes several longitudinal and transverse canals.

The structure of the longitudinal and circular muscles of the body wall is unusual (Fig. 7.4). The muscles are hollow and tubular, with a number of anastomosing connections, and the lumen of each muscle fiber is continuous with the lacunar system. The interplay of the lacunar system and hollow muscle fibers probably supports the efficient transport of nutrients and waste products throughout the organism, while simultaneously serving as a hydrostatic skeleton. The muscles of the body wall also are closely associated with the **rete network** (Fig. 7.4A). The rete system is a network of branching tubules lying on the medial surface of the longitudinal muscles or between the longitudinal and circular muscle layers. The rete network may be involved in the initiation of muscle contraction.

7.2.2 Osmoregulatory and excretory system

Most acanthocephalans lack an excretory system and waste materials are eliminated by diffusion through the body wall, probably through the pores in the tegument (Dunagan & Miller, 1986). Acanthocephalans are poor osmoregulators and the osmotic pressure of the fluid in their body cavity is close to that of the host's intestine.

7.2.3 Reproductive system

Acanthocephalans are dioecious and exhibit sexual dimorphism in size, with females usually being larger than males (Figs. 7.5, 7.6). Both sexes have one or two ligament sacs that extend from the proboscis receptacle to near the genital pore, forming an envelope that surrounds the gonads and accessory organs. The ligament sacs are permanent in some species, whereas in others they disappear as the individuals mature sexually.

Males have two testes, each with a vas efferens through which spermatozoa travel to a common vas deferens, or to a penis, or both. There are two important accessory organs in males, the **cement glands** and the copulatory **bursa** (Figs. 7.5, 7.7). The cement glands

may or may not be syncytial; if they are not syncytial, they range from one to eight in number. The cells produce copulatory cement, which, in some species, may be stored in a cement reservoir. The number and shape of the cement glands are important characters used in the classification of acanthocephalans. The copulatory bursa is a bell-shaped structure that is everted only during copulation and is used to hold the female in place during insemination (Fig. 7.7).

The female reproductive system (Fig. 7.5) consists of an ovary which, early in development and usually before the acanthocephalan reaches the definitive host, fragments into ovarian balls that float freely within the ligament sac. The posterior end of the ligament sac is attached to a unique acanthocephalan structure called the **uterine bell**, a muscular, funnel-shaped organ that allows mature fertilized eggs to pass through into the uterus and vagina and to the outside via the gonopore. Immature eggs are returned to the ligament sac. The selection process appears to be size-dependent, with only the mature, larger eggs passing into the uterus (Whitfield, 1970).

The male's copulatory bursa (Figs. 7.5, 7.7) is everted during copulation by the hydrostatic pressure of fluid forced into its lacunar system by a muscular sac called the **Saefftigen's pouch** (Fig. 7.5). The bursa is then wrapped around the posterior end of the female and the penis is introduced into the female gonopore. The released spermatozoa migrate through the vagina, the uterus, and the uterine bell into the ligament sac where fertilization of the oocytes occurs in the ovarian balls. After the transfer of sperm is completed, the male plugs the posterior end of the female with cement from the cement glands, forming a **copulatory cap** (Fig. 7.8). The copulatory cap appears to last only for a few days, but may act as a temporary occlusion to prevent loss of sperm. The cap may also be used as a kind of 'chastity belt,' ensuring that sperm from a given male will not encounter sperm from other males. In a bizarre twist, male acanthocephalans have been observed to also have copulatory plugs (see Box 7.2).

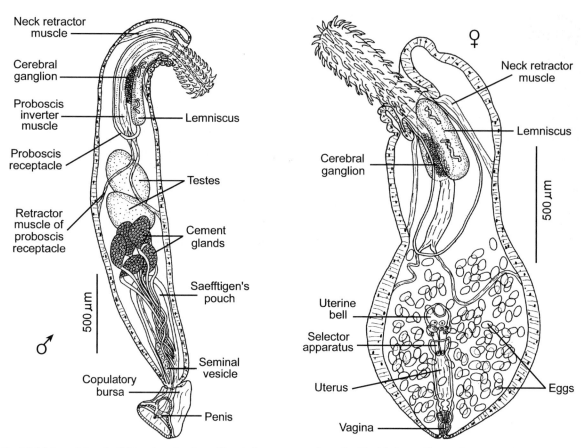

Fig. 7.5 Male and female *Echinorhynchus lageniformis* from a starry flounder *Platichthys stellatus*, illustrating the basic morphology of acanthocephalans. (Modified from Olson & Pratt, 1971, with permission, *Journal of Parasitology*, **57**, 143–149.)

7.2.4 Nervous system and sensory input

The nervous system of acanthocephalans is relatively simple. Male worms have three different types of ganglia, i.e., cerebral, genital, and bursal ganglia. Females have only a cerebral ganglion located in the body wall. Nerves from the cerebral ganglion reach the two lateral sense organs of the neck and the apical sense organ of the proboscis, if present. In males, the two genital ganglia are located along the bursal muscle and appear to control the musculature associated with the reproductive system. Several nerve tracts connect the cerebral ganglion with the genital ganglia. The bursal ganglion, also found

only in males, innervates the muscles of the bursa and the adjacent body wall. Near the cerebral ganglion there is a large multinucleated cell with processes that extend to the lateral and apical sensory organs (Miller & Dunagan, 1983). Although its function is not known, it may be part of the sensory network.

7.3 Nutrient uptake and metabolism

Most of the knowledge about the feeding physiology and metabolism of acanthocephalans comes from

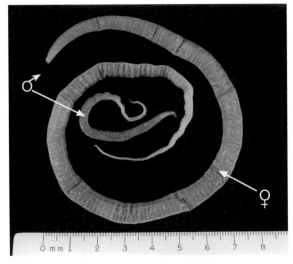

Fig. 7.6 Male and female individuals of *Oligacanthorhynchus tortuosa* from a Virginia opossum *Didelphis virginiana*, showing sexual dimorphism in size. The flat appearance is the normal condition in the host. The lacunar canals can be seen in the female. (Photograph courtesy of Tom Dunagan.)

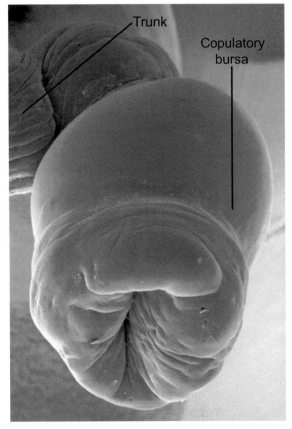

Fig. 7.7 Scanning electron micrograph of the posterior region of male of *Serrasentis sagittifer* from a Japanese threadfin bream *Nemipterus japonicus*, with the copulatory bursa everted. (Micrograph courtesy of Tom Dunagan.)

research on three species: *Moniliformis moniliformis*, a rat parasite and an excellent research model easily maintained in the laboratory; *Polymorphus minutus*, a parasite of ducks and chickens; and *Macracanthorhynchus hirudinaceus*, a large parasite of pigs. Because acanthocephalans lack a digestive system, nutrients are acquired through the body surface from the contents of the host intestine. Some sugars, nucleotides, amino acids, and some triglycerides are regularly absorbed. *Moniliformis moniliformis*, for example, absorbs glucose, mannose, fructose, and galactose. The transport of sugars is accomplished by facilitated diffusion. Immediately after absorption, sugars are phosphorylated into hexose phosphate, and although some of it is used for general metabolism, e.g., glycogenesis and other synthetic processes, much is converted into the disaccharide trehalose.

The tegument of *M. moniliformis* contains aminopeptidases that can cleave small peptides into amino acids, which are then absorbed (Uglem *et al.*, 1973). The mechanism of lipid absorption is poorly known, although absorption of triglycerides begins at the anterior half of the proboscis (Taraschewski &

Mackenstedt, 1991). Adult worms accumulate large quantities of neutral lipids, but they are not used as energy sources. It appears that acanthocephalans can control the absorption of lipids but do not have selective mechanisms to pick and choose specific ones. Nucleotides such as thymine are also absorbed and incorporated into mitochondrial DNA in the body wall, and nuclear DNA in ovarian balls and testes.

Most metabolism research has been directed at the acanthocephalan's intermediary carbohydrate metabolism and related activities. It appears that their glycogen content ranges between 22% and 24% of the dry weight in adult worms removed from unfasted hosts. Moreover, glycogen is apparently the primary

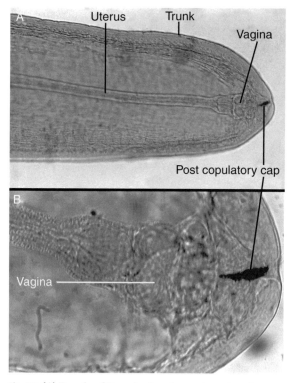

Fig. 7.8 (A) Female of *Pomphorhynchus patagonicus* from a South American silverside *Patagonina hatcheri*, showing the posterior end of the reproductive system and a copulatory cap; (B) higher magnification showing the vagina and post copulatory cap. (Modified from Semenas *et al.*, 1992, with permission, *Research and Reviews in Parasitology*, 52, 89–93.)

energy resource, with up to 75% being lost during a single 24-hour fasting period. Trehalose is a non-reducing disaccharide that comprises 2%–4% of the solid tissues of *M. moniliformis*. The function of trehalose in acanthocephalans is problematic. It has been suggested that trehalose may be involved in maintaining internal osmolality, but there is also evidence that it may function as a 'shuttle' in the transfer of glucose to glycogen.

A functional TCA cycle is absent in acanthocephalans, as is the case in a number of other intestinal helminths, and the worms probably operate as facultative anaerobes. In some acanthocephalans, the main wastes of carbohydrate metabolism are lactate and succinate, the latter being produced by carbon dioxide fixation, and the former as a product of glycolysis. In other species, various combinations and quantities of ethanol, lactate, butyrate, and acetate are excreted. Lipid metabolism as an energy resource is apparently unimportant. For additional information in this area, see Starling (1985).

7.4 Development and general life cycle

Following copulation, the fertilized eggs undergo early development in the ligament sac or body cavity of the female worm. Early in cell division, the cell

---

Box 7.2 | **Sexual selection in the Acanthocephala**

When Charles Darwin proposed his theory of sexual selection to help explain the evolution of extravagant and costly ornaments, and/or bizarre behaviors of males of some animals, he was probably not thinking about acanthocephalans! Yet, there are several lines of evidence suggesting that sexual selection has been a driving force in acanthocephalan evolution. First, in many dioecious animals there is strong selection for females to be large in order to maximize fecundity, while in males there is typically no reproductive advantage to attaining large body size, at least from a fertilization perspective. This process can lead to pronounced sexual dimorphism as is seen in many acanthocephalan species. Second, among many acanthocephalan species there is evidence for intense male–male competition for access to females and this can

Box 7.2 | (continued)

favor large male body size in some species (Sinisalo *et al.*, 2004). Third, male testis size has been shown to vary as a function of sperm competition; species in which sperm competition is unimportant (because they employ copulatory caps) tend to have smaller testes (Poulin & Morand, 2000). Fourth, male acanthocephalans have an unusual reproductive adaptation (cement glands) and behavior (copulatory caps). During copulation in acanthocephalans, after the transfer of sperm is completed, cement from the cement glands of the male plugs the vagina and posterior end of the female forming a copulatory cap. The external cap is transient, but the cement packed into the genital tract of the female may last for weeks. It is possible then, that cement glands and capping behavior evolved in response to sexual selection (Abele & Gilchrist, 1977). The capping of females would prevent subsequent insemination by other males at least for a few days, a sort of 'chastity belt,' ensuring that sperm from the capping male will encounter and fertilize mature oocytes without competition from sperm from another male.

Copulatory caps have also been found on the posterior end of male acanthocephalans (Crompton, 1985), and they may represent another manifestation of male–male competition for the limited female resource. The capping of males, effectively, albeit temporarily, eliminates other males from the pool competing for the female resource. Although it is possible that male capping may be the result of poor sex recognition, no sperm from the capping males has been found in the capped males, indicating that capping occurred without previous insemination. Abele & Gilchrist (1977) propose that "sperm competition may have led to the evolution of cement glands and capping behavior and that this may represent a preadaptation that under sexual selection may have assumed the additional function of removing male competitors from the reproductive pool."

Sexual selection may influence more than female and male acanthocephalan body size and mating success. It turns out that sexual selection may also influence the spatial distribution of acanthocephalans in the intestine of their host. For example, the distribution of males of *Corynosoma magdaleni* in a natural population of Saimaa ringed seals was not random. Instead, larger males congregated around non-mated females, providing evidence that male–male competition for access to females can be intense (Sinisalo *et al.*, 2004). Finally, sexual selection may also influence growth of male and female cystacanths in arthropod intermediate hosts. Banesh and Valtonen (2007) demonstrated that, at least for some acanthocephalans, larval life histories diverge between males and females, and that adult and cystacanth dimorphism was positively correlated. These authors conclude that "future studies should consider the type and size of the intermediate host, relative sexual development achieved by cystacanths, and phylogenetic history in order to reach more generalized conclusions about the impact of sexual selection on intermediate host use by male versus female acanthocephalans."

membranes disappear and the embryo becomes syncytial. When egg development is completed, the embryo, or **acanthor**, is surrounded by three or four membranes. The mature eggs (Fig. 7.9A) leave the female's body through the gonopore and reach the external environment with the host feces. The acanthor inside the eggs is the infective stage for the next host in the life cycle and will not develop any further until the appropriate arthropod intermediate host ingests it. The acanthor is also a resting, resistant stage. Under normal conditions, and depending on the species, an acanthor inside an egg can remain viable for several months. The acanthor is generally elongated and armed at the anterior end with six to eight hooks

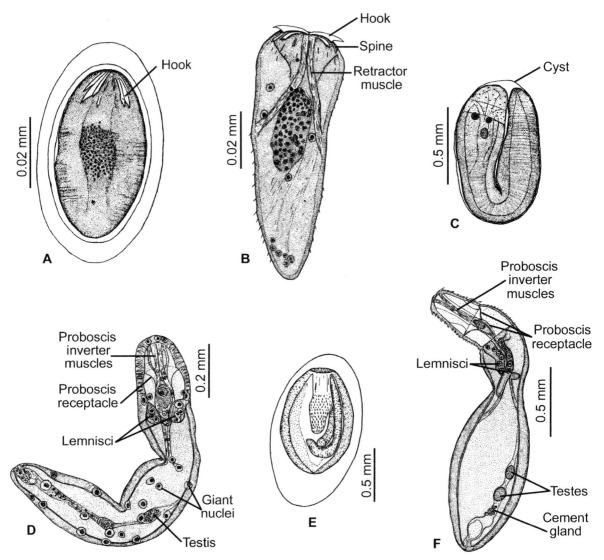

Fig. 7.9 Developmental stages of *Mediorhynchus grandis*, an acanthocephalan of birds, including grackles, crows, and robins. (A) Mature egg containing hooked acanthor; (B) mature acanthor removed from the egg; (C) acanthella, age 24 days; (D) acanthella removed from the cyst, age 25 days; (E) cystacanth, age 27–30 days; (F) cystacanth removed from the cyst, age 27–30 days showing the male reproductive structures. (Modified from Moore, 1962, with permission, *Journal of Parasitology*, **48**, 76–86.)

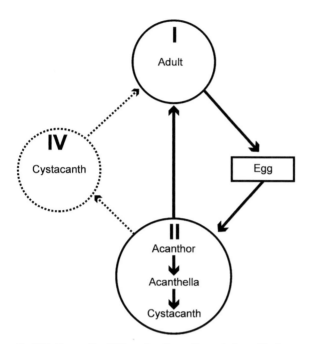

Fig. 7.10 Generalized life cycle of acanthocephalans. Host types are represented by roman numerals (I = definitive host, II = intermediate host, IV = paratenic host). Solid lines represent a 'typical life cycle,' whereas dotted lines represent the incorporation of a paratenic host. Stages within hosts are enclosed in circles while the free-living egg stage in the environment is enclosed within a rectangle. (Figure courtesy of Jackie Fernández.)

or spines (Fig. 7.9B). This spinose structure is used to penetrate the intestine of the arthropod intermediate host. When a suitable arthropod intermediate host ingests an egg, the acanthor is released and bores through the intestinal wall until it reaches the host hemocoel. In most cases, the acanthor stops under the serosa of the intestine. In either location, the acanthor begins to absorb nutrients from the host, grows, and develops primordia of all organs present in the adult. This growing stage is known as the **acanthella** (Fig. 7.9C, D). At the end of the acanthella stage, when development is completed, and a cyst wall is secreted, the worm is known as a **cystacanth** (Fig. 7.9E, F). This is the infective stage for the paratenic or definitive host in the life cycle of acanthocephalans. When the intermediate host infected with the cystacanth is eaten by a suitable vertebrate definitive host, the worm excysts,

attaches to the intestinal wall, and matures sexually. This generalized life cycle is shown in Fig. 7.10. Infection of both the intermediate and the definitive hosts is passive; the mature egg must be eaten by a specific arthropod host and the cystacanth-infected arthropod must be ingested by the paratenic or vertebrate definitive host.

7.5 Biodiversity and life-cycle variation

Acanthocephalans infect a wide variety of terrestrial and aquatic (both marine and freshwater) vertebrates. Their success in exploiting such a host range is related to their complex life cycles involving predator–prey interactions featuring obligatory arthropod intermediate hosts. Thus, much like the trematodes being evolutionarily linked to the molluscs, the evolutionary history of acanthocephalans is tied to the arthropods, primarily the two most speciose groups – the mostly aquatic crustaceans and the primarily terrestrial insects. Given the importance of crustaceans in the diet of freshwater vertebrates it is not surprising that a diversity of acanthocephalans infect primarily aquatic vertebrates, especially fish and waterfowl.

In the marine environment, decapod crustaceans, e.g., shore crabs, are also intermediate hosts for many species of acanthocephalans, e.g., *Profilicollis* spp. and *Hexaglandula* spp., that mature primarily in shorebirds (Nickol *et al.*, 2002). *Bolbosoma* spp. (Fig. 7.11) infect filter-feeding whales and use planktonic copepods and euphausid crustaceans (krill) as intermediate hosts.

To exploit terrestrial vertebrates many acanthocephalans have evolved life cycles involving insects as intermediate hosts. For example, cockroaches are intermediate hosts of the rat acanthocephalan, *Moniliformis dubius*. In addition, isopod crustaceans (pill bugs) serve as intermediate hosts for some acanthocephalan species of terrestrial birds such as starlings, e.g., *Plagiorhynchus cylindraceus* (Box 7.1).

Several species of acanthocephalans have diversified to infect vertebrates that do not directly prey upon arthropods. For example, species within the speciose genus *Corynosoma* (Fig. 7.1B) have been reported from

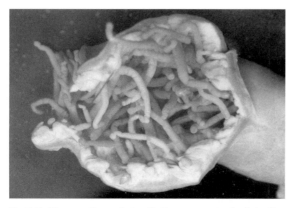

Fig. 7.11 The acanthocephalan *Bolbosoma* sp. in the intestine of a minke whale *Balaenoptera acurostrata* found stranded off the coast of Alaska. Each centimeter of intestine contained ≈{ts}144 acanthocephalans. (Photograph courtesy of Robert Rausch.)

a variety of piscivorous marine mammals and seabirds worldwide, and use amphipods as intermediate hosts (Color plate Fig. 8.2). Cystacanths of *Corynosoma* spp. have also been reported from a range of paratenic fish hosts. Hoberg (1986) suggested that the use of fish paratenic hosts by *Corynosoma* spp. led to opportunities for host switching (from seals to seabirds, for example) and resultant diversification within the genus. Another speciose genus, *Centrorhynchus*, infects several species of avian and mammalian carnivores (Fig. 7.1A). Ewald *et al.* (1991) found cystacanths at high abundances in shrews; these voracious insectivores were likely important paratenic hosts for *Centrorhynchus aluconis* in tawny owls in Britain. In Australia, *Oncicola pomastomi* parasitizes dingoes and feral cats, with leopard cats serving as definitive hosts in Borneo, Malaysia, and probably the Philippines. Schmidt (1983) speculated that leopard cats from Malaysia, Borneo, and the Philippines were the original hosts and that migrating birds, which serve as paratenic hosts for *O. pomastomi*, were responsible for introducing the parasite into Australia. In these instances, vertebrate paratenic hosts act to bridge the trophic link between the intermediate host and definitive host and may act as a significant transmission adaptation, enhancing spatial and temporal dispersal. Paratenic hosts also provide a very important means

for acanthocephalans to overcome the rigidity of a two-host life cycle (Kennedy, 2006).

7.6 Phylogenetic relationships and classification

Molecular phylogenetic analyses based on 18S ribosomal RNA gene sequence data are consistent with the traditional classification based upon morphological characteristics, intermediate and definitive host specificity, and host ecology. Thus, the Archiacanthocephala, Palaeacanthocephala, and the Eoacanthocephala are currently recognized as monophyletic (Fig. 7.12; Box 7.4; Near *et al.*, 1998; Garcia-Varela *et al.*, 2000; Garcia-Varela & Nadler, 2005). Near *et al.* (1998) suggest that acanthocephalans have a strict pattern of historical association with their arthropod intermediate hosts, meaning that the arthropod–acanthocephalan parasitic relationship is an ancient one. Conversely, vertebrate host plasticity is pronounced among acanthocephalans; for example, birds and mammals are definitive hosts in the Archiacanthocephala as well as in the Polymorphidae in the Class Palaeacanthocephala (see below and Boxes 7.3, 7.4).

Adult archiacanthocephalans parasitize birds and mammals, while their intermediate hosts are insects and myriapods, e.g., *Moniliformis* and *Oligacanthorhynchus* (Box 7.3). Palaeacanthocephalans are the most diverse of the acanthocephalans in terms of their definitive hosts, parasitizing all vertebrate classes and using malacostracan crustaceans as intermediate hosts. Another common feature among the Palaeacanthocephala is that most of them infect hosts such as fishes or migratory waterfowl that are linked to aquatic habitats, and employ aquatic crustaceans as intermediate hosts (e.g., *Pomphorhynchus* and *Polymorphus*). One unusual palaeacanthocephalan is *Megapriapus ungriai*, a parasite of South American potamotrygonid freshwater stingrays, and, intriguingly, the only acanthocephalan known from elasmobranchs. Potamotrygonid stingrays originally inhabited the Pacific Ocean, but after the rising of the

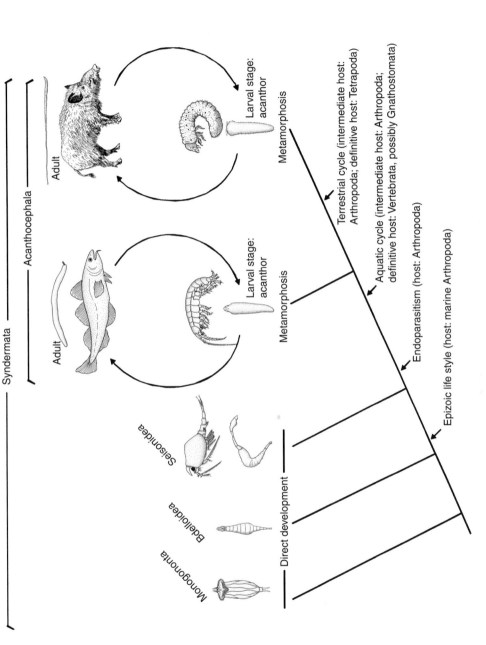

Fig 7.12 Phylogeny of the Syndermata as inferred from 18S rDNA and EST analyses, showing the relationships of the Acanthocephala to Monogononta, Bdelloidea, and Seisonidea. Also illustrated are the proposed steps leading to the evolution of endoparasitism, and aquatic and terrestrial life cycles in the Acanthocephala. The representative acanthocephalans include the aquatic palaeacanthocephalan *Echinorhynchus gadi* and the terrestrial archiacanthocephalan *Macracanthorhynchus hirudinaceus*. (Figure courtesy of Bernd Baumgart, Göttingen, Germany.)

Box 7.3 | Classification of the Acanthocephala

Current classification schemes for the acanthocephalans are in a state of change as more species are subject to molecular-based phylogenetic analyses. For example, Garcia-Varela & Nadler (2005) found that neither of the diverse palaeacanthocephalan orders Polymorphida or Echinorhynchida are monophyletic. Their molecular phylogenetic data also indicate that family-level classification within the class needs revision.

Class Archiacanthocephala
  Order Moniliformida
  Representative genus: *Moniliformis*
Order Gigantorhynchida
  Representative genera: *Gigantorhynchus, Mediorhynchus*
Order Oligacanthorhynchida
  Representative genera: *Macracanthorhynchus, Oligacanthorhynchus, Oncicola*
Order Apororhynchida
  Representative genus: *Apororhynchus*
Class Palaeacanthocephala
  Order Echinorhynchida
  Representative genera: *Acanthocephalus, Echinorhynchus, Fessisentis,*
    Heteracanthocephalus, Hypoechinorhynchus, Pomphorhynchus, Polyacanthorynchus,
    Rhadinorhynchus
Order Polymorphida
  Representative genera: *Bolbosoma, Centrorhynchus, Corynosoma, Hexaglandula,*
    Plagiorhynchus, Polymorphus, Profilicollis
Class Eoacanthocephala
  Order Gyracanthocephalida
  Representative genera: *Acanthogyrus, Quadrigyrus*
Order Neoechinorhynchida
  Representative genera: *Neoechinorhynchus, Tenuisentis, Dendronucleata*
Class Polyacanthocephala
  Order Polyacanthorhynchida
  Representative genus: *Polyacanthorhynchus*

Andes mountain chain, they apparently became isolated in freshwater ecosystems (Brooks *et al.*, 1981; Brooks, 1992). The phylogenetic position of the genus *Megapriapus* and its single species remains unclear, as well as how, when, and where it became a parasite of freshwater stingrays (see Chapter 13).

Another palaeacanthocephalan that has been the focus of increased attention by ecologists is *Profilicollis altmani*. This species matures in shore

Box 7.4 | Phylogeny of the Acanthocephala and the evolution of parasitism

Deducing the evolutionary relationships of invertebrates has proven enormously difficult and has been the subject of considerable debate among taxonomists and evolutionary biologists over the past few decades. This is especially true for parasitic taxa, since many potentially taxonomically valuable morphological features have been lost on their evolutionary path to becoming parasites. Such is the case for the Acanthocephala, a most intriguing situation indeed, as there is now overwhelming morphological and molecular phylogenetic evidence that the Acanthocephala and Rotifera share a common ancestry (Sörensen & Giribet, 2006). At first glance, it is difficult to conceive of such a scenario since the two taxa cannot be more different morphologically, physiologically, or ecologically. Most rotifers are ciliated, free-swimming inhabitants of marine or freshwater habitats. Many rotifers are renowned for spectacular reproductive and ecological adaptations for life in unstable freshwater habitats, including the evolution of parthenogenesis and anhydrobiosis. In one group (the bdelloid rotifers), no males have ever been found and in another (the monogont rotifers) the males are short-lived dwarfs! This is in marked contrast to a group of dieocious obligate parasites that never reproduce by parthenogenesis, have a retractable spined holdfast, mature in the intestine of vertebrate animals, and have evolved an indirect life cycle.

And yet, there are similarities when one looks closer. First, both taxa have cement glands. As mentioned earlier, in male acanthocephalans the adhesive product of cement glands is used by the male to seal the female's reproductive tract after insemination. Pedal or cement glands are also found in rotifers, in both males and females, where they function to produce an adhesive that is used for temporary substratum attachment. The pedal glands of bdelloid rotifers are morphologically and functionally similar to the cement glands of archiacanthocephalans, suggesting that the structures are homologous. Four other shared morphological features hint to a monophyly of the rotifers and acanthocephalans: syncytial epidermis; intracytoplasmic lamina; sperm cells with an anterior flagella; and, epidermal cells with apical crypts (review in Near, 2002). Current molecular phylogenetic analyses also support a close evolutionary relationship and, therefore, the taxon Syndermata has been proposed to accommodate both the Rotifera and the Acanthocephala (Herlyn *et al.*, 2003: see Fig. 7.12). The key question is how can we explain the current host distribution and radiation of the Acanthocephala and how did they become endoparasites?

Near's (2002) 18S inferred rRNA phylogeny provides a framework to investigate the diversification of features associated with parasitism in acanthocephalans. First and foremost, the class of arthropod used as intermediate host is highly conserved (=congruent) on the phylogeny. Palaeacanthocephalans use malacostracan crustaceans, e.g., isopods and amphipods, as intermediate hosts, while maxillipod crustaceans, e.g., ostracods and copepods, are used by the Eoacanthocephala. Uniramians, e.g., insects, are used as intermediate hosts by the Archiacanthocephala. Such remarkable conservation of intermediate hosts may be driven by

Box 7.4 | (continued)

physiological constraints, due to the more intimate relationship between acanthocephalans and their arthropod hosts (Near, 2002). In contrast, acanthocephalans exhibit wide variation in the use of vertebrate definitive hosts. For example, mapping definitive hosts on their phylogeny reveals that acanthocephalans have independently invaded avian and mammalian hosts in the Archiacanthocephala and the polymorphid palaeacanthocephalans. Such independent evolution must be strongly influenced by trophic interactions among intermediate and definitive hosts. The evolution of life cycles in the acanthocephalans is also characterized by shifts between aquatic and terrestrial environments. For example, such shifts were likely promoted by the availability of malacostracans, e.g., isopods, in both terrestrial and aquatic habitats, leading to the diversity of habitat utilization of the Palaeacanthocephala. The radiation of the eoacantho-cephalans to terrestrial hosts is prevented by the lack of appropriate intermediate hosts; all eoacanthocephalans require entirely aquatic maxillipod crustaceans (Near, 2002).

Herlyn *et al.* (2003) propose an evolutionary scenario for acanthocephalan endoparasitism (Fig. 7.12). They propose that the common ancestor of *Seison* (a small genus of rotifer that today lives as ectocommensals of crustaceans) and acanthocephalans developed an epizoic (or possibly ectoparasitic) life style on a marine crustacean ancestor. This 'stem species' then invaded the body cavity of the crustacean host and established an endoparasitic life style. Vertebrates subsequently became part of the life cycle by feeding on infected aquatic crustaceans. With the radiation of the Insecta and the terrestrial isopod crustaceans, terrestrial life cycles for the acanthocephalans could become established and acanthocephalans radiated to exploit terrestrial vertebrates as discussed above. Witek *et al.* (2008) mapped morphological character evolution onto molecular phylogeny. They concluded that the reduction of the corona and the emergence of a retractable anterior end (rostrum, proboscis) occurred before the separation of the Acanthocephala (Fig. 7.12). In particular, the evolution of the rostrum might have been a key event leading to the later evolution of acanthocephalan endoparasitism, especially given the significance of the proboscis for anchoring of the adults to their definitive host's intestinal wall (Witek *et al.*, 2008).

birds such as surf scoters in California and uses decapod crustaceans, including sand and mole crabs, as intermediate hosts. This acanthocephalan has been implicated as an agent of mortality of endangered southern sea otters along the west coast of North America. Normally, sea otters forage for macro-invertebrates in rocky reef and kelp forest habitats for a variety of molluscs as well as sea urchins. When their natural prey is scarce they turn to alternative prey such as sand crabs in sandy substrata habitats. When sea otters eat infected shore crabs, *P. altmani* often perforate the intestine causing bacterial infection of the abdominal cavity and a condition known as septic or acanthocephalan peritonitis (Mayer *et al.*, 2003). Lafferty and Gerber (2002) found that years in which *P. altmani* infections were common were followed by years with high sea otter mortality. They speculated that human-induced changes in the environment may be responsible for the southern sea otter's shifts in diet, which lead to heavy infections and acanthocephalan-induced peritonitis. In short, this may be an emerging disease of sea otters and the parasite may be a contributing factor limiting southern sea otter populations.

The eoacanthocephalans primarily use fish as definitive hosts and aquatic crustaceans as intermediate hosts. There has also been a radiation of *Neoechinorhynchus* in North American freshwater turtles.

The phylogenetic relationships of the Acanthocephala with other animal groups are not absolutely clear, although recent analyses support the hypothesis that the Acanthocephala and Rotifera are closely related. Conway-Morris & Crompton (1982) originally suggested that Priapulida (a small group of free-living, marine, benthic worms called penis worms) was the most likely sister group of the Acanthocephala. These authors indicated that the immediate ancestor of acanthocephalans might have been a marine interstitial meiofaunal worm, similar to some of the priapulid

worms found in the fossils that mark the Cambrian explosion. Because the first vertebrates, i.e., jawless fish, did not appear until the end of the Cambrian, it is thought that the well-represented arthropods may have been the first hosts used by acanthocephalans in that period. The incorporation of a vertebrate host to produce the present two-host life cycle was probably a later development, accomplished by the mid Paleozoic, when fishes and amphibians evolved.

Phylogenetic analyses of morphological data (Lorenzen, 1985) and 18S ribosomal RNA (Winnepenninckx *et al.*, 1995; Garey *et al.*, 1996; Near *et al.*, 1998) indicate that acanthocephalans and rotifers in the class Bdelloidea are sister groups and are more closely related to each other than to any other group of organisms. Herlyn *et al.* (2003) suggest that the commensal rotifer genus *Seison* is the acanthocephalan sister group. As pointed out by Near (2002), "the identification of a free-living sister taxon to the entirely parasitic Acanthocephala offers an unprecedented opportunity to study the evolution of obligate parasitism." In short, the acanthocephalans provide a model system to investigate the processes involved in the evolution of parasitism (see Box 7.4).

References

Abele, L. G. & Gilchrist, S. (1977) Homosexual rape and sexual selection in acanthocephalan worms. *Science*, **197**, 81–83.

Banesh, D. P. & Valtonen, E. T. (2007) Sexual differences in larval life history traits of acanthocephalan cystacanths. *International Journal for Parasitology*, **37**, 191–198.

Bethel, W. M. & Holmes, J. C. (1973) Altered evasive behavior and responses to light in amphipods harboring acanthocephalan cystacanths. *Journal of Parasitology*, **59**, 945–956.

Brooks, D. R. (1992) Origins, diversification, and historical structure of the helminth fauna inhabiting neotropical freshwater stingrays (Potamotrygonidae). *Journal of Parasitology*, **78**, 588–595.

Brooks, D. R., Thorson, T. B. & Mayes, M. A. (1981) Freshwater stingrays (Potamotrygonidae) and their helminth parasites: testing hypothesis of evolution and coevolution. In *Advances in Cladistics: Proceedings of the First Meeting of the Willi Henig Society*, ed. V. A. Funk & D. R. Brooks, pp. 147–175. New York: New York Botanical Garden.

Conway-Morris, S. & Crompton, D. W. T. (1982) The origins and evolution of the Acanthocephala. *Biological Reviews of the Cambridge Philosophical Society*, **57**, 85–115.

Crompton, D. W. T. (1985) Reproduction. In *Biology of the Acanthocephala*, ed. D. W. T. Crompton & B. B. Nickol, pp. 213–271. Cambridge: Cambridge University Press.

Crompton, D. W. T. & Nickol, B. B. (1985) *Biology of the Acanthocephala*. Cambridge: Cambridge University Press.

Dunagan, T. T. & Miller, D. M. (1986) A review of protonephridial excretory systems in Acanthocephala. *Journal of Parasitology*, **72**, 621–632.

Ewald, J. A., Crompton, D. W. T., Johnson, I. & Stoddart, R. C. (1991) The occurrence of *Centrorhynchus* (Acanthocephala) in shrews (*Sorex araneus* and *Sorex minutus*) in the United Kingdom. *Journal of Parasitology*, **77**, 485–487.

Garcia-Varela, M. & Nadler, S. A. (2005) Phylogenetic relationships of Palaeacanthocepha (Acanthocephala) inferred from SSU and LSU rDNA gene sequences. *Journal of Parasitology*, **91**, 1401–1409.

Garcia-Varela, M., Pérez-Ponce de Leon, G., de la Torre, P., *et al.* (2000) Phylogenetic relationships of Acanthocephala based on analysis of 18S ribosomal RNA gene sequences. *Journal of Molecular Evolution*, **50**, 532–540.

Garey, J. R., Near, T. J., Nonnemacher, M. R. & Nadler, S. A. (1996) Molecular evidence for Acanthocephala as a subtaxon of Rotifera. *Journal of Molecular Evolution*, **43**, 287–292.

Herlyn, H., Piskurek, O., Schmitz, J., *et al.* (2003) The syndermatan phylogeny and the evolution of acanthocephalan endoparasitism as inferred from 18S rDNA sequences. *Molecular Phylogenetics and Evolution*, **26**, 155–164.

Hoberg, E. (1986) Aspects of ecology and biogeography of Acanthocephala in Antarctic seabirds. *Annales de Parasitologie Humaine et Comparée*, **61**, 199–214.

Kennedy, C. R. (2006) *Ecology of the Acanthocephala*. Cambridge: Cambridge University Press.

Lafferty, K. D. & Gerber, L. R. (2002) Good medicine for conservation biology: the intersection of epidemiology and conservation theory. *Conservation Biology*, **16**, 593–604.

Lorenzen, S. (1985) Phylogenetic aspects of pseudocoelomate evolution. In *The Origins and Relationships of Lower Invertebrates*, ed. S. Conway-Morris, J. D. George, R. Gibson & H. M. Platt, pp. 210–223. Oxford: Clarendon Press.

Mayer, K. A., Dailey, M. D. & Miller, M. A. (2003) Helminth parasites of the southern sea otter *Enhydra lutris nereis* in central California: abundance, distribution and pathology. *Diseases of Aquatic Organisms*, **53**, 77–88.

Miller, D. M. & Dunagan, T. T. (1983) A support cell to the apical and lateral sensory organs in *Macracanthorhynchus hirudinaceus* (Acanthocephala). *Journal of Parasitology*, **69**, 534–538.

Moore, J. (1983) Responses of an avian predator and its isopod prey to an acanthocephalan parasite. *Ecology*, **64**, 1000–1015.

Moore, J. (2002) *Parasites and the Behavior of Animals*. Oxford: Oxford University Press.

Near, T. J. (2002) Acanthocephalan phylogeny and the evolution of parasitism. *Integrative and Comparative Biology*, **42**, 668–677.

Near, T. J., Garey, J. R. & Nadler, S. A. (1998) Phylogenetic relationships of the Acanthocephala inferred from 18S ribosomal DNA sequences. *Molecular Phylogenetics and Evolution*, **10**, 287–298.

Nickol, B. B., Hear, R. W. & Smith, N. F. (2002) Acanthocephalans from crabs in the southeastern U.S., with the first intermediate hosts known for *Arhythmorhynchus frassoni* and *Hexaglandula corynosoma*. *Journal of Parasitology*, **88**, 79–83.

Poulin, R. (2007) Evolution in attachment: evolution of anchoring structures in acanthocephalan parasites. *Biological Journal of the Linnean Society*, **90**, 637–645.

Poulin, R. & Morand, S. (2000) Testes size and male–male competition in acanthocephalan parasites. *Journal of Zoology*, **250**, 551–558.

Poulin, R., Wise, M. & Moore, J. (2003) A comparative analysis of adult body size and its correlates in acanthocephalan parasites. *International Journal for Parasitology*, **33**, 799–805.

Schmidt, G. D. (1983) What is *Echinorhynchus pomatostomi* Johnston and Cleland, 1912? *Journal of Parasitology*, **69**, 397–399.

Sinisalo, T., Poulin, R., Hogmander, H., *et al.* (2004) The impact of sexual selection on *Corynosoma magdaleni* (Acanthocephala) infrapopulations in Saimaa ringed seals (*Phoca hispida saimensis*). *Parasitology*, **128**, 179–185.

Sörensen, M. V. & Giribet, G. (2006) A modern approach to rotiferan phylogeny: combining morphological and molecular data. *Molecular Phylogenetics and Evolution*, **40**, 585–608.

Starling, J. A. (1985) Feeding, nutrition, and metabolism. In *Biology of Acanthocephala*, ed. D. W. T. Crompton & B. B. Nickol, pp. 125–212. Cambridge: Cambridge University Press.

Taraschewski, H. & Mackenstedt, U. (1991) Autoradiographic and morphological studies on the uptake of the triglyceride [3H]-glyceroltriolate by acanthocephalans. *Parasitological Research*, **77**, 247–254.

Uglem, G. L., Pappas, P. W. & Read, C. P. (1973) Surface aminopeptidase in *Moniliformis dubius* and its relation to amino acid uptake. *Parasitology*, **67**, 185–195.

Whitfield, P. J. (1970) The egg sorting function of the uterine bell of *Polymorphus minutus* (Acanthocephala). *Parasitology*, **61**, 111–126.

Winnepenninckx, B., Backeljau, T., Mackey, L. Y., *et al.* (1995) 18S rRNA data indicate that aschelminthes are polyphyletic in origin and consist of at least three distinct clades. *Molecular Biology and Evolution*, **12**, 1132–1137.

Witek, A., Herlyn, H., Meyer, A., *et al.* (2008) EST based phylogenomics of Syndermata questions monophyly of Eurotatoria. *BMC Evolutionary Biology*, DOI:10.1186/1471-2148-8-345

Nematoda: the roundworms

8.1 General considerations

The Nematoda (Greek, *nema*thread), or roundworms, are perhaps the most abundant and diverse group of multicellular animals on earth. Free-living nematode densities can exceed over 1 million individuals per square meter in some shallow-water marine sediments. Furthermore, free-living nematodes exploit a greater array of ecological habitats than any other metazoan. Many are significant detritivores or decomposers that play a disproportionately large role in recycling chemicals and organic nutrients in aquatic and terrestrial ecosystems. Others feed on bacteria and other microorganisms and are important in food web relationships. Approximately 20 000 nematode species have been described. However, it is certain that this is an underestimate of total biodiversity, with perhaps as many as one million species. The difficulty in describing nematodes, in part, is related to their small sizes and their notorious uniformity in internal and external morphology. However, it is clear that their small sizes and cylindrical, tapered shapes have contributed to their extraordinary adaptive radiation within diverse ecosystems, as well as their exploitation of both plants and animals as hosts.

Nematodes are typically dioecious, exhibit sexual dimorphism, and vary in size from 1 mm to well over 1 m in length. The world's largest nematode species is the massive *Placentonema gigantissima* in the placenta of sperm whales; female worms are approximately 8 m long! Nematodes have a fluid-filled pseudocoelom and are bilaterally symmetrical. All possess a complex multilayered cuticle that is molted four times before reaching sexual maturity. Like the acanthocephalans, nematodes also exhibit **eutely** whereby most adult tissues are composed of a constant number of cells and growth is due to increased cell size rather than an increase in cell number.

Adults possess a complete, but simple, digestive system. Their reproductive systems are also relatively simple, but their life cycles are exceptionally varied.

No other group of metazoan parasites has received as much attention as the nematodes. Given their diverse associations with humans and their domesticated animals, it is not difficult to understand why this is so. As plant parasites, for example, they can be significant agricultural pests for many types of food crops. In addition, nematodes may act as vectors for pathogenic plant viruses (Taylor & Brown, 1997). As parasites of vertebrates, they are the causative agents of some of the most debilitating diseases of humans, our livestock, and our pets, as well as wildlife. Moreover, many of the so-called entomopathogenic nematodes are used for the biological control of several insect agricultural and forest pests (Smart, 1995; Stock, 2005; Box 8.1). Given the significant attention that the nematodes have received from parasitologists, it is no surprise that nematode–host interactions have had tremendous impact on diverse subfields such as parasite population ecology (Chapter 12), biogeography (Chapter 14), evolution (Chapter 16), and coevolution (Chapter 16). In this chapter, we restrict our focus to a generalized description of the nematode body plan, with subsequent emphasis on nematode life-cycle and life-history variation, and classification. Detailed coverage of this important group is provided in Anderson (2000), Kennedy & Harnett (2001), and Roberts & Janovy (2009).

8.2 Form and function

8.2.1 Cuticular structures

The body shape of most nematodes is a naked cylinder, i.e., fusiform (spindle-like) or filiform (thread-like). Typically, there are no distinct divisions of the body and arbitrary reference must be made to a head and a tail (Fig. 8.1). Among the nematode parasites of vertebrates, there are only a few exceptions to this general uniform cylindrical body shape (Fig. 8.2). Much more common are variations in the external morphology of the anterior or posterior (or both) regions (Figs. 8.3, 8.4).

Anteriorly, the primitive number of lips is thought to be six, but in extant nematodes, the number may range from zero to six. In addition, lateral lips, called pseudolabia, may be present, particularly on those with a reduced number of lips. The lips open internally to a **buccal capsule** lined by hardened (sclerotized) cuticle. Hookworms have a highly modified buccal capsule containing lateral cutting plates, sometimes with teeth. Often, it is some combination of these 'teeth,' lips, or the buccal capsule itself (Figs. 8.3A, D) that nematodes use as temporary holdfasts to attach to the host. Rather prominent anterior structures on some nematodes are cuticular projections, called **cordons** (Fig. 8.3B). Finally, some nematodes have lateral projections, called cephalic **alae**, associated with the anterior end (Fig. 8.3C).

Also found anteriorly are sensory structures that may include some combination of glands, papillae, or bristles. **Amphids** are paired sensory organs, derived from cilia. These occur in pairs and open internally into the amphidial glands, the largest of the anterior sense organs. Though often reduced in zooparasitic nematodes, amphids nonetheless are thought to be important as chemosensory structures. Amphids are also secretory, although the function of the secretory proteins seems variable (Riga *et al.*, 1995). Associated with the lips are labial papillae, but these are often fused, accompanying a reduction in the number of lips. At about the same level as the amphids, are cephalic papillae or bristles. Slightly posterior to these is a pair of cervical papillae known as **deirids**. Papillae or bristles are thought to serve a tactile function and their number and location is an important taxonomic character.

Behind the cephalic structures, some nematodes may have some combination of spines, lateral alae, additional sensory papillae, or other modifications of the body. This region may also be important for taxonomic purposes, at least in the trichostrongyles, as this

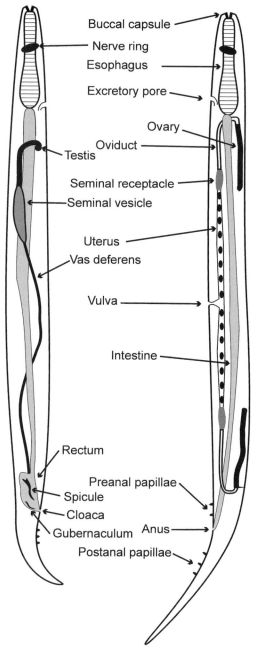

Fig. 8.1 Schematic representation of the body plan of male (left) and female (right) nematodes. The reproductive systems are reduced for clarity; typically, structures such as the ovary and testis are many times longer than the body and are wrapped around the intestine. (Figure courtesy of Al Bush.)

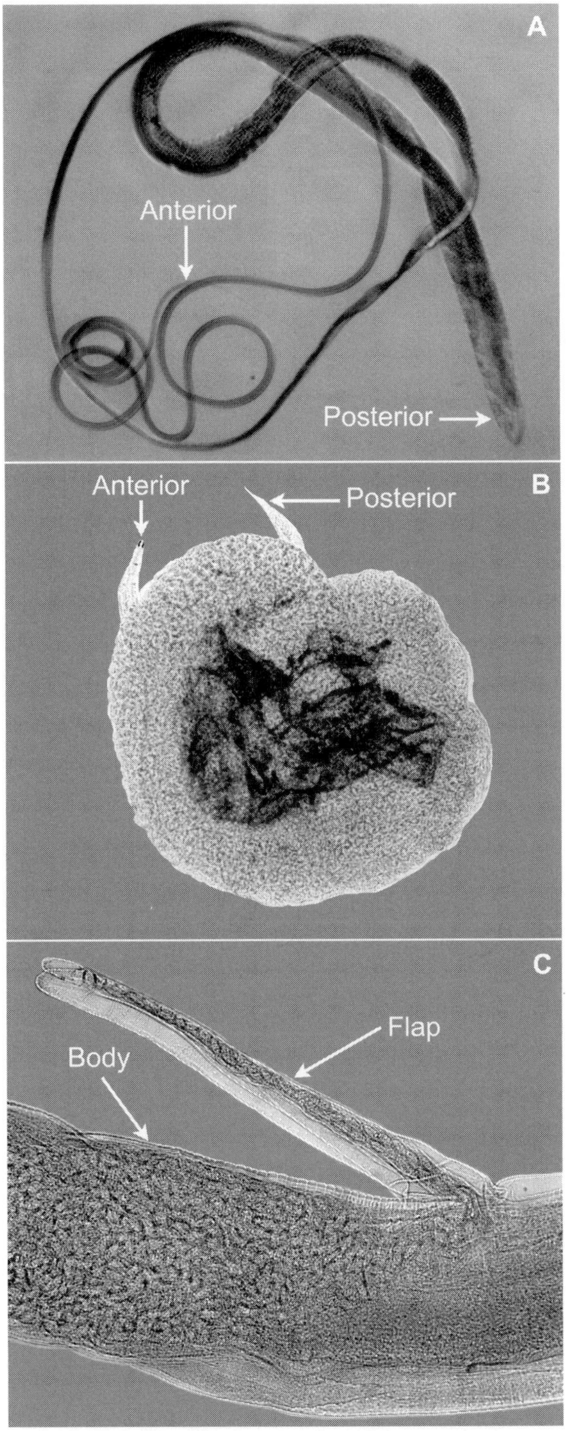

Fig. 8.2 Diversity of body shapes in selected nematodes. (A) *Trichuris*; (B) female *Tetrameres*; (C) vulvar flap of *Evaginurius*. (Photographs courtesy of Al Bush.)

is where the cuticular ridges (**synlophe**) are most prominent. For most nematodes, however, this is a region largely devoid of any cuticular elaborations.

Posteriorly may be found pore-like **phasmids**, caudal papillae, and, on some males, caudal alae or an elaborate copulatory bursa (Fig. 8.4). Phasmids are important for taxonomic purposes. They characterize the Rhabditea, which includes many nematode parasites of animals. Structurally, phasmids are similar to amphids, though the internal glands to which they connect are often smaller and, exceptionally, absent. Caudal papillae are also important taxonomically and, though more prominent in males, may be found on both sexes. Caudal alae, smaller than, but otherwise similar to, cephalic alae, are lateral extensions of the cuticle (Fig. 8.4A). Copulatory bursae are found on the males of two unrelated groups of nematodes. A fleshy bursa with no supporting rays is characteristic of the Dioctophymatida, while males in the extremely diverse Strongylida have well-developed bursae supported by finger-like muscular extensions, called rays (Figs. 8.4B, C). The males use the copulatory bursae to grasp the female during insemination.

8.2.1.1 Holdfasts

Specialized holdfast structures are uncommon in the parasitic nematodes. Exceptions generally involve nematodes parasitic within the gut tube that face peristalsis. For example, the oral plates and teeth of the hookworms act as impressive holdfast structures (Fig. 8.3A). Others, such as *Trichuris* spp., thread the narrow anterior portion of the body through the intestinal mucosa to maintain their position (Fig. 8.2A). Still others, such as female *Tetrameres* (Fig. 8.2B), which live in the glands of the proventriculus of birds, may rely simply on their massive girth (attained after entering the glands) to hold them in place. *Spinitectus carolini* in centrarchid sunfishes possess a series of posteriorly oriented, toothed ridges that are large towards the anterior end of the body, then decline in size and disappear before the first third of the body. The part of the parasite with the ridges is buried in the mucosa, while the rest

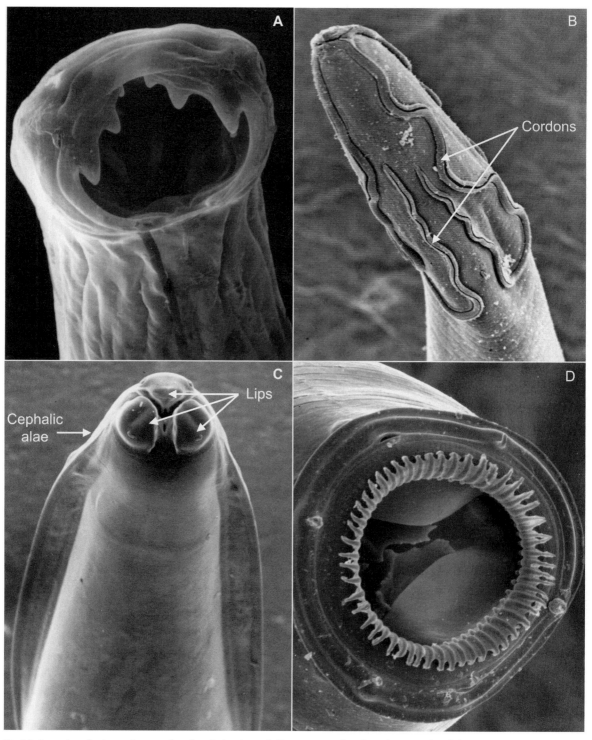

Fig. 8.3 Scanning electron micrographs of variation in cephalic structures in selected nematodes. (A) Buccal capsule of *Ancylostoma*; (B) cordons of *Dispharynx*; (C) lips and cephalic alae of *Toxocara*; (D) buccal capsule of *Triodontophorus*. (Micrographs courtesy of Russ Hobbs.)

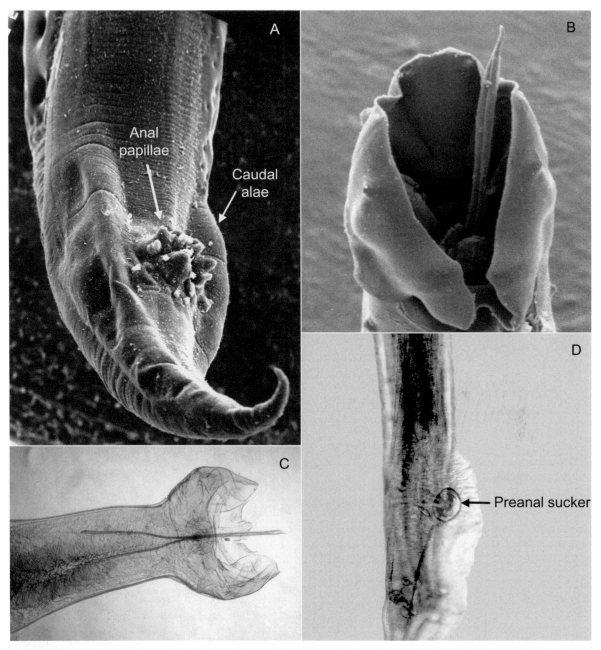

Fig. 8.4 Scanning electron micrographs of variation in caudal structures in selected nematodes. (A) Caudal alae and anal papillae on *Eugenurus*; (B) copulatory bursa on male *Chabertia* with extended spicule; (C) light microscopy photograph showing bursa on *Uncinaria* with extended spicule; (D) preanal sucker on *Heterakis*. ((A,B) Photographs courtesy of Russ Hobbs; (C) photograph courtesy of Toni Raga.)

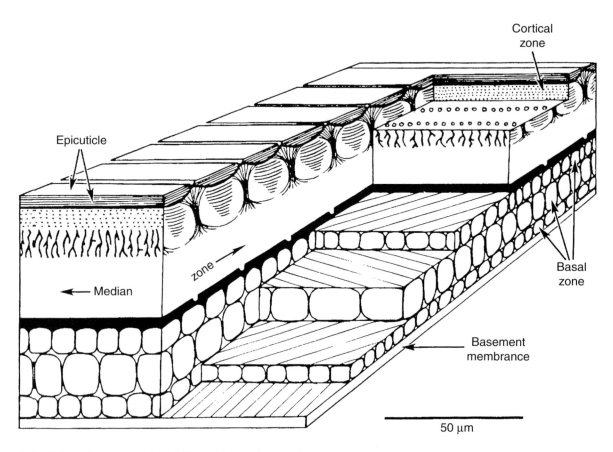

Fig. 8.5 Schematic representation of the complex, multilayered cuticle of the nematode *Ascaris suum*. (Modified from Bird & Bird, 1991, *The Structure of Nematodes*, 2nd edition, with permission, Elsevier.)

hangs in the lumen of the intestine. Secretions from this nematode have been speculated to act as a 'biochemical holdfast' by reducing intestinal peristalsis. Lee (1996) suggests that the mechanism likely involves a vasoactive intestinal polypeptide-like protein that reduces gastrointestinal motility.

8.2.1.2 Body wall and cuticle

The nematode cuticle is a multilayered exoskeleton that provides environmental protection and allows growth by molting. The complex and highly conserved structure of the cuticle has played an important role in the success and diversification of the nematodes, just as in other ecdysozoan taxa such as the arthropods.

The architecture of the cuticle (Fig. 8.5) provides a strong, yet flexible exoskeleton that permits locomotion via attachment to muscle and provides structural support for the hydrostatic skeleton. The impervious cuticle also protects against desiccation and digestion, and it may play a role in nutrition and excretion in some smaller taxa. It is the first line of defense against pathogens and, for parasitic nematodes, other environmental stressors such as anthelmintics. Blaxter *et al.* (1992) suggest a secretory role for the cuticle that may be involved in antigenic activity.

Much of what is known regarding the structure, synthesis, and development of the nematode cuticle comes from ultrastructural and molecular studies involving *Caenorhabditis elegans* (review in Page &

Johnstone, 2007). Politz & Philipp (1992) highlight the application of this enormous research effort to the parasitic nematodes. The cuticle is a collagenous extracellular matrix, secreted by underlying syncytial epidermal cells, that envelops the entire body (Fig. 8.5). Material is secreted by the epidermis and polymerizes along the outer surface, remaining in place until molting.

Ultrastructurally, the outermost layer of the nematode body wall is the lipid-rich epicuticle (Fig. 8.5). This is followed internally by the cuticle, which is a non-cellular layer. Chemically, it is composed of proteins, most notably collagen. In some species, a highly complex network of collagenous fibers is layered in a lattice-like arrangement on the epidermis that permits extensive bending, stretching, and shrinking. In cross-section, there are three main layers in addition to the epicuticle, each of which is collagen-based: an external cortical zone, an intervening median zone, and a basal fibrous zone. The number and substructure of layers varies between species and between different developmental stages within a species (Neuhaus *et al.*, 1996; Martínez & de Souza, 1997). Overlaying the epicuticle is a thin, protein-rich coating that is synthesized from the excretory system and specific gland cells. This surface coat has been associated with immune evasion in a number of parasitic nematodes (Blaxter *et al.*, 1992).

A basement membrane marks the division between the cuticle and the underlying epidermis, internal to which are longitudinal muscles. The muscle cells have two distinct parts: a contractile U-shaped projection (composed of striated fibers) inserting on the epidermis and the other a non-contractile **myocyton** (or cell body) containing the nucleus. Arising from the myocyton are muscle cell arms, terminating in non-contractile, digit-like projections that insert directly on the nerve cords (Figs. 8.6, 8.7). While not unique to the Nematoda, this relationship between muscle and nerve is unusual.

8.2.1.3 The pseudocoelom

Between the somatic musculature and the digestive system is the fluid-filled **pseudocoelom** (Fig. 8.7). The non-compressible pseudocoelom, in combination with the longitudinal muscles and rigid exoskeleton, are

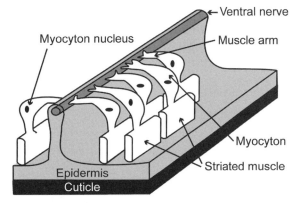

Fig. 8.6 Schematic representation of somatic musculature in nematodes showing the arrangement of the contractile and non-contractile elements of individual muscle cells. All muscle cells in the lower hemisphere attach to the ventral nerve while all muscle cells in the upper hemisphere attach to the dorsal nerve. (Diagram courtesy of Al Bush.)

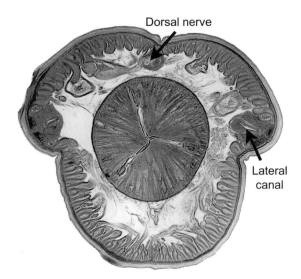

Fig. 8.7 Cross-section through the esophageal region of *Ascaris suum* indicating the powerful musculature associated with the esophagus and its triradiate lumen. The lateral canals, embedded in epidermis, are clearly visible as is the dorsal nerve, also embedded in epidermis, and several muscle cells. (Photograph courtesy of Al Bush.)

main components of the hydrostatic skeleton. In the absence of a circulatory system, hemolymph in the pseudocoelom serves to transfer solutes, e.g., trehalose, from tissue to tissue (Behm, 1997). Although the fluid in the pseudocoelom is mostly cell-free, small cells called **coelomocytes** are thought to play diverse roles in synthesis, storage, and secretion.

8.2.1.4 Digestive system

Unlike any of the helminths discussed thus far, the Nematoda have a complete digestive system, with mouth and anus. The nematode digestive system can be thought of as a tube within a tube, beginning anteriorly with a mouth (stomodaeum) and ending posteriorly with a subterminal proctodaeum (cloaca in males). The mouth leads directly into a buccal capsule that may be simple (most nematodes) or complex (hookworms) and this, in turn, connects with the pharynx (= esophagus). These structures are lined by cuticle that is continuous with the cuticle covering their body. Therefore, when nematodes molt their body covering, they also molt the lining of their foregut and hindgut. The esophagus is triradiate in cross-section and is often muscular (Fig. 8.7). It sucks food into the worm and propels it into, and along, the intestine. Variable numbers of glands are interspersed with the muscles of the esophagus and, in nematodes such as the spirurids, it is often divided into a muscular anterior portion and a glandular posterior portion. Esophageal glands are associated with the digestive system in that their secretions may be involved in such functions as anticoagulation (Pritchard, 1995). The morphology of the buccal capsule, esophagus, and glands associated with the esophagus (Fig 8.8) are important taxonomically.

The foregut terminates with a valve and is followed by a relatively unspecialized, uncuticularized midgut composed of a single layer of columnar epithelial cells bearing microvilli. In some nematodes such as *Ancylostoma*, parts of the midgut are syncytial and in others, such as *Haemonchus*, the entire intestinal epithelium is syncytial. Being pseudocoelomates, nematodes lack any muscle associated with the midgut and propulsion of food is due to the muscular activity of their pharynx, coupled with their high internal hydrostatic pressure and possibly movement of their body. The high internal pressure seems to obviate

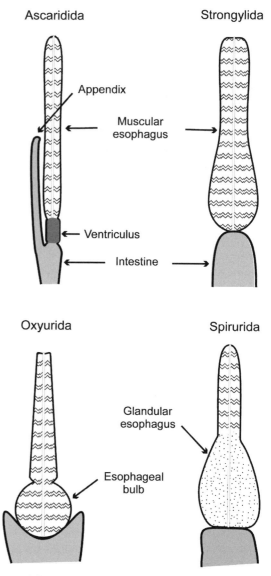

Fig. 8.8 Schematic representation of the variation in esophageal patterns seen in four major groups of nematodes. Although the glands are most prominent in the spirurids, all nematodes have glands associated with the muscular portion of the esophagus. (Figure courtesy of Al Bush.)

regional specialization along the gut; for example, under experimental conditions, *Ascaris lumbricoides* empties its gut about every 3 minutes (Crofton, 1966), and the free-living nematode *Caenorhabditis elegans*

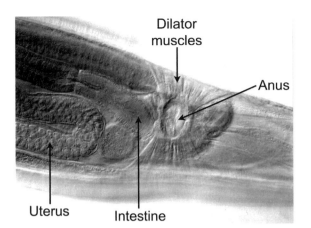

Dilator muscles

Anus

Uterus   Intestine

Fig. 8.9 Posterior region of a female of *Oesophagostomum* sp. showing the rectal dilator muscles surrounding the anus, as well as the intestine and egg-filled uterus. (Photograph courtesy of Al Bush.)

empties its gut about every 45 seconds (Bird & Bird, 1991). That same high internal pressure requires rectal dilator muscles to open the anus/cloaca (Fig. 8.9). The hindgut is essentially a cuticularized tube connecting the posterior intestine with the anus in females or the cloaca in males. Rectal glands, if present, open into the hindgut.

8.2.1.5 Osmoregulatory/excretory system

Osmoregulatory function in nematodes appears correlated with their environment. Free-living forms (and this would include the larvae of parasitic forms) appear capable of moving significant amounts of water across their cuticle. Adult nematodes are often found to be osmoconformers, but Davey (1995) suggests that short-term osmoregulation may occur in some species.

Some nematodes possess a secretory–excretory system composed of one or two **renette cells** that are unique to the phylum. The secretory–excretory system is typically some variation on an H-shape with two longitudinal canals lying in lateral epidermal cords, connected transversely near the anterior end, and extending to the **excretory pore** via an **excretory canal**. Variations include, for example, additional glandular cells, or the extent of the distribution of the lateral canals.

Several nematodes have been shown to excrete organic acids across their cuticles and Sims *et al.* (1996) suggest that this may be common in nematodes. The intestine, however, probably serves as the primary means for removal of nitrogenous waste products, principally ammonia.

8.2.1.6 Reproductive system

Animal-parasitic nematodes are typically dioecious and often show sexual dimorphism. Generally, the female is larger than the male and the male tends to have a curved tail in contrast to the straight posterior end found in females. Genetic sex determination based upon the XX (female)/XO (male) chromosome system is widespread among nematodes and is probably ancestral. Environmental sex determination appears to be restricted to only a handful of groups. Mermithid nematodes (see below) in lightly infected arthropod hosts produce only female offspring, whereas males are favored in heavily infected hosts (Harlos *et al.*, 1980). Similarly, sex determination of some species of *Strongyloides* is determined by the immune status of their vertebrate host (Gemmill *et al.*, 1997). Haag (2005) provides a comprehensive review of variation in genetic and molecular mechanisms underlying nematode sex determination.

Males (Fig. 8.1A) typically have a single testis that is solid and thread-like. This continues as a seminal vesicle that becomes a vas deferens before terminating in the cloaca as an ejaculatory duct. Accessory structures often include spicules, a gubernaculum, and, in the Dioctophymatida and Strongylida, copulatory bursae (Fig. 8.4 B, C). **Spicules** are chitinized structures often surrounded by a sheath. They vary in number (one or two), length, and shape and are used to hold the female's vulva open (against the high hydrostatic pressure) while the non-flagellated sperm pass from male to female. The **gubernaculum**, if present, is a sclerotized portion of the rectum and it appears to act as a guide for the spicules as they pass through the cloaca. Bursae function solely to bind the male tightly to the female during copulation. The size and shape of the spicules, bursae, and gubernaculum are important

taxonomic characters in distinguishing between various taxa of nematodes.

Unlike males, the female reproductive system
(Fig. 8.1) is usually paired, a condition known as
didelphic. Proximally, a thread-like ovary merges with
a slightly larger oviduct and this grades into the uterus.
The uterus is muscular, particularly at the distal ends
where it may be so muscular that it forms a distinct
structure known as the **ovijector**. Ovijectors from the
two uteri unite to form a short vagina that opens to the
exterior through a ventral (usually) vulva. An important, but structurally obscure, region is the junction of
the oviduct and uterus. This area serves as a seminal
receptacle and is the site where fertilization occurs.
After fertilization, eggshell formation begins. The
eggshell of most nematodes is composed of three
layers. Eggs of some groups, e.g., the oxyurids and
trichurids, have one or two opercula, thought to facilitate hatching. In the filarids, the females are ovoviviparous and give birth to live larvae known as
microfilariae. Eggshells in these worms are usually
delicate, composed of a vitelline membrane and a thin,
chitinous shell. In many cases, the microfilariae retain
an egg membrane as a sheath and, in a sense, they
represent highly developed embryos.

8.2.1.7 Nervous system and sensory input

A dominant structure in the nervous system of nematodes is the circumesophageal nerve ring (or 'brain').
As the name implies, it is always located in the anterior
portion of the body in conjunction with the pharynx or
esophagus. It is composed of few cells (four neural and
four glial cells in *Ascaris*) that give rise anteriorly and
posteriorly to a series of nerves. Anteriorly directed
nerves connect with the amphids and the cephalic
papillae. Posteriorly directed are the prominent dorsal
and ventral nerve cords along with less-prominent
lateral nerves. There may be variable numbers of
ganglia associated with these cords as well as commissures connecting the dorsal and ventral cords. The
most posteriad one in the ventral cord is called the
preanal ganglion. It gives rise to additional anteriorly
directed nerves that innervate the rectum and form

what is called a posterior nerve ring. The posterior
nerves also service the caudal papillae and phasmids (if
present). In the much-studied free-living nematode,
Caenorhabditis elegans, a number of behavioral
mutants have been identified and these appear to be
associated with specific nerve cells.

As noted previously, sensory structures for nematodes are mainly papillae and glands. Papillae have
sensory endings that are modified cilia and whose
function is probably tactile. Glands are either amphids,
or phasmids, or both; their sensory endings are also
cilia and their function is largely chemoreception and
secretion. Where muscles are innervated, acetylcholine
is thought to be the primary neurotransmitter. Various
other neurotransmitters have been found in the
nematodes, but their function remains unknown.

8.3 Nutrient uptake and metabolism

Among nematodes, there exists a variety of mechanisms for the acquisition of nutrients. As free-living
juveniles, they feed on bacteria and possibly other
microorganisms. As third-stage juveniles, enclosed
within the second-stage molt, they are seemingly
incapable of feeding, although Ogbogu & Storey
(1996) suggest that L_3 larvae of the filarial nematode
Litosomoides carinii may feed by a transcuticular
route. Adults parasitize almost all organs and tissues
and there is, therefore, an assortment of nutrient
resources available to them, ranging from various tissues and tissue fluids to that which is ingested by the
host. Tremendous variation in mouth structure reflects
the variety of foods exploited by nematodes.

Food is taken in through the mouth, which may be
relatively simple, as might be expected in those that
graze, or quite complex in those that attach and feed
on body fluids. The nematode pharynx is composed
of a syncytium of radial muscle. The strong musculature associated with the pharynx (Figs. 8.7, 8.8) is
necessary to 'pump' food against the high internal
pressure of the worm. It is also the pumping of the
pharynx that helps propel the food through the
remainder of the worm's gut.

Given the diversity of nematodes and the habitats they occupy, metabolic challenges are great. For example, many adult nematodes live in oxygen-poor environments, yet have eggs or larvae that occur in comparatively oxygen-rich environments. The eggs of many parasitic nematodes that are shed in feces have been examined and all require oxygen for continued development. When the pO_2 is reduced, development slows; under anaerobic conditions, development ceases. Interestingly however, although development may be curtailed, the eggs (and even the infective larvae) of some species that infect the host by an oral route are resistant to oxygen deprivation, a possible hedge against poor environmental conditions. On the other hand, the larvae of some skin-penetrating species such as *Strongyloides* are very sensitive to a lack of oxygen. Little is known about the intermediary metabolism of nematode larvae. There are reports that glycolysis, the TCA cycle, and cytochrome systems may be present and functional (Kita *et al.*, 1997).

Adult nematodes have two options with regard to intermediary carbohydrate metabolism. The first is to process glucose solely through glycolysis. This is a catabolic process, independent of oxygen, and occurs in the cytosol. In energetic terms, it is extremely wasteful since glycolysis alone yields a net of only two ATP from a molecule of glucose. (Aerobically, when glucose is metabolized through glycolysis, the TCA cycle, and the electron transport system resulting in CO_2 and H_2O, 36 to 38 ATP are produced. For this complete degradation to occur, pyruvate must enter the mitochondria and oxygen must be present.) However, many nematodes live in glucose-rich environments and we have already noted that they pass food through their guts rapidly. Thus, the 'wasteful' (this process extracts only about 2% of the energy present in glucose) use of glucose may pose no problems for them. The last intermediate in glycolysis is pyruvate, which can be reduced to lactate (and $NADH_2$ oxidized). Nematodes employing this form of catabolism are called homolactic fermenters.

An alternative has been to develop different metabolic pathways, to become what are called CO_2 fixers. The route followed is typical for glycolysis to the point of phosphoenolpyruvate. Instead of being converted to pyruvate, however, the phosphoenolpyruvate is carboxylated to oxaloacetate. The oxaloacetate is reduced to malate, which then enters the mitochondria where malate is metabolized to succinate and pyruvate. This yields additional ATP beyond simple glycolysis. The end products are succinate, which is excreted as an acid end product, and pyruvate, which is rarely excreted. In addition, other organic acids such as lactate and proprionate may be excreted. Several nematodes have some of the enzymes characteristic of the TCA cycle and still others have elements of the cytochrome system.

Protein catabolism is not well understood in the nematodes, except that ammonotelic, ureotelic, and even uricotelic excretion are all known to occur. Since excess ammonia is toxic, it would only be expected where suitable quantities of water are available. For example, *Ascaris lumbricoides* is predominantly ammonotelic; interestingly, however, when water becomes limiting, it switches to ureotelic excretion.

Anabolism, or synthesis, is the other side of the story to catabolism. Anabolic pathways have received little study in the nematodes. Glycogen is the major energy reserve and it is synthesized from carbohydrates ingested during feeding. Typically, higher glycogen reserves are found in those nematodes that do not have direct access to the host's carbohydrate reserves, e.g., *Ascaris* sp., compared to those with more direct access, e.g., *Ancylostoma* sp. When provided with suitable substrates, e.g., acetate, glycine, and glucose, some nematodes can synthesize a wide variety of amino acids. Others are known to synthesize fatty acids.

8.4 Development and general life cycle

The study of nematode development has a long history. Indeed, the first elucidation of fertilization and meiosis in animals was provided by studies on the ascarids of horses. Nematode zygotes undergo deterministic cleavage. Many, but not all, nematode eggs are laid in the single- or two-cell stage. Exceptions include eggs being laid with fully developed larvae or even ovoviviparity. The first cell division (resulting in two blastomeres) is important because one of the daughter cells will go on to produce only somatic tissue, while the other will produce some somatic tissue and all germinal tissue. In some species, the blastomere whose fate is to produce only somatic tissue undergoes the process of chromosome diminution, or more properly, chromatin diminution. Here, segments of chromosomes are programmed for removal within particular cells (review in Muller *et al.*, 1996). In this progression, parts of the chromosomes fragment, resulting in the loss of much genetic information. Chromatin diminution occurs several times so that, by the 64-cell stage, only two cells retain the complete genetic code. These two cells will produce the gametes for the next generation. All of the other cells, which may have lost 25–85% of their genetic coding through chromatin diminution, will produce all of the somatic tissue. Molecular studies involving *Ascaris suum* have shown that this enigmatic process involves highly regulated genome rearrangement involving attack by degrading enzymes at breakage points on precise regions of DNA.

Embryogenesis progresses through a typical morula and blastula stage prior to gastrulation. Once organogenesis is complete, mitosis ceases and, with the rare exception of some intestinal or epidermal cells, or both, there is nuclear constancy throughout life (eutely).

The eggshell of nematodes is important because it often acts as a barrier both within the host and in the free-living environment. Typically, the nematode eggshell is divided into three layers produced by the egg itself, i.e., an inner lipid layer, a middle chitinous layer, and an outer vitelline layer. Added to these egg secretions may be one or two uterine layers (secreted by the uterus). The uterine layers often include pores, spaces, proteins, or some combination. Although the layers of nematode eggs are permeable to gases and lipid solvents, they are nonetheless quite resistant to many substances. For example, the eggs of *Ascaris* spp. can survive in solutions such as 9% sulfuric acid, 14% hydrochloric acid, and even 12% formalin (Bird & Bird, 1991). They are obviously highly resistant to

environmental extremes. For example, a high concentration of trehalose (Behm, 1997), the same sugar involved in anhydrobiosis in taxa as diverse as the rotifers and tardigrades, is thought to provide nematode eggs with the ability to survive environmental stresses such as cold and desiccation.

Excepting those that are ovoviviparous, and a very few others, most parasitic nematodes produce an egg that will be deposited in the host's feces. Once in the environment, and after embryonation (if necessary), the egg may hatch into a first-stage larva or L_1 (such as occurs in strongyles), or it may remain as a resistant stage (as in some ascarids). Hatching in the environment may be spontaneous, but some environmental factors serve as triggers. In any case, the actual process of hatching involves the breakdown of the lipid layer (possibly enzymatic) and chitinous layer (due to the secretion of chitinases). The secretion of other enzymes also helps to degrade the shell to the point where the larvae can eclose. A rhabditiform larva (L_1) hatches from the egg and is fully free-living, and ready to feed on microorganisms. It will molt to become a rhabditiform larva (L_2) that is also free-living and and microbotrophic. At some point, the juvenile will molt and the cuticle will remain in place as a sheath surrounding the third-stage filariform larva (L_3). Although freeliving, this stage is incapable of feeding and must survive on stored energy reserves, primarily lipids (Medica & Sukhdeo, 1997). This is usually the infective stage and, depending on the parasite's life-cycle strategy, this stage must penetrate or be eaten by a host. Rarely, for example in some metastrongyles, the L_1 will penetrate an intermediate host and develop to the infective stage within that host.

Alternatively, an egg may remain as a resistant stage in the environment until it is ingested. Once ingested by an appropriate host, it will require a hostspecific stimulus to induce hatching. For example, *Ascaris* eggs require a fairly specific temperature, redox potential, and CO_2 concentration to induce hatching. Hatching under such conditions may result in an infective L_3 for those parasites having a direct life cycle or it may result in a first-stage juvenile, which will develop to the infective L_3 in those forms that use an intermediate host.

Some filariform larvae penetrate the final host directly and lose their sheaths in the process; others, ingested directly or via an intermediate or paratenic host, must exsheath. Like the hatching of ingested eggs, exsheathment often requires specific stimuli and, in fact, the stimuli to exsheath are often the same as those required for egg hatching.

Developmental arrest is common to many nematodes and may involve either the egg or the sheathed filariform juvenile. Functionally, it acts as a hedge against mortality since the appropriate stage is resistant to environmental vagaries or physiological degradation (Fetterer & Rhoads, 1996). An excellent review of the extent and significance of developmental arrest in the parasitic nematodes of animals is provided in Anderson (2000). We will also see that arrest enables several mammalian nematodes to utilize transmammary or transplacental transmission routes.

8.5 Biodiversity and life-cycle variation

The notion that animal parasites display astonishing variability in their life cycles and life histories has been emphasized in earlier chapters, especially in our coverage of groups such as the protists and platyhelminths. Yet, in terms of life cycle and life history diversity, there is no question that the nematodes rival these other groups. The simplified phylogenetic tree provided at the end of this chapter includes parasitic nematode taxa that reproduce via dioecy, facultative or obligate **parthenogenesis, haplodiploidy**, or hermaphroditism. The tree also includes groups that have direct life cycles involving the ingestion of eggs or larvae, others that involve the penetration of larvae through the hosts' epidermis, and still others that incorporate one or two intermediate hosts or vectors. This sequence of examples does not include the bizarre types that seem to 'do their own thing,' such as some pinworms that can autoinfect without ever leaving their hosts, the trichinellids that use a single animal as both final and intermediate host, and some rhabditids

that produce larvae that can either develop into dioecious adults in the soil, or into parthenogenetic parasitic females. Arguably, no single group of related organisms possesses greater life cycle and life history variability than the nematodes. Some of that complexity and versatility is illustrated in Figure 8.10 and in Box 8.1. The figure is complex, yet it still lacks some variants. For example, omitted are the metastrongyles, which have L_1's that penetrate an intermediate host. Also omitted are species of Cephalobidae, i.e., *Daubaylia potomaca*, in which adult females are the infective stage for their molluscan definitive hosts (Zimmermann *et al.*, 2011). Nonetheless, the illustration serves to emphasize the diversity and life-cycle variation encountered within the nematodes.

Whether free-living or parasitic in plants, invertebrates, or vertebrates, nematodes must be regarded as among the most diverse creatures on earth. Indeed, it is an interesting exercise to consider a single other taxa on the Tree of Life that is successful within the three main habitats: water, land, and hosts. If 'success' is further indicated by the diversity of hosts infected, then all vertebrates that have been studied to date, regardless of their habitat, are host to at least one nematode species. Lastly, unlike some parasites, e.g., the cestodes and the acanthocephalans, where the adults are restricted to the small intestine, nematodes exploit virtually any tissue in the vertebrate body.

In this section, we cover selected examples that span the diversity of life cycles and life histories of the nematode parasites of animals. We first introduce the diversity of nematodes whose adults utilize invertebrate hosts. Although we recognize that a few nematodes are found as adults in invertebrates such as earthworms and molluscs, our focus is on the nematodes of insects. We then shift our focus onto selected examples that span a portion of the nematode diversity within the vertebrates.

8.5.1 Diversity of nematodes in invertebrate hosts

The utilization of invertebrates as hosts for adult nematodes has arisen independently in four main

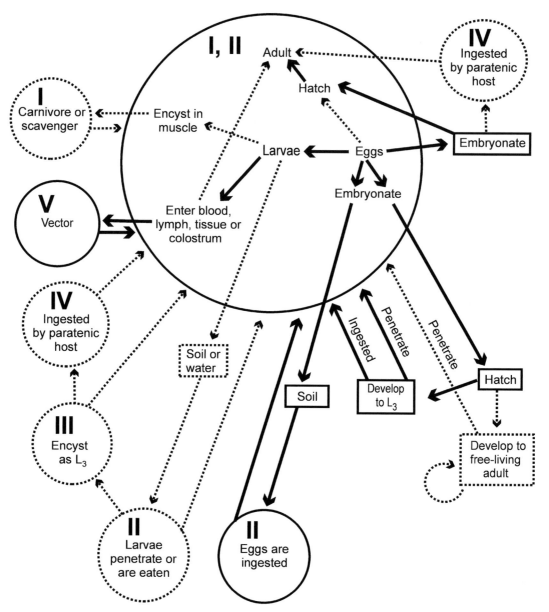

Fig. 8.10 Alternative pathways found in the life cycles of the parasitic Nematoda. Host types are represented by roman numerals (I=definitive host, II=first intermediate host, III=second intermediate host, IV=paratenic host, V=haematophagous or tissue-eating micropredator or 'vector'). Stages within circles are found within hosts and stages in rectangles are found in the environment. Despite the obvious complexity, there are really four basic patterns of interest (solid lines) and all begin with the fate of the egg. If the egg leaves the body unembryonated, it is ingested by the host and the larva hatches and passes through all larval stages within the final host. If the egg embryonates (at least partially) prior to leaving the definitive host, there are two possible fates. First, it can hatch in the external environment and develop to the infective L3. This may then penetrate the definitive host directly or be eaten by the definitive host. Second, an intermediate host may ingest the embryonated egg. In this case, the juveniles develop to the infective stage in the intermediate host. Finally, the eggs may hatch inside the definitive host and larvae will be ingested by a micropredator; after a period of development, the larvae will be infective to a definitive host when the micropredator feeds again. (Figure courtesy of Al Bush.)

Box 8.1 | **Evolution of nematode life-cycle plasticity: developmental switching in *Strongyloides* spp.**

In an ideal world, we might expect organisms to have evolved that could alternate between a completely free-living generation and a completely parasitic generation. Such organisms could enjoy the benefits of the parasitic life style, such as high fecundity and fast generation time, yet avoid costs such as host immunity. The closest we come to this hypothetical scenario lies in the parasitic rhabditid nematodes of animals, many of which incorporate a soil-dwelling, dioecious stage into their life cycles.

The nematode *Strongyloides ratti* is an intestinal parasite of rats. Similar to other members of the genus, this species has a remarkable life cycle outside the host that switches between homogonic and heterogonic development. In homogonic development, the eggs from parthenogenetic females pass into the feces and the larvae develop directly through two molts into infective 3rd stage larvae. In heterogonic development, larvae molt four times into dioecious, free-living adults. Following mating in the soil, eggs from the female hatch and proceed through two further molts into infective L_3's.

Over the course of an adult female's lifetime, her eggs may develop homogonically, heterogonically, or she may adopt a mixed strategy and produce both. Identification of the mechanisms that determine such developmental plasticity has vexed parasitologists for many years (review in Hawdon & Schad, 1991). Results from a series of clever laboratory experiments with various *Strongyloides* model systems have provided some intriguing answers, some of which give tantalizing insights into the evolution of nematode parasitism itself (Ashton *et al.*, 1998; Viney, 2009; Ogawa *et al.*, 2009).

Results from Mark Viney's initial experiments confirmed prior studies that identified two general classes of factors that influence larval development. Extra-host factors, especially high soil temperatures and high soil moisture, favor heterogonic development. Yet intra-host factors were also important, chief among these being host immunity. Gemmill *et al.* (1997) manipulated host immune status, showing that larvae from rats with acquired immunity against *S. ratti* tend to develop heterogonically. In contrast, hypothymic rats and those with suppressed immune systems (via hormone treatment) tended to produce homogonic larvae. These effects occurred independently of potential confounds such as worm intensity and worm age. In an interesting twist, the production of heterogonic larvae was also favored by immune responses to nematodes other than *S. ratti* (West *et al.*, 2001). Taken together, these results indicate that heterogony, i.e. sexual reproduction, is selected for when eggs or newly hatched L_1's detect stressful environmental conditions, either directly in the soil or indirectly via the host's immune status.

Subsequent studies have determined the precise physical and chemical mechanisms behind the detection of environmental stressors by recently hatched L_1's. Chemosensory

Box 8.1 | (continued)

neurons located in the lateral amphids of *S. stercoralis* L_1's appear to play a key role in the detection of environmental cues in the soil. Ashton *et al.* (1998) switched larval development from homogonic to heterogonic by surgically ablating the cell bodies of two classes of amphidial neuron. They suggest that sensory input into both of these groups of neurons controlled the homogonic/heterogonic developmental switch. At least one endocrine pathway has been identified that links the detection of environmental stressors and developmental switching in *Strongyloides* (Ogawa *et al.*, 2009). One nuclear receptor, known as DAF-12, binds to ligands associated with the production of numerous growth factors. Ligands for DAF-12 are the dafochronic acids (DAs). Ogawa *et al.* (2009) have shown that soil conditions that are conducive to the production of heterogonic larvae generate DA under the action of various pathways. When DA was added directly to newly hatched larvae of *Strongyloides papillosus* (a parasite of rabbits), development switched from homogonic to heterogonic. These experiments provide compelling evidence for a chemosensory mechanism that links environmental cues with life-cycle versatility – a key element in the success of the parasitic nematodes.

These results have important implications for the evolution of nematode parasitism. Intriguingly, the two classes of amphidial neurons in *S. stercoralis* that control the homogonic/heterogonic switch are homologs of those present in *Caenorhabditis elegans*. In *C. elegans*, ablation studies have shown that these neurons control the switch leading to the **dauer juvenile** versus non-dauer juvenile stages of the life cycle. Further, the mechanism for the dauer–non-dauer developmental switch involves DA/DAF-12 signaling (Ogawa *et al.*, 2009). Thus, both the signaling pathway and the structure and position of the amphidial neurons seem to be deeply conserved in nematodes. Viney (2009) proposes the intriguing hypothesis that minor tinkering of key developmental switches at the dauer–non-dauer interface was a key step in the evolution of nematode parasitism. He proposes that environmental cues linked to food availability (and the production of DAs) lead to the resumption of development of *C. elegans* dauers, whereas environmental cues linked to host suitability lead to resumption of development of infective L_3's.

groups. Phylogenetic studies indicate that there is an association between invertebrate and vertebrate parasitism, and that invertebrate parasitic clades are basal to major vertebrate parasitic ones (Dorris *et al.*, 1999). Three families of primarily insect nematodes, in particular, have been the focus of considerable research effort. These include species within the Steinernematidae, Heterorhabditidae, and Mermithidae. Their potential as biological control agents of insect pests is an important reason why they have been the subject of intense applied research (Gaugler, 2002; Stock, 2005). The steinernematatids have also been the focus of biodiversity surveys and phylogenetic studies (Stock & Gress, 2006).

8.5.1.1 Steinernematidae and Heterorhabditidae

Entomopathogenic nematodes within these families are used as biological control agents for significant insect pests (Gaugler, 2002; Stock, 2005; Box 8.2). Like other Rhabditida, these nematodes are bacteriovores. The infective L_3 of the ubiquitous entomopathogenic nematodes *Steinernema* and *Heterorhabditis* carry bacterial symbionts in their gut. In *Steinernema* species, the symbiotic bacteria *Xenorhabdus* are located in a small intestinal vesicle whereas in *Heterorhabditis* the bacterial symbionts *Photorhabdus* are located in the anterior region of the infective juvenile's gut. Each nematode species is associated with only one species of bacterial symbiont, almost all of which are found only in the intestines of entomopathogenic nematodes and the body cavities of dead and dying insects that are infected by one of these nematode species. The bacteria produce anti-microbial metabolites, some of which are novel, and show considerable promise in the development of antibacterial products and other bioactive compounds that may benefit humans (Webster *et al.*, 1998, 2002; Box 8.2).

The life cycle varies depending on the insect–nematode system. Generally, following entry into the insect, the infective L_3 penetrates into the hemocoel and releases its symbionts. Once in the hemocoel, the symbiotic bacteria multiply rapidly and the insect dies from septicemia, usually within 24–48 hours. The nematodes feed on the increasing number of bacteria as rhabditiform L_3's, growing and molting into adults, and completing 1–3 generations in the host cadaver (Stock & Gress, 2006). In the steinernamatids, mated females typically produce eggs that develop into filariform L_3's (Smart, 1995). Eventually, as food resources are depleted, the filariform L_3's exit the insect corpse and enter the soil, then disperse in search of new hosts. They carry in their intestine their respective species of bacterial symbiont for transmission to another insect. The infective juveniles do not feed in the soil, but can survive for several weeks on stored reserves in a state of anhydrobiosis (Smart, 1995).

8.5.1.2 Mermithidae

Mermithid nematodes are unusual in that they are parasitic only as juveniles and must leave the host before reaching the adult stage, which requires water for further development and reproduction. This life history characteristic is found in only one other phylum, the Nematomorpha (Chapter 9). Like the nematomorphs, mermithids are primarily parasites of the body cavity of insects; however, a variety of spiders and crustaceans also host mermithid nematodes (Poinar, 1983, 1985). Moreover, both mermithid and nematomorph juveniles can reach extraordinarily large sizes relative to their hosts (Color plate Figs. 2.1, 8.3). Adult mermithids do not feed and thus rely on stored nutrients retained during the juvenile stage within the host's hemocoel. Nutrients available in the insect's hemolymph are transported across the rapidly developing worm's thin cuticle. The nutrients (mostly lipids) are then stored in a structure known as the **trophosome** for subsequent utilization by free-living aquatic adults.

Mermithids impact their hosts in a variety of ways, ranging from fat body depletion to castration, and modification of behavior to sex reversal. Mermithids, like insect parasitoids, typically kill their host upon emergence. It is for this reason that several species have been examined for their biological control possibilities against insect pest populations. In this regard, the most famous mermithid is probably *Romanomermis culicivorax*, a parasite of mosquito larvae. In the 1970s, this species was briefly commercialized and experimental field trials were conducted to control mosquito populations. The product went by the catchy name, *Skeeter-Doom*! Intriguingly, this mermithid has also been recently shown to secrete and then shed an extracellular surface coat as an adaptation to avoid the mosquito immune response (Shamseldean *et al.*, 2007).

There is increasing evidence that mermithid-infected hosts induce a behavior change that enhances the probability that their terrestrial arthropod host

Box 8.2 | **Entomopathogenic nematodes and their microbial symbionts: tapping their chemotherapeutic potential**

There is no doubt that parasitic nematodes have developed a bad reputation. This comes as no surprise, for they can cause important diseases of plants, humans, livestock, and our pets, as well as wildlife. Yet, within this extraordinarily diverse group of parasites, there are the so-called entomopathogenic nematodes, several species of which have been raised commercially and used successfully as environmentally friendly biological control agents for significant insect pests. For example, a biological insecticide formulated with the nematode *Steinernema carpocapsae* is used in the biological control of the black vine weevil, an important pest of cranberry crops. The cranberry industry was the first in North America to employ beneficial nematodes as biological control agents on a commercial basis (Webster *et al.*, 1998; 2002).

These nematodes have unique intestinal bacterial symbionts and the metabolites that they produce have potent antibacterial and antifungal properties that delay the putrefaction of the dead insect. As stressed by Webster *et al.* (1998; 2002), this enables the continued growth, development, and reproduction of the bacterial symbionts in the relative absence of competition from the soil and insect gut microflora.

Xenorhabdins and a range of indole derivatives have been isolated from various strains and species of *Xenorhabdus*, all of which have antibiotic and/or antifungal activities (Paul *et al.*, 1981). They appear to function by inhibiting microbe RNA synthesis. *Photorhabdus* species also produce indoles that have antibacterial properties. These diverse metabolites have been shown to also have both nematicidal and insecticidal activities (Webster *et al.*, 2002). The nematicidal activities in particular may be ecologically significant. By repelling other nematode species that compete for food and space, the entomopathogenic nematodes increase their potential for survival and transmission to new insect hosts.

The properties of these metabolites suggest that they might be effective against drug-resistant human bacterial pathogens. The activity of nematophin (Li *et al.*, 1997) produced by *Xenorhabdus nematophilus* is of particular significance in this regard. It has been shown to have specific antibacterial activity against drug-resistant strains of the human pathogen *Staphylococcus aureus*. Another major advantage of some of these new antibiotics over conventional ones is that they are not structurally related to currently used antibiotics. This lessens the probability that the bacteria are already resistant to the antibiotic action of these novel metabolites. Another of these promising compounds, originally extracted as a metabolite from a culture of one of the bacterial symbionts, and later synthesized, has demonstrated anti-cancer activity (Li *et al.*, 2007). Yet another is presently in clinical trials as an anti-inflammatory agent with potential for use as a topical therapeutic against psoriasis (John Webster, personal communication).

will come into contact with water. Poulin & Latham (2002) found that sand-inhabiting beach hoppers (amphipods) containing the mermithid nematode *Thaumamermis zealandica*, burrow more deeply into moist sand substrata than do uninfected hosts. This behavior can be viewed as a parasite adaptation since adults require this water-saturated sand to complete the life cycle and reproduce. Williams *et al.* (2004) propose an intriguing potential mechanism for this behavioral change. They suggest that the mermithid increases hemolymph osmolality, which in turn induces a need for water in the host, and the observed water-seeking behavior. Thus, it appears that mermithids may induce a 'thirst response' in their arthropod hosts.

The life cycles of mermithids can either be direct or indirect. Infective larvae of direct life cycle mermithids hatch from eggs in the aquatic environment and second-stage larvae penetrate the cuticle of aquatic arthropod hosts using a stylet apparatus. They then develop rapidly into the large third-stage juvenile in the hemocoel. Conversely, some mermithids with entirely terrestrial hosts incorporate paratenic hosts. For these types, infective juveniles penetrate aquatic insect larvae then enter a dormant state within the hemocoel, effectively bridging the aquatic–terrestrial transmission barrier. They remain in arrest, even surviving the host's metamorphosis. The mermithid-infected paratenic hosts are then eaten by predatory arthropods and the parasite develops into third-stage juveniles in the hemocoel. A post-parasitic juvenile then emerges from the host into water (or moist habitat) to mature, mate, and oviposit.

Both direct and indirect life cycles occur in mermithids infecting spiders (Poinar, 1985; Penney & Bennett, 2006). For example, Poinar & Early (1990) demonstrated that aquatic invertebrates are paratenic hosts for a New Zealand species of trap door spider mermithid *Aranimermis giganteus* (Color plate Fig. 2.1). Paratenic hosts are also used by those mermithids that parasitize predatory social insects such as ants (Poinar *et al.*, 2006). Further, the mermithid *Pheromermis pachysoma* exploits the complex social behavior of its predatory hymenopteran hosts, familiar yellow jacket wasps. Females oviposit eggs in water;

L_1's hatch and penetrate aquatic insect larvae, in which they coil up in diapause. These paratenic hosts are captured by worker female wasps and fed to developing yellow jacket larvae in the wasp colony. Once in the wasp, the larvae begin development, growing enormously and surviving wasp metamorphosis. When the worker wasp leaves the hive and comes into contact with water the juvenile parasite emerges and searches for a mate to begin the life cycle anew.

Several species of fossil mermithid nematodes have been described. These were in the process of emerging from their insect hosts in tree resin millions of years ago, trapped, and then 'frozen in time' in amber (Fig. 8.11). George Poinar and his associates have pioneered the study of these amber fossils (Poinar, 2002a; 2002b; Poinar *et al.*, 2006; Poinar & Monteys, 2008), some dating back 120–135 million years. Poinar *et al.* (2006) described a fossil mermithid emerging from an ant in amber collected from the Dominican Republic. They concluded that "ant infection by mermithids in the Neotropics is widespread and has occurred for at least 20–30 million years."

Another family of nematodes in the order Mermithida is the Tetradonematidae. One recently described species, *Myrmeconema neotropicum*, infects arboreal ants in the rainforest canopies of Panama and Peru. The abdomens of infected ants turn a bright red and they elevate their gaster region (Color plate Fig. 7.4). This dramatic change in appearance and behavior represents another fascinating case of phenotypic manipulation (see Chapter 15). It is a special case of parasite-induced fruit mimicry, an adaptation to facilitate predation by fruit-eating birds and enhance dispersal of the nematode's eggs (Yanoviak *et al.*, 2008; see Box 15.1).

8.5.2 Diversity of nematodes in vertebrate hosts

Thousands of species of nematodes infect vertebrates. Here, we explore vertebrate–nematode diversity and life history variation by considering selected examples within seven important orders, i.e., Trichocephalida, Dioctophymatida, Rhabditida, Strongylida, Ascaridida, Oxyurida, and Spirurida. Phylogenetic analyses support

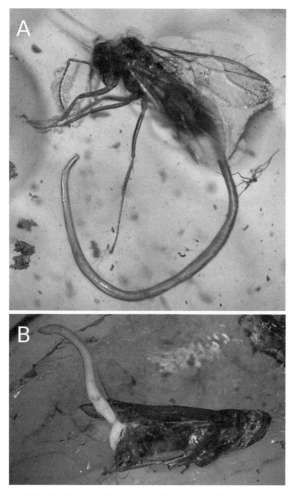

Fig. 8.11 Fossil mermithid nematodes preserved in 40 million-year-old Baltic amber. (A) *Heydenius formicarius* emerging from the ant *Prenolepis henschei*; (B) *Heydenius brownii* emerging from an unidentified planthopper (Achilidae). (Photographs courtesy of George Poinar, Jr.)

the hypotheses that these orders, with the exception of the Rhabditida (and possibly the Spirurida), are monophyletic (Blaxter *et al.*, 1998; Smythe *et al.*, 2006; see Box 8.5).

8.5.2.1 Trichocephalida

The major characteristic of the trichocephalids is the possession of a unique esophageal structure known as

a **stichosome** (Fig. 8.12). It is composed of large glandular cells called stichocytes, which surround the thin-walled esophagus. Other characteristics are bi-operculate eggs, a single gonad in both sexes, and males having one, or no, spicule or spicule sheath. The three most diverse and potentially medically important genera within the order are *Trichurus*, *Capillaria*, and *Trichinella*.

Trichuris is worldwide (but tropical and subtropical) in distribution. Various species are found in the large intestine and cecum of terrestrial mammals, especially wild and domestic ruminants. Transmission is direct and requires ingestion of fully embryonated eggs by the definitive mammalian host. Their common name, whipworm, derives from their rather distinctive body shape in which the anterior end is thread-like (composed mostly of stichosome) compared to a more robust posterior (Fig. 8.2A). The thread-like anterior end embeds in the gut mucosa. *Trichuris trichiura* is the species found in humans and other primates. Pathology due to this parasite can be severe. In heavy infections, it can cause intestinal epithelium damage, leading to hemorrhage and anemia. It is one of the major **geohelminths** of humans in tropical and subtropical countries. Like many direct life cycle nematodes, the combination of poor sanitation and a humid climate are favorable for egg survival and development.

The 300 described species of *Capillaria* are also worldwide in distribution and can parasitize all classes of vertebrates, including humans. Most systematists feel that there are probably several genera contained within '*Capillaria*' but, for convenience sake, we will refer to the single genus. Unlike *Trichuris*, they are found in various parts of the gut, urinary tracts, and organs such as the liver and spleen. Most *Capillaria* species are transmitted directly by ingestion of eggs. One exception is *C. caudinflata*, a potentially serious pathogen in several species of birds, notably chickens and turkeys, which requires an earthworm intermediate host. Adults of *C. hepatica* occur in the livers of microtine rodents, but females never release eggs into the gastrointestinal tract. Instead, the livers of infected

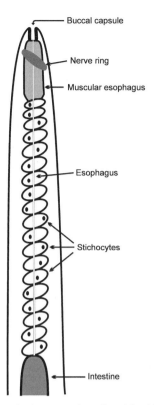

Fig. 8.12 Schematic representation of a trichurid stichosome composed of individual stichocytes. (Figure courtesy of Al Bush.)

Fig. 8.13 Larva of *Trichinella spiralis* in a nurse cell within skeletal muscle. The outer layer is the collagenous capsule. (Photograph courtesy of John Sullivan.)

rodents must be consumed by a scavenger or predator. The resistant eggs pass through the predator, followed by embryonation in the soil.

Species in the genus *Trichinella* are the best known of the trichocephalids. Since the discovery of the life cycle of *T. spiralis* in 1860, it has been regarded as one of the most geographically widespread and clinically important parasites of humans in the world. The genus has undergone major taxonomic revision over the past decade. Molecular analyses currently distinguish at least 11 morphologically indistinguishable taxa in mammals, birds, crocodilians, and saurians (Pozio & Zarlenga, 2005). The role of *T. spiralis* as a human pathogen has led to extensive research on its pathology, epidemiology, systematics, and evolution. Characterization of its complete genome is forthcoming. In an authoritative

review, Dickson Despommier (2004), a research leader on *Trichinella*, notes: "*Trichinella spiralis* continues to hold our interest, both as a nematode parasite with a complex life style and as a thing of utter beauty."

Trichinella spiralis is a small worm (1–3 mm) with an atypical life cycle, in that the same individual serves as both intermediate and definitive host. Transmission takes place by ingestion (via carnivory or scavenging) of raw or undercooked muscle tissue that contains infective larvae (Fig. 8.13). Site selection occurs within columnar epithelium cells along the upper part of the small intestine. Immature adults occur within rows of these cells, making them 'intra-multicellular' in terms of site selection (Despommier, 1993). Mating occurs within these cells, leading to the production of live L_1's. These larvae possess an oral stylet that penetrates through host cells, allowing entry into the general circulatory system. Adult females continue to produce eggs for 4–16 weeks, with the duration of patency determined primarily by acquired host immunity.

Migrating *T. spiralis* L_1's leave the circulatory system and enter host cells. Those that do not enter skeletal muscle cells die, or they exit the cell and re-enter the circulatory system. The penetration of skeletal muscle cells by L_1's initiates a process leading to the

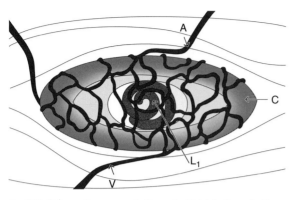

Fig. 8.14 Schematic representation of a *Trichinella spiralis*–nurse cell complex within mammalian host skeletal muscle. Arterioles (A) and venules (V) form a circulatory rete to nourish the developing juvenile (L_1) within a collagenous capsule (C). (Figure courtesy of Danielle Morrison; modified from Despommier, 1990, *Parasitology Today*, **6**, 193–196.)

formation of the characteristic *Trichinella* **nurse cell** (review by Despommier, 1998). This remarkable structure (Fig. 8.14) nourishes and protects individual larvae for up to 30 years following the penetration of individual cells, a process that neatly resolves the problem of 'staying in' (see Chapter 1) for this intra-cellular macroparasite.

The fascinating process of nurse cell formation starts with de-differentiation of a muscle cell that has been penetrated by an L_1. From 1–5 days after penetration, contractile elements within the cell disappear, as do most organelles. The permanent loss of host mitochondria is a critical step at this early stage, as the larvae are functionally anaerobic. The process of de-differentiation of muscle cells is followed by re-differentiation of the muscle cell that is controlled and regulated by the developing worm. By Day 8, host cell nuclei have expanded in size. At this time, unique proteins are produced and secreted by stichocytes on developing larvae. By Day 8 (and permanently thereafter) segments of these proteins are present within host nuclei. These secretions are thought to be involved in the process of nuclear expansion and, importantly, in the regulation of both host and parasite gene expression. For example, secretions from

stichocytes are implicated in the up-regulation of genes for collagen synthesis and also for vascular endothelial growth factor (VEGF). The latter is known to regulate angiogenesis that leads to the formation of the characteristic rete circulatory network that develops adjacent to the collagen capsule (Fig. 8.14). The rete is composed of numerous flattened blood vessels that regulate movement of nutrients and wastes in and out of the nurse cell.

Trichinella species can be serious pathogens of humans. Much of the pathology includes tissue destruction caused by invasive newborn juveniles, accompanied by a strong inflammatory reaction. Intense muscular and joint pain are typical symptoms of trichinosis. For unknown reasons, juveniles have a predilection for the maseter, diaphragm, and inter-costal muscles associated with chewing, swallowing, and breathing. Difficulties in breathing caused by heavy infection can lead to pneumonia and death. Curiously, unlike many nematodes that cause human disease in tropical countries, trichinosis is usually associated with more temperate regions. The epidemiology of the disease is complex, depending on the species involved and the geographical region in which it occurs (Campbell, 1988). Several major epidemiological patterns are recognized. Among these are included domestic trichinosis, a zoonotic disease involving *T. spiralis*, which is sustained in nature by passage back and forth between rats and pigs. Humans become involved when poorly cooked pork containing trichinae (L_1's) is ingested. In contrast, sylvatic trichinosis involves wild animals. This form of the disease has a wide geographical distribution, with species such as *T. nativa* in northern regions, *T. nelsoni* in Africa, and *T. britovi* in the temperate areas of eastern Russia. *Trichinella murrelli* was first described in Pennsylvania. *Trichinella pseudospiralis* is a non-encapsulating form that occurs in birds. Its geographical range is unsettled. Pozio *et al.* (2004) described two more non-encapsulating species, *T. papuae* from wild pigs and saltwater crocodiles in southeast Asia and *T. zimbabwensis* from crocodiles (and probably mammals) in Africa. These authors speculate that the trichinellids are an ancient group that originated in poikilothermic reptiles and evolved with the homio-thermic birds and mammals.

There have been several fairly recent outbreaks of trichinellosis, worldwide. Some of the most serious have occurred in France and Italy where at least three species of *Trichinella* were involved. The source of the infection was horse meat, some of which was imported from Serbia (Pozio *et al.*, 2001). Chaos in the former Yugoslavia has led to the re-emergence of serious problems with trichinosis. Djordjevic *et al.* (2003) cited data from several sources that the prevalence of the disease in Serbia was as low as 0.009% in the 1980s. However, after the break-up of the former Yugoslav Federation, prevalence of *T. spiralis* in inspected pigs jumped by nearly 300% and the number of cases of trichinosis in humans increased two- to four-fold. The description by Djordjevic *et al.* (2003) of a more recent endemic outbreak of trichinosis in Serbia is a classic case illustrating how the decline in high standards of meat inspection and other control measures are directly related to the disintegration of what they refer to as "underlying political, socio-economic, and demographic factors."

8.5.2.2 Dioctophymatida

A major characteristic of dioctophymatids is the production of thick-shelled eggs with one or two polar opercula. Although they are not of medical importance, they are common parasites in a range of aquatic birds and terrestrial mammals. They include among the largest and most colorful (bright red) of the nematodes. An intriguing, unifying feature is that all use oligochaete annelids as intermediate hosts. Three of the most commonly encountered genera include *Dioctyophyme*, *Eustrongylides*, and *Soboliphyme*.

Dioctophyme renale normally infects the kidneys of mustelid carnivores such as mink and ferrets. Females are enormous, reaching 100 cm in size; such large worms effectively turn the kidney into an empty, non-functional capsule. Thus, *D. renale* can cause renal pathology in wild and domestic mammalian hosts. Eggs are shed in the urine and hatch when ingested by

aquatic oligochaete annelid worms in which they develop into third-stage juveniles. Infected annelids are either eaten directly by the definitive mammalian host, or the worms are ingested by fish paratenic hosts, where they encyst in the musculature or viscera. The swallowed juvenile worms undergo a complex migration to develop and mature in the kidney of the mammalian host.

Eustrongylides spp. are commonly encountered as large, red juveniles on the mesenteries inside the body cavity of many fish species (Color plate Fig. 2.2). The parasite matures in the proventriculus of several species of piscivorous birds. The life cycle of *E. tubifex* is complex and involves two predator–prey pathways. Eggs containing first-stage juveniles are released in the feces and hatch into second-stage juveniles when eaten by freshwater oligochaetes. They develop into third-stage juveniles within the ventral blood vessel of these annelids (Measures, 1988a). Fish ingest these infected worms and act as second intermediate hosts for the membrane-encapsulated juveniles. They molt into infective fourth-stage juveniles within these fish hosts. The life cycle is completed when birds eat the fishes containing the infective fourth-stage juveniles. Measures (1988b) proposed that the ability of *E. tubifex* to develop to the advanced fourth stage in fish second intermediate hosts represents an adaptation to facilitate transmission to transient migratory waterfowl hosts. Thus, *E. tubifex* matures rapidly in waterfowl that visit a water body only briefly during a given year, ensuring a short-term pulse of eggs that, in turn, can infect annelids.

Soboliphyme spp. are dioctophymatid parasites in the stomachs of shrews and mustelid carnivores. *Soboliphyme baturini*, for example, is a stomach-dwelling species maturing primarily in mustelids such as marten and sable. Eggs are shed from mustelids into the terrestrial environment and ingested by enchytraeid oligochaetes. Infective third-stage juveniles develop in these annelids, which, in turn, may be ingested by shrew paratenic hosts (Karpenko *et al.*, 2007). Shrew paratenic hosts increase spatial dispersal, thus representing a significant transmission adaptation for a parasite that exploits wide-ranging carnivores as final host.

8.5.2.3 Rhabditida

A key characteristic of the worms in this order is their bewildering diversity of life-cycle strategies. Thus, within this single order, there exists species that are exclusively free-living, some that are facultative parasites, some that are obligate parasites, and some that incorporate mixed strategies involving both parasitic and free-living stages. Evolutionary biologists have become keenly interested in understanding how natural selection favors and maintains such phenomenal variation in life-cycle strategies of the parasitic rhabditids (Box 8.1).

Among the diverse rhabditid nematodes of vertebrates are the lung-inhabiting *Rhabdias* spp. in amphibians and reptiles. These lungworms occur in most adult amphibians (and some snakes), often in large numbers. Infection is via direct penetration of L_3 larvae present in the soil. Following a brief period within muscles, L_4's penetrate the lungs and develop into robust, blood-feeding hermaphroditic adults. Hermaphroditism is via protandry. Thus, the male system develops first, producing sperm that are stored in a seminal receptacle. The male system then disappears, except for the seminal receptacle, to be replaced by a female system that uses sperm stored in the seminal receptacle to fertilize ova. Embryonated eggs hatch in the lungs, and then the L_1's are coughed into the mouth, swallowed, and excreted in feces. The life cycle then takes an unusual twist. Under some environmental conditions, the L_1's molt to the L_2 stage, then molt again into infective L_3's that penetrate host tissues. The larvae undergo a final molt within the musculature and then migrate to the lungs, where they mature to the adult stage. This development pattern is homogonic (see Box 8.1). Alternatively, the L_1's become free-living males or females, i.e., heterogonic development. Following mating in the soil, the males die, but the female broods her larvae *in utero*. After a short period of development, these larvae break out of the uterus and devour their mother – an astonishing

example of parental care! The filariform larvae that exit their mothers' protective cuticle are infective to their final host.

Strongyloides spp. have a cosmopolitan distribution, occurring in the gut mucosa of most vertebrates. Like *Rhabdias*, *Strongyloides* spp. have life cycles involving both homogonic and heterogonic development (Box 8.1). The tiny parasitic females produce genotypically female eggs by mitotic parthenogenesis (but both male and female larvae are produced). Via homogony, eggs (or first-stage larvae) passed in the feces may molt twice to become infective filariform larvae (L_3's). Following skin penetration, they are picked up in the circulatory system and carried to the lungs. There, they emerge through capillaries into the alveoli. They molt a third time to the L_4 stage before moving up the 'bronchial escalator,' to the throat where they are then coughed up and swallowed. On reaching the final site of infection in the intestine, they molt a fourth time to become parthenogenetic females.

The life cycles of *Rhabdias* spp. and *Strongyloides* spp. differ in details, but they are similar in that embryos of most species develop either homogonically or heterogonically. The 'decision' to produce dioecious adults must represent a key life history choice for these parasites. What mechanism determines this key developmental alternative? What are the fitness costs and benefits of homogonic versus heterogonic development? What factors promote the evolution of hermaphroditic adults in one family of rhabditids, but parthenogenetic females in the other? Although these questions have not been addressed for any *Rhabdias* spp., empirical and molecular approaches using the *S. ratti*–rat model system have provided some intriguing insights (Box 8.1).

8.5.2.4 Strongylida

These are commonly called the bursate nematodes referring to the conspicuous copulatory bursa found on the posterior end of the males (Figs. 8.4B, C). Most strongyle nematodes have a direct life cycle, although some species may use intermediate and/or paratenic hosts. Certain nematodes within the Strongylida are of great economic and human public health importance throughout the world.

This is a very large and diverse group. We will focus on three of the most well-studied strongyloid families. Members of the Ancylostomatidae, the hookworms, typically infect their host by direct penetration across the host's skin and exhibit an extensive migration within the host's body. Species of Trichostrongylidae are most often directly ingested as infective third-stage juveniles and do not migrate extensively through tissues. Lungworms in the Protostrongylidae generally use terrestrial gastropods as intermediate hosts, and migrate extensively in tissues.

Ancylostomatidae

As adults, hookworms are found in the small intestine of a wide variety of mammals. The life cycle is direct, with thin-shelled eggs being passed in the host's feces. Females shed up to 30 000 eggs per day (Hotez et al., 2005). Under suitable environmental conditions, the eggs hatch and pass through a sequence of free-living juvenile stages resulting in infective filariform L_3's.

The juveniles lose their sheaths when penetrating the mammal and make their way to the lymphatic system. Ultimately they arrive in the lungs where they penetrate into the alveoli and molt to the fourth stage. They then migrate via the trachea (using the 'bronchial escalator'), then are coughed up and swallowed to the intestine. A final molt results in adults in the small intestine where they attach to the mucosa with their buccal capsule (Fig. 8.3A). The buccal capsule often contains broad cutting plates, and teeth, used to erode the intestinal lining and access the host's blood. Unless the infection is massive, little pathology seems to be associated with the migration of the larvae through the body. Instead, the significant pathology is attributed to blood-feeding of the adults; each worm can ingest 0.2 to 0.3 ml every day. There is a correlation between hookworm intensity and host intestinal blood loss. Silent blood loss from heavy hookworm infection may result in iron-deficiency anemia and protein malnutrition (Hotez & Pritchard, 1995; Hotez et al., 2005).

The main hookworms of humans, *Ancylostoma duodenale* and *Necator americanus*, are included among the important geohelminths. Approximately 1.3 billion people, living primarily in warm, humid climates, are infected with one, or both, of these two species, making hookworm disease one of the most globally significant. Children are at greatest risk of hookworm infection; chronic heavy infections are associated with growth retardation and cognitive impairment (Hotez et al., 2005). The World Health Organization estimates that over 40 million pregnant women are infected and, because there is no consensus on teratogenic effects that antihelminthic drugs may have on the fetus, they are mostly left untreated. Since women in developing countries spend approximately half their lives either pregnant or lactating, anemia in these individuals is an important health consideration.

Soil conditions are important for the transmission of hookworms. A combination of poor sanitation and warm, moist habitats are factors important in hookworm epidemiology. Not surprisingly, in all regions of the developing world where hookworm disease is present, there is a striking relationship between *N. americanus* prevalence and low socio-economic status (Hotez et al., 2005). Lilley et al. (1997) attribute a significant increase in human hookworm infection in Haiti to deforestation along riverbanks. They postulate that flooding caused by deforestation and the associated silting provides an ideal microhabitat for the development of hookworm larvae.

Hookworm disease has proven to be notoriously difficult to eliminate or eradicate in poverty-stricken endemic areas. Control efforts to reduce morbidity among the world's impoverished people have focused on 'deworming' with antihelmintic drugs, especially benzimidazoles (Hotez et al., 2005). Those at great risk of acquiring heavy infections, i.e., school-age children, have been the focus of these deworming programs. However, preschool children and pregnant mothers are also at high risk. Moreover, there is concern that drug resistance is emerging due to the frequent and repeated use of drugs directed at the soil-transmitted helminth. Thus, the Human Hookworm Vaccine Initiative (HHVI) is currently developing a safe, reliable, and cost-effective vaccine in the fight against hookworm disease. The recombinant hookworm larval antigen ASP-2 (ancylostoma secreted protein-2) vaccine has shown considerable promise, and clinical trials are underway (Hotez *et al.*, 2005). The aim is to reduce the numbers of infective larvae entering the intestine and establishing as blood-feeding adults.

Despite the significant pathogenesis of hookworm in humans, paradoxically, a counter-intuitive relationship between humans and hookworms has been suggested (Pritchard & Brown, 2001). In the developed world, an increase in the number of autoimmune diseases has occurred, which has been linked to improved hygiene (Croese & Speare, 2006). Thus, the absence of exposure to certain parasites, including bacteria and viruses, appears to be playing a significant role in the development of autoimmune diseases in more 'sterile' industrialized countries (see also Chapter 2). There is increasing evidence that infections with intestinal nematodes such as *N. americanus* can protect infected hosts against autoimmune diseases such as Crohn's disease (Croese & Speare, 2006).

Canine (*Ancylostoma caninum*, *A. braziliense*) and feline (*A. braziliense*, *A. tubaeformae*) hookworms can also infect humans. *Ancylostoma caninum* is one of the reasons that it is important to deworm dogs at a young age. Puppies can acquire the parasite via transmammary transmission. Filariform L_3's enter a state of arrested development in female dog tissues and then become activated and enter the mammary glands, ready for transmission by suckling puppies (Box 8.3). A condition called cutaneous larval migrans, or creeping eruption, can occur when infective juveniles of *A. braziliense* penetrate human skin. In this case, the infective juveniles penetrate the epidermal layer of the human, but do not proceed further. They migrate through subcutaneous tissues causing extreme itching until they are finally killed by the human's immune response.

Trichostrongylidae

This is a diverse family of nematodes in all vertebrates except fishes, reaching their greatest diversity in mammals. Although mainly parasitic in the mammalian gastrointestinal tract, these small worms may occupy other diverse habitats ranging from lungs to bile ducts. They are of massive economic importance because of the diseases they cause in domestic animals, especially ruminants such as sheep and cattle. The normal mode of infection is by the ingestion of infective third-stage juveniles. Like other strongyles, however, cutaneous, transplacental, and transmammary strategies are common. No intermediate hosts are required. The most typical life-cycle pattern found in those that parasitize herbivorous mammals is for the adult female to deposit eggs that pass out with the host's feces. With appropriate environmental conditions, the egg hatches to a rhabditiform L_1. This stage and the L_2 stage are free-living. The filariform L_3's are also free-living, but non-feeding. These infective stages employ questing behavior to increase their chances of being ingested by grazing herbivores. Once eaten, cues from the host, e.g., CO_2 tension and pH, cause the larvae to secrete an exsheathing fluid that acts to liberate the worm from the sheath. Typically, the exsheathed larvae invade the mucosa, develop to the L_4 stage, and return to the lumen where they undergo a final molt to become adults.

Haemonchus contortus is regarded as one of the most pathogenic parasites of domesticated animals. It is primarily a parasite of sheep, but has been reported to survive and reproduce in other ruminants as well. It is a parasite with a worldwide distribution, being found wherever sheep occur. Throughout its distribution, it is a source of major economic loss. As an adult, it inhabits the abomasum (the 'true,' or fourth, stomach) and, like the hookworms, the major pathology is attributed to the blood-feeding habits of the adults.

Free-living stages of *H. contortus* seem to have fairly specific environmental requirements for survival. Perhaps for this reason, arrested development (early in the fourth stage) is very common (up to 100%) in this species. The arrested larvae may act as a hedge against poor environmental conditions (and thus

Box 8.3 | Transmammary transmission in hookworms

A much-heralded feature of parasites is the phenomenal diversity of ways that parasites have evolved to facilitate their transmission to new hosts. Perhaps the surest way to ensure an individual parasite's long-term survival is to never enter the free-living environment, with all its inherent risks. Such is the case for parasites that use transmammary and transplacental transmission, remarkable life-cycle strategies common to several nematode species, such as certain hookworms, ascarids, and protostrongyles, but also to alariid digeneans (see Chapter 6). In transplacental transmission, infective stages pass directly into the offspring across the placenta while fetuses are still *in utero*. In transmammary transmission, the offspring are born uninfected, but acquire infective stages from mother's milk during nursing.

Both transmission strategies depend on the juvenile stages encysting within host tissues in a state of arrested development and then resuming development in synchrony with its host's reproductive biology. *Uncinaria lucasi*, a hookworm parasite of northern fur seals and northern sea lions, is an example of an ecological situation that favors transmammary transmission. The hosts spend most of their lives at sea, coming ashore at rookeries on islands in the Bering Sea only during a short summer breeding season. When adults arrive at the rookeries they may have filariform larvae in their blubber and other tissues, but will not have adults in their intestines. At this point, filariform larvae that have survived the winter in the soil of the rookery penetrate the host's flippers. These are distributed to various parts of the body, accumulating in the blubber and, in females, the mammary glands as well. Newborn pups are infected during nursing. After migrating through the lungs and being coughed up and swallowed, these juveniles develop to adults in the small intestine of the pups, thereby recontaminating the rookery. The L_3's that have survived in the rookeries over winter will also infect the pups. However, these filariform larvae, as in the adult hosts, will remain in host tissues in arrested development until the hosts mature sexually, at which time the females are ready to pass larvae to a new cohort of seal pups.

Pregnancy is somehow associated with reactivation of latent, tissue-dwelling infections of third-stage juveniles in parasites such as the canine hookworm, *Ancylostoma caninum*. Experiments with *A. caninum* using a mouse model have shown that the hormonal effects of estrogen and prolactin may mediate reactivation, resulting in transmammary transmission of infection to nursing puppies. More specifically, host-derived transforming growth factor-beta (which is up-regulated by estrogen and prolactin) is hypothesized to signal a parasite-encoded receptor to trigger the reactivation of tissue-arrested juveniles (Arasu, 2001).

equate to an 'overwintering' population) in temperate climates. It is, at present, unclear what stimulates the arrest. Both immunological and environmental stimuli have been suggested; perhaps both are involved.

Heavily infected hosts suffer anemia and emaciation and often die. Those that survive often develop an immunological resistance known as a 'self-cure'. This phenomenon, in which the extant infrapopulation is spontaneously expelled, is thought to be elicited by the recruitment of a new cohort of juveniles. Sometimes, all worms (resident adults and newly ingested larvae) are expelled. At other times, only the adults are lost and the juveniles re-establish the infrapopulation. Finally, in some instances, the adults are lost, but the juveniles are retained in the arrested state. The ability to elicit the 'self-cure' phenomenon may be related to the strain of sheep, the strain of the worms, or both.

Another trichostrongylid, *Ostertagia ostertagi*, is regarded as the most economically important parasitic helminth of cattle in temperate parts of the world. Infections cause a reduction in feed intake, which is an important factor in pathogenesis. Thus, there is a marked change in energy metabolism, and parasitized calves exhibit an increase in fatty acid levels due to mobilization of adipose tissue and reduction in digestive efficiency (Fox, 1993). Diverse suites of other ostertagiines are dominant parasites of the gastrointestinal tract (especially the abomassum) of wild and domestic grazing animals, especially cervids (deer, caribou, moose, caribou). It is typical to find 100% of a deer population infected, with individual hosts containing thousands of worms. Abomassal ostertagiines have been implicated in the declines of populations of introduced reindeer in the Norwegian Svalbard archipelago (Albon *et al.*, 2002).

Trichostrongylus spp. are the smallest species in the family and infect a variety of mammals and birds. One species, *Trichostrongylus tenuis*, inhabits the ceca of primarily terrestrial birds such as pheasants, chickens, and grouse, including the red grouse in the UK; red grouse are an important sport-hunting resource in Britain. Consequently, extensive long-term research has been conducted on the population dynamics of red grouse and the roles of its predators and parasites, especially *T. tenuis* (Hudson *et al.*, 2003). Grouse are infected with the nematode when they consume vegetation containing infective L_3's. Adults of *T. tenuis* burrow into the cecal mucosa, causing internal bleeding. Heavy infections are responsible for significant mortality and the parasite has been shown to reduce red grouse fecundity (Hudson *et al.*, 2003; see Chapter 15).

The trichostrongylid *Heligmosomoides polygyrus* infects mice. Due to the ease of maintaining this host–parasite system in the laboratory it is a renowned model system in many parasitological disciplines, especially in ecological parasitology (see Chapters 12, 13, 15).

Protostrongylidae

Unlike the other strongyles discussed, the protostrongyles have an indirect life cycle. They are found primarily in ruminant mammals and most are 'lungworms.' In a typical life cycle, unembryonated eggs are shed into the lungs where they develop to first-stage juveniles. These make their way up the respiratory system, are swallowed and pass out of the host in the feces. Once in the environment, the first-stage juveniles penetrate terrestrial molluscs in which they will develop and molt into the L_3 stages. The life cycle is completed when a ruminant accidentally ingests the infected mollusc as it grazes.

Protostrongyles are important to wildlife the world over, although some of the lungworms of domestic animals, e.g., *Metastrongylus apri* and *M. pudendotectus* in swine, may have an economic impact on humans. *Protostrongylus stilesi* (Fig. 8.15) is a lung nematode of Rocky Mountain bighorn sheep, as well as Stone's and Dall's sheep, across their geographical ranges. The parasite has also been found in muskoxen, a situation that may represent a host switch from sympatric Dall's sheep in the high Arctic (Kutz *et al.*, 2004). In heavy infections, lungworm-induced pneumonia may develop. In bighorn sheep, the parasite is considered a predisposing factor leading to pneumonia outbreaks (review in Monello *et al.*, 2001). *Protostrongylus* spp. initiate an immune response in infected adult hosts, resulting in newly ingested juveniles being sequestered in the lungs. Transplacental transmission occurs in species of this genus. In a pregnant ewe, these larvae can migrate across the placenta and lodge in the liver of the developing fetus. At lamb birth, the juveniles migrate to the lungs where they develop to adults. Eggs deposited by the females, and the juveniles that hatch from them, produce a substantial inflammatory response ultimately resulting in granuloma formation. Various species of bacteria and viruses invade this diseased tissue resulting

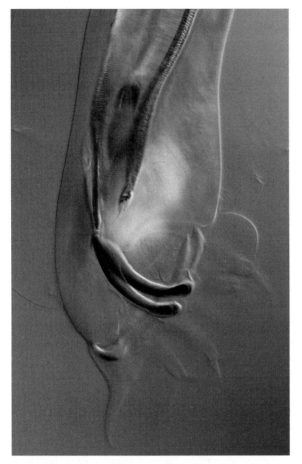

Fig. 8.15 Lateral view of the posterior region of a male of *Protostrongylus stilesi* from lungs of a Dall's sheep, showing the two spicules, characteristic gubernaculum, and the rays of the copulatory bursa. (Photograph courtesy of Eric Hoberg.)

in bronchopneumonia and potential death of the young lamb (e.g., Jenkins *et al.*, 2007).

The meningeal worm *Parelaphostrongylus tenuis* matures in the meninges adjacent to the venous sinuses of the brain of white-tailed deer and other cervid ungulates, such as moose. Females deposit eggs into the sinuses and the eggs are then carried via the venous blood to the heart, then to the lungs where first-stage juveniles hatch. These migrate to the pharynx via the bronchial escalator, are swallowed, and then voided with the feces. These L_1's penetrate

terrestrial snails and slugs, in which they develop into filariform L_3's. When infected molluscs are eaten accidentally by browsing cervids, the third-stage juveniles undergo an astonishing migration. The juvenile worms leave the stomach and then invade the spinal cord, migrating to the brain meninges where development to the adult occurs. In white-tailed deer the parasite is benign; however, in moose, it can cause a crippling neurological paralysis, or so-called 'moose disease.' In regions where white-tailed deer are sympatric with moose, the parasite may be one of the ecological factors influencing current distributions and abundances of moose (see Chapter 15).

8.5.2.5 Oxyurida

All oxyurids have a direct life cycle and live in the posterior intestine of their hosts. Oxyurids infect all vertebrate classes although they are not particularly common in fishes. A prominent diagnostic feature is the presence of a large esophageal bulb (Figs. 8.7, 8.8). The oxyurids are of little importance as agents of morbidity or mortality but, because of their life cycles, they are interesting biologically. A typical life cycle involves a female crawling out of the anus (usually nocturnally) and depositing eggs on the perianal region of the host. These eggs develop (two molts within the egg) within hours. During the course of grooming, the eggs are ingested, hatch in the small intestine, undergo two additional molts, and migrate to the appendix or sigmoidal-rectal region of the large intestine where they reside as adults. It has been suggested that retroinfection may occur whereby L_3's hatch from the eggs in the perianal region and then crawl back through the anus to continue development. This has not been demonstrated conclusively.

A reproductive adaptation seen in some oxyurids is **poecilogyny**, a process where two different types of females are produced. One form of female produces thin-shelled eggs containing fully developed larvae. These may play a role in autoinfection. The other kind of female produces thick-shelled, unembryonated eggs that pass out with the feces and embryonate in the environment. These play a role in colonization of new hosts. An elaboration on this theme is found in *Gyrinicola batrachiensis*, an oxyurid of frogs. In this species, female worms are didelphic. One branch of the uterus produces the thin-shelled, autoinfective eggs, while the other branch produces thick-shelled, unembryonated eggs that pass out into the environment.

Enterobius vermicularis, perhaps the best known of the pinworms, seems to be especially well adapted for group transmission into humans. When females crawl out of the anus to deposit eggs, they cause intense itching. Frequently, especially in young children, this leads to scratching and the transfer of the eggs to the mouth. The tiny eggs can easily become airborne. This feature, together with very high female fecundity, can lead to the notoriously high rates of transmission within school groups and within families. Other than the discomfort of itching, and a mother's psychological distress of having a child infected with 'worms,' pinworms are of little significance to human health. Diagnosis of pinworms in humans remains 'low tech' whereby a piece of 'Scotch tape' (a so-called NIH swab) is applied in the perianal region and, when removed, it is then placed on a microscope slide. Diagnosis is made when either eggs, or the worms themselves, are observed on the tape.

8.5.2.6 Ascarida

This is a very large and diverse order of nematodes. Adults are parasitic in the digestive tract (mostly the small intestine) in all vertebrate groups. The majority of ascarids are large and stout worms with three prominent lips. Some have direct life cycles; however, using an intermediate or paratenic (or both) host is common. In fact, some noted authorities (e.g., Anderson, 2000) suggest that the use of an intermediate host in the ascarids is the norm. Intermediate hosts may be either invertebrates or vertebrates. Some ascarids employ transplacental or transmammary transmission and, although **progenesis** has not been reported, some nematodes, particularly among the ascarids, exhibit precocious development of juveniles while in the intermediate host.

Heterakidae

The members of this ascarid family infect reptiles, amphibians, and birds. They are characterized by having a large preanal sucker possessing a cuticularized rim (Fig. 8.4D). *Heterakis gallinarum* is a ubiquitous parasite living in the ceca of poultry and wild partridges and pheasants. Eggs are passed in the feces and may remain viable for very long periods of time. When ingested by the definitive host, the eggs hatch in the small intestine and the larvae migrate to the ceca where they become mature adults. It is also common for paratenic hosts (annelids and arthropods) to be included in the life cycle. This is an important parasite in the poultry industry because it also transmits the flagellated protozoan *Histomonas meleagridis*, the etiological agent of a disease called 'turkey blackhead,' a serious pathogen for turkey poults. The protist is acquired by the worm and transmitted to eggs via the worm's ovaries. *Heterakis gallinarum* has been implicated as an agent of parasite-mediated competition among gamebirds in the United Kingdom (see Chapter 15).

Ascarididae

Members of this family are mainly parasites in the small intestine of terrestrial vertebrates and their transmission may be direct or, more commonly, indirect. Other than being comparatively large and robust (Fig. 8.16), they have few distinguishing features. *Ascaris lumbricoides*, a serious parasite in

Fig. 8.16 Adults of the ascarid nematode *Ascaris suum* from the intestine of a pig. (Photograph courtesy of Russ Hobbs.)

humans, has been studied extensively. This species is among the most common helminth parasites reported in humans, with an estimated worldwide prevalence of 25% and 1.4 billion. Worldwide, heavy *A. lumbricoides* infections cause approximately 60 000 deaths per year, mainly in children. It is most common in developing subtropical and tropical countries, where 100% of the inhabitants in some villages may be infected. It frequently co-occurs with *Trichuris trichiura* and hookworms, perhaps because of similar environmental requirements for development outside the host. Poor sanitation and hygiene are the factors most important in *Ascaris* transmission to humans.

The life cycle of *A. lumbricoides* is direct, involving the ingestion of tiny embryonated eggs. The eggs hatch as sheathed L_3's that penetrate the intestine and enter the hepatic portal system prior to passing on to the liver, where the sheath is shed. These larvae migrate to the lungs and molt to the fourth stage. The L_4's migrate up the trachea, are coughed up, and swallowed. Upon their return to the gut, the larvae molt again to the adult stage. Paradoxically then, the migration phase starts and ends at the same site. Tissue migration strategies of this type, which are obligate features in the life cycles of many intestinal nematodes, e.g., hookworms and *Strongylus*, have long vexed parasitologists. Migrating larvae must incur costs in terms of energy expenditure, delay in reaching reproductive maturity, exposure to host defenses, and host immunopathology. Intuitively, natural selection should favor individuals that skip the migratory phase.

The traditional explanation of this seemingly pointless migratory behavior is that it is an obligate developmental requirement that repeats what had originally occurred in an intermediate host. *Toxascaris leonina*, another member of the family, bears some evidence for the loss of a required intermediate host. Dogs and cats can become infected directly by ingesting embryonated eggs. In this species, there is then a tissue phase before the worms enter the small intestine and mature. Yet, if L_3's are infected via the ingestion of an infected intermediate host (a rodent),

the tissue phase is eliminated. An alternative explanation is that the costs of migration are compensated by benefits. Read and Skorping (1995) showed that species of nematodes with an obligate migration phase tend to be larger than their closest relatives that develop directly in the gut, after controlling for confounds such as site, host size, generation time, and diet. They suggest that migrating larvae obtain key nutrients that are not available in the gut, supporting relatively rapid growth to maturity, large body sizes, and high worm fecundity. Their costs/benefit analysis appears to neatly resolve the long-standing paradox of tissue migration in some species of nematode.

The migratory phase is one source of pathology in ascariasis. Sometimes, juveniles seem to get lost when leaving the lungs. They may migrate into the right side of the heart and then via the arterial system be distributed to other tissues and organs causing localized host reactions. Even when the juveniles follow the conventional route, pathology associated with their molt will occur in the lungs. When large numbers of juveniles penetrate the lungs, secondary infection and clogging of air passages can occur. When the juveniles molt in the lungs, the shed cuticle and residual molting hormone can induce localized hypersensitivity resulting in edema, followed by pneumonia. Large numbers of large ascarid adults in the intestine can also be pathogenic by causing intestinal blockage, or by migrating into the pancreatic or bile ducts (Figs. 8.17A, B).

Toxocara canis and *Toxocara cati* (found mainly in canids and felids, respectively) also have direct life cycles with pronounced tissue migrations. Both are capable of using paratenic hosts and *T. canis*, at least, is capable of transplacental and transmammary transmission. This explains why it is important to deworm a puppy at 8 weeks of age, even if it has never been out of doors. Because their life cycles are direct and they undergo extensive tissue migrations, they can be a concern to humans. When humans ingest eggs, the larvae begin their typical migration across the small intestine. They do not proceed through normal development, however, and wander

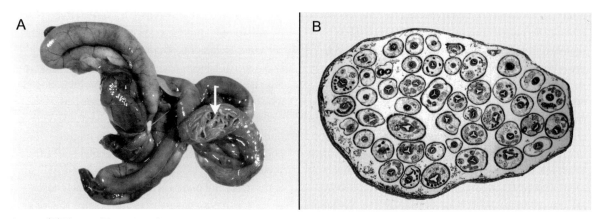

Fig. 8.17 (A) The small intestine of a pet regent parrot, cut open to reveal numerous individuals of the ascarid nematode *Ascaridia platyceri* (arrow); (B) cross-section of the same parrot's intestine showing the large number of nematodes *in situ*. (Photographs courtesy of Russ Hobbs.)

through the body causing visceral larval migrans. Most commonly, these wandering larvae infect the liver, but any organ, including the brain, can be invaded. The degree of pathology in humans is related to the number of juvenile worms present and their final site of infection. Ultimately, the larvae will be killed by a strong host reaction.

Anisakidae

Several anisakid nematodes mature in marine birds and mammals. Two of the most extensively researched anisakids are found in the stomachs of marine mammals. These are the sealworm, *Pseudoterranova decipiens* and the whaleworm, *Anisakis simplex*; the former uses pinnipeds, especially gray seals, as definitive hosts, while *A. simplex* mature primarily in cetaceans. The life cycles of both nematode species are complex, involving primarily crustaceans as intermediate hosts and fish as paratenic hosts. Transmission occurs by diverse predator–prey pathways. Indeed, *P. decipiens* has an extraordinarily wide host distribution, and the use of paratenic hosts can be extreme (Fig. 8.18). A diversity of invertebrates, ranging from copepods to mysid, isopod, decapod, and amphipod crustaceans, as well as benthic polychaetes, are known to be naturally infected with *P. decipiens* (Marcogliese, 2001). Similarly, at least 60 species of fish have been reported with sealworm infections in the North Atlantic. Microcrustaceans

ingest hatched L_2's or L_3's. These infected intermediate hosts may then be ingested by macroinvertebrates such as amphipods and mysids (Fig. 8.18; McClelland *et al.*, 1990), which in turn can be ingested by small benthic fishes such as smelt, sculpin, and American plaice. Piscivorous fish, including Atlantic cod and eelpout, act as further paratenic hosts, potentially accumulating large numbers of L_3's in the musculature prior to the fish eventually being eaten by a seal (see Chapter 12). Interestingly, *P. decipiens* is a precocious nematode and most of the growth takes place as L_3's within fish. Thus, when the mammal becomes infected, maturation of the worms is rapid.

Since anisakids can be found in the flesh of many commercially important fish species, they can affect marketability of the product and thus be costly to the fishing industry. Thus, it comes as no surprise that a tremendous amount of basic and applied research has been devoted to these nematodes (McClelland *et al.*, 1990; Marcogliese, 2001). Because so many fish species of commercial importance are infected, it is estimated that the increased processing cost attributable to *P. decipiens* infections (due to detection and manual removal of the parasite) in Atlantic Canada in 1984 was approximately $30 million (Malouf, 1986). Adding to the problem is that anisakids are notorious for undergoing migrations from internal tissues into the flesh upon death of the fish host.

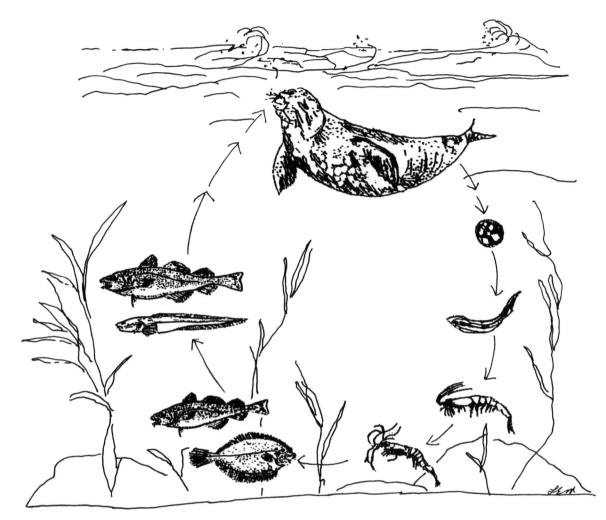

Fig. 8.18 Life cycle plasticity of the sealworm, *Pseudoterranova decipiens*, showing diversity of potential invertebrate intermediate and fish paratenic hosts. Eggs pass into water from seal feces and hatch into a second-stage or third-stage larva. These are ingested by benthic microcrustaceans such as harpacticoid copepods. Macroinvertebrates (e.g., mysid shrimp) then ingest these infected microcrustaceans. Small fish such as juvenile cod or American plaice become infected when infected macroinvertebrates are eaten. Large piscivorous fish, including Atlantic cod or eelpout, eat planktivorous fish or macroinvertebrates containing infective third-stage larvae. The life cycle is completed when seals eat the infected fish and the larva undergoes its final molt to an adult in the seal's stomach. (Artwork courtesy of Lisa Esch McCall; modified from Marcogliese & Price, 1997, *Global Biodiversity*, **7**, 7–15; and McClelland *et al.*, 1990, *Canadian Bulletin of Fisheries and Aquatic Sciences*, **222**, 83–118.)

With an increase in the popularity of consumption of raw fish, e.g. sushi and cerviche, the likelihood of humans acquiring anisakid nematodes is on the rise. It appears that the main cause of anisakiasis is *Anisakis simplex* (Fig. 8.19), which uses euphausid crustaceans (krill) as intermediate hosts, and Pacific hake and herring as paratenic hosts. Adequate cooking and freezing kills the juvenile nematodes. This is one reason why reputable sushi bars freeze their product before passing it on to consumers. The worms do not mature in humans and the pathology is mostly associated with L_3's penetrating the small intestine or

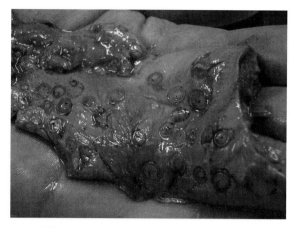

Fig. 8.19 Heavy infection of third-stage juveniles of the anisakid nematode *Anisakis simplex* in the liver of a widow rockfish *Sebastes entomelas*. (Photograph courtesy of David Bush and Vanessa Phillips.)

stomach, causing acute and severe abdominal pain and nausea. The degree of discomfort is likely a consequence of the number of juveniles penetrating. Invasive anisakiasis can occur when the juvenile worms migrate out of the gastrointestinal tract and invade organs such as the liver and pancreas. Be careful when eating raw seafood delicacies!

8.5.2.7 Spirurida

Spirurids produce eggs containing fully developed L_1's. In addition, all spirurids have an indirect life cycle, requiring an arthropod (including hematophagous insects) as an intermediate host. The spirurids are the most morphologically diverse of the nematode orders. Thus, it is among the spirurids that one finds the greatest disparity in body sizes, the greatest diversity of body ornamentation, and the greatest range of microhabitats inhabited by the adult parasites in their vertebrate hosts. Certain of the spirurids cause considerable disfigurement in humans; some of these have been targeted for eradication by the World Health Organization in developing countries. Other spirurids are important parasites of wild animals. *Anguillicoloides crassus* for example, is a serious pathogen of European eels (Color plate Fig. 2.3).

Dracunculidae

Members of this nematode family inhabit the tissues of reptiles, birds, and mammals. One of these, *Dracunculus medinensis*, known since antiquity, and of great public health importance, is commonly called the 'fiery serpent' or 'guinea worm.' It is the causative agent of the dreaded disease known as dracunculiasis. Whereas infection is usually not fatal, it causes intense pain and suffering. A milestone in parasitology was the 1870 discovery by the Russian biologist, Fedchenko, that a cyclopoid copepod (*Cyclops* sp.) was the intermediate host for the guinea worm. Humans acquire the infection by consumption of infected copepods in drinking water. The parasite, in which the female may be almost a meter in length, lives under the skin. Dracunculids are ovoviviparous, with the uterus possessing over 1 000 000 juveniles. Gravid females, most frequently found under the skin of the feet and legs, initially induce the formation of a painful blister. When the blister bursts, and the open lesion is immersed in water, the worm protrudes and releases juvenile worms via a ruptured uterus (interestingly, the vulva is located along the mid-body and actually atrophies, making it non-functional for release of L_1's). A copepod ingests the free-swimming L_1, which subsequently penetrates into the hemocoel and develops to the filariform L_3 stage (Fig. 8.20). Humans become infected by drinking water containing infected copepods. Although the migration of the adult worm causes discomfort, it is the open blister in the skin that causes intense pain. The result is partial or complete disability for up to several months. Once all larvae are shed from the female worm, she will die and must be removed carefully, a mechanical process that can take several weeks (Fig. 8.21).

A global initiative for the control of dracunculiasis began in 1982 and was specifically targeted for eradication by the World Health Organization in 1995. This lofty goal is made possible by the potential ease of effectively breaking the parasite's life cycle. The predominant environmental feature in the areas where guinea worm disease is endemic is that clean drinking water tends to be scarce,

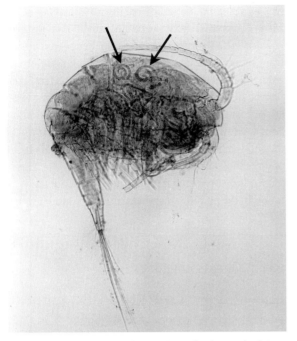

Fig. 8.20 Two infective third-stage juveniles (arrows) of the nematode *Dracunculus medinensis* in the hemocoel of *Cyclops* sp. (Photograph courtesy of the Armed Forces Institute of Pathology.)

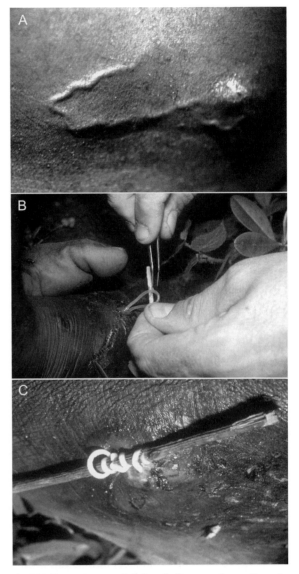

Fig. 8.21 Removing an adult female *Dracunculus medinensis*. Once a blister is ruptured (A) the worm is grasped carefully (B) and gently wound onto a stick (C). ((A, C) Photographs courtesy of Jean-Phillipe Chippaux; (B) photograph courtesy of Daniel Heuclin.)

especially during the dry season. The primary method of eradication is water filtration to remove infected copepods. As of 2003, reported dracunculiasis cases have been reduced by 96% since 1989. A total of 168 countries that were endemic in the 1990s have stopped guinea worm transmission and have been certified free of the nematode. Dracunculiasis transmission is now confined to just 12 African countries, with civil war-torn Sudan being responsible for approximately 80% of cases of dracunculiasis in 2006. Global eradication seems a strong possibility.

Onchocercidae

Like the dracunculids, onchocercid nematodes are tissue-dwelling parasites that use arthropods as intermediate hosts. One important group, the so-called filariids (or filarial worms), are unique among macroparasites in that transmission to vertebrates depends upon blood-feeding vectors. Gravid females are ovoviviparous, and produce juveniles known as **microfilariae** (mf) that are released into the blood or lymph of the vertebrate host and thus available for transmission to insects such as mosquitoes. Filariids are parasites of all classes of vertebrates except fish.

However, by far the best known are those species that cause pathology and death in domestic animals and those that cause horrific infections in humans.

Many filariids have evolved fascinating adaptations facilitating transmission and survival. Notable among these is the phenomenon of **periodicity** in which mf undergo periodic migrations into and out of the peripheral blood, coinciding with the timing of blood-feeding by the specific vector. Filariid nematodes also have a mutualistic relationship with bacteria in the genus *Wolbachia*. These bacterial endosymbionts play important roles in survival and reproduction of their hosts, as well as in filarial immunity and disease (Bandy *et al.*, 1999; Taylor *et al.*, 2001; Fenn & Blaxter, 2006; Box 8.4).

Dirofilaria immitis, or heartworm, has been reported from several mammalian hosts, including a variety of wild and domestic canids and bears, sea lions, and cats. Canine heartworm is one of the most significant veterinary parasites as it can be a serious pathogen in dogs worldwide. The large adults are found mainly in the right ventricle and pulmonary arteries of the heart of dogs (Figs. 8.22A, B; Color plate Fig. 2.4). Microfilariae released from the gravid female are found in the peripheral and visceral circulatory system

---

Box 8.4 | *Wolbachia* spp.: endosymbiotic bacteria in filariid nematodes

Wolbachia spp. are endosymbiotic gram-negative bacteria, which means that their cell walls cannot be stained with a crystal violet dye. Most gram-negative bacteria are pathogens. It is estimated that up to 75% of all insect species possess *Wolbachia* symbionts, as well as a range of other arthropods that includes isopods, mites, and spiders. A number of parasitic nematodes, including species that cause river blindness and elephantiasis, are also 'home' to the bacteria. There is growing evidence that lateral (horizontal) gene transfer of virtually the entire bacterial genome to the host has occurred in several host/*Wolbachia* systems (Hotopp *et al.*, 2007). *Wolbachia* present in insects have phages that possess ankyrin-repeat proteins; these may "be exported to the host cell to manipulate reproduction" (Fenn & Blaxter, 2006). Oddly, *Wolbachia* in nematode hosts are without phages.

Wolbachia is acquired via egg transfer (transovarian transmission) and, therefore, is typically associated with the female reproductive system. Males cannot transfer the bacteria to their offspring. In arthropods, the effects of *Wolbachia* are varied. For example, male killing has been reported, accompanied by induced parthenogenesis in remaining females, thereby resulting in female-only populations. Feminization of males may also occur, even to the point that sperm production ceases and ova are produced. Gametic cytoplasmic incompatability may be generated whereby *Wolbachia*-infected males are unable to successfully reproduce with uninfected females or with females infected with a different strain of *Wolbachia*. The upshot of the latter reproductive combinations is that one of the 'competing' *Wolbachia* strains gains an advantage over others.

As mentioned, *Wolbachia* is also an endosymbiont of several filarial worms, where they can be found in the paired lateral chords of males and females, plus the oocytes of females. The bacteria have not been found in any other nematodes (or in nematomorphs). When present, they are associated with 100% of a nematode population, which implies they are not accidental partners and that in all likelihood there is some sort of positive reciprocity occurring between the bacteria

Box 8.4 | (continued)

and the host. The resulting relationship between filarial worms and their endosymbiotic *Wolbachia* is, thus, quite unusual. If the *Wolbachia* population is removed from an individual *Brugia malayi*, *Onchocerca volvulus*, or *Wuchereria bancrofti*, the worm dies or becomes infertile.

The immune response of humans to *Wolbachia* is unclear at the present time. Brattig *et al.* (2001) described the presence of *Wolbachia* in human **onchocercomas**. They observed host inflammatory cells that stained positive for *Wolbachia* in nodules containing dead and degenerated microfilariae. They concluded the latter result was a definitive signal that *Wolbachia* was stimulating the host's immune system. Subsequently, Simon *et al.* (2003) reported evidence that humans infected with *Dirofilaria immitis* possessed antibodies specific for the endosymbiotic bacteria.

Molecular evidence has shown a strong phylogenetic relationship between *Wolbachia* and several rickettsial bacteria, including species of *Rickettsia*, *Erlichia*, *Anaplasma*, and *Neorickettsia*. Interestingly, a number of chemotherapeutic drugs effective against the latter species are also effective against *Wolbachia*. Studies by Fenollar *et al.* (2003) indicate the greatest effects on *Wolbachia* are produced by doxycycline and rifampin. Strong mutualistic tendencies between *Wolbachia* and several of the filarial worms infecting humans have suggested that these two drugs may be useful in the treatment of filariasis and onchocerciasis, i.e., destroy the bacteria and the *Wolbachia*-dependent helminths cannot survive. In summary, a great deal of research has been conducted on *Wolbachia* in both arthropods and nematodes, but much remains to be learned about the unusual mutualistic association between these odd bacteria and their hosts.

(Fig 8.23). The mf are ingested by a variety of mosquito species. Within the mosquito, the juveniles undergo two molts and develop to infective filariform L_3's in the insect's Malpighian tubules. The L_3's migrate to the salivary glands and are available for inoculation when the mosquito again feeds. The precise route taken by the L_3's in the mammalian host is unclear. Approximately 6 months after injection of mf by an infected mosquito, *D. immitis* matures within the heart. Infected dogs have an unmistakable chronic cough and typically suffer from shortness of breath due to blockage of the pulmonary blood vessels. The disease can lead to heart failure and is frequently fatal. It is very difficult to remove adult worms from the heart. Ivermectin is an effective prophylactic drug in the treatment of heartworm disease since it works against pre-adult stages and mf.

Lymphatic filariasis is caused by two of the most horribly disfiguring parasites in humans, i.e., *Brugia malayi* (Brugian filariasis) and *Wuchereria bancrofti* (Bancroftian filariasis). It is estimated that 120 million people are infected and that 20% of the world's population is at risk. As adults, they occur in the lymphatic system causing blockage that often leads to the disfiguring condition known as elephantiasis. Although generally similar, there are differences (other than morphological) between the two species. Brugian filariasis is most common in southeast Asia, and mainly causes elephantiasis in the human extremities (Fig. 8.24). It also occurs in a diversity of wild and domestic animals. Bancroftian filariasis is

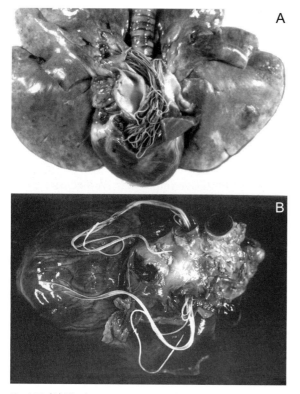

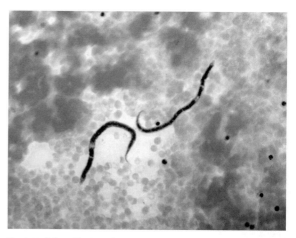

Fig. 8.23 Microfilariae of *Dirofilaria immitis* from the peripheral circulation of a dog. (Photograph courtesy of Al Bush.)

Fig. 8.22 (A) The heartworm nematode *Dirofilaria immitis* in the heart of a domestic dog; (B) *Dirofilaria immitis* in the pulmonary arteries of the heart of a red wolf *Canis rufus*. ((A) Photograph courtesy of Russ Hobbs; (B) photograph courtesy of Danny Pence.)

circumtropical in distribution, may also cause elephantiasis of human extremities, and is restricted to humans. The life cycles of the species are similar. Microfilariae, circulating in the blood, are ingested by several species of mosquitoes. The larvae develop to the infective third stage in the thoracic muscles, migrate to the mouthparts, and are inoculated when the vector takes a blood meal. Once in the human host, the juveniles enter the lymphatic vessels or lymph nodes and mature.

An intriguing feature about the circulating mf of *B. malayi* and *W. bancrofti* is that several strains show a marked diurnal periodicity. In one strain, mf are found in the peripheral circulation during the day. Accordingly, day-biting mosquitoes transmit that strain. At night, mf are absent or rare in the peripheral

circulation. In another strain, mf are most common in the peripheral circulation at night and are transmitted by night-biting mosquitoes. Although grossly disfiguring, the parasites are not particularly pathogenic. Further, 'elephantiasis' results from repeated infections over long periods of time and is the product of a complex immunogenic reaction. There is evidence that, in the absence of new infections, the symptoms will disappear after the adults die.

Filariasis is another parasitic disease targeted for elimination by the World Health Organization, but will not be nearly as easy to solve as *D. medinensis* since the former will require drug therapy. In 1998, SmithKline Beecham agreed to make their drug Albendazole®, shown to be 99% effective against the parasites, available free of charge. This is an impressive commitment since the treatment regime calls for all people in infected areas to receive a dose each year for about 5 years. The WHO estimates the program will run for at least 20 years and that we might hope for elimination of lymphatic filariasis by about 2020. The genome of *B. malayi* was the first parasitic nematode to be sequenced (Ghedin *et al.*, 2004). As a result, the search and identification of new targets of therapeutic interest will be enhanced in the years to come.

Adults of *Onchocerca volvulus*, unlike *B. malayi* and *W. bancrofti*, occur in subcutaneous nodules known as onchocercomas (Fig. 8.25), where they survive for up

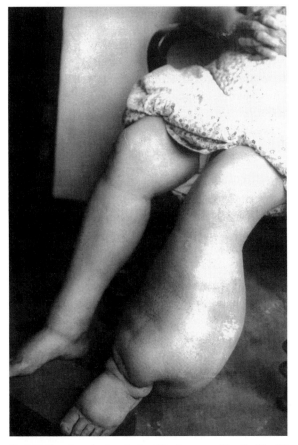

Fig. 8.24 Brugian elephantiasis of the leg caused by the filariid nematode *Brugia malayi*. (Photograph courtesy of the Armed Forces Institute of Pathology.)

Fig. 8.25 Adult *Onchocerca volvulus* in an onchocercoma. (Photograph courtesy of WHO and OCP.)

8.6 Phylogenetic relationships and classification

Advances in molecular systematics have become increasingly useful in clarifying the phylogenetic relationships of the nematodes. Recognizing that 18S rDNA sequences evolve rapidly in the rhabditid nematodes, Aguinaldo *et al.* (1997) employed only the slowest evolving sequences from representative taxa (for nematodes, e.g., *Trichinella spiralis*). Based on their results, they identified a clade called the Ecdysozoa, comprising all animals that molt (see also Chapter 11). This viewpoint has radically altered our current view of the Tree of Life. Not only does it encompass the two most biodiverse metazoan clades on earth, the arthropods and nematodes, but it also incorporates lesser-known phyla such as the Kinorhyncha, Priapulida, Nematomorpha, Onychophora, and Tardigrada. The inclusion of the Ecdysozoan clade into the Tree of Life is now widely accepted by systematists, underscoring the importance of the exoskeleton and the process of molting as a key evolutionary innovation in the history of life on earth.

Traditionally, the Nematoda have been divided into two classes, the primarily terrestrial Secernentea and the mostly aquatic Adenophorea (Inglis, 1983; Adamson, 1987). These taxonomic distinctions were based on few morphological characteristics. One serious complication of the Secernentean/Adenophorean distinction is that many important morphological features are now

to 10 years. Onchocerciasis is widely distributed in tropical Africa. It is also reported in Brazil, Mexico, and Guatemala. Approximately 18 million people are infected with *O. volvulus*. The most important source of pathology is the mf, which, unlike *B. malayi* and *W. bancrofti*, are found mainly in the host's subcutaneous tissues. When the mf invade the cornea, iris, and optic nerve, however, the result is often permanent blindness. The vectors are various species of black flies (*Simulium*) whose larvae and pupae require fast-moving streams for development. Combining the major pathology to humans with the breeding requirements of the vectors results in the common name 'river blindness.'

thought to have arisen independently during nematode evolution. Thus, there remains considerable doubt that this dichotomy reflected true evolutionary relationships. More recently, molecular phylogenetics has dramatically changed nematode systematics (Kampfer *et al.*, 1998; Blaxter *et al.*, 1998; De Ley & Blaxter, 2004; De Ley, 2006; Smythe *et al.*, 2006). Blaxter *et al.* (1998) examined small subunit rDNA sequences from 53 nematode species. Their analyses, as well as those of Smythe *et al.* (2006), indicate that nematode parasitism evolved independently at least four times (Fig. 8.26). Their phylogenetic studies also support the hypothesis that plant, arthropod, and vertebrate–nematode parasites have evolved multiple times from free-living nematodes. The small subunit rDNA sequences of over 1800 species of nematodes are now available, making for a more robust and comprehensive view of nematode phylogenetic relationships. As stated by De Ley (2006), these analyses

have, in fact, confirmed that "one of the two traditional classes (Secernentea) is deeply phylogenetically embedded within the other (Adenophorea)." A recent classification scheme recognizes that three major lineages exist within the Nematoda: Chromadoria, Enoplia, and Dorylaimia (see De Ley, 2006).

Even at lower levels of resolution, the phylogeny of the nematodes is controversial. Nadler (1995), for example, used full-length sequences of 18S rDNA and 300 nucleotides of cytochrome oxidase II to resolve 13 ascaridoid species. His analyses were mostly consistent with current taxonomic assignments at lower ranks, but "inconsistent with most proposed arrangements at higher taxonomic levels." Nadler & Hudspeth (1998) argue that "some key features emphasized by previous workers represent ancestral states or highly homoplastic characters." It is highly likely that the same argument applies to other groups of nematodes (see Box 8.5).

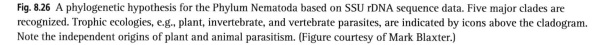

Fig. 8.26 A phylogenetic hypothesis for the Phylum Nematoda based on SSU rDNA sequence data. Five major clades are recognized. Trophic ecologies, e.g., plant, invertebrate, and vertebrate parasites, are indicated by icons above the cladogram. Note the independent origins of plant and animal parasitism. (Figure courtesy of Mark Blaxter.)

Box 8.5 | **Classification of the Nematoda**

The classification scheme we use here is one of convenience and may not reflect the 'true' evolutionary relationships of the nematodes parasitic in animals. It is based upon the five-clade model of Blaxter *et al.* (1998) (Fig. 8.26), as modified and systematized by De Ley and Blaxter (2004), and supported by Meldal *et al.* (2007) and Holterman *et al.* (2006; 2008). The Subclass Tylenchia and the Orders Drilonematida, Mermithida, and Rhigonematida (identified with asterisks) are parasitic exclusively in invertebrates. It is interesting to note that the latter group of insect parasites have been shown to be related to the vertebrate parasites within the Ascaridida, demonstrating that this major parasitic group is paraphyletic (Smythe *et al.*, 2006). There is much to learn about the evolutionary relationships amongst nematodes.

Class Enoplea
  Subclass Dorylaimia
    Order Dioctophymatida
      Representative genera: *Dioctophyme,*
        Eustrongylides, Soboliphyme
    *Order Mermithida
      Representative genera: *Aranimermis, Mermis,*
        Myrmeconema, Romanomermis
    Order Muspiceida
    Order Trichinellida
      Representative genera: *Capillaria, Trichinella,*
        Trichuris
Class Chromadorea
[Order Rhabditida]
  *Suborder Tylenchina
    InfraOrder Panagrolaimomorpha
      Representative genera: *Rhabdias, Steinernema,*
        Strongyloides
    *InfraOrder Drilonematomorpha
    *InfraOrder Tylenchomorpha
      Representative genus: *Sphaerularia*
  Suborder Spirurina
    *InfraOrder Rhigonematomorpha
    InfraOrder Oxyuridomorpha
      Representative genera: *Enterobius,*
        Pharyngodon, Thelastoma
    InfraOrder Ascaridomorpha

Box 8.5 | (continued)

Representative genera: *Anisakis, Ascaris,*
Cosmocerca, Cucullanus, Heterakis, Kathlania, Maupasina, Quimperia, Skrjabinura,
Toxocara
InfraOrder Spiruromorpha
Representative genera: *Anguillicoides, Aprocta, Ascarophis, Brugia, Camallanus,*
Cystidicoloides, Dirofilaria, Dracunculus, Hedruris, Onchocerca, Philometra,
Physaloptera, Spiroxis, Spirura, Tetrameres, Thelazia
SubOrder Rhabditina
InfraOrder Rhabditomorpha
Including Superfamily Strongyloidea:
Representative genera: *Heterorhabditis,*
Ancylostoma, Angiostrongylus, Haemonchus, Heligmosomoides, Metastrongylus,
Necator, Ostertagia, Protostrongylus, Parelaphostrongylus, Strongylus, Syngamus,
Trichostrongylus, Uncinaria

References

Adamson, M. L. (1987) Phylogenetic analysis of the higher classification of the Nematoda. *Canadian Journal of Zoology,* 65, 1478–1482.

Aguinaldo, A. M. A., Turbeville, J. M., Linford, L. S., *et al.* (1997) Evidence for a clade of nematodes, arthropods and other moulting animals. *Nature,* 387, 489–493.

Albon, S. D., Stein, A., Irvine, R. J., *et al.* (2002) The role of parasites in the dynamics of a reindeer population. *Proceedings of the Royal Society of London Series B, Biological Sciences,* 269, 1625–1632.

Anderson, R. C. (2000) *Nematode Parasites of Vertebrates: Their Development and Transmission,* 2nd edition. Wallingford: CAB International.

Arasu, P. (2001) In vitro reactivation of *Anyclostoma caninum* tissue-arrested third-stage larvae by transforming growth factor-beta. *Journal of Parasitology,* 87, 733–738.

Ashton, F. T., Bhopale, V. M., Holt, D., *et al.* (1998). Developmental switching in the parasitic nematode *Strongyloides stercoralis* is controlled by the ASF and ASI amphidial neurons. *Journal of Parasitology,* 84, 691–695.

Bandy, C. McCall, J. W., Genchi, C., *et al.* (1999) Effects of tetracycline on the filarial worms *Brugia pahangi* and *Dirofilaria immitis* and their bacterial endosymbionts, *Wolbachia. International Journal for Parasitology,* 29, 357–364.

Behm, C. A. (1997) The role of trehalose in the physiology of nematodes. *International Journal for Parasitology,* 27, 215–229.

Bird, A. F. & Bird, J. (1991) *The Structure of Nematodes,* 2nd edition. San Diego: Academic Press.

Blaxter, M. L., Page, A. P., Rudin, W. & Maizels, R. M. (1992) Nematode surface coats: actively evading immunity. *Parasitology Today,* 8, 243–247.

Blaxter, M. L., De Ley, P., Garey, J. R., *et al.* (1998) A molecular evolutionary framework for the phylum Nematoda. *Nature,* 392, 71–75.

Brattig, N. W., Buttner, D. W. & Hoerauf, A. (2001) Neutrophil accumulation around *Onchocerca* worms and chemotaxis of neutrophils are dependent on *Wolbachia* endobacteria. *Microbes and Infection*, **3**, 439–446.

Campbell, W. C. (1988) Trichinosis revisited: another look at modes of transmission. *Parasitology Today*, **4**, 83–86.

Croese, J. & Speare, R. (2006) Intestinal allergy expels hookworms: seeing is believing. *Trends in Parasitology*, **22**, 547–550.

Crofton, H. D. (1966) *Nematodes*. London: Hutchinson.

Davey, K. G. (1995) Water, water compartments and water regulation in some nematodes parasitic in vertebrates. *Journal of Nematology*, **27**, 433–440.

De Ley, P. (2006) A quick tour of nematode diversity and the backbone of nematode phylogeny (January 25, 2006). *WormBook*, ed. The *C. elegans* Research Community, WormBook, DOI/10.1895/wormbook.1.41.1, www.wormbook.org.

De Ley, P. & Blaxter, M. L. (2004) A new system for Nematoda: combining morphological characters with molecular trees, and translating clades into ranks and taxa. In *Nematology Monographs and Perspectives*, ed. R. Cook, D. J. Hunt, pp. 633–653. Leiden: E. J. Brill.

Despommier, D. D. (1993) *Trichinella spiralis* and the concept of niche. *Journal of Parasitology*, **79**, 472–482.

Despommier, D. D. (1998) How does *Trichinella spiralis* make itself at home? *Parasitology Today*, **14**, 318–323.

Despommier, D. D. (2004) www.trichinella.org. Columbia University.

Djordjevic, M., Bacic, M., Petricevic, M., *et al.* (2003) Social, political, and economic factors responsible for the reemergence of trichinellosis in Serbia: a case study. *Journal of Parasitology*, **89**, 226–231.

Dorris, M., De Ley, P. & Blaxter, M. L. (1999) Molecular analysis of nematode diversity and the evolution of parasitism. *Parasitology Today*, **15**, 188–193.

Fenn, K. & Blaxter, M. L. (2006) *Wolbachia* genomes: revealing the biology of parasitism and mutualism. *Trends in Parasitology*, **22**, 60–65.

Fenollar, F., Maurin, M. & Raoult, G. (2003) *Wolbachia pipientis* growth kinetics and susceptibilities to 13 antibiotics determined by immunofluorescence staining and real-time PCR. *Antimicrobial Agents and Chemotherapy*, **47**, 1665–1671.

Fetterer, R. H. & Rhoads, M. L. (1996) The role of the sheath in resistance of *Haemonchus contortus* infective-stage larvae to proteolytic digestion. *Veterinary Parasitology*, **64**, 267–276.

Fox, M. T. (1993) Pathophysiology of infection with *Ostertagia ostertagi* in cattle. *Veterinary Parasitology*, **46**, 143–158.

Gaugler, R. (2002) *Entomopathogenic Nematology*. Wallingford: CAB International.

Gemmill, A. W., Viney, M. E. & Read, A. F. (1997) Host immune status determines sexuality in a parasitic nematode. *Evolution*, **51**, 393–401.

Ghedin, E., Wang, S., Foster, J. M. & Slatko, B. E. (2004) First sequenced genome of a parasitic nematode. *Trends in Parasitology*, **20**, 151–153.

Haag, E. S. (2005) The evolution of nematode sex determination: *C. elegans* as a reference point for comparative biology. In *WormBook*, ed. The *C. elegans* Research Community, WormBook. DOI/10.1895/wormbook.1.120.1.

Harlos, J., Brust, R. & Galloway, T. (1980) Observations on a nematode parasite of *Aedes vexans* (Diptera: Culicidae) in Manitoba. *Canadian Journal of Zoology*, **58**, 215–220.

Hawdon, J. M. & Schad, G. A. (1991) Developmental adaptations in nematodes. In *Parasite–Host Associations: Coexistence or Conflict?*, ed. C. H. Toft, L. Aeschliman, & L. Bolas, pp. 274–298. Oxford: Oxford University Press.

Holterman, M., van der Wurff, A., van den Elsen, S., *et al.* (2006) Phylum-wide analysis of SSU rDNA reveals deep phylogenetic relationships among nematodes and accelerated evolution toward crown clades. *Molecular Biology and Evolution*, **23**, 1792–1800.

Holterman, M., Holovachov, O., van den Elsen, S., *et al.* (2008) Small subunit ribosomal DNA-based phylogeny of basal Chromadoria (Nematoda) suggests that transitions from marine to terrestrial habitats (and vice versa) require relatively simple adaptations. *Molecular Phylogenetics and Evolution*, **48**, 758–763.

Hotez, P. J. & Pritchard, D. I. (1995) Hookworm infection. *Scientific American*, **272**, 68–74.

Hotez, P. J. Bethony, J., Bottazzi, M. E., *et al.* (2005) Hookworm: the great infection of mankind. *PLoS Medicine* **2**, e67, 187–191.

Hotopp, J. C. D., Clark, M. E., Oliveira, D. C. S. G., *et al.* (2007) Widespread lateral gene transfer from intracellular bacteria to multicellular eukaryotes. *Science*, **317**, 1753–1756.

Hudson, P. J., Dobson, A. P. & Newborn, D. (2003) Parasitic worms and the population cycles of Red grouse (*Lagopus lagopus scoticus*). In *Population Cycles*. ed. A. Berryman, pp. 109–129. Oxford: Oxford University Press.

Inglis, W. G. (1983) An outline classification of the phylum Nematoda. *Australian Journal of Zoology*, **31**, 243–255.

Jenkins, E. J., Vietch, A. M., Kutz, S. J., *et al.* (2007) Protostrongylid parasites and pneumonia in captive and wild thinhorn sheep (*Ovis dalli*). *Journal of Wildlife Diseases*, **43**, 189–205.

Kampfer, S., Sturmbauer, C. & Ott, J. (1998) Phylogenetic analysis of rDNA sequences from adenophorean nematodes and implications for the Adenophorea-Secernentea controversy. *Invertebrate Biology*, **117**, 29–36.

Karpenko, S. V., Dokuchaev, N. E. & Hoberg, E. P. (2007) Nearctic shrews, *Sorex* spp., as paratenic hosts of

Soboliphyme baturini (Nematoda: Soboliphymidae). *Comparative Parasitology*, **74**, 81–87.

Kennedy, M. W. & Harnett, W. (2001) *Parasitic Nematodes: Molecular Biology, Biochemistry, and Immunology*. Wallingford: CAB International.

Kita, K., Hirawake, H. & Takamiya, S. (1997) Cytochromes in the respiratory chain of helminth mitochondria. *International Journal for Parasitology*, **27**, 617–630.

Kutz, S. J., Hoberg, E. P., Nagy, J., *et al.* (2004) "Emerging" parasitic infection in Arctic ungulates. *Integrative Comparative Biology*, **44**, 109–118.

Lee, D. L. (1996) Why do some nematode parasites of the alimentary tract secrete acetylcholinesterase? *International Journal for Parasitology*, **26**, 499–505.

Li, J., Chen, G. & Webster, J. M. (1997) Nematophin, a novel antimicrobial substance produced by *Xenorhabdus nematophilus* (Enterobactereaceae). *Canadian Journal of Microbiology*, **43**, 770–773.

Li, B., Lyle, M. P. A., Chen, G., *et al.* (2007) Substituted 6-amino-4H-[1,2]dithiolo[4,3-b]pyrrol-5-ones: synthesis, structure-activity relationships, and cytotoxic activity on selected human cancer cell lines. *Bioorganic and Medicinal Chemistry*, **15**, 4601–4608.

Lilley, B., Lammie, P., Dickerson, J. & Eberhard, M. (1997) An increase in hookworm infection temporally associated with ecologic change. *Emerging Infectious Disease*, **3**, 391–393.

Malouf, A. H. (1986) Seals and sealing in Canada. *Report of the Royal Commission*, vol. 3. Ottawa: Canadian Government Publishing Centre.

Marcogliese, D. J. (2001) Review of experimental and natural invertebrate hosts of sealworm (*Pseudoterranova decipiens*) and its distribution and abundance in macroinvertebrates in eastern Canada. *NAMMCO Scientific Publications*, **3**, 27–38.

Martinez, A. M. B. & de Souza, W. (1997) A freeze-fracture and deep-etch study of the cuticle and hypodermis of infective larvae of *Strongyloides venezuelensis* (Nematoda). *International Journal for Parasitology*, **27**, 289–297.

McClelland, G., Misra, R. K. & Martell, D. J. (1990) Larval anisakine nematodes in various fish species from Sable Island Bank and vicinity. In W. D. Bowen (ed.) Population biology of sealworm (*Pseudoterranova decipiens*) in relation to its intermediate and seal hosts. *Canadian Bulletin of Fisheries and Aquatic Sciences*, **222**, 83–118.

Measures, L. (1988a) The development of *Eustrongylides tubifex* (Nematoda: Dioctophymatoidea) in oligochaetes. *Journal of Parasitology*, **74**, 294–304.

Measures, L. (1988b) Epizootiology, pathology, and description of *Eustrongylides tubifex* (Nematoda: Dioctophymatoidea) in fish. *Canadian Journal of Zoology*, **66**, 2212–2222.

Medica, D. L. & Sukhdeo, M. V. K. (1997) Role of lipids in the transmission of the infective stage (L3) of *Strongylus*

vulgaris (Nematoda: Strongylida). *Journal of Parasitology*, **83**, 775–779.

Meldal, B. H., Debenham, N. J., De Ley, P., *et al.* (2007) An improved molecular phylogeny of the Nematoda with special emphasis on marine taxa. *Molecular Phylogenetics and Evolution*, **42**, 622–636.

Monello, R. J., Murray, D. L. & Cassirer, E. F. (2001) Ecological correlates of pneumonia epizootics in bighorn sheep herds. *Canadian Journal of Zoology*, **79**, 1423–1432.

Muller, F., Bernand, V. & Tobler, H. (1996) Chromatin diminution in nematodes. *BioEssays*, **18**, 133–138.

Nadler, S. A. (1995) Advantages and disadvantages of molecular phylogenetics: a case study of ascaridoid nematodes. *Journal of Nematology*, **27**, 423–432.

Nadler, S. A. & Hudspeth, D. S. S. (1998) Ribosomal DNA and phylogeny of the Ascaridoidea (Nemata: Secernentea): implications for morphological evolution and classification. *Molecular Phylogenetics and Evolution*, **10**, 221–236.

Neuhaus, B., Bresciani, J., Christensen, C. & Frandsen, F. (1996) Ultrastructure and development of the body cuticle of *Oesophagostomum dentatum* (Strongylida, Nematoda). *Journal of Parasitology*, **82**, 820–828.

Ogawa, A., Streit, A., Antebi, A. & Sommer, R. J. (2009) A conserved endocrine mechanism controls the formation of dauer and infective larvae in nematodes. *Current Biology*, **19**, 67–71.

Ogbogu, V. C. & Storey, D. M. (1996) Ultrastructure of the alimentary tract of third-stage larvae of *Litomosoides carinii*. *Journal of Helminthology*, **70**, 223–229.

Page, A. P. & Johnstone, I. L. (2007) The cuticle. WormBook, ed. The *C. elegans* Research Community, WormBook, DOI/10.1895/wormbook.1.138.1, http://www.wormbook.org.

Paul, V. J., Frautschy, S., Fenical, W., & Nealson, K. H. (1981) Antibiotics in microbial ecology, isolation and structure assignment of several new antibacterial compounds from the insect-symbiotic bacteria *Xenorhabdus* spp. *Journal of Chemical Ecology*, **7**, 589–597.

Penney D. & Bennett, S. P. (2006) First unequivocal mermithid–linyphiid (Araneae) parasite–host association. *Journal of Arachnology*, **34**, 273–278.

Poinar, Jr., G. O. (1983) *The Natural History of Nematodes*. Englewood Cliffs: Prentice-Hall.

Poinar, Jr., G. O. (1985) Mermithid (Nematoda) parasites of spiders and harvestmen. *Journal of Arachnology*, **13**, 121–128.

Poinar, Jr., G. O. (2002a) *Heydenius brownii* sp. n. (Nematoda: Mermithidae) parasitizing a planthopper (Homoptera: Achilidae) in Baltic amber. *Nematology*, **3**, 753–757.

Poinar, Jr., G. O. (2002b) First fossil record of nematode parasitism of ants; a 40 million year tale. *Parasitology*, **125**, 457–459.

Poinar, Jr., G. O. & Early, J. W. (1990) *Aranimermis giganteus* n. sp. (Mermithidae: Nematoda), a parasite of New Zealand mygalomorph spiders (Araneae: Arachnida). *Revue de Nématololologie*, 13, 403–410.

Poinar, Jr., G. O. & Monteys, V. S. (2008) Mermithids (Nematoda: Mermithidae) of biting midges (Diptera: Ceratopogonidae): *Heleidomermis cataloniensis* n. sp. from *Culicoides circumscriptus* Kieffer in Spain and a species of *Cretacimermis* Poinar, 2001 from a ceratopogonid in Burmese amber. *Systematic Parasitology*, 69, 13–21.

Poinar, Jr., G. O., Lachaud, J.-P., Castillo, A. & Infante, F. (2006) Recent and fossil nematode parasites (Nematoda: Mermithidae) of Neotropical ants. *Journal of Invertebrate Pathology*, 91, 19–26.

Politz, S. M. & Philipp, M. (1992) *Caenorhabditis elegans* as a model for parasitic nematodes: a focus on the cuticle. *Parasitology Today*, 8, 6–12.

Poulin, R. & Latham, A. D. M. (2002) Parasitism and the burrowing depth of the beach hopper *Talorchestia quoyana* (Amphipoda: Talitridae). *Animal Behaviour*, 63, 269–275.

Pozio, E. & Zarlenga, D. S. (2005) Recent advances on the taxonomy, systematics and epidemiology of *Trichinella*. *International Journal for Parasitology*, 35, 1191–1204.

Pozio, E., Tamburrini, A., & La Rosa, G. (2001) Horse trichinellosis: an unresolved puzzle. *Parasite*, 8, 263–265.

Pozio, E., Marucci, G., Casulli, A., *et al.* (2004) *Trichinella papuae* and *Trichinella zimbabwensis* induce infection in experimentally infected varans, caimans, pythons, and turtles. *Parasitology*, 128, 333–342.

Pritchard, D. (1995) The survival strategies of hookworms. *Parasitology Today*, 11, 255–259.

Pritchard, D. J. & Brown, A. (2001) Is *Necator americanus* approaching a mutualistic symbiotic relationship with humans? *Trends in Parasitology*, 17, 169–172.

Read, A. F., & Skorping, A. (1995) The evolution of tissue migration by parasitic nematode larvae. *Parasitology*, 111, 359–371.

Riga, E., Perry, R. N., Barrett, J. & Johnston, M. R. L. (1995) Biochemical analyses on single amphidial glands, excretory–secretory gland cells, pharyngeal glands and their secretions from the avian nematode *Syngamus trachea*. *International Journal for Parasitology*, 25, 1151–1158.

Roberts, L. S. & Janovy, Jr., J. (2009) *Foundations of Parasitology*, 8th edition. New York: McGraw Hill.

Shamseldean, M. M., Platzer, E. G. & Gaugler, R. (2007) Role of the surface coat of *Romanomermis culicivorax* in immune evasion. *Nematology*, 9, 17–24.

Simon, F., Prieto, G., Morchon, R., *et al.* (2003) Immunoglobulin G antibodies against endosymbionts of filarial nematodes (*Wolbachia*) in patients with pulmonary dirofilariasis. *Clinical Diagnostic and Laboratory Immunology*, 10, 180–181.

Sims, S. M., Ho, N. F. H., Geary, T. G., *et al.* (1996) Influence of organic acid excretion on cuticle pH and drug absorption by *Haemonchus contortus*. *International Journal for Parasitology*, 26, 25–35.

Smart, G. C. (1995) Entomopathogenic nematodes for the biological control of insects. *Journal of Nematology*, 27, 529–534.

Smythe, A. B., Sanderson, M. J. & Nadler, S. A. (2006) Nematode small subunit phylogeny correlates with alignment parameters. *Systematic Biology*, 55, 972–992.

Stock, S. P. (2005) Insect-parasitic nematodes: From lab curiosities to model organisms. *Journal of Invertebrate Pathology*, 89, 7–66.

Stock, S. P. & Gress, J. C. (2006) Diversity and phylogenetic relationships of entomopathogenic nematodes (Steinernematidae and Heterorhabditidae) from the Sky Islands of southern Arizona. *Journal of Invertebrate Pathology*, 92, 66–72.

Taylor, C. E. & Brown, D. J. F. (1997) *Nematode Vectors of Plant Viruses*. Wallingford: CAB International.

Taylor, M. J., Cross, H. F., Ford, L., *et al.* (2001) *Wolbachia* bacteria in filarial immunity and disease. *Parasite Immunology*, 23, 401–409.

Viney, M. E. (2009) How do parasitic worms evolve? *BioEssays*, 31, 496–499.

Webster, J. M., Chen, G., & Li, J. (1998) Parasitic worms: an ally in the war against the superbugs. *Parasitology Today*, 14, 161–163.

Webster, J. M., Chen, G., Hu, K. & Li, J. (2002) Bacterial metabolites. In *Entomopathogenic Nematology*, ed. R. Gaugler, pp. 99–114. Wallingford: CAB International.

West, S. A., Gemmill, A. W., Graham, A., *et al.* (2001) Immune stress and facultative sex in a parasitic nematode. *Journal of Evolutionary Biology*, 14, 333–337.

Williams, C. M., Poulin, R. & Sinclair, B. J. (2004) Increased haemolymph osmolality suggests a new route for behavioural manipulation of *Talorchestia quoyana*. (Amphipoda: Talitridae) by its mermithid parasite. *Functional Ecology*, 18, 685–691.

Yanoviak, S. P., Kaspari, M., Dudley, B. & Poinar, Jr., G. O. (2008) Parasite-induced fruit mimicry in a tropical canopy ant. *The American Naturalist*, 171, 536–544.

Zimmermann, M. R., Luth, K. E. & Esch, G. W. (2011) The unusual life cycle of *Daubaylia potomaca*, a nematode parasite of *Helisoma anceps*. *Journal of Parasitology*, 97, 430–434.

Nematomorpha: the hairworms

9.1 General considerations

The common name of the nematomorphs, 'horsehair worm' or 'hairworm,' arises from their long filariform, cylindrical morphology. Adult body size among the approximately 350 described species varies considerably, ranging from a few centimeters in length to over 2 m. Nematomorphs are dioecious and the large adults (Fig. 9.1) are free-living in aquatic habitats, mostly permanent freshwater lakes, ephemeral ponds, and streams. In contrast, juvenile nematomorphs are obligate parasites within the hemocoel of arthropods, a characteristic they share with the mermithid nematodes (see Chapter 8). The juveniles of almost all described nematomorphs are parasitic in terrestrial arthropods (the gordiids), whereas the remainder (the nectonematids) are parasites of marine invertebrates, especially crustaceans. Nematomorphs are often referred to as gordiids or Gordian worms on account of the tangled mass of swarming adults (**Gordian knots**) that are frequently observed in shallow aquatic habitats. Compared to the mermithids and their potential for biological control of insect pests and vectors, the nematomorphs are a poorly studied group. Following the first completion of a nematomorph life cycle under laboratory conditions (Hanelt & Janovy, 2004b), significant advances have been made in our understanding of the ecology, systematics, and life cycles of this enigmatic taxon (review in Hanelt *et al.*, 2005). The facilitation of nematomorph transmission between its parasitic larval stage and its free-living adult stage in water is now a well-known case of parasite-induced alteration in host behavior (see Chapter 15, Color plate Fig. 8.3).

9.2 Form and function

9.2.1 Body wall

Gross and ultrastructural studies of general body morphology have been restricted to a handful of model nematomorph/host interactions, especially involving the adult stages of the gordiids *Paragordius varius* and *Gordius aquaticus*. Overall, the morphology of juvenile and adult stages appears to be uniform and nematode-like. Thus, body shape is generally cylindrical and filariform, with tapered ends. Macroscopic elaborations to the body wall, if present, are restricted to reproductive structures at the posterior end of a few groups. Mature worms tend to be brownish in color, often with a lightly shaded area at the anterior end.

Schmidt-Rhaesa (2005) described the complex ontogenetic changes in cuticle ultrastructure for the parasitic phase of *Paragordius varius* in experimentally infected crickets *Gryllus firmus*. At approximately 10 days post-infection (p.i.), the juvenile cuticle overlaying the epidermis is homogenous and thin. During this period, the epidermis contains a high density of mitochondria and rough endoplasmic reticulum, indicating metabolically active cells. Further, the distal edge of the epidermis is completely enveloped by a complex network of microvilli. Little is known regarding the biochemical structure of the juvenile cuticle, nor is it known to what extent the unusual extensions to the epidermis play a role in the absorption of host material across the body wall. Given the transient nature of both structures, and their elaboration during the rapid growth phase in the arthropod host, some role in absorption seems likely.

A rudimentary adult cuticle is visible by approximately 20 days p.i., corresponding with the

Fig. 9.1 Free-living adults of the nematomorph *Gordius robustus* in a mating mass, or Gordian knot. (Photograph courtesy of Ben Hanelt.)

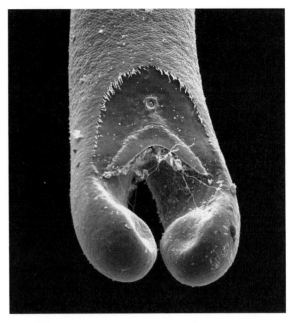

Fig. 9.2 Scanning electron micrograph of posterior region of the male of *Gordius difficilis*, showing the inverted U-shaped postcloacal crescent and row of hairs anterior to the cloacal opening. (Micrograph courtesy of Matthew Bolek.)

disappearance of the microvilli surrounding the epidermis. By 25 days p.i., the juvenile cuticle is molted and the adult cuticle, comprising multiple layers of proteinaceous fibers, is complete. The complex, laminar ultrastructure of the adult cuticle has been described for several freshwater and marine nematomorphs (Schmidt-Rhaesa, 2005). Presumably, the fibrous nature of the cuticle provides protection during the free-living phase in water and also provides tensile strength to support the action of the hydrostatic skeleton. High rates of metabolism within the juvenile epidermis would support the secretion of the adult cuticle as would secretions of the microvilli that cover the epidermis. Studies designed to clarify the absorptive vs. secretory role of the transient layer of microvilli that envelope the epidermis between 10 and 25 days p.i. are needed.

Elaborations to the adult cuticle are important in nematomorph taxonomy especially those present around the cloacal opening in males. In *Gordius* spp., for example, there is a semi-circular fold called the **postcloacal crescent** (Fig. 9.2). The cloacal opening of males may be ringed by characteristic spines or bristles. In most species, the surface of the cuticle is enveloped by minute ornate protrusions called **areoles**, which may form characteristic patterns and be useful in species determination. However, patterns of expression of these structures vary intra-specifically, geographically, and also along the anterior/posterior axis of individual worms (Bolek & Coggins, 2002).

In contrast to the uniform morphology of juvenile and adult gordiids, larval morphology is complex. SEM imagery indicates that the body is typically divided into two distinct regions, the preseptum and the postseptum, each bearing superficial annuli (Fig. 9.3). The preseptum contains a complex, retractile proboscis that is ringed by at least three rows of characteristic spines. The number, configuration, and shape of the spines vary between species. The proboscis is dorso-ventrally flattened and bears its own elaborate spinous projections (Marchiori *et al.*, 2009). The postseptum is often curled away from the preseptum to give the larvae a characteristic J-shape. Larval hooks possess long intracuticular roots and the stylets have apical teeth (Jochmann & Schmidt-Rhaesa, 2007). The morphological adaptations of the preseptum presumably assist in penetration of the gut of the intermediate and/or paratenic arthropod host.

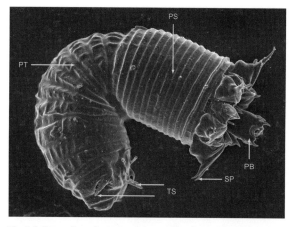

Fig. 9.3 Scanning electron micrograph of a newly hatched larva of *Paragordius obamai*, showing preseptum (PS), postseptum (PT), spines (SP), proboscis (PB) with stylets, and terminal spines (TS). (Micrograph courtesy of Andreas Schmidt-Rhaesa.)

9.2.2 Metabolism and nutrition

Adult nematomorphs lack a mouth and the digestive system is greatly reduced. Thus, the short life span of the adult is supported by stored energy reserves (mainly glycogen) obtained as a juvenile within the insect host. Skaling & MacKinnon (1988) report esterases and acid and alkaline phosphatases associated with active absorption of nutrients across both the body wall and intestine of juvenile *Nectonema agile*. The relative roles of ingestion versus absorption in supporting the enormous growth rates of juvenile nematomorphs are not known.

9.2.3 Reproductive system

Nematomorphs are dioecious and sexually dimorphic, with females longer and thicker than males. Males can also be distinguished from females by the shape of the posterior end, which may have two tail lobes (Fig. 9.2), and also the position of the cloacal opening. The males' lobes are used to attach to a female during copulation. In males, the posterior end curls inwards and the cloacal opening is ventrally located (Fig. 9.2), while in females the posterior region is typically rounded with a terminal cloaca. Most of the body cavity of mature

gordiid adults is filled with paired gonads. Testes and ovaries join the cloaca through ducts. In females, a seminal receptacle is located near the oviduct and the cloacal aperture and serves to store sperm. The numbers of eggs contained within adult female nematomorphs is phenomenal, among the highest of any metazoan species. Intriguingly, a newly described nematomorph species, *Paragordius obamai*, has been shown to reproduce asexually via parthenogenesis; a life history adaptation which may be more common in gordiids than previously thought (Hanelt *et al.*, 2012).

9.2.4 Nervous system

The gordiid nervous system consists of a cerebral ganglion and mid-ventral nerve cord. A peripheral system of nerves connects the underlying musculature. The anterior region and posterior end of nematomorphs is well supplied by nerves, suggesting probable sensory functions. However, little is known about the sensory cells of nematomorphs.

9.3 Development and general life cycle

Development from a microscopic larva into a large adult requires 1 to 3 months, depending on species. Based on ultrastructural studies of *Paragordius varius*, ontogenetic changes in organ systems generally parallel the striking changes in the development of the cuticle. Thus, longitudinal muscles grow continuously through the juvenile stage and the ventral nerve chord shifts to a submuscular position with sustained growth. Gonads develop from paired dorso-lateral compact strands, extending in size until they fill the body cavity. The gonads carry gametes during the final developmental period within the insect host (Hanelt *et al.*, 2005).

The hairworm life cycle has four distinct phases. Included are a free-living aquatic adult and larva (Figs. 9.3, 9.4), a parasitic cyst stage (Fig. 9.5), and a large, parasitic juvenile (Color plate Fig. 8.3). Nematomorphs attain sexual maturity shortly after leaving the arthropod host and reaching freshwater. In

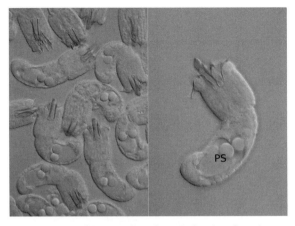

Fig. 9.4 Larvae of *Paragordius obamai*, showing the spines, evertable proboscis with stylets, and complex pseudointestine (PS). The pseudointestine secretes the cyst wall. (Photograph courtesy of Matthew Bolek.)

Fig. 9.5 Cyst of *Chordodes* sp. in tissues of the snail *Physa gyrina*. (Photograph courtesy of Ben Hanelt.)

Paragordius varius and other temperate species, copulation typically involves several adult males and females congregating in a writhing Gordian knot (Fig. 9.1). Several million fertilized eggs, in long gelatinous strings, are released into the water or among plant roots along the shore of ponds or streams. Adults are short-lived and seasonal. Males die after copulation and, after laying their eggs, females die soon thereafter. Masses of larvae emerge from the egg sacs.

Until recently, it was unclear how the larvae, after emerging from the egg, reach their terrestrial insect host. Infection trials completed under laboratory conditions indicate that larvae of *P. varius* can encyst upon a variety of surfaces, e.g., grass blades, and, following ingestion by crickets, initiate juvenile development (Hanelt & Janovy, 2004a). Perhaps more surprisingly, invertebrate paratenic hosts may play an important role in the completion of some life cycles (Fig. 9.6). When cysts of *Paragordius varius* were removed from physid snails and fed to crickets, a nematomorph life cycle was completed experimentally for the first time (Hanelt & Janovy, 2004b). The complete life cycle of this hairworm took approximately 45 days and could be maintained in the lab for several generations.

Perhaps paratenic hosts help bridge the aquatic to terrestrial transmission barrier (Schmidt-Rhaesa, 2001). For example, larvae of the common North American gordiid *Gordius robustus* encyst in various freshwater invertebrate taxa, including pulmonate snails, aquatic crustaceans, annelids, and larvae or naiads of aquatic insects. Hanelt *et al.* (2001) found high infection levels of *G. robustus* cysts in the common pulmonate snail *Physa gyrina* (Fig. 9.5). The emergence of infected insects such as mayflies and midges from the water, upon metamorphosis, may also disperse the infective hairworm cysts into the terrestrial habitat (Hanelt & Janovy, 2003). Presumably, when the paratenic host is ingested by the definitive host, the larva excysts, bores through the intestine, and enters the host's hemocoel (Schmidt-Rhaesa, 2005; Fig. 9.6).

9.4 Biodiversity and ecology

Most nematomorph species have been found worldwide as juvenile parasites in beetles and orthopteran insects, especially crickets and grasshoppers. Other insect taxa, including praying mantids, and also centipedes, millipedes, and leeches have been reported as hosts for nematomorph species. While 326 hairworm species have been described, a conservative estimate

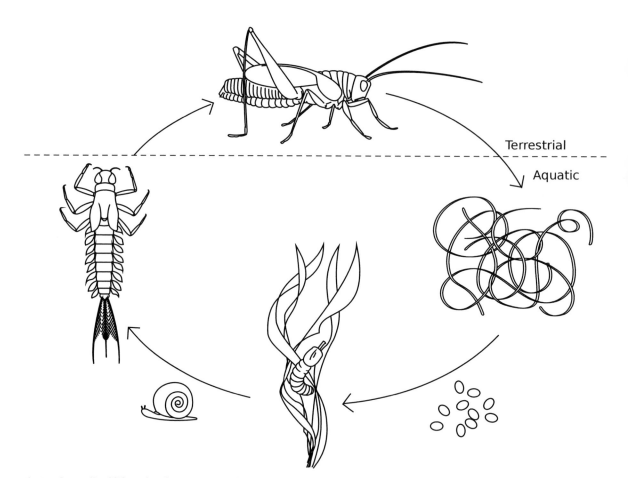

Fig. 9.6 Generalized life cycle of a nematomorph, e.g., *Paragordius varius*. Adults emerge from crickets in fresh water and mate in a Gordian knot. Eggs are deposited on vegetation and hatch into larvae which are eaten by paratenic hosts such as insects (e.g., mayfly nymphs) and snails, in which they encyst. The life cycle is completed when infected paratenic hosts are eaten by terrestrial insect definitive hosts, in which the cysts excyst. Larvae penetrate and then develop within the hemocoel. Upon maturation, hairworms manipulate host behavior causing the insect to jump into water. (Figure courtesy of Danielle Morrison.)

for global species diversity may exceed 2000 species (Poinar, 2008). The most species-rich group of insects, the carabid beetles, are host to several species of nematomorphs (Poinar *et al.*, 2004). Targeting this single host group should reveal many more nemato-morph species new to science.

Species of *Nectonema* are exclusively marine. Adults are free-living, pelagic, and their large juvenile stages parasitize crustaceans such as pelagic shrimps and benthic crabs (Nielsen, 1969). The larvae have been rarely observed and the mode of infection for *Nectonema* spp. is unknown. To date, only the

decapod crustacean hosts containing the juvenile stages have been identified. *Nectonema agile*, for example, is found in the hemocoel of shrimps along the northeastern coast of North America, where it reaches at least 10 cm in length. Copulation in *Nectonema* takes place near the ocean surface; plank-tonic larvae may then infect the crustacean hosts. In the case of *Nectonema munidae*, which parasitizes crabs in deep Norwegian fjords, infection probably occurs in the upper layers of the water. It has been speculated that when the infected zoea larvae descend and settle out of the plankton, the nematomorph

Plate 1. Host and site specificity in parasitic flatworms.

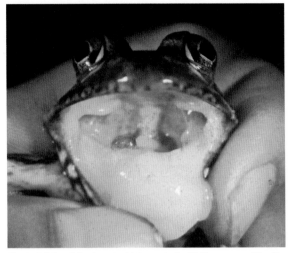

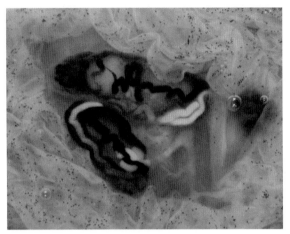

Fig. 2 Adults of the plagiorchiid trematode *Haematoloechus longiplexus* in the lung of a bull frog *Lithobates catesbeianus*. (Photograph courtesy of Tyrel Pinnegar.)

Fig. 1 Adult of the hemiurid trematode *Halipegus occidualis* attached under the tongue of a green frog *Lithobates clamitans*.

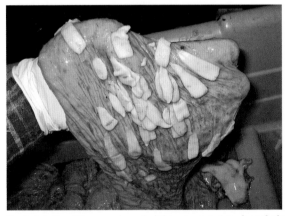

Fig. 4 Adults of the anoplocephalid tapeworm *Anoplocephala perfoliata* attached to the iliocecal junction of a horse. (Photograph courtesy of Sara Skotarek.)

Fig. 3 Hydatid cysts of the taeniid tapeworm *Echinococcus granulosus* in the liver of a moose *Alces alces*. (Photograph courtesy of Blake Parker.)

Plate 2. Host and site specificity in parasitic nematodes.

Fig. 1 Coils of a third-stage juvenile of the mermithid *Aranimermis giganteus* in the abdominal cavity of a trap door spider *Cantuaria borealis* (Photograph courtesy of George Poinar, Jr.)

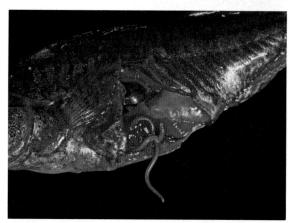

Fig. 2 A fourth-stage juvenile of *Eustrongylides tubifex* protruding from the body cavity of a threespine stickleback *Gasterosteus aculeatus*. (Photograph courtesy of Bill Pennell.)

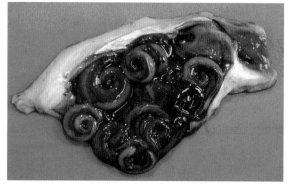

Fig. 3 Adults of *Anguillicoloides crassus* in the swim bladder of a European eel *Anguilla anguilla*. (Photograph courtesy of Eva Jakob.)

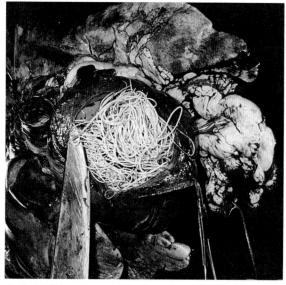

Fig. 4 Adult heartworms *Dirofilaria immitis* occluding the right ventricle of the heart of a domestic dog. (Photograph courtesy of Gordon Mackenzie.)

Plate 3. Morphological variation and diversity of parasitic crustaceans.

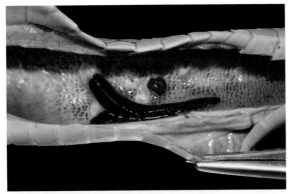

Fig. 1 The pentastome *Waddycephalus punctulatus* on the lung of a common tree snake *Dendrelaphis punctulatus*. (Photograph courtesy of Tim Portas.)

Fig. 3 Female bopyrid isopod *Orthione griffenis* in the branchial chamber of a mud shrimp *Upogebia putettensis*. (Photograph courtesy of Jason Williams.)

Fig. 2 Juveniles and adults of the cyamid amphipod *Cyamis scammoni* on the skin of a beach-stranded gray whale *Eschrichtius robustus*. (Photograph courtesy of Graeme Ellis.)

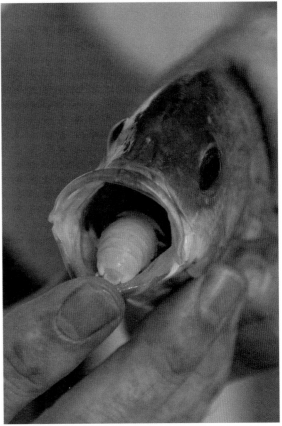

Fig. 4 Female of the cymothoid isopod *Cymothoa exigua* inside the mouth of a yellowtail snapper *Ocyurus chrysurus*. The parasite first attaches to the tongue of the host, and then replaces it. (Photograph courtesy of Eric Demers.)

Plate 4. Morphological variation, site, and host specificity among marine parasitic copepods.

Fig. 1 Females of *Lepeophtheirus salmonis* (Caligidae) on the skin of a Chinook salmon *Oncorhynchus tshawytscha*. (Photograph courtesy of Bill Pennell.)

Fig. 2 Two transformed females of *Phrixocephalus cincinnatus* (Pennellidae) attached to the eye of an arrowtooth flounder *Atheresthes stomias*. (Photograph courtesy of Dane Stabel.)

Fig. 3 Transformed female of *Haemobaphes diceraus* (Pennellidae) in the gill cavity of a shiner perch *Cymatogaster aggregata*. The parasite's holdfast is inserted into the bulbus arteriosus of the heart of the fish host.

Fig. 4 Females of *Ismaila* sp. (Splanchnotrophidae) in the nudibranch *Janolus fuscus*. The egg sacs protrude from the dorsal surface between the nudibranch's brightly colored cerata. (Photograph courtesy of Michael Miller.)

Plate 5. Diversity of rhizocephalan barnacles of decapod crustaceans.

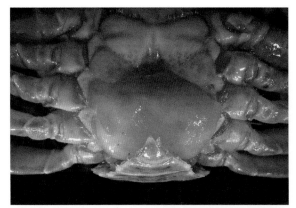

Fig. 1 Externa of the rhizocephalan *Sacculina carcini* on the abdomen of the green crab *Carcinus maenas*. (Photograph courtesy of Jens Høeg.)

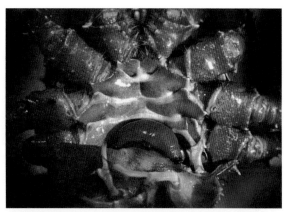

Fig. 2 Externa of the rhizocephalan *Briarosaccus callosus* on the abdomen of the king crab *Lithodes aequispina*, illustrating the phenomenon of egg mimicry. (Photograph courtesy of Susan Bower.)

Fig. 3 Externa of the rhizocephalan *Peltogaster paguri* on the abdomen of the hermit crab *Pagurus bernhardus*. (Photograph courtesy of Jens Høeg.)

Fig. 4 Several externae of the rhizocephalan *Peltogasterella sulcata* on the hermit crab *Pagurus cuanensis*. (Photograph courtesy of Jens Høeg.)

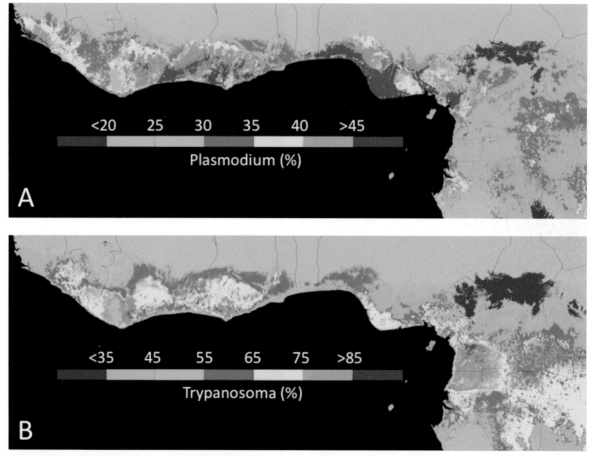

Plate 6. Risk maps depicting geographical variation in the prevalence of avian blood parasites in the olive sunbird *Cyanomitra olivacea* across West Africa and Cameroon. (A) *Plasmodium* spp.; (B) *Trypanosoma* spp. Colors from blue to red indicate increasing prevalence of blood parasites in birds. Regions shaded in gray correspond to absence of the host–parasite system. Predictive spatial patterns were developed with GIS-based modeling tools that associated parasite prevalence with ground-based and remotely sensed environmental variables (see Chapter 14). (Figure courtesy of Ravinder Sehgal; modified from Sehgal *et al.*, 2010, *Proceedings of the Royal Society, Series B*, **278**, 1025–1023, with permission, The Royal Society.)

Plate 7. Parasite-induced alterations to host morphology.

Fig. 1 The snail *Succinea putris* infected with the trematode *Leucochloridium macrostomum*. The striped, brightly colored tentacles contain branched sporocysts which are filled with metacercariae. (Photograph courtesy of Josef Hlásek.)

Fig. 2 Supernumerary limb deformity of the Pacific treefrog *Pseudacris regilla* caused by the trematode *Ribeiroia ondatrae*. Metacercariae encyst in the developing hind limbs of tadpoles, causing the limb malformations when the tadpole metamorphoses into a frog. (Photograph courtesy of David Herasimtschuk.)

Fig. 3 Plerocercoids of the tapeworm *Schistocephalus solidus* removed from the body cavity of the threespine stickleback *Gasterosteus aculeatus*. The large sizes of the larvae cause a distended abdomen. (Photograph courtesy of David Marcogliese.)

Fig. 4 A South American ant *Cephalotes atratus* infected with the tetradonematid nematode *Myrmeconema neotropicum*. This is the only known case of parasite-induced fruit mimicry. The red color of the ant's abdomen is due to the presence of gravid female nematodes. (Photograph courtesy of Steven Yanoviak.)

Plate 8. Diversity of parasite-induced host behavioral manipulations.

Fig. 1 Ants (*Formica* sp.) clinging to a flower. These ants are infected with metacercariae of the trematode *Dicrocoelium dendriticum*. The behavior of infected ants is altered, such that they are induced to cling to vegetation during times of the day when their herbivorous definitive hosts are most likely to be feeding. (Photograph courtesy of Andy Hurly.)

Fig. 2 Cystacanth larvae of the acanthocephalan *Corynosoma* sp. in the body cavity of the amphipod *Hyallella azteca*. The phototaxic behavior of infected amphipods is altered, increasing vulnerability to predation by specific aquatic definitive hosts. (Photograph courtesy of Matthew Bolek.)

Fig. 3 The gordiid nematomorph *Paragordius varius* emerging from a cricket *Aceta domestica* into water. The behavior of infected crickets is modified so that infected hosts expel their worms into water. (Photograph courtesy of Ben Hanelt.)

Fig. 4 Larva of the ichneumonid parasitoid *Hymenoepimecis argyraphaga* on the orb-weaving spider *Plesiometa argyra*. Infected spiders' normal web-building behavior is altered and the spider is manipulated to build a special 'cocoon web' for the parasitoid's larva. (Photograph courtesy of William Eberhard.)

survives its crab host's metamorphosis and then grows, along with its host.

The growth and development of the large juveniles have been demonstrated to reduce the growth and reproduction of infected arthropod hosts. The ovaries of *Nectonema*-infected female shrimp atrophy and become opaque. Crickets infected with *G. robustus* lacked gonads and fat bodies (Hanelt & Janovy, 1999). Male crickets infected with *Paragordius tricuspidatus* never developed testes and female fecundity was sharply reduced (Biron *et al.*, 2005). Further, the courtship and oviposition behaviors of infected crickets differed from uninfected ones. However, contrary to popularized reports, hairworm emergence is not always lethal for arthropod hosts and the negative impact of infection on individuals is not always clear-cut. Crickets (especially females) can survive several months following hairworm emergence. No studies have evaluated the population-level impacts of hairworm infection on their hosts.

It is easy to imagine that a hairworm-infected insect that jumps into water would be vulnerable to surface-feeding aquatic predators. Due to the large size of hairworms, emergence from the host at the water's surface can take several minutes, further increasing the risk of predation. Indeed, Ponton *et al.* (2006) observed that crickets harboring, or expelling, hairworms were often eaten by generalist fish and amphibian predators. These authors observed that the worms escaped such predation by wriggling out of the mouth, nose, or gills of the predator that had eaten the infected cricket.

As mentioned above, gordiid nematomorphs copulate in freshwater. Yet, the stage prior to copulation often occurs within hosts (e.g., crickets) that rarely, if ever, enter aquatic habitats. This presents a significant transmission barrier. Presumably, natural selection would strongly disfavor the emergence of mature hairworms on land but would favor mechanisms that would increase an infected host's contact with water. Experiments completed by Frédéric Thomas and his colleagues have demonstrated that infected hosts display a completely novel and sometimes lethal behavior, one that is originally absent from the host's repertoire. In effect, hairworms manipulate the behavior of their host by causing them to seek water and then 'commit suicide' by jumping into an appropriate freshwater habitat. Thomas *et al.* (2002) observed suicidal behaviors in nine insect species following emergence of hairworms. They compared the behavior of *Paragordius tricuspidatus*-infected and uninfected hosts and showed that infected insects were more likely to jump into water then uninfected ones. Furthermore, results from behavior assays showed that infected insects first display an erratic behavior that eventually brings them close to water, followed by a behavioral change that causes them to enter it, typically at night. Nematomorph–insect model systems now play a central role in tests of hypotheses regarding the molecular and physiological mechanisms of phenotypic manipulation of infected hosts (Thomas *et al.*, 2003; Biron *et al.*, 2005, 2006; see Chapter 15).

Nematomorph-induced alteration in arthropod behavior can have dramatic effects on the overall flow of energy through aquatic ecosystems (Sato *et al.*, 2011). Camel crickets and grasshoppers infected with the gordiid, *Gordionus chinensis*, are 20 times more likely to enter streams than uninfected hosts. An astonishing 60% of the annual energy budget of Japanese trout is accounted for by the ingestion of these infected orthopterans. These results provide some of the clearest evidence of the ecosystem-level effects of a manipulating parasite (see Chapter 15).

9.5 Phylogenetic relationships and classification

Fossilized nematomorphs have been found exquisitely preserved as amber depositions. Poinar (1999), for example, describes the hairworm *Paleochordodes protus* from a fossil cockroach trapped in amber and estimated to be between 15 and 45 million years old. It is the first unequivocal record of a fossil nematomorph. Recently, Poinar & Buckley (2006) described the hairworm *Cretachordodes burmitis* from early Cretaceous Burmese amber. This fossil, complete with clearly discernible areole patterns, was dated at 100–110 million years.

As is the case for other 'minor' invertebrate phyla, determining evolutionary relationships of the nematomorphs has been difficult. Morphological and molecular studies consistently support the view that nematomorphs are monophyletic and that they share a sister-group relationship with the Nematoda. Two orders of nematomorphs are currently recognized and, based on molecular and morphological phylogenetic studies, they are also monophyletic and are considered sister groups (Bleidorn *et al.*, 2002). The two groups are united in the clade Nematoidea. The nematomorph order Nectonematida comprises the five species within the single marine genus, *Nectonema*. The Order Gordiida comprises the three most speciose freshwater families of hairworms, the Gordiidae, e.g., *Gordius, Gordionus, Paragordius*, Chordodidae, e.g., *Chordodes*, and the Spinochordodidae, e.g., *Spinochordodes*.

References

Bleidorn, C., Schmidt-Rhaesa, A., & Garey, J R. (2002) Systematic relationships of Nematomorpha based on molecular and morphological data. *Invertebrate Biology*, 121, 357–364.

Biron, D. G., Ponton, F., Joly, C., *et al.* (2005) Water-seeking behaviour in insects harboring hairworms: should the host collaborate? *Behavioral Ecology*, 16, 656–660.

Biron, D. G., Ponton, F., Marché, L., *et al.* (2006) Suicide of crickets harbouring hairworms: a proteomics investigation. *Insect Molecular Biology*, 15, 731–742.

Bolek, M. G. & Coggins, J. R. (2002) Seasonal occurrence, morphology, and observations on the life history of *Gordius difficilis* (Nematomorpha: Gordioidea) from southeastern Wisconsin, United States. *Journal of Parasitology*, 88, 287–294.

Hanelt, B. & Janovy, Jr., J. (1999) The life cycle of a horsehair worm, *Gordius robustus* (Nematomorpha: Gordioidea). *Journal of Parasitology*, 85, 139–141.

Hanelt, B. & Janovy, Jr., J. (2003) Spanning the gap: identification of natural paratenic hosts of horsehair worms (Nematomorpha: Gordioidea) by experimental determination of paratenic host specificity. *Invertebrate Biology*, 122, 12–18.

Hanelt, B. & Janovy, Jr., J. (2004a) Life cycle and paratenesis of American gordiids (Nematomorpha: Gordiidae). *Journal of Parasitology*, 90, 240–244.

Hanelt, B. & Janovy, Jr., J. (2004b) Untying a Gordian knot: the domestication and laboratory maintenance of a Gordian worm, *Paragordius varius* (Nematomorpha: Gordiida). *Journal of Natural History*, 38, 939–950.

Hanelt, B., Grothier, L. E. & Janovy, Jr., J. (2001) Physid snails as sentinels of freshwater Nematomorphs. *Journal of Parasitology*, 87, 1049–1053.

Hanelt, B., Thomas, F., & Schmidt-Rhaesa, A. (2005) Biology of the Phylum Nematomorpha. *Advances in Parasitology*, 59, 243–305.

Hanelt, B., Bolek, M. G. & Schmidt-Rhaesa, A. (2012) Going solo: discovery of the first parthenogenetic gordiid (Nematomorpha: Gordiida). *PLoS One*, 7, e34472.

Jochmann, R., & Schmidt-Rhaesa, A. (2007) New ultrastructural data from the larvae of *Paragordius varius* (Nematomorpha). *Acta Zoologica*, 88, 137–144.

Marchiori, N. C., Periera, J., & Castro, L. A. S. (2009) Morphology of larval *Gordius dimorphus* (Nematomorpha: Gordiida). *Journal of Parasitology*, 95, 1218–1220.

Nielsen, S.-O. (1969). *Nectonema munidae* (Brinkman) (Nematomorpha) parasitizing *Muida tenuimana* G.O. Sars (Crust. Dec.) with notes on host parasite relations and new host species. *Sarsia*, 38, 91–110.

Poinar, Jr., G. O. (1999) *Paleochordodes protus* n. g., n. sp. (Nematomorpha, Chordodidae), parasites of a fossil cockroach, with a critical examination of other fossil hairworms and helminths of extant cockroaches (Insecta: Blattaria). *Invertebrate Biology*, 118, 109–115.

Poinar, Jr., G. O. (2008) Global diversity of hairworms (Nematomorpha: Gordiaceae) in fresh water. *Hydrobiologia*, 595, 79–83.

Poinar, Jr., G. O., & Buckley, R. (2006) Nematode (Nematoda: Mermithidae) and hairworm (Nematomorpha: Chordodidae) parasites in early Cretaceous amber. *Journal of Invertebrate Pathology*, 93, 36–41.

Poinar, Jr., G. O., Rykken, J. & LaBonte, J. (2004) *Parachordodes tegonotus* n. sp. (Gordioidea: Nematomorpha), a hairworm parasite of ground beetles (Carabidae: Coleoptera), with a summary of gordiid parasites of carabids. *Systematic Parasitology*, **59**, 139–148.

Ponton, F. Lebarbenchon, C., Lefèvre, R., *et al.* (2006) Parasite survives predation on its host. *Nature*, **440**, 756.

Sato, T., Watanabe, K., Kanaiwa, M., *et al.* (2011) Nematomorph parasites drive energy flow through a riparian ecosystem. *Ecology*, **91**, 201–207.

Schmidt-Rhaesa, A. (2001) The life cycle of horsehair worms (Nematomorpha). *Acta Parasitologica*, **46**, 151–158.

Schmidt-Rhaesa, A. (2005) Morphogenesis of *Paragordius varius* (Nematomorpha) during the parasitic phase. *Zoomorphology*, **124**, 33–46.

Skaling, B. & MacKinnon, B. M. (1988) The absorptive surfaces of *Nectonema* sp. (Nematomorpha: Nectonematoidea) from *Pandalus mantagui*: histology, ultrastructure, and absorptive capabilities of the body wall and intestine. *Canadian Journal of Zoology*, **66**, 289–295.

Thomas, F., Schmidt-Rhaesa, A., Martin, G., *et al.* (2002) Do hairworms (Nematomorpha) manipulate the water seeking behaviour of their terrestrial hosts? *Journal of Evolutionary Biology*, **15**, 356–361.

Thomas, F., Ulitsky, P., Augier, R., *et al.* (2003) Biochemical and histological changes in the brain of the cricket *Nemobius sylvestris* infected by the manipulative parasite *Paragordius tricuspidatus* (Nematomorpha). *International Journal for Parasitology*, **33**, 435–443.

Pentastomida: the tongue worms

10.1 General considerations

The pentastomids (Greek, *penta*five, *stoma*mouth), or tongue worms, are a small group of obligatory parasites that includes about 130 species. The taxonomic name was erroneously coined in the belief that each of the hooked appendages that flank the true mouth had a mouth. Some species supposedly resemble a miniature vertebrate tongue. Adult pentastomids are found primarily in the respiratory passages of terrestrial vertebrates, mostly reptiles. They range in size from a few millimeters to 15 cm in length. Approximately 70% of the definitive hosts for pentastomes are snakes; several pentastome species have also been described from lizards and freshwater turtles and crocodilians. Relatively few adult pentastomes have been described from amphibians, birds, or mammals, although some species are reported from such unusual sites and hosts as the air sacs of marine birds, the trachea of vultures, and the nasopharynx and sinuses of canines and felines. *Raillietiella* is the most speciose pentastome genus and is the only one known to mature in amphibian hosts. The unique site specificity of pentastomes, coupled with their hematophagus feeding habit, large body size, and long-lived nature have inspired fascinating studies in parasite ecology and evolution (review in Riley, 1986). Particular focus has been on examining the mechanisms by which these large parasites evade their vertebrate host's immune response (reviews in Riley, 1992; Riley and Henderson, 1999; see Box 10.1).

10.2 Form and function

10.2.1 Head and body wall

Adult pentastomids are elongated and cylindrical, but occasionally flattened dorsoventrally. The body appears to be externally segmented, forming distinct **annuli**, or false segments (Figs. 10.1, 10.2, 10.3). The anterior end of the body, also called a head or cephalothorax, bears frontal and dorsal sensory papillae, frontal glands, hook glands, the mouth, and four openings with curved, retractile hooks (Figs. 10.4, 10.5). In the more primitive group (the Cephalobaenida), the retractile hooks are located at the tip of four appendages and the mouth is located at the tip of a fifth projection. The mouth is constantly held open because its lining is hardened or sclerotized. The shape of this sclerotized lining is an important morphological character used in pentastomid classification. The retractile hooks are controlled by muscles and are used both to tear host tissues and to attach to the host.

The body surface is a tegument composed of a chitinous arthropodan cuticle overlying a thin epidermis. In pentastomids the cuticle is soft, transparent, and flexible, similar to the cuticle of the larvae of holometabolous insects. These characteristics of the

Fig. 10.1 Adult of the pentastomid *Porocephalus crotali* in the lung of a rattlesnake *Crotalus atrox*. The uterine coils of the females are clearly visible through the transparent body wall. (Photograph courtesy of John Riley.)

Fig. 10.2 Adult of the pentastomid *Armillifer armillatus* in the lung of a Gaboon viper *Bitis gabonica*. (Photograph courtesy of John Riley.)

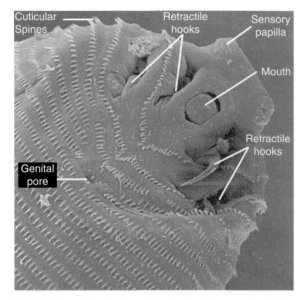

Fig. 10.3 Scanning electron micrograph of the nymph of the pentastomid *Linguatula serrata* from the lungs and liver of an intermediate host, the Chilean deer *Pudu pudu*. (Modified from Fernández & Villalba, 1986, with permission, *Parasitología al Día*, 10, 29–30.)

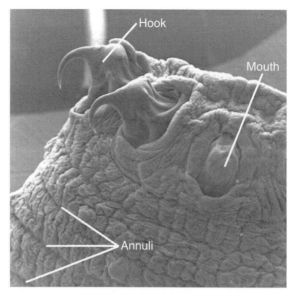

Fig. 10.4 Scanning electron micrograph of the nymph of the pentastomid *Porocephalus crotali* from a rat intermediate host. (Micrograph courtesy of John Riley.)

Fig. 10.5 Adult male and female *Reighardia sternae* from the air sacs of a herring gull *Larus argentatus*. Note the marked sexual dimorphism in size between the sexes. (Photograph courtesy of John Riley.)

cuticle allow the parasites to move through restricted spaces during their complex migration inside the host. Growth requires periodic shedding of the cuticle. The body surface may be annulated, appear serrated, or have spines. It also has a large number of **subparietal glands**. These glands, together with the frontal and hook glands of the head, produce a secretion that plays a protective role against the host immune system (Box 10.1; Riley *et al.*, 1979; Jones *et al.*, 1992; Riley, 1992). Pentastomids have a well-developed muscular system, consisting of longitudinal and circular muscle fibers that are loosely attached to the cuticle. However, because the cuticle is soft and flexible, it does not provide a rigid skeleton upon which muscles can act.

Box 10.1 | Pentastomids: masters of immunoevasion

Adult pentastomids are large, long-lived macroparasites. Typically, they are associated with the lungs or air sacs of vertebrate hosts, sites that should be relatively straightforward for the sophisticated host immune system to access. Yet pentastome-induced inflammatory responses and immunopathology of the pulmonary system are rarely reported. Thus, adult pentastomids must possess effective mechanisms to protect themselves from the attack of their host's immune system.

One of the striking characteristics of pentastomids is a profusion of glands, e.g., subparietal, hook, and frontal, associated with the cuticle. The large cells in the subparietal glands, for example, constantly secrete a lamellate material onto the cuticle that seems to protect it from host immune responses. Studies of the pentastomid *Porocephalus crotali* (a parasite of rattle-snakes and an important model system) indicate that this lamellate coating is acquired during growth of the parasite in its mouse intermediate host (Ambrose & Riley, 1988).

The lungs of all tetrapods are covered by a membranous surfactant secreted by alveolar cells that seems to have a double purpose. First, the membranous surfactant lowers the alveolar surface tension at the air–water interface and facilitates the diffusion of gases. Second, the surfactant functions as an immunosuppressor, protecting the delicate gas exchanging regions of the lungs from detrimental immune reactions to inhaled particles and microbes. Also, at least in mammals, pulmonary surfactant plays a critical role in regulating the activity of alveolar macrophages; these are the most important phagocytes of the bronchioles and alveoli (Riley & Henderson, 1999).

Remarkably, it appears that the overall lipid composition of the lamellate secretions produced by pentastomid glands is very similar to that of the vertebrate pulmonary surfactant, although the proportion of the various components is different. Pentastomid surfactant is channeled to the cuticlar surface via chitin-lined ducts, and every part of the cuticle is membrane-coated. Moreover, ultrastructural studies cannot distinguish between membranes produced by the lamellate secretions of the parasites and the pulmonary membranous surfactant produced by the lungs of the vertebrate host. Thus, it is likely that the host immune system cannot distinguish between its own membranous surfactant and the lamellate secretions of the pentastomids. It appears then that lipids in the lamellate secretions of pentastomids mimic the pulmonary surfactant of the host and suppress the potential host immune response against the parasite (Riley, 1992). In effect, lung-dwelling pentastomes evade immune surveillance and also reduce immunopathology by coating their chitinous cuticle with their own surfactant.

The coelomic fluid within the body, however, provides a hydrostatic skeleton for supporting internal organs and in locomotion via waves of peristalsis. Movement, then, is accomplished by localized changes in the shape of the body, using the fluid skeleton as the basis for muscle antagonism. Because there is no circulatory or respiratory system, peristaltic movements of the body wall agitate the coelomic fluid, and aid in the

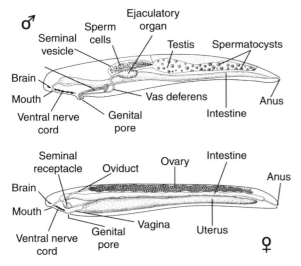

Fig. 10.6 Body plans of a generalized male and female cephalobaenid pentastome. (Modified from Riley, 1983, with permission, *Parasitology*, 86, 59–83, Cambridge University Press.)

transport of gases. Gas exchange probably occurs through the permeable cuticle.

10.2.2 Digestive system

The digestive system is divided into a foregut, midgut, and hindgut (Fig. 10.6). Consistent with other arthropods, the foregut and hindgut are lined with cuticle. The foregut includes the buccal cavity, pharynx, and esophagus and is separated from the midgut by a valve that prevents the backflow of ingested food, primarily blood. The midgut is a straight tube without compartments and is where extracellular digestion occurs.

10.2.3 Osmoregulatory system

Pentastomids lack an excretory system, but they have specialized, epidermal ion-transporting cells called **ionocytes**, or **chloride cells**, that are responsible for osmotic and ionic regulation. These cells are presumably involved in the secretion of excess electrolytes (mainly sodium and chloride) contained in the blood and lymph they consume. These ion-transporting cells

are protected from the host immune system by the secretions of the subparietal, frontal, and hook glands. Pentastomids are likely to regulate hypoosmotically, maintaining their ion concentration lower than that of the host. Other waste products, e.g., nitrogen, diffuse across the cuticle of the parasite probably in the form of ammonium ions.

10.2.4 Reproductive system

Pentastomids are dioecious and exhibit some degree of sexual dimorphism, with males being smaller than females (Fig. 10.5). The male has a single tubular testis (except *Linguatula* which has two) that occupies about 50% of the body cavity and connects to the seminal vesicle, which stores the sperm until copulation. The seminal vesicle is continuous with a pair of muscular ejaculatory organs, each of which has a duct or vas deferens that reaches into a terminal penis (Fig. 10.6). The penis fits into a sclerotized **dilator organ**. The dilator organ may serve to guide the thread-like penis during copulation (Storch, 1993).

Females have a single ovary that occupies most of the body cavity. The ovary leads to one or two oviducts that continue into the uterus. There is a short muscular vagina that opens to the exterior through the gonopore. Attached to the uterus is a pair of seminal receptacles (spermatheca) where the male deposits sperm during copulation (Fig. 10.6).

Females apparently copulate only once in their lifetime, whereas males probably copulate more than once and with different females. Copulation occurs when both sexes are approximately the same size, which means that females are not yet fully mature, and their uterus is still undeveloped. After copulation, the sperm are stored in the seminal receptacle for several months until her reproductive structures are fully developed, at which time the seminal receptacle provides sperm for the fertilization of oocytes. Since females only copulate once in their lifetime, the total number of sperm held in the seminal receptacle is the limiting factor for egg production. Sufficient sperm must be stored to fertilize millions of eggs over many years in some cases. In the Porocephalida, the uterus is tubular, elongated, extensively folded, and occupies most of the hemocoel. In this group, egg production is massive and continuous, with the uterus containing eggs in different stages of development; eggs mature as they travel down the convoluted uterus. In females of *Linguatula serrata* (Fig. 10.4), about 500 000 eggs are stored in the uterus at any one time, and several million eggs can be produced in their lifetime. In the cephalobaenid pentastomes, however, the uterus is saccate and eggs are accumulated and stored before being released. In the cephalobaenids, the vagina is modified and forms a unique sieving function allowing only large, fully embryonated eggs to pass into the host, while retaining immature ones. Females of *Reighardia sternae* (Fig. 10.5), for example, have about 2900 eggs in the uterus, the total number of eggs produced in their lifetime. Thus, the pentastomes have two fundamentally different mechanisms by which mature eggs pass into their host's respiratory tract.

10.2.5 Nervous system and sensory input

The organization of the nervous system differs between the two groups of pentastomids. The more derived Porocephalida have a pair of ventral nerve cords, and all the ganglia are fused into a subesophageal or cerebral ganglion. In the Cephalobaenida, three separate ganglia form the cerebral ganglion, with two more located anteriorly along the ventral nerve cord, a pattern similar to the primitive arthropod nervous system. The only apparent sensory structures in pentastomids are the frontal and dorsal papillae of the cephalothorax, and the chains of lateral papillae located in between the abdominal annuli.

10.3 Nutrient uptake and metabolism

Adult pentastomids living in the lungs and air sacs of reptiles and birds feed on blood sucked in from ruptured capillaries and lymph, whereas those living in the nasopharyngeal cavities of mammals feed on mucus and sloughed cells. Encysted nymphs in the intermediate hosts feed on blood, lymph, and lymphoid cells. Early larval stages feed predominantly on eosinophils recruited to the infection site by the inflammatory response of the host. Digestion in adult pentastomids seems to be mostly extracellular because hematin, the end product of hemoglobin digestion, accumulates in the intestinal lumen, causing it to darken. Some intracellular digestion also must occur, however, because iron accumulates in certain gastrodermal cells that are shed periodically into the intestinal lumen.

Very little is known regarding the biochemistry and metabolism of pentastomids. It appears that glycogen is the main storage form for carbohydrates, being sequestered mainly within the striated muscles and the gastrodermis. Although details regarding the catabolism by *Kiricephalus pattoni* are unknown, it seems that this species, at least, is capable of some form of oxidative metabolism.

10.4 Development and general life cycle

Within the uterus of a female, the fertilized oocyte develops into a **primary larva**, which is enclosed by three layers in the Cephalobaenida and four in the Porocephalida. The innermost layer is chitinous, covered by mucus, and has a characteristic opening called the **facette**. The mucus of the innermost layers is produced by the embryo's **dorsal organ** and flows out through the facette, surrounding the chitinous stratum. The middle layer is the chorion followed by a third layer that becomes the eggshell. The fourth layer present in the Porocephalida is secreted by the reproductive tract of the female and forms a hyaline capsule when exposed to water (Storch, 1993).

By the time the egg is shed by the female pentastomid into the host's respiratory tract, it contains a fully developed primary larva that is infective to the next host in the life cycle (Fig. 10.7A). This larva is ovoid, has a short bifurcate tail, four stumpy legs each with a pair of claws, and a penetration apparatus at the anterior end (Fig. 10.7B). The penetration apparatus, together with the clawed legs, is used to tear through

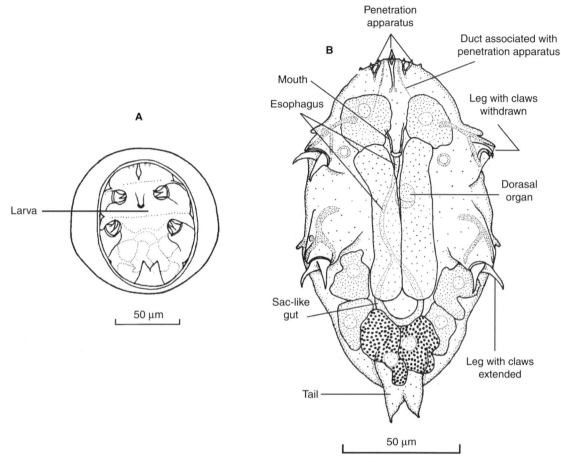

Fig. 10.7 Schematic representation of the egg and primary larva of the pentastomid *Porocephalus crotali*. (A) Mature egg containing a fully developed primary larva; (B) ventral view of the primary larva. (Modified from Esslinger, 1962, with permission, *Journal of Parasitology*, 48, 457–462.)

the tissues of the intermediate host during the migration of the primary larva from the host intestine to the abdomen. Two ducts open within the penetration apparatus and are thought to secrete histolytic enzymes that aid in the migration process. Between the anterior pair of legs, there is a simple mouth that continues into an esophagus ending in a blind sac. In some species, a thin hindgut is present. An important characteristic of all the pentastomid larvae is the presence of a dorsal organ (Fig. 10.7B), which secretes a mucoid substance that becomes part of the first protective layer surrounding the larva. The dorsal organ is formed by a number of gland cells

surrounding a central, hollow vesicle that opens to the surface by a dorsal pore.

Although just a few life cycles are known for pentastomids, most species require an intermediate and a definitive host. The intermediate host in most cases is a vertebrate (fish, amphibian, reptile, or mammal), or an insect in some species of *Raillietiella*. Intermediate hosts become infected by consuming the eggs. In pentastomes of the lungs, eggs pass up the trachea of the host, are swallowed, and eliminated with the host feces. It is also likely that some eggs may reach the outside with the host sputum. In those species that inhabit the nasal passages and sinuses of the host, eggs are discharged with nasal secretions and while sneezing.

When a suitable intermediate host ingests an infective egg, the primary larva hatches in the host's intestine, penetrates the intestinal wall using the penetration apparatus, and migrates into the abdominal cavity. After a number of molts in the host's abdomen, the primary larva becomes quiescent and metamorphoses into a **nymph**, which is a miniature version of the adult. The nymph is the infective stage for the definitive host. In most pentastomes the infective nymph is protected within the last cuticle molt. When the intermediate host is eaten by a suitable definitive host, the nymph is freed in the digestive tract, penetrates the intestinal wall, and bores into the lung where it matures. In the case of *Linguatula*, the freed nymph migrates to the nasopharyngeal cavity directly up the esophagus from the stomach, without boring through tissues.

Pentastomid nymphs, unlike the adults, are more diverse in terms of site selection. For example, the larvae of *Linguatula* (Fig. 10.4) prefers mesenteric and lymph nodes; *Porocephalus crotali* in mice and rat intermediate hosts prefers the fatty tissue around the intestine and reproductive organs. The nymphs of *Sebekia* are found free in the body cavity in the viscera and *Subtriquetra* prefer the swim bladder of the fish intermediate host. The nymphs of species of *Raillietiella* that use insects as intermediate hosts are found on the surface of the viscera or in the fat bodies (for more details, see Riley, 1986). There is evidence that nymphs may cause pathology in some intermediate hosts. For example, Boyce *et al.* (1987) noted that some fish species, e.g., swordtails, suffer an extensive granulomatous inflammatory reaction and prominent lesions when infected with nymphs of *Sebekia mississipiensis* (see also Self, 1972).

10.5 Biodiversity and life-cycle variation

Approximately 90% of adult pentastomids parasitize reptiles and 70% of these species are found in snakes (see Boxes 10.2, 10.3). *Armillifer* is a common parasite in the lungs of snakes (Fig. 10.2). *Armillifer armillatus* and *A. grandis* infect pythons and vipers in Africa, whereas *A. moniliformis* infects pythons in southeast Asia. *Waddycephalus punctatus* is a parasite in the lung of the common tree snake in Australia (Color plate Fig. 3.1). A wide range of mammals, including rats, cats, monkeys, giraffes, antelopes, and sometimes humans, serve as intermediate hosts, although humans and big game animals represent dead-end hosts. Several species within the genus *Porocephalus* parasitize snakes in North, Central, and South America; all of these species use mammals as intermediate hosts. *Porocephalus crotali* is probably the best known of these species (Figs. 10.1, 10.4). It is commonly found in the lungs of crotalid snakes, especially rattlesnakes in North America, where rodents are the intermediate hosts most important in transmission.

Species of *Sebekia* and *Alofia* parasitize crocodilian species worldwide and are thought to use a variety of fishes as intermediate hosts. Fourteen of the currently 23 crocodilian species have been recorded as being host to one or more pentastomes; the African crocodile harbors eight pentastome species, while the Indo-Pacific crocodile has seven (Boomker, 2006). One noteworthy genus found only in this latter host is *Selfia*, named after the great pioneer and guru of pentastomes, Professor J. Teague Self. *Sebekia mississippiensis* is found in the bronchi of the American alligator and nymphs of this pentastome species have been found mostly in several species of freshwater fishes, as well as snakes, turtles, and mammals (Overstreet *et al.*, 1985). Two monospecific genera of chelonians, *Pelonia* and *Diesingia*, have been reported from terrapin species, paradoxically, in South Africa and South America, respectively (e.g., Boomker, 2006).

Linguatula is found in the nasal airways, frontal sinuses, and tympanic cavity of carnivorous mammals such as canids, felids, and hyaenids (Fig. 10.3). Large grazing herbivores are intermediate hosts. *Linguatula serrata* is a cosmopolitan parasite of dogs, wolves, and foxes; cattle, sheep, goats, and other ruminants serve as intermediate hosts. This parasite can cause hemorrhage and breathing difficulties. Humans can also be accidental hosts and pathogenesis can occur.

Box 10.2 | **Fossil pentastomes from the Cambrian era suggest an alternative phylogenetic hypothesis**

Comparative morphology of fossil organisms has contributed significantly to deciphering macroevolutionary patterns of speciation and extinction, and to understanding the history of multicellular life on earth. For soft-bodied taxa, such an approach is limited to certain fossil sites such as the Burgess Shale in British Columbia, Canada. Not surprisingly, the fossil record is virtually non-existent for most soft-bodied parasitic organisms. In most cases, both the host and parasite must also be fossilized together – a rare event indeed! It is no small wonder, then, that the fossil pentastomes discovered by Walossek and Müller (1994) from upper Cambrian fossil beds in Sweden generated much interest. Stephen Jay Gould detailed the striking morphological and embryological similarities of these fossils with extant pentastomes when he wrote "Of tongue worms, velvet worms and water bears" (Gould, 1995).

Molecular and morphological studies convinced many biologists that pentastomes were crustaceans closely related to the Branchiura (e.g., Abele *et al.*, 1989; Storch & Jamieson, 1992). This hypothesis favors the idea that pentastomes are a recently derived group that arose later in earth's history when true crustaceans were present (about 280 million years ago). These fossils have altered this notion and suggest a much earlier divergence from a proto-arthropod ancestor stem group, 530 million years ago and millions of years before most of their present-day hosts (reptiles) had evolved. In other words, these Cambrian fossils completely alter our view of pentastome origins. Walossek and Müller (1994) conclude: "The long history of the group and its remarkable morphological stasis invalidates any hypothesis of their evolution from terrestrial arthropods."

The existence of Cambrian pentastomes raises the obvious question – what were their original hosts? An enigmatic and prominent group of microscopic fossils known as conodonts (tooth elements) may provide a clue (Gould, 1995). Conodont fossils are common in all Cambrian localities from which pentastomes have been found. Conodonts are now thought to be the fossil teeth of early fish-like vertebrates. Thus, the earliest pentastomes may have been parasites of these early vertebrates. They may have made their critical 'jump' onto land by first infecting amphibian lungs and then reptiles during the Mesozoic era. Pentastome-reptilian associations could also have originated either in the Paleozoic among the freshwater reptiles of the Carboniferous and Permian, or in the Mesozoic, when a considerable proportion of the then-existing reptiles became secondarily adapted to a marine existence.

There are exceptions to the general pentastome life-cycle pattern. These include *Reighardia sternae*, *R. lomviae*, and *Subtriquetra subtriquetra*. *Reighardia sternae* (Fig. 10.5), a parasite of the air sacs of gulls and terns, has a direct life cycle. It appears that females induce coughing and vomiting by the infected host and infection of other hosts occurs when they ingest the egg-infected vomit of a parasitized bird. *Raillietiella trachea*, a terrestrial pentastome inhabiting the trachea of the white-backed vulture (Riley *et al.*, 2003), as well as *Rileyella petauri* from the lungs and nasal sinus of an Australian petaurid marsupial

(Spratt, 2003), also appear to have direct life cycles. *Reighardia lomviae*, a parasite of guillemots, has a similar life cycle, but the pattern of prevalence (very high in young birds with a sharp decline in adults by the end of the breeding season) suggests that transmission probably occurs between infected parents and their nestlings. Infection is likely to occur when the parents feed their young with fish that were held in their mouths and thereby contaminated with pentastomid eggs (Banaja *et al.*, 1975; 1976). *Subtriquetra subtriquetra*, a parasite in the nasopharynx of South American crocodiles, is also unusual. It is the only pentastome species known to have a free-living larva that actively searches for its fish intermediate host (Winch & Riley, 1986).

Although most pentastomids occupy the lumen of the host lungs and a few occupy the sinuses and nasal passages, they are normally able to move freely, changing their feeding site. However, certain porocephalid females of snakes are permanently buried in the epithelium of the lungs and are unable to move. A host reaction at the attachment site of these pentastomids further encloses the anterior end with a fibrotic capsule, sequestering the parasite.

Although pentastomids are not common parasites of humans, they can become a zoonotic problem in some parts of the world. Visceral pentastomiasis occurs when humans ingest infective eggs of species such as *Linguatula serrata*, *Armillifer armillatus*, and *A. moniliformis*. Most cases of visceral pentastomiasis are caused by *A. armillatus* in Central and West Africa and *A. moniliformis* in southeast Asia. Infections are probably acquired by the consumption of food or water contaminated with snake feces containing eggs, or by eating inadequately cleaned and undercooked snake meat. Among Malaysian aborigines who consume snake meat, for example, the prevalence of visceral pentastomiasis by *A. moniliformis* can reach up to 45%. Although the pathology is restricted to inflammatory and granulomatous responses, an infection may sensitize a person in such a way that a subsequent infection may elicit a strong allergic response. Nymphs develop and encapsulate in a thin-walled cyst, mainly in the liver, intestinal wall, and mesenteries.

Nasopharyngeal pentastomiasis results from the ingestion of nymphs of *L. serrata* that, upon ingestion, migrate and establish in the nasal passages of humans. This condition, also known as nasopharyngeal linguatulosis, can have dramatic effects. Infections are acquired by eating raw or undercooked meat, or viscera, of domestic herbivores harboring encapsulated nymphs. In rural areas of Lebanon, for example, consumption of raw liver from goats and sheep is relatively common; in Sudan, a similar dish consists of raw stomach, liver, lung, and trachea of sheep, goats, cattle, and camels. Nasal and lachrymal discharges, together with violent coughing and sneezing, may help dislodge the nymphs and produce immediate relief of the symptoms. If complications arise, they may include abscesses of the auditory canal, facial swelling, paralysis, and, sometimes, asphyxiation and death.

10.6 Phylogenetic relationships and classification

Pentastomes have previously been allied as distinct phyla linked to the Tardigrada and Onycophora, and even the Annelida. Current consensus based on morphological (primarily on the basis of cuticular structure and comparative sperm morphology) and molecular (based on 18S ribosomal nucleotide sequences) phylogenetic analyses suggest an alliance between pentastomes and arthropods, specifically the branchiuran crustaceans (Wingstrand, 1972; Riley *et al.*, 1978; Riley, 1986; Abele *et al.*, 1989; Storch & Jamieson, 1992). Currently, pentastomids are considered as highly modified crustacean endoparasites. As stated by Martin & Davis (2001), "the inclusion of pentastomids among the Crustacea takes the known morphological diversity and life style extremes of the Crustacea – already greater than for any other taxon on earth – to new heights." Pentastomes may not resemble 'typical' crustaceans, but neither do several groups of highly modified parasitic crustaceans such as the pennellid copepods, rhizocephalan barnacles, or entoniscid isopods (see Chapter 11). It is just that these

Box 10.3 | **Classification of the Pentastomida**

We treat the pentastomids as an independent group in this chapter, since a consensus regarding their precise linkages to the crustaceans has not been reached (e.g. Almeida & Christofferson, 1999). The classification scheme proposed by Riley (1986, 2001) is presented here and modified to include the genera *Rileyella* and *Pelonia*. Almeida & Christoffersen (1999) analyzed the relationships among pentastomid groups using cladistic methods. Their scheme differs from that of Riley's in a few ways. They propose a new order Raillietiellida to contain the family Raillietellidae and the recognition of the Reighardiida as a new order to contain the family Reighardiidae. As well, they proposed to remove the family Sambonidae. These differences will be resolved as more pentastomes are subject to morphological and molecular phylogenetic studies in the future.

Order Cephalobaenida
  Family Cephalobaenidae
    Representative genera: *Cephalobaena,*
      Raillietiella, Rileyella
  Family Reighardiidae
    Representative genus: *Reighardia*
Order Porocephalida
  Family Sebekiidae
    Representative genera: *Sebekia, Alofia, Pelonia,*
      Leiperia, Diesingia, Selfia, Agema
  Family Subtriquetridae
    Representative genus: *Subtriquetra*
  Family Sambonidae
    Representative genera: *Sambonia, Elenia,*
      Waddycephalus, Parasambonia
  Family Porocephalidae
    Representative genera: *Porocephalus,*
      Kiricephalus
  Family Armilliferidae
    Representative genera: *Armillifer, Cubirea,*
      Gigliolella
  Family Linguatulidae
    Representative genus: *Linguatula*

groups reveal their crustacean affinities more readily as they are aquatic, have free-swimming nauplius larvae, and do not live a terrestrial life in snake lungs! Further, the discovery of fossil pentastomes preserved in detail in Swedish limestone deposits of the middle Cambrian period has exacerbated the phylogenetic debate (Walossek & Müller, 1994; Riley, 2001; see Box 10.3).

References

Abele, L. G., Kim, W. & Felgenhauer, B. E. (1989) Molecular evidence for inclusion of the phylum Pentastomida in the Crustacea. *Molecular Biology and Evolution*, **6**, 685–691.

Almeida, W. O & Christoffersen, M. L. (1999) Evolutionary pathways in Pentastomida. *Journal of Parasitology*, **85**, 695–704.

Ambrose, N. C. & Riley, J. (1988) Fine structural aspects of secretory processes in a pentastomid arthropod parasite in its mouse and rattlesnake hosts. *Tissue Cell*, **20**, 381–404.

Banaja, A. A., James, J. L. & Riley, J. (1975) An experimental investigation of a direct life-cycle in *Reighardia sternae* (Diesing, 1864) a pentastomid parasite of the herring gull (*Larus argentatus*). *Parasitology*, **71**, 493–503.

Banaja, A. A., James, J. L. & Riley J. (1976) Some observations on egg production and autoreinfection of *Reighardia sternae* (Diesing, 1864) a pentastomid parasite of the herring gull (*Larus argentatus*). *Parasitology*, **72**, 81–91.

Boomker, J. K (2006) Check-list of the pentastomid parasites of crocodilians and freshwater chelonians. *Ondersteport Journal of Veterinary Research*, **73**, 27–36.

Boyce, W. M., Kazacos, E. A., Kazacos, K. R. & Engelhardt, J. A. (1987) Pathology of pentastomid infections (*Sebekia mississippiensis*) in fish. *Journal of Wildlife Diseases*, **23**, 689–692.

Gould, S. J. (1995) Of tongue worms, velvet worms, and water bears. *Natural History*, **104**, 323–330.

Jones, D. A. C., Henderson, R. J. & Riley, J. (1992) Preliminary characterization of the lipid and protein components of the protective surface membranes of a pentastomid *Porocephalus crotali*. *Parasitology*, **104**, 469–478.

Martin, J. W. & Davis, G. E. (2001) *An Updated Classification of the Recent Crustacea*. Los Angeles: Natural History Museum of Los Angeles County.

Overstreet, R. M., Self, J. T. & Vliet, K. A. (1985) The pentastomid *Sebekia mississippiensis* sp. n. in the American alligator and other hosts. *Proceedings of the Helminthological Society of Washington*, **52**, 266–277.

Riley, J. (1986) The biology of pentastomids. *Advances in Parasitology*, **25**, 48–128.

Riley, J. (1992) Pentastomids and the immune response. *Parasitology Today*, **8**, 133–137.

Riley, J. (2001) *Pentastomida. Encylopedia of Life Sciences*. Hoboken: John Wiley & Sons.

Riley J. & Henderson, R. J. (1999) Pentastomids and the tetrapod lung. *Parasitology*, **119**, S89–105.

Riley, J., Banaja, A. A. & James, J. L. (1978) The phylogenetic relationships of the Pentastomida: the case for their inclusion within the Crustacea. *International Journal for Parasitology*, **8**, 245–254.

Riley, J., James, J. L. & Banaja, A. A. (1979) The possible role of the frontal and sub-parietal gland systems of the pentastomid *Reighardia sternae* (Diesing, 1864) in the evasion of the host immune response. *Parasitology*, **78**, 53–66.

Riley, J., Oaks, J. L. & Gilbert, M. (2003) *Raillietiella trachea* n. sp., a pentastomid from the trachea of an oriental white-backed vulture *Gyps bengalensis* taken in Pakistan, with speculation about its life-cycle. *Systematic Parasitology*, **56**, 155–161.

Self, J. T. (1972) Pentastomiasis: host responses to larval and nymphal infections. *Transactions of the American Society of Microscopy*, **91**, 2–8.

Spratt, D. M. (2003) *Rileyella petauri* gen. nov., sp. nov. (Pentastomida: Cephalobaenidae) from the lungs and nasal sinus of *Petaurus breviceps* (Marsupialia: Petauridae) in Australia. *Parasite*, **10**, 235–241.

Storch, V. (1993) Pentastomida. In *Microscopic Anatomy of Invertebrates*, vol. 12, *Onycophora, Chilopoda, and Lesser Protostomata*, ed. F. W. Harrison & M. E. Rice, pp. 115–142. New York: Wiley-Liss.

Storch, V. & Jamieson, B. G. M. (1992) Further spermatological evidence for including the Pentastomida (tongue worms) in the Crustacea. *International Journal for Parasitology*, **22**, 95–108.

Walossek, D. & Müller, K. J. (1994) Pentastomid parasites from the lower Paleozoic of Sweden. *Transactions of the Royal Society of Edinburgh, Earth Sciences*, **85**, 1–37.

Winch, J. M. & Riley, J. (1986) Studies on the behavior and development in fish of *Subtriquetra subtriquetra*: a uniquely free-living pentastomid larva from a crocodilian. *Parasitology*, **93**, 81–98.

Wingstrand, K. G. (1972) Comparative spermatology of a pentastomid, *Raillietella hemidactyli*, and a branchiuran crustacean, *Argulus foliaceus*, with a discussion of pentastomid relationships. *Kongelige Danske Videnskab Selskab Biologiske Skrifter*, **19**, 1–72.

11 Arthropoda: the joint-legged animals

11.1 General considerations

The phylum Arthropoda (Greek, *arthro*jointed, *poda*foot) comprises the most abundant and diverse of all animals, with more species than all other animal groups combined. One group alone, the insects, likely includes well over 5 000 000 species, although 'only' about 1 000 000 have been described (Gullan & Cranston, 2010). Arthropods have an extraordinarily rich fossil record that extends back to the Cambrian period. Since then, arthropods have undergone a phenomenal adaptive radiation and exploit virtually every conceivable habitat on earth. Within these habitats, arthopods are of ecological significance, playing many integral ecosystem functions. As free-living animals, as micropredators, and as mutualists and parasites, they are of massive evolutionary importance, driving the evolution of countless other organisms. Arthropods are also of economic importance to humans. Many are sources of food, while others are direct competitors for our food or destroy valuable products, and can be devastating pests. Life on earth would not exist as we know it, were it not for arthropods.

Arthopods feature prominently in parasitology. Thus, we have learned in previous chapters that arthropods can act as intermediate hosts for a variety of parasites, demonstrating their importance in food web relationships and predator–prey interactions. We have also discovered that many arthropods are micropredators and act as vectors for many species of protist and nematode parasites infecting vertebrates. Given their diversity, it should come as no surprise that a substantial number of arthropods have also evolved a parasitic life style. Most are ectoparasitic, with exquisite morphological adaptations for attachment to a host. Others are endoparasitic and live within a host.

Our focus in this chapter is to present an overview of the biology of ectoparasitic and endoparasitic arthropods, especially the crustaceans, arachnids, and insects. These three groups include over 95% of all arthropod species (Pechenik, 2010). While some of these parasites resemble their free-living ancestors, we will encounter others that rival the most bizarre and highly specialized of any of the parasitic animals that we have discussed up to now.

Before introducing arthropods as parasites, we consider the general arthropod body plan that helps to explain their overwhelming success. This 'winning combination' of features sets the stage for considering the evolution of parasitism and the diversity of parasites in the phylum. The first concerns the relatively small body sizes and rapid generation times characteristic of many arthropods. As we will see in this chapter, hosts offer a diverse array of ecological niches for these small arthropods to undergo speciation and adaptive radiations.

The second characteristic gives the phylum the name – jointed appendages. These appendages have been variously modified during arthropod evolution to serve many functions, including diverse modes of locomotion, food handling, as well as sensory and reproductive functions. The diversity of mouthpart appendages, for example, helps explain the many types of foods and feeding life styles of arthropods. The versatility of mouthpart and locomotory appendages is realized when we find that they can be modified to serve as tenacious holdfasts in many ectoparasitic arthropods.

The third important feature of arthropods concerns segmentation of the body, or **metamerism**. Thus, metameric segmentation refers to the arthropod body being divided into similar parts, arranged in a linear series along the antero-posterior axis. In most arthropods, the underlying metameric segmentation is masked because the body segments are combined into functional groups forming the different regions of the body. This process is known as **tagmosis** or **tagmatization**. The different regions of the body formed by the fusion of segments are called **tagmata**. The body of arthropods usually is formed of between 10 and 25 segments, which are grouped into two or three tagmata, such as the head, thorax, and abdomen of insects. Tagmatization has often been lauded as instrumental in the overall success of arthropods because it enables the appendages of all its segments to perform efficiently and to operate as a unit.

A continuous chitinous exoskeleton, or cuticle, covers the entire surface of the body and appendages, providing protection and support. The cuticle provides an extensive surface area for the jointed appendages. The exoskeleton consists of hardened regions, or plates, connected by thinner, and flexible, articular membranes that permit the movement of the jointed appendages. The presence of this versatile exoskeleton plays a significant role in the success of arthropods. During arthropod evolution, it was molded into a tremendous diversity of structures with many feeding, sensory, locomotory, and physiological functions. Despite its protective advantages and functional versatility, all arthropods must shed or molt this external armour to grow. Thus, a complex, hormonally mediated process of **ecdysis** or **molting** is a significant characteristic of arthropods.

Arthropods exhibit protostome development and show pronounced cephalization, with a brain and a complex central nervous system coordinating numerous sensory structures. They possess a complete digestive system, a reduced coelomic cavity (hemocoel), open circulatory system, and varied respiratory and osmoregulatory structures. For the most part, the sexes are separate and fertilization is internal. Life cycles are extraordinarily diverse and postembryonic development typically involves larval or nymphal stages that undergo a series of molts prior to **metamorphosis** into the adult. Some parasitic crustaceans, e.g., rhizocephalan barnacles and pennellid copepods, and many parasitic insects undergo a truly spectacular

metamorphosis as part of their complex life cycles (Box 11.1).

Determining the phylogenetic relationships of arthropods is a daunting challenge. Indeed, whether the phylum Arthropoda is monophyletic has long been controversial. A combination of embryological, morphological, and molecular comparative evidence is currently used to interpret evolutionary relationships and construct robust phylogenies for the Arthropoda (e.g., Regier *et al.*, 2005; Mallatt & Giribet, 2006). As we mention in Chapter 8, the new taxonomic grouping Ecdysozoa has been proposed to accommodate the arthropods and other animals that molt, including the nematodes and nematomorphs (Aguinaldo *et al.*, 1997; Mallatt *et al.*, 2004). This would mean that arthropods are more closely related to the nematodes and other molting animals than they are to the annelids. It appears now that metamerism in the Annelida and Arthropoda reflects evolutionary convergence.

A widely accepted classification system recognizes five main evolutionary lines of arthropods. These include the extinct trilobytes (Trilobyta), the Myriapoda (millipedes, centipedes), the Crustacea (shrimps, crabs, copepods, barnacles, isopods), Chelicerata (ticks and mites), and the Hexapoda (insects). Of course, the jury is still out and not everyone agrees that these subphyla are monophyletic (see Box 11.6). These taxa do, however, provide a framework in which to explore the crustaceans, chelicerates, and insects as parasitic animals.

11.2 Crustacea

11.2.1 General considerations

As we mention in the last chapter (the Pentastomida), the Crustacea are the most morphologically diverse of all arthropodan taxa, and include familiar forms such as crabs, shrimp, barnacles, and copepods. Most crustaceans are aquatic and marine. Free-living species can play vital ecological roles, e.g., planktonic copepods as integral constituents of food webs linking phytoplankton with higher trophic levels. Crustaceans also display a variety of life styles, including parasitism. Indeed, perhaps no other group displays such an astonishing array of morphological and life history adaptations to parasitism. Some of these crustacean parasites are so highly adapted to parasitism that, as adults, they bear no resemblance to arthropods whatsoever!

11.2.2 Form and function

11.2.2.1 Morphology

A 'typical' crustacean has a head, a thorax, and an abdomen, although often one, or several, thoracic segments are fused to the head, forming a **cephalothorax**, which may be covered by a **carapace**. The head bears five segments with five pairs of appendages, including two pairs of chemosensory **antennae**; however, in some parasitic species these are modified for attachment to the host. Next, the **mandibles** and the first and second **maxillae** are used as feeding appendages. All of these appendages can be variously modified as holdfasts or to exploit different diets in parasitic crustaceans. Behind the head is the thorax and the thoracic appendages, or **pereopods**. Most thoracic appendages are used in locomotion, although one, or more, of them may be incorporated into the mouthparts and used for food manipulation. When this happens, they are called **maxillipeds**. The abdomen bearing appendages, or **pleopods**, follows the thorax, and it ends in a **telson**. The telson is flanked by the most posterior abdominal appendages, known as **uropods**. Pereopods and pleopods can be modified for walking, swimming, or copulation, or as tenacious holdfasts in some parasites, e.g., isopods.

When considering other morphological adaptations of crustacean parasites, three key trends are apparent. First, there is a reduction in locomotory appendages. Second, there is typically an increase in the size of parasitic crustacean species relative to their free-living relatives. Often, sexual dimorphism is extreme, with male 'dwarfism' common. The trends towards gigantism and morphological modification are particularly evident in parasitic females of many crustaceans. Further, the larger body sizes, especially of females, are often accompanied by dramatic changes in their body

proportions. Third, there is a noticeable trend towards fusion of body segments, often with a complete loss of segmentation. In the parasitic Copepoda, for example, these characteristics can be seen with reference to the three parasitic species shown in Fig. 11.1. At one end of the spectrum are the ergasilid copepods, with pronounced segmentation reminiscent of free-living copepods. At the other extreme are extraordinarily modified lernaeopodid copepods that lack segmentation and locomotory appendages, and possess remarkable holdfast modifications such as *Ommatokoita elongata* (Fig. 11.1). Many of these same trends will be apparent when we consider the diversity of other parasitic crustaceans.

11.2.2.2 Body wall

The chitinous exoskeleton of crustaceans consists of several layers containing protein, lipid, and polysaccharides, all secreted by the underlying epidermis. The outermost layer of the exoskeleton is the epicuticle, which includes mostly chemically inert proteins. Beneath the epicuticle lies the endocuticle, which is formed by three layers. The outer layer is calcified and pigmented with tanned proteins; the middle layer is relatively thin, calcified, untanned, and unpigmented; the inner layer is thin, uncalcified, and untanned. Beneath the endocuticle is the epidermis that secretes all the overlying cuticular layers. Under the epidermis are clusters of tegumental glands that open via ducts into the surface of the exoskeleton. The secretions from these glands are involved in the production of the epicuticle. The middle and inner layers of the endocuticle contain chitin and protein, but the protein is not sclerotized, i.e., hardened. These layers are membranous and flexible, and become the articular membranes of the junctions.

Little is known about the nature of the cuticle in parasitic crustaceans. In the least-modified forms, the cuticle probably conforms to the basic pattern found in most free-living crustaceans but, in the highly modified parasitic crustaceans, it is rarely clear what kind of changes occur to adapt to their particular life styles. During the earlier life cycle stages, parasitic copepods

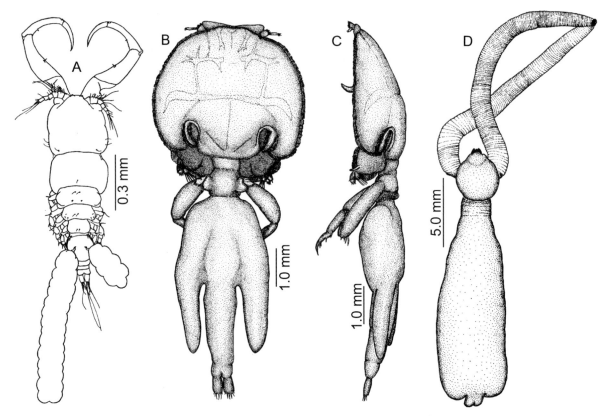

Fig. 11.1 Morphological variability in parasitic copepods, illustrating highly modified holdfasts, as well as trends towards reduced appendages and segmentation. (A) Female of *Ergasilus versicolor* from the gills of a yellow bullhead *Ameiurus natalis*, dorsal view; (B) female of *Paralebion elongatus* from the mouth of a whitetip reef shark *Triaenodon obesus*, dorsal view; (C) female of *P. elongatus*, lateral view showing dorsoventrally compressed body; (D) female of *Ommatokoita elongata* from the eye of a Pacific sleeper shark *Somniosus pacificus*, dorsal view, showing the highly modified second maxillae the parasite uses to attach to its host's eye. (A) Modified from Roberts, 1969, with permission, *Journal of the Fisheries Research Board of Canada*, 26, 997–1011, NRC Research Press; (B, C) Modified from Benz *et al.*, 1992, with permission, *Journal of Parasitology*, 78, 1027–1035; (D) Modified from Benz *et al.*, 1998, with permission, *Journal of Parasitology*, 84, 1271–1274.)

grow like all other crustaceans, by stepwise increase in size through a series of molts. However, once they reach sexual maturity, copepods stop molting, but many continue to grow gradually. It is not quite clear how this is accomplished, but the epicuticle of at least one species, *Pennella elegans*, does not contain amino acids with sulfur or aromatic rings, which are normally involved in hardening. The proteins that make up the cuticle of *P. elegans* also contain many di- and tri-tyrosine links. It appears that these links help stabilize resilin, a rubber-like compound present in the cuticle of arthropods, that helps keep the cuticle flexible

(Kannupandi, 1976a, b). In other highly specialized copepods, the cuticle has been modified in response to the particular habitat they occupy.

11.2.2.3 Digestive system

In most Crustacea, the digestive tract is a fairly straight tube with few modifications. As in all arthropods, it is divided into foregut, midgut, and hindgut. The foregut includes the esophagus and stomach. The stomach may be equipped with chitinous ridges, teeth, or calcareous ossicles that help grind food. The midgut varies in size and contains a variable number of ceca or diverticula where enzymes are secreted and food is absorbed. Sometimes the ceca are modified to form a large solid digestive gland, or hepatopancreas. The hindgut or intestine is a single narrow tube that terminates in an anus. The gut is equipped with muscles used in defecation and, in some cases, there are dilator muscles used for anal pumping. These muscles also seem to aid in intestinal peristalsis and in the circulation of fluids through the body. This generalized digestive system describes the basic pattern found in many parasitic crustaceans. In the most modified groups, such as the rhizocephalan barnacles, a digestive system is absent and nutrients are absorbed through the body wall.

11.2.2.4 Osmoregulatory and excretory systems

The organs responsible for excretion in crustaceans are the antennal glands located near the antennae, and the maxillary glands located near the maxillae. Both types of glands are found in larval crustaceans, but only one type persists in the adults. Nitrogenous wastes from the blood are filtered into the excretory gland and secreted. Ammonia is the main product of excretion, although amines, urea, and uric acid are also excreted. The urine produced is isosmotic with the blood of the crustacean and it seems that these glands do not function in osmoregulation. In large, free-living crustaceans, the gills usually function as osmoregulatory organs, secreting salts in marine species and absorbing salts in freshwater ones. Crustaceans also have nephrocytes, specialized cells capable of picking up and accumulating waste materials. Nephrocytes are particularly common in the gills and in the bases of the legs.

11.2.2.5 Circulatory system

The circulatory system in crustaceans is open. A simple dorsal heart is surrounded by a pericardial sinus. Blood enters the dorsal heart through lateral openings or ostia, and leaves through one or more arteries. The blood flows from the arteries into a system of open spaces called sinuses where it bathes the tissues directly. The blood then returns to the pericardial sinus and heart aided by the movements of the body. Parasitic copepods do not have a heart, and body movements and the gyrations of the gut move their blood. Similarly, highly modified crustaceans like the rhizocephalans do not have a circulatory system. The blood often has one, or more, respiratory pigments such as hemoglobin or hemocyanin, and one, or more, types of motile cells involved in clotting and phagocytosis.

11.2.2.6 Respiratory system

In small crustaceans and most parasitic species, gas exchange occurs through the body surface or by way of foliacean appendages that increase the body surface area. Branchiurans, for example, have cuticular extensions on the edge of the carapace that facilitate gas exchange. Cyamid amphipods, which are ectoparasites of marine mammals, have tubular gills (Fig. 11.2) to facilitate gas exchange.

11.2.2.7 Reproductive system

Most crustaceans are dioecious and many are also sexually dimorphic. Sexual dimorphism normally involves differences in body size and the degree of development of certain appendages. Males often have various appendages modified as clasping organs and, in some, the males are dwarfs that live permanently attached to the female.

In males, paired testes empty into sperm ducts that may have special seminal vesicles to store sperm. Females have paired ovaries that continue into the uterus and empty into oviducts and may, or may not, have seminal receptacles to store sperm. During copulation, the male clasps the female using modified appendages such as antennae, mouthparts, or thoracic appendages. A penis, or other modified abdominal appendage, is used to deposit the sperm within the seminal receptacle of the female or on the surface near the female gonopore. The sperm of most crustaceans lack flagella and, therefore, cannot swim. In some groups, the sperm are packed into a **spermatophore** that is delivered to the female.

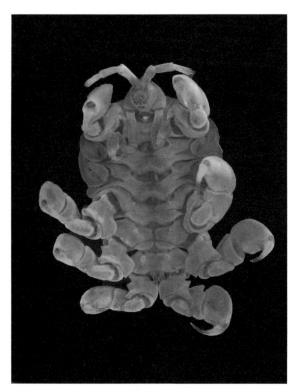

Fig. 11.2 The cyamid amphipod *Cyamus kessleri* from a gray whale *Eschrichtius robustus* showing the two pairs of tubular gills and the prehensile thoracic appendages or pereopods. (Photograph courtesy of Bill Pennell.)

11.2.2.8 Nervous system and sensory structures

The nervous system of crustaceans displays a tendency towards concentration and fusion of elements. In the primitive condition, there is a supraesophageal ganglion or brain, nerves that supply the cephalic sense organs, a pair of nerve trunks that connect the brain with a pair of subesophageal ganglia, and a pair of ventral cords that lie beneath the digestive system connecting the segmental ganglia. In most advanced crustaceans, including many of the parasitic species, there is a tendency towards fusion of the segmental ganglia, and of the two ventral cords.

The sense organs of most Crustacea include eyes, sensory hairs, proprioceptors, and statocysts. The eyes can be naupliar or compound. A naupliar eye is a characteristic feature of the nauplius larva and it may degenerate or persist in the adult. It contains a few photoreceptors and is probably used for orientation, enabling the animal to determine the direction of light. Compound eyes are common in adult crustaceans and possess a variable number of ommatidia, depending on the species. Adult copepods have only a naupliar eye and some of the parasitic species lack eyes completely. Ectoparasitic isopods and amphipods retain their compound eyes, and branchiurans such as *Argulus* have a pair of compound eyes and a median naupliar eye.

Sensory hairs are located over the body surface, especially the appendages. **Aesthetases** for example, are chemoreceptive hairs usually present on the antennae and are important in food and mate recognition. In parasitic copepods, they may be important in host recognition. Caligid copepods have a little-known sensory organ located in the middle of the anterior margin of the carapace. Its structure suggests a chemosensory function, but it also may be involved in host recognition (Kabata, 1974). Proprioceptors and stretch receptors are found in muscles, tendons, or articular membranes, and may be important in reflexes associated with body control during movement.

11.2.3 Nutrient uptake

What is ingested by the different parasitic crustaceans, and how, depends mostly on the type of oral appendages present and the location of the parasite on the host. Caligid-like copepods have a tubular mouth with a pair of mandibles that bear a sharp blade on one side and teeth on the other. The mandibles pierce and tear off pieces of the host tissues, which are then sucked up by the tubular mouth. *Ergasilus sieboldi*, a copepod parasite on the gills of fishes, ingests gill epithelium, mucus, erythrocytes, and white blood cells, in that order of preference.

In pennellid copepods, the holdfast of transformed adult females is embedded in major blood vessels, trunk musculature, or the visceral cavity of the host and serves to anchor the copepods while they feed on blood or tissue fluids. The holdfast of *Cardiodectes medusaeus*, a parasite of lanternfishes, reaches into the

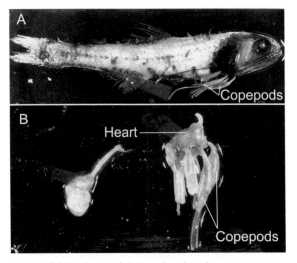

Fig. 11.3 (A) The lanternfish *Stenobrachius leucopsarus* parasitized by four transformed females of the copepod *Cardiodectes medusaeus*. The trunk region and egg sacs of the parasite project from the gill cavity, and the head is embedded in the bulbus arteriosus of the host's heart. (B) Heart removed from infected lanternfish showing the four attached females. Note the larger size of the bulbus arteriosus when compared to that of an uninfected heart on the left. (Photographs courtesy of Mike Moser.)

bulbus arteriosus of the host's heart (Fig. 11.3). The holdfast not only functions as an anchoring device, but also acts in the digestion of host erythrocytes, as well as in the detoxification and storage of iron freed from the catabolism of hemoglobin (Perkins, 1985). Females of *Phrixocephalus*, on the other hand, attach to the eyes of fishes. In *P. cincinnatus* (Color plate Fig. 4.2), a highly modified pennellid copepod of the eyes of certain flatfishes, the holdfast penetrates the cornea until it reaches the back of the eye and becomes embedded in the choroid, eventually establishing a network of multi-branched, intertwined rootlets (see Box 11.1). Development of the elaborate holdfast causes the formation of a large hematoma on which the parasite feeds. However, unlike *C. medusaeus*, the holdfast of *P. cincinnatus* does not function in digestion, and the blood is digested primarily in the intestine (Perkins, 1994). Interestingly, the epicuticle of highly modified copepods, including the holdfast of these two pennellid copepods, has abundant microvilli and it

has been suggested that they may function in the absorption of nutrients from the tissues of the host (Bresciani, 1986).

Most parasitic isopods feed by sucking blood and other body fluids. Rhizocephalans, as well as other endoparasitic crustaceans that have lost most of their appendages and internal organs, probably absorb nutrients directly through their body wall. The cuticle of rhizocephalans shows acid phosphatase activity, suggesting that nutrients are probably absorbed through it (Bresciani & Dantzer, 1980). Branchiurans generally feed by rasping at the host integument with the serrated mandibles and burying the proboscis into a blood vessel to suck blood. Other crustacean species, such as amphipods, however, feed primarily on mucus, epithelial cells, and extracellular fluid obtained from the surfaces of their hosts.

11.2.4 Development and general life cycle

Most crustaceans retain their fertilized eggs during embryonation within a specialized brood chamber or within an egg sac. Isopods and amphipods, for example, brood their eggs in a ventral brood chamber, or **marsupium**. Copepods retain and brood their eggs in an egg sac produced by secretions from the oviduct as the eggs pass by. The egg sacs remain attached to the female genital segment. Female branchiurans, on the other hand, do not retain their eggs. They leave the host temporarily and deposit their eggs on objects found on the substratum. In other highly modified crustaceans, such as the sarcotacid copepods, the body of the enormous female becomes a brooding chamber.

Development in crustaceans is indirect. Usually, a planktonic, free-living larva hatches from the eggs. This larva, known as a **nauplius**, is the most basic and easily recognized larval stage among crustaceans, even among those parasite species that are highly modified as adults (Fig. 11.4). It has three pairs of appendages (antennules, antennae, and mandibles), a naupliar eye, and no apparent segmentation. As the nauplius molts, segmentation becomes apparent and more appendages are developed. When a full complement of functional appendages has been acquired, the

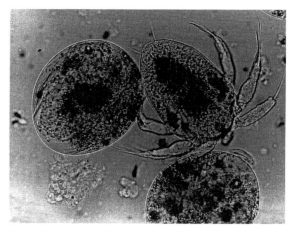

Fig. 11.4 Embryonated eggs and nauplius larvae of the endoparasitic copepod *Sarcotaces verrucosus*. (Photograph courtesy of Raúl González.)

nauplius becomes a postlarva. The postlarva is usually quite similar to the adult in its general appearance and increases in size through successive molts until it reaches sexual maturity. In most parasitic crustaceans, the postlarva is also parasitic.

Although many crustaceans conform to the basic nauplius, postlarva, and adult pattern described above, there are large differences among, and within, the groups regarding the stage of development at which the larva hatches from the egg and the number of molts in each stage. This variation is particularly apparent in the parasitic crustaceans as we see in the next section; indeed, the diversity of life cycles and the wealth of larval stages in the various parasitic crustaceans rival that of most other parasite groups, including the Platyhelminthes.

11.2.5 Biodiversity and life-cycle variation

The morphological and life-cycle diversity within the parasitic crustaceans requires that each taxonomic group be considered separately. Each taxon has unique parasitic adaptations, and representative species within each impacts its hosts in different ways. Here, we adopt the taxonomic classification scheme of Martin & Davis (2001). These authors recognize the class Maxillopoda, which includes the Copepoda,

Branchiura, Cirripedia, and the Tantulocarida, all of which contain parasitic representatives. Our focus is on the former three subclasses. For information on the enigmatic Tantulocarida, tiny ectoparasites of other crustaceans such as copepods and isopods, see Boxshall & Lincoln, (1983), Huys *et al.* (1993), and Raibaut & Trilles (1993). The class Malacostraca includes the orders Amphipoda, Isopoda, and the Decapoda (see Box 11.6). Our focus here will be on the former two orders.

11.2.5.1 Copepoda

The parasitic life style has evolved independently many times among copepods. As a result, a bewildering diversity of body forms, habitats, development patterns, and life cycles is evident, making generalizations on aspects of parasitic copepod biology almost impossible. Copepods parasitize a wide variety of hosts including cnidarians, annelids, molluscs, arthropods, echinoderms, ascidians, fishes, and even marine mammals. Most of the species parasitize marine fishes, but some are parasitic on freshwater fishes. This diversity of hosts and the habitats they offer has led to an extraordinary variety of adaptations to parasitism. Thus, parasitic copepods illustrate varying degrees of specialization to a parasitic mode of life. Indeed, they can be thought of as spanning a continuum, ranging from those that resemble their free-living planktonic counterparts, to those that are among the most highly specialized of parasitic animals (Fig. 11.1).

Kabata (1979) grouped parasitic copepods based on the degree of intimacy of their host–parasite association. At one extreme are the ectoparasites, such as the many species in the Trebiidae and Caligidae. These possess locomotory appendages, are segmented, and are able to move freely over the surface of their hosts. Other ectoparasites are permanently attached to the gill regions of their hosts and display incredible modifications to the general copepod body plan. Two diverse families are the Lernaeopodidae and the Chondracanthidae. At the other extreme, Kabata identifies mesoparasites as those that are partly external, yet are highly invasive. Thus, in the

Sphyriidae and the Pennellidae, for example, the transformed female's reproductive structures protrude from the host and their elaborate holdfasts extend deeply into their host's tissues. Endoparasitic copepods live inside their hosts. They include highly modified species in the Philichthyidae and Sarcotacidae. This continuum emphasizes the fascinating parasitic adaptations evolved by these families of parasitic copepods. Thus, with the evolution towards mesoparasitism and endoparasitism, dramatic appendage modifications have occurred, as well as reductions in segmentation. Moreover, the degree of sexual dimorphism and metamorphosis, and the potential for parasite-induced host damage accompanies these pronounced morphological specializations.

There are two major orders of parasitic copepods, distinguished primarily on the basis of their mouthparts. Species of the more speciose order, the Siphonostomatoida, have a mouth forming a short cylindrical tube called the **siphonostome**. Many of the siphonostomatoid copepods possess a **frontal filament**. This is a larval attachment organ that assures contact with the host when the copepod undergoes extensive morphological changes during attachment. The Poecilostomatoida comprise those copepods having slit-like mouths and sickle-shaped mandibles.

Members of the Ergasilidae are the least specialized of the ectoparasitic copepods. These have retained pronounced segmentation and thoracic appendages. In this group, only the females are parasitic. Males remain free-swimming throughout their life cycle and mating occurs in water. The females search and attach to the gills and exposed surfaces of their host. The one obvious parasitic modification is their powerful, prehensile second antennae (Fig. 11.1A). There is a tendency to permanent attachment to the gills in the family, although most are able to change position on their host. *Ergasilus* is by far the most speciose genus within the family.

Caligid and trebiid copepods parasitize the surface of many fishes (Figs. 11.1B, C; 11.5; Color plate Fig. 4.1). They are able to move freely, feeding on mucus and sloughed tissues. Their bodies are flattened dorsoventrally, using their modified cephalothorax as a suction

Fig. 11.5 The copepod *Trebius shiinoi* on the surface of an embryo of the Japanese angel shark *Squatina japonica*. Trebiid copepods are normally found on the surface of the body or branchial chambers of the host. This species is unusual in that it parasitizes the surface of shark embryos while still in the uterus. (Modified from Nagasawa *et al.*, 1998, with permission, *Journal of Parasitology*, **84**, 1218–1230.)

cup for attachment to their host. The cephalothorax is concave on the ventral side and has a membrane on its perimeter that, when applied to the surface of the host, seals the ventral cavity (Figs. 11.1B, C). This cavity is then partially evacuated of water creating a vacuum that firmly presses the copepod against the surface of the host. As further holdfast adaptations, they have modified their second antennae and maxillipeds as prehensile structures (Fig. 11.6).

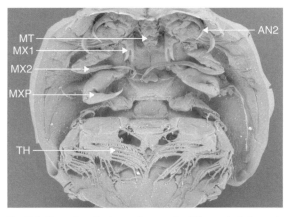

Fig. 11.6 Scanning electron micrograph of the cephalothoracic region of the caligid copepod *Lepeophtheirus salmonis* from Chinook salmon *Oncorhynchus tshawytscha*. The second antennae (AN2) and maxillipeds (MXP) are modified holdfast appendages. The mouth tube (MT), first maxillae (MX1), and thoracic appendages (TH) are also shown. (Micrograph courtesy of Brad van Paridon.)

Caligid genera, such as *Lepeophtheirus* (Color plate Fig. 4.1) and *Caligus*, are the ones most often referred to as sea lice. They are among the most notorious pests affecting cultured marine fishes (see Chapter 15), with infestations costing millions of dollars in salmon farms, worldwide (Todd, 2007). In nature, these skin ectoparasites may not be pathogenic; however, at high intensities they may inflict damage, often leading to secondary bacterial infections, especially under crowded aquaculture conditions. Unlike most other parasitic copepods, many caligids are capable of moving over their hosts and even swimming and switching hosts. This ability of caligid copepods to transfer between hosts enables them to act as mechanical vectors in the spread of pathogenic viruses, such as infectious salmon anemia virus (Overstreet *et al.*, 2009) and infectious haematopoietic necrosis virus (Jacob *et al.*, 2011), to fish. Barker *et al.* (2009) have also isolated a variety of bacterial pathogens, such as *Aeromonas salmonicida* from *Lepeophtheirus salmonis*. The comprehensive reviews of sea lice infecting salmonids by Pike & Wadsworth (1999) and Jones & Beamish (2011) are recommended.

The Cecropidae comprise a small family of copepods. One representative, *Cecrops latreilli*, is found on the

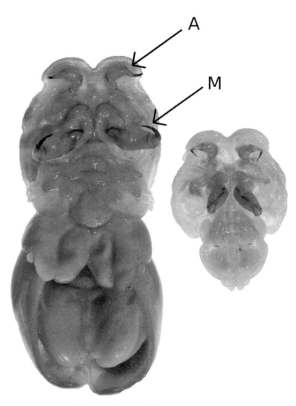

Fig. 11.7 Sexual dimorphism of the cecropid copepod *Cecrops latreilli* from the ocean sunfish *Mola mola*. Note the strongly clawed second antennae (A) and maxillipeds (M) on both the female (left) and male (right). (Photograph courtesy of Bill Pennell.)

gills of the ocean sunfish, *Mola mola*. This is a large parasite with females reaching 5 cm in size. Sexual dimorphism is also apparent with males smaller than females (Fig. 11.7). They have a pronounced cephalo-thoracic shield and a much reduced abdominal region. Unlike the caligids with their long egg sacs extending off the genital complex, egg sacs are held ventrally and covered by membranous plates (Fig. 11.7). The second antennae and maxillipeds of cecropids are prehensile and powerfully clawed (Fig. 11.7).

Perhaps the most striking of all the siphostomatoid copepods are the mesoparasites, such as those in the Pennellidae. Sexual dimorphism is extreme, with males being tiny dwarfs and possessing segmentation and motile appendages (Fig. 11.8). Due to their highly invasive nature, many are implicated as causative

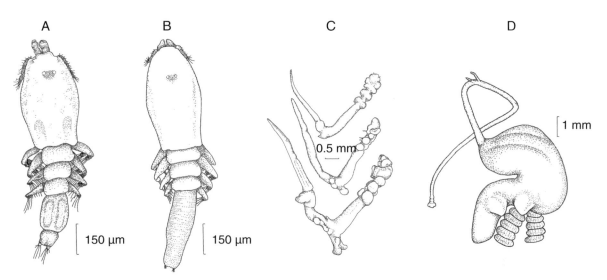

Fig. 11.8 Sexual dimorphism and dramatic metamorphosis of pennellid copepods. (A) Male of *Lernaeocera lusci*; (B) female of *Lernaeocera lusci*; (C) three stages in the metamorphosis of adult female of *Haemobaphes ambiguus*; (D) fully transformed female of *Haemobaphes diceraus* from the gill cavity of shiner perch, *Cymatogaster aggregata*, showing the highly modified cephalothorax and trunk region. The cephalothorax extends into the bulbus arteriosus of the heart of the fish host. (Figure courtesy of Danielle Morrison; modified from Kabata, 1979, *Parasitic Copepoda of British Fishes*, London: The Ray Society.)

agents of disease of various fish species (Kabata, 1979). Female pennellids undergo an astonishing metamorphosis during their development, taking on the most unusual shapes, and losing all traces of segmentation and motile appendages (Figs. 11.8C, D; see Box 11.1). Large sizes are common among female pennellids. Species of *Pennella* are the most abundant in the family. All are extremely modified, with complex holdfasts extending deeply into the flesh of their hosts. *Pennella balaenopterae* penetrates deeply into the skin of whales and reaches 30 cm in size, taking the evolution of gigantism to an extreme and making it one of the largest of all arthropods. All *Pennella* spp. have slender, cylindrical trunks and abdominal regions; lateral rows of fine processes forming abdominal brushes project from the latter.

Phrixocephalus cincinnatus (Color plate Fig. 4.2), and *Haemobaphes diceraus*, with their characteristic bright red trunks and spiral egg sacs (Color plate Fig. 4.3), illustrate further the dramatic morphologies and site specificity exhibited by transformed female pennellids. In both species, the enlarged trunk region and egg sacs remain largely outside the host, but the cephalothorax transforms into an elaborate holdfast that penetrates deeply into host tissue. In *H. diceraus*, for example, the transformed females reside within the gill cavity. The neck region of the cephalothorax becomes highly modified into a long tubular structure (Fig. 11.8D), which then traverses the gills and the lobed holdfast is inserted into the bulbus arteriosus of the heart of shiner perch, *Cymatogaster aggregata*. In the case of *P. cincinnatus*, females develop luxuriant holdfasts consisting of branched cephalic antlers and numerous processes, extending deeply into the choroid of the eye of arrowtooth flounder, *Atheresthes stomias* (Fig. 11.9; see Box 11.1).

Lernaeocera spp. also develop complex holdfasts that reach the aorta, branchial vessels, and even the heart. *Lernaeocera branchialis*, for example (Fig. 11.10A), a parasite of the Atlantic cod, *Gadus morhua*, attaches to the branchial region. The neck elongates and penetrates the fish body until the cephalothorax/holdfast region reaches the bulbus arteriosus of the heart. The large trunk, bearing the reproductive organs and the egg sacs, remains on the surface of the fish. The parasite severely affects the

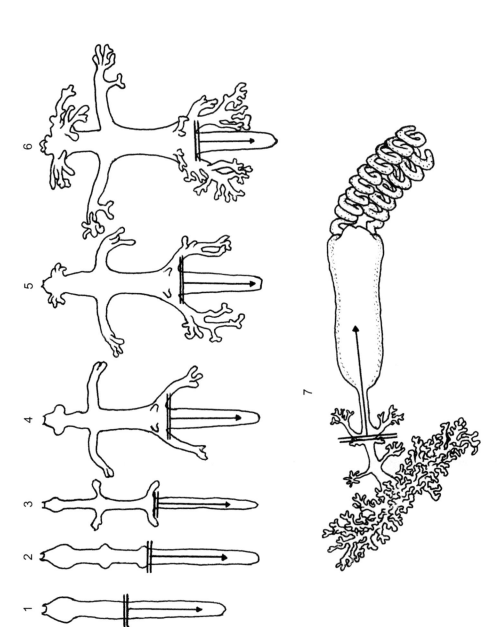

Fig. 11.9 Stages in the complex metamorphosis of the female pennellid copepod *Phrirocephalus cincinnatus* from the eye of the arrowtooth flounder *Atheresthes stomias*. The genital–abdominal region of each of the seven stages is marked off by two transverse lines and an arrow. Upon attachment to the eye of the fish, cephalothoracic swellings form (stage 2), enlarge (3, 4), and then branch extensively, forming anterior cephalic antlers (5, 6) that act as holdfasts. The primary anterior holdfast branches extensively into the eye, and the genital–abdominal region expands greatly (7). Metamorphosis is complete when the cylindrical trunk of the mature female erupts out of the eye, eventually producing egg sacs (7). (Figure courtesy of Nicole Maguire; modified from Kabata, 1979, *Parasitic Copepoda of British Fishes*, London: The Ray Society.)

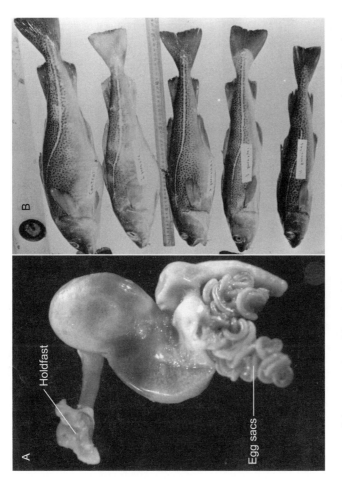

Fig. 11.10 (A) Metamorphosed female of the pennellid copepod *Lernaeocera branchialis* from the Atlantic cod *Gadus morhua*; (B) Effects of *L. branchialis* on the growth of parasitized cod 9 months post-infection. All fish were approximately the same size when the infection started. The fish on top was not infected (control); the fishes below had had one, two, three, and four copepods, respectively. (Modified from Khan, 1988, with permission, *Journal of Parasitology*, **74**, 586–599.)

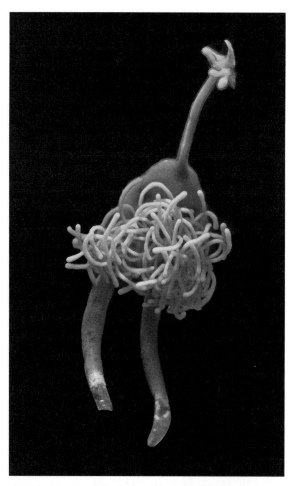

Fig. 11.11 Female of the lophourid copepod *Lophoura* sp. removed from the musculature of the rattail, *Coryphaenoides armatus*. The tubular neck region ends with an elaborate holdfast. Numerous profusely branched processes and two egg sacs project from the enlarged genital complex. (Photograph courtesy of Bill Pennell.)

growth, reproduction, and survival of its host (Fig. 11.10B). The presence of the copepod is associated with anemia, weight loss, reduction in fat content and liver weight, and a decrease in reproductive potential (Khan, 1988). Khan *et al.* (1990) found that up to 33% of cod infected with *L. branchialis* died over a period of 4 years. The holdfast of *Cardiodectes medusaeus* also reaches the bulbus arteriosus of the lanternfish host and feeds on blood. Like *Haemobaphes* spp., egg sacs and the trunk region project from the gill region. In

multiple infections, the heart becomes enlarged (Fig. 11.3B), and the flow of blood through it may be affected (Moser & Taylor, 1978).

Two other mesoparasitic families, typically associated with deep-water marine fishes, are the Sphyriidae and the Lophuridae. An elaborate, lobed holdfast and a long neck region connect to a large trunk, from which egg sacs and many branching respiratory cylinders extend (Figs. 11.11, 11.12A). The holdfast extends deeply into the flesh of their hosts, while the trunk and egg sacs protrude (Fig. 11.12B).

The Lernaeopodidae is another highly modified and diverse family of ectoparasitic copepods of fishes. The success of this family is the result of the evolution of unique modes of attachment and feeding. Three morphological structures on transformed female lernaeopodids are used to distinguish species. These include the relative sizes and positions of the cephalothorax and the second maxillae, as well as the trunk morphology. Female lernaepodid copepods typically attach via a modification of the second maxillae and a unique attachment structure called the **bulla**. *Salmincola* spp. (Fig. 11.13) comprise among the most diverse lernaeopodids of various freshwater salmonid fishes, worldwide. The second maxillae are normally fused to the bulla, which is securely embedded in the host's tissues. *Ommatokoita elongata*, a lernaeopodid infecting the eyes of Pacific sleeper and Greenland sharks (Figs. 11.1D, 11.14), has a pair of long, tubular second maxillae terminating in the bulla, which then attaches to the cornea of the eye (Benz *et al.*, 1998). Transformed females of *Ommatokoita elongata* also possess a pair of clawed maxillipeds for attachment to the host's eye (Fig. 11.15).

Several other lernaepodid species have a relatively long cephalothorax and short maxillae. The mouth is located at the tip of the cylindrical cephalothorax. Despite the permanent attachment, the cephalothorax/mouth region is mobile and the transformed adult female can acquire nutrients from a region around the point of attachment that is equal to the length of the cephalothorax (Fig. 11.16). Thus, while they are permanently secured to a host, they have an 'extended grazing range,' an adaptive feature contributing to the success of the lernaeopodid copepods (Kabata, 1979).

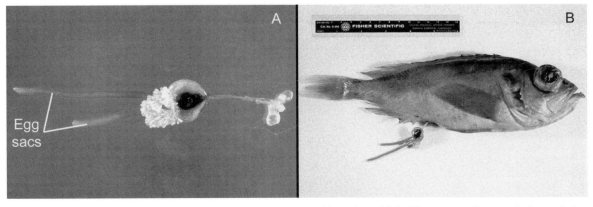

Fig. 11.12 (A) Female of the sphyriid copepod *Sphyrion lumpi* removed from the red fish *Sebastes mentella*; note the long tubular neck culminating in an elaborate holdfast with two lateral lobes. Numerous filamentous processes project off the genital complex. (B) Female of *S. lumpi* embedded in the musculature of the redfish. A portion of the neck, posterior processes, and egg sacs project from the host. (Photographs courtesy of Jonathan Moran.)

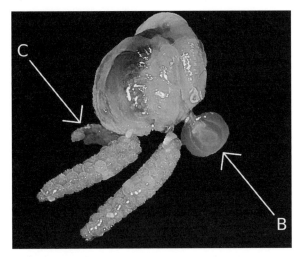

Fig. 11.13 Transformed female of the lernaeopodid copepod *Salmincola californiensis* from the gills of a rainbow trout *Oncorhynchus mykiss*, showing the positions of the second maxilla with the holdfast bulla (B) and the cephalothorax (C). (Photograph courtesy of Bill Pennell.)

Fig. 11.14 Transformed female of the lernaeopodid copepod *Ommatokoita elongata* attached to the eye of a Greenland shark *Somniosus microcephalus*. (Micrograph courtesy of George Benz.)

Two families within the Poecilostomatoida include the Chondracanthidae and the Philichthyidae. The former comprise the most diverse of the poecilostomes commonly infecting the gill region of marine fishes. The metamorphosed females are highly modified morphologically, often with extremely bizarre appearances. The unusual shapes of chondracanthid females, such as *Chondracanthus zei*, are due to prominent outgrowths or processes extending from their large trunk region (Fig. 11.17). The genito-abdominal region is greatly reduced. All traces of locomotory appendages and segmentation are lost on females following metamorphosis. The major attachment structures are the hooked second antennae (Fig. 11.17). Females eventually become permanently attached because a host tissue reaction completely

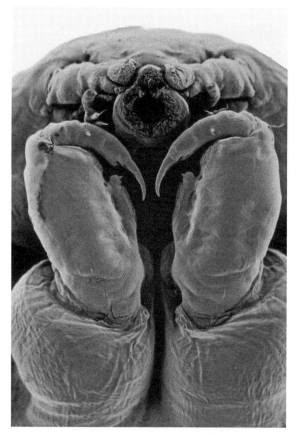

Fig. 11.15 Scanning electron micrograph of the mouth region of a transformed female of the lernaeopodid copepod *Ommatokoita elongata* from the eye of a Greenland shark *Somniosus microcephalus*. (Micrograph courtesy of George Benz.)

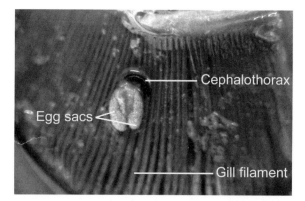

Fig. 11.16 Female of the lernaeopodid copepod *Naobranchia lizae* from the gill filaments of a gray mullet *Mugil cephalus*. The copepod attaches permanently to a gill filament using the bulla, while the flexible cephalothorax–mouth region 'sweeps' the surface of the filaments. (Photograph courtesy of Jackie Fernández.)

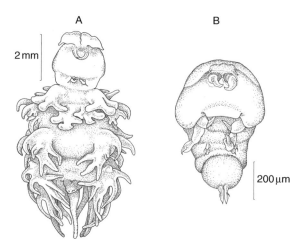

Fig. 11.17 Female (A) and male (B) of *Chondracanthus zei* from a John Dory, *Zeus faber*, showing the pronounced sexual dimorphism (note scale bars) characteristic of the Chondracanthidae. The presence of outgrowths on metamorphosed females is a feature of chondracanthid copepods, especially numerous and prominent on this species. (Figure courtesy of Danielle Morrison; modified from Kabata, 1979, *Parasitic Copepoda of British Fishes*, London: Ray Society.)

surrounds the cephalothorax of the parasite. The extrabuccal digestion carried out by the female copepod at the point of attachment irritates the tissues around the attachment area, leading to encapsulation of the anterior end of the copepod. Sexual dimorphism among chondracanthids is pronounced, with the dwarf males approximately 10 000 times smaller than females (Fig. 11.17). Chondracanthid males lack the outgrowths and attach via their prehensile second antennae to the genital segment of the female.

Small body cavities that communicate to the outside, such as the pores of the lateral line, the frontal mucus ducts, and sinuses of teleosts and elasmobranchs, are the preferred sites of endoparasitic philichthyid copepods, such as *Colobomatus* spp (Fig. 11.18). Because of this protected habitat, these copepods do not require specialized attachment structures and most of their appendages have been lost. However, females have

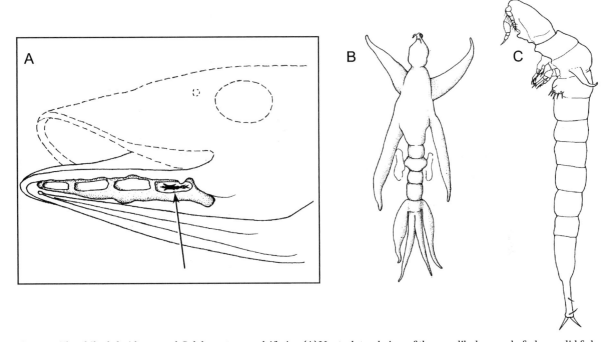

Fig. 11.18 The philychthyid copepod *Colobomatus quadrifarius*. (A) Ventrolateral view of the mandibular canal of a haemulid fish showing the position of the copepod; (B) ventral view of a female copepod with prominent lateral outgrowths projecting from thoracic region; (C) lateral view of male, showing the segmented abdomen. (Modified from Cressey & Schotte, 1983, with permission, *Proceedings of the Biological Society of Washington*, **96**, 189–201, Allen Press Publishing Services.)

developed prominent lateral body processes off their thoracic region that may help maintain their position inside the cavities they occupy (Fig. 11.18B). Smaller males are less modified than females and retain most of their appendages and segmentation (Fig. 11.18C).

Many copepods are also parasites of invertebrates. Species of Lichomolgidae infect a wide variety of marine invertebrates, including corals, anemones, polychaetes, molluscs, echinoderms, and ascidians. These copepods have retained most of their free-living characteristics, including segmentation and swimming legs, although in some species, the latter are reduced. Species of *Ismaila* and *Splanchnotrophus* include highly modified poecilostomatoid copepods in the Splanchnotrophidae. These copepods are found in a diversity of marine opisthobranch gastropods, especially nudibranchs (Huys, 2001). In the case of *Ismaila* spp., large egg sacs protrude from the nudibranch's body while the rest of the female copepod develops within (Color plate Fig. 4.4).

The basic life cycle of most free-living copepods involves a gradual metamorphosis and includes five or six naupliar stages and six copepodid (or chalimus) stages before reaching adulthood (Kabata, 1981). Parasitic copepods follow this basic plan, with a tendency to reduce the number of larval stages. However, the number of each of these stages varies greatly among parasitic species. In general, mature females release nauplii larvae, which then become copepodids. The latter then initiate contact with a host and attachment occurs via second antennae and maxillipeds. At this stage, they are a chalimus and typically attach even more securely via their frontal filament. Post-chalimus stages of many copepod parasites, such as the caligids, may give rise to pre-adults resembling the adults in many cases. However, as we have seen, females of many copepods undergo further complex development and a profound metamorphosis (see Box 11.1).

Perhaps the most unusual life cycle pattern among crustacean parasites is that exhibited by some members

Box 11.1 | **Astonishing metamorphosis in pennellid copepods and rhizocephalan barnacles**

The evolution of complex life cycles and the phenomenon of metamorphosis are among the most intriguing of adaptations. Metamorphosis typically involves abrupt changes in morphology, physiology, ecology, and behavior between larva and adult stages. It is a distinctive life history feature common to an array of animals, ranging from most marine invertebrates, to insects, to many fish and amphibians. As we have seen, complex life cycles featuring dramatic metamorphoses are associated with many parasitic animals. The ecological and evolutionary advantages of maintaining complex life cycles, and the many environmental, genetic, and physiological factors controlling metamorphosis, have been extensively studied in this diverse range of taxa. The pennellid copepods and the rhizocephalan barnacles, in particular, provide among the most stunning examples of morphological transformations to be encountered in the animal kingdom.

The succession of free-living nauplius, copepod, and attached chalimus larval stages, coupled with the evolution of two-host life cycles, are intriguing features of pennellids. However, it is the metamorphosis that follows attachment of the free-swimming female to a host for which the pennellids are renowned. What sequence of events transforms a segmented, motile stage into the bizarre metamorphosed egg-producing female of *Phrixocephalus cincinnatus* inhabiting the eyes of flounder, for example? Bob Kabata, the world's expert on parasitic copepods, meticulously worked out the details of this bizarre animal's development (Kabata, 1979). He identified six major developmental stages (refer to Fig. 11.9), prior to becoming the fully transformed, mature female parasite. The mated juvenile female pierces the eye and travels through the cornea. She undergoes two molts and reaches the choroid of the eye and attaches. Cephalothoracic swellings form (stage 2) and then enlarge (stages 3 and 4). Stage 4 is marked by an increased growth rate of the anterior region. The cephalic antlers branch extensively (5) and the mouth tube forms (6). At this point, she feeds on blood within the choroid of the flatfish's eye. Metamorphosis is complete when the genito-abdomen lengthens and the cylindrical trunk erupts out of the eye. At this stage, the primary holdfast is enormous, consisting of highly branching and intertwined processes (stage 6), ensuring permanent attachment of the adult female. Eventually, the spiral egg sacs form and sexual maturity is attained (see Fig. 11.9, stage 7; Color plate Fig. 4.2).

Another even more spectacular transformation occurs when a rhizocephalan barnacle's female cypris larva contacts its specific crustacean host. This free-swimming cypris resembles the larva of free-living barnacles. However, the resemblance to barnacles ends when the host is contacted. The sequence of developmental events that transforms this female larva into a large external gonad (=externa) located on the ventral abdominal surface of the crab host is now understood precisely. This observation is due primarily to the patience and expertise of Henrik Glenner and Jens Høeg, two scientists who have contributed much to our understanding of these extraordinary parasites.

The female cypris settles on the cuticle of its crustacean host and undergoes a metamorphosis into a bizarre stage called the **kentrogon**. This stage possesses a hollow stylet, adapted for

Box 11.1 | (continued)

piercing the host's cuticle and acting as a tube for the injection of 'parasite material' into the hemocoel. The nature of this injected material and the internal development of rhizocephalans remained a mystery for many years, until Glenner & Høeg (1995) and Glenner *et al.* (2000), working with the rhizocephalan *Loxothylacus panopaei* in mud crabs, demonstrated that the injected substance was a motile, multicellular, and worm-shaped structure. They named this highly simplified endoparasitic stage the **vermigon**, closing the rhizocephalan life cycle once and for all. The vermigon migrates intact through the hemolymph of the host and attaches near the site where the externa will eventually emerge and begin to grow out of the highly invasive rootlet system (=**interna**). Glenner *et al.* (2000) determined that the vermigon included the most important cellular components forming the adult parasite, including the cuticle, epidermis, and a body of cells that later mature into the ovary. The ultrastructural changes accompanying formation of each stage and the precise timing of each are presented in meticulous detail in Glenner (2001).

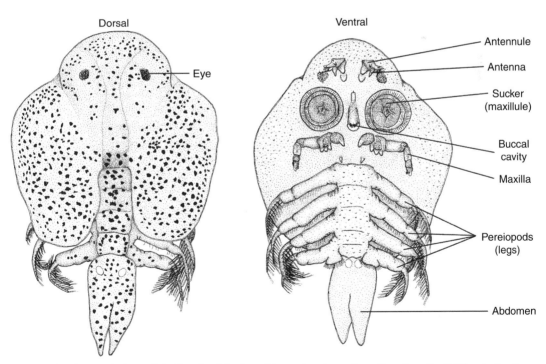

Fig. 11.19 Dorsal and ventral views of the female of the branchiuran *Argulus melanosticus* from the California grunion *Leuresthes tenuis*. The first maxillae (maxillules) are modified into sucker-like holdfasts in this genus. (Modified from Benz *et al.*, 1995, with permission, *Journal of Parasitology*, 81, 754–761.)

of the Pennellidae. Remarkably, several pennellids have evolved a two-host life cycle. *Lernaeocera branchialis*, located in the gill regions of Atlantic cod definitive hosts, releases nauplius larvae. These metamorphose into infective copepodids that seek and attach to an intermediate host, in this case flounder, sculpin, or lumpfish (Khan, 1988). On these fish hosts, the copepodid molts into a chalimus, and male and female chalimi mate in the gill cavity. Fertilized females then leave the intermediate host in search of cod, attaching to the gills and undergoing the massive growth and transformation characteristic of female pennellids. A similar life cycle has been suggested for *Haemobaphes diceraus* from shiner perch gill cavities. In this case, bay pipefish are the proposed intermediate hosts (Goater & Jepps, 2002). *Pennella* uses cephalopods and fishes as intermediate hosts, whereas the hosts for *Cardiodectes* are pelagic gastropods and fishes.

11.2.5.2 Branchiura

Branchiurans are a relatively small group of crustaceans parasitic on marine and freshwater fishes. Morphologically, they have a broad dorsal shield that makes them superficially resemble caligid copepods. The prominent dorsal shield (Fig. 11.19) is formed by lateral outgrowths of the head and thorax. Although the bilobed abdomen of the branchiurans has lost all traces of segmentation, the thorax still bears four pairs of legs used for swimming. They attach to the host by their hooked second antennae and modified first maxillae (maxillules) that form cup-like suckers in *Argulus* (Fig. 11.19). Argulids feed by piercing the host's skin with a **preoral stylet**; they then ingest blood or tissue fluids via a mouth tube. Some species are known to be pathogenic in freshwater fish aquaculture, where they can reach large numbers and cause mortality. One effect is the production of ulcerated bloody lesions on the fish's body. This can lead to secondary infections of the wounds by fungi and bacteria, contributing to mortality. In addition to this damage, proboscis glands of *Argulus* spp. produce toxins, leading to inflammation.

Unlike most parasitic crustaceans, branchiurans like *Argulus* spp. exhibit low host specificity, and are highly mobile and able to move freely over the surface of the host. They are able to leave the host to find mates, to locate new hosts, and to lay eggs in the aquatic substratum. Nauplius stages occur within the egg. When the egg hatches, it closely resembles the adult and the juvenile swims, locates a host, and the second antennae and first maxillae become modified as anchors. This ability to switch hosts means new fish can be colonized rapidly, especially under hatchery conditions. Moreover, due to their mobility and blood-feeding habits, argulids are known to act as vectors of certain pathogenic viruses of fish (Overstreet *et al.*, 2009).

11.2.5.3 Cirripedia

Barnacles are exclusively marine and are the only totally sessile group of crustaceans as adults. All barnacles start life as pelagic larvae. Normally, the larvae are released as nauplii, always recognizeable from other crustacean nauplii by having a pair of conspicuous horns. While the nauplii are adapted for growth and dispersal, the terminal larval stage, the **cyprid** (or **cypris**) has the critical task of locating a site suitable for permanent attachment and, then, for metamorphosis into the attached juvenile barnacle. The sessile life style has led to the evolution of unique features in the adult, including protective shell plates and six pairs of thoracic suspension-feeding appendages known as **cirri**. Species of barnacles range from completely free-living as in many of the familiar intertidal species, to commensalism, and to parasitism. Several host-specific commensal/parasitic barnacle species in the Order Thoracica have been described from sharks, marine turtles, and baleen whales. *Cryptolepas rachianecta*, for example, infects gray whales (Fig. 11.20). Highly modified calcareous shell plates grow into the whale's skin for permanent attachment.

Perhaps the most curious and highly modified of all parasitic animals are the barnacles parasitic on decapod crustaceans, members of the Order Rhizocephala (='root-headed'). The rhizocephalans are often used as prime examples of extreme adaptations and modifications to parasitism. In fact, it is only their planktonic cyprid larva that betrays the fact that they are even

Fig. 11.20 The thoracican barnacle *Cryptolepas rachianecta* embedded in the skin of a gray whale *Eschrichtius robustus*. (Photograph courtesy of Bill Pennell.)

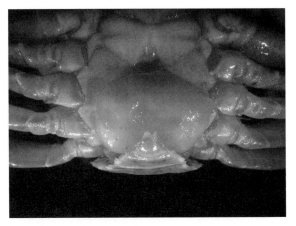

Fig. 11.21 Externa of the rhizocephalan barnacle *Sacculina carcini* on the ventral abdomen of the green crab *Carcinus maenas*. (Photograph courtesy of Jens Høeg.)

Fig. 11.22 Multiple externae of the rhizocephalan barnacle *Thompsonia* sp. on the abdomen of an unidentified alpheid (snapper) shrimp. (Photograph courtesy of Jens Høeg.)

barnacles! Adult rhizocephalans have no gut or other internal organs; moreover, they lack appendages and have lost all traces of segmentation. The adult parasite consists of an external female reproductive portion, the externa, connected to a network of nutrient-absorbing processes, the interna, or rootlet system. The large externa is located on the ventral surface of crabs or shrimp, always mimicking the position, size, and color of the female host's egg mass. For example, the large externa of the well-known rhizocephalan *Sacculina carcini* of shore crabs is located under the host's abdomen, in the precise location where the crab's egg mass would normally be located (Fig. 11.21; Color plate Fig. 5.1). Similarly, the colors and positions of the orange externa of *Briarosaccus callosus* of king crabs (Color plate Fig. 5.2) and the bright red externa of *Peltogaster paguri* of hermit crabs (Color plate Fig. 5.3) nicely illustrate the dramatic egg mimicry found in some rhizocephalans.

The externa is the site where male fertilization occurs and where eggs and larvae are brooded. It contains the female's gonad, a large brood chamber, and a pair of phenomenally modified organs that contain the males. The parasite is anchored to its host by a stalk, from which the root-like interna ramifies and penetrates through virtually all host tissues. Some rhizocephalans, such as *Thompsonia* spp. and *Peltogasterella* spp. of hermit crabs and shrimp, produce gregarious infections and several externae are produced asexually by budding from the root system (Fig. 11.22; Color plate Fig. 5.4). The multiple externae originate from a single cypris larva. Other sacculinid rhizocephalans have also been shown to reproduce in this manner. Glenner *et al.* (2003) conducted a molecular phylogenetic study and erected the new genus, *Polyascus*, to accommodate these species.

The life cycles of so-called kentrogonid rhizocephalans are complex and the manner by which they infect their decapod host is fascinating (refer to Fig. 11.23). The egg hatches into a nauplius larva that undergoes a series of molts, eventually metamorphosing into a male or female cypris larva. The female cypris attaches to a suitable crab host via its antennules and sheds most of its appendages. It then undergoes an astonishing metamorphosis into a most bizarre stage, the kentrogon (see Box 11.1). The kentrogon has a hollow, cuticular stylet that is specialized for piercing and penetrating specific sites of the crustacean host's cuticle. The kentrogon injects a motile, multicellular worm-shaped body called the vermigon into the crab's hemocoel. This vermigon is an instar that forms a direct link between the kentrogon and the maturing internal parasite, eventually forming the absorptive and highly invasive interna of the female parasite (Glenner & Høeg, 1995; Glenner *et al.*, 2000; Glenner, 2001). The developing parasite eventually emerges through the ventral abdomen, forming the virgin externa. Virgin externa attract male cyprids; these enter the mantle cavity of the externa and extrude a mass of male cells transforming into a **trichogon** larva, the male, parallel to the female vermigon. The trichogon then migrates via a narrow duct into one of two special chambers, the receptacle, where they become permanently implanted (Fig. 11.23). Only a single male trichogon can enter each of the two receptacles; remarkably, they shed their cuticle in the duct and thus block entrance of late arriving males, a marvelous extension to conventional male–male competition. The trichogons normally remain in the male receptacles throughout the life of the female parasite, producing spermatozoa to fertilize the eggs produced in the mantle cavity of the female externa (Ritchie & Høeg,

1981; Raibaut & Trilles, 1993; Høeg, 1995). In most kentrogonids, the externa liberates nauplii that develop into the morphologically distinct male and female cyprids. Interestingly, the male cells present in the rhizocephalan females were thought to be testes and the rhizocephalans were thought to be hermaphrodites. The discovery of the hyperparasitic dwarf males inside the enormous female makes the rhizocephalans a classic example of **cryptogonochorism**, where sexes are separate, but hidden.

Many rhizocephalans are well known for causing diverse pathological effects, including alterations in host physiology and behavior. For example, rhizocephalans have been shown to affect the hormonal and reproductive activities of their decapod hosts. In many rhizocephalan-infected hosts, molting is inhibited and growth rates are severely impacted. It is likely that rhizocephalans influence the molting patterns of their hosts by one, or more, hormonal mechanisms. Thus, rhizocephalans may interfere with production of host hormones, or may sequester active hormones produced by the host, maintaining a subthreshold titer in the host. In addition, reproduction is inhibited. In fact, several rhizocephalans are known for inducing loss of gonad function and causing parasitic castration. The host is not able to reproduce because the gonads atrophy and some sex characteristics change due to parasite-induced hormonal changes.

Rhizocephalans basically take full control of the host, to the point that the host is manipulated morphologically, physiologically, and behaviorally in astonishing ways. Thus, castrated female crabs are induced to protect, groom, and ventilate the parasite's egg mass as if it were its own. Remarkably, infected male crabs are feminized morphologically and display the appropriate maternal behaviors. Male crabs develop a more female-like morphology, with a broader, segmented abdomen. Furthermore, they may develop pleopods that resemble the female's egg-tending appendages. The male crab then cares for the parasite's externa as if it were a female looking after its

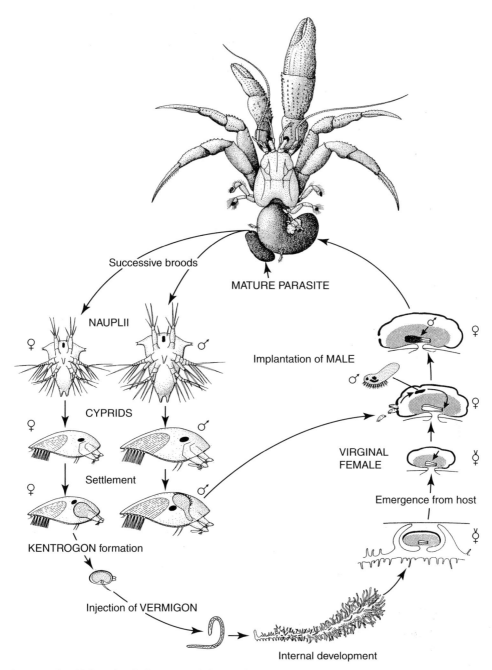

Fig. 11.23 The generalized life cycle of a kentrogonid rhizocephalan, e.g., *Peltogaster paguri*. Externa of the mature female releases nauplii, which molt into cyprid larvae. Female cyprids settle on potential hosts, while male cyprids settle only on juvenile virgin externa. The settled female cyprid metamorphoses into a kentrogon and injects the vermigon stage into the host. The vermigon, once in its permanent site, undergoes internal development and forms a branched trophic root system. The parasite emerges on the abdomen of the host as a virgin externa with empty male receptacles (arrow). Male cyprids settle on the externa, implant, and metamorphose into trichogons and migrate into the two male receptacles. The implanted trichogons develop into hyperparasitic dwarf males. (Figure courtesy of Jens Høeg.)

Labels in figure:

Successive broods
NAUPLII
MATURE PARASITE
Implantation of MALE
CYPRIDS
Settlement
VIRGINAL FEMALE
KENTROGON formation
Emergence from host
Injection of VERMIGON
Internal development

own eggs! Female crabs normally exhibit phero-monally mediated spawning behavior. A vigorous pumping response is exhibited to create currents to dispense larvae into the water column. It turns out that some rhizocephalan-infected crabs induce their hosts to perform this spawning behavior, even though they are castrated and have no eggs of their own. Rhizocephalans may produce chemicals that induce the larval release behavior in order to maximize release of the parasite's nauplii into water. The castration and hormonal interference in males is so pronounced that they also exhibit this spawning behavior.

11.2.5.4 Amphipoda

Although many amphipods develop symbiotic associations with marine organisms, true parasitism is surprisingly rare. Hyperiideans are pelagic amphipods that have established symbi-otic relationships with other pelagic organisms such as jellyfishes, ctenophores, molluscs, and tunicates.

Cyamid amphipods are a unique group of crusta-ceans parasitic on whales and dolphins. Most of the approximately 50 species in the family exhibit host specificity; few cetaceans are host to more than one species of cyamid (Martin & Heyning, 1999; Kaliszewska *et al.*, 2005). Cyamids, commonly called 'whale lice,' have several morphological adaptations to ectoparasitism. Their body is flattened dorsoven-trally, not laterally like other amphipods, and they possess powerful, prehensile appendages called per-eopods (Figs. 11.2; 11.24). They also possess a pair of sharply clawed, recurved **gnathopods** on the ventral surface, near the mouth region. These appendage modifications are clearly adaptations to prevent dislodgment from their marine mammal hosts and ensure tenacious attachment onto such mobile hosts.

Cyamids are unable to swim at any stage of their life cycle. Most attach to crevices, or to natural openings such as the mammary and genital slits, or to wounds or

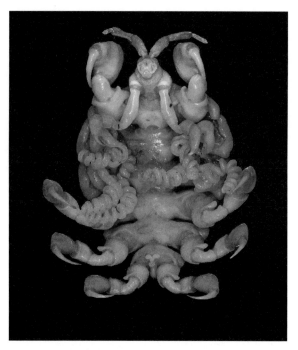

Fig. 11.24 Male of the cyamid amphipod *Cyamis scammoni* from the skin of a gray whale *Eschrichtius robustus*. The coiled structures are gills. Clawed pereopods serve as tenacious holdfasts. (Photograph courtesy of Bill Pennell.)

other areas that provide some shelter from the currents produced by the swimming of the host. Attachment is such a critical factor that the abundance of cyamids on a host is inversely proportional to its swimming speed. In slower-swimming baleen whales, for example, dense populations consisting of several thousand individuals of different ages can be found clinging to one host (Color plate Fig. 3.2). Moreover, right whales and gray whales are unusual in that each carries large populations of at least three different *Cyamis* species, which do not occur on other cetaceans (Kaliszewska *et al.*, 2005). In faster-swimming cetaceans such as dolphins, however, population densities are much smaller (Raga, 1997).

Cyamid amphipods feed mainly on host skin. There is evidence that even juvenile cyamids still present in the marsupium or brood chamber of the female might leave to feed on the host's skin and then return to the brood chamber (Rowntree, 1996). Since cyamids cannot swim and they lack a larval stage, dispersal capabilities are limited. Juveniles settle near their parents. Transfer to new hosts can only occur between whales that contact each other, such as during lactation and during mating. Even though cyamids cannot swim, they are able to roam widely over the surface of their hosts, feeding on mucus or bacteria growing on the surface of their host's skin (Rowntree, 1996).

Kaliszewska *et al.* (2005) studied the mitochondrial DNA sequence variation of the three species of whale lice occurring on right whales, *Eubalaena* spp., in order to decipher their host's biogeographical and evolutionary histories. Intriguingly, their molecular clock calibrations and the genetic differences between the three species indicated that a single right whale species diverged into three species (North Atlantic, North Pacific, Southern Ocean right whales) five to six million years ago. This coincided with the movements of the Earth's tectonic plates, which formed the isthmus of Panama, separating the Atlantic and Pacific Oceans. Warm equatorial currents were created, effectively isolating the right whales since they cannot tolerate warm water. In short, the cyamids acted as 'replicated evolutionary experiments' that enabled these scientists to decipher the population histories of their right whale hosts (Kaliszewska *et al.*, 2005).

11.2.5.5 Isopoda

Parasitic isopods are morphologically diverse and their adaptations to parasitism are extensive. Three suborders are recognized, i.e., the Gnathiidea, Flabellifera, and Epicaridea. Gnathiid isopods parasitize marine fishes only during their larval stages. A parasitic larva known as a **praniza** attaches to a fish host, sucks blood, and molts until it becomes an adult. The adults leave the host, settle in the benthos and do not feed, subsisting on the reserve nutrients accumulated during their larval stage, until mating and larval production occurs.

A diverse group of flabelliferan isopods includes the cymothoids. They are permanent ectoparasites of primarily marine fishes (Bunkley-Williams & Williams,

1998). Some species such as *Lironeca amurensis* burrow under a scale of the host and, as the isopod grows, it becomes completely surrounded by host tissue, communicating to the exterior by a small hole. Most cymothoids, however, have pereopods that are armed with strong, prehensile claws (Figs. 11.25, 11.27), used to cling to the skin, gills, operculum, and buccal cavity of the host, where they feed on blood or plasma from wounds. Most cymothoids are highly host and site-specific. In many cases, the isopods cause pressure atrophy of the structures to which they attach, such as the gills and the tongue. The dramatic 'tongue-eating isopods,' such as *Cymothoa exigua* (Figs. 11.26, 11.27, Color plate Fig. 3.4) are the only parasites known to completely replace an entire organ!

A distinctive feature of cymothoids is the phenomenon of **protandrous hermaphroditism** or **protandry** where juveniles develop first into males and later into females. Isopods normally brood their eggs in a

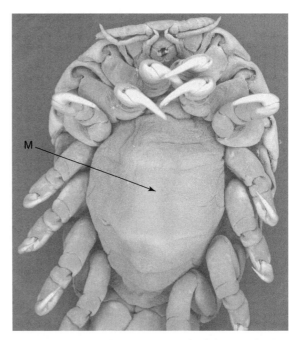

Fig. 11.25 Scanning electron micrograph of the cymothoid isopod *Lironeca californica* from the gill cavity of shiner perch *Cymatogaster aggregata*, showing the clawed pereopods and the marsupium (M). (Micrograph courtesy of Brad van Paradon.)

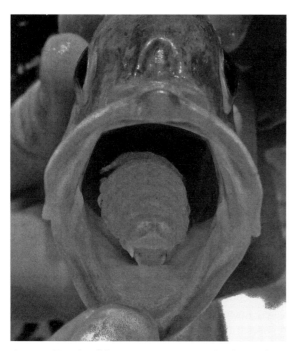

Fig. 11.26 Female of the tongue-eating cymothoid isopod *Cymothoa exigia* within the buccal cavity of a yellowtail snapper *Ocyurus chrysurus* from Belize. (Photograph courtesy of Sabiha Kamani.)

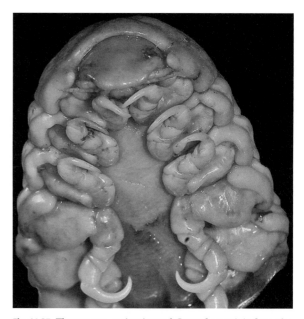

Fig. 11.27 The tongue-eating isopod *Cymothoa exigia* from the buccal cavity of a yellowtail snapper *Ocyurus chrysurus*, showing the prehensile pereopods which attach the parasite to the host's tongue remains. (Photograph courtesy of Bill Pennell.)

marsupium (Fig. 11.25) until late in development when a larva known as a **manca** hatches. The manca larva leaves the brood pouch or marsupium and swims until it finds a specific host. As soon as the larva locates a host, it loses its swimming setae and begins development into a male. Later, this male will transform into a mature, egg-producing female.

Epicaridean isopods are parasites of crustaceans during their entire lives. A highly modified and diverse epicaridean family, the Bopyridae, are ectoparasites of the gill chamber of marine crustaceans, including crabs and shrimp (Markham, 2001). Bopyrids comprise the most diverse taxa of isopods; those that parasitize the gill chamber may cause a large swelling of the host carapace (Fig. 11.28A). This bulge is due to the presence of the enormous and flattened female developing in the gill cavity. Male bopyrids are tiny, segmented dwarfs (Fig. 11.28B) that attach to the much larger female. The bulged carapace induced by bopyrids is so distinctive that it has also been observed in many fossil

decapod crustaceans. Fossil decapods with a swollen carapace induced by epicarideans are known from the Jurassic and Cretaceous, making these isopods one of the very few parasites to leave a fossil history (Rasmussen *et al.*, 2008).

Bopyrids have a complex life cycle with three distinct larval stages and a two-host life cycle (Fig. 11.29). Females produce an enormous number of eggs in their marsupium. An **epicaridium** larva hatches from an egg and seeks out a copepod intermediate host. On this host, it transforms into a feeding **microniscus** larva, eventually becoming a **cryptoniscus** larva. The cryptoniscus larva detaches, becomes free-swimming, and settles on its specific crustacean host (Fig. 11.29). A recently settled juvenile is known as the bopyridium. Often, if the host is a new one, the bopyrid juvenile becomes a female; all subsequent individuals become males.

Bopyrids are blood-feeders and induce a form of castration of their crustacean hosts. The bopyrid *Orthione griffenis* infects a variety of burrowing mud shrimp species. In western North America, this parasite

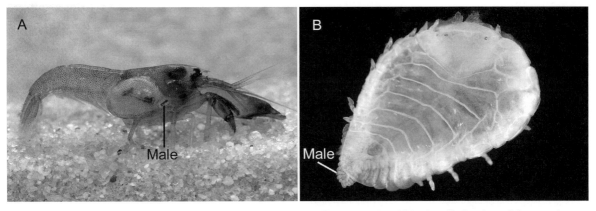

Fig. 11.28 (A) Swelling of the carapace induced by the isopod *Bopyrella macginitiei* within the gill chamber of the snapping shrimp *Alpheopsis equidactylus*. Note the presence of the dwarf male. (B) Flattened female of *B. macginitiei* removed from the gill chamber of the shrimp. Note the location of the segmented male on the female. (Photographs courtesy of Todd Huspeni.)

is now considered an invasive species, likely introduced as larvae released in ship ballast water from Asia (Williams & An, 2009). This introduced parasite is now a common parasite of the mud shrimp, *Upogebia pugettensis* (Color plate Fig. 3.3). Prevalences of greater than 80% have been reported for *O. griffenis* in mud shrimp from some estuaries. Significantly, 100% of the hosts are castrated, potentially leading to population declines of *U. pugettensis* (Smith *et al.*, 2008). For this reason, *O. griffenis* can be considered a 'keystone parasite,' one having an impact on a free-living species of ecological importance and causing cascading ecological effects (see Chapter 15). The host occurs in high density in muddy estuaries along the Pacific coast, and is considered a key biogeochemical ecosystem engineer due to the variety of ecosystem functions they perform. It is no wonder that this relatively new host–parasite interaction has been the subject of considerable research focus (Williams & An, 2009).

Isopods in the Entoniscidae are also extensively modified parasites and, although they are technically ectoparasites, they are located inside the body of their crustacean host (Fig. 11.30). The female feeds on blood and becomes very large, whereas the males remain small and live on the surface of the female. The female broods the eggs and the larvae leave through a pore on the host's gill cavity.

Isopods such as *Danalia* and *Liriopsis* extend the concept of parasitism one step further. They are hyperparasites of rhizocephalan barnacles parasitic on various crabs. *Liriopsis pygmaeae*, for example, is an epicaridean isopod hyperparasitic on the rhizocephalan *Briarosaccus callosus* of king crabs (Peresan & Roccatagliata, 2005). Such complex host–parasite systems serve to stress the biodiversity and adaptability of parasitic crustaceans, and the biological wonders of their diverse and often bizarre morphologies and life cycles.

11.3 Chelicerata

11.3.1 General considerations

The Chelicerata include the Merostomata (horseshoe crabs), Pycnogonida (sea spiders), and Arachnida. The Arachnida is, by far, the largest chelicerate class and includes spiders, scorpions, mites, and ticks. Most arachnids are terrestrial except for a few groups that have become secondarily aquatic. The ticks and mites, within the subclass Acari, often referred to as acarines, include many parasitic organisms.

Mites are exceptionally diverse and include both free-living and parasitic species, whereas all ticks are parasitic. They perform vital ecological roles in aquatic

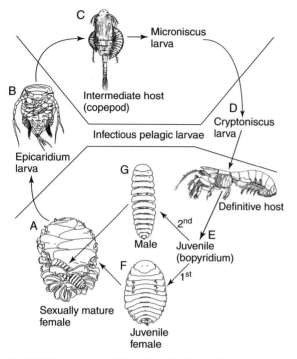

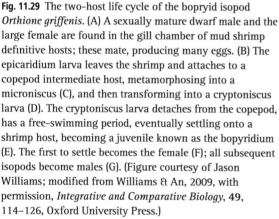

Fig. 11.29 The two-host life cycle of the bopryid isopod *Orthione griffenis*. (A) A sexually mature dwarf male and the large female are found in the gill chamber of mud shrimp definitive hosts; these mate, producing many eggs. (B) The epicaridium larva leaves the shrimp and attaches to a copepod intermediate host, metamorphosing into a microniscus (C), and then transforming into a cryptoniscus larva (D). The cryptoniscus larva detaches from the copepod, has a free-swimming period, eventually settling onto a shrimp host, becoming a juvenile known as the bopyridium (E). The first to settle becomes the female (F); all subsequent isopods become males (G). (Figure courtesy of Jason Williams; modified from Williams & An, 2009, with permission, *Integrative and Comparative Biology*, **49**, 114–126, Oxford University Press.)

and terrestrial ecosystems, including nutrient recycling and in predator–prey interactions. Many plant parasitic mite species are significant agricultural pests. Many other species of mites are phoretic on a variety of insects, e.g., dung and carrion beetles, using these hosts to 'hitch a ride' to rich feeding grounds. Some parasitic mites (*Varroa destructor*, *Acarapis* spp.) are devastating pests of ecologically important animals such as honey bees, thus acting as keystone parasites (see Box 11.3).

Most of the parasitic mites are ectoparasites on both vertebrates and invertebrates, although some have become endoparasites through the infection of respiratory passages. Some mites are parasitic only as larvae, whereas others are parasitic during their entire life. Attachment, however, is not permanent and many are found on the host only when feeding. Ticks parasitize terrestrial vertebrates throughout their development and, in most cases, they remain attached to the host only when feeding.

The pathology caused by parasitic ticks and mites can be of three general types, i.e., local damage, systemic damage, and via transmission of important pathogens. Local damage at, or around, the site of infection may include inflammation, swelling, ulceration, and itching. Sometimes, secondary infection by bacteria may occur. Ticks sometimes cause systemic damage. When ticks bite humans or other animals near the base of the skull, toxic secretions by the tick might cause a problem known as tick paralysis. Heavy infections of ticks can also result in high blood loss, with the debilitating effects of anemia. Ticks also are important vectors of significant infectious organisms such as viruses, bacteria, rickettsias, spirochaetes, protozoa, and microfilariae, including those causing Lyme disease (see Box 11.2) and Rocky Mountain spotted fever. Some free-living mites are intermediate hosts for a number of helminths (see Chapter 6).

The classification and taxonomy of ticks and mites is extraordinarily complex and dynamic. In particular, attempts to identify monophyletic higher taxa of mites have proven most difficult. The most up-to-date and comprehensive acarine resource is the *Manual of Acarology* edited by Krantz & Walter (2009). Walter & Proctor (1999) provide a fascinating exploration into the biodiversity, ecology, evolution, and behavior of mites.

11.3.2 Form and function

11.3.2.1 Morphology

Most mites are tiny, measuring between 0.2 and 0.8 mm; ticks, on the other hand, can reach up to 30 mm. The bodies of most arachnids can be divided into a prosoma

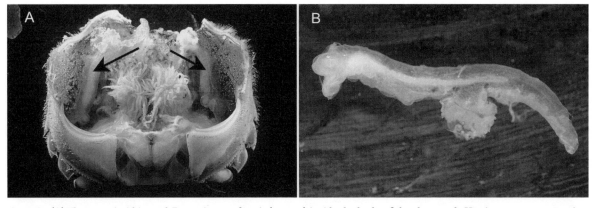

Fig. 11.30 (A) The entoniscid isopod *Portunion conformis* (arrows) inside the body of the shore crab *Hemigrapsus oregonensis*, after removal of the crab's carapace; (B) young female of *P. conformis* removed from the crab. (Photographs courtesy of Todd Huspeni.)

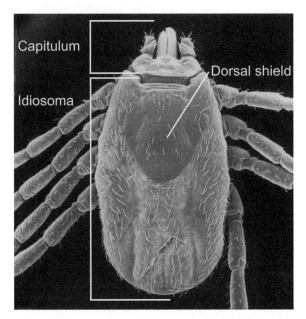

Fig. 11.31 Scanning electron micrograph of the tick *Boophilus microplus*, showing the two main body regions characteristic of ticks and mites: the gnathosoma or capitulum and the idiosoma. The large dorsal shield or scutum of this female individual is also shown. (Micrograph courtesy of the US National Tick Collection, Georgia Southern University.)

or cephalothorax, and an opisthosoma or abdomen. In ticks and mites, most segmentation has disappeared and the abdomen has fused with the cephalothorax. In ticks and mites, the body can be divided into two regions, the gnathosoma or **capitulum** and the **idiosoma** (Fig. 11.31). The capitulum carries all the mouthparts and feeding appendages (Fig. 11.32A, B). These include a pair of **chelicerae**, commonly used in piercing, tearing, or gripping host tissues, and a pair of **pedipalps** also used in feeding. The coxae (first segment) of the pedipalps are fused and extend forward to form the **hypostome** (Fig 11.32B). The hypostome, together with a labrum, forms the **buccal cone**, which, in some species, can be retracted. In other species, the pedipalps are greatly reduced and function as sensory organs. Most of the remaining body is formed by the idiosoma. It contains most internal organs and legs, and is covered dorsally with a single carapace or shield. Ticks and mites, like other arachnids, have four pairs of legs, but some may have fewer. A pair of claws is present at the tip of the legs in most ectoparasitic Acari to aid in attachment.

The anus is near the posterior end of the body and the gonopore is normally located ventrally, in a genital plate, between the last two pairs of legs. Hairs, or setae of various shapes, many of which are sensory, cover the body and legs of most species.

11.3.2.2 Body wall

The cuticle of arachnids, as in other arthropods, is secreted by the epidermis and is formed by several

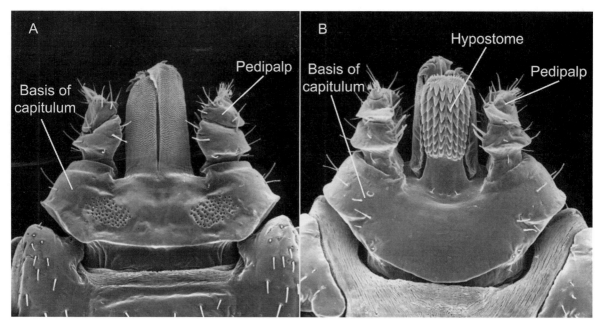

Fig. 11.32 Scanning electron micrographs of the capitulum of *Boophilus microplus*. (A) Dorsal view; (B) ventral view showing the toothed hypostome. (Micrographs courtesy of the US National Tick Collection, Georgia Southern University.)

layers, containing proteins, polysaccharides, and lipids. The epicuticle, the outermost layer of the cuticle, is made of cuticulin, a protein. Cuticulin is normally covered by a lipoidal layer that prevents water loss and by a kind of 'varnish' that protects the impermeable layer from abrasion. Immediately under the epicuticle is the procuticle that, in different species, has different degrees of sclerotization.

11.3.2.3 Digestive system

As in other arthropods, the digestive system is divided into a foregut, midgut, and hindgut. The mouth continues into a muscular, sucking pharynx, and an esophagus. There is a pair of salivary glands that open to the buccal cone. The esophagus leads into a stomach with several ceca and continues into a sacculate intestine that terminates in the anus. In many acarines, the stomach becomes separated from the intestine and ends in a blind sac. In these species, the indigestible food is stored in specialized gut cells that accumulate in the ceca. When one of these ceca is full,

it detaches and is eliminated through a split in the dorsal cuticle.

11.3.2.4 Osmoregulatory and excretory system

The excretory organs in most ticks and mites are **coxal glands** and **Malpighian tubules**. Coxal glands open to the outside at the base (coxa) of certain legs. The Malpighian tubules lie free in the hemocoel and empty to the terminal part of the midgut. In those Acari whose stomach ends in a blind sac, the Malpighian tubules are connected to the hindgut and waste is still excreted through the anus. Uric acid, amino acids, and salts are taken up by the Malpighian tubules and secreted into their lumen. The main nitrogenous waste in ticks is guanine.

11.3.2.5 Circulatory system

The circulatory system of Acari is reduced in most groups and consists of a network of sinuses with colorless blood, although in some, a heart may be present.

Movement of the blood probably is accomplished by the contraction of body muscles.

11.3.2.6 Respiratory system

Most ticks and mites possess **tracheae**, similar to those present in insects, for gas exchange. The tracheae are simple, chitin-lined tubes, branched or unbranched, through which gas diffuses. The tracheae open to the outside in one to four pairs of **spiracles**, all located on the anterior half of the body. Astigmatid mites, however, lack a tracheal system, have no spiracles, and respiration occurs through the cuticle.

11.3.2.7 Reproductive system

Acari are dioecious and fertilization is internal. The male reproductive system consists of a pair of testes located in the mid region of the body. A vas deferens arises from each testis and both may join together to open through a gonopore, or through a chitinous penis projecting to the outside through the gonopore. In those species with a penis, transfer of sperm to the female is direct. In most Acari, however, sperm transfer is indirect. Sperm can be packed into a spermatophore that is delivered to the female or injected by the chelicerae into special openings on the female body.

The female reproductive system consists of a single ovary that continues into an oviduct and uterus to open in the gonopore. During copulation, the spermatozoa are introduced into the uterus, although in some mites, a vagina is present. A seminal receptacle and accessory glands are also present.

11.3.2.8 Nervous system and sensory input

The nervous system of acarines is greatly compressed and includes a brain that is located above the esophagus. The brain is a ganglionic mass formed by two pairs of lobes, the protocerebrum and tritocerebrum, which innervate the eyes and anterior gut, respectively. Nerve trunks from the brain connect with the subesophageal ganglion forming a ring around the esophagus. Most, or all, of the ganglia originally present in the abdomen are fused with the subesophageal ganglia. Nerves arising from this anterior complex or central nerve mass innervate the appendages, organs, and sensory structures.

Sensory setae are important sensory structures in parasitic Acari. Innervated pits and slit sense organs are also present. Although many mites are blind, some possess simple eyes. The Ixodida (all ticks) have a large set of different types of receptive sensilla clumped together on the first tarsi. This structure, known as **Haller's organ**, has an important aggregate of receptors, including mechano-, thermo-, hygro-, and chemoreceptors, which are used for host and mate location. Mites have a similar structure, but it lacks the modifications that characterize the Haller's organ in Ixodida (Lompen & Oliver, 1993).

11.3.3 Nutrient uptake

Ticks and mites ingest mostly fluids, but, when feeding on solid foods, the food is digested externally and liquified, then absorbed by the action of the sucking pharynx. Some feed by sucking blood and lymph released from tissues cut or damaged with the chelicerae. Others bite the skin of the host and feed on dermal tissue that is hydrolyzed externally by proteolytic enzymes. Ticks feed by pushing the hypostome into the skin of the host. A number of barbs present in the hypostome anchor the tick to the skin (Fig. 11.32B), although some species also produce a cement-like substance to help with anchoring. Sharp 'teeth' in the hypostome cut blood vessels under the skin, causing the blood to form a pool. The tick then sucks this blood into its gut through the hypostome.

Tick saliva contains a number of compounds that aid in the feeding process, as well as in avoiding or neutralizing the host immune system. There are three mechanisms by which the host can stop bleeding. It can plug a hole in the capillaries using platelets, it can form a blood coagule, or it can pinch off the vessel via vasoconstriction. The tick saliva contains compounds able to block all three mechanisms. In addition, tick saliva contains the tick version of prostaglandins, which can relax blood vessels and also suppress the host immune response (review in Bowman *et al.*, 1997). It also has compounds that can block the host chemicals that cause inflammatory responses, such as histamines and thromboxane. Still another compound in tick saliva breaks down bradykinin, a host chemical that causes pain at the bite site, allowing the tick to feed undetected (Ribeiro *et al.*, 1985; Ramachandra & Wikel, 1992).

11.3.4 Development and general life cycle

Development of acarines is direct, although their life cycle may include one or more hosts. After mating, the female of most tick species requires a blood meal to initiate egg production. Following an incubation period of 2 to 6 weeks, a sexually immature larva hatches from the egg. The larva has only three pairs of legs and looks like a miniature version of the adult. In most acarines, the fourth pair of legs is acquired after the first molt, when the larva changes into a **proto-nymph**. In many acarines, successive molts transform the protonymph into a **deutonymph**, a **tritonymph**, and, finally, an adult. During these changes, adult structures are gradually acquired. This pattern varies among the different groups. In aquatic mites, for example, the deutonymph may enter a quiescent stage called the **imagochrysalis**. In the Ixodidae, or hard ticks, there is only one nymph; in the Argasidae, or soft ticks, there may be up to eight nymphal stages. Both males and females require a blood meal in every developmental stage in order to molt into the next developmental stage. In mites, the number is highly variable, ranging from zero to several.

The life cycle of ticks (Ixodida) may include one to many hosts, depending on the species. After mating and obtaining a blood meal, female ticks leave the host and lay the eggs in the soil. A six-legged larva hatches from the egg, climbs into the vegetation, exhibits questing behavior by waving its front legs, and waits for a suitable host. Once a host is found, the larva feeds, molts, and becomes a nymph. Some ticks, like *Boophilus*, complete their development with just one host. Others may require two, three, or even more

hosts. Most ixodids require three hosts, one for the larva, one for the nymph, and one for the adult, whereas most argasids use more than three hosts because they have more than one nymphal stage.

The life cycle of parasitic mites is similar to that described for ticks, with a larva, one or more nymphal stages, and an adult. But there are enormous variations on this theme. In many species, the life cycle includes free-living and parasitic stages. Some prostigmatid mites, like the chiggers, are parasitic only during their larval stage, and their nymphal and adult stages are free-living. Water mites in the genus *Unionicola* have life cycles that involve freshwater molluscs or sponges. *Unionicola formosa*, for example, are commensals in the mantle cavity of freshwater mussels, while larvae are ectoparasites of chironomid dipterans.

Host specificity is variable in ticks and mites, ranging from highly specific to opportunistic. Both host specificity and the number of hosts needed to complete the life cycle have implications for their dispersal capabilities and for the potential to spread disease in those species that act as vectors. For example, in one-host ticks, the transmission of disease organisms from one animal to another is very low, but in multiple-host ticks, transmission among animals is very likely greatly complicating the epidemiology of tick-transmitted diseases (see Box 11.2). Nonetheless, one-host ticks still carry disease-producing pathogens and, in this case, transmission is vertical, from the mother tick to her progeny, which, in turn, infect new hosts.

11.3.5 Biodiversity and life-cycle variation

Walter & Proctor (1999) and Krantz & Walter (2009) highlight the phenomenal biodiversity of mites and the interdisciplinary field of acarology. There are approximately 540 described mite families, ranking them as among the most species-rich groups of organisms. They are ubiquitous in virtually every imaginable habitat. In terms of those using animals as hosts, there are many ectocommensals and ectoparasites of both aquatic and terrestrial animals. In terms of endo-parasitic mites, there are species that occur in the respiratory systems of diverse hosts, ranging from snakes, to seals, to honey bees. Some live in bizarre places, including on the feet and antennae of army ants, the ears of noctuid moths, the nostrils of hummingbirds, and even in the facial pores and eyebrow follicles in humans.

The diversity of Acari will be approached on a group-by-group basis following the taxonomic system proposed by Evans (1992), and modified recently by Krantz & Walter (2009). These authors present the most current and comprehensive treatment of mite taxonomy and phylogeny. While still hotly debated, they adopt a system of two superorders (Parasitiformes and Acariformes) and six orders. Only the groups of Acari that have parasitic species in their ranks will be considered (see Box 11.6).

11.3.5.1 Ixodida

These ticks are highly specialized obligate parasites of a wide variety of mostly terrestrial vertebrates, including amphibians, reptiles, birds, and mammals, plus a few marine snakes and lizards. About 800 species have been described, but most of the attention has been focused on the 80 or so species of medical or veterinary significance, either because of the direct pathology they cause, or because of the pathogens they carry. Two kinds of ticks are generally recognized, i.e., hard ticks and soft ticks.

Hard ticks

Hard ticks belong to the Ixodidae. In hard ticks, the capitulum is terminal and can be seen from above. They also have a sclerotized dorsal plate, or **scutum**, that covers the entire dorsum of the body in males, but only the leg-bearing portion of females (Fig. 11.31). In females, the cuticle of the abdomen is folded, allowing the abdomen to expand when feeding and engorging on blood from the host. Most hard ticks have a single nymphal stage between the larva and the adult. During their life span, hard ticks attach and feed on one, two, or more often, three hosts, which can be the same or different species. These ticks are called one-host, two-host, or three-host ticks. Hard ticks feed for an extended period. Each blood meal, depending on the

developmental stage, can take 4–7 days to ingest. For most tick species, host specificity is low. In spite of their low host specificity, many species of tick show developmentally related patterns, where larvae and nymphs are found on small mammals and birds, whereas the adults are most frequently found on larger mammals (Fig. 11.33). After mating and engorging on the last host, the adult female leaves the host, lays a large clutch of eggs on the substratum, often in the thousands, and then dies.

Although most hard ticks have three-host life cycles, species of *Boophilus*, including the American cattle tick *B. annulatus*, are one-host ticks, with all the developmental stages occurring on the same individual host. Several species of *Boophilus* are vectors for a number of viruses, rickettsias, and protozoans, including *Babesia bigemina*, the protozoan responsible for Texas cattle fever, also known as red-water fever. Although one-host ticks are, in general, poor vectors, pathogens can be passed vertically, from mother to offspring, by transovarian transmission. Newly hatched ticks thus are already infected and ready to pass the disease onto new hosts when they disperse.

Hard ticks in *Ixodes* are ubiquitous. Although many species of *Ixodes* do not cause extensive damage to their vertebrate hosts, larvae and nymphs do grow considerably while feeding during their development, and as they become fully engorged with blood (Fig. 11.34). Several are of medical significance since they can act as vectors of the spirochete bacterium responsible for Lyme disease, *Borrelia burgdorferi* (see Box 11.2; Bennett, 1995; Oliver, 1996). *Ixodes scapularis*, the black-legged tick (Figs. 11.35, 11.36), is a common vector of Lyme disease in eastern North America, whereas *I. pacificus* is the vector in western North America, *I. ricinus* in Europe, and *I. persulcatus* (Fig. 11.34) in Japan. Several species of *Dermacentor*, such as the Rocky Mountain wood tick *D. andersoni*, the American dog tick *D. variabilis*, and the Pacific Coast tick *D. occidentalis*, are vectors for the pathogen causing Rocky Mountain spotted fever and for a number of other viral and bacterial diseases.

The lone star tick *Amblyomma americanum* is also a vector for the organism that causes Rocky

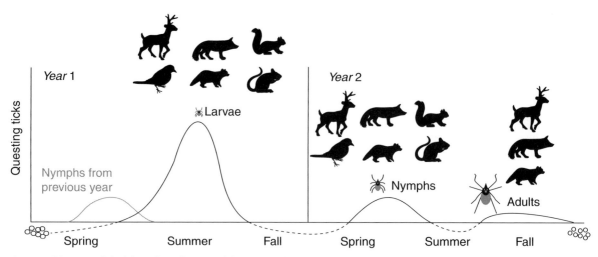

Fig. 11.33 Diagram of the life cycle and seasonal dynamics of the black-legged tick *Ixodes scapularis* in eastern North America. Displacement above the horizontal line represents relative abundance of host-seeking (questing) ticks, whereas displacement below the horizontal line (dashed part of curve) represents post-feeding quiescence on the forest floor. In mid-summer of Year 1, eggs hatch into larvae, which take their single blood meal from one of many different species of mammals and birds. After their 3- to 5-day blood meal, larvae drop off the host, molt into nymphs, and remain quiescent until taking their single blood meal the following late spring or early summer of Year 2. Fed nymphs molt into adults and undergo a several month quiescent period before seeking their final blood meal. Mating typically takes place while on the adult host and egg-laying occurs the following spring. (Figure courtesy of Richard Ostfeld; modified from a diagram by Jesse Brunner in Ostfeld, 2011, *Lyme Disease: The Ecology of a Complex System*. New York: Oxford University Press).

Mountain spotted fever. Larvae and nymphs of this species usually feed on birds and almost any small terrestrial mammal available, whereas the adults, again, prefer larger hosts. This lack of specificity reaches as far as humans and, unlike most other ticks, all three developmental stages of *A. americanum* attack humans.

Dermacentor albipictus, also known as the winter tick, overwinters on wild ungulates (moose, elk, white-tailed and mule deer) throughout the northern USA and southern Canada. Moose seem to be the preferred host, with some individuals harboring more than 100 000 ticks. Moose that are heavily infected can be anemic, have low fat stores, and perform extensive grooming, which leads to the premature loss of winter hair and a condition known as 'ghost moose syndrome' (Samuel & Welch, 1991). The loss of hair, especially in late winter, decreases thermoregulatory capacity, leading to reduced survival and marked reductions in host population size (Chapter 15).

Although ticks do not parasitize many amphibians, *Amblyomma rotundatum* is common on the marine toad *Bufo marinus*, in Central and South America. This tick was introduced to Florida in the last 50 to 60 years with the toad host, and is now widely spread along both coasts of Florida (Oliver *et al.*, 1993). Unlike other ticks, *A. rotundatum* is a parthenogenetic species, a factor that aids in the colonization of new areas. Species of *Amblyomma* have invaded the marine environment and are parasites of certain reptiles. For example, *Amblyomma nitidium* is a tick found on certain sea snake species. The famously named tick, *Amblyomma darwini* parasitizes the marine iguana of the Galapagos Islands.

The ticks' ability to endure environmental harshness is exemplified by species of *Hyalomma*. These hard ticks are found in desert areas of the Middle East, Asia, and Africa, where there is little shelter, very few hosts, and the ticks must endure extensive periods of starvation. A possible adaptation to the scarcity of hosts is

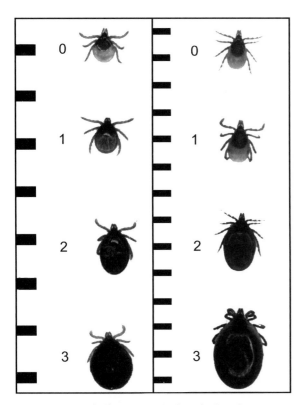

Fig. 11.34 Larvae (left) and nymphs (right) of *Ixodes persulcatus* at different stages of feeding. (0) Unfed; (1) partially fed for 1 day; (2) partially fed for 2 days; (3) fully fed for 3 days (engorged). Note scales in mm. (Modified from Nakao & Sato, 1996, with permission, *Journal of Parasitology*, **82**, 669–672.)

a reduction in the number of hosts required to complete their life cycle, and thus *Hyalomma* is a two-host tick. Many species of *Hyalomma* are vectors for a number of viruses, rickettsias, and protists, many of which are carried over long distances by migrating birds infected with ticks.

Soft ticks

Soft ticks belong to the Argasidae. In soft ticks, the capitulum is subterminal and cannot be seen in dorsal view because it lies in a depression that forms a hood over it. Soft ticks have several nymphal stages and are called many-host ticks. Unlike hard ticks, soft ticks do not engorge but feed repeatedly, mainly at night, with each feeding session lasting from minutes to 1 hour. After feeding, they leave the host and rest in their hide-out between meals. They are normally found in nests, burrows, crevices, caves, loose soil of resting or rolling areas of big game, and other places often visited or inhabited by the hosts. In most species, the larvae remain on the host until molting, whereas the nymphs and adults feed quickly and return to their resting place. Soft ticks repeatedly parasitize the same individual or family members. Adult females deposit eggs in their hiding places several times between feedings, and, unlike hard ticks, clutch sizes are small, with a few hundred eggs or less. Argasids prefer habitats with low humidity and are

---

Box 11.2 | On acorns, ticks, and mice: the epidemiology of Lyme disease

Understanding the complex epidemiology of microbial and parasitic diseases can be extremely difficult. It often requires an interdisciplinary approach, linking parasite life-cycle details with host ecology and behavior. This is especially the case when the microbial disease in question is transmitted by a multi-host vector (Fig. 11.33). Such is the case for Lyme disease, a serious illness of humans caused by the spirochaete bacterium *Borrelia burgdorferi* and transmitted by the tick *Ixodes scapularis* in eastern North America. When *Ixodes* ticks take a blood meal from an infected host, the spirochaetes multiply in its midgut and then enter the tick's circulation. Transmission to humans occurs during tick feeding.

The pathogenesis of Lyme disease represents a complex of symptoms. An initital symptom is the characteristic, expanding bulls-eye shaped rash around the tick's bite. This is typically

Box 11.2 | (continued)

followed by flu-like symptoms of fever, chills, and muscle aches. Severe late-stage symptoms can arise when the bacteria migrate to the nervous system and joints, leading to painful and debilitating arthritic and neurological disorders. Other disease manifestations include cardiac pathology, encephalitis, and memory loss. Chronic pathogenesis is the result of immunologically based inflammatory responses to bacteria in target tissues.

Lyme disease is on the increase. Climate change is predicted to increase the rate of spread, leading to a substantial increase in Lyme disease risk in humans (Ogden *et al.*, 2006; Leighton *et al.*, 2012). In some high-risk regions, it has reached epidemic status and thousands of new cases are reported each year, making it the most common vector-borne disease in the USA. In some endemic areas, very high prevalences (50–70%) of infected ticks are reported; it is in these areas that humans are at greatest risk. What factors account for so many infected ticks? Can ecologists and epidemiologists make predictions as to what factors are most important? The answer is yes to both questions, and thanks to the extensive field and modeling studies of Richard Ostfeld and his colleagues, it turns out that a complex array of ecological interactions influences the transmission biology and risk of this zoonotic disease (review in Ostfeld, 2011).

The epidemiology of Lyme disease in northeastern USA has been tied to crucial details of oak tree ecology! Indeed, in the northeastern USA, Lyme disease is virtually restricted to forested landscapes (Ostfeld, 1997). This is because of the life cycle of the vector, the black-legged tick *I. scapularis* (Figs. 11.35, 11.36), and the parasite's dependence on forest-dwelling natural reservoirs of infection, primarily the white-footed mouse (flying and gray squirrels, chipmunks and ground-inhabiting birds like robins and blue jays can be hosts as well) and the white-tailed deer (Fig. 11.33). Eggs are deposited in the spring and larvae hatch in the summer, then infect white-footed mice. By the following summer, larvae have overwintered in diapause, dropped off the host, and molted

Fig. 11.35 The black-legged tick *Ixodes scapularis*, the vector of Lyme disease in eastern North America. (Photograph courtesy of Scott Bauer, USDA Agricultural Research Service, Bugwood.org.)

Box 11.2   (continued)

Fig. 11.36 A fully engorged female of the black-legged tick *Ixodes scapularis*. (Photograph courtesy of Scott Bauer, USDA Agricultural Research Service, Bugwood.org.)

into nymphs, which seek out and feed on the mice again. By fall, adults emerge, locating and feeding a third time, usually on white-tailed deer. This is an important host on which *I. scapularis* mate, although other mammals such as racoons and red foxes can act as hosts for the adults (Fig. 11.33). It is primarily the larvae and nymphs in white-footed mouse populations that maintain the *B. burgdorferi* bacteria in nature. The bacterium is most often transmitted to humans by nymphs that 'step out' of this natural life cycle and bite people instead, typically in mid-summer when humans are most likely in a wilderness setting.

Thus, it is primarily the distribution and abundance of rodents and white-tailed deer that determine the human risk of acquiring Lyme disease. Factors influencing the population dynamics of these hosts may interact to determine Lyme disease outbreaks. In particular, when white-footed mouse populations are high, increased transmission of the bacterium in the mouse population occurs. A particularly important predictor of high rodent numbers and thus increased numbers of *B. burgdorferi*-infected ticks is the episodic production of acorns in autumn (Ostfeld, 1997; Ostfeld *et al.*, 2006). Genetic and climatic factors cause oak trees to exhibit so-called masting behavior, leading to bumper crops of acorns in some years. It is this mast crop of acorns that sets off an ecological cascade or chain reaction that influences Lyme disease transmission to humans. Rodent populations increase dramatically the summer after a mast year due to the increase in availability of acorn food reserves. More rodents lead to an increase in the abundance of infected nymphal ticks. More infected ticks means increased risk of humans being bitten, especially in the second summer following a mast year in oak forests. By monitoring annual acorn production, the number of Lyme disease cases can be predicted almost 2 years in advance (Ostfeld *et al.*, 2006).

To further complicate matters, migrating songbirds contribute to the long-distance dispersal of *B. burgdorferi*-infected ticks (Ogden *et al.*, 2008). For example, several migratory songbird species have recently been shown to disperse *B. burgdorferi*-infected *I. scapularis* ticks across Canada, with the potential for starting new tick populations endemic for Lyme disease (Scott *et al.*, 2012). In short, complex interactions between climate, several reservoir-competent vertebrate hosts, and human activities determine the risk of exposure to this expanding disease.

common in desert areas, where they are able to aestivate for months, or even years, without food.

Since most species of soft tick are found in areas of very low humidity, whenever they happen to occur in humid or wetter climates, they seek dry microhabitats. Several species of *Ornithodoros* parasitize rodents and other non-flying mammals that inhabit burrows, dens, and caves. Other species infect a number of bat species and a few are found on colonial, ground-nesting marine birds. *Ornithodoros savignyi*, the sand tampan, is the most widely distributed species in this genus. It is common in dry regions of Africa, the Near East, and India where it feeds on camels, livestock, humans, and wildlife. It is a nocturnal species and most of its victims are asleep when the tick attacks. Although it appears not to vector any disease-causing organisms, the bite is painful. The extreme pain associated with the bite of *O. savignyi* seems to be a rather common phenomenon among most soft ticks. *Ornithodoros hermsi* is a common species found in forests up to 2500 m in altitude between the Rocky Mountains and the Pacific Coast of North America. The tick infests nests of chipmunks, wood rats, pine squirrels, and other rodents in hollow logs, tree stumps, and wood cabins, where tourists, woodsmen, hikers, and residents are frequently bitten and may develop a severe relapsing fever caused by *Borrelia hermsi*.

A few species of *Ornithodoros* are associated with ground-nesting marine birds, including the Galapagos and Humboldt penguins. The Humboldt penguin, *Spheniscus humboldti*, breeds on the barren coasts and offshore islands of Peru and Chile, where they normally cohabit with guano-producing bird colonies. *Ornithodoros spheniscus* is ubiquitous among penguins, but they eagerly attack humans causing slow-healing blisters, fever, and headaches. The tick is annoying to the host, and the irritation produced by feeding ticks may induce the penguins to abandon their nests (Hoogstraal *et al.*, 1985). Other related tick species also are known to cause nest desertion by terns and brown pelicans. Some species of Peruvian guano birds are prone to remain in flight longer than needed, staying away from their breeding grounds to avoid tick bites. Eventually, they still may desert their nests (Feare,

1976; King *et al.*, 1977; Duffy, 1983). *Ornithodoros transversus* infects Galapagos tortoises. This tick is unusual in that it oviposits directly on its host, making it the only ixodid to spend its entire life cycle on its host (Walter & Proctor, 1999). The host specificity of ticks such as *O. transversus* on endangered reptiles prompted Durden & Keirans (1996) to write their intriguing paper entitled "Host–parasite coextinction and the plight of tick conservation." We do not often think of the potential extinction of hosts leading to the extinction of their parasites.

Soft ticks, including species of *Argas*, are almost exclusively parasites of birds, both marine and terrestrial. Most are nest parasites, feeding at night and hiding during the day in crevices, cracks, litter, or any other hiding place in, or near, the nest of the host. The fowl tick, *A. persicus*, is native to central Asia where it attacks arboreal nesting birds such as sparrows and crows, but has spread to many parts of the world, where it has successfully colonized domestic chickens, turkeys, and other fowl. Under favorable conditions, this tick can form large populations in henhouses and become a serious pest. Humans that visit chicken coops at night are usually attacked. *Argas cucumerinus* is the exception to this common nocturnal behavior. This species is found on barren cliffs facing the Pacific Ocean along the Peruvian coast and feeds on birds and humans that happen to stop and rest on the rock ledges where the ticks shelter. Because *A. cucumerinus* is not associated with a nest, host encounters are random and scarce. When a host is detected during the daytime, the tick races towards the host using its long, spider-like legs and feeds very rapidly. The racing legs, daytime behavior, and fast feeding are unique characteristics of this species, very different from the typical tick, and clearly constitute adaptations to the scarce and irregular presence of hosts.

Otobius megnini (Fig. 11.37) is a highly specialized soft tick that lives and feeds inside the ears of the host. Deer, mountain sheep, and pronghorn antelope are natural hosts, but the tick has adapted easily to cattle, other domestic animals, and humans. Unlike other soft ticks, *O. megnini* has a one-host life cycle, adult ticks do not feed, and mating occurs in the ground.

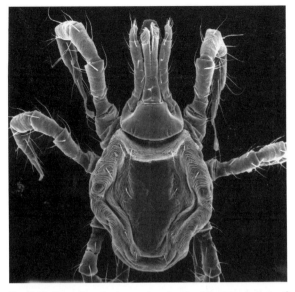

Fig. 11.37 Scanning electron micrograph of the larval stage of the ear tick *Otobius megnini*. (Note that larval stages have only three pairs of legs.) (Micrograph courtesy of the US National Tick Collection, Georgia Southern University.)

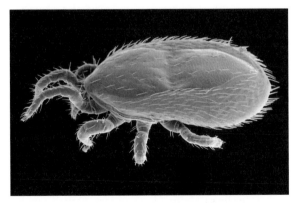

Fig. 11.39 Scanning electron micrograph of *Ornithoynyssus bacoti* from a captive rock rat *Zyzomys pedunculatus*. (Micrograph courtesy of David Walter.)

11.3.5.4 Mesostigmata

Mesostigmatid mites are parasites of a variety of vertebrates and invertebrates, and many of them parasitize the respiratory system of their hosts. Many mesostigmatid mites are associated with the nests and bodies of birds. Several bird nest mites can become an irritating nuisance to humans. The chicken mite,

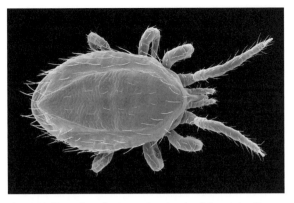

Fig. 11.38 Scanning electron micrograph of the nest mite *Ornithonyssus bursa* from an unknown bird host in Australia. (Micrograph courtesy of David Walter.)

Dermanyssus gallinae, parasitizes fowl and many wild birds worldwide, and causes severe dermatitis in humans. The mites live near roosting places and feed at night, laying eggs during the day in nests. In heavy infections, they become so annoying that hens leave their nests and young chickens are likely to die. Pigeons may carry these mites near humans when roosting in the roofs or eaves of buildings. The northern fowl mite, *Ornithonyssus sylviarum*, is a blood-feeding pest of poultry and wild birds throughout the world. Heavy infestations can give the birds' plumage a blackish appearance (Walter & Proctor, 1999). *Ornithonyssus bursa* (Fig. 11.38) is a common nest mite infecting various bird species, including starlings and swallows. These, and other avian nest mite–bird systems, have been used extensively as research models in evolutionary biology, notably in testing hypotheses of sexual selection by female choice (reviewed by Walter & Proctor, 1999; Proctor & Owens, 2000).

Echinolaelaps muris, the common rat mite, is the vector for a malaria-like protist among rats, whereas *Liponyssus sanguineus*, the house mouse mite, is the vector for the microorganism responsible for rickettsial pox in humans, a mild febrile condition that lasts about 2 weeks. The tropical rat mite, *Ornithonyssus bacoti* (Fig. 11.39), can become a serious nuisance in research facilities where mouse colonies are maintained. In heavy infections, it can retard growth and

Fig. 11.40 Female of the mite *Varroa destructor* (arrow) on a western honey bee *Apis mellifera*, showing the large size of this mite species relative to its host. (Photograph courtesy of Scott Bauer, USDA Agricultural Research Service, Bugwood.org.)

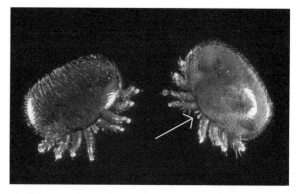

Fig. 11.41 Females of *Varroa destructor* from a western honey bee *Apis mellifera*. Note the chelicerae (C) and the dorsoventrally flattened body. (Photograph courtesy of Scott Bauer, USDA Agricultural Research Service, Bugwood.org.)

even kill young mice, whereas in humans it may cause severe dermatitis in addition to the sharp pain produced by the bite.

Rhynonyssid mites parasitize the respiratory tracts of birds, feeding on blood and other tissues, and can be a significant agent of disease among wild birds. Halarachnids, on the other hand, are parasites of the respiratory system of other vertebrates. Species of *Orthohalarachne* parasitize the respiratory tract of marine mammals, such as fur seals, seal lions, and walruses. Males and females are relatively non-motile and have well-developed claws to maintain their position in the nares of the host. Transmission from host to host may occur by active larvae that crawl, or more likely, by larvae that are sneezed from one animal to another.

A notorious mesostigmatid mite of honey bees is *Varroa destructor*. These relatively large ectoparasitic mites (Figs. 11.40, 11.41) feed on the hemolymph of worker and drone honey bees, weakening hives and eventually leading to their death if their populations are not controlled. They are known worldwide as the most devastating parasites of honey bees. *Varroa* mites have also been implicated as one of the causes of colony collapse disorder. This is a recent, widespread phenomenon leading to the mysterious, catastrophic losses of honey bee colonies (Box. 11.3).

11.3.5.5 Prostigmata

The Prostigmata includes both free-living and parasitic mites. The familiar chiggers, follicular mites, and several mites that infect insects belong to this group. Chiggers are the larval forms of trombiculid mites. They are unusual in that only the larvae are parasitic, whereas the nymphs and adults are free-living and prey on small terrestrial invertebrates and their eggs. Chiggers parasitize every major group of terrestrial vertebrates, and some aquatic ones too, but they rarely parasitize invertebrates.

Most chiggers cause some type of skin problem and some are vectors for pathogens. Contrary to popular belief, the skin condition caused by most chiggers is not due to burrowing of the larvae under the skin, but to their feeding activity. Chiggers do not burrow under the skin; instead, after the mouthparts penetrate the skin, the larva injects a proteolytic salivary secretion that digests host cells, which are then sucked up, together with interstitial fluids. A specialized hollow feeding tube, or **stylostome**, made from hardened host and larva secretions, facilitates feeding, working as a drinking straw. The larvae of *Blankaartia sinnamaryi* parasitize young Florida sandhill cranes and other birds and mammals in tropical America. These chiggers are bright orange and can be seen with the naked eye in the central depression of firm, raised papules.

Box 11.3 | A mite-y ecological and economic problem: *Varroa destructor* of honey bees

Bees are host to many species of symbiotic mites. As a testament to this diversity, at least 30 lineages of mites have independently colonized the bee habitat (Oldroyd, 1999; Walter & Proctor, 1999). Of these, one particularly nasty mite species has received the greatest attention, since it is a parasite of European honey bees. Given their ecological importance as pollinators and their economic importance in the production of honey, this is not surprising. Honey bees have been referred to as the glue that holds terrestrial ecosystems together. Thus, the 'Varroa mite,' *Varroa destructor*, can be considered one of the world's most significant keystone parasites.

Varroa destructor (formerly *jacobsoni*) (Anderson & Trueman, 2000) entered the USA in 1987; it was originally a parasite of the Asian honey bee, *Apis cerana*. The parasite causes little harm in Asian honey bees, since the host evolved so-called hygienic behaviors to control mite population sizes. When the European honey bee, *Apis mellifera*, a superior producer of honey, was introduced to Asia, the Varroa mite found a new and suitable host to colonize and a host shift occurred (Oldroyd, 1999). Since then, due to the widespread dispersal of hives by humans, *Varroa* mites have spread throughout the world, with devastating consequences. *Varroa* mites are relatively large, external parasites of larvae, pupa, and adult honey bees (Fig. 11.40). The mites particularly target the developing pupa (especially the drone brood) inside the wax cells of the brood comb in the honey bee colony. When mites are observed on flying worker bees, the bee keeper knows that the colony is seriously infected, likely in a severely weakened state, and may not survive. The mites pierce the body wall and feed on hemolymph, causing the disease known as varroatosis. A variety of pathological impacts have been noted, including declines in adult worker emergence weights, and reductions in water and metabolic reserves (Bowen-Walker & Gunn, 2001). Some *Varroa* strains are particularly virulent. In addition to the direct pathogenesis, *Varroa* mites act as vectors for significant honey bee virus pathogens, including the deformed wing virus. Finally, the mites have been implicated as a contributor in the widespread, mysterious losses of honey bee colonies, a phenomenon known as colony collapse disorder. More and more evidence indicates that there may be a link between *Varroa* mites and transmission of another viral pathogen – the Israeli acute paralysis virus (Martin, 2001).

The adult female mites are reddish-brown in color and are dorsoventrally flattened (Fig. 11.41). The flattened shape enables females to attach and hide between the bee's segments, preventing grooming by bees while the mite feeds. The mite's life cycle and reproductive biology are finely tuned and synchronized with crucial aspects of the honey bee social system and development (see Oldroyd, 1999 for precise timing). Mated female mites, ready to reproduce, abandon the bee inside the hive and enter a wax cell containing a bee larva just before the cell is capped by worker bees. The 'mother' mite hides in the larval food and is then sealed with the larva. She feeds on the developing larval bee and then commences to lay eggs, which hatch rapidly. Like their bee hosts, *Varroa* is haplodiploid; unfertilized eggs result in males and

Box 11.3 | (continued)

fertilized eggs hatch into females. Her first egg is unfertilized and hatches into a male proto-nymph. The next four or five eggs are fertilized and hatch into daughter protonymphs. Once they become deutonymphs and then adults, the male mates with his sisters in the mite fecal pile accumulating at the bottom of the wax cell. By the time the honey bee pupa is mature, it will emerge with the original mother and several mated daughters. These are now free to roam inside the hive and repeat the cycle. Mites spread from colony to colony by drifting workers and drones, or during the swarming process. Honey bees can also acquire *Varroa* when robbing other colonies that happen to be infected. Infected colonies are weakened and are more likely to be robbed, facilitating the spread of *Varroa* mites to infect new colonies.

A solution to the honey bee crisis is not yet in sight. Currently, a few acaricides are available to control mite infections in hives but, unfortunately, the mites are already showing signs of resistance. Several biological strategies also are being developed to fight these mites. One is to let natural selection runs its course until resistance increases in the bee population. It may take some time, but it has happened before with honey bees in Europe and South America, some of the earlier locations of the Asian mite invasion. Other potential strategies include the development of natural chemical attractants that lure the mite away from their hosts, artificial selection of honey bees with appropriate hygiene behaviors, introduction of benign strains of *Varroa*, and the introduction of strains of honey bees with genetic resistance to mites. The development of 'Varroa-sensitive hygienic' strains of bees is particularly promising. Hygienic behavior is controlled by two recessive genes. This allows the bees to detect the presence of *Varroa* in cells, leading to uncapping of the cell and removal of the mite-infested brood, thus reducing the mite population size in the hive over time. In 2006, the entire honey bee genome was sequenced. This milestone will likely contribute significantly to understanding the mechanisms of genetic resistance and immunity in response to honey bee pathogens, such as *Varroa destructor* and the microsporidian *Nosema ceranae* (see Box 4.1).

The papules are the result of a host reaction to the chigger, specifically to its stylostome, which triggers a localized granulomatous dermatitis (Spalding *et al.*, 1997).

A variety of prostigmatid mites infect lizards. *Geckobia bataviensis*, for example, is a beautifully ornamented species found on the toes of the Asian house gecko *Hemidactylus frenatus* (Fig. 11.42). Some pterygosomatid lizard mite species are reported to act as vectors of haemogregarine blood parasites to their reptilian hosts.

Demodicid mites, commonly known as follicular mites, are minute, elongated parasites, with short stumpy legs. Follicular mites parasitize the hair follicles and sebaceous glands in the skin of mammals, where they feed on their secretions. Often, all the developmental stages, from egg to adult, are found in a single follicle. Demodicid mites are common in every group of mammals, including marsupials, and exhibit strict host specificity. *Demodex canis* is a common parasite of the hair follicles of dogs. Sometimes, however, the mite population proliferates beyond normal

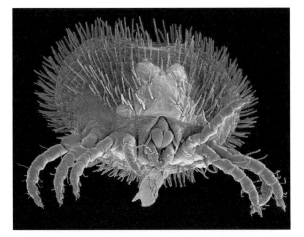

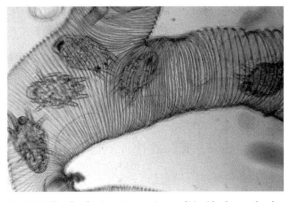

Fig. 11.43 Tracheal mites *Acarapis woodi* inside the tracheal tubes of a honey bee. (Photograph courtesy of the Agricultural Research Service, US Department of Agriculture.)

Fig. 11.42 Scanning electron micrograph of the lizard mite *Geckobia bataviensis* from the toes of the Asian house gecko *Hemidactylus frenatus*, showing the exquisite structuring of the cuticle, characteristic of many mites. A piece of gecko skin is still attached to tips of the mite's chelicerae. (Micrograph courtesy of David Walter.)

levels, causing a clinical condition known as red mange. Heavily infected dogs have dry, scaly patches of hair loss, and may develop pustular lesions covered with a foul-smelling exudate, probably as a result of secondary infection by *Staphylococcus*. Young puppies or older dogs suffering from malnutrition, parasitic infections, or debilitating diseases are most susceptible; symptoms disappear gradually in older, healthy dogs, probably due to acquired immunity. There is some evidence that the host immune response is responsible for controlling the mite population.

Humans are no strangers to mites and are usually infected (unknowingly) by two species common in the follicles and sebaceous glands of facial hair. *Demodex folliculorum* measures almost 0.3 mm and inhabits the follicles of hairs above the level of the sebaceous gland, where three or more specimens can be found in a single follicle. *Demodex brevis* is slightly smaller, about 0.2 mm, and lives in the sebaceous glands of hairs, where a single individual, or at the most two (a female and her offspring), can be found (Desch & Nutting, 1972). *Demodex folliculorum* consumes cells of the follicular epithelium, whereas *D. brevis* feeds on cells from the sebaceous gland, a nice example of

habitat segregation and resource partitioning. Both species are rather benign, although some forms of rosacea and juvenile acne may be related to their presence.

Pyemotid mites are ubiquitous parasites of insects. Unlike other mites, the female retains the eggs in her abdomen and they complete all their developmental stages inside. Males emerge first and then wait to copulate immediately with the emerging females. *Pyemotes tritici*, the straw itch mite, is a common parasite of the grain moth, but it also feeds readily on humans, especially granary workers. The mites are so abundant that, in most cases, they kill their moth host. For this reason, this mite has been used for the biological control of several insects.

Species of *Acarapis* are highly host-specific parasites of bees. *Acarapis woodi*, the dreaded tracheal mite of honey bees, entered the USA from Mexico in 1984. The mite, however, has been recognized as a problem since 1917 in other parts of the world. *Acarapis woodi* live in the tracheal system (Fig. 11.43) where they perforate the wall of the trachea, reach into the body cavity of the honey bee, and feed on hemolymph. In high numbers, the mites also reduce the rate of air exchange in the tracheae, affecting the bee's capacity for flight. The bee may eventually die of starvation or suffocation. Under adverse environmental conditions, the bee colony may be severely weakened. In warmer climates, where brood rearing is continuous, the constant production of new bees allows the colony to survive. In more temperate climates, however, the seasonal breeding is not enough to maintain the functionality of the hive and the colony may not survive.

Water mites are among the most abundant and diverse arthropods in various aquatic habitats. The most species-rich taxa, the Hydracarina, includes about 5000 mostly freshwater species. Many water mites have a complex life cycle in which larvae are ectoparasitic on a diversity of fresh water insects and post-larval stages that are predatory. Several species of *Arrenurus* parasitize mosquito pupae while in the water; when the adult mosquito emerges, they transfer and parasitize the adults. In natural populations, female mosquitoes harbor more parasitic mites than males. The aquatic mites seem to detect and attach preferentially to female mosquito pupae, although males are also parasitized (Lanciani, 1988). Although several hypotheses may explain this preference, it seems that because the mites must return to the water to complete their life cycle, they are more likely to do so when parasitizing a female mosquito that also needs to return to the water to lay her eggs. The larva of the water mite, *Hydrachnella virella*, parasitizes the back-swimmer, *Buenoa scimitra*. It attaches to the external surface of the insect, punctures the cuticle, and feeds on hemolymph. Infections by several mites may induce host mortality, but also, the number of mites on a host is inversely correlated with mite size (Lanciani, 1984). Larvae of *Hydrachna* sp., an ectoparasite of the predaceous water scorpion *Ranatra drakei*, like many other aquatic mites are often brightly colored; their vivid red colors make them conspicuous on their host.

11.3.5.6 Astigmata

Astigmatid mites lack a respiratory system. They are common parasites of mammals and birds, causing several types of mange and scabies. Mange is a generic term used to describe the dermatitis caused by various species of mites. Scabies, or sarcoptic mange, on the other hand, is the allergic condition resulting from the burrowing of *Sarcoptes* spp. in the epidermis.

Fig. 11.44 *Sarcoptes scabiei*, the itch mite responsible for scabies or sarcoptic mange. (Photograph courtesy of Jackie Fernández.)

Sarcoptes scabiei (Fig. 11.44) is one of the best known mites and is responsible for the scabies of humans and many wild and domestic animals. *Sarcoptes scabiei* burrows in the skin of the host, particularly in moist areas between fingers, behind knees and elbows, ankles, toes, and genital regions. Transmission is only by contact. Males live on the surface of the skin where they move seeking mates. Mature females, however, burrow and tunnel under the skin, where they live for about 2 months, during which time their tunnel advances between 0.5 and 5 mm per day. After 1 or 2 months of tunneling, humans develop an allergic reaction to eggs, mites, and mite droppings present in the tunnels, which results in rashes and intense itching.

Sarcoptic mange in coyotes (also caused by *S. scabiei)*, however, becomes a chronic debilitating infection as evidenced by significant decreases in fat deposits and total body weight. Even though coyotes mount a significant humoral response, as indicated by marked increases in gamma globulin, the percentage of coyotes that recover completely from the infection is very low (Pence *et al.*, 1983). Secondary infections by fungi are common, and it is not clear to what extent the fungi contribute to the pathology of mange in wild animals.

Scabies and the mite causing it have been known for a long time. Aristotle and Cicero were familiar with both. Because of the accurate record-keeping on this parasite in recent times, it seems that scabies follows epidemic cycles among humans, ranging from 20 to 30 years between peak levels of infection. The mechanism behind this cycling phenomenon is not clear. Some correlate these peaks with periods of social unrest, such as wars. Others attribute them to the inability of physicians to recognize the disease because of lack of experience in periods of low prevalence, which results in a subsequent high level of transmission in the following years. A third explanation attributes the epidemic cycling to the development of 'herd immunity.' When herd immunity occurs, the proportion of the population that has been infected and overcomes the infection (the immune population) increases until it becomes so large that transmission decreases. Reduced transmission then results in a decline of prevalence and the disease may disappear for some time. Once a new susceptible population has grown up (a new host generation), a new wave of infection begins and prevalence increases. For further details regarding scabies, see Burgess (1994).

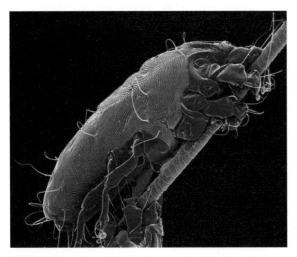

Fig. 11.45 Scanning electron micrograph of the fur mite *Koalachirus perkinsi* attached to the hair shaft of the koala *Phascolarctos cinereus*. The first two pairs of legs are modified for grasping hairs. (Micrograph courtesy of David Walter.)

Psoroptid mites, in particular species of *Psoroptes*, are responsible for a condition known as psoroptic mange or scabs. These mites do not burrow under the skin; instead, they pierce the skin at the base of hairs and suck on lymph and other exudates. The excess fluids left on the skin solidify, forming the characteristic scabs seen in infected animals. The scabs provide a protective surface for the mites, under which they reproduce and proliferate very rapidly. A related group of species in *Otodectes* attacks the ears and head of several canids and cats.

Several other site and host-specific astigmatid mites are known from a diversity of mammals. The Gastronyssidae, for example, are ectoparasites of the gastric mucosa, corneas, and nasal cavities of bats. The so-called 'fur mites' in the Listrophoridae and Chirodiscidae are permanently attached to the hair shafts of their hosts (Walter & Proctor, 1999). The fur mite, *Koalachirus perkinsi*, attaches to the koala's hair shaft via modified first and second pairs of legs (Fig. 11.45).

Birds are also hosts for a huge diversity of astigmatid mites. Well over 2000 species of 'feather mites' in approximately 30 families live in, and among, the feathers of birds. They can inhabit the surface of feathers, the skin under the feathers, and even the interior of feathers. Feather mites are among the most bizarrely shaped and spectacularly ornamented acarines (Fig. 11.46). Several mite species can be found on the same bird individual; partitioning into the many

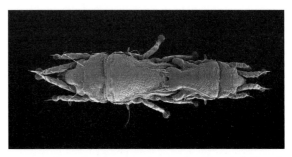

Fig. 11.46 Scanning electron micrograph of female (left) and male of the feather mite *Rhytidelasma punctata* in copula from an Australian parrot, the pale-headed rosella *Platycercus adscitus*. (Micrograph courtesy of David Walter.)

microhabitats offered by a bird's plumage may allow several species to co-occur on the same bird host (Walter & Proctor, 1999).

11.4 Hexapoda

11.4.1 General considerations

The Hexapoda includes the most common and familiar terrestrial invertebrates on earth, the insects. The Insecta are of massive ecological importance performing vital ecosystem functions, including decomposition and nutrient recycling, and as mutualists in pollination of plants, for example. The class includes about 75% of all animal species. The huge radiation of the insects is attributable to several of the arthropod adaptations already discussed. In addition to these, two other adaptations have contributed significantly to the success of the insects. The first and most obvious is the evolution of the two pairs of thoracic wings and associated neuromusculature, enabling powered flight and dispersal in most insects. The second is the evolution of a unique egg-laying structure – the **ovipositor**. This enabled insects to lay their eggs in protected and inaccessible places, including inside hosts, an adaptation that helps to further understand their phenomenal adaptive radiation.

Most insects are free-living and terrestrial, although several have evolved a parasitic life style and have numerous adaptations to life in, or on, a specific living host. The versatility of the arthropod cuticle becomes immediately apparent in these insects. Thus, many of the ectoparasitic insects have highly modified cuticular holdfast modifications for attachment to a host. The feeding adaptations and life histories encountered among the parasitic insects are also extremely diverse. Some insects are permanently associated with the host, while others are only temporary. Many species are free-living as adults and are endoparasitic only during their larval stage. Recall that many insects are vectors for a number of disease-causing organisms, including those responsible for sleeping sickness, malaria, elephantiasis, and leishmaniasis. Some ectoparasitic insects are vectors of other microbial diseases including yellow fever, typhus, plague, and typhoid fever.

11.4.2 Form and function

11.4.2.1 Morphology

The body of all insects is divided into a head, thorax, and abdomen. The head bears one pair of compound eyes, one pair of antennae, and a mouth with three pairs of appendages that aid in feeding. Between the eyes and the antenna, one or more ocelli, or simple eyes, may be present. The mouth and mouthparts include a pair of mandibles, a pair of maxillae, and a single **labium**. The mandibles are the primary feeding appendages; the maxillae and labium may also have palps that aid in food handling. The mouth is covered by an extension of the head, forming an upper lip or **labrum**. Some insects also may have a tongue-like structure, or **hypopharynx**, in the floor of the mouth. Depending on the group of insects and their feeding behavior, these mouthparts may be greatly modified or even lost. In chewing lice, the mouthparts conform to the basic pattern. In blood-sucking, micropredatory insects, however, some of the mouthparts are modified into stylets, used to penetrate the skin of the host in search of a blood vessel. In sucking lice, the stylets are retracted into the body when not being used. Although biting flies also feed on blood, they do not have stylets. Instead, the mandibles produce a wound in the skin and blood is collected by a sponge-like labium and conveyed to the mouth.

The thorax consists of three segments, each bearing a pair of legs. The second and third segments normally bear a pair of wings each. If wings are missing, their loss has been secondary. Some groups, such as the parasitic lice and fleas, lack wings completely, and others, like ants and termites, have wings only during certain periods of their life cycle. In the Diptera, which includes mosquitoes and flies, the second pair of wings is reduced to knobs, called **halteres**, which function as balancing organs during flight.

The abdomen contains most of the internal organs. Externally, it may bear a pair of terminal sensory cerci that function as receptors for air currents, and the genital appendages or genitalia. The ovipositor on female insects is frequently the most obvious external posterior abdominal structure. Several cuticular processes such as setae, scales, spines, etc., may be present on the surface of the exoskeleton.

11.4.2.2 Body wall

The body wall of insects is similar to that of arachnids. The tri-layered morphology is composed of an outer cuticle that contains chitin, polysaccharides, and proteins, an epidermis that lies beneath the cuticle and secretes it, and a thin, non-cellular layer beneath the epidermis, called the basement membrane. The cuticle can be divided into epicuticle and procuticle. The epicuticle, the thin outermost layer of the cuticle, is made of cuticulin, a protein. Cuticulin is normally covered by a wax layer that prevents water loss, and by a cement layer on top that prevents the abrasion of the wax and subsequent loss of water. Immediately under the epicuticle is the procuticle. The outer part of the procuticle is often dark and hard and is called the exocuticle; the remaining basal section is called the endocuticle.

The body wall has a number of internal processes or infoldings that strengthen the exoskeleton and serve as attachment points for the muscles.

11.4.2.3 Digestive system

The insect gut is divided into three sections, i.e., foregut, midgut, and hindgut. Food taken into the mouth continues into the foregut, which is commonly divided into an anterior pharynx, an esophagus, a crop, and a proventriculus. In insects that suck fluid, the pharynx is modified as a pumping organ. The crop is used for storage and may be absent. The function of the proventriculus depends on the type of food ingested. In insects that eat solid food, the proventriculus is a gizzard for triturating food, but in sucking insects it becomes a valve to regulate the movement of food into the midgut. Most insects have a pair of salivary glands that open into the mouth. The saliva produced by

insects has a different composition depending on their feeding habits. Anticoagulants, for example, are present in the saliva from blood-sucking insects.

The midgut, or stomach, is where enzyme production, digestion, and absorption occur. In many insects, the midgut secretes a thin membrane made of protein and chitin that surround the food as it moves along the gut. This structure, called the **peritrophic membrane**, protects the gut from abrasion by the food mass while it remains permeable to enzymes and the products of digestion. Several gastric ceca located in the anterior end of the midgut are important in food and water absorption. Following the midgut is the hindgut, divided into intestine and rectum, which are responsible for the elimination of waste and for regulating water and salts.

11.4.2.4 Osmoregulatory and excretory system

Malpighian tubules are the main structures used for excretion in insects. They lie free in the hemocoel and open into the digestive system at the junction of the midgut and hindgut. Uric acid, amino acids, and salts are actively transported into the Malpighian tubules, but water enters passively. Water, some salts, and other compounds are later reabsorbed and are transferred back to the hemolymph in the body cavity. The uric acid within the Malpighian tubules is deposited into the hindgut where it crystallizes as water is reabsorbed. The crystallization of uric acid allows most terrestrial insects to excrete solid waste with the feces, saving valuable water in the process.

11.4.2.5 Circulatory system

The circulatory system of insects is open and the body cavity functions as a hemocoel that is divided into several sinuses or compartments by perforated diaphragms. A dorsal and a ventral diaphragm divide the body into three compartments, i.e., a pericardial sinus, a perivisceral sinus, and a perineural sinus. A tubular heart that extends through most of the abdomen pumps blood from the posterior to the anterior end. The blood returns to the heart through the sinuses, aided by body movements. Blood enters the heart from the pericardial sinus through lateral openings, or ostia. The ostia are one-way valves that close when the heart contracts to force the blood anteriorly.

The blood, or hemolymph, is a clear fluid, usually colorless, but because of certain pigments it may be green, blue, or yellow. The red hemolymph of certain insects such as the gasterophilid bot flies, owes its red color to hemoglobin. Hemolymph contains a number of distinct cells, or hemocytes, that play a variety of functions in insects, such as phagocytosis, wound healing and cell-mediated immunity via encapsulation of foreign bodies (see Chapter 2). Insect hemolymph is also critical in the storage of nutrients and in the transport of hormones for coordinating many physiological functions, including feeding, reproduction, molting, and metamorphosis.

11.4.2.6 Respiratory system

Gas exchange between the tissues and the environment occurs through internal, air-filled tracheae that ramify throughout the body. The respiratory system of insects is geared to reducing evaporative water loss. The smallest branches, or tracheoles, contact all internal organs and tissues, and are especially numerous in tissues with high oxygen requirements. The tracheal system opens to the outside via spiracles. In many endoparasitic dipteran larvae, however, the spiracles are absent; in these species, the tracheae divide peripherally forming a network that covers the body surface and allows for cutaneous gas exchange within the host. Bot flies, such as *Dermatobia hominis*, extend a respiratory siphon through a boil-like lesion in the flesh to access air while within vertebrate tissues.

11.4.2.7 Reproductive system

The gonads of insects are located in the abdomen and their ducts open to the outside near the posterior end of the abdomen. The female reproductive system includes two ovaries with several ovarioles, and two oviducts that unite and continue into a vagina that opens to the outside. A seminal receptacle, or **spermatheca**, in which sperm are stored, plus paired accessory glands are connected to the vagina. The accessory glands secrete either an adhesive that cements the eggs to the substratum when they are laid or a material to form an egg capsule. In many insects, most of the eggs mature inside the female's body before they are laid.

The male reproductive system consists of a pair of testes that continue into a pair of vas deferens that fuse into a vas efferens or ejaculatory duct. The ejaculatory duct opens to the outside through a penis or **aedeagus**. A section of each vas deferens may be dilated into a seminal vesicle, where sperm is stored. Also associated with the ejaculatory duct is a pair of accessory glands that secrete nutrient-rich seminal fluid to carry the sperm, or to form a sperm-containing capsule, the spermatophore. Many insects use spermatophores to transfer sperm, which are deposited directly into the female vagina during copulation.

11.4.2.8 Nervous system and sensory input

The nervous system of insects consists of a brain located in the head above the esophagus, a subesophageal ganglion connected to the brain by two commissures, and a ventral nerve cord that runs the length of the body. Three pairs of lobes, the protocerebrum, deuterocerebrum, and tritocerebrum, which innervate the eyes, antennae, and anterior gut, respectively, form the brain. The ventral nerve cord is double and connects the ganglia of the different body segments. In some insects, these ganglia fuse and there are fewer ganglia than segments.

Sensory organs are abundant and extremely well developed in insects. A pair of compound eyes containing many **ommatidia** is typically responsible for photoreception. Sensilla are sensory organs scattered over the body designed to monitor specific types of signals. In general, they function as mechanoreceptors, proprioceptors, chemoreceptors, and auditory receptors. Mechanoreceptors are used primarily for touch, orientation in space, and vibration reception. Most proprioceptors work as stretch receptors, and provide information about the tension of a particular muscle, or the forces being applied to the exoskeleton.

Chemoreceptors are concentrated in the antennae, mouth, and around the ovipositor, and are involved in feeding behavior, mating, and habitat selection, including host selection. Auditory receptors, e.g., tympana, are especially developed in sound-producing insects.

11.4.3 Nutrient uptake

Insects have adapted to practically all types of food resources. Parasitic insects, however, feed mainly on blood, body secretions, feathers, skin, and flesh. Most chewing lice feed by chewing the feathers, hair, and skin of their hosts (Fig. 11.47). The keratin present in these structures is digested with the aid of mutualistic bacteria. Some, however, may ingest blood, if

Fig. 11.47 Feather louse from the breast feathers of the clapper rail *Rallus longirostris*. (Modified from Overstreet, 1978, with permission from the Mississippi-Alabama Sea Grant Consortium.)

available. Scratching by the host or intense chewing of the skin may form small wounds, which provide chewing lice with blood. Some endoparasitic dipteran larvae are able to absorb food through their body surface, although most of them have fully functional digestive systems and consume tissues, blood, or other body fluids.

There are two basic types of blood-feeding insects: pool feeders and vessel feeders. Pool feeders use their strong mouthparts to rip and tear through tissue, causing bleeding. The insect then just laps up the blood as it flows out of the wound. Vessel-feeding insects, on the other hand, take blood directly from small blood vessels. Their mouthparts are highly modified and function on the principle of a hypodermic syringe. The mouthparts of mosquitoes, for example, include six stylets. Two of them are used to pierce the skin and the blood vessel, and the remaining four stylets form a straw-like tube through which blood is drawn. The saliva of vessel feeders usually contains anticoagulants, anticlotting agents, anesthetics, and even painkillers, so that the insect can feed without being disturbed by the host.

Blood-sucking insects must contend with the problem of the excess liquid that is ingested with their meal and that needs to be removed from the food before it comes in contact with the digestive enzymes. In blood-sucking Hemiptera, for example, the blood meal is temporarily stored in the crop; the water is absorbed through the crop wall into the hemolymph and passes to the hindgut by way of the Malpighian tubules.

Although blood is nutritious, it does not provide all the necessary compounds needed by these parasites to survive. Strict blood-feeders cannot supplement their diet with other foods. Instead, they have mutualistic microorganisms that produce the compounds that are in short supply. These symbiotic microbes are housed in specialized organs called **mycetomes**. Parents normally transmit these symbionts to their offspring in several ways. In some species, the eggs are smeared with feces containing the symbionts and the shells are consumed by the hatching parasite. In sucking lice, the symbionts are incorporated directly into the egg shell, and in the viviparous tsetse flies, the growing larva inside the female feeds through a special gland that provides nutrients and the needed symbionts.

11.4.4 Development and general life cycle

Most insects are oviparous, i.e., a larva or nymph hatches from the egg after being laid by the female. A few insects, however, some of them parasitic, are ovoviviparous. In these species, the developing eggs are retained and hatched within the body of the female. The number of eggs produced in this case may be small, but the protection offered by the female, and the fact that the young can commence feeding immediately, offsets the reduction in egg number. The sheep bot fly, for example, is ovoviviparous and the emerging larvae are deposited directly into the nostrils of sheep, increasing colonization success. A few other insects, such as tsetse flies and louse flies, are viviparous. In viviparity, development also occurs within the female body, but the embryos are not inside an egg. The female feeds the embryos inside her body until their development is complete, at which time they are born.

Once they hatch, insects exhibit one of three patterns of development, i.e., **ametabolous, hemimetabolous,** or **holometabolous**. In the ametabolous pattern, typical of the free-living Thysanura and Collembola, the newly hatched larva is a miniature of the adult. Growth occurs by successive molts and no significant morphological changes occur.

In hemimetabolous development, the newly hatched larva resembles the adult, with well-developed appendages, compound eyes, and the rudiments of the external genitalia. Only the wings and reproductive organs are undeveloped. The larval stages of hemimetabolous insects are called nymphs and have external wing buds that develop gradually through molting, until they reach the adult form. With some exceptions, the habits of the nymphs are usually similar to those of the adult, especially feeding. Several orders of insects, including some totally or partially parasitic such as Phthiraptera and Hemiptera, have hemimetabolous development.

In holometabolous development, there are three developmental stages: a feeding larva, a non-feeding **pupa**, and the adult. The newly hatched larva has no wings, and the antenna and eyes are often rudimentary. The larva usually goes through several instars, or developmental stages, before entering a non-feeding and quiescent stage, called a pupa. The pupal stage is normally passed in a protective location during which time adult structures are developed from embryonic buds called **imaginal discs** and the organism is completely reorganized. When development is completed, metamorphosis occurs and a fully developed adult emerges from the pupa. Many common groups that include parasitic species, such as Diptera, Coleoptera, Hymenoptera, and Siphonaptera, have holometabolous development. In holometabolous insects, the larva and the adult normally occupy different ecological niches, especially regarding their feeding preferences, an adaptation to minimize intraspecific competition. In mosquitoes, for example, the larvae are aquatic and filter feed on microorganisms, whereas adult females suck blood and adult males feed on plant juices. In fleas, the larva is a scavenger with chewing mouthparts, whereas the adults feed on blood using piercing and sucking mouthparts.

Most parasitic insects require only one host to complete their life cycle. Some insects need a host during their larval stages, others only as adults, and still others need a host throughout their lives. Regardless of their host requirements, the developmental pattern remains basically the same. Hemimetabolous insects go from egg to nymph to adult, whereas holometabolous insects follow the egg, larva, pupa, and adult sequence.

11.4.5 Biodiversity and life-cycle variation

Although most insects are free-living, it comes as no surprise that many of them, about 15%, have adapted to a parasitic mode of life. The Insecta are divided into about 30 orders, a few more or a few less, depending on whose systematic scheme is followed. The Psocodea (Phthiraptera) and Siphonaptera comprise only parasitic insects; the Diptera, Strepsiptera, and Hymenoptera have a significant number of parasitic and parasitoid species; and the Hemiptera, Coleoptera, Lepidoptera, and Dermaptera, are mostly free-living, with very few parasitic species in their ranks.

In a group as diverse as the Insecta, parasitism as a life style is also diverse, with many life history strategies. Some insects are parasites throughout their lives, some only as adults, and some only during their larval stages. In those that are parasitic throughout their lives, some stay on the host at all times, whereas others visit the host only when feeding. Among those insects that are parasitic during their adult stage, only the females are parasites in some species, whereas both sexes are parasitic in other species.

The species parasitic only during their larval stages can be of two kinds, i.e., parasites in the traditional sense and parasitoids (see Chapter 1). Traditional parasites live and feed in, or on, the host and, under normal circumstances, they do not kill the host. Parasitoids, on the other hand, eventually kill their host. These specialized organisms begin as typical parasites in the sense that they are much smaller than the host and live in intimate association with it. However, they eventually grow, sometimes as much as the host, which is eventually killed. At this point, the original host–parasite association changes and becomes more like a prey–predator relationship. Because they are neither parasites in the traditional sense nor predators, they are referred to as parasitoids (see Chapter 1). We provide a brief overview of dipteran and hymenopteran parasitoids after first considering the ecto- and endoparasitic insects.

11.4.5.1 Psocodea (Phthiraptera)

Lice are tiny ectoparasites of birds and mammals, rarely exceeding a few millimeters in length. Their body is flattened dorso-ventrally; they have reduced eyes, lack wings, and have well-developed grasping legs to attach to the host's hair or feathers (Fig. 11.46). Lice are parasitic throughout their life cycles and do not survive away from the host. Development is hemimetabolous and the life cycle is completed on one individual host. The female glues the eggs to the feathers or hairs of the host and dispersal or transfer from one host to another occurs by direct contact during host mating, brooding, roosting, or lactation. Lice have traditionally been divided into chewing lice (Mallophaga) and sucking lice (Anoplura) based on the structure of their mouthparts and the resulting feeding strategy.

The chewing lice are ectoparasites of birds and mammals. Most chewing lice feed on feathers or hair, skin, and secretions from sebaceous glands. A few, however, ingest blood and serum. Most species of chewing lice are highly host specific, parasitizing only one species of host. One individual host, however, can host several species of chewing lice. When this occurs, the lice are often site-specific and their morphology correlates with the sites preferred. Chewing lice found on the head and neck of birds have round bodies, large heads, and are slow moving. If they wander into other regions of the bird's body, they are likely to be removed during preening. Species living on the back and wings, however, are flattened and elongated, and capable of moving very fast across the broad feathers of their habitat to avoid removal during preening. Chewing lice living in other body regions are normally intermediate between these forms. An interesting and highly site-specific louse is *Actornithophilus patellatus*. It lives inside the quill or shaft of the primary and secondary feathers of curlews, feeding on the pith of the shaft. It is rarely found on the outside, and all the developmental stages, from egg to adult, can be found in a single feather.

Several species of chewing lice are common in birds associated with humans. *Columbicola columbae* is the pigeon louse, *Anaticola crasicornis* and *A. anseris* are common in ducks, *Goniocotes gallinae* lives in the fluff at the base of the feathers in chickens, and *Menacanthus stramineus* is the yellow body louse of chickens and turkeys. Pelicans and cormorants are parasitized by species of *Piagetiella*. They live in the throat pouch of the birds, and remain in place by

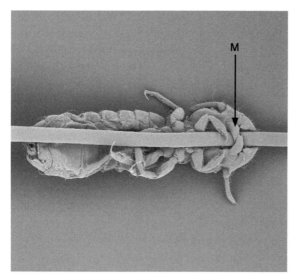

Fig. 11.48 Scanning electron micrograph of a trichodectid chewing louse *Bovicola tibialis* attached to a hair shaft from the black-tailed deer *Odocoileus hemionus columbianus*. The louse's heavily chitinized mandibles (M) completely 'embrace' the deer's hair which, in turn, is cradled in the ventral median groove of the head capsule. (Micrograph courtesy of Brad van Paridon.)

attaching to the mucous membrane with their mandibles, where they feed on blood and mucus. They only leave the pouch temporarily to lay eggs in the feathers of the same bird. The adults and nymphs are very mobile and, during the breeding season, they move rapidly from the adult birds to the newly hatched chicks (Samuel *et al.*, 1982).

Species of *Bovicola* are common trichodectid chewing lice on horses, cattle, goats, sheep, deer, and other ruminants. *Bovicola tibialis*, for example, normally infests European fallow deer, but in North America, it infests Columbian black-tailed deer (Fig. 11.48), often reaching very high intensities. This louse is an invasive species that may have been introduced to black-tailed deer in western North America when fallow deer were introduced (Bildfell *et al.*, 2004; Mertins *et al.*, 2011). Parasitized animals often try to eliminate these lice by biting the skin and rubbing themselves against

hard substrates. Such behavior may lead to a condition in cervids known as hair loss syndrome, resulting in poor body condition and possible mortality.

The sucking lice are exclusively ectoparasites of mammals. Their mouthparts are specialized for piercing the skin and sucking blood. Host and site specificity are also relatively high among sucking lice. Lice in the Enderleinellidae, for example, are parasites of squirrels, the Hematopinidae parasitize ungulates, and the Echithiriidae are found in the fur of seals and walruses. The latter are well adapted to the diving habits of their hosts; when in the water, the lice remain in the underfur and use the air trapped there as an oxygen source.

Humans are also host to three different types of sucking lice, i.e., the body louse, the head louse, and the crab louse. Body and head lice are closely related and usually are considered to be the same species, but different subspecies or varieties. Most people know the adults by some of their common names, e.g., cooties, graybacks, and 'mechanized dandruff,' whereas the eggs that are laid and cemented to hair are commonly called nits. The head louse, *Pediculus humanus capitis*, is found mainly in the hair behind the ears and back of the neck. Transmission from one host to another occurs by physical contact and by stray hairs carrying lice or their eggs.

The body louse, *P. humanus humanus*, is rather unusual in that it spends most of its time, and even lays its eggs, in the clothing of the host, visiting the host's body only to feed. Despite this habit, the lice must stay in close contact with the body of the host and are extremely sensitive to changes in temperature. If, during their nymphal stages, clothing is removed, development slows and the lice may even die if the host does not wear the contaminated clothes for a few days. Although the species status of these two human lice is still under debate, it seems likely that the body louse is a relatively new form that evolved from a head louse ancestor when humans began wearing clothes (see also Chapter 14).

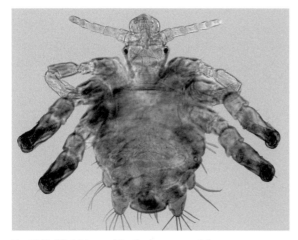

Fig. 11.49 *Phthirius pubis*, the human crab louse. (Photograph courtesy of Jackie Fernández.)

The third species found in humans, *Phthirius pubis*, the crab louse (Fig. 11.49), is generally found in the pubic region. They normally insert their mouthparts in the skin and, unlike the body and head lice, they remain in the same position for some time, causing intense itching. Transmission from host to host is by direct contact with the affected area. For a comprehensive review of human lice and their management, see Burgess (1995).

Lice are vectors for many prokaryotic pathogens. In humans, the body louse transmits at least three important diseases, i.e., epidemic typhus and trench fever caused by a rickettsial organism, and relapsing fever caused by a spirochaete. Epidemics of these diseases correlate with increases in lice populations, which, in turn, are favored when living quarters are crowded, sanitation is at a minimum, and people go for long periods without changing clothes. These conditions are commonly seen in times of social unrest.

11.4.2.2 Hemiptera

Most Hemiptera, also known as true bugs, are free-living, hemimetabolous insects. All have mouthpart modifications for piercing plant or animal tissues and sucking fluids. In the blood-sucking hemipterans, these mouthparts form a needle-like stylet through which blood can be ingested and saliva injected. Three families, the Cimicidae, or bedbugs, the Triatomidae, or kissing bugs, and the Polyctenidae are of importance from public health or parasitological perspectives. The former two families can be considered as temporary ectoparasites or as micropredators (see Chapter 1) since they feed on host blood, but leave the host after each feeding. The polyctenids, however, are totally dependent on their bat hosts and die within 6–12 hours after removal from the host.

Bedbugs are wingless, temporary parasites of birds and mammals, mainly bats. Most bedbugs are host specific. *Cimex lectularis*, *C. hemipterus*, and *Leptocimex boueti* are found in human habitations; bedbug outbreaks can occur and they can become an irritating nuisance.

The kissing bugs (Reduviidae) feed from a diversity of mammals, including humans. Most kissing bugs are not host specific, and feed on multiple-host species, mainly at night. In the case of humans, they preferentially bite the face around the mouth; this site preference gives them their common name, kissing bugs. Several species are vectors of *Trypanosoma cruzi*, the protist that causes Chagas' disease (see Chapter 3). Although all kissing bugs are suitable hosts for *T. cruzi*, their susceptibility varies. *Triatoma infestans*, *T. dimidiata*, *Rhodnius prolixus*, and *Panstrongylus megistus* are the most common and important vectors of *Trypanosoma cruzi* in the Americas.

The Polyctenidae are found exclusively in the micropteran bats of the tropics and subtropics. They are very well adapted to a totally parasitic life style. They lack eyes, the wings are reduced to a pair of large lobes, and the body has a number of holdfast combs similar to those present in fleas. Most of the bat hemipterans form small assemblages in tree holes and caves, which facilitates the transfer of these ectoparasites between bats (Marshall, 1982).

11.4.2.3 Siphonaptera

The Siphonaptera, or fleas, are blood-sucking ectoparasites as adults. Fleas have laterally flattened

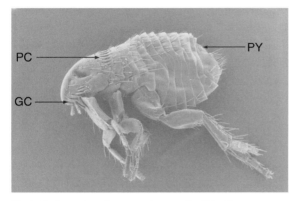

Fig. 11.50 Scanning electron micrograph of the flea *Ctenocephalides canis* from a dog, *Canis familiaris*, illustrating the genal ctenidia (GT), pronotal ctenidia (PT) and posterior pygidium (PY). Note the powerful femora of the hind legs. (Micrograph courtesy of Brad van Paridon.)

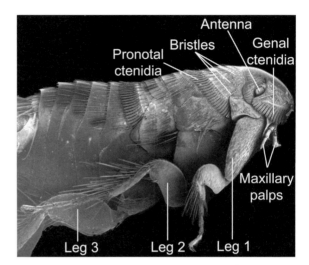

Fig. 11.51 Scanning electron micrograph of the flea *Stenoponia americana* (male) from the white-footed mouse *Peromyscus leucopus*, showing details of the external morphology. (Micrograph courtesy of David Salmon, Ralph Eckerlin, & Sherman Hendrix.)

bodies, have piercing-sucking mouthparts, and lack wings, an adaptation to facilitate their movement between hairs on their hosts. Most fleas have characteristic bristles or spines directed backward, called **ctenidia**, for snagging hairs when the host attempts to remove them (Figs. 11.50, 11.51). Development is holometabolous. Fleas lay their eggs in the nest of the host, or on the host, but because the eggs are not cemented to the host as in lice, they normally drop off and end up on the ground or in the host's nest. The larvae are maggot-like and feed on organic debris and flea feces accumulated in the host's nest. The pupae can remain dormant for a long period, until a host appears. The hind legs of most fleas are modified for jumping, and some can leap more than 100 times their body length (Rothschild *et al.*, 1973). The jumping ability of most fleas is correlated with the type of host they parasitize. In general, fleas that parasitize nesting hosts are poor jumpers because the nest provides a restricted and safe place for transmission. Fleas that parasitize non-nesting hosts, however, have large hind femora and are superb jumpers, being able to reach a host from the ground. Vibrations, heat, and high concentrations of carbon dioxide trigger the jumping or host-searching behavior of fleas. A unique sense organ, the **pygidium** (Fig. 11.50), is found on the abdomen of fleas. This structure functions in

detection of air currents and vibrations, thus aiding in host detection.

Most adult fleas parasitize mammals, but birds are infected too. Some fleas are not very host specific, but they do have preferred hosts. For example, despite their common names, *Ctenocephalides canis* (Fig. 11.50) and *C. felis*, the dog and the cat flea, respectively, are widespread and parasitize cats, dogs, humans, and other mammals. Similarly, the human flea, *Pulex irritans*, has been found on a number of other hosts, including pigs, dogs, coyotes, and squirrels.

Although most fleas can feed on a variety of hosts, some species require specific hosts to produce viable eggs. In other cases, such as the European rabbit flea, *Spilopsyllus cuniculi*, and other related species parasitizing rabbits and hares, the host's hormones control reproduction in the flea (Rothschild & Ford, 1972). The European rabbit flea cannot mature sexually until the flea feeds on a pregnant rabbit. When the rabbits are born, the now sexually mature fleas move from the mother to the newborn where they feed, mate, and lay eggs. Then, they return to the mother rabbit, leaving the newborn rabbits with a bumper crop of flea eggs. Hormonal changes in the mother rabbit during pregnancy trigger the sexual development of the fleas, and growth hormones in the newborn rabbits stimulate mating and egg laying.

Fleas are also vectors for a number of infectious diseases. A number of species, including the human flea *Pulex irritans*, but in particular the oriental rat flea *Xenopsylla cheopis*, are the main vectors for plague and murine typhus. Plague, also known as black death, is caused by the bacterium *Yersinia pestis* and can affect humans, rats, mice, rabbits, and dogs. Epidemics of plague have had a profound effect on the history of humanity. The Black Death that swept Europe during the fourteenth century killed 25 million people. Similar epidemics in the sixth century and the Great Plague of London in the seventeenth century also caused a heavy toll in human lives. Sporadic epidemics continued well into the twentieth century. About 7000 cases per year are reported, but large-scale outbreaks are now prevented by insecticides (to limit flea proliferation) and antibiotics (for prompt treatment of infected individuals). Murine typhus, caused by *Rickettsia mooseri*, is a much less fatal disease affecting humans and other animals. A third disease transmitted by fleas is myxomatosis, a viral disease of rabbits transmitted by several blood-sucking arthropods, including the European rabbit flea, *Spilopsyllus cuniculi*. The flea and the myxomatosis virus have been introduced in Australia to control the rabbit population (see Chapter 16).

Several flea species associated with humans and domestic animals can be intermediate hosts for cestodes such as *Vampirolepis nana*, *Hymenolepis diminuta*, and *Dipylidium caninum*. Transmission of these tapeworms occurs during grooming behavior; the host ingests infected fleas when attempting to remove the irritating pests with their teeth.

11.4.2.4 Diptera

The Diptera comprises an extremely diverse assemblage of insects with several parasitic and many parasitoid species. The Diptera exhibit holometabolous development; the larva and the adult usually occupy radically different ecological niches. Some are parasitic only during their larval stages, others only as adults, and some others throughout their lives. Furthermore, in many, only the females are parasitic, living permanently on the host, or just temporarily while feeding. As we have learned in previous chapters, many blood-feeding dipterans are of huge parasitological significance as vectors of many parasites, including filariid nematodes and protists. Ceratopogonids, or biting midges, and the culcids, or mosquitoes, are particularly noteworthy, as are the Glossinidae or tsetse flies. Herein, we include only those dipterans that can be considered as true ectoparasites, or endoparasites, or as endoparasitoids.

Hippoboscid flies are blood-sucking ectoparasites of birds and mammals; only a few species are host specific. In some cases, certain ectoparasitic lice can reach their specific hosts by attaching via their mandible to flying hippoboscid flies, a classic case of phoresy. Keds and louse flies are the most common hippoboscids. Typically, louse flies lose their wings when they locate a host (Fig. 11.52B). The sheep ked, *Melanophagus ovinus*, spends its entire life associated with the host and, consequently, has lost its wings. The deer ked, *Lipoptena cervi*, is a common louse fly of various deer species as well as elk, horses, and cattle. This parasite has been implicated in causing a condition known as deer ked dermatitis if humans are bitten by *L. cervi*, and as a vector in the transmission of a pathogenic bacterium, *Bartonella schoenbuchensis*, within ruminant populations (Dehio *et al.*, 2004). *Lipoptena depressa* is the common louse fly of black-tailed and mule deer of western North America (Fig. 11.52).

The Streblidae and Nycteribiidae are highly specialized, blood-sucking flies that feed exclusively on bats. These obligate ectoparasites live in the fur or on the wing membranes of diverse bat hosts. Both families are worldwide in distribution and are much more diverse in the tropics. Many bat flies are strikingly host specific and are restricted to a single bat species (Dick & Patterson, 2006). Several bat fly species can coexist on a single host, typically partitioning the bat's fur and wing membranes (Patterson *et al.*, 2008). They are morphologically diverse and show varying degrees of lateral versus dorsoventral body compression, as

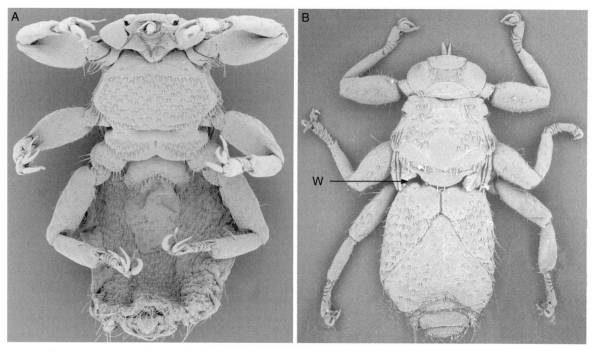

Fig. 11.52 Scanning electron micrographs of the hippoboscid louse fly *Lipoptena depressa* from the black-tailed deer *Odocoileus hemionus columbianus*. (A) Ventral view showing the three pairs of prominent thoracic appendages, each equipped with powerful holdfast claws. (B) Dorsal view showing the tiny wings; the wings break off when a host is colonized. (Micrographs courtesy of Brad van Paridon.)

well as variation in wing and eye reduction. Streblids are more speciose in the western hemisphere. Streblids, in particular, have evolved radically different morphologies. Some have a flea-like ctenidium along the posteroventral margin of the head. Several are extremely long-legged and spider-like in appearance and all possess wings, although the wings may be reduced and dispersal via flight is limited (Fig. 11.53). Strongly prehensile tarsal claws (Fig. 11.54) enable streblids to firmly grasp their bat host's fur or wing membranes. Nycteribiids are more common in the eastern hemisphere and comprise the most modified of the bat flies. They have holdfast ctenidia, are completely wingless, and have reduced eyes. The thorax is reduced as well; the small head and the insertion of the legs are displaced into a dorsal position, also giving these flies a spider-like appearance. Many bat flies in both families are extremely mobile on their hosts, an adaptation that enables them to avoid host grooming.

Fig. 11.53 Scanning electron micrograph of the streblid bat fly *Megistopoda aranea* from the Neotropical fruit bat *Artibeus* sp., illustrating the elongated, spider-like legs, as well as the reduced wings. (Micrograph courtesy of Katharina Dittmar.)

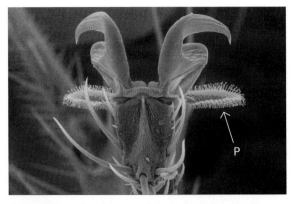

Fig. 11.54 Scanning electron micrograph of the paired protarsal claws and pulvilli (P) of the tarsi of the streblid bat fly *Stylidia* sp. from the roundleaf bat *Hipposideros* sp. (Micrograph courtesy of Katharina Dittmar.)

Their diverse morphological adaptations and the ecological diversity of bats make bat flies promising models for research in deciphering host–parasite phylogenies, biogeographical patterns, and coevolution (Dick & Patterson, 2006; Dittmar *et al.*, 2006).

Bat flies and louse flies demonstrate a type of curious form of viviparous development termed adenotrophic viviparity. Larvae of these parasites feed from accessory gland secretions within the mother's reproductive system. The full-grown larva is deposited and pupates immediately. In the case of bat flies, larvae molt twice inside the female and gravid females leave the bat host and deposit a single third-instar larva (prepupa). Once deposited, the larva immediately pupates. Adult bat flies emerge from the pupiparum and must rapidly locate and colonize a host (Dick & Patterson, 2006). In some cases, pupal deposition occurs some distance away from roosting cave-inhabiting bats, such as in the cave passageways (Dittmar *et al.*, 2009). Such spatial segregation of pupa and adults suggests that the host-seeking adaptations, e.g., mechanisms of sensory perception, of bat flies are extraordinarily complex.

In a number of dipteran families, the adult flies are free-living, whereas the larva is the parasitic stage. The infection of vertebrate hosts by fly larvae, or maggots, is referred to as **myiasis**. Parasitic larvae can feed on host tissue, body fluids, secretions, or food ingested by the host. The three most important families with larval forms endoparasitic in vertebrates are the Calliphoridae, Sarcophagidae, and Oestridae. A fascinating and detailed account of the systematics, ecology, and epidemiology of the oestrid flies is provided in Colwell *et al.* (2006).

Adult calliphorids, known as blow flies, are about the size of a house fly and frequently are metallic green or blue. The screwworm fly, *Cochliomyia hominivorax*, is a common species that lays eggs in the wounds of mammals. After hatching, the screwworm larva feeds on living tissue and, when fully developed, drops to the ground and pupates.

The Oestridae include the bot flies as well as the terrifying warble flies. The adult flies are robust, 'hairy,' and resemble bees. Unlike the screwworm flies, the larvae of oestrids do not require a pre-existing wound to lay eggs and gain entrance to the host. Skin bot flies, such as *Cuterebra* spp., are covered in peg-like spines (Fig. 11.55); these lay their eggs on or near natural orifices of the mammalian host. After hatching, the larva enters the body, tunnels under the skin, and lives in a cyst that communicates to the exterior through a small hole. The large, fully mature larva then leaves the cyst and pupates on the ground. Head maggots such as *Oestrus ovis*, the sheep bot fly, and *Cephenemyia* sp., a bot fly of deer, deposit their larvae in the nostrils. The larvae then enter into the sinuses and nasal passages where they attach to the mucosa and feed. If present in large numbers, the nasal bot flies can cause considerable damage and pain; they may also be agents of mortality.

Warble flies are primarily parasites of bovids and cervids. The presence of adult flies terrifies cattle, which initiate intense evasive maneuvers attempting to avoid the flies. The female lays eggs in the hairs of the host and, on hatching, the larvae penetrate the skin. What terrifies the host is probably the extensive subcutaneous migration that lasts several months and that takes the larvae first to the front end of the host and finally to the dorsal area, where they open a small hole in the skin.

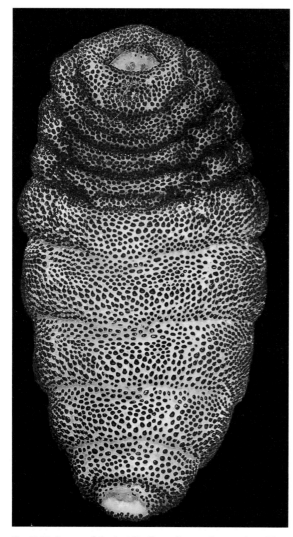

Fig. 11.55 Larva of the bot fly *Cuterebra* sp. from a dog. The entire cuticle is covered in small peg-like spines. (Photograph courtesy of Bill Pennell.)

Fig. 11.56 Larva of the bot fly *Gasterophilis intestinalis* from the stomach of a horse. The cuticle has many rows of robust holdfast spines. Note also the sharp mouthparts for feeding and attachment. (Photograph courtesy of Bill Pennell.)

Several species of bot flies parasitize the stomach of horses, elephants, and rhinoceroses, causing gastric myiasis. Each stomach bot fly has a different infection strategy and occupies different sites in the alimentary canal of the host. Species of *Gasterophilus* deposit their eggs on grass or on the skin of horses from where they are ingested when eating or licking themselves, respectively. The larva hatches in the mouth, penetrates the mucosa, and migrates through the host's body to its preferred site; depending on the species, this may be the stomach, intestine, or rectum. *Gasterophilis intestinalis*, the stomach bot fly of horses, attaches to the stomach with large mouth hooks and powerful cuticular spines (Fig. 11.56). Eventually, the larva detaches, passes out with the feces, and pupates in the soil.

Not all myiasis-causing flies deposit their eggs directly on the host (or on the grass the host eats). The human skin bot fly, *Dermatobia hominis*, (Fig. 11.57), is a common species found in the rainforests of Central and South America. The life cycle is unique in that a vector is involved in transmission. An egg-laden female bot fly captures a mosquito or other blood-feeding arthropod and glues fertile eggs onto its abdomen and then releases it. When the mosquito next takes a blood meal, the body heat of the mammalian host triggers the hatching of the bot fly egg and the microscopic larva immediately burrows into the host. The 'bot' creates its own burrow in the host's skin and develops and grows substantially through a series of molts for 6 to 8 weeks. At the posterior end of the maggot is a snorkel-like siphon with a respiratory spiracle, which it pokes through the skin. After their 2-month period of development, the larva exits the host, pupates in the soil, and metamorphoses into adults to reproduce and begin the life cycle again.

Typically, human bots are difficult to squeeze out alive because they have exquisite morphological adaptations, including a series of rows of strong, backwards-pointing spines on their cuticle (Fig. 11.57). These function as holdfasts. They hold the maggot tenaciously in place in their burrows and prevent the host from manually removing them so that they can continue their larval development. Often, a tried and true technique for bot fly removal can be used without the need for surgical intervention. Nail polish can be liberally applied to the bot fly wound and then it can be sealed over tightly with duct tape in order to poison and suffocate the larva. After 24 hours of this nail polish/duct tape treatment the maggot is dead and can be very gently, but firmly, squeezed out, intact (Fig. 11.58).

Fig. 11.57 Larva of the human skin bot fly *Dermatobia hominis* removed from human flesh, showing the five rows of backwards-pointing cuticular spines and the pair of sharp, curved mouthparts. (Photograph courtesy of Bill Pennell.)

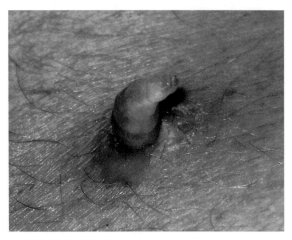

Fig. 11.58 A larva of the bot fly *Dermatobia hominis* has been gently squeezed out of human skin. Note the posterior respiratory siphon and a row of holdfast cuticular spines.

The human skin bot can also develop in the skin of many other mammals. For an in-depth discussion of myiasis in humans and domestic animals, see Hall & Wall (1995).

Some dipterans are agents of so-called lethal myiasis; these cause such extensive tissue and organ destruction that they can be agents of mortality of various vertebrates in nature (Box 11.4).

Box 11.4 | Fly maggots as agents of lethal parasitism

The infection of live vertebrate hosts by dipteran larvae is referred to as myiasis. There is no doubt that the human bot fly, *Dermatobia hominis*, and many of those other species of dipterans infecting livestock and wild vertebrates causes discomfort as the maggots feed and grow within the host's tissues. However, in some species, the myiasis-causing flies can induce destructive tissue damage, causing so-called lethal myiasis. Several species cause significant mortality and may impact the population ecology of vertebrates in nature. Such lethal parasitism is known from a diversity of vertebrates, ranging from frogs, to rodents, to birds. *Notochaeta* spp. infect various leptodactylid and dendrobatid frogs in Central and South America. Crump & Pounds (1985) discovered an extraordinary case of lethal myiasis in harlequin frogs in Costa Rica caused by a sarcophagid fly, *Notochaeta bufonivora*. The fly maggots consume the frog's musculature first and then the internal organs. Consequently, they were a source of density-dependent mortality, especially in female harlequin frogs. For this reason, the parasite was implicated as one explanation for the mating system of these frogs and the observed highly skewed sex ratio towards males. Other species of flies, such as *Bufolucilia* spp., cause lethal myiasis in North American frogs and toads. Larvae of the sarcophagid fly *Neobellieria citellivora* cause massive tissue damage

(Fig. 11.59) and fatal cutaneous myiasis in Richardson's ground squirrels; in particular, the parasite is a major source of mortality for juveniles during the summer months (Michener, 1993).

Philornis downsi is perhaps the most notorious of the flies causing lethal myiasis in birds. This dipteran was introduced to the Galapagos archipelago and has emerged as a threat to Darwin's finches. This parasite kills finch nestlings and has had significant impact on the reproductive success of Darwin's finch populations on several islands in the Galapagos (Fessl & Tebbich, 2002; Fessl *et al.*, 2006).

Fig. 11.59 Lethal myiasis of juvenile Richardson's ground squirrel, caused by larvae of the sarcophagid fly *Neobellieria citellivora*. Inset is a close-up of the larvae within the wound. (Photograph courtesy of Gail Michener.)

Box 11.4   (continued)

A particularly intriguing natural history lesson involving lethal myiasis stresses the complexity of ecological interactions in the Neotropics. This extraordinary story, which was worked out by a renowned tropical biologist Neil Smith, revolves around four key players: a species of dipteran fly (*Philornis* sp.), the giant cowbird, birds known as oropendolas, and venomous and aggressive social bees and wasps (Smith, 1968). *Philornis* is a devastating parasite and can be a significant source of mortality among oropendola nestlings. Oropendolas can avoid bot fly parasitism, as well as mammalian and snake predation of their young, by building their pendulous nests in trees in proximity to the aggressive wasp or bee colonies. Any disturbance by bot flies or predators triggers attack by the vicious wasps, and the nestlings are protected until they fledge. A remarkable feature of this system is that the oropendolas that nest in waspless trees and are thus subject to bot fly attack are protected if a giant cowbird **brood parasite** nestling is present. In other words, the host oropendola bird benefits from being parasitized! Avian brood parasites are those birds that lay their eggs in other bird species' nests; the foster host parent birds raise the brood parasite offspring as if they were their own. Usually the brood parasite is truly parasitic, reducing the fitness of the host bird. However, in this case, paradoxically, when a cowbird nestling is present in the nest more oropendola nestlings fledge. This is because the cowbird nestling preens and eats the bot fly, or its eggs and/or larvae, on its host nest mates before they invade the host and cause irreversible tissue damage. Thus, in waspless trees when the risk of bot fly parasitism is high, oropendola behavior is altered and they do not chase cowbirds away. Moreover, they accept the cowbird brood parasite's eggs even though the eggs do not resemble their own. In effect, the brood parasite acts as a mutualist in some ecological circumstances because of its impact on controlling another parasite (Smith, 1968). The giant cowbird provides its avian hosts with parasitic fly protection in exchange for foster care!

Another diverse and ecologically important family of Diptera is the Tachinidae. These are endoparasitoids, typically occurring singly in caterpillars. Female flies oviposit on specific hosts and the larva feeds internally, developing to a relatively large size prior to exiting the host and pupating in soil. Many tachinids are important regulators of terrestrial insect populations in nature and have been used in the biological control of specific lepidopteran pests. Stireman *et al.* (2006) review the biodiversity, ecology, behavior, and evolution of tachinid parasitoids.

11.4.2.5 Strepsiptera

Strepsipterans are minute endoparasites with extreme sexual dimorphism and remarkable life cycles. They are parasites of a diverse range of other insects, especially the Orthoptera, Hemiptera, and Hymenoptera. The males are free-living and winged, whereas the females are wingless and never leave the host. The parasitic females are highly modified; they usually lack legs, antennae, and eyes, and the head and thorax are fused, resulting in a bizarre, vermiform body. Females inside the host produce large numbers of larvae called triungulins, which escape from her body and the body of the host. In strepsipterans that parasitize eusocial hymenopterans, triungulins leave their hosts while on flowers, and seek out and penetrate the specific wasp or bee host. In this way, they can 'hitch a ride' to the nest where they can then enter a host egg or larva (Gullan & Cranston, 2010).

The triungulin larvae have well-developed eyes and thoracic legs and actively seek and enter the body of the specific insect host, where they feed and develop into pupae. When development is completed, the adult males emerge from the host, whereas the females remain inside, with only the anterior part of their body protruding through the abdominal segments of the host. The virgin female releases pheromones to attract recently emerged males and copulation occurs. Males usually cause extensive damage to the insect host when emerging and, although the host is not typically killed, it may be castrated, show unusual coloration patterns, or other morphogical and physiological abnormalities.

11.4.2.6 Coleoptera

Despite the phenomenal adaptive radiation of the beetles, there are surprisingly few parasitic representatives. The most highly modified ectoparasites are a group of small beetles in the Leodidae, specifically those in the Platypsyllinae. Three genera include *Leptinus*, ectoparasitic on small rodents, moles and shrews; *Leptinillus* on rodents including beavers; and *Platypsyllus*, a parasite of beavers. All genera are wingless and are eyeless or have reduced eyes, and are markedly dorsoventrally flattened. The beaver beetle, *Platypsyllus castoris* (Fig. 11.60), is an ectoparasite of American and European beavers across their geographical distributions. Holdfast adaptations include stout setae and flea-like ctenidia projecting from the head (Fig. 11.60A, B). The beetle is blind and is a permanent ectoparasite of beavers as a larva and adult, feeding on epidermal tissue and the fatty skin secretions produced by the host. The pupal stage is the only life stage not spent on the host. It is passed in a pupal

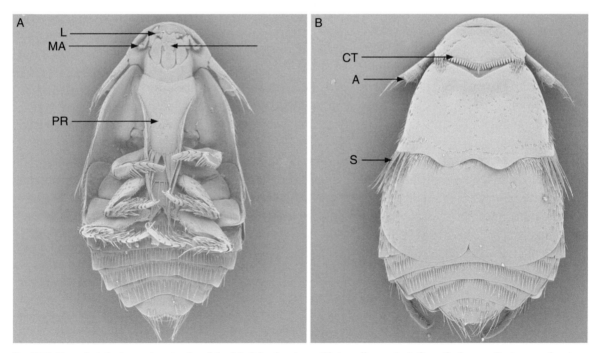

Fig. 11.60 Scanning electron micrographs of the leiodid coleopteran *Platypsyllus castoris* from the beaver *Castor canadensis*. (A) Ventral view showing the mouthparts, including the labium (L) maxilla (M) and mentum (ME), and the large pronotal sternum or prosternum (PR). (B) Dorsal view showing the wingless condition and the holdfast ctenidia (CT), antennae (A), and stout setae (S). (Micrographs courtesy of Brad van Paridon.)

chamber formed in substrate located within the beaver's lodge. This species is one of the few holometabolous insects that are parasitic on the same host both as larva and adult. The extreme modifications to parasitism in *Platypsyllus* suggest that it had mammal hosts more ancient than beavers, possibly a burrow-inhabiting small rodent or shrew (Peck, 2006).

11.4.2.7 Hymenoptera

The Hymenoptera comprises one of the largest and ecologically most important orders of insects and includes the familiar bees, ants, and wasps. Among the wasps, a huge adaptive radiation occurred and, in addition to the many solitary and eusocial species, there are thousands of species of parasitoids. Most belong to the extraordinarily species-rich families Ichneumonidae, Braconidae, and Chalcidae, which parasitize the eggs or larvae of diverse terrestrial insects, notably the Lepidoptera, Coleoptera, Diptera, and Hymenoptera, as well as spiders. Many species are hyperparasitoids, parasitizing the parasitoid of the insect host. Thus, a significant component of our planet's biodiversity consists of the herbivorous insects and the parasitoids that infect them. Parasitoids are important regulators of insect populations in nature; several are used in the biological control of economically devastating insect pests.

Hymenopteran parasitoid females have highly specialized ovipositors that are used both to lay eggs and to inject venom to paralyze the host. The venom from the sting temporarily (and sometimes permanently) immobilizes the host victim, making it easier for the female parasitoid to lay her eggs, either on the surface of the host or in the hemocoel. Parasitoids that attack insects living in wood, or in galls, or underground, or in other concealed places often have extremely long ovipositors (Fig. 11.61). Cutting ridges on the ovipositor allow many parasitoid wasps to penetrate through plant tissues, to locate and lay eggs in the hidden insect larvae within.

Typically, the adult female locates a host and lays her eggs directly on, or in, it. Some parasitoid larvae develop within the body of their hosts, feeding from the inside,

Fig. 11.61 The ichneumonid wasp pupal ectoparasitoid *Diapetimorpha introita* is preparing to lay an egg in the pupal tunnel of a corn earworm pupa *Heliothis zea*. Note the long ovipositor by which the female wasp lays her eggs. (Photograph courtesy of the Agricultural Research Service, US Department of Agriculture.)

and are called endoparasitoids. They may feed on hemolymph and/or host tissues within the hemocoel, eventually emerging from the host, spinning a silken cocoon, and pupating. Invariably, the host is killed soon after emergence. Ectoparasitoids on the other hand, live externally, but their mouthparts are buried in the body of the host (Color plate Fig. 8.4). Parasitoids can be termed gregarious or solitary. Gregarious species inject multiple eggs during oviposition, while solitary species inject a single egg per host.

Some parasitoids are highly specific, attacking one or very few species, while others are generalists and infect a variety of hosts. Host specificity is higher among endoparasitoids, where a most intimate host–parasitoid relationship exists. Parasitoids are renowned for their sophisticated interactions with their host's immune, endocrine, and nervous systems. Host molting and metamorphosis are frequently impacted prior to death. In addition, it turns out that female ichneumonid and braconid endoparasitoids also introduce unique, mutualistic **polydnaviruses** (PDVs) along with the eggs and venom. These PDVs are potent regulators of the insect immune response, inducing a profound immunosuppression (Beckage, 1998). PDVs interfere with the structure and function of the host's hemocytes, preventing encapsulation, and allowing the compatible parasitoid's larvae to develop (Lavine & Beckage, 1995; see Chapter 2).

Often, parasitoids modify the normal development and behavior of their hosts in dramatic ways. For this reason, parasitoids have featured prominently in the parasite-induced behavioral manipulation literature. Perhaps the most famous of the parasitoid-induced behavioral manipulations involve the induction of bizarre web-building behaviors in infected neotropical spiders (Eberhard, 2000; 2010; Box 11.5). Another spectacular example occurs when a braconid parasitoid (*Glyptapanteles* sp.) induces its geometrid caterpillar host to behave as a bodyguard of its pupae. After the parasitoid's larvae exit the host to pupate nearby, the host remains alive, stops feeding, and stays close to the pupae. Remarkably, the infected caterpillar responds to disturbance from predators with violent head-swings (Grosman *et al.*, 2008). The caterpillar dies soon after the adult parasitoids emerge from the pupae. Moreover, field experiments demonstrated that the presence of the bodyguard hosts resulted in a two-fold reduction in mortality of this parasitoid's pupae. This confirms that the modified behavior is beneficial to the parasitoid under certain ecological situations (Janssen *et al.*, 2009).

Box 11.5 | **Parasite's web of death! Manipulation of spider web-building behavior by an insect parasitoid**

Insect parasitoids alter the ecology, physiology, and/or behavior of their hosts in a diversity of ways, spectacularly so, in some cases. A particularly fascinating example comes from the Costa Rican rainforest (Eberhard, 2000; 2001). The story revolves around an ichneumonid wasp parasitoid, *Hymenoepimecis argyraphaga* and its host, an orb-weaving spider, *Plesiometa argyra* (Color plate Fig. 8.4). William Eberhard is a renowned evolutionary biologist and neotropical spider expert at the Smithsonian Tropical Research Institute and the University of Costa Rica. While examining spider webs in the field he came across some very peculiar ones. In addition to the normal, symmetrical prey-capture orb webs, he found smaller webs attached with unusually sturdy silken cables. Attached to these strong cables and lying in the center of the unusual web, Eberhard witnessed these spiders building a bright orange cocoon. It turned out that *H. argyraphaga* was using the spiders as a 'nursery' for its larvae by manipulating its normal web-building behavior. In other words, the wasp forces its spider host to weave a special web for its own ends (Eberhard, 2000).

The details are fascinating. A female wasp attacks the spider at the hub of its orb. It stings the spider to temporarily paralyze it and then lays an egg on the spider's abdomen. The spider then resumes normal behavior for the next week or two, constructing normal orb webs to capture insect prey. During this time, the wasp's egg hatches and the attached larva punctures the

Box 11.5 | (continued)

abdomen and drinks hemolymph, growing to a large size (Color plate Fig. 8.4). On the evening that it kills its spider host, the larval wasp induces the spider to spin the bizarre, unusually strong web. Once this so-called cocoon web is built, the wasp larva then molts, kills, and sucks the spider dry over the next 12 hours. The larva sits in the hub of this special web until the next evening when it spins its own orange-colored pupal cocoon that hangs from the cocoon web by a silk-reinforced line. Within the cocoon, the wasp pupates and then undergoes metamorphosis into an adult. Adult wasps then leave the pupal cocoon to search for a mate, eventually returning to search for new spider victims and beginning the life cycle again.

Eberhard determined that the cocoon webs were highly modified orbs, with greatly strengthened radial lines that attach to a substrate; the sticky spiral lines of the normal prey-capture orb were also absent. Furthermore, several aspects of normal web-building routines such as reeling up and replacing lines were absent. Essentially, the spider is induced to leave out some critical steps and repeat others to build the cocoon web. Why manipulate spider behavior to produce this strange cocoon web? Eberhard believes that by doing so, the web acts as a strong and durable support or platform for the wasp's cocoon. This is especially required during torrential rains when the pupae are particularly vulnerable to dislodgement.

It appears that this complex spider–parasitoid interaction is not unique. Several other poly-sphinctine wasps have now been shown to induce their spider hosts to build resistant, protective structures that shelter the larval cocoon from predators and/or the elements. For example, *Polysphincta gutfreundi* also induces its orb-weaving spider, *Allocyclosa bifurca*, to build a highly modified, stable web, on which the wasp's pupal cocoon is found (Eberhard, 2010). In this case, the modified web design may function in camouflaging the wasp's cocoon. Eberhard found that the behavioral effects induced by this wasp were different than those of *H. argyraphaga*; the differences represent adaptive adjustments to differences between the natural histories of their hosts. In other words, different species of polysphinctine wasps induce behavioral effects fine tuned to details of each spider host's natural history (Eberhard, 2010).

What is the mechanism for such an exquisitely timed behavioral change? Eberhard believes it to be chemically mediated as the parasitoid-induced effects are rapid and long-lasting. Furthermore, the modifications in the spider's web are reversed when the wasp's larva was experimentally removed. It is hypothesized that the wasp larva injects a psychotropic chemical (or several chemicals) that manipulates the spider's nervous system, causing it to dramatically modify normal web-building behaviors. The effects appear to be dose-dependent. Effects on webs were more pronounced when the larva reached its penultimate instar, indicating that greater concentrations of psychotropic chemicals were injected at this time (Eberhard, 2010). Eberhard (2000) concludes, "*Hymenoepimecis*'s manipulation of its spider host is probably the most finely directed alteration of behavior ever attributed to an insect parasitoid."

For comprehensive and fascinating reviews of parasitoids from ecological, behavioral, and evolutionary perspectives, see Godfray (1994), Hawkins & Sheenan (1994), Brodeur & Boivin (2004), and Stireman *et al.* (2006).

11.5 Phylogenetic relationships and classification

Despite the rich biodiversity of the arthropods, their rich fossil record going back to the Cambrian, and much intensive study, their phylogenetic relationships remain unclear. This is true for both the associations among the different arthropod groups, and between the arthropods and other phyla. Moreover, as we mention at the outset of this chapter, the basic question regarding the origin of arthropods is still debated. It is not clear if all the arthropod groups derive from a common ancestor or if the arthropod groups derive from more than one common ancestor.

Historically, monophyly has been the prevailing view because all arthropods share a number of traits, such as a cuticle made of chitin and proteins, segmentation with at least some segments bearing paired, articulated limbs, and similar patterns of cephalization. However, extensive pioneering work by Sidney Manton on the comparative morphology and embryology of the different groups of arthropods suggested that the three major groups, Chelicerata, Crustacea, and Uniramia, should be recognized as separate phyla having no common ancestor (Manton, 1973). Supporters of the polyphyletic origin indicate that the similarities between the groups should be regarded as extensive convergence, in which the possession of an exoskeleton leads, by necessity, to an array of changes that always result in an organism recognized as an arthropod, regardless of the starting point. It all boils down to a sort of chain reaction that results in what Willmer (1990) calls a 'syndrome of arthropodization' or what Gullan & Cranston (2010) refer to as 'monophyly of arthropodization.'

Currently, most authors support the hypothesis regarding a monophyletic origin of arthropods. They base their views on the fact that all arthropods are more closely related to each other than to members of other phyla, that their nearest common ancestor itself looked like an arthropod, and that their similarities are too great to be accounted for by anything but shared ancestry. Most importantly, perhaps, cladistic analyses of arthropods, including Cambrian fossils, strongly indicate that the primitive crustaceans are the living forms that come closest to the ancestral arthropod (Briggs & Fortey, 1989; Nielsen, 1995). Compelling data from molecular embryology, especially similarities in developmental expression of certain regulatory genes in crustaceans and hexapods, adds further support (Gullan & Cranston, 2010). This evidence is also supported by the fact that most crustaceans are aquatic, whereas the hexapods and chelicerates are, for the most part, terrestrial, and many of their embryological and anatomical specializations are related to the change in habitat, from aquatic to terrestrial. Gullan & Cranston (2010) present an overview of the phylogenetic relationships of the six-legged animals (Hexapoda) with other arthropods (see also Regier *et al.*, 2005).

The phylogenetic relationships of the arthropods with other phyla are not clear either. Evidence based on the nature of the nervous system and the segmentation of the body suggested that arthropods derived from one or more ancestral annelids. The Annelida, then, was thought to be the sister group of arthropods. However, a number of studies using morphological, paleontological, and molecular data question the arthropod–annelid relationship. Phylogenetic analysis of 18S and 28S ribosomal DNA sequences indicates a close relationship between arthropods, nematodes, tardigrades, onycophorans, nematomorphs, kinorhynchs, and priapulids, but not with annelids. The one unifying thread between the arthropods and their close allies is that they all undergo cuticle molting during their growth and development, leading to erection of the taxon Ecdysozoa to encompass this vast group of animals (Aguinaldo *et al.*, 1997; Mallatt *et al.*, 2004; Mallatt & Giribet, 2006).

Box 11.6 | Classification of the Arthropoda

A widely accepted scheme of classification for the arthropods does not exist. Several authors have proposed classifications for the many groups of arthropods but there is no consensus yet as to which one is a better reflection of their phylogeny. Despite their differences, most classifications are similar and differ mainly in the hierarchical level of the groups. For example, some consider the Crustacea, Chelicerata, and Uniramia as classes within the phylum Arthropoda, whereas others assign them to subphylum level. Adding to the complexity and controversy, it is unlikely that the Uniramia is a monophyletic taxon and other classification schemes have been proposed as a result, including placing the Insecta in the Subphylum Hexapoda (although other classifications refer to the superclass Hexapoda). A recent proposed classification refers to the Subphylum Mandibulata that includes the hexapods and crustaceans (Mallatt & Giribet, 2006; Pechenik, 2010).

The classification below is a simplified one that is similar to that presented in Roberts & Janovy (2009) and Pechenik (2010). Only the groups of arthropods that include parasitic species and are mentioned in this chapter are considered here.

Subphylum Crustacea
  Class Maxillopoda
  Subclass Copepoda
  Order Poecilostomatoida
  Representative genera: *Bomolochus, Chondracanthus, Colobomatus, Ergasilus, Lichomolgus, Ismaila, Sarcotaces*
  Order Siphonostomatoida
  Representative genera: *Caligus, Cecrops, Clavella, Haemobaphes, Lepeophtheirus, Lophoura, Pennella, Phrixocephalus, Salmincola, Sphyrion*
  Subclass Tantulocarida
  Representative genera: *Basipodella, Deoterthron*
  Subclass Branchiura
  Representative genera: *Argulus, Dolops*
  Subclass Cirripedia
  Order Rhizocephala
  Representative genera: *Briarosaccus, Chthamalophilus, Clistosaccus, Lernaeodiscus, Peltogaster, Peltogasterella, Polyascus, Ptychascus, Sacculina, Thompsonia*
  Order Thoracica
  Representative genera: *Conchoderma, Coronula, Cryptolepas*
 Class Malacostraca
  Order Amphipoda
  Suborder Hyperiidea
  Representative genera: *Phronima, Hyperia, Primmo*
  Suborder Caprellidea
  Representative genera: *Aeginella, Caprella, Cercops, Cyamus, Paracyamus, Syncyamus*

Box 11.6 | (continued)

Order Isopoda
  Suborder Gnathiidea
  Representative genus: *Gnathia*
  Suborder Flabellifera
  Representative genera: *Cymothoa, Anilocra, Lironeca*
  Suborder Epicaridea
  Representative genera: *Bopyrus, Danalia, Entoniscus, Orthione, Pinnotherion, Portunion, Liriopsis*
Subphylum Chelicerata
  Class Arachnida
  Subclass Acari
  Order Ixodida (Metastigmata)
  Representative genera: *Amblyomma, Argas, Boophilus, Dermacentor, Hyalomma, Ixodes,*
    Nuttalliella, Ornithodoros, Otobius, Rhipicephalus
  Order Mesostigmata
  Representative genera: *Dermanyssus, Echinolaelaps, Lyponyssus, Ornithonyssus,*
    Orthohalarachne, Pneumonyssus, Sternostoma, Varroa
  Order Astigmata (Sarcoptiformes)
  Representative genera: *Knemidocoptes, Koalachirus, Megninia, Otodectes, Psoroptes, Sarcoptes*
  Order Prostigmata (Trombidiformes)
  Representative genera: *Acarapis, Arrenurus, Blankaartia, Demodex, Hydrachna,*
    Hydrachnella, Pyemotes, Trombicula
Subphylum Hexapoda (Superclass to some)
  Class Insecta
  Order Psocodea (Phthiraptera)
  Representative genera: *Actornithophilus, Anaticola, Bovicola, Columbicola, Haematopinus,*
    Menacanthus, Menopon, Pediculus, Phthirius, Piagetiella, Trichodectes
  Order Hemiptera
  Representative genus: *Cimex*
  Order Siphonaptera
  Representative genera: *Ctenocephalides, Echidnophaga, Pulex*
  Order Strepsiptera
  Representative genera: *Corioxenos, Elenchus, Eoxenos, Stylops*
  Order Coleoptera
  Representative genera: *Platypsyllus, Leptinus*
  Order Diptera
  Representative genera: *Cuterebra, Dermatobia, Gasterophilus, Hippelates, Hippobosca, Lucilia,*
    Lutzomyia, Melophagus, Oestrus
  Order Hymenoptera
  Representative genera: *Aphelinus, Coeloides, Eupelmus, Macrocentrus, Phytodietus, Rhyssella,*
    Tetrastichus

References

Aguinaldo, A. M. A., Turbeville, J. M., Linford, L. S., *et al.* (1997) Evidence for a clade of nematodes, arthropods and other moulting animals. *Nature*, **387**, 489–493.

Anderson, D. L. & Trueman, J. W. H. (2000) *Varroa jacobsoni* (Acari: Varroidae) is more than one species. *Experimental and Applied Acarology*, **24**, 165–189.

Barker, D. E., Braeden, L. M., Coombs, M. P. & Boyce, B. (2009) Preliminary studies on the isolation of bacteria from sea lice, *Lepeophtheirus salmonis*, infecting salmon in British Columbia, Canada. *Parasitology Research*, **105**, 1173–1177.

Beckage, N. E. (1998) Parasitoids and polydnaviruses. *BioScience*, **48**, 305–311.

Bennett, C. E. (1995) Ticks and Lyme disease. *Advances in Parasitology*, **36**, 343–405.

Benz, G. W., Lucas, Z. & Lowry, L. L. (1998) New host and ocean records for the copepod *Ommatokoita elongata* (Siphonostomatoida: Lernaeopodidae), a parasite of the eyes of sleeper sharks. *Journal of Parasitology*, **84**, 1271–1274.

Bildfell, R. J., Mertins, J. W., Mortenson, J. A. & Cottam, D. F. (2004) Hair-loss syndrome in black-tailed deer of the Pacific Northwest. *Journal of Wildlife Diseases*, **40**, 670–681.

Bowen-Walker, P. L. & Gunn, A. (2001) The effect of the ectoparasitic mite, *Varroa destructor* on adult worker honeybee (*Apis mellifera*) emergence weights, water, protein, carbohydrate, and lipid levels. *Entomologia Experimentalis et Applicata*, **101**, 207–217.

Bowman, A. S., Coons, L. B., Needham, G. R. & Sauer, J. R. (1997) Tick saliva: recent advances and implications for vector competence. *Medical and Veterinary Entomology*, **11**, 277–285.

Boxshall, G. A. & Lincoln, R. J. (1983) Tantulocarida, a new class of Crustacea ectoparasitic on other crustaceans. *Journal of Crustacean Biology*, **3**, 1–16.

Bresciani, J. (1986) The fine structure of the integument of free-living and parasitic copepods: a review. *Acta Zoologica (Stockholm)*, **67**, 125–145.

Bresciani, J. & Dantzer, V. (1980) Fine structural localization of acid phosphatase in the root system of the parasite *Clistosaccus paguri* (Crustacea, Rhizocephala). *Electron Microscopy*, **2**, 290–291.

Briggs, D. E. & Fortey, R. A. (1989) The early radiation and relationships of the major arthropod groups. *Science*, **246**, 241–243.

Brodeur, J. & Boivin, G. (2004) Functional ecology of immature parasitoids. *Annual Review of Entomology*, **49**, 27–49.

Bunkley-Williams, L. & Williams, Jr., E. H. (1998) Isopods associated with fishes: a synopsis and corrections. *Journal of Parasitology*, **84**, 893–896.

Burgess, I. (1994) *Sarcoptes scabiei* and scabies. *Advances in Parasitology*, **33**, 235–292.

Burgess, I. F. (1995) Human lice and their management. *Advances in Parasitology*, **36**, 271–342.

Colwell, D. D., Hall, M. J. R. & Scholl, P. J. (2006) *The Oestrid Flies: Biology, Host–Parasite Relationships, Impact and Management*. Wallingford: CAB International.

Crump, M. L. & Pounds, J. A. (1985) Lethal parasitism of an aposematic anuran (*Atelopus varius*) by *Notochaeta bufonivora* (Diptera: Sarcophagidae). *Journal of Parasitology*, **71**, 588–591.

Dehio, C., Sauder, U. & Hiestand, R. (2004) Isolation of *Bartonella schoenbuchensis* from *Lipoptena cervi*, a bloodsucking arthropod causing deer ked dermatitis. *Journal of Clinical Microbiology*, **42**, 5320–5323.

Desch, C. & Nutting, W. B. (1972) *Demodex folliculorum* (Simon) and *D. brevis* (Akbulatova) of man: redescription and reevaluation. *Journal of Parasitology*, **58**, 169–177.

Dick, C. W. & Patterson, B. D. (2006) Bat flies: obligate ectoparasites of bats. In: *Micromammals and Macroparasites: From Evolutionary Ecology to Management*, ed. Morand, S., Krasnov, B., Poulin, R., pp. 179–194. Tokyo: Springer.

Dittmar, K., Porter, M., Murray, S. & Whiting, M. F. (2006) Molecular phylogenetic analysis of nycteribiid and streblid batflies (Diptera: Brachycera, Calyptratae): implications for host association and phylogeographic origins. *Molecular Phylogenetics and Evolution*, **38**, 155–170.

Dittmar, K., Dick, C. W., Patterson, B. D., *et al.* (2009) Pupal deposition and ecology of bat flies (Diptera: Streblidae): *Trichobius* sp. (caecus group) in a Mexican cave habitat. *Journal of Parasitology*, **95**, 308–314.

Duffy, D. C. (1983) The ecology of tick parasitism in densely nesting Peruvian seabirds. *Ecology*, **64**, 110–119.

Durden, L. A. & Keirans, J. E. (1996) Host–parasite coextinction and the plight of tick conservation. *American Entomologist*, **42**, 87–91.

Eberhard, W. G. (2000) Spider manipulation by a wasp larva. *Nature*, **406**, 255–256.

Eberhard, W. G. (2001) Under the influence: webs and building behavior of *Plesiometa argyra* (Araneae: Tetragnathidae) when parasitized by *Hymenoepimecis argyraphaga* (Hymenoptera: Ichneumonidae). *Journal of Arachnology*, 29, 354–366.

Eberhard, W. G. (2010) Recovery of spiders from the effects of parasitic wasps: implications for fine-tuned mechanisms of manipulation. *Animal Behaviour*, 79, 375–383.

Evans, G. O. (1992) *Principles of Acarology*. Wallingford: CAB International.

Feare, C. J. (1976) Desertion and abnormal development in a colony of sooty terns *Sterna fuscata* infested by virus-infected ticks. *Ecology*, 118, 112–115.

Fessl, B. & Tebbich, S. (2002) *Philornis downsi*: a recently discovered parasite on the Galapagos archipelago – a threat for Darwin's finches? *Ibis*, 144, 445–451.

Fessl, B., Kleindorfer, S. & Tebbich, S. (2006) An experimental study on the effects of an introduced parasite in Darwin's finches. *Biological Conservation*, 127, 55–61.

Glenner, H. (2001) Cypris metamorphosis, injection and earliest internal development of the rhizocephalan *Loxothylacus panopaei* (Gissler). Crustacea: Cirripedia: Rhizocephala: Sacculinidae. *Journal of Morphology*, 249, 43–75.

Glenner, H. & Høeg, J. (1995) A new motile, multicellular stage involved in host invasion by parasitic barnacles (Rhizocephala). *Nature*, 377, 147–150.

Glenner, H., Høeg, J., O'Brien, J. J. & Sherman, T. D. (2000) Invasive vermigon stage in the parasitic barnacles *Loxothylacus texanus* and *L. panopaei* (Sacculinidae): closing of the rhizocephalan life-cycle. *Marine Biology*, 136, 249–257.

Glenner, H., Lützen, J. & Takahashi, T. (2003) Molecular and morphological evidence of a monophyletic clade of asexually reproducing Rhizocephala: *Polyascus*, new genus (Cirripedia). *Journal of Crustacean Biology*, 23, 548–557.

Goater, T. M. & Jepps, S. F. (2002) Prevalence and intensity of *Haemobaphes diceraus* (Copepoda: Pennellidae) from shiner perch, *Cymatogaster aggregata* (Embiotocidae) *Journal of Parasitology*, 88, 194–197.

Godfray, H. C. J. (1994) *Parasitoids: Behavioral and Evolutionary Ecology*. Princeton: Princeton University Press.

Grosman, A. H., Janssen, A., De Brito, E. F., *et al.* (2008). Parasitoid increases survival of its pupae by inducing hosts to fight predators. *PLoS One*, 3: e2276. DOI: 10.1371/journal.pone.0002276.

Gullan, P. J. & Cranston, P. S. (2010) *The Insects: An Outline of Entomology*, 4th edition. Oxford: Wiley-Blackwell.

Hall, M. & Wall, R. (1995) Myiasis of humans and domestic animals. *Advances in Parasitology*, 35, 257–334.

Hawkins, B. A. & Sheenan, W. (1994) *Parasitoid Community Ecology*. New York: Oxford University Press.

Høeg, J. (1995) The biology and life cycle of the Cirripedia Rhizocephala. *Journal of the Marine Biological Association UK*, 75, 517–550.

Hoogstraal, H., Wassef, H. Y., Hays, C. & Keirans, J. E. (1985) *Ornithodoros (Alectorobius) spheniscus* n. sp. [Acarina: Ixodoidea: Argasidae: *Ornithodoros (Alectorobius) capensis* group], a tick parasite of the Humboldt penguin in Peru. *Journal of Parasitology*, 71, 635–644.

Huys, R. (2001) Splanchnotrophid systematics: a case of polyphyly and taxonomic myopia. *Journal of Crustacean Biology*, 21, 106–156.

Huys, R., Boxshall, G. A. & Lincoln, R. J. (1993) The tantulocaridan life cycle: the circle closed? *Journal of Crustacean Biology*, 13, 432–442.

Jacob, E., Barker, D. E. & Garver, K. A. (2011) Vector potential of the salmon louse *Lepeophtheirus salmonis* in the transmission of infectious haematopoietic necrosis virus (IHNV). *Diseases of Aquatic Animals*, 97, 155–165.

Janssen, A., Grosman, A. H., Cordeiro, E. G., *et al.* (2009) Context-dependent fitness effects of behavioral manipulation by a parasitoid. *Behavioral Ecology*, 21, 33–36.

Jones, S. R. M. & Beamish, R. J. (2011) *Salmon Lice: An Integrated Approach to Understanding Parasite Abundance and Distribution*. Chichester: Wiley-Blackwell.

Kabata, Z. (1974) Two new features in the morphology of Caligidae (Copepoda). *Proceedings of the Third International Congress of Parasitology*, 3, 1635–1636.

Kabata, Z (1979) *Parasitic Copepoda of British Fishes*. London: Ray Society.

Kabata, Z. (1981) Copepoda (Crustacea) parasitic on fishes: problems and perspectives. *Advances in Parasitology*, 19, 2–71.

Kaliszewska, Z. A., Seger, J., Rowntree, V. J., *et al.* (2005) Population histories of right whales (Cetacea: *Eubalaena*) inferred from mitochondrial sequence diversities and divergences of their whale lice (Amphipoda: *Cyamus*). *Molecular Ecology*, 14, 3439–3456.

Kannupandi, T. (1976a) Cuticular adaptations in two parasitic copepods in relation to their mode of life. *Journal of Experimental Marine Biology and Ecology*, 22, 235–248.

Kannupandi, T. (1976b) Occurrence of resilin and its significance in the cuticle of *Pennella elegans*, a copepod parasite. *Acta Histochemica*, 56, 73–79.

Khan, R. A. (1988) Experimental transmission, development, and effects of a parasitic copepod, *Lernaeocera branchialis*, on Atlantic cod, *Gadus morhua*. *Journal of Parasitology*, 74, 586–599.

Khan, R. A., Lee, E. M. & Barker, D. (1990) *Lernaeocera branchialis*: a potential pathogen to cod ranching. *Journal of Parasitology*, 76, 913–917.

King, K. A., Keith, J. O., Mitchell, C. A. & Keirans, J. E. (1977) Ticks as a factor in nest desertion of California brown pelicans. *Condor*, 79, 507–509.

Krantz, G. W. & Walter, D. E. (2009) *A Manual of Acarology*, 3rd edition. Lubbock: Texas Tech University Press.

Lanciani, C. A. (1984) Crowding in the parasitic stage of the water mite *Hydrachna virella* (Acari: Hydrachnidae). *Journal of Parasitology*, **70**, 270–272.

Lanciani, C. A. (1988) Sexual bias in host selection by parasitic mites of the mosquito *Anopheles crucians* (Diptera: Culicidae). *Journal of Parasitology*, **74**, 768–773.

Lavine, M. D. & Beckage, N. E. (1995) Polydnaviruses: potent mediators of host insect immune dysfunction. *Parasitology Today*, **11**, 368–378.

Leighton, P. A., Koffi, J. K., Pelcat, Y., *et al.* (2012) Predicting the spread of tick invasion: an empirical model of range expansion for the Lyme disease vector *Ixodes scapularis* in Canada. *Journal of Applied Ecology*, **49**, 457–464.

Lompen, J. S. H. & Oliver, J. H. Jr. (1993) Haller's organ in the tick family Argasidae (Acari: Parasitiformes: Ixodida). *Journal of Parasitology*, **79**, 591–603.

Mallatt, J. M. & Giribet, G. (2006) Further use of nearly complete 28S and 18S rRNA genes to classify Ecdysozoa: 37 more arthropods and a kinorhynch. *Molecular Phylogenetics and Evolution*, **40**, 772–794.

Mallatt, J. M., Garey, J. R. & Schultz, J. W. (2004) Ecdysozoan phylogeny and Bayesian inference: First use of nearly complete 28S and 18S rRNA gene sequences to classify the arthropods and their kin. *Molecular Phylogenetics and Evolution*, **31**, 178–191.

Manton, S. M. (1973) Arthropod phylogeny: a modern synthesis. *Journal of Zoology, London*, **171**, 111–130.

Markham, J. C. (2001) A review of the bopyrid isopods parasitic on thalassinidean decapods. *Crustacean Issues*, **13**, 195–204.

Marshall, A. G. (1982) The ecology of the bat ectoparasite *Eoctenes spasmae* (Hemiptera: Polyctenidae) in Malaysia. *Biotropica*, **14**, 50–55.

Martin, S. J. (2001) The role of *Varroa* and viral pathogens in the collapse of honeybee colonies: a modelling approach. *Journal of Applied Ecology*, **38**, 1082–1093.

Martin, J. W. & Davis, G. E. (2001) *An Updated Classification of the Recent Crustacea*. Los Angeles: Natural History Museum of Los Angeles County.

Martin, J. W. & Heyning, J. E. (1999) First record of *Isocyamus kogiae* Sedlak-Weinstein, 1992 (Crustacea, Amphipoda, Cyamidae) from the Eastern Pacific, with comments on morphological characters, a key to the genera of the Cyamidae, and a checklist of cyamids and their hosts. *Bulletin of the Southern California Academy of Science*, **98**, 26–38.

Mertins, J. W., Mortenson, J. A., Bernatowicz, J. A. & Briggs Hall, P. (2011) *Bovicola tibialis* (Phthiraptera: Trichodectidae): occurrence of an exotic chewing louse on cervids in North America. *Journal of Medical Entomology*, **48**, 1–12.

Michener, G. R. (1993) Lethal myiasis of Richardson's ground squirrels by the sarcophagid fly *Neobellieria citellivora*. *Journal of Mammology*, **74**, 148–155.

Moser, M. & Taylor, S. (1978) Effects of the copepod *Cardiodectes medusaeus* on the lanternfish *Stenobrachius leucopsarus* with notes on hypercastration by the hydroid *Hydrichthys* sp. *Canadian Journal of Zoology*, **56**, 2372–2376.

Nielsen, C. (1995) *Animal Evolution: Interrelationships of the Living Phyla*. Oxford: Oxford University Press.

Ogden, N. H., Maarouf, A., Barker, I. K., *et al.* (2006) Climate change and the potential for range expansion of the Lyme disease vector *Ixodes scapularis* in Canada. *International Journal for Parasitology*, **36**, 63–70.

Ogden, N. H., Lindsay, L. R., Hanincová, K., *et al.* (2008) Role of migratory birds in introduction and range expansion of *Ixodes scapularis* ticks and of *Borrelia burgdorferi* and *Anaplasma phagocytophilum* in Canada. *Applied and Environmental Microbiology*, **74**, 1780–1790.

Oldroyd, B. P. (1999) Coevolution while you wait: *Varroa jacobsoni*, a new parasite of western honeybees. *Trends in Ecology and Evolution*, **14**, 312–315.

Oliver, Jr., J. H. (1996) Lyme borreliosis in the southern United States: a review. *Journal of Parasitology*, **82**, 926–935.

Oliver, Jr., J. H., Hayes, M. P., Keirans, J. E. & Lavender, D. R. (1993) Establishment of the foreign parthenogenetic tick *Amblyomma rotundatum* (Acari: Ixodidae) in Florida. *Journal of Parasitology*, **79**, 786–790.

Ostfeld, R. S. (1997) The ecology of Lyme-disease risk. *American Scientist*, **85**, 338–346.

Ostfeld, R. S. (2011) *Lyme Disease: The Ecology of a Complex System*. New York: Oxford University Press.

Ostfeld, R. S., Canham, C. D., Oggenfuss, K., *et al.* (2006) Climate, deer, rodents, and acorns as determinants of variation in Lyme-disease risk. *PLoS Biology*, **4**: e145. DOI:10.1371/journal.pbio.0040145.

Overstreet, R. M., Jovonovich, J. & Ma, H. (2009) Parasitic crustaceans as vectors of viruses, with an emphasis on three penaeid viruses. *Integrative and Comparative Biology*, **49**, 127–141.

Patterson, B. D., Dick, C. W. & Dittmar, K. (2008) Parasitism by bat flies (Diptera: Streblidae) on neoptropical bats: effects of host body size, distribution, and abundance. *Parasitology Research*, **103**, 1091–1100.

Pechenik, J. A. (2010) *Biology of the Invertebrates*, 6th edition. New York: McGraw-Hill.

Peck, S. B. (2006) Distribution and biology of the ectoparasitic beaver beetle *Platypsyllus castoris* Ritsema in North America (Coleoptera: Leioididae: Platypsyllinae). *Insecta Mundi*, **20**, 85–94.

Pence, D. B., Windberg, L. A., Pence, B. C. & Sprowls, R. (1983) The epizootiology and pathology of sarcoptic mange in coyotes, *Canis latrans*, from South Texas. *Journal of Parasitology*, **69**, 1100–1115.

Peresan, L. & Roccatagliata, D. (2005) First record of the hyperparasite *Liriopsis pygmaea* (Cryptoniscidae, Isopoda) from a rhizocephalan parasite of the false king crab *Paralomis granulosa* from the Beagle Channel (Argentina), with a redescription. *Journal of Natural History*, **39**, 311–324.

Perkins, P. S. (1985) Iron crystals in the attachment organ of the erythrophagous copepod *Cardiodectes medusaeus* (Pennellidae). *Journal of Crustacean Biology*, **5**, 591–605.

Perkins, P. S. (1994) Ultrastructure of the holdfast of *Phrixocephalus cincinnatus* (Wilson), a blood-feeding parasitic copepod of flatfishes. *Journal of Parasitology*, **80**, 797–804.

Pike, A. W. & Wadsworth, S. (1999) Sealice on salmonids, their biology and control. *Advances in Parasitology*, **44**, 233–337.

Proctor, H. & Owens, I. (2000) Mites and birds: diversity, parasitism and coevolution. *Trends in Ecology and Evolution*, **15**, 358–365.

Raga, J. A. (1997) Parasitology of marine mammals. In *Marine Mammals, Seabirds and Pollution of Marine Systems*, ed. T. Jauniaux, J. M. Bouquegneau & F. Coignoul, pp. 67–90. Liège: Presses de la Faculté de Médecine Vétérinaire de l'Université de Liège.

Raibaut, A. & Trilles, J. P. (1993) The sexuality of parasitic crustaceans. *Advances in Parasitology*, **32**, 367–444.

Ramachandra, R. N. & Wikel, S. K. (1992) Modulation of host immune response by ticks (Acari: Ixodidae): effect of salivary gland extracts on host macrophages and lymphocyte cytokine production. *Journal of Medical Entomology*, **29**, 818–826.

Rasmussen, H. W., Jacobsen, S. L. & Collins, J. S. H. (2008) Raninidae infested by parasitic Isopoda (Epicaridea). *Bulletin of the Mizunami Fossil Museum*, **34**, 31–49.

Regier, J. C., Schultz, J. W. & Kambic, R. E. (2005) Pancrustacean phylogeny: Hexapods are terrestrial crustaceans and maxillipods are not monophyletic. *Proceedings of the Royal Society of London, series B, Biological Sciences*, **272**, 395–401.

Ribeiro, J. M. C., Makoul, G. T., Levine, J., *et al.* (1985) Antihemostatic, antiinflammatory, and immunosuppressive properties of the saliva of a tick, *Ixodes dammini. Journal of Experimental Medicine*, **161**, 332–344.

Ritchie, L. E. & Høeg, J. T. (1981) The life history of *Lernaeodiscus porcellanae* (Cirripedia: Rhizocephala) and co-evolution with its porcellanid host. *Journal of Crustacean Biology*, **1**, 334–347.

Roberts, L. S. & Janovy, Jr., J. (2009) *Foundations of Parasitology*, 8th edition. New York: McGraw-Hill.

Rothschild, M. & Ford, B. (1972) Breeding cycle of the flea *Cediopsylla simplex* is controlled by breeding cycle of host. *Science*, **178**, 625–626.

Rothschild, M., Schlein, Y., Parker, K., *et al.* (1973) The flying leap of the flea. *Scientific American*, **229**, 92–100.

Rowntree, V. J. (1996) Feeding, distribution, and reproductive behavior of cyamids (Crustacea: Amphipoda) living on humpback and right whales. *Canadian Journal of Zoology*, **74**, 103–109.

Samuel, W. M. & Welch, D. A. (1991) Winter ticks on moose and other ungulates: factors influencing their population size. *Alces*, **27**, 169–182.

Samuel, W. M., Williams, E. S. & Rippin, A. B. (1982) Infestations of *Piagetiella peralis* (Mallophaga: Menopodidae) on juvenile white pelicans. *Canadian Journal of Zoology*, **60**, 951–953.

Scott, J. D., Anderson, J. F. & Durden, L. A. (2012) Widespread dispersal of *Borrelia burgdorferi*-infected ticks collected from songbirds across Canada. *Journal of Parasitology*, **98**, 49–59.

Smith, N. G. (1968) The advantage of being parasitized. *Nature*, **219**, 690–694.

Smith, A. E., Chapman, J. W. & Dumbault, B. R. (2008) Population structure and energetics of the bopyrid isopod parasite *Orthione griffenis* in mud shrimp *Upogebia pugettensis. Journal of Crustacean Biology*, **28**, 228–233.

Spalding, M. G., Wrenn, W. J., Schwikert, S. T. & Schmidt, J. A. (1997) Dermatitis in young Florida sandhill cranes (*Grus canadensis pratensis*) due to infestation by the chigger, *Blankaartia sinnamaryi. Journal of Parasitology*, **83**, 768–771.

Stireman, J. O., III, O'Hara, J. E. & Wood, D. M. (2006) Tachinidae: evolution, behavior, and ecology. *Annual Review of Entomology*, **51**, 525–555.

Todd, C. D. (2007) The copepod parasite (*Lepeophtheirus salmonis* (Kroyer), *Caligus elongatus* Nordmann) interactions between wild and farmed Atlantic salmon (*Salmo salar* L.) and wild sea trout (*Salmo trutta* L.): a mini review. *Journal of Plankton Research*, **29**, 161–171.

Walter, D. E. & Proctor, H. C. (1999) *Mites: Ecology, Evolution and Behaviour*. Wallingford: CAB International.

Williams, J.D. & An, J. (2009) The cryptogenic parasitic isopod *Orthione griffenis* Markham, 2004 from the eastern and western Pacific. *Integrative and Comparative Biology*, **49**, 114–126.

Willmer, P. (1990) *Invertebrate Relationships: Patterns in Animal Evolution*. Cambridge: Cambridge University Press.

Parasite population ecology

12.1 General considerations

Our previous chapters focused on the functional morphology, life cycles and ecology of a wide range of animal parasites. One of our aims in these chapters was to highlight the complexity and fascination of the parasitic life style. Another was to introduce the idea that the seemingly infinite diversity of parasite life cycles and adaptations could be interpreted under an ecological umbrella. Armed with this background knowledge, we now consider unifying principles of ecology and evolution that can be applied to the phenomenon of parasitism.

We begin our transition with a consideration of the complex nature of parasite populations and the general ecological characteristics of the individuals that comprise them. We highlight the nature of enquiry at this level by first considering two examples. In a field survey, Cornwell & Cowan (1963) monitored the transmission of gut helminths into 180 canvasback ducks, *Aythya valisineria*, sampled from a small wetland in western Canada. The authors controlled for sampling heterogeneity by restricting their collections to ducklings within individual clutches. Their results showed that even within a single clutch, individual siblings harbored between 90 and 6000 worms! In another field survey, Valtonen *et al.* (2004) censused adult acanthocephalan populations in individual ringed seals, *Phoca hispida*, collected from the Baltic Sea. Although the collections spanned a 22-year period when the seal and intermediate host populations varied extensively, the prevalence of acanthocephalans was always 100% and the mean number of worms fluctuated within a single order of magnitude. Thus, on the one hand, mean parasite infrapopulation sizes can vary tremendously, even within very narrow spatial and temporal scales. On the other hand,

population sizes can be remarkably stable, barely fluctuating around an equilibrium value. Characterizing this extreme variation at both narrow and broad temporal and spatial scales and understanding the underlying mechanisms that determine it are the central objectives of parasite population ecologists and ecological epidemiologists.

There exists a strong theoretical foundation for the study of parasite population dynamics. Following the seminal work of Crofton (1971a, b), Roy Anderson and Robert May spearheaded theoretical developments that continue to define the direction of enquiry into the complex nature of parasite population dynamics (Anderson & May, 1979; May & Anderson, 1979). Both authors were knighted in recognition of their enormous contributions to science. It is remarkable that Sir Roy Anderson's research roots lie in describing the population dynamics of a primitive caryophyllidean cestode in freshwater fish (see Chapter 6) collected from a small English pond (Anderson, 1974). Perhaps his early fascination with the ecology and epidemiology of somewhat obscure parasites parallels your own! Roy Anderson's research trajectory from the guts of freshwater fish to one of the world's leading parasite population ecologists and human epidemiologists is described in Esch (2007).

The conceptual foundation led by Anderson and May is now closely allied to theoretical developments in the broader field of epidemiology (reviews in Grenfell & Dobson, 1995; Hudson *et al.*, 2001). In the strictest sense, most of us consider epidemiology as the study of disease dynamics in human populations, including diseases associated, for example, with cancer and heart conditions. However, ecologists extend the term to include populations of any host, while also restricting it to parasites and other diseases. Thus, ecological epidemiologists and parasite population ecologists are interested in the dynamics of parasitism within host populations. A further important distinction is that ecological epidemiologists acknowledge and assume that the dynamics of host and parasite populations interact, such that population dynamics of the parasites can be strongly dependent on the population dynamics of the host, and vice versa.

In this chapter, we introduce the broad subdiscipline of parasite population ecology. We begin with an overview of terminology and general approaches. What exactly constitutes a parasite population? How are parasite population densities determined, and how do we monitor changes in population size? We then introduce the general theoretical foundation of parasite population ecology and epidemiology championed by Roy Anderson and Robert May, concluding with empirical studies aimed at understanding the complex nature of parasite population dynamics, parasite aggregation within host populations, and regulation of parasite populations.

12.2 Terminology and general approaches

A population can be defined as a group of organisms of the same species occupying the same space in time and comprising a unique gene pool. Each population, whether free-living or parasitic, can be characterized by well-known parameters that are described in introductory ecology texts. The most important of these is density, i.e., a measure of the number of organisms occurring within a defined area, e.g., a square meter, or a square centimeter of tissue. Four basic biological rates determine how population densities change over time – birth, death, immigration, and emigration. Depending on the nature of the questions being considered, populations can also be described by features such as age distribution, reproductive potential, biomass, sex ratio, and genetic composition. All of these terms can be applied to the study of micro- and macroparasite populations.

However, the nature of parasite life cycles imposes special constraints on how we define, and evaluate, their populations. Indeed, similar constraints exist for all organisms with complex and/or indirect life cycles. Consider the nematodes in the genus *Trichinella* that we highlighted in Chapter 8. How might we define a population of a species of *Trichinella*? Perhaps interest is in the population of dioecious adults living in the gut of a single host. Or, the focus may be on population-level characteristics of a different life cycle stage, such as the larvae encysted within a carnivore's striated muscle. Often, the focus of epidemiological studies rests in the assessment of differences in parasite populations between individual hosts, or between populations of hosts. The nature of enquiry becomes increasingly complex at this point, particularly in the case of generalist parasites such as *Trichinella* that can infect multiple species of hosts. The key point is that even in a fairly simple case involving a parasite with only one obligate intermediate host, and no free-living larval stage, defining the scale at which questions are asked, and the precise unit of study (individual parasite, individual host, and so on) is a critical feature of any population-level study involving parasites.

In an effort to clarify these issues, Esch *et al.* (1975) proposed that parasites within a single host and those within all hosts in an ecosystem be considered independently. They developed the concepts of the **infrapopulation** and **suprapopulation** to address these issues. An infrapopulation includes all of the parasites of a single species in one host individual, whereas a suprapopulation includes all of the parasites of a given species, in all stages of development, within all hosts in an ecosystem. Subsequently, Bush *et al.* (1997) extended the terminology to define **component population** as all of the infrapopulations of a species of parasite within all hosts of a given species in an ecosystem. Many population ecologists define the term metapopulation in the same way. Thus, an infrapopulation of *Trichinella* is comprised of all adults in the intestine of an individual host, or all of the encysted larvae located within its striated muscles. A component population of adult *Trichinella* is all individuals counted within a sampled host population. The *Trichinella* suprapopulation is all of the component and infrapopulations of adults and larvae in all its hosts! Given the logistic complexity, comprehensive studies of parasite suprapopulations are rare.

Most epidemiological and population-level studies involving parasites require an accurate assessment of the numbers of parasites in an individual host and the extent of variation within a host population. Although the determination of parasite count data can be straightforward for some parasites, the collection of these data is difficult in others. Thus, traditional kill and count methods can provide a direct population census of many species of macroparasite in, or on, a host, such as those in the gut and lungs or on the skin. However, these methods are rarely appropriate for macroparasites that are located within host tissues, or for microparasites. Moreover, such methods are often associated with potentially biased samples of hosts that are collected opportunistically, perhaps from hunters or fishermen. To address the practical and ethical constraints of destructive host sampling, indirect assessments of worm burden are commonly used. One of the most common for adult parasites is the quantification of the numbers of eggs or larvae released in the host's feces, e.g., eggs per gram feces or eggs shed per day. Immunodiagnostic tools have also become widely available to detect antibodies located within host sera, or parasite antigens present within host feces. In many cases, the association between these indirect assessments and actual worm counts is poor, requiring extensive validation on a species-by-species basis (Wilson *et al.*, 2002).

Despite these challenges, the direct or indirect assessment of the numbers of parasites in a host is a cornerstone of parasite population ecology and epidemiology. Intensity is among the most widely used terms in ecological parasitology, referring to the number of parasites of a given species in a host. Prevalence is the percentage of hosts in a sample that is infected with a given parasite species. Mean intensity is defined as the mean number of parasites of a given species in a sample of infected hosts. Mean abundance is defined as the mean number of parasites in a sample of infected and uninfected hosts. These means must always include a measure of variation in parasite counts within a sample. This terminology is especially appropriate for most macroparasites, although some measure of intensity is also useful for studies involving microparasites, e.g., parasitemia (see Chapter 3). Bush *et al.* (1997) elaborate on the terminology of parasite population ecology.

12.3 Introduction to parasite population ecology

12.3.1 Theoretical considerations

In their extensions to the original models developed by H. D. Crofton (1971a; 1971b), Roy Anderson and Robert May provided a theoretical and conceptual foundation for studies involving the coupled dynamics of parasite and host populations (Anderson & May, 1979; May & Anderson, 1979). As we will see, the foundation that they developed is of great significance for classical epidemiologists, with interests in human, livestock, and crop parasites and diseases, and for ecological epidemiologists interested in interactions between host and parasite populations. In our brief introduction to this vast discipline, we restrict our focus to their simplest mathematical models involving direct life cycle macroparasites. A more complete synopsis of the history, development, and application of their models can be found in Anderson (1993) and McCallum (2000).

As an introduction to this approach, consider the changes in size of a macroparasite population within a host population. For simplicity, consider that the parasite has a direct life cycle and that infective larvae do not undergo extensive development outside the host. Examples include the direct life cycle nematodes (e.g., *Ascaris, Ancylostoma, Trichuris, Trichostrongylus*) and monogenean trematodes, and many of the ticks, mites, lice, and copepods that we covered in earlier chapters. In the interest of even greater simplicity, assume that density-dependent parasite mortality and fecundity is absent and that parasite-induced host mortality is linearly tied to the number of parasites in, or on, a host. When these assumptions are made, Anderson & May (1979) expressed the dynamics of parasite population (P) growth within a host population (H) by the following differential equation:

$$\frac{dp}{dt} = \frac{\lambda HP}{\left[\frac{\mu}{\beta} + H\right]} - P(\gamma + \alpha + b) - \frac{\alpha(k+1)P^2}{kH} \quad (12.1)$$

Although this equation looks formidable, it can be distilled down to three key rates – one that increases the numbers of parasites in a host, followed by two that decrease parasite numbers. The gain term at the left-hand side of the equation reflects the numbers of infective stages that actually survive to infect another host. The numerator of this term is the product of the total number of parasites in the host population (HP) times their *per capita* birth rate (λ). However, only a fraction of those infective stages will survive to infect a new host, at a rate determined by their natural *per capita* death rate (μ), and a transmission coefficient (β) that determines how many surviving infective stages actually reach a naïve host. The middle term reflects the loss of parasites due to natural parasite mortality (γ), or from the loss of parasites from hosts that die naturally (b), or from the parasites themselves (α). The last term is the loss of parasites arising from the rate of parasite-induced host mortality, i.e., parasite virulence, and the extent to which the parasite population is aggregated within the host population (k). For example, loss of parasites in the system will be high when α is high and parasite aggregation is low, i.e., where most hosts in a population are infected with virulent parasites. Figure 12.1 is a visual representation of the gain and loss of adult parasites and their infective larvae in this model.

The parameters in the equation and in Fig. 12.1 can be rearranged to develop an expression for the **Basic Reproductive Rate** for a macroparasite in a host population (Anderson & May, 1991):

$$Ro = \frac{\lambda H}{\left[\frac{\mu}{\beta + H}\right][\lambda + \alpha + b]} \quad (12.2)$$

For macroparasites, Ro is the average number of infective offspring produced over the lifetime of an adult parasite that infect and reproduce in a new host.

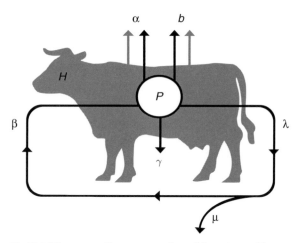

Fig. 12.1 Diagrammatic representation of the gains and losses of a direct life cycle parasite in a hypothetical host. Parameters are defined in the text. (Figure courtesy of Lori Goater.)

Thus, if the average individual in a parasite infrapopulation replaces itself with at least one successful offspring in the subsequent generation (i.e., $Ro > 1$), the parasite population will increase in size. If $Ro < 1$, and the parasite does not replace itself, then population size will decrease. The case where $Ro = 1$ is known as the **Transmission Threshold**, a key epidemiological parameter depicting a 'tipping point' that must be crossed if a parasite is to spread within a host population. Because Ro provides an estimate of the reproductive potential of an individual macroparasite, and thus its potential to spread in a host population, it is a cornerstone of parasite population ecology and epidemiology. A complete discussion of methods used to estimate Ro for macro- and microparasites, and its significance in ecological epidemiology, is provided in Anderson & May (1991) and Tompkins et al. (2002).

All models are abstractions of reality. Yet, over the 30 years since they were introduced, epidemiological models of the type described in Fig. 12.1 have provided a useful tool to enhance our understanding of the complex processes that determine parasite population dynamics. One advantage is that they focus attention on a set of key parameters that can be quantified. The relative importance of individual parameters can then be determined by varying their magnitude while keeping the others constant. A second advantage is that extensions to the Basic Model (Fig. 12.1) can include more complex scenarios involving one or two intermediate hosts, and phenomena such as density-dependent parasite survival, immunity, and inter-specific competition. Perhaps most important, the models can be used to predict future outcomes. Examples include predicting the impact of climate change on β and Ro for human and wildlife parasites (Lafferty, 2009), predicting the proportion of a host population required to be vaccinated for parasite control (Hudson et al., 1998), or forecasting the role of parasites in the regulation of host populations (Hudson et al., 1998). Of course, for the predictions to be meaningful, the parameters must be validated. Much of the day-to-day work undertaken by parasite population ecologists and epidemiologists involves collecting data from laboratory and field experiments designed to estimate rates such as β and α, together with the various rates linked to parasite reproduction and mortality (Fig. 12.1).

12.3.2 Distribution of parasites in host populations

Imagine a scenario in which you and 30 of your friends were instantaneously exposed to the infective stages of a parasite. Perhaps this group lunched from a plate of sushi that contained raw fish infected with larval anisakid nematodes. Or perhaps this group waded into water containing the bird schistosome cercariae that causes swimmer's itch. After a period of time, imagine that you had the opportunity to evaluate parasite intensity in each of your friends. How might these parasites be distributed among this sample of hosts? One prediction is that the overall population of parasites would be distributed at random. You could easily test this by evaluating whether the distribution of worm counts fit a random (or Poisson) distribution. An alternative is that the parasites would be distributed uniformly among your friends, such that each student had roughly the same number. A further prediction is that the parasite counts would be aggregated so that most of your classmates would have no or very few parasites, but some would have many.

Based upon the results of countless empirical surveys, parasite ecologists and epidemiologists know that it is the third outcome – aggregation – that invariably arises in natural host–parasite combinations (review in Wilson et al., 2002). Indeed, not only is a pattern of aggregated distribution remarkably consistent for many micro- and macroparasites of animals, but the magnitude of heterogeneity within a natural host population can be staggering. The example from the beginning of this chapter involving helminths of canvasback ducks provides an extreme case showing that even when features such as host genetics, host age, season, and habitat are controlled, parasite intensities often vary tremendously among individuals within a host population. In fact, even when conditions of exposure and host genetics are tightly controlled in laboratory settings, e.g., Trichinella in congenic strains of mice, considerable variation among individuals remains (review in Goater & Holmes, 1997).

Ever since Crofton's seminal work, and the extensions by Anderson and May, the concept of aggregation has become a central paradigm in parasite population biology and ecological epidemiology. From an applied and economic perspective, the diagnosis and treatment of the heavily infected hosts at the 'tail' of the parasite frequency distribution is key in terms of parasite control. From a theoretical perspective, we saw in the previous section that the magnitude of parasite aggregation plays a key role in the regulation and determination of parasite population sizes. Parasite aggregation also influences the methods we use to study and analyze parasite populations. Because so many parasites are aggregated in their host populations, we almost always need large samples of hosts to obtain an accurate picture of true patterns of parasite prevalence and intensity in a natural host population. Further, because aggregated parasite distributions typically depart from standard normal distributions, the analyses involving comparisons among samples of hosts often require special statistical tools (Wilson & Grenfell, 1997; Rózsa et al., 2000).

The frequency distributions of parasite counts within a host population follow one of the three basic patterns, represented in graphic form in Fig. 12.2 (Anderson & Gordon, 1982). Random distributions occur when the

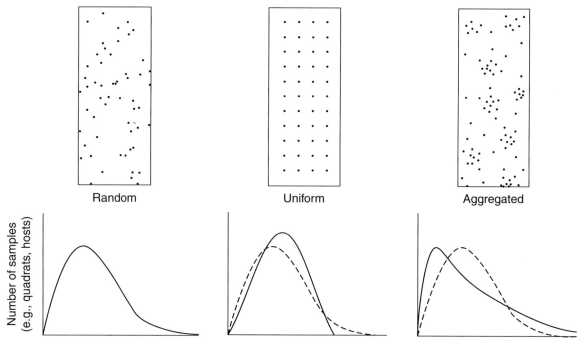

Fig. 12.2 Distributional patterns in space (above) for populations that are random, regular, and aggregated. Below are the frequency curves that describe these distributions; the dashed line is the same random distribution for comparison.

position of one individual is independent of any other and when each segment of the habitat has the same probability of being colonized. As such, random distributions are an appropriate 'null model' against which to test observed patterns. Uniform distributions are exceedingly rare in nature, typically only occurring where interactions between individuals are very strong and consistent. As we indicated in the hypothetical example, the most common form of frequency distribution for the majority of macroparasites is one in which the parasites are aggregated among the available hosts in a population. Thus, most hosts in a population harbor zero or few parasites, while relatively few hosts harbor many parasites.

The magnitude of parasite aggregation within a host population can be estimated a number of ways (review in Poulin, 2007). Crofton (1971a, b) showed that the negative binomial model could be used as a statistical representation of aggregation, similar to the way the Poisson distribution represents a random distribution.

It is especially convenient because it is defined by only two parameters, the mean of the distribution and the exponent k. As k approaches zero, aggregation increases; as k increases, the parasite distribution becomes more random. Most mathematical models use k as the index of aggregation. The ratio of the variance in parasite counts to the mean provides another estimate of parasite aggregation within a sample of hosts. This is the so-called variance-to-mean ratio. If the variance and mean are approximately the same, the parasite population is randomly distributed in the host population (Fig. 12.2). As the variance/mean ratio increases above one, the parasite population becomes increasingly aggregated. In those rare instances when the variance/mean ratio is less than one, the population is uniformly distributed. Factors such as sample size, parasite transmission strategies, and whether interest is on the right-hand side of the distribution, i.e., heavily infected hosts, or left-hand side of the distribution, i.e.,

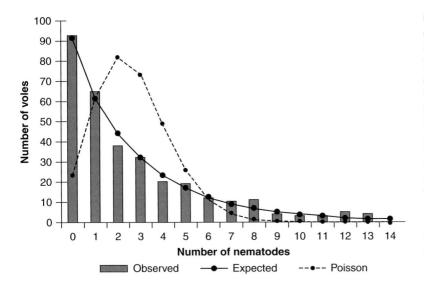

Fig. 12.3 Aggregated frequency distribution of the intestinal nematode *Heligmosomum mixtum* from a population of bank voles *Myodes glareolus* in Poland, showing the observed distribution and the predictions of the best-fit negative binomial model. The observed distribution closely corresponds to the expected curve of the negative binomial distribution, but does not conform to the Poisson distribution. The mean intensity and variance/mean ratio was 2.7 and 5.8 respectively (n = 319; k = 0.9). (Figure courtesy of Danielle Morrison; based on data provided by Jerzy Behnke.)

lightly-infected hosts, play a central role in determining which index of aggregation to use (Scott, 1987; Poulin, 2007).

A summary of these general ideas regarding parasite aggregation is described in Fig. 12.3. In a series of studies over several years, Jerzy Behnke and colleagues studied the helminth communities of bank voles, *Myodes glareolus*, from a site in Poland (e.g., Behnke *et al.*, 2008). The frequency distribution of their most common intestinal nematode, *Heligmosomum mixtum*, in the bank voles was determined. Their results showed that about 60% of over 850 nematodes present in the sample of over 300 hosts were found in 20% of the voles. Most hosts contained 0–2 worms, whereas a small number of hosts contained over 10 (Fig. 12.3). The negative binomial distribution closely described the observed distribution of the nematode counts, whereas the Poisson distribution did not. In this example, k = 0.9 and the variance/mean ratio = 5.6, indicating moderate aggregation. The ubiquity of this general pattern is demonstrated in Fig. 12.4, involving four very different host–parasite interactions. In each case, the overall form of the frequency distribution is similar, despite the wide variation in the natural histories of the hosts and parasites, variation in sample sizes, and variation in infrapopulation intensities.

It should not be surprising that parasites of humans are also aggregated. Epidemiological studies relying on fecal egg counts can be revealing, despite the notoriously high variation that invariably arises from this approach. Evidence from studies involving the soil-transmitted, enteric nematodes such as *Trichuris trichiura*, *Ascaris lumbricoides*, *Ancylostoma duodenale*, and *Necator americanus* that we covered in Chapter 8 are almost always aggregated within their human populations (review in Holland & Kennedy, 2002). Most of these studies are based upon counts of parasite eggs or larvae in fecal samples. However, intensity can also be measured directly by counting the number of worms passed in the feces immediately following anthelmintic, e.g., ivermectin chemotherapy. Holland *et al.* (1989) used this method to assess intensities of *A. lumbricoides* in over 800 children in Ile-Ife, Nigeria following chemotherapy. In this case, only about 8% of the children contained 40 or more of the large roundworms (Fig. 12.5). Taken together, epidemiological data on a wide range of parasites of humans are consistent with the patterns demonstrated for parasites of wildlife and other animals.

This consistent pattern of aggregation in so many host–parasite interactions has inspired a large number of empirical and field-based studies designed to uncover underlying mechanisms. Anderson & Gordon

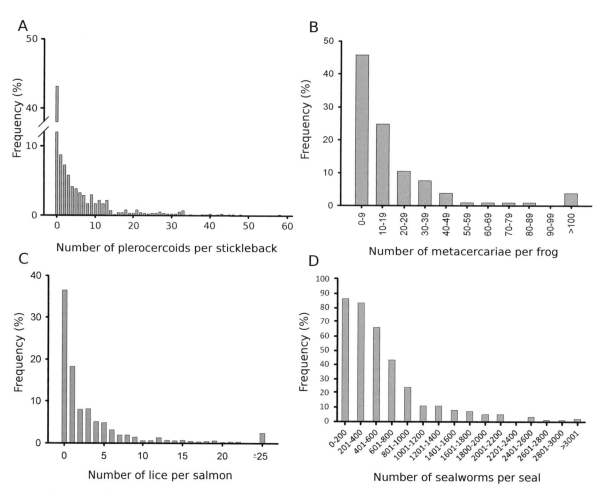

Fig. 12.4 Frequency distributions of parasites from a number of wild host species, illustrating that many macroparasites exhibit an aggregated distribution within the host population. (A) Distribution of plerocercoids of *Schistocephalus solidus* in a population of threespine stickleback *Gasterosteus aculeatus* (*n* = 1022) from Scout Lake, Alaska. (B) Distribution of metacercariae of *Ribeiroia ondatrae* in a population of metamorphs of leopard frogs *Lithobates pipiens* (*n* =105) from a wetland in Minnesota, USA. A total of 245 metacercariae were found in one frog. (C) Distribution of sea lice *Lepeophtheirus salmonis* on juvenile pink salmon *Oncorhynchus gorbuscha* (*n* = 818) from the Broughton Archipelago, British Columbia, Canada. The maximum number of lice on a pink salmon was 79. (D) Distribution of nematodes *Pseudoterranova decipiens* (juveniles and adults) in gray seals *Halichoerus grypus* (*n* = 357) from Nova Scotia, Canada. The most heavily infected seal contained 6243 sealworms. (Plate courtesy of Danielle Morrison; based on data provided by David Heins (A), Pieter Johnson (B), Simon Jones (C), and Gary McClelland (D).)

(1982) showed that observed patterns of aggregation could be attributed to a mix of both host and parasite factors; whereas some of these factors increase aggregation, others tend to make distributions more random or uniform. Using models similar to the one represented in Fig. 12.1, they showed that aggregation can arise by imposing heterogeneity in the rates that individual hosts are exposed to infective stages. They also showed that even if rates of exposure were completely random, the resulting distribution of parasites in the host population was still aggregated. Indeed, the infective stages of many parasites are probably not acquired by hosts randomly, but are themselves aggregated within their microhabitats and over time.

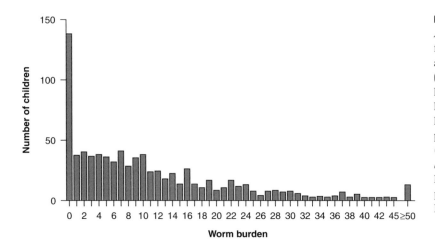

Fig. 12.5 Frequency distribution of *Ascaris lumbricoides* in children from Ile-Ife, Nigeria following anthelmintic chemotherapy ($n = 808$). (Figure courtesy of Danielle Morrison; modified from Holland, C. V. & Boes, J., 2002, Distributions and predisposition: people and pigs. In *The Geohelminths: Ascaris, Trichuris and Hookworm*, ed. C.V. Holland & M. W. Kennedy, pp. 1–24, with permission, Springer Science and Business Media.)

This means that for parasites with resting stages in intermediate hosts, aggregation in the subsequent host can arise from a single predator–prey interaction. Likewise, aggregation can arise in grazing mammals that are exposed to larval ticks that 'quest' together in large groups on the tips of vegetation, or to nematode larvae that are non-randomly distributed on pasture (see Chapter 8). These results show that when the distribution of infective stages is non-random, aggregation levels in the next host are magnified (Keymer & Anderson, 1979; Leung, 1998). Thus, aggregation can arise from simple stochastic transmission processes linked to heterogeneity in rates of exposure and parasite recruitment.

Heterogeneity in rates of exposure can also arise from variation in host traits such as age, gender, diet, and body size. For parasites that transmit via random contact with eggs or larvae, or those with active infective stages, small differences in host body size can lead to large differences in exposure. Differences among individuals in host behavior can be especially important when certain activities, e.g., rate of movement, foraging, hygiene, and vagility, are associated with increased risk of exposure to infective stages. The aggregation of *A. lumbricoides* among children in villages in southeast Madagascar is influenced by gender-specific behaviors (Kightlinger *et al.*, 1998). It is customary in this culture for boys to tend cattle and to participate in the cultivation of rice, away from the

primary source of parasite eggs within the villages. In contrast, girls tend to stay at home where they are continuously exposed. For other parasites of humans, variation in socio-economic status and access to health care facilities can play a role in determining risk of exposure to micro- and macroparasites. A close linkage between patterns of aggregation and behaviors that determine risk of exposure exists in many natural and anthropogenic host–parasite interactions.

For each of the aggregated distributions illustrated in Fig. 12.4, it would be straightforward to describe a range of factors that could lead to heterogeneity in rates of exposure. Thus, aggregation of larval *S. solidus* in populations of sticklebacks could arise from the stochastic ingestion of infected copepods. Age- and gender-dependent differences in foraging behavior and microhabitat selection may also play a role in determining which individuals are most likely to ingest infected copepods. Likewise, tadpoles with the highest intensities of *Ribeiroia ondatrae* metacercariae might be the larger, more active ones that happened to encounter a cloud of cercariae released from an infected snail. The key point is that understanding the various underlying causes of aggregation is not a simple task, particularly for those host–parasite interactions that are not amenable to experimental manipulation. Clearly, understanding the causes of aggregation requires extensive background knowledge on the natural history of both hosts and parasites.

In addition to heterogeneity in exposure, heterogeneity in susceptibility can also lead to aggregated patterns of distribution in host populations. Often, but not always, such susceptibility is linked to genetically based immune defenses (see Chapter 2). Thus, the characteristically high variation in immune competence that exists within host populations can, in theory, be sufficient to produce aggregated parasite distributions (Wilson *et al.*, 2002; Poulin, 2007). Wassom *et al.* (1986) show that observed patterns of aggregation of a cestode in wild mice populations could be explained by the presence of a few susceptible individuals that were homozygous recessive for a gene that controlled the rate of worm expulsion (see Section 12.3.3). However, cases in which resistance and susceptibility are determined by single genes are probably rare in most host–parasite interactions.

Despite extensive efforts, the role of host genetics in determining patterns of aggregation is unclear. One problem is that distinguishing genetic versus ecological hypotheses for the determination of variation in parasite numbers is not a simple task. The difficulty is perhaps best illustrated by studies involving human parasites. One common observation arising from detailed studies involving certain geohelminths is that following anthelmintic chemotherapy, individuals tend to acquire worm burdens similar to those they harbored before treatment. Thus, 'wormy persons' can be predisposed to heavy infection, perhaps on account of their genetically based immunity (review in Holland & Kennedy, 2002). Indeed, evidence from some epidemiological surveys supports the idea that high worm counts tend to be clustered within family units, indicative of genetic predisposition. The problem is that individuals within family units are also likely to share hygiene behaviors and other patterns of activity that could lead to high rates of exposure (e.g., Bundy & Medley, 1992). Confounding effects of this type are magnified in non-model and non-anthropogenic systems. Thus, resolution of the relative roles of host ecology versus host genes requires experimental approaches involving genetically characterized hosts and/or epidemiology surveys that incorporate modern genetic fingerprinting tools.

As we have seen, the consequences of aggregation span most areas of parasite population ecology and epidemiology, and also evolutionary biology. It should not be surprising that the parameter k is present in virtually all models that seek to understand parasite population dynamics and the role of parasites on their host populations (see Chapter 15). For parasites such as *A. lumbricoides* in humans, the practical implications of aggregation are enormous, since the diagnosis and selective treatment of 'wormy persons' will not only reduce morbidity, but also decrease overall rates of transmission (Anderson & May, 1991). Yet, aside from these obvious applications, the magnitude of aggregation will also influence the mate-finding opportunities for these dioecious worms and will influence the opportunities for density-dependent regulation of adult worm size, fecundity, and mortality. Levels of aggregation for *A. lumbricoides* will also impact the probability of encounter with potential interspecific competitors such as hookworms and whipworms, and will impact the magnitude of parasite-induced host mortality. Since parasite-mediated natural selection for enhanced host immunity or tolerance (see Chapter 15) is likely to be highest in heavily infected hosts, parasite aggregation also has evolutionary and coevolutionary implications for both the hosts and parasites. We return to the phenomenon of parasite aggregation in the remaining chapters.

12.3.3 Parasite population dynamics

One of the main aims of enquiry in parasite population ecology is to characterize the dynamic nature of parasite populations. A second is to understand the underlying factors leading to the loss and gain of parasites in infrapopulations. Earlier in this chapter, and also in prior chapters, we have hinted at the complexity and fascination of population-level patterns. Consider again the various rates that are involved in determining the loss and gain of parasites in a simple, direct life cycle macroparasite (Fig. 12.1). For example, λ (*per capita* parasite fecundity) is influenced by host-related factors such as immunity, size, diet, and gender, parasite-related factors such as size, age, and genotype, and, for poikilothermic hosts, environmental factors such as temperature. The following case studies emphasize this complexity, and emphasize the diversity of approaches that have been used to characterize and understand it.

12.3.3.4 The cestode *Caryophyllaeus laticeps* in fish

It may come as a surprise that the roots of quantitative parasite population ecology and ecological epidemiology can be traced back to this obscure group of cestodes that mature in freshwater fish. As we described earlier, studies on the population dynamics of *C. laticeps* in fish launched the career of Sir Roy Anderson. Parallel studies on this system at other locations in England by Clive Kennedy provided much of the background data in support of the first comprehensive text on parasite ecology (Kennedy, 1975). The general biology of the caryophyllidean cestodes was described in Chapter 6. Adult *C. laticeps* mature within the intestines of several species of freshwater

fish throughout the Northern Hemisphere. Eggs are shed from definitive hosts and ingested by a tubificid oligochaete. An oncosphere emerges from the egg and then develops into a procercoid within the worm's hemocoel. When a bottom-feeding fish eats the infected worm, the parasite develops to maturity in the gut.

Anderson (1974, 1976) monitored changes in the size and structure of bream *Abramis brama* and *C. laticeps* populations over 1 year. His results showed that the parasites were aggregated within samples of bream, likely arising from periodic waves of ingestion of procercoids that were themselves aggregated within the oligochaete population. His results also indicated a seasonal pattern of transmission from intermediate hosts to final hosts as well as a seasonal pattern of worm growth and survival. Transmission occurred throughout the year, but peaked in winter and spring, particularly within older fish. Worm intensity peaked in early summer, then declined sharply as water temperature increased (Fig. 12.6). Although worm survival dropped in summer, most of the survivors were gravid.

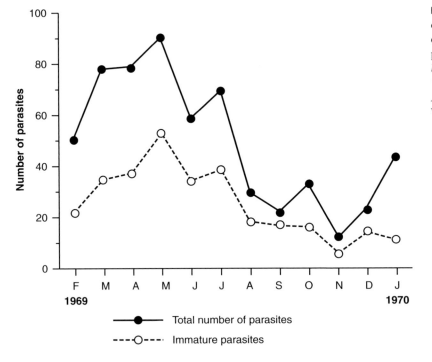

Fig. 12.6 Seasonal variation in total cestode numbers and immature cestodes *Caryophyllaeus laticeps* per sample of 30 bream *Abramis brama*. (Modified from Anderson, 1974, with permission, *Journal of Animal Ecology*, **43**, 305–321, John Wiley & Sons.)

It was these few survivors that contributed most to the overall contamination of the substrate with eggs that then became available for ingestion by oligochaetes (Anderson, 1974). An annual cycle of transmission occurred when fully developed procercoids became available for ingestion the following spring.

The population dynamics of C. laticeps in a river population of dace Leuciscus leuciscus was also strongly seasonal (Kennedy, 1968). Transmission of procercoids into dace began in December, peaked in spring and early summer, then stopped. As in the pond study, transmission of procercoids, adult worm maturation, and adult worm mortality were closely linked to rising temperatures. Kennedy (1969) also found that oligochaetes were infected throughout the year and that the feeding behavior of dace was consistent from one season to the next. The implication was that dace were constantly being exposed to procercoids in infected intermediate hosts but, during certain times of the year, many parasites were incapable of becoming established. The results from these two studies show that important differences in the details of transmission occur for the same parasites in different definitive hosts, and between habitats. Yet, despite differences in patterns of transmission and worm reproduction, a temperature-dependent decline in worm numbers occurred each summer in both study systems.

12.3.3.5 The monogenean *Pseudodiplorchis americanus* in desert toads

Ecologists often study extreme cases as a tool to understand general phenomena. Indeed, 'extreme' is one way to characterize the potentially insurmountable challenges facing an aquatic parasite that only infects a desert host. Surely, the transmission coefficient that is so prominent in the Anderson & May models will be zero! Yet, in the deserts of southeast Arizona, USA, the urinary bladders of most spadefoot toads Scaphiopus couchii contain the polystomatid monogenean Pseudodiplorchis americanus.

The transmission dynamics and population ecology of P. americanus are described in a series of studies completed by Richard Tinsley and his students (review in

Tinsley & Jackson, 2002). The hosts are explosive breeders, only visiting their ephemeral breeding ponds during a 1-3 day period following extreme rainfall events (Fig. 12.7). For the rest of the year, the toads are foraging on land, or are estivating at least 1 m beneath the surface. Thus, transmission of oncomiricidia is only possible during this brief spawning period. Remarkably, the transmission window is narrowed further because

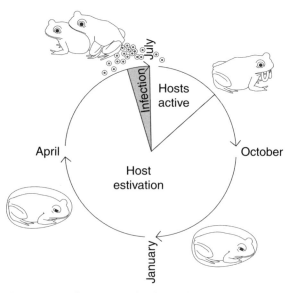

Fig. 12.7 Annual activity cycle of spadefoot toads *Scaphiopus couchii* precisely determines life cycle events and transmission opportunities for the monogenean *Pseudodiplorchis americanus* in Arizona. Toads estivate below ground for 10–11 months of the year. They emerge during torrential rains and males call and amplex females, and spawning occurs at night in temporary desert pools in July. The toads feed extensively during evenings in July and August. Transmission of *P. americanus* oncomiracidia and infection of toads occurs only during host spawning, and is restricted to only a maximum of 7 hours during 1 to 3 nights throughout the year. Ciliated oncomiracidia invade via the nostrils while toads are submerged and spawning in water. Juvenile worms develop in the lungs and migrate internally while the toads are actively feeding. Maturation of the parasite occurs in the urinary bladder and larvae are retained *in utero* during host estivation, ready for the next opportunity for transmission 1 year after infection. As toads enter water to spawn the following July, encapsulated larvae are released, hatching into immediately infective oncomiracidia. (Figure courtesy of Danielle Morrison; modified from Tinsley, 1989, *Parasitology Today*, 5, 190–195.)

the breeding toads are strictly nocturnal. Tinsley (1990) has shown that the transmission of oncomiricidia is restricted to approximately 7 hours per night during a 3-night period per year. At one study site, every male toad was infected in most years, with mean intensities remarkably stable at about 100 worms/host (Tinsley & Jackson, 1988). This transmission success arises from adaptations that lead to precise integration of host and parasite reproduction (reviews in Tinsley, 1989; 1990). Further, adult worms are ovoviviparous, releasing fully developed, relatively long-lived oncomiricidia the instant the worms detect a rise in host reproductive hormones. These infective offspring pass out of the toad's urinary bladder into water. Approximately 30% of individual oncomiracidia hatch and infect a new host (Tinsley & Jackson, 1988).

Invading oncomiracidia invade the nostrils and then migrate into the lungs where they remain for about 1 month (Fig. 12.7). While in the lungs, the worms acquire the protective characteristics of the adult tegument that provide for their survival during migration through the hostile stomach environment. Eventually, the worms pass into the urinary tract and end up in the urinary bladder, where they mature and begin to accumulate the infective larvae over the next 11 months, despite estivation by their host.

Worm intensity appears to be strongly regulated by the toad's immune system. In a comparable monogenean/toad interaction, the results of laboratory exposures showed that hosts develop a strong acquired immune response that reduces worm survival, time to reproduction, and fecundity (Tinsley et al., 2012). A response of this type can explain the decline in P. americanus prevalence to about 50% and the decline in mean intensity to about six adults/host (Tinsley & Jackson, 2002). Juvenile worms in the lungs and adults in the urinary bladder are blood-feeders. As a consequence, they are a chronic drain on host resources and a potential cause of density-dependent, parasite-induced host mortality (Tinsley et al., 2002). Toads require an adequate energy reserve in the form of fat in order to survive the 11 months of estivation. Tinsley and his colleagues estimated that an average parasite load of six worms consumed approximately 7% of a toad's body fat over 11 months (Tocque & Tinsley, 1994). The eventual outcome is dependent on the success of host feeding during the previous summer and the duration of the estivation period when the toads are not able to feed (Tinsley et al., 2002).

12.3.3.6 The cestode *Hymenolepis citelli* in mice

Adults of *Hymenolepis citelli* are found within the intestines of deer mice, *Peromyscus maniculatus*, where they produce eggs that are shed in feces. When camel crickets, *Ceuthophilus utahensis*, accidentally ingest the eggs, a larva emerges, penetrates the gut wall, and then migrates to the hemocoel to develop into a cysticercoid. Following ingestion by a mouse, the cysticercoid develops into an adult.

The ease with which deer mice can be raised in the laboratory makes this a suitable model for combined field and empirical approaches. Initial laboratory studies showed that young deer mice are 100% susceptible to *H. citelli*, but that most mice eliminate the worms before they become mature (see Wassom et al., 1986). When these mice are subsequently challenged with *H. citelli*, most are resistant. But a few mice in the lab population remained susceptible to both the stimulating and the challenge infection. This resistance is density dependent, being directly related to the dose of cysticercoids presented to the mice in the challenge infection. It is now known that acquired resistance to *H. citelli* is regulated by a single autosomal dominant gene and that susceptible hosts are homozygous for the recessive gene. Moreover, the immunity could be transferred to uninfected hosts via cells originating from the lymph of resistant mice. The results of field surveys showed that the distribution of *H. citelli* was restricted to small foci of susceptible hosts that represent about 25% of the total population. The authors concluded that host genetics, especially genetically based resistance/susceptibility, played a key role in determining population structure, transmission dynamics, and patterns of aggregation of this cestode in natural populations of mice (Wassom et al., 1986).

The results of these early studies involving mice confirmed prior studies involving easily manipulated crop plants and domestic and laboratory animals. Thus, it became clear that the genetic structure of a host population could determine key aspects of parasite population dynamics (review in Wilson *et al.*, 2002). This is a key advance from an applied perspective, since the identification of genes that determine host susceptibility can be used in targeted breeding programs to improve resistance in domestic animals and to target treatment and control strategies for human parasites. However, it has become equally clear that the complex cascade of immune recognition, rejection, and control of parasites that we covered in Chapter 2 is regulated by a vast network of genes. Thus, single gene control of parasite populations such as that observed in the *H. citelli*-mouse model is probably rare. Currently, the availability of microsatellite markers and other molecular tools makes it possible to identify gene loci responsible for resistance/susceptible phenotypes (see also Chapters 2 and 15). In the case of the mouse nematode *Heligmosomoides polygyrus*, at least four loci have been identified that link worm egg production with host immunity (Behnke *et al.*, 2003). Likewise, egg production of certain nematodes of Soay sheep, *Ovis aries*, on isolated islands off the Scottish coast (see Chapter 15) are determined by specific genotypes at a number of microsatellite loci, some of which are located within the MHC (Paterson *et al.*, 1998).

12.3.3.7 The trematode *Halipegus* spp. in frogs

Recall that these hemiurid flukes mature in green frogs, *Lithobates clamitans*, and use pulmonate snails, ostracods, and odonates in their obligate four-host life cycle (Goater *et al.*, 1990; see Fig. 6.24). For parasite population ecologists, site selection of the adult flukes within the frog's buccal cavity (see Color plate Fig. 1.1) or eustachian tubes means that infrapopulations of the adult flukes can be enumerated over time without killing the host (review in Esch et al., 1997). Moreover, individual adult worms can be added to, or removed from, individual marked hosts (and these frogs

subsequently recaptured) to assess the roles of density dependence, predisposition to infection, and host characteristics in determining population-level patterns. In Charlie's Pond, North Carolina, extensive field studies have shown year-to-year consistency in the population dynamics of *H. occidualis*. Intensities vary from one to about 40 in frogs, and the distribution of the adult flukes within the frog population is aggregated (Wetzel & Esch, 1997). Metacercariae transmission into frogs is seasonal, with a spring/early summer peak similar to the pattern observed for *C. laticeps*. An almost instantaneous rise in mean infrapopulation intensities in spring is most likely due to chance predation on a small number of heavily infected odonates. Maturation of immature flukes occurs within about 7 days. Frog recapture data further revealed that adult flukes can overwinter in their hosts and that the longevity of *H. occidualis* is at least 9 months (Wetzel & Esch, 1997). Recaptures of infected frogs showed that rapid increases in metacercariae transmission in spring/summer were followed by equally rapid and sharp decreases in infrapopulation intensities. Wetzel & Esch (1997) suggested that density- and temperature-dependent immunological responses were the cause of worm mortality in individual hosts.

Wetzel & Esch (1997) showed that the rate of transmission of *Haligegus* spp. metacercariae into frogs was related to the patterns of emergence of odonate paratenic hosts. An observed decline in adult infrapopulation sizes in fall coincided with a decline in the ingestion of odonates by green frogs. In contrast, the summer breeding season of green frogs coincided with maximum *H. occidualis* transmission. During this time, males are highly territorial and female frogs show strong site fidelity. This limited host mobility enabled Zelmer *et al.* (1999) to assess the suitability of particular microhabitats within the pond for metacercariae transmission. By plotting the maximum *H. occidualis* abundance of each frog against its capture site over a 4-year period, the authors found that mean abundance was highest in frogs that foraged within four 'hotspots' of transmission (Fig. 12.8). These four transmission foci occurred in areas with

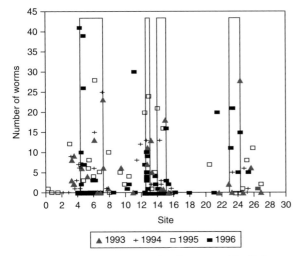

Fig. 12.8 Infrapopulation sizes of *Halipegus occidualis* in green frogs *Lithobates clamitans* captured from 30 microhabitats in Charlie's Pond, North Carolina over a 4-year period. Rectangles delineate four foci of infection where mean abundances were significantly higher than in the rest of the pond. (Modified from Zelmer *et al.*, 1999, with permission, *Journal of Parasitology*, **85**, 19–24.)

large littoral zones and patches of emergent vegetation that likely favored transmission of all life-cycle stages. In this case, microhabitat selection and host breeding behavior predisposed a few frogs to high rates of cercariae transmission (Zelmer *et al.*, 1999).

During one field season, Zelmer & Esch (2000) manipulated the intensities of *H. occidualis* in frogs by removing all but one adult fluke from heavily infected hosts (those with ≥ 15 flukes). In effect, Zelmer was testing the 'selective treatment' notion proposed for treating humans infected with parasites, i.e., concentrating therapy on heavily infected individuals. While overall prevalence was not altered, mean intensity and aggregation were decreased by 85% and 63%, respectively (Zelmer & Esch, 2000). Data collected the following season from marked frogs revealed that mean intensity and aggregation returned to pretreatment levels, indicating no noticeable effects of worm removal and long-term stability of the frog–*H. occidualis* system. Zelmer & Esch (2000) propose that the observed stability was

independent of parasite intensity and aggregation. They suggested that parasite prevalence is of critical importance because a frog infected with a single gravid parasite will release large numbers of infective eggs to infect snails.

These case studies illustrate a number of key points. First, when natural parasite populations are monitored over time and space, parameters such as rate of transmission, rate of mortality, and rate of parasite reproduction are highly variable. Whereas variation in these rates arises from combinations of host, parasite, and seasonal habitat characteristics, much of it is due to stochastic environmental factors that act at various stages within parasite life cycles. Second, despite enormous levels of background variation and case-specific details, clear patterns of parasite loss and gain can be observed. But the identification of key underlying processes requires experimental manipulations. For instance, each example involves a characteristic period during which parasites are added to infrapopulations, followed at some time later by a period of loss. In the case of *C. laticeps* in fish, that loss could be due to a temperature-dependent immune response, density-dependent parasite survival, or annual worm mortality. Distinguishing among these alternatives is rarely possible with field data alone.

Fortunately, there is a large literature on the population dynamics of parasites of model and non-model hosts raised under laboratory and natural conditions. The general approach is to expose individual hosts to known numbers of infective larvae and then evaluate as many of the rates described in Fig. 12.1 as possible. In this way, the contribution of host, parasite, and environmental factors to overall variation in targeted rate parameters can be assessed. A logical extension to this process is to expose hosts multiple times to known numbers of infective stages, a scenario that realistically parallels continuous exposure in nature. Empirical approaches of this sort have clarified the processes that underlie the patterns observed in the case studies described above:

temperature-dependent growth and survival of macro- and microparasites of fish and amphibians (e.g., Scott & Anderson, 1984), temperature- and density-dependent immunity in poikilotherms (e.g., Tinsley *et al.*, 2012), density-dependent parasite growth and survival (see Section 12.3.4), and the relative roles of parasite and host genetics (e.g., Sayles & Wassom, 1988).

12.3.4 Regulation of parasite populations

Our examples up to now have demonstrated that numerous factors can be identified that act solely and interactively to determine the wax and wane of natural parasite populations. Indeed, factors such as temperature, moisture, and humidity are probably the key determinants of infrapopulation sizes in some systems, particularly those involving poikilothermic hosts. However, ecologists are keenly interested in distinguishing those factors that *determine* population size from those that *regulate* population size. In the latter case, ecologists focus their attention on factors such as predation, competition, and parasitism that reduce the *per capita* survival and/or reproduction of individuals in a population as the density of the population rises. Thus, population regulation requires the influence of density-dependent processes, which typically lead to population sizes that fluctuate over time around a fairly stable value. Such long-term stability is a hallmark of some parasite populations, inspiring numerous studies aimed to detect density-dependent processes (review in Poulin, 2007).

Theoretically, the regulation of parasite populations is closely tied to the phenomenon of aggregation. Depending on the species, parasites that are not highly aggregated in their host populations will exist within infrapopulations that rarely reach intensities where density-dependent effects on survival and growth would occur. On the other hand, if the parasites are highly aggregated, then most parasites will occur in the small number of infrapopulations where density-dependent effects will be strong. Thus, depending on the precise nature of each host–

parasite interaction, the manner in which parasites are aggregated in their host population may be a key determinant of regulation (Anderson & Gordon, 1982).

Although it is possible to detect density-dependent effects with field-collected data, the interpretation of underlying processes can be difficult (Shostak & Scott, 1993). In contrast, experimental approaches have provided solid evidence for the importance of density-dependent effects. Goater (1992), for example, demonstrated density-dependent effects on worm establishment, growth, and fecundity in toads experimentally infected with the lung nematode *Rhabdias bufonis*. Adult worm establishment in the lungs declined as the number of worms in the lungs increased, and heavily infected toads produced fewer larvae (Fig. 12.9). Thus, there was a threshold for the number of worms maturing in the lungs, and this in turn affected growth of individual worms, the larval output per host, and *per capita* fecundity. For these lungworms, space constraints and/or nutrient limitation limited worm performance. Prevalence of lungworms is often 100% in natural populations of adult toads, and intensities typically span the range depicted in Fig. 12.9. Thus, density-dependent worm growth and survival may regulate adult lungworm populations in natural populations of toads (Goater, 1992). These results are consistent with the results of a large number of similar studies, tracing back to the earliest demonstrations of the 'crowding effect' in cestode and acanthocephalan infrapopulations (Read, 1951; Bush & Lotz, 2000). Although it is usually difficult to distinguish the relative roles of competition for nutrients versus space availability as underlying mechanisms, the consequences on population regulation are the same.

Density dependence can also occur in intermediate hosts, with important consequences for the overall regulation of parasite populations. In a classic example, the notoriously high variation in cercariae output from individual snails infected with trematode larvae tends to be independent of the

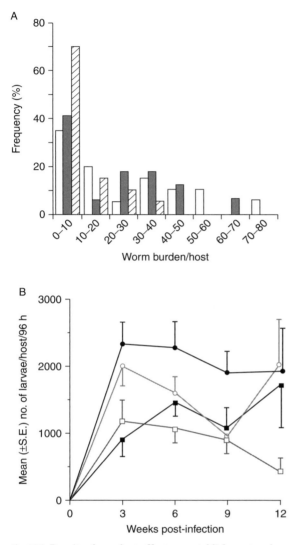

Fig. 12.9 Density-dependent effects on establishment and fecundity of *Rhabdias bufonis* in lungs of juvenile toads *Bufo bufo*. (A) Frequency distribution of the numbers of adult *R. bufonis* recovered from toads at necropsy after 3 weeks (□), 6 weeks (■), and 12 weeks (▨). (B) Temporal changes in the mean numbers of *R. bufonis* larvae recovered/host from toads exposed to 10 (●), 40 (○), 80 (■), or 160 (□) infective larvae. (Modified from Goater, 1992, with permission, *Parasitology*, 104, 179–187, Cambridge University Press.)

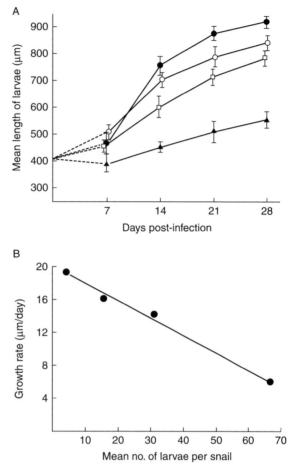

Fig. 12.10 Density-dependent effects on larval length and growth rate of *Elaphostrongylus rangifera* in terrestrial snails *Arianta arbustorum*. (A) Growth in length of *E. rangifera* at different larval densities. Mean larval densities for each of the four snail groups are 3.8 (●), 15.5 (○), 30.8 (□), and 66.8 (▲). Vertical bars are 95% confidence intervals. (B) The growth rate (mean growth in length per day) of *E. rangifera* at different larval densities (r = 0.99). (Modified from Skorping, 1984, with permission, *Oecologia*, 64, 34–40, Springer Science and Business Media.)

numbers of penetrating miracidia (e. g., Gerard *et al.*, 1993). Density-dependent parasite performance has also been demonstrated for the resting stages of parasites with indirect life cycles. Skorping (1984)

showed that the development rates of larvae of the reindeer nematode, *Elaphostrongylus rangiferi*, in the terrestrial snails it uses as intermediate host depended on infrapopulation density (Figs. 12.10, 12.11). Whereas about 90% of the larvae within low-exposure snails reached the infective stage, only 2%

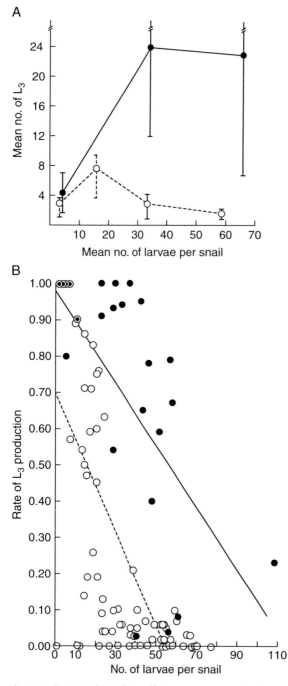

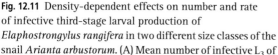

Fig. 12.11 Density-dependent effects on number and rate of infective third-stage larval production of *Elaphostrongylus rangifera* in two different size classes of the snail *Arianta arbustorum*. (A) Mean number of infective L_3 of

of larvae in high-exposure snails did so. Larvae were also severely stunted in the latter group (Fig. 12.10A) and they grew at a slower rate, especially in small snails (Fig. 12.11B). Intraspecific competition for limited food resources within snails was proposed as the potential mechanism (Skorping, 1984). Density dependence has also been demonstrated for larval trematodes, cestodes, and acanthocephalans in their second intermediate hosts, and reduced *per capita* performance during the resting phase can carry over to reduced performance in the adult phase (Poulin, 2007). This means that for parasites with indirect life cycles, the consequences of regulation occurring at one stage can be transferred to other stages.

Regulation of parasite infrapopulations can also arise when the magnitude of a hosts' defense response is determined by parasite infrapopulation size. Although there is little doubt regarding the key role of host immunity in determining key characteristics of infrapopulations (see Chapter 2), less is known regarding its role in population regulation. In an early example, variation in the duration of sterilizing immunity that mice develop following exposure to the cestode *Hymenolepis nana* is thought to be linked to a host's past history of exposure to infective stages (Esch, 1983). Likewise, the magnitude of the response of laboratory rodents to a second exposure to schistosome cercariae is dependent upon the numbers of cercariae they were exposed to in their first exposure (Terry, 1994). For rats exposed to the nematode, *Strongyloides ratti*, density-dependent effects on establishment, survivorship, and

Caption for Fig. 12.11 (cont.)
E. rangifera at 28 days post-infection in small (snail diameter = 3.4 mm) (○), and large (snail diameter = 6.0 mm) (●) snails; (B) The rate of L_3 production of *E. rangifera* with increasing larval density in the two size classes of snails. (Modified from Skorping, 1984, with permission, *Oecologia*, **64**, 34–40, Springer Science and Business Media.)

fecundity are mediated by the host immune response, not by space or nutrient limitation. Moreover, the effects are absent in rats with compromised immune systems (Paterson & Viney, 2002). Finally, the consistent decline to a small and stable number of adult bladder flukes in toads (see Section 12.3.3) is determined by host immunity (Tinsley *et al.*, 2012). Taken together, these results indicate that density-dependent immunity can act singularly, or in combination with other forms of density dependence, to regulate parasite infrapopulation sizes.

References

Anderson, R. M. (1974) Population dynamics of the cestode *Caryophyllaeus laticeps* (Pallas, 1781) in bream (*Abramis brama* L.). *Journal of Animal Ecology*, **43**, 305–321.

Anderson, R. M. (1976) Seasonal variation in the population dynamics of *Caryophyllaeus laticeps*. *Parasitology*, **72**, 281–305.

Anderson, R. M. (1993) Epidemiology. In *Modern Parasitology*, 2nd edition, ed. F. E. G. Cox, pp. 75–116. Oxford: Oxford University Press.

Anderson, R. M. & Gordon, D. M. (1982) Processes influencing the distribution of parasite numbers within host populations with special emphasis on parasite-induced mortalities. *Parasitology*, **85**, 373–398.

Anderson, R. M. & May, R. M. (1979) Population biology of infectious diseases. I. *Nature*, **280**, 361–367.

Anderson, R. M. & May, R. M. (1991) *Infectious Diseases of Humans*. Oxford: Oxford University Press.

Behnke, J. M., Iraqi, F., Menge, D., *et al.* (2003) Chasing the genes that control resistance to gastrointestinal nematodes. *Journal of Helminthology*, **77**, 99–109.

Behnke, J. M., Bajer, A., Harris, P. D., *et al.* (2008) Temporal and between-site variation in helminths of bank voles (*Myodes glareolus*) from N.E. Poland. 2. The infracommunity level. *Parasitology*, **135**, 999–1018.

Bundy, D. A. P. and Medley, G. F. (1992) Immunoepidemiology of human geohelminthiasis: ecological and immunological determinants of worm burden. *Parasitology*, **104**, S105–S119.

Bush, A. O. & Lotz, J. M. (2000) The ecology of crowding. *Journal of Parasitology*, **86**, 212–213.

Bush, A. O., Lafferty, K. D., Lotz, J. M. & Shostak, A. W. (1997) Parasitology meets ecology on its own terms: Margolis *et al.* revisited. *Journal of Parasitology*, **83**, 575–583.

Cornwell, G. W. & Cowan, A. B. (1963) Helminth populations of the canvasback (*Aythya valisineria*) and host–parasite environmental interrelationships. *Transactions of the North American Wildlife Conference*, **28**, 172–199.

Crofton, H. D. (1971a) A quantitative approach to parasitism. *Parasitology*, **62**, 179–193.

Crofton, H. D. (1971b) A model for host–parasite relationships. *Parasitology*, **63**, 343–364.

Esch, G. W. (1983) The population and community ecology of cestodes. In *Biology of the Eucestoda*, ed. P. Pappas & C. Arme, pp. 81–137. New York: Academic Press.

Esch, G. W. (2007) *Parasites and Infectious Disease: Discovery by Serendipity, and Otherwise*. Cambridge: Cambridge University Press.

Esch, G. W., Gibbons, J. W. & Bourque, J. E. (1975) An analysis of the relationship between stress and parasitism. *American Midland Naturalist*, **93**, 339–353.

Esch, G. W., Wetzel, E. J., Zelmer, D. A. & Schotthoefer, A. M. (1997) Long-term changes in parasite population and community structure: a case history. *American Midland Naturalist*, **137**, 369–387.

Gerard, C., Mone, H. & Theron, A. (1993) *Schistosoma mansoni–Biomphalaria glabrata*: dynamics of the sporocyst population in relation to the miricidial dose and the host size. *Canadian Journal of Zoology*, **71**, 1880–1885.

Goater, C. P. (1992) Experimental population dynamics of *Rhabdias bufonis* (Nematoda) in toads (*Bufo bufo*): density dependence in the primary infection. *Parasitology*, **104**, 179–187.

Goater, C. P. & Holmes, J. C. (1997) Parasite-mediated natural selection. In: *Host–Parasite Evolution: General Principles and Avian Models*, ed. D. Clayton & J. Moore, pp. 9–29. Oxford: Oxford University Press.

Goater, T. M., Browne, C. L. & Esch, G. W. (1990) On the life history and functional morphology of *Halipegus occidualis* (Trematoda: Hemiuridae), with emphasis on the cystophorous cercariae stage. *International Journal for Parasitology*, **20**, 923–934.

Grenfell, B. T. & Dobson, A. P. (1995) *Ecology of Infectious Diseases in Natural Populations.* Cambridge: Cambridge University Press.

Holland, C. V. & Kennedy, M. W. (2002) *The Geohelminths: Ascaris, Trichuris and Hookworm.* Norwell: Kluwer Academic Publishers.

Holland, C. V., Asaolu, S. A., Crompton, D. W. T., *et al.* (1989) The epidemiology of *Ascaris lumbricoides* and other soil-transmitted helminths in primary school children from Ile-Ife, Nigeria. *Parasitology,* **99,** 275–285.

Hudson, P. J., Dobson, A. P. & Newborn, D. (1998) Prevention of population cycles by parasite removal. *Science,* **282,** 2256–2258.

Hudson, P. J., Rizzoli A., Grenfell, B. T., *et al.* (2001) *The Ecology of Wildlife Diseases.* Oxford: Oxford University Press.

Kennedy, C. R. (1968) Population biology of the cestode *Caryphyllaeus laticeps* (Pallus, 1781) in dace, *Leuciscus leuciscus* L., of the River Avon. *Journal of Parasitology,* **54,** 538–543.

Kennedy, C. R. (1969) Seasonal incidence and development of the cestode *Caryophyllaeus laticeps* (Pallus) in the River Avon. *Parasitology,* **59,** 783–794.

Kennedy, C. R. (1975) *Ecological Animal Parasitology.* Oxford: Blackwell Scientific Publications.

Keymer, A. E. & Anderson, R. M. (1979) The dynamics of infection of *Tribolium confusum* by *Hymenolepis diminuta*: the influence of infective-stage density and spatial distribution. *Parasitology,* **79,** 195–207.

Kightlinger, L. B., Seed, J. R. & Kightlinger, M. B. (1998) *Ascaris lumbricoides* intensity in relation to environmental, socioeconomic, and behavioral determinants of exposure to infection in children from southeast Madagascar. *Journal of Parasitology,* **84,** 480–484.

Lafferty, K. D. (2009) The ecology of climate change and infectious disease. *Ecology,* **90,** 888–900.

Leung, B. (1998) Aggregated parasite distributions on hosts in a homogeneous environment: examining the Poisson null model. *International Journal for Parasitology,* **28,** 1709–1712.

May, R. M. & Anderson, R. M. (1979) Population biology of infectious diseases. II. *Nature,* **280,** 455–461.

McCallum, H. (2000) *Population Parameters: Estimation for Ecological Models.* Oxford: Blackwell Scientific Publications.

Paterson, S. & Viney, M. E. (2002) Host immune responses are necessary for density dependence in nematode infections. *Parasitology,* **125,** 283–292.

Paterson, S., Wilson, K. & Pemberton, J. M. (1998) Major histocompatibility complex variation associated with juvenile survival and parasite resistance in a large unmanaged ungulate population (*Ovis aries*). *Proceedings of the National Academy of Sciences USA,* **95,** 3714–3719.

Poulin, R. (2007) *Evolutionary Ecology of Parasites,* 2nd edition. Princeton: Princeton University Press.

Read, C. P. (1951) The 'crowding effect' in tapeworm infections. *Journal of Parasitology,* **37,** 174–178.

Rózsa, L., Reiczigel J. & Majoros, G. (2000) Quantifying parasites in samples of hosts. *Journal of Parasitology,* **86,** 228–232.

Sayles, P. C. & Wassom, D. L. (1988) Immunoregulation in murine malaria: susceptibility of inbred mice to infection with *Plasmodium yoelii* depends on the dynamic interplay of host and parasite genes. *Journal of Immunology,* **141,** 241–248.

Scott, M. E. (1987) Temporal changes in aggregation: a laboratory study. *Parasitology,* **94,** 583–595.

Scott, M. E. & Anderson, R. M. (1984) The population dynamics of *Gyrodactylus bullatarudis* (Monogenea) within laboratory populations of the fish host *Poecilia reticulata. Parasitology,* **89,** 159–194.

Shostak, A. W. & Scott, M. E. (1993) Detection of density-dependent growth and fecundity of helminths in natural populations. *Parasitology,* **106,** 527–539.

Skorping, A. (1984) Density-dependent effects in a parasitic nematode, *Elaphostrongylus rangiferi*, in the snail intermediate host. *Oecologia,* **64,** 34–40.

Terry, R. J. (1994) Human immunity to schistosomes: concomitant immunity? *Parasitology Today,* **10,** 377–378.

Tinsley, R. C. (1989) The effects of host sex on transmission success. *Parasitology Today,* **5,** 190–195.

Tinsley, R. C. (1990) Host behaviour and opportunism in parasite life cycles. In *Parasitism and Host Behaviour,* ed. C. J. Barnard & J. M. Behnke, pp. 158–192. London: Taylor & Francis.

Tinsley, R. C. & Jackson, H. C. (1988) Pulsed transmission of *Pseudodiplorchis americanus* between desert hosts (*Scaphiopus couchii*). *Parasitology,* **97,** 437–452.

Tinsley R. C. & Jackson, J. A. (2002) Host factors limiting monogenean infections: a case study. *International Journal for Parasitology,* **32,** 353–365.

Tinsley, R. C., Cable, J. & Porter, R. (2002) Pathological effects of *Pseudodiplorchis americanus* (Monogenea: Polystomatidae) on the lung epithelium of its host, *Scaphiopus couchii. Parasitology,* **125,** 143–153.

Tinsley, R. C., Stott, L., York, J., *et al.* (2012) Acquired immunity protects against helminth infection in a natural host population: long-term field and laboratory evidence. *International Journal for Parasitology,* **42,** 931–938.

Tocque, K. & Tinsley, R. C. (1994) The relationship between *Pseudodiplorchis americanus* (Monogenea) density and host resources under controlled environmental conditions. *Parasitology,* **108,** 175–183.

Tompkins, D. M., Dobson, A. P., Arneberg, P., *et al.* (2002). Parasites and host population dynamics. In *The Ecology of Wildlife Diseases*, ed. P. J. Hudson, A. Rizzoli, B. T. Grenfell, *et al.*, pp. 45–62. Oxford: Oxford University Press.

Valtonen, E. T., Helle, E. & Poulin, R. (2004) Stability of *Corynosoma* populations with fluctuating population densities of the seal definitive host. *Parasitology*, **129**, 635–642.

Wassom, D. L., Dick, T. A., Arnason, N., *et al.* (1986) Host genetics: a key factor in regulating the distribution of parasites in natural host populations. *Journal of Parasitology*, **72**, 334–337.

Wetzel, E. J. & Esch, G. W. (1997) Infrapopulation dynamics of *Halipegus occidualis* and *Halipegus eccentricus* (Digenea: Hemiuridae): temporal changes within individual hosts. *Journal of Parasitology*, 83, 1019–1024.

Wilson, K. & Grenfell, B. T. (1997) Generalized linear modelling for parasitologists. *Parasitology Today*, 13, 33–38.

Wilson, K., Bjornstad, O. N., Dobson, A. P., *et al.* (2002) Heterogeneities in macroparasite infections: patterns and processes. In *The Ecology of Wildlife Diseases*, ed. P. J. Hudson, A. Rizzoli, B. T. Grenfell, *et al.*, pp. 6–44. Oxford: Oxford University Press.

Zelmer, D. A. & Esch, G. W. (2000) Relationship between structure and stability of a *Halipegus occidualis* component population in green frogs: a test of selective treatment. *Journal of Parasitology*, 86, 233–240.

Zelmer, D. A., Wetzel, E. J. & Esch, G. W. (1999) The role of habitat in structuring *Halipegus occidualis* metapopulations in the green frog. *Journal of Parasitology*, 85, 19–24.

13 Parasite community ecology

13.1 General considerations

The lesser scaup, *Aythya affinis*, is a common duck of the western North American prairies. Even the etymology of its genus name, *Aythya* (=water bird) suggests it is an unremarkable species of duck. Yet, for parasite ecologists it is far from unremarkable. In a classic study of parasite biodiversity, Albert O. Bush counted almost 1 000 000 parasites in a sample of 45 scaup, representing an astonishing 52 species of gut helminths alone (Bush & Holmes, 1986a, b). Although parasite biodiversity of this magnitude was foreshadowed by earlier studies on bird–helminth interactions by the fathers of parasite ecology in the former Soviet Union (Dogiel, 1964) and Poland (Wisniewski, 1958), Bush's work was the first to rigorously quantify the occurrence and intensity of individual species within individual hosts. His study was also the first to place parasite communities into the conceptual framework of mainstream community ecology.

We highlight the scaup–helminth example to emphasize a point that lies at the heart of parasite community ecology, i.e., individual hosts almost always contain more than one species of parasite. This basic tenet prompts several intriguing questions. Do the species compete for limiting resources and then assort themselves in a non-random manner to reduce interspecific overlap? Does the presence of one species impose fitness costs on co-occurring species? Do species assort themselves to increase the probability of intraspecific mating or to avoid interspecific mating? Are biodiverse parasite communities less likely to be invaded than depauperate communities? Similar questions can be asked at different scales. Why, for example, is parasite biodiversity higher in scaup than in other species of bird, even sympatric ones with which they share bodies of water? What is the role of host phylogeny versus host ecology in determining interspecific variation in parasite biodiversity? It is questions such as these that have inspired parasite ecologists since the earlier studies by Dogiel and Wisniewski. Such questions provide the framework for this chapter.

Esch *et al.* (1990) provided the first synthetic review of parasite community ecology, summarizing the field and setting the conceptual framework for its development over the subsequent two decades. Reviews regarding the structure of parasite communities in and on individual hosts, especially the role of interspecific competition in determining restricted distributions, are covered in Combes (2001), Janovy (2002), and Poulin (2007). Poulin & Morand (2004) review patterns of parasite species richness within, and among, species of hosts, paralleling the interest by ecologists in broad patterns of animal biodiversity that began in the mid 1990s. Theoretical and applied aspects of parasite community ecology are reviewed by Roberts *et al.* (2002). In this chapter, we provide an overview of selected aspects of parasite community ecology, with focus on two key features that have defined the subdiscipline. We first provide a general introduction to parasite community ecology, covering terminology and general approaches. We then characterize the manner in which multiple species of parasites are distributed within host individuals, and then describe the proposed mechanisms that underlie the phenomenon of 'restricted niches'. Next, we describe the factors that underlie the notoriously high variation in parasite biodiversity that exists between host taxa and between host populations.

13.2 Introduction to parasite community ecology

13.2.1 Terminology

Parasite community ecologists are interested in all aspects of parasite biodiversity. This all-encompassing term deals with all variation in our planet's life forms, from genetic variation among individuals to variation in the characteristics of species assemblages between populations and regions. Yet one fundamental component of studies in the biodiversity sciences, and parasite community ecologists are no exception, lies in the characterization of **species diversity**. Thus, a central and unifying focus lies in the description of the numbers (**species richness**) or types of species that exist within a defined region, be it a locality, a host, or a site within a host.

The terminology of parasite community ecology extends logically from the hierarchical structure that was described at the population level (Chapter 12). Bush & Holmes (1986a; 1986b) coined the term **infracommunity** to include all of the parasite infra-populations that exploit a single host individual. Infracommunity data are typically structured as a complete census of the presence and intensity of each parasite species in an individual host. When the procedure is repeated for additional hosts, variation between species of parasite in occurrence and abundance can be estimated (Table 13.1). Data arising from replicated infracommunities typically provide the raw material for statistical treatment of questions centered on interspecific interactions, dominance, evenness, departure from null models of species assembly, and comparisons of average species richness among samples of hosts. These data are at the lowest level of

Table 13.1. **Helminth species from the small intestine of 45 lesser scaup from Alberta, Canada**

Parasite[a]	Status A[b]	Status B[c]	Prevalence	Mean intensity±s.d.
Hymenolepis pusilla[C]	Co	S	98	4786 ± 6490
H. spinocirrosa[C]	Co	S	96	7868 ± 10177
H. abortiva[C]	Co	S	96	6690 ± 8697
Polymorphus marilis[A]	Co	S	93	35 ± 61
H. tuvensis[C]	Co	S	89	1590 ± 1996
Fimbriaria fasciolaris[C]	Co	G	89	98 ± 151
Apatemon gracilis[T]	Co	G	84	52 ± 119
Hymenolepis sp.[C]	Se	U	60	2194 ± 3692
H. recurvata[C]	Se	S	51	206 ± 335
Dicranotaenia coronula[C]	Se	G	51	19 ± 15

The list of species included here does not include the 38 species that were present in fewer than 50% of hosts.
[a] T=trematode; C=cestode; A=acanthocephalan.
[b] Co=core species; Se=secondary species.
[c] S=specialist; G=generalist; U=unknown.
(Modified from Bush & Holmes, 1986a, with permission, *Canadian Journal of Zoology*, **64**, 132–141.)

the hierarchy and they typically represent the source of data for higher levels.

The procedures used to collect infracommunity data extend from the population-level approach (Chapter 12) to include species that co-occur in or on hosts. Traditionally, 'kill and count' methods are used to characterize the abundance of individuals within particular sites in individual hosts, ultimately leading to the cross-sectional, 'snapshot' data shown in Table 13.1. This approach is more challenging for microparasite communities where emphasis is necessarily on crude estimates of occurrence and/or abundance based upon histological sections (e.g., myxozoans in fish), blood smears (e.g., *Plasmodium* in birds), or fecal smears (e.g., *Eimeria* in mammals). While the cross-sectional approach continues to dominate parasite community ecology, advancements in molecular diagnostics show promise for species-specific characterization of parasite communities (e.g., Gasser, 1999). These developments are important because they allow researchers to repeatedly sample individual hosts over time (longitudinal sampling) in order to characterize temporal changes in community assembly and structure.

Holmes & Price (1986) extended the hierarchical terminology by adopting ideas proposed for free-living communities. This terminology parallels that used for parasite populations. Thus, the next highest level is the **component community**, which includes all parasite species exploiting a host population. A component community is the sum of infracommunities within a sample of a single species of host. Figure 13.1 is a schematic representation of a hypothetical component community. Here, the 12 black ellipses represent an entire infracommunity and each white rectangle represents the gastrointestinal tract within that host. An examination of each gastrointestinal community indicates that species C and species D co-occur with each other and with either species A or B. However, species A and species B never co-occur. Extending this idea, consider species E and F, both of which live outside of the gastrointestinal tract. Species F can co-occur with any of the other species, while species E can co-occur with all but species B. One explanation

for this pattern is that the presence of A in the intestine precludes the presence of B (or vice versa), perhaps via interference or exploitative competition for limited nutrients or space in the gut. Likewise, the presence of species E may preclude the presence of species B, perhaps via host immunity.

Summed species lists originating from replicated component communities provide an estimate of all parasite species that exploit a host species. Thus, a **parasite fauna** is the list of parasites found in populations of hosts sampled across its distributional range (Poulin & Morand, 2004). Parasite faunas represent the most encompassing level of organization of parasite assemblages. Researchers interested in understanding broad patterns of parasite biodiversity among species of hosts often focus at this level.

13.2.2 Delineation of parasite communities

The search for general patterns in parasite communities often begins with the delineation of the parasites that comprise a community into certain types. Typically, the goal is to identify those parasites that share common attributes, such as life history characteristics, e.g., size, biomass, life cycle stage, common sites of occupation, and/or common food resources.

One of the most common distinctions is the degree to which constituent species are host- and site-specific. As was true at the population level (Chapter 12), characterization of species as host generalist or specialist can be useful in community-level analyses. For the rich intestinal infracommunities in lesser scaup ducks (Table 13.1), Bush & Holmes (1986b) showed that the relatively small subset of host specialists (those found exclusively in scaup, or in sympatric ducks and waders) formed the predictable component of the overall community. The host generalists were unpredictable in their overall occurrence and assorted themselves randomly along the small intestine. Poulin (1997a) showed that the richest parasite communities in 116 species of fish were composed of both host specialists and host generalists, but species-poor communities were composed mostly of host generalists. These empirical and survey results

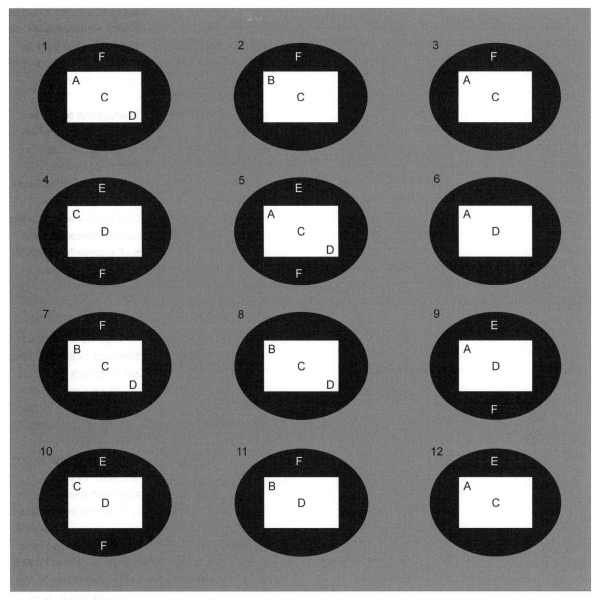

Fig. 13.1 A stylized component community (the large square) containing 12 infracommunities (ellipses). The white rectangles within the infracommunities represent the small intestine. Letters differentiate different species of parasites. Species A–D are found in the gastrointestinal tract whereas species E and F are found elsewhere in the body. Note that species C, D, and F appear to co-occur with any other parasite species. Species A and E can co-occur with each other and with any other species but neither co-occurs with species B. (Figure courtesy of Al Bush.)

indicate that the delineation of both infra- and component communities into specialists and generalists provides important insights into general patterns of parasite community organization.

Another general distinction in the types of parasites that make up infracommunities lies in their patterns of occurrence and abundance. In the parasites of lesser scaup shown in Table 13.1, it is clear that some species are more common, and also more numerous, than others. While these are relative and subjective terms, evidence from many studies supports the idea that individual hosts tend to harbor parasites that can be classified according to their frequency of occurrence (review in Poulin, 1999). Hanski (1982) proposed the core/satellite species dichotomy to describe these contrasting patterns of distribution and abundance within free-living communities. His hypothesis makes two predictions: there is a positive relationship between the occurrence of a species within a community and its abundance, and there is a bimodal distribution of species within a community. In effect, core species are regionally common and locally abundant, i.e., they colonize most appropriate habitats and are found in high numbers. Satellite species, in contrast, are regionally uncommon and locally rare, i.e., they colonize few appropriate habitats and are found in low numbers.

Bush & Holmes (1986a) tested the core–satellite species hypothesis using data from the parasites of lesser scaup collected from western Canada. They used prevalence to estimate regional dispersion and mean intensity to estimate abundance. They tested for a positive correlation between prevalence and intensity and examined the modes of distribution before assigning species to categories. Based on these criteria, they described a trimodal, not bimodal, pattern. Eight of 52 species were core, representing over 90% of all individual parasites. Another group of eight 'secondary species' represented approximately 7% of all individual parasites. The remaining 36 were described as satellite species, representing only 1% of the total. In this case, the predictions of the core-satellite hypothesis were generally met. Further, the distinction between the two (in this case, three) groups of species allowed for a more detailed understanding of potential mechanisms underlying parasite community structure (see below). Complete tests of the core-satellite species hypothesis are rare in the parasite community literature, although the idea is frequently used in a semi-quantitative manner to categorize species in a community into common or rare types. In general, the first prediction of the

core-satellite species hypothesis seems to be met across a range of community types. However, supportive evidence for the bimodal (versus continuous) nature of species distributions within communities is uncommon.

Depending on the precise questions being asked, delineation of parasite species into functional groups or **guilds** (Bush *et al.*, 1997) can also be useful. This concept is applied frequently to the study of potential competitive interactions between parasites along an environmental gradient. For example, a hypothetical gut community in a vertebrate may be composed of two species of nematode, two cestodes, and an acanthocephalan, each of which is similar in size. If the questions address the use of space along some linear axis, the functional group might be considered a single enteric guild and would include all five species. If the questions address acquisition of nutrients, two guilds would be meaningful, i.e., an engulfing guild (the two nematode species) and an absorbing guild (the two cestodes and the acanthocephalan). Finally, if the questions address niche restriction based on mating processes, two guilds would be meaningful, with a dioecious guild (the two nematode species and the acanthocephalan) and a monoecious guild (the two cestodes).

13.2.3 General approaches

Parasite community ecologists tend towards either descriptive or mechanistic approaches. In the former, comparisons are typically made between different populations of hosts, usually evaluating some form of similarity in community attributes between different component communities. This descriptive approach allows many questions to be addressed, ranging from the relative importance of parasite/host phylogeny and ecology in determining patterns of species richness to assessing the impact of pollution on parasite species richness. Minimally, for a descriptive study, we need a list of the parasites present, and some indication of the mean and variance in the numbers of worms of each species present in a sample of hosts (e.g., Table 13.1). From these data, ecologists may also calculate a measure of species diversity in a single datum incorporating not only the actual number of species present, but also a measure of their evenness in distribution across the sample of hosts (Poulin & Morand, 2004). The intent is to weight those species having more individuals present. The relative merit of various diversity indices, estimators of species richness, and different measures of abundance, as they apply to parasite communities, are described in Bush et al. (1997) and Poulin & Morand (2004).

Mechanistic approaches focus on how communities are organized. Here, emphasis is usually on comparisons between infracommunities, i.e., the infracommunities are used as replicates. Questions at this level focus on numerical and functional responses of one or more species of parasite on another. In the case of numerical responses, interest is on the manner in which the abundance of one species affects the abundance, biomass, and/or reproductive status of another species. In the case of functional responses, focus is on the mechanism by which one species affects the way another species utilizes resources, typically in the form of a shift in location away from an optimal site. It is in mechanistic studies that one can ask whether biotic forces such as competition, reproductive assurance, or host immunity structure communities.

If interest is in functional responses, data are required on the distribution of each parasite's population with respect to some resource axis (the parasite's **realized niche**). We need to know how much of that resource axis can be used by the parasite in the absence of other parasite species (the parasite's **fundamental niche**). Determining the fundamental niche is best accomplished experimentally by infecting an appropriate host with varying numbers of a particular parasite and determining the range of the resource axis used. Often, however, such experimental infections are intractable and the fundamental niche must be inferred. Bush & Holmes (1986b) suggested that, by summing the realized niches of a parasite species from a sample of infracommunities, a measure of the fundamental niche could be derived (Fig. 13.2).

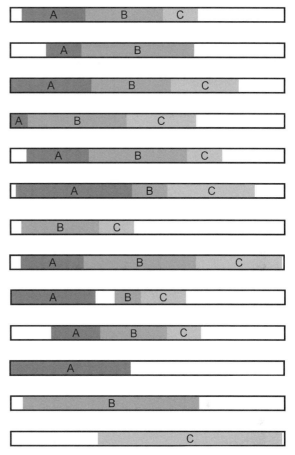

Fig. 13.2 Realized and fundamental niches of three parasites along a linear gradient in infracommunities. The first 10 infracommunities represent the actual (realized) observed distributions of each of three different parasites along a linear gradient. The bottom three artificial infracommunities show the fundamental (potential) distributions of each species. (Figure courtesy of Al Bush.)

13.3 The structure of parasite infracommunities: restricted niches

Ecologists recognize that no two species, even if they are closely related, utilize precisely the same habitats. Thus, the niches of virtually all species are restricted to some extent via the utilization of different food resources, different seasons, different geographical areas, and so on. Although there is often much interspecific overlap along these and other niche

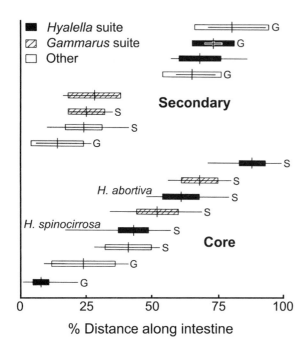

Fig. 13.3 The average realized distribution of core and secondary parasites along the small intestine of lesser scaup ducks. The vertical lines represent the mean of the median distributions of each parasite in each infracommunity; boxes enclose one standard deviation of the mean, and horizontal lines connect the mean anterior and mean posterior endpoints in each infracommunity. Note the even distribution of core species; secondary species are similar although none regularly occurs in the middle of the gut. S and G refer, respectively, to parasites that are specialists in lesser scaup ducks and parasites that are generalists in waterfowl. The status of two species, with respect to this dichotomy, is uncertain. (Modified from Bush & Holmes, 1986, with permission, *Canadian Journal of Zoology*, **64**, 142–152, NRC Research Press.)

dimensions, niche restriction and niche segregation seems to be a universal pattern. As we discussed in Chapter 12, all parasites exhibit various degrees of specialization, and no form of specialization is more apparent than specialization to a specific site in, or on, a host. The distribution of parasites A, B, and C along the length of the intestine in a hypothetical host in Figure 13.3 provides an example. In this case, the three species are roughly segregated into the anterior, middle, and posterior regions of the intestine. Relative to this example, we can ask two questions. What

mechanism leads to the restriction of species A to the anterior regions of the host intestine? Furthermore, what mechanism leads to the contraction of species A into particular regions of the anterior intestine in some hosts? These questions raise important issues in evolutionary ecology in general, and in parasite community ecology in particular. In this section, we discuss the main hypotheses explaining niche restriction and niche segregation in communities of parasites. For complete coverage of these phenomena, including detailed treatment of numerical responses, see Lotz & Font (1994), Combes (2001), and Poulin (2007).

13.3.1 Niche restriction mediated by interspecific competition

Studies in 'classical' community ecology have emphasized the role of interspecific competition in determining restricted niches. Under this paradigm, niches are restricted because co-occurring species with similar ecological requirements (especially food resources) restrict expansion into adjacent niches, or lead to contracted niches as species assort themselves to avoid each other. Not surprisingly, understanding the role of interspecific competition in determining restricted niches has received considerable attention from parasite ecologists.

Competitive exclusion can be considered an extreme example of niche restriction and segregation. Direct evidence for competition-mediated exclusion comes from studies that combine field and laboratory-based approaches in 'simple' two species assemblages. For instance, two species of polystome monogenean (*Protopolystoma* spp.) are common within the urinary bladders of the anuran frog *Xenopus laevis* collected from isolated wetlands in eastern Africa (Jackson *et al.*, 2006). Although both species are almost always present at each site, they never co-occurred in the same individual host. Via laboratory exposures, Jackson *et al.* (2006) tested the competition hypothesis, confirming that adults of the two species competed to such an extent that they did not co-occur in the same individual. Competitive exclusion of this magnitude is probably restricted to cases where worms compete for space and/or nutrients within a restricted site in a host. Further supportive evidence for exclusion of this type comes from several studies involving the larval stages of trematodes in their snail first intermediate hosts (review in Kuris & Lafferty, 1994). In this case, strong interference competition (or outright predation) from developing larval stages (e.g., rediae) often leads to the exclusion of smaller, or later-arriving, species. These results indicate the potential for interspecific competition to play an important role in determining restricted niches and overall patterns of parasite community structure.

Further direct evidence for the role of interspecific competition in determining restricted niches comes from experimental studies involving manipulation of the occurrence and intensity of potential competitors. Again, focus here has been on fairly simple, two species assemblages of enteric or gill helminths. The studies by John Holmes (Holmes, 1961, 1962a; 1962b) are of this type. He showed that when adults of the cestode *Hymenolepis diminuta* are present alone in rats, they attach within the anterior 35% of the gut. Likewise, adults of the acanthocephalan *Moniliformis dubius* also attached within the first 35% of the gut when they were present alone. When the two species were introduced simultaneously, *M. dubius* continued to occupy the same site, but *H. diminuta* shifted its attachment position further down the gut. It is important to note that the posterior niche shift by *H. diminuta* was associated with a decrease in average *per capita* biomass and worm length, respectively (Holmes, 1962a). Thus, competition-mediated 'interactive site selection' (Holmes & Price, 1986) was associated with fitness costs, at least for the cestode.

In a clever twist of this experimental paradigm, Patrick (1991) demonstrated ecological release by one nematode parasite when interspecific competitors were removed. He evaluated the occurrence and distribution of helminth parasites in the intestine of naturally infected flying squirrels, *Glaucomys volans*. Three species of nematode, *Strongyloides robustus*, *Capillaria americana*, and *Citellinema bifurcatum*, co-occurred in the anterior 30% of the small intestine

of most infected squirrels. Squirrels were then live-trapped, dosed with an anthelmintic drug to remove all nematodes, and then re-exposed to controlled numbers of *S. robustus* larvae. In the absence of competitors, *S. robustus* expanded its niche significantly. Further, as worm population size increased, the mean linear range of *S. robustus* along the gut increased and, in natural infections, there was no overlap in the distribution between parasite species. In combination, these results provide strong evidence for competition between *S. robustus* and the other two nematodes.

Indirect support for a role for interspecific competition comes from field studies involving a wide range of hosts. One of the most detailed studies invoking interspecific exploitative competition from field data is that of Bush & Holmes (1986b) on the biodiverse helminth communities of lesser scaup ducks. As noted earlier, they used the core–satellite species hypothesis to identify those parasites that were regionally common and locally abundant. The core species and secondary species, collectively, occupied all parts of the small intestine and exhibited an even dispersion along the small intestine (Fig. 13.3). Note from Figure 13.3 that six of the core and secondary species use the same amphipod intermediate host, *Hyalella azteca*, and that another four species use the sympatric amphipod *Gammarus lacustris*. Further, species identified as host specialists occupied all but the very anterior of the small intestine. Focusing on two species of small absorbers, e.g., members of the same guild both in their use of intermediate hosts and in their mode of using the small intestine, Bush & Holmes (1986b) provided evidence that the niche of *Hymenolepis abortiva* increased symmetrically in response to higher populations, but that under the same conditions, the niche of *H. spinocirrosa* shifted anteriorly (Fig. 13.4). The results from other studies involving communities of absorber helminths in water birds are consistent with these findings (Goater & Bush, 1988; Stock & Holmes, 1988), indicating that both inter- and intraspecific exploitative competition appear to be important in niche restriction.

Other kinds of field studies revolve around the assumption that the gut of a host is a linear, or radial,

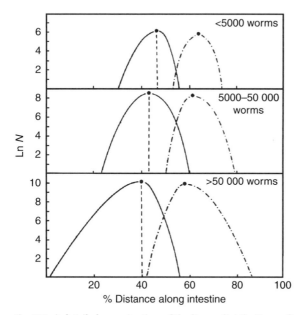

Fig. 13.4 A detailed examination of the linear distributions of two purported exploitative competitors, *Hymenolepis spinocirrosa* (solid line on the left) and *H. abortiva* (dot/dashed line on the right) along the small intestine of lesser scaup ducks. Each of the three panels represents the average distribution of the two species under different total numbers of all helminths present. The distributions are based on the mean anterior, median, and posterior individuals. The mean of the medians is indicated by a dot along each curve. The dashed line within the distribution of *H. spinocirrosa* shows the significant anteriad movement of the mean medianth individual. (Modified from Bush & Holmes, 1986, with permission, *Canadian Journal of Zoology*, **64**, 142–152, NRC Research Press.)

gradient and that it represents a meaningful niche axis for a parasite. Such studies usually demonstrate that the observed distributions of the parasites differ from the distributions expected from a model of simple random assembly. Bush & Holmes (1983) used this approach to demonstrate that the parasites of scaup were concordant in their distribution across infra-communities, that they co-occurred frequently, and that their realized niches were significantly reduced from their fundamental niches in the presence of potential competitors. The assumption of this latter observation is that, in the absence of competitors, a parasite's linear, or radial, distribution would equal its

fundamental distribution, i.e., ecological release would occur, as has been demonstrated in experimental studies.

The evidence from both experimental and field studies indicates that exploitative interspecific competition can lead to restricted niches, be they along the gut, gills, bladder, or on the skin of mammals or birds. The evidence is strongest for habitat shifts due to co-occurring, potential competitors – sometimes it is symmetrical, sometimes asymmetrical. There is some support, although from a restricted number of studies, that such shifts are associated with reductions in *per capita* biomass and egg production (e.g., Holmes, 1962a). Thus, there is a preferred site where parasite fitness is highest. The presence of other species shifts some members of an infrapopulation away from this optimal site, where fitness costs can be substantial.

Although there is little doubt that contemporary interactions between species of parasite are important determinants of parasite community structure, such interactions are not universal. There exists little supportive evidence for competition-mediated numerical or functional responses for the gut helminths of fish or mammals, for example (reviews in Kennedy, 1990; Behnke *et al.*, 2005), or for the specialized communities of monogeneans on the gills of fish (see below). In such cases, evidence supports the idea that the fundamental and realized niches of each species are almost identical, and that their niche dimensions are independent of the presence of co-occurring species. Holmes & Price (1986) recognized this dichotomy in community type by proposing an 'interactive versus isolationist continuum,' and urged researchers to focus on understanding the factors, especially relative colonization rates and relative probabilities of co-occurrence, that lead to different positions along the continuum.

13.3.2 Niche restriction mediated by other factors

A parasite's niche in contemporary time might simply reflect evolutionary descent. Thus, a parasite's present-day niche can be considered a consequence of its ancestor's niche in an ancestral host species

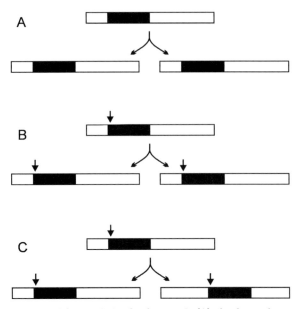

Fig. 13.5 Niche restriction by descent. In (A), the descendant parasites mirror the ancestral niche position in the descendant species. In (B), the descendant parasites mirror the ancestral niche position in the descendant species but here, the arrow indicates the location of a cue (perhaps the entry of the bile duct?). In (C), the niche position of the descendant parasite in one descendant daughter host differs from that in the ancestral host, but it seems to track the same cue (which has changed position in the daughter host). (Figure courtesy of Al Bush.)

(e.g., Brooks, 1980). A hypothetical example of this scenario is depicted in Figure 13.5A. Here, the distributions of the descendant parasites in the descendant hosts simply mirror the distribution of the ancestral parasite in the ancestral host. In this case, long-term specialization to particular host resources and/or long-term coevolution with host defense mechanisms may provide an explanation for niche restriction in the ancestral host (e.g., Price, 1980). Alternatively, the quantity or quality of an 'establishment cue' may differ following host speciation (indicated by the arrow pointing to the resource gradient in Fig. 13.5B), providing a focal point for the development of the parasites. In this case, niche restriction is mediated by orientation to a morphological or physiological parameter. In Figure 13.5C, we extend this hypothetical example to suggest that the precise location of the

parameter may differ in descendants of the ancestral host leading to a niche shift by the parasite in that descendant lineage.

Niche restriction might also enhance opportunities for sexual reproduction and/or reinforce reproductive barriers between co-occurring species. This mechanism for restricted niches is favored by Rohde (1994; 2005) following from the results of his studies involving monogenean trematode communities of fish. These communities are often dominated by several congeneric host specialists, each of which tends to be specific for a precise microhabitat on the gills (e.g., Simková et al., 2000). Although the requirements for competitive interactions are present in these highly diverse communities, evidence for interactive site selection is weak (Rohde, 2005). Instead, monogeneans with the narrowest host ranges and microhabitats tend to have increased intraspecific contact, and presumably higher reproductive success. Further, species coexistence is facilitated by highly specialized adaptations to attachment and copulatory structures that are, in part, determined by historical contingency (Morand et al., 2002). Detailed studies on the rich monogenean communities of European freshwater fish also support the notion that coexistence is facilitated by reproductive isolation (e.g., Simková et al., 2002). In this case, isolation and coexistence is mediated by the complex morphology of copulatory organs, such that species with the most similar reproductive morphologies tend to be segregated into different microhabitats. Although evidence for the reproductive assurance hypothesis is necessarily indirect (multi-species monogenean communities are difficult to mimic in the laboratory), there exists strong support for its importance in some groups.

13.3.3 Summary: restricted niches

Parasite ecologists have made significant progress in describing, and understanding, the phenomenon of restricted and segregated parasite niches. We can conclude that contemporary interspecific interactions are important under certain conditions, leading to concordant, predictable, and non-random assortment of species along meaningful resource gradients. We can also conclude that under some conditions, restricted niches increase parasite reproductive success and will be strongly favored by natural selection.

However, there is still much that we do not know. Two key obstacles remain. First, most parasite communities continue to be intractable from an experimental point of view. Biodiverse systems such as those found in lesser scaup, for example, would be impossible to mimic under laboratory or field conditions. Thus, experimental model systems are by necessity highly simplified. Jackson *et al.* (2006) caution that the seemingly simple competitive exclusion that occurs between two species of monogenean in African clawed frogs depends upon factors such as the immune status of the host and the sequence in which the host is exposed to the two competitors. Extending this 'simple' system to multi-species assemblages, such as those found in scaup, is currently not possible. The problem with the intractable nature of many host–parasite systems is that experiments designed to

distinguish, for example, the competition versus reproductive assurance hypotheses are non-existent. By default then, we remain uncomfortably reliant on observational data, an approach "bedevilled by pitfalls" (Behnke *et al.*, 2005). The second main obstacle is that we continue to face interpreting community-level data in the absence of information on available nutrients, the cornerstone of competition-mediated community structure.

Yet, despite the challenges associated with these important obstacles, it could be argued that we are entering a most exciting phase in parasite community ecology. Evidence from laboratory model systems, field surveys, and mathematical models is pointing to the important role of host defenses in mediating interactions between parasite species (see Box 13.1). While emphasis here has been on the evaluation of linkages between host immunity and numerical responses within parasite communities, the future is bright for the application of similar approaches to the phenomenon of restricted niches.

Box 13.1 | **Host immunity and parasite infracommunity structure: new evidence for an old hypothesis**

We might consider the 1980s as a 'heyday' of parasite community ecology. During this period, several laboratories around the world were developing 'simple' model systems, perhaps in the hopes of making fundamental contributions to general community ecology. After all, parasite communities had clearly defined boundaries that were relatively easily replicated. Further, powerful structuring forces in free-living systems, such as predation and habitat disturbance, were generally absent. Thus, any detectable non-random structure within these 'simple' communities must be due to other factors. Of course, we now know that the 'simplicity' of parasite communities is deceptive. One reason for this is the potential role of host defenses, especially immunity, in mediating interspecific interactions in or on hosts. This idea has a long history (e.g., Schad, 1966), but only recently have empirical studies provided supportive evidence. The notion that interspecific interactions between parasites may be mediated by host immunity is an important current focus in parasite community ecology (e.g., Karvonen *et al.*, 2009).

Remarkable results from a field study involving a community of monogenean trematodes on the gills of carp provided initial evidence for host-mediated parasite community structure (Paperna, 1964). Four species of *Dactylogyrus* were present on the gills, three of which could

Box 13.1 (continued)

co-occur. However, an immunopathological response to a fourth species, *D. vastator*, caused enough damage to the host's gills to exclude the other species. Ultimately, damage by *D. vastator* to the gills was so severe that it too could no longer survive. The now parasite-free gills recovered gradually and were recolonized, but not by *D. vastator*! In this case, parasite community structure is strongly affected by interspecific interactions. However, the interactions appear to be independent of host resources, and are, instead, mediated by species-specific host immunity.

Given the intricacies of the vertebrate host immune response (see Chapter 2), it should not be surprising that community structure mediated by host defenses presents a bewildering complexity of potential interactions. In general, we can classify the role of host immunity into two broad categories. In the first, antagonistic interactions between species of parasite may arise from cross-reactive responses leading to elimination (or reduction) of one or more species. Here, one species of parasite leads to a strong but generalized immune response that causes changes in features such as the structure and/or function of the gut (or the gills, in the case of the example described above). Loss of the cestode *Hymenolepis diminuta* from *Trichinella*-infected mice provides one example of exclusion via host immunity, in this case mediated by a generalized inflammatory response (Behnke *et al.*, 1977). In the second category, synergistic interactions between species may arise if one species depresses host immunity to the extent that other species are able to grow and reproduce more successfully. Enhanced survival of *Trichinella* in mice infected with the nematode *Nematospiroides dubius* is an example (Behnke *et al.*, 1978).

Laboratory and field-based studies completed over the last decade provide support for both antagonistic and synergistic interactions mediated by host defenses. The evidence comes from a range of community types, including protists in rodents and helminths of rabbits, mice, and fish. In a study involving laboratory mice, for example, the extent of competitive suppression of one clone of *Plasmodium chabaudi* by other clones was determined by the immune status of the host (Råberg *et al.*, 2006). In contrast, results from a field study, also involving mice as host, provided evidence for synergistic interactions mediated by host immunity. Jackson *et al.* (2009) showed that the common nematode (*H. polygyrus*), and especially a louse (*Polyplax serrata*), lowered the innate immune response of field-collected wood mice (*Apodemus sylvaticus*). Indeed, immunosuppression mediated by *H. polygrus* provides a key mechanism to explain the observation that helminth species richness increases in wood mice that are infected with the nematode (Behnke *et al.*, 2009). These results highlight the importance of subsets of species within a community in determining overall parasite community structure, in this case, via their modulation of immune function.

It is not difficult to imagine scenarios where both categories of immunity modulation could exist within an individual host to structure parasite communities. Such complexity appears to be a feature within at least two sharply contrasting host–parasite systems. In a study involving the gut helminths of field-collected rabbits, interspecific interactions covered a broad spectrum, ranging from competition, to cross-immunity, to immunosuppression (Lello *et al.*, 2004).

Box 13.1 | (continued)

Similarly, the biodiverse microparasite communities of voles (*Microtus agrestis*) were structured by a complex network of interactions that included both cross-reactive immunity and immunosuppression. Here, variation in the probability of a parasite colonizing a host was best explained by infection with other species, and not by factors such as host age, host condition, and season of the year (Telfer *et al.*, 2010). Results from studies such as these provide compelling evidence that community structure is much more complex, and interesting, than was ever envisioned.

13.4 The structure of parasite communities: species richness

We now shift our focus to the composition of parasite communities. Here, we explore general patterns of variation in community composition between samples of hosts and describe proposed underlying factors. The question of parasite community composition between samples of hosts, and its notoriously high variation between individuals, host species, and between host populations, lies at the root of parasite community ecology (e.g., Wisniewski, 1958; Dogiel, 1964). More recently, the question parallels the general interest by macroecologists in characterizing patterns of animal biodiversity, and in the general interest by evolutionary ecologists in searching for patterns in trait characteristics across species. Combes (2001) and Poulin & Morand (2004) provide reviews of theoretical and empirical advancements in this area.

Questions regarding comparative patterns in community composition can be evaluated between any two samples of host. In the previous section, we were concerned with patterns at the level of infracommunities, where features such as interspecific interactions, host specificity, and host immunity could determine the suite of species that colonize and establish within individual hosts. In that case, infracommunities are assembled within local regions, e.g., a lake, a pasture, etc., and within ecological time frames. In this section, our main interest is in community-level patterns at

higher hierarchical scales, especially at the level of component communities and at the level of entire parasite faunas (see Section 13.2 for definitions). At these hierarchical scales, we can expect that combinations of processes occurring over evolutionary time frames, such as species invasion, extinction, colonization, and speciation, will potentially play a role.

The related ideas of supply and screens (or filters) provide a useful context to begin a discussion of parasite community composition at each hierarchical level (Fig. 13.6; Combes, 2001). The idea of 'supply' derives from Connell (1985) and is expressed as 'supply-side' ecology by Lewin (1986). Simply stated for parasites, supply means that a parasite, with any and all of its required hosts, is available in the system under study. Holmes (1987) was the first to propose the notion that the mere presence of a parasite in a system does not mean that any host is exposed, or ultimately, infected. Indeed, he notes, "The parasite communities in individual hosts are non-random samples of the parasites available in the environment they occupy . . ." Subsequently, Combes (2001) has championed the position that characterizing various filters at each hierarchical level is necessary to accurately interpret the composition of natural parasite communities.

In Figure 13.6, we depict how supply and screens might influence patterns of species richness. The hypothetical Global Pool contains 15 parasites (each species is represented by a different letter) that can

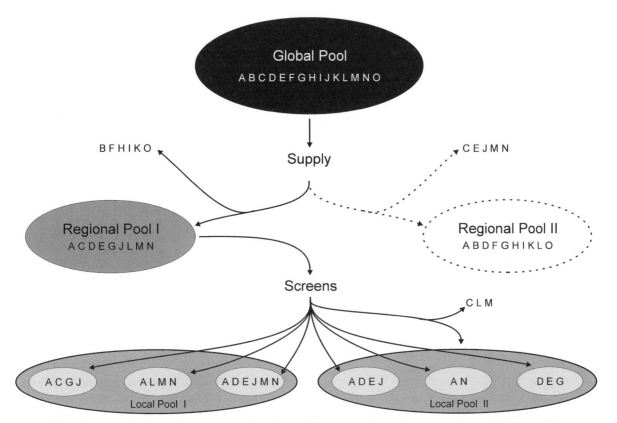

Fig. 13.6 The relationship between the presence of parasites in global, regional, and local pools. The small ellipses within local pools represent component communities and the letters designate different species of parasites. Supply will determine the subset of parasites available to different regional pools, whereas screens will determine what actually is available to local pools and, therefore, to component communities. (Figure courtesy of Al Bush.)

potentially infect one, or more, host species. Because of supply, hypothetical Regional Pool I will contain only nine species, whereas the second hypothetical Regional Pool will contain 10 species, i.e., supply determines what is actually available at the regional level. Some species are common to both regional pools, but Regional Pool I can never have species B, F, H, I, K, or O. Often, restriction due to supply would equate to historical/zoogeographical events, e.g., a glaciation. Alternatively, anthropogenic events, e.g., introductions, may alter supply in a contemporary time frame.

It is at the Local Pool level that a variety of potential screens become especially important. Screens determine what species are actually found in any particular component community (the smallest ellipses in

Fig. 13.6). For example, the screens may reflect biotic features such as host phylogeny (including both descent and cospeciation, coupled with host specificity), interactions between species, host immunology, or host food habits. Alternatively, screens may reflect abiotic features such as acid rain, episodic droughts, or extinctions. In Figure 13.6, no screen effectively prevents any potential parasite from being available at Local Pool I; some may be important at the component community level as implied by the three arrows pointing to different component communities. However, some powerful screen(s) does exclude parasites C, L, and M from being available to hosts that draw their parasites from Local Pool II. Therefore, just as parasites B, F, H, I, K, and O were unavailable to

Regional Pool I, parasites C, L, and M are unavailable to Local Pool II. And, just as in Local Pool I, some other screen(s), implied by the three arrows, may be important in determining what parasites infect different component communities. Furthermore, screens are not necessarily important in determining the species found in component communities; the occurrence of parasites in the component communities drawing their parasites from Local Pool I could simply reflect random colonization. Likewise, once the screen(s) excluding parasites C, L, and M is considered, the component communities drawing their parasites from Local Pool II might represent stochastic colonization.

13.4.1 Variation in parasite species richness between host species

When the composition of parasite communities is compared between two host species, extensive variation in species richness is the rule. The pattern is one of the most striking features of results reported by field-based parasitologists, regardless of host taxa or parasite taxa. Thus, the parasite fauna of lesser scaup collected from their breeding ponds and lakes located in western Canada comprises 52 species of intestinal helminths; on average, individual infra-communities contain about 14 species (Bush & Holmes, 1986a). On the one hand, the helminth faunas of sympatric white-winged scoters and canvasbacks are comparable to those in scaup (43 and 48 species, respectively), although their infracommunities have, on average, 17 and 19 species (Butterworth, 1982). On the other hand, helminth faunas in sympatric buffle-heads and gadwalls comprise about 20 species and individual hosts rarely harbor over 10 species (Butterworth, 1982). Community metrics from a handful of other species of duck fell between these extremes, as did the helminth communities in sympatric waders (Bush, 1990). Differences in parasite biodiversity of this magnitude are commonly reported in studies involving samples of hosts from other taxonomic groups.

A large number of factors can explain variation in community composition, or entire parasite faunas, between host species. For instance, it is straightforward to envision that quantitative differences in community composition among the sympatric water-fowl on prairie ponds and lakes could be due to inter-specific variation in features such as host size, geographical range, migration behavior, diet, bill dimensions, and so on. A standard approach is to use correlation analyses to evaluate variation in these ecological factors and variation in parasite community composition. The identification of such associations is important because they can indicate why some host species contain significantly more parasite species than others. Indeed, this semi-quantitative approach has been used to characterize patterns in composition within a wide variety of host groups infected by a wide variety of parasite taxa (reviews in Esch et al., 1990; Poulin & Morand, 2004).

Yet, while the identification of interspecific patterns in community composition and/or parasite faunas is straightforward, the interpretation of underlying causation is not. One problem lies in the sheer numbers of potential confounding variables, many of which are inter-correlated (e.g., host age, size, and longevity) that can severely restrict the ability to identify important single factors. Another confounding feature of stand-ard comparative approaches lies in the linkage between sampling effort and species richness. Thus, as more hosts and/or more populations of hosts are sampled, the numbers of rare species tends to rise non-linearly to a maximum value (Walther et al., 1995). The problem of unequal sampling effort becomes even more acute if sampling bias is towards larger or more wide-ranging hosts. It is for this reason that tests for patterns in species richness between two or more host species must control for variation in sampling effort.

Another methodological problem that plagues the comparative approach is the difficulty in disentangling phylogenetic (historic) from ecological (contemporary) effects. This is a general problem in comparative biology (Felsenstein, 1985; Harvey & Pagel, 1991), and is especially problematic for parasites (Bush et al., 1990). Just as ecological factors such as host feeding rate and host size must influence the opportunities for colonization of individual hosts by parasites (Fig. 13.6), so too must host phylogeny and other historical features influence the pool of parasites available for colonization (Fig. 13.6, and see Chapter 14). In the case of sympatric waterfowl on western Canadian ponds and lakes, to what extent are differences in parasite component communities or faunas attributable to factors such as feeding strategy (diving versus dabbling) or to phylogenetic relatedness among hosts? Both factors are undoubtedly important in most parasite communities. The problem from a methodological point of view is that due to relatedness, data points involving individual hosts are not statistically independent, and are not appropriate for standard parametric correlation analyses. Thus, tests for patterns in species richness between two or more host species must control for phylogenetic relatedness, typically by using phylogenetically independent contrasts (review in Poulin & Morand, 2004).

Poulin & Morand (2004) reviewed the results of a large number of comparative studies that reported between-host differences in parasite species richness. Their aim was to seek general patterns in the significance and direction of correlations between parasite species richness and host characteristics such as host body size, host diet, and host geographical range. In their analysis, they distinguished those studies that corrected for sampling effort and phylogenetic influences from earlier studies that did not. Their analyses provided two important results. First, earlier results that pointed to significant associations between ecological factors and species richness were found to no longer be significant when sampling biases were considered. This pattern seemed to be especially evident for tests of species richness and host geographical range. Second, studies that corrected for both unequal sampling and phylogenetic effects provided mixed results. Thus, one study concluded that host body size correlated strongly with species richness in vertebrates (Gregory et al., 1996), whereas another did not (Poulin, 1995). Two studies involving monogenean species richness and geographical range also generated contrasting results (Poulin & Rohde, 1997; Morand et al., 1999). Poulin & Morand (2004) concluded that while factors such as host size and geographical range are

associated with species richness in some circumstances, generalizations regarding the overriding determinants of parasite species richness among hosts are not possible.

Not all parasite ecologists support this lack of overall pattern, and not all support the overriding importance of host phylogeny. In an analysis that aimed to disentangle ecological versus phylogenetic effects in the parasite component communities of birds, mammals, and herptiles (amphibians and reptiles), Bush et al. (1990) showed that species richness was significantly higher in aquatic compared to terrestrial hosts. The magnitude of this effect was most striking in birds, where, on average, aquatic species had twice as many component species as their terrestrial counterparts. Although this study did not control for variation in sampling effort, or for phylogenetic effects, the pattern seems striking. Explanations for these strong differences also seem to make intuitive sense, whereby terrestrial hosts simply are not exposed to the many species (often as 'flocks') of cestode, trematode, and acanthocephalan that have obligate aquatic stages. The same argument can be made for the striking differences in species richness between birds (and one semi-aquatic mammal, the rice rat) and fish described by Kennedy et al. (1986). In this case, component community richness in the selected aquatic birds (and the mammal) was always orders of magnitude higher than in any component community described from fish.

13.4.2 Historical and ecological determinants of parasite species richness

In the next chapter, we emphasize how large-scale, historical events such as glaciation can influence parasite radiations, rates of host colonization, and extinction. It is clear that each of these processes alone, and in combination, can have a major impact on the composition of contemporary communities. It is equally conceivable that antagonistic interactions in the past between parasite species (the ghost of competition past) and between hosts and parasites (via host immunity) could have an effect. Although we cover some of these issues in the following chapter, and several are reviewed by Poulin & Morand (2004) and Poulin (2007), we wish to make two key points relative to historical effects.

The community of gut helminths in lesser scaup is dominated by a guild of small cestodes in the genus *Hymenolepis* (now *Microsomacanthus*; Table 13.1). Six of the 10 most common species belong to this guild, and of those species that reach intensities >1000 in individual hosts, all belong to members of this genus. Such 'species flocks' are commonly reported – dilepidid cestodes in shorebirds, larval diplostomatid trematodes in freshwater fish, dactylogyrid monogeneans in freshwater fish – and, when present, they tend to dominate their communities. Kennedy & Bush (1992) investigated the occurrence of 'species flocks' from published accounts. They were unable to identify a pattern in the distribution of these 'flocks' among host groups, but their occurrence alone highlights the importance of host phylogeny and historic events in determining current patterns of species richness. What ecological and/or historic factors promoted the radiation of these groups into their hosts? What factors restrict their colonization into sympatric species of host? We currently do not know the answers to these questions, but it is clear that the identification of the conditions favoring the occurrence of species flocks will increase our understanding of variation in community composition among hosts.

We have noted previously that all parasites show some degree of specificity, whether it is to a host species, a host genus, a host family, or to an even higher level of host taxon. Yet, specificity on its own can be a key determinant of species richness, as we see in Table 13.1. Poulin (1997a) proposed a novel idea based on his investigation of parasite species richness in Canadian freshwater fishes. His meta-analysis showed that the parasites found in species-rich communities occurred in fewer host species, while the parasites found in depauperate communities occurred in many host species. In other words, rich communities contained specialists and generalists, whereas depauperate communities were composed of generalists. Similarly, the biodiverse communities of ectoparasitic

arthropod on mammals are dominated by specialists, whereas depauperate communities are dominated by generalists (Vazquez et al., 2005). A study by Herreras et al. (1997) on the parasites of harbor porpoises (*Phocoena phocoena*) shows this latter feature, as do almost all studies on marine mammals. Taken together, these results point to the disproportionate role of specialists in the assembly of species-rich communities. Yet, while the pattern seems to be increasingly clear across a variety of systems, underlying mechanisms remain unknown.

Many interesting questions remain to be addressed. For example, it is clear that some parasite communities in closely related host species appear mostly to be descendants of an ancestral community. In these communities, what factors constrain parasites to retain similarity to an ancestral type? In contrast, if the parasite communities diverge to some extent between closely related host species, what is, or was, the driving force? It is this latter question that drives the idea that ecology can influence the diversity of parasite communities.

Parasite community ecologists have evaluated countless ecological characteristics for their associations with parasite community composition. Factors such as host behavior, host metabolism (e.g., endothermy vs. ectothermy), host genetics, host life histories, and so on have each been evaluated for their effects on species richness and faunal diversity (review in Poulin & Morand, 2004). Although individual factors have been shown to play a role in single systems, few generalizations can be made. One exception seems to arise from the many studies designed to understand the factors underlying variation in parasite communities between host populations (component communities). Studies involving fish collected from adjacent lakes, or oceanic reefs, have been especially prominent. Here, factors such as lake size, depth, and habitat heterogeneity, together with distance between lakes have been shown to play a role under some conditions (Marcogliese & Cone 1991; Marcogliese, 2001). Limnological characteristics of the lakes such as pH, calcium concentration, temperature, and productivity have also been shown to play key roles.

Some ecological determinants of parasite species richness can be considered surrogates for host feeding, especially for helminths. The key role of host diet was emphasized in early studies (Wisniewski, 1958; Dogiel, 1964), although empirical tests have only come in the last few decades. For example, several authors (Gregory, 1990; Poulin, 1997b) note that species richness increases with host geographical range (but see discussion above). An embedded implication is that hosts will be exposed to more parasites because they will encounter a wider array of intermediate hosts over their wide ranges. Bush (1990) compared the parasite communities of willets on their breeding grounds in several freshwater environments, and on their wintering grounds in several saltwater environments. No significant difference in species richness was found, but there was a radical change in the kinds of parasites, from mostly cestodes in freshwater to mostly digeneans in saltwater. It was clear that marked differences in host feeding patterns, and thus ingestion of different intermediate hosts, were a critical factor that produced the dramatic differences in parasite faunas. Edwards & Bush (1989) showed that ecological events could override phylogenetic events in determining species richness in avocets, *Recurvirostra americana*. Under certain environmental conditions, avocets fed on amphipods infected with several species of parasite that were normally found only in lesser scaup ducks. Clearly, contemporary ecological factors can play a strong role in determining patterns of richness.

Some anthropogenic factors are related to host feeding as well. One of the earliest to address this was Wisniewski (1958) who examined hosts from eutrophic Lake Druzno and mesotrophic Lake Goldapivo in Poland. After sampling many thousands of potential vertebrate and invertebrate hosts, Wisniewski concluded that parasites were moving primarily from fishes, amphibians, and certain invertebrates, to birds. Thus, parasites were following food chains (see also Marcogliese, 1991). Esch (1971) speculated that eutrophic systems were 'open' and subject to more opportunities for aquatic–terrestrial interaction such as may occur between piscivorous birds. Oligotrophic systems, on the other hand, were described as more

'closed,' with less aquatic–terrestrial interaction and less opportunity for the transmission of parasites through piscine intermediate hosts to avian definitive hosts. The speculation was, therefore, essentially linked to the nature of predator–prey relationships; according to Esch (1971), this was the primary determinant of parasite community richness in oligotrophic versus eutrophic ecosystems. Marcogliese & Cone (1996) examined the parasite communities in eels taken from Nova Scotian (Canada) rivers with acidic and more neutral pHs. They observed lower species richness in rivers with reduced pH. They attributed these lower diversities to reduced complexities in food chains that would typically involve parasite transmission (see Chapter 17).

Ecological determinants of species richness need not be related directly to feeding. For example, the impact of anthropogenic factors may be independent from what a host eats. This is particularly true of digeneans that have molluscs as obligate first intermediate hosts. Curtis & Rau (1980) showed that the presence of strigeids in their second intermediate hosts, fish, was related to calcium ion concentrations in lakes. Lakes with low concentrations did not support snails, the obligate first intermediate host. Valtonen *et al.* (1997) concluded that the low prevalence of digeneans and glochidia in Lake Vatia, Finland was due to chemical pollutants, which caused a depauperate molluscan fauna. Following several years of reduced organochlorine input into the lake, the prevalences of digeneans and glochidia approached levels of unpolluted lakes. The same authors noted that the immune system, which may be important in regulating populations of monogeneans, was compromised in fish under pollution stress. Sentinel fish from unpolluted lakes introduced into a cage in Lake Vatia showed a rapid increase in populations of monogeneans, but caged control fish in the unpolluted lake did not show a similar increase. The use of parasitological data to monitor environmental health (see Chapter 17) attests further to the importance that anthropogenic factors may have as a determinant of parasite species richness.

Epidemiologists have long been aware of a positive relationship between rates of parasite transmission within local host populations and host density. It is straightforward to extend this idea from the population level to the community level (Roberts *et al.*, 2002), with the prediction that parasite species richness will be positively associated with host population density. Morand & Poulin (1998) examined parasites from 79 species of terrestrial mammals and found a significant relationship between parasite species richness and host density, once various confounds were controlled. They concluded that colonization by parasites is based on how many hosts there are in a particular region. Further empirical support for the importance of host density comes from studies on nematodes and ecto-parasites of mammals (Arneberg, 2002; Stanko *et al.*, 2002) and endoparasites of fish (Takemoto *et al.*, 2005). Thus, current evidence supports the idea that high host densities are associated with increased opportunities for colonization of component communities by a range of parasites over ecological and evolutionary time.

Just as the structure of host populations can play a role in determining patterns of species richness, so too can the structure of the host community. The highly diverse communities of gastrointestinal nematodes in grazing mammals are determined, in part, by the exchange of worms between sympatric species with which they share pasture. Similarly, the composition of the gastrointestinal community in American avocets is determined by the exchange of generalist parasites from sympatric lesser scaup ducks (Edwards & Bush, 1989). In these two examples, the structure of the host community determined the exchange of generalists between sympatric hosts. The structure of the sympatric host community was also identified as a key determinant of the composition of parasite communities in whitefish, *Coregonus clupeaformes*, collected from pristine lakes in northern Canada (Goater *et al.*, 2005). In this case, the host community did not contribute generalists, but rather permitted the completion of the life cycles of the dominant members of the community, all of which were host specialists.

For parasites with complex life cycles, the role of the host community in determining species composition

can be extended to the structure of the intermediate host community. Bush *et al.* (1993) noted that some intermediate hosts might be likened to 'source communities.' Thus, the community seen in a particular definitive host may simply mirror what was ingested in an intermediate host. For example, piscivorous birds that forage on small-bodied 'minnows' on their northern breeding sites, may ingest an 'instant' community comprised of up to 15 species of larval helminth (Sandland *et al.*, 2001). Likewise, oyster-catchers, *Haematopus ostralegus*, are exposed to up to eight species of larval helminth with each ingested shellfish (Goater *et al.*, 1995). Under this scenario, the identification of 'pattern' in definitive hosts becomes more challenging because it requires that transferred versus post-transmission processes be clearly identified (review in Lotz *et al.*, 1995). The key point to make here is that one determinant of the composition and structure of some parasite communities may simply lie in the structure of the intermediate host community, and the manner in which processes occurring within intermediate hosts are transferred to definitive hosts via ingestion.

We conclude this section by re-iterating the exciting possibility that a host's physiological or immune status can determine species richness (review in Pederson & Fenton, 2006). In Box 13.1, we outlined the mechanisms by which host immunity could both increase and decrease species richness. To date, the best evidence comes from the latter scenario, where the immunosuppressive effect of the common gut nematode *Heligmosomoides polygyrus* led to increased colonization of mice by three other helminth species (Fig. 13.7; Behnke *et al.*, 2009).

13.4.3 Summary: species richness

Parasite ecologists have a long history of characterizing variation in parasite biodiversity, be it between host species, between host populations, and/or between geographical regions. Over the past few decades, we have clarified our methods of describing that variation, we have sharpened our terminology, and have adopted analyses developed by mainstream

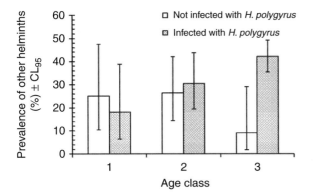

Fig. 13.7 Prevalences of other species of helminths in three age classes of wood mice *Apodemus sylvaticus* with or without the nematode *Heligmosomoides polygyrus*. Differences in prevalence of other helminths were approximately four-fold higher in mice with, compared to those without, *H. polygyrus* in age-3 mice. (Figure courtesy of Jerzy Behnke; modified from Behnke *et al.*, 2009, with permission, *Parasitology*, **136**, 793–804, Cambridge University Press.)

ecologists to help interpret general patterns. Yet, while extreme variation in biodiversity of infra- and component communities is the clear pattern, our search for determinants of that variation continues to be fraught with difficulty. While it is certain that key factors such as host diet, host demography, and host community structure (in both definitive and intermediate hosts), together with a wide range of other ecological factors, are important in some hosts, there is little scope for generalization relative to mechanism(s). There is little scope for generalizations regarding the relative roles of host phylogeny versus host ecology within particular host–parasite interactions.

The lack of macroecological patterns, which seems often to contrast results from free-living systems, has led some to cast doubt on the very existence of structure at the host population or component community levels. This perspective mirrors earlier queries by Simberloff (1990) on whether parasite communities exist at all or whether they are merely collections of species that colonize their hosts independently, and at random. The debate continues and is unlikely to be resolved soon. For example, among studies of monogenean parasites, there are studies invoking random colonization (Bagge & Valtonen, 1999), deterministic colonization (Gutiérrez & Martorelli, 1999), and both forms of colonization within the same community (Jackson *et al.*, 2006).

Poulin (1997b) used a null model developed by Janovy *et al.* (1995) to test for departures from random assembly in published studies involving a range of parasite infracommunities. Here, the expected frequencies of all combinations of species (based upon known prevalences) are computed to build artificial communities with presumed random structure and then compared to those in real communities. Poulin (1997b) concluded that random assembly was a common feature in natural communities. Exceptions occurred in about a third of cases, likely reflecting the importance of interspecific interactions in a limited subset of communities. Gotelli & Rhode (2002) used an alternative null model to reach similar conclusions relative to monogeneans of marine fish.

References

Arneberg, P. (2002) Host population density and body mass as determinants of species richness in parasite communities: comparative analyses of directly transmitted nematodes of mammals. *Ecography*, **25**, 88–94.

Bagge, A. M. & Valtonen, E. T. (1999) Development of monogenean communities on the gills of roach fry (*Rutilus rutilus*). *Parasitology*, **118**, 479–487.

Behnke, J. M., Bland, P. W. & Wakelin, D. (1977) The effect of the expulsion phase of *Trichinella spiralis* on *Hymenolepis diminuta* in mice. *Parasitology*, **75**, 79–88.

Behnke, J. M., Wakelin, D. & Wilson, M. M. (1978) *Trichinella spiralis*: delayed rejection in mice concurrently infected with *Nematospiroides dubius*. *Experimental Parasitology*, **46**, 121–130.

Behnke, J. M., Gilbert, F. S., Abu-Madi, M. A. & Lewis, J. W. (2005) Do the helminth parasites of wood mouse interact? *Journal of Animal Ecology*, **74**, 982–993.

Behnke, J. M., Eira, C., Rogan, M., *et al.* (2009) Helminth species richness in wild wood mice, *Apodemus sylvaticus*, is enhanced by the presence of the intestinal nematode *Heligmosomoides polygyrus*. *Parasitology*, **136**, 793–804.

Brooks, D. R. (1980) Allopatric speciation and non-interactive parasite community structure. *Systematic Zoology*, **29**, 192–203.

Bush, A. O. (1990) Helminth communities in avian hosts: determinants of pattern. In *Parasite Communities: Patterns and Processes*, ed. G. W. Esch, A. O. Bush & J. M. Aho, pp. 197–232. London: Chapman & Hall.

Bush, A. O. & Holmes, J. C. (1983) Niche separation and the broken-stick model: use with multiple assemblages. *The American Naturalist*, **122**, 849–855.

Bush, A. O. & Holmes, J. C. (1986a) Intestinal helminths of lesser scaup ducks: patterns of association. *Canadian Journal of Zoology*, **64**, 132–141.

Bush, A. O. & Holmes, J. C. (1986b) Intestinal helminths of lesser scaup ducks: an interactive community. *Canadian Journal of Zoology*, **64**, 142–152.

Bush, A. O., Aho, J. M. & Kennedy, C. R. (1990) Ecological versus phylogenetic determinants of helminth parasite community richness. *Evolutionary Ecology*, **4**, 1–20.

Bush, A. O., Heard, R. W., Jr. & Overstreet, R. M. (1993) Intermediate hosts as source communities. *Canadian Journal of Zoology*, **71**, 1358–1363.

Bush, A. O., Lafferty, K. D., Lotz, J. M. & Shostak, A. W. (1997) Parasitology meets ecology on its own terms: Margolis *et al.* revisited. *Journal of Parasitology*, **83**, 565–583.

Butterworth, E. W. (1982) A study of the structure and organization of intestinal helminth communities in ten species of waterfowl (Anatinae). Ph.D. Thesis, University of Alberta, Edmonton.

Connell, J. H. (1985) The consequences of variation in initial settlement vs. post-settlement mortality in rocky intertidal communities. *Journal of Experimental Marine Biology*, **93**, 11–45.

Combes, C. (2001) *Parasitism: The Ecology and Evolution of Intimate Interactions*. Chicago: University of Chicago Press.

Curtis, M. A. & Rau, M. E. (1980) The geographical distribution of diplostomiasis (Trematoda: Strigeidae) in fishes from northern Quebec, Canada, in relation to the calcium ion concentrations of lakes. *Canadian Journal of Zoology*, **58**, 1390–1394.

Dogiel, V. A. (1964) *General Parasitology*. Edinburgh: Oliver Boyd.

Edwards, D. D. & Bush, A. O. (1989) Helminth communities in avocets: importance of the compound community. *Journal of Parasitology*, **75**, 225–238.

Esch, G. W. (1971) Impact of ecological succession on the parasite fauna in centrarchids from oligotrophic and eutrophic ecosystems. *American Midland Naturalist*, **86**, 160–168.

Esch, G. W., Bush, A. O. & Kennedy, C. R. (1990) *Parasite Communities: Patterns and Processes*. London: Chapman & Hall.

Felsenstein, J. (1985) Phylogenies and the comparative method. *The American Naturalist*, **125**, 1–15.

Gasser, R. B. (1999) PCR-based technology in veterinary parasitology. *Veterinary Parasitology*, **84**, 229–258.

Goater, C. P. & Bush, A. O. (1988) Intestinal helminth communities in long-billed curlews: the importance of congeneric host-specialists. *Holarctic Ecology*, **11**, 140–145.

Goater, C. P., Goss-Custard, J. D. & Kennedy, C. R. (1995) Population biology of two species of helminth in oyster-catchers, *Haematopus ostralegus*. *Canadian Journal of Zoology*, **73**, 296–308.

Goater, C. P., Baldwin, R. E. & Scrimgeour, G. (2005) Physico-chemical determinants of helminth community structure in whitefish (*Coregonus clupeoformis*) from adjacent lakes in northern Alberta, Canada. *Parasitology*, **131**, 713–722.

Gotelli, N. J. & Rhode, K. (2002) Co-occurrence of ectopara-sites of marine fishes: a null model analysis. *Ecology Letters*, **5**, 86–94.

Gregory, R. D. (1990) Parasites and host geographic range as illustrated by waterfowl. *Functional Ecology*, **4**, 645–654.

Gregory, R. D., Keymer, A. E. & Harvey, P. H. (1996) Helminth parasite richness among vertebrates. *Biodiversity and Conservation*, **5**, 985–997.

Gutiérrez, P. A. & Martorelli, S. R. (1999) The structure of the monogenean community on the gills of *Pimelodus maculatus* in Rio de la Plata (Argentina). *Parasitology*, **119**, 177–182.

Hanski, I. (1982) Dynamics of regional distribution: the core and satellite species hypothesis. *Oikos*, **38**, 210–221.

Harvey, P. H. & Pagel, M. D. (1991) *The Comparative Method in Evolutionary Biology*. Oxford: Oxford University Press.

Herreras, M. V., Kaarstad, S. E., Balbuena, J. A., *et al.* (1997) Helminth parasites of the digestive tract of the harbour porpoise *Phocoena phocoena* in Danish waters: a compar-ative geographical analysis. *Diseases of Aquatic Organisms*, **28**, 163–167.

Holmes, J. C. (1961) Effects of concurrent infections on *Hymenolepis diminuta* (Cestoda) and *Moniliformis dubius* (Acanthocephala). I. General effects and comparison with crowding. *Journal of Parasitology*, **47**, 209–216.

Holmes, J. C. (1962a) Effects of concurrent infections on *Hymenolepis diminuta* (Cestoda) and *Moniliformis dubius* (Acanthocephala). II. Effects on growth. *Journal of Parasitology*, 48, 87–96.

Holmes, J. C. (1962b) Effects of concurrent infections on *Hymenolepis diminuta* (Cestoda) and *Moniliformis dubius* (Acanthocephala). III. Effects in hamsters. *Journal of Parasitology*, 48, 97–100.

Holmes, J. C. (1987) The structure of helminth communities. *International Journal for Parasitology*, 17, 203–208.

Holmes, J. C. & Price, P. W. (1986) Communities of parasites. In *Community Ecology: Pattern and Process*, ed. D. J. Anderson & J. Kikkawa, pp. 187–213. Oxford: Blackwell Scientific Publications.

Jackson, J. A., Pleass, R. J., Cable, J., *et al.* (2006) Heterogenous interspecific interactions in a host–parasite system. *International Journal for Parasitology*, 36, 1341–1349.

Jackson, J. A., Friberg, I. M., Bolch, L., *et al.* (2009) Immunomodulatory parasites and toll-like receptor-mediated tumour necrosis factor alpha responsiveness in wild mammals. *BMC Biology*, 7, 16–29.

Janovy, Jr., J. (2002) Concurrent infections and the community ecology of helminth parasites. *Journal of Parasitology*, 88, 440–445.

Janovy, Jr., J., Clopton, R. E., Clopton, D. A., *et al.* (1995) Species density distributions as null models for ecologically significant interactions of parasite species in an assemblage. *Ecological Modelling*, 77, 189–196.

Karvonen, A., Seppälä, O. & Valtonen, E. T. (2009) Host immunization shapes interspecific associations in trematode parasites. *Journal of Animal Ecology*, 78, 945–952.

Kennedy, C. R. (1990) Helminth communities in freshwater fish: structured communities or stochastic assemblages? In *Parasite Communities: Patterns and Processes*, ed. G. W. Esch, A. O Bush & J. M. Aho, pp. 130–156. London: Chapman & Hall.

Kennedy, C. R. & Bush, A. O. (1992) Species richness in helminth communities: the importance of multiple congeners. *Parasitology*, 104, 189–197.

Kennedy, C. R., Bush, A. O. & Aho, J. M. (1986). Patterns in helminth communities: why are birds and fish different? *Parasitology*, 93, 205–215.

Kuris, A. M. & Lafferty, K. D. (1994) Community structure: larval trematodes in snail hosts. *Annual Review of Ecology and Systematics*, 25, 189–217.

Lello, J., Boag, B., Fenton, A., *et al.* (2004) Competition and mutualism among the gut helminths of a mammalian host. *Nature*, 428, 840–844.

Lewin, R. (1986) Supply-side ecology. *Science*, 234, 25–27.

Lotz, J. M. & Font, W. F. (1994) Excess positive associations in communities of intestinal helminths of bats: a refined null hypothesis and a test of the facilitation hypothesis. *Journal of Parasitology*, 80, 398–413.

Lotz, J. M., Bush, A. O. & Font, W. F. (1995) Recruitment-driven, spatially discontinuous communities: a null model for transferred patterns in target communities of intestinal helminths. *Journal of Parasitology*, 81, 12–24.

Marcogliese, D. J. (1991) Pursuing parasites up the food chain: implications of food web structure and function on parasite communities in aquatic systems. *Acta Parasitologica*, 46, 82–93.

Marcogliese, D. J. (2001) Implications of climate change for parasitism of animals in the aquatic environment. *Canadian Journal of Zoology*, 79, 1331–1352.

Marcogliese, D. J. & Cone, D. K. (1996) On the distribution and abundance of eel parasites in Nova Scotia: influence of pH. *Journal of Parasitology*, 82, 389–399.

Morand, S. & Poulin, R. (1998) Density, body mass and parasite species richness of terrestrial mammals. *Evolutionary Ecology*, 12, 717–727.

Morand, S., Poulin, R., Rohde, K. & Hayward, C. (1999) Aggregation and species coexistence of ectoparasites of marine fishes. *International Journal for Parasitology*, 29, 663–672.

Morand, S., Simková, A., Matejusová, I., *et al.* (2002) Investigating patterns may reveal the processes: evolutionary ecology of ectoparasitic monogeneans. *International Journal for Parasitology*, 32, 111–119.

Paperna, I. (1964) Competitive exclusion of *Dactylogyrus extensus* by *Dactylogyrus vastator* (Trematoda: Monogenea) on the gills of reared carp. *Journal of Parasitology*, 50, 94–98.

Patrick, M. J. (1991) Distribution of enteric helminths in *Glaucomys volans* L. (Sciuridae): a test for competition. *Ecology*, 72, 755–758.

Pederson, A. B. & Fenton, A. (2006) Emphasizing the ecology in parasite community ecology. *Trends in Ecology and Evolution*, 22, 133–139.

Poulin, R. (1995) Phylogeny, ecology and the richness of parasite communities in vertebrates. *Ecological Monographs*, 65, 283–302.

Poulin, R. (1997a) Parasite faunas of freshwater fish: the relationship between richness and the specificity of parasites. *International Journal for Parasitology*, 27, 1091–1098.

Poulin, R. (1997b) Species richness of parasite assemblages: evolution and patterns. *Annual Review of Ecology and Systematics*, 28, 341–358.

Poulin, R. (1999) The intra- and interspecific relationships between abundance and distribution in helminth parasites of birds. *Journal of Animal Ecology*, 68, 719–725.

Poulin, R. (2007) *Evolutionary Ecology of Parasites*, 2nd edition. Princeton: Princeton University Press.

Poulin, R. & Rohde, K. (1997) Comparing the richness of metazoan ectoparasite communities of marine fishes: controlling for host phylogeny. *Oecologia*, 110, 278–283.

Poulin, R. & Morand, S. (2004) *Parasite Biodiversity*. Washington, D.C.: Smithsonian Institution Press.

Price, P. W. (1980) *Evolutionary Biology of Parasites*. Princeton: Princeton University Press.

Råberg, L., de Roode, J. C., Bell, A. S., *et al.* (2006) The role of immune-mediated apparent competition in genetically diverse malaria infections. *The American Naturalist*, 168, 41–53.

Roberts, M. G., Dobson, A. P., Arneberg, P., *et al.* (2002). Parasite community ecology and biodiversity. In *The Ecology of Wildlife Diseases*, ed. P. J. Hudson, A. Rizzoli, B. T. Grenfell, H. Heesterbeek & A. P. Dobson, pp. 63–82. Oxford: Oxford University Press.

Rohde, K. (1994) Niche restriction in parasites: proximate and ultimate causes. *Parasitology*, 109, S69–S84.

Rohde, K. (2005) *Nonequilibrium Ecology*. Cambridge: Cambridge University Press.

Sandland, G. J., Goater, C. P. & Danylchuk, A. J. (2001) Population dynamics of *Ornithodiplostomum ptychocheilus* metacercariae in fathead minnows (*Pimephales promelas*) from four northern Alberta lakes. *Journal of Parasitology*, 87, 744–748.

Schad, G. A. (1966) Immunity, competition, and the natural regulation of parasite populations. *The American Naturalist*, 100, 359–364.

Simberloff, D. (1990) Free-living communities and alimentary tract helminths: hypotheses and pattern analyses. In *Parasite Communities: Patterns and Processes*, ed. G. W. Esch, A. O. Bush & J. M. Aho, pp. 289–320. London: Chapman & Hall.

Simková, A., Desdevises, Y., Gelnar, M. & Morand, S. (2000) Co-existence of nine gill ectoparasites (*Dactylogyrus*: Monogenea) parasitizing the roach, *Rutilis rutilis* (L.): history and present ecology. *International Journal for Parasitology*, 30, 1077–1083.

Simková, A., Ondracková, M., Gelnar, M. & Morand, S. (2002) Morphology and coexistence of congeneric ectoparasite species: reinforcement of reproductive isolation? *Biological Journal of the Linnean Society*, 76, 125–135.

Stanko, M., Miklisová, D., Gouy de Bellocq, J. & Morand, S. (2002) Mammal density and patterns of ectoparasite species richness and abundance. *Oecologia*, 131, 289–295.

Stock, T. M. & Holmes, J. C. (1988) Functional relationships and microhabitat distributions of enteric helminthes of grebes (Podicipedidae): the evidence for interactive communities. *Journal of Parasitology*, 74, 214–227.

Sukhdeo, M. V. K. & Sukhdeo, S. C. (1994) Optimal habitat selection by helminths within the host environment. *Parasitology*, 109, S41–S55.

Takemoto, R. M., Pavanelli, G. C., Lizama, M. A. P., *et al.* (2005) Host population density as the major determinant of endoparasite species richness in floodplain fishes of the upper Paraná River, Brazil. *Journal of Helminthology*, 79, 75–84.

Telfer, S., Lambin, X., Birtles, R., *et al.* (2010) Species interactions in a parasite community drive infection risk in a wildlife population. *Science*, 330, 243–246.

Valtonen, E. T., Holmes, J. C. & Koskivaara, M. (1997) Eutrophication, pollution, and fragmentation effects on parasite communities in roach (*Rutilus rutilus*) and perch (*Perca fluviatilis*) in four lakes in central Finland. *Canadian Journal of Fisheries and Aquatic Sciences*, 54, 572–585.

Vazquez, D. P., Poulin, R., Krasnov, B. R. & Shenbrot, G. I. (2005) Species abundance and the distribution of specialization in host–parasite interaction networks. *Journal of Animal Ecology*, 74, 946–955.

Walther, B. A., Cotgreave, P., Price, R. D., *et al.* (1995) Sampling effort and parasite species richness. *Parasitology Today*, 11, 306–310.

Wisniewski, W. L. (1958) Characterization of the parasitofauna of a eutrophic lake (parasitofauna of the biocoenosis of Druzno lake, part I). *Acta Parasitologica Polonica*, 6, 1–64.

14 Parasite biogeography and phylogeography

14.1 General considerations

We now shift our attention to the geographical distributions of parasites. Although spatial aspects of parasite distributions were considered in the previous chapter, our focus here is on parasite distributions at broader scales, typically on the order of continents or regions. This perspective is the realm of parasite biogeography. In general, biogeographers aim to characterize patterns in the geographical distribution of organisms and to understand the role of historical events in determining present-day distributions of populations, species, and entire biotas (Brown & Lomolino, 1998). This fundamental aim seems straightforward, but the subject matter of biogeography is notoriously complex. Modern studies in biogeography combine aspects of phylogenetics, ecology, geographical information sciences, paleontology, and geology to understand present-day distributions, and to reconstruct the sequence of events leading to the assembly of faunas in space and time. For parasite biogeographers, the task is especially complex, and interesting, because understanding parasite distributions requires an intimate knowledge of the ecological and evolutionary factors that determine the distributions of all of their hosts.

For convenience, biogeographers recognize two broad research traditions. Historical biogeographers seek to understand the origin, dispersal, and extinction of species relative to geological events such as continental drift, glaciation, and the emergence and submergence of landmasses. A key paradigm that underlies historical biogeography is the notion that the geographical range of a species, or an entire biota, can be split into isolated parts by physical barriers to dispersal or gene flow (e.g., the isthmus of Panama, the Beringia land bridge) that results in the formation of sister species or sister faunas. This research arm is known as **vicariant biogeography**. In contrast, **ecological biogeography** tends to focus on the extent to which the distributions of species result from current ecological processes. While the former focuses on parasite distributions at global or continental scales, the latter tends to focus on narrower temporal and spatial scales. In reality, the two research traditions are probably not so discrete. This is because an understanding of history provides the unifying backbone for deciphering biotic structure across wide spatial and temporal scales. Thus, ecological factors undoubtedly play key roles in determining current distributions, but they do so within a deeper historical context.

Over the past decade, advances in molecular methods have integrated the once disparate fields of molecular population genetics and biogeography. From this combination of approaches, **phylogeography** has arisen as a subset of historical biogeography (Avise, 2000). Although phylogeographers tend to place focus at the population and landscape scales, they explore the same conceptual issues as historical biogeographers who tend to focus on phylogenies and species. Thus, the questions regarding the biogeographical distributions of species and biota are similar for phylogeographers and biogeographers, just at different spatial and temporal scales (review in Hoberg & Brooks, 2010).

In this chapter, we review developments in parasite biogeography and phylogeography. Our emphasis is on how modern approaches in biogeography, phylogeography, and in the geographical information sciences have transformed our understanding of parasite distributions at global, regional, and local scales (reviews in Brooks & McLennan, 2002; Morand & Krasnov, 2010). In the first section on historical biogeography, we focus on how the integration of molecular tools with traditional vicariance biogeography has led to important new insights regarding the distribution of parasite biodiversity across broad scales, and its origins. In this section, we necessarily emphasize the role of macroevolutionary phenomena such as host geographical isolation, speciation, and host dispersal. This is because the patterns we observe in present-day parasite distributions tend often to result from complex events that occurred over evolutionary time scales. We return to a more detailed discussion of certain macroevolutionary phenomena in Chapter 16. In the second section on ecological biogeography, we cover patterns of parasite distribution at finer spatial and temporal scales. Here, focus is on the extent to which variation in parasite distributions across a landscape is associated with variation in key ecological variables. Emphasis in the last section is on applied aspects of parasite biogeography.

14.2 Historical biogeography

We begin our coverage of historical biogeography with an example from prehistoric humans and the parasites they left behind. Indeed, paleo-parasitological traces of parasites found in 'fecal fossils' or in preserved mummies provide remarkable signatures of historical events leading to the current distributions of human parasites (Dittmar, 2010). For instance, the eggs or larvae of hookworm (*Necator* and *Ancylostoma*), whipworm (*Trichuris*), threadworm (*Strongyloides stercoralis*), pinworm (*Enterobius*), and head lice (*Pediculus*) have been recovered from archeological sites in North and South America dated at 7000–10 000 years (Araujo *et al.*, 2008). The eggs of these species had to be passed by humans that were present in these regions thousands of years prior to the arrival of Columbus. Given that each of these species is a strict host specialist in hominids (see Chapters 8, 11), it is most unlikely that they originated from sympatric vertebrates. Instead, these specialists must have 'hitchhiked' into the western hemisphere with their

hosts, perhaps when climate conditions created the Beringia land bridge about 13 000 years ago. As this example shows, the present-day distributions of these human parasites are best understood within an historical context.

The tools of modern phylogenetics and cladistics can be used to evaluate the origin and radiation of these and other human parasites much further back in evolutionary time. The radiation of species of pinworms is almost perfectly congruent with the radiation of their primate hosts (Hugot, 1999). This is an important result from a biogeographical perspective, indicating that as humans and other primates speciated and radiated into isolated geographical regions, their parasites speciated and dispersed with them. Similarly tight co-phylogenetic patterns have been demonstrated for the highly host-specific species of lice on humans and other mammals (e.g., Reed *et al.*, 2007). In humans, the louse *P. humanus* occurs in two forms (the head and body lice, see Chapter 11) that are morphologically similar, but ecologically distinct. Reed *et al.* (2004) combined morphological and molecular data to reconstruct the evolutionary history of these and related lice, and also tested for co-phylogenetic patterns with their mammalian hosts. Their results indicated that the two forms of *Pediculus* found in modern humans represent ancient lineages, one with a worldwide distribution and the other restricted to the New World (Fig. 14.1). This biogeographical pattern can be explained by the divergence of the two lineages about 1 000 000 years ago coinciding with the divergence of archaic humans, followed by a further 1 000 000 years of isolation. The current distributions of the two lineages are best explained by direct physical contact (i.e., a host switch) between *Homo sapiens* and a sympatric species (probably *H. erectus*) that was restricted to the Old World. In a remarkably parallel example to the scenario depicted in Fig. 14.1, the two species of *Taenia* that infect humans likely also diverged around 1 000 000 years ago, one of which is currently restricted to Asia (Hoberg *et al.*, 2001). The key point to arise from these examples of human parasites is that explanations for the current geographical distributions of parasites involve complex

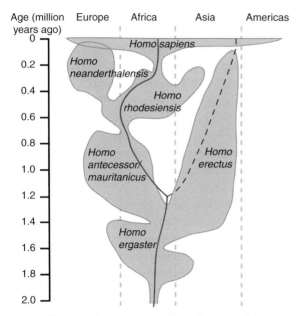

Fig. 14.1 The evolutionary history of two lineages of the human louse *Pediculus humanus* superimposed on a view of the temporal and geographical distribution of the hominids. This schematic demonstrates the parallel divergence of the two louse lineages (NW and WW) with an ancient divergence of ancient *Homo* spp. The contemporary restriction of the NW clade (broken line) to humans in Asia reflects a recent host switching event from *H. erectus*, whereas the worldwide distribution of the WW clade (bold line) coincides with human population expansion into the New World. (Modified from Reed *et al.*, 2004, *PLoS Biology*, **2**, 1972–1983.)

interactions among vicariance, coevolutionary history, host specificity, and ecological factors that promote host switching.

Although we emphasize human examples to introduce ideas in historical biogeography, similar patterns have been recognized for many years in a wide range of host–parasite combinations. Early in the development of the discipline, researchers observed close congruence between the geographical distribution of hosts and some of their parasites. Indeed, years before geologists had accepted the Theory of Continental Drift, Cameron (1929) postulated the close coevolutionary and biogeographical ties between primates and their pinworms. Likewise, Metcalf (1940) considered that certain families of frogs from South America and

Australia were closely related based on the similarity of their protist parasites. Next, a series of ground-breaking studies by Harold Manter (e.g., Manter, 1963) showed that vicariant events associated with continental drift could explain the geographical distributions and affinities of digeneans along the Atlantic and Pacific coasts of Central America, digeneans in marine fishes globally, and digeneans in freshwater fishes of South America. Intercontinental relationships normally included related genera present in fishes in South America, Africa, and India, all components of the now widely accepted Gondwana supercontinent (Fig. 14.2).

Starting in the late 1970s, Daniel Brooks combined the Theory of Continental Drift with early studies on parasite phylogenetic systematics to lead a rigorous program of research that defined the direction of historical parasite biogeography (reviews in Brooks & McLennan, 1993, 2002). One of the first studies that combined biogeographical and phylogenetic approaches of both hosts and parasites involved the evolutionary relationships between crocodiles and their digeneans (Brooks, 1979). In this case, evidence supported the notion that intercontinental relationships of the digeneans were concordant with the patterns of continental fragmentation since the Cretaceous. These relationships were also congruent with the biogeographical affinities of the crocodiles. Over the past decade, Brooks' approach has expanded to embrace the conceptual and theoretical framework of molecular-based phylogeography to help interpret the geographical origins, and subsequent dispersal, of a wide range of parasites. Klassen (1992) and Hoberg & Klassen (2002) provide accounts of the history of parasite biogeography.

These examples are consistent with the idea that vicariance is a key process leading to the current global and continent-wide distributions of many parasites. The general pattern is one where a geographical barrier promotes the origin of sister species of hosts and their parasites, followed by long periods of isolation and cospeciation. Often, but certainly not always (see Chapter 16), vicariance scenarios will be consistent

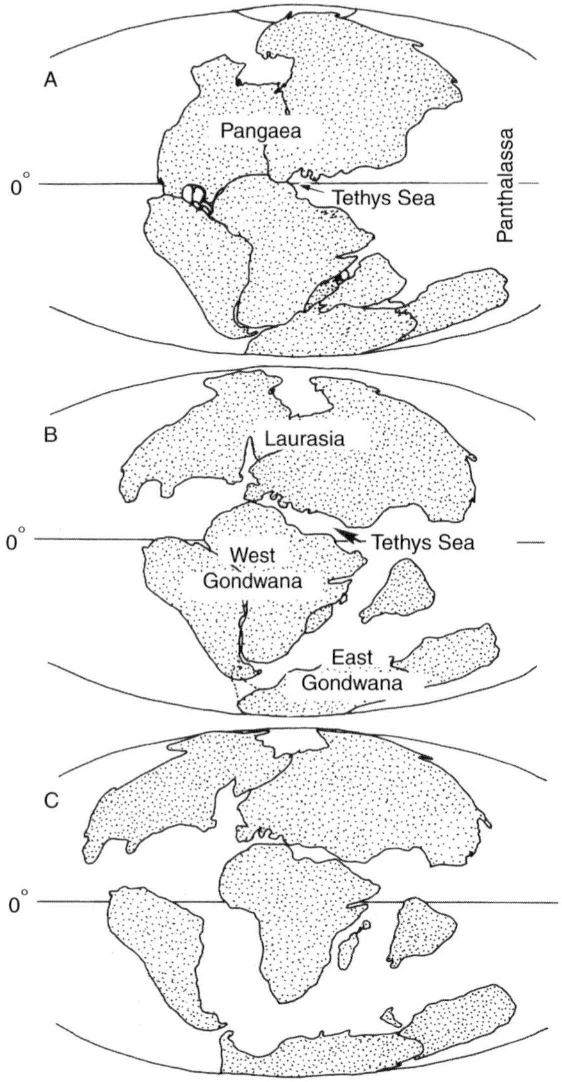

Fig. 14.2 Past configuration of the landmasses and oceans at different times based on current views of plate tectonics and Continental Drift. (A) Supercontinent Pangaea in the Triassic, 200 million ybp. Panthalassa was the ancestral Pacific Ocean and the Tethys Sea the ancestral Mediterranean. (B) In the late Jurassic or early Cretaceous, after 65 million years of drift (135 million ybp) the Northern Pangaea became Laurasia; the Southern Pangaea, after becoming Gondwana, split into East Gondwana, West Gondwana, and India. The Atlantic and Indian Oceans are already opened. (C) At the end of the Cretaceous, after 135 million years of drift (65 million ybp), the Mediterranean is recognizable; the Atlantic has become a major ocean separating South America from Africa, while Australia still remains attached to Antarctica. (Figure courtesy of Jackie Fernández.)

with congruent host–parasite phylogenies indicative of tight coevolution (Hoberg & Brooks, 2010). The biogeographical extension of this general process may lead to a pattern whereby the distributions of parasites are congruent with those of their hosts. Complex patterns of geographical distribution of entire parasite biotas can then be explained as products of vicariance and long-term coevolutionary dynamics between hosts and parasites. This fundamental paradigm of host–parasite biogeography is conceptualized on the left side of Fig. 14.3.

The results from comparative phylogenetic and historical biogeographical studies involving the parasite faunas of arctic vertebrates were among the first to indicate that interpretations based solely upon single vicariant events, followed by host dispersal, were too simplistic. Surveys involving the cestode faunas of alcid birds (puffins, murres, auks) and pinniped mammals (seals, sea lions, walruses) provided initial examples (Hoberg, 1992, 1995). One challenge to the fundamental paradigm was the observation that the host and parasite phylogenies for both alcids and their cestodes, and pinnipeds and their cestodes, were highly incongruent. The pattern indicated little coevolution within these host–parasite combinations, but extensive host switching followed by subsequent radiation. Instances of host switching occurred mainly among hosts of similar trophic and ecological requirements. Further, the patterns of speciation of cestodes in alcids and cestodes in pinnipeds were approximately congruent and synchronized. Based upon these observations, Hoberg (1995) concluded that the host–parasite assemblages of arctic vertebrates were not remnants of a previous biota, but were new assemblages structured by extensive host switching, geographical colonization, and allopatric speciation (Fig. 14.4). He proposed that during periods of glacial maxima during the Pliocene–Pleistocene, sea levels dropped, exposing the continental shelf and trapping populations of hosts and their parasites into restricted refugia throughout the Holarctic region. These conditions favored cestode speciation through both vicariance and host switching. During the episodic retreat of the glaciers, the ocean levels rose

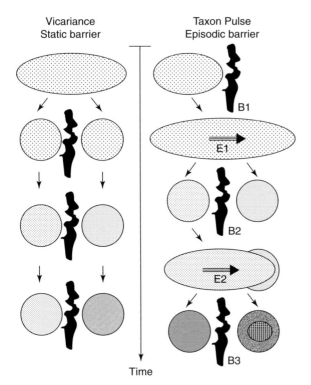

Vicariance
Static barrier

Taxon Pulse
Episodic barrier

B1

E1

B2

E2

B3

Time

Fig. 14.3 Schematic representation of vicariance and taxon pulse models of historical parasite biogeography and the contrasting consequences for biotic structure. In the former, emphasis is on the development of a single or unique static barrier leading to isolation. In the latter, there is episodic development and breakdown of geographical barriers. Shown here is the simple scenario for taxon pulses based on episodic periods of development and breakdown of barriers (B1–B3), followed by periods of expansion (E1–E2). (Figure courtesy of Eric Hoberg; modified from Hoberg & Brooks, 2010, Chapter 1. Beyond vicariance: integrating taxon pulses, ecological fitting, and oscillation in evolution and historical biogeography. In *The Biogeography of Host–Parasite Interactions*, ed. S. Morand & B. R. Krasnov, Fig. 1.1, p. 9, with permission, Oxford University Press.)

development of a mountain range) is replaced by recognition of barriers that develop and break down at episodic intervals (Fig. 14.3). For example, radical changes in climate during the Pleistocene period were associated with up to 20 glaciation events, each of which consisted of periods of glacier advance, stabilization, and retreat (Martinson *et al.*, 1987). Each glacial period in North America, Eurasia, and Antarctica lasted about 100 000 years, with inter-glacial periods spanning about 10 000 years. These periodic geological upheavals likely enhanced opportunities for the colonization of parasites into new areas and hosts, followed by long periods of isolation during which coevolutionary dynamics could proceed.

In their **taxon pulse model** of historical parasite biogeography, Hoberg & Brooks (2010) propose a two-stage process leading to the development of complex parasite distributions (Fig. 14.3). Initially, there is a period of host and parasite isolation that follows the formation of a barrier (i.e., vicariance), leading to the development of pairs of adjacent sister species. This period of isolation is then followed by a second stage that involves a 'pulse' of biodiversity expansion into new regions and hosts that is associated with the breakdown of the original barrier. Thus, extensive opportunities for range expansion and host switching arise from the repeated but episodic perturbations that have been so pervasive throughout earth's history. According to their model, vicariance alone cannot result in the complex mosaics of parasite populations, species, and biotas so commonly observed in the present-day faunas.

Supportive evidence for their taxon pulse model comes from studies involving marine, aquatic, and terrestrial systems (review in Hoberg & Brooks, 2010). One example involves the contemporary worldwide distribution of the nematode *Trichinella*. The life cycle and biology of this important genus was described in Chapter 8. Molecular phylogeographical analyses involving comparisons of DNA sequences of two mitochondrial genes in isolates collected from around the world showed patterns consistent with the taxon pulse model (Zarlenga *et al.*, 2006; Pozio *et al.*, 2009).

and allowed the confined species to disperse and extend their geographical range, further increasing host switching events (Fig. 14.4).

Results from a series of phylogeographical studies are consistent with a modified view of parasite biogeography that emphasizes the episodic nature of historical upheavals over geological time. Here, the notion of a static geological barrier (e.g.,

Biogeographical history

Fig. 14.4 Summary of historical biogeography for species of *Anophryocephalus* among phocids and otariids and for species of *Alcataenia* among alcids. Map shows the extent of exposed continental shelf (shaded area) during a reduction in sea level to 100 m during glacial maxima. Partitioning of the North Pacific and Arctic basin into regional (e.g., Arctic Basin, Pacific Basin) and insular refugia (Aleutian Islands and Kurile Islands) was influenced by fluctuations in sea level. (1) The postulated origin of *Anophryocephalus* in *Phoca* sp. and *Alcataenia* in larids (gulls) was in the North Atlantic about 3.0–3.5 million ybp. (2) Range expansion occurred through the Arctic Basin and resulted in the development of early Holarctic distributions for hosts and parasites about 2.5–3.0 million ybp. (3) Initial entry to the North Pacific Basin through the Bering Strait occurred soon after the submergence of Beringia, about 2.5–3.0 million ybp. *Alcataenia* diversified following colonization of puffins, through sequential colonization and radiation in auklets, murres, and guillemots. (4) Secondary Holarctic ranges were attained later by species of *Alcataenia* among murres and guillemots during the Quaternary period. (5) *Anophryocephalus* colonized otariid mammals less than 2.0 million ybp. (Modified from Hoberg, 1992, with permission, *Journal of Parasitology*, **78**, 601–615.)

Their results indicated that the *Trichinella* 'homeland' was in central Asia, from which the ancestoral taxa radiated (Fig. 14.5). An initial radiation occurred within Eurasia, followed by two or three independent expansions into Africa, three into the Nearctic, and at least one into the Neotropics from Eurasia. The authors consider that the contemporary worldwide distribution of *Trichinella* species arose from the episodic development of land bridges and dispersal corridors, followed by periodic breakdowns in geographical isolation.

Examples involving Beringian parasite assemblages in vertebrates are also consistent with the taxon pulse model of historical parasite biogeography over recent evolutionary time. The bridge that connected Eurasian and Nearctic faunas was undoubtedly a primary

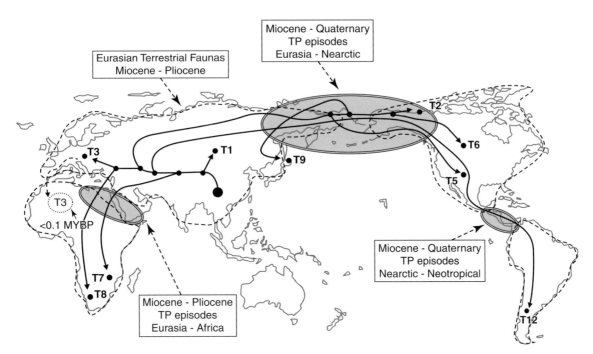

Fig. 14.5 Phylogeny and colonization of the encapsulated species of *Trichinella* superimposed onto their contemporary worldwide distributions. The schematic demonstrates the timing of taxon pulses from their origin in Eurasia, followed by expansion into Africa and Nearctic/Neotropical regions. Taxa are designated as T1 (*T. spiralis*), T2 (*T. nativa*), T3 (*T. britovi*), T5 (*T. murrelli*), T7 (*T. nelsoni*), and four unnamed species represented as T6, T8, T9, and T12. Arrows indicate the main terrestrial pathways for expansion that linked Eurasia with Africa, and the Nearctic regions with Neotropical regions. (Figure courtesy of Eric Hoberg; based on Zarlenga *et al.*, 2006, Pozio *et al.*, 2009; modified from Hoberg & Brooks, 2010, Chapter 1. Beyond vicariance: integrating taxon pulses, ecological fitting, and oscillation in evolution and historical biogeography. In *The Biogeography of Host–Parasite Interactions*, ed. S. Morand & B. R. Krasnov, Fig. 1.2, p.10, with permission, Oxford University Press.)

pathway for the dispersal of terrestrial hosts and their parasites. Indeed, the expansion of species ranges during numerous interglacial periods, following by range contraction into glacial refugia, likely provided ecological opportunities for host switching and diversification that have been unparalleled in the geological history of this region. Thus, the phylogeography of the rich nematode faunas of grazing mammals, the nematodes of both large and medium-sized carnivores (e.g., bears and martens), and the cestodes of rodents demonstrate patterns consistent with highly complex episodic expansion into the Nearctic followed by periods of isolation (Waltari *et al.*, 2007; Hoberg *et al.*, 2012). The results of studies involving parasites of vertebrates in adjacent marine areas also demonstrate

a key role for taxon pulses that link North Atlantic, North Pacific, and arctic host and parasite faunas (Hoberg, 1995).

There are two important messages to draw from these examples. One is the key role that methodological advances have made to help us understand the present-day distributions of parasites. Studies that combine modern phylogenetic reconstructions of hosts and parasites, molecular clock technologies, and knowledge of host and parasite geographical distributions have firmly positioned the study of parasite phylogeography into the framework of classical historical biogeography. The second is recognition that biogeographical patterns of many parasite distributions, especially if we consider entire parasite faunas, are more complex than we originally thought. Thus, some patterns can be interpreted to arise from single vicariant events, followed by tight host–parasite cospeciation and tight geographical matching between hosts and their parasites (e.g., ectoparasites of mammals). Yet, many others are consistent with multiple vicariant events, leading to episodic periods of parasite faunal expansion and contraction.

14.3 Ecological biogeography

When the characteristics of parasite populations, communities, or faunas are compared between hosts sampled from two localities, extensive variation is almost always present (see also Chapters 12, 13). A pattern of extensive geographical variation holds even among homogeneous samples of hosts, i.e., when confounding factors such as host age, gender, size, and so on, are held constant. The pattern is most striking when samples of hosts are compared between adjacent and perhaps interconnected sites. The prevalence and average abundances of *Fasciola hepatica* in cattle in the southeastern USA can differ 100-fold between adjacent ranches (Malone, 2005). The average infection characteristics for the trematode *Schistosoma japonicum* in 20 Chinese villages separated by only a few kilometers differ by orders of magnitude (Spear *et al.*, 2004). Similar spatial patterns occur for waterfowl and fish collected from adjacent, interconnected waterbodies (Edwards & Bush, 1989; Goater *et al.*, 2005). These examples indicate that geographical variation in infection characteristics is a fundamental feature of host–parasite interactions. Understanding the factors that underlie this variation has obvious consequences for human, livestock, and wildlife health, and for the design of parasite monitoring and control programs.

We emphasized the role of climatic factors, especially temperature and moisture, in determining parasite population sizes in Chapter 12. Thus, evidence from countless empirical studies demonstrate a role for factors such as temperature and humidity in determining population sizes, either directly on parasite

development and transmission rates or indirectly on parasite-induced host mortality, host immunity, or host population sizes. We contend that in any well-studied host–parasite system, ecological factors such as temperature can be seen to affect any or all of the parameters (e.g., B, k, α) that mathematically define parasite population dynamics (see Chapter 12). As we saw in Chapter 12, ecological factors can often mask factors such as host immunity and host behavior to determine parasite population sizes. Given the pervasive role of ecological factors in determining parasite population sizes, we should expect that they have a strong impact on the geographical distributions of parasites.

Early studies that linked geographical variation in parasite distributions with variation in environmental factors involved painstaking field surveys, often in combination with laboratory studies on the effects of factors such as temperature on parasite development. In a classic study, Ollerenshaw & Rowlands (1959) combined meteorological data from weather stations between 1948 and 1957 (precipitation, number of rainy days, and potential evapotranspiration) with prevalence data on individual ranches to forecast *Fasciola hepatica* outbreaks on English rangelands. The Ollerenshaw Index is currently used to provide ranchers in the United Kingdom with short-term forecasts of fascioliasis risk, upon which they base treatment strategies. Extensions of these early methods are being used to predict future fascioliasis outbreaks under conditions of potential global warming (Fox *et al.*, 2011).

Another striking example of the effect of temperature on parasite distribution involves a suite of hemiurid trematodes in marine fish. Manter (1934) found these trematodes in fishes collected at depths ranging from 270 to 550 m off the coast of Florida. However, in more northern latitudes, these species are abundantly distributed in fishes from much shallower waters. One species, *Derogenes varicus*, has been found in well over 100 species of fishes from near the Arctic to near the Antarctic, at a wide range of depths, and from coastal to deep-sea fishes. This type of circumtropical distribution has often been reported for free-living benthic species that follow isothermal bands of water across the ocean floor. Cold water (4 °C) is closer to the surface at the poles than in the tropics. These masses of polar water move from pole to pole across the Atlantic and Pacific Oceans following their submergence at about 75 °S, an oceanic frontal system delimiting the Antarctic from the sub-Antarctic. Manter (1934) proposed that the distribution of *D. varicus* in tropical waters is determined by its association with masses of cold water that move through the tropics at greater depths.

The results of numerous geographical surveys of hosts ranging from snails, to arthropod vectors, to humans, have identified a wide range of ecological and socio-economic factors, in addition to temperature and humidity, which determine the spatial component of parasite distributions. In terrestrial host–parasite interactions, factors such as soil type and composition, moisture content, vegetation, and altitude can be biogeographical determinants. In freshwater habitats, water flow rates, substrate type, depth, habitat area, and limnological characteristics such as pH, dissolved oxygen concentration, calcium concentration, and trophic status have been shown to affect parasite distributions (see Chapter 17). In marine systems, temperature, salinity, depth, physiography of the substrata, and currents (as distinctive bodies of water) are important factors affecting distributions. In many cases, these factors limit the distributions of the intermediate hosts used in the life cycle.

Over the past 20 years, advances in geographical information systems (GIS) and satellite-based remote sensing (RS) techniques have provided important tools for the characterization of parasite distributions and for the assessment of underlying environmental determinants (review in Hay *et al.*, 2006). A fundamental paradigm of this overall approach is that the development and survival of the free-living stages of parasites, and of larval stages within intermediate hosts and vectors, are highly sensitive to climatic factors, especially temperature. Thus, the spatial distributions of many parasites are amenable to mapping based upon environmental data that is readily available from satellite sources. The general approach proceeds as follows (Simoonga *et al.*, 2009). First, parasite prevalence and abundance data are collected from representative samples of hosts. Depending on the questions being considered, the scale of enquiry can range from global, to regional, to individual villages or farms. Next, environmental data are collected at the appropriate scale, either from satellite sources or from ground-based field surveys. Most typically, these data are imported as individual 'layers' into appropriate GIS software, allowing each to be visualized on a preliminary map. Finally, various modeling tools are then used to evaluate statistical significance between parasitological parameters and independent environmental covariates.

One key advance of this overall approach lies in the ability of spatial risk models to predict patterns of prevalence and abundance at sites that have not been sampled. Another lies in the diversity of potential environmental predictors that are made available via RS, some of which are appropriate surrogates for ground-based meteorological data. Various RS platforms are available that provide absolute data on temperature and moisture conditions at geographical scales ranging from about 10 m to 5 km, depending on temporal resolution (Hay *et al.*, 1996). Yet, depending on the particular satellite, derived data are also available at the same resolutions that characterize vegetation cover, precipitation, and overall habitat 'wetness.' Lastly, it is also straightforward to incorporate socio-economic indicators such as GDP, host demographic characteristics, availability of health care, and so on, as layers in various GIS platforms. Taken together, the integration of modern geospatial tools into parasitology provides a powerful tool for understanding the geographical distribution of parasites over a range of scales.

At a global scale, GIS and RS methodology was used to assess the role of habitat characteristics in determining the distributions of the major soil-transmitted nematodes of humans (Pullan & Brooker, 2012). These authors used data available in the Global Atlas of Helminth Infection (www.thiswormyworld.org) to estimate the prevalence of *A. lumbricoides*, *T.*

trichiura, and hookworm (*N. americanus* and *A. duodenale*) at almost 5000 sites around the world. The geographical distribution of each species was unique, but overall, high and low temperature extremes and highly arid environments limited their transmission. At a national level, nematode prevalence was statistically associated with the rural versus urban nature of a particular site and with *per capita* gross domestic product. These methods were used to define the locations where an estimated 5.3 billion people are at risk of transmission of at least one of these soil-transmitted nematodes, including one billion school-aged children. These results highlight the value of GIS methodology in determining regions where risk of infection is highest and, therefore, where control strategies should be targeted. They also identify some of the ecological and socio-economic conditions associated with transmission risk. Their results also provide a baseline for monitoring studies that can track future changes in global disease burden.

On a regional scale, Sehgal *et al.* (2010) used GIS methods to characterize the geographical distributions of *Plasmodium* sp. and *Trypanosoma* sp. in its host, the olive sunbird, *Cyanomitra olivacea*. This is a non-migratory and sedentary bird, common within a wide range of habitats in central and western Africa. The authors combined microscopy and molecular diagnoses to estimate parasite prevalence from blood samples of birds collected from a range of sites. Environmental data were collected remotely via satellite and from ground-based field studies. Their analyses showed that geographical variation in the prevalence of both taxa was associated with geographical variation in temperature, canopy humidity, and canopy structure (Color plate Fig. 6A, B). Maximum prevalence of *Plasmodium* spp. was associated with the maximum temperature of the warmest month of the year. In this case, high temperatures likely promoted optimal rates of development within mosquito vectors. In contrast, variation in the humidity of the rainforest canopy was negatively associated with variation in prevalence of *Trypanosoma*

spp. Sehgal *et al.* (2010) suggest that extended periods of high humidity within the canopy were required for trypanosome development within the vector and for transmission to birds. These results demonstrate how satellite-based meteorological data can explain variation in taxa-specific distributions across a landscape.

From a biogeographical perspective, a key outcome of modern geospatial approaches is the development of 'risk maps' that highlight areas where parasite prevalences or abundances are higher relative to other areas (Color plate Fig. 6). The maps are especially informative when they are used to predict transmission at sites that have not been sampled. In the ovenbird–blood parasites system, results from intensive sampling of birds in central Africa were used to predict regions of high prevalence in unsampled regions of coastal Nigeria (Sehgal *et al.*, 2010). GIS and RS approaches are now used to map risk of infection for important human and livestock parasites, at a range of scales. Simoonga *et al.* (2009) review studies that use GIS and RS technology to map the spatial risk of schistosomiasis, particularly in Africa. In a handful of these cases, predictions of spatial risk based upon features such as temperature, vegetation cover, and socio-economics were used to guide mass control interventions with praziquantal. Similar approaches have been used for *Fasciola hepatica* in the southeastern USA (Malone, 2005), for soil-transmitted helminths in sub-Saharan Africa (Brooker *et al.*, 2000) and for parasites that are vectored by arthropods, especially *Plasmodium*, *Trypanosoma*, and some filariid nematodes (Hay *et al.*, 2006).

In conclusion, understanding where, when, and why hosts are most at risk of parasite exposure provides a critical piece of information for understanding parasite biogeography. Yet, the applications of 'medical GIS' to host–parasite interactions are especially intriguing (Malone, 2005). Perhaps in the near future, a virtual globe may be available that will depict current (to the day?) risk zones for malaria, schistosomiasis, filariasis, and other important human parasites. Modern geospatial tools could then link current risk maps with various models of habitat change (e.g., global temperature increases, habitat loss) to predict future areas of concern (e.g., Fox *et al.*, 2011). Lastly, these tools could be used to estimate risk at various scales, such that the risk of exposure to *Schistosoma* sp. cercariae for example, could be zoomed from a global scale down to a scale appropriate to individuals living within a particular village. Each of these applications can equally be envisioned for parasites of other animals, especially in the context of tracking and controlling the spread of important invasive species. For students interested in expanding the ongoing, interdisciplinary integration between the geographical information sciences and parasitology, the future is bright indeed.

14.4 Applied aspects of parasite biogeography and phylogeography

14.4.1 Parasite distributions as 'biological tags' of host distributions

The presence of a parasite in a host provides unambiguous evidence that precise ecological events occurred in that host's recent past. The occurrence of malaria-causing *Plasmodium* spp. in a human host indicates a previous blood meal by a female mosquito, just as the presence of *Schistosoma* spp. indicates the host came into contact with water containing cercariae-releasing snails. Further, the presence of *Schistocephalus solidus* in a piscivorous bird collected along the coast of California means the bird had foraged within a lake in Canada within the past few weeks. The idea that parasite distributions can be used to trace the activities of their host in ecological time provides the foundation for biogeographical studies that utilize parasites as 'biological tags.' In Chapter 11, for instance, we indicated that ectoparasitic amphipods were valuable in deciphering the biogeographical and population histories of their right whale hosts. However, most common are studies that use parasites

to assign individual hosts, usually fish, to their population of origin (review in MacKenzie, 2002).

From a conservation perspective, such assignments can inform wildlife and fisheries managers about animal dispersal patterns. From an economic perspective, the assignment of individuals to 'source' populations can have profound implications relative to the management of resources. When the resources cross international or national political jurisdictions, the accurate assignment of individuals to their source is a key piece of information, upon which evidence-based harvesting decisions can be made. The best examples occur for global fisheries stocks, many of which involve species that have migratory life histories and/or occur in mixed stocks. In the case of wild salmonids, the determination of international harvesting quotas often rests on the accurate assignment of individuals to a particular river of origin.

Parasites were first used as biomarkers in fisheries research more than 50 years ago. When the offshore salmon fishing industry began in the early 1950s in the Pacific Northwest, it was not known if salmon from Pacific and Asian rivers were intermixing in the ocean or if they remained as distinct populations. Similarly, the geographical origin of the salmon caught by the different countries was unknown. These shortcomings had political ramifications that affected international fisheries policy for many years. Since then, the migrations of salmon in the North Pacific have been well studied using parasites as markers. Margolis (1963) was able to distinguish between sockeye salmon, *Oncorhynchus nerka*, of Pacific and Asian origin based on two of their parasites. Whereas sockeye salmon from Alaska were infected by the plerocercoid larvae of the cestode *Triaenophorus crassus*, those from Kamchatka were infected by the nematode *Dacnitis truttae* (Fig. 14.6). The cestode was acquired through the ingestion of infected zooplankton in lakes of western Alaska, while nematode transmission into juvenile sockeye occurred within rivers of Kamchatka. Further, sampling of salmon at sea showed that salmon from both Alaska and Kamchatka dispersed up to

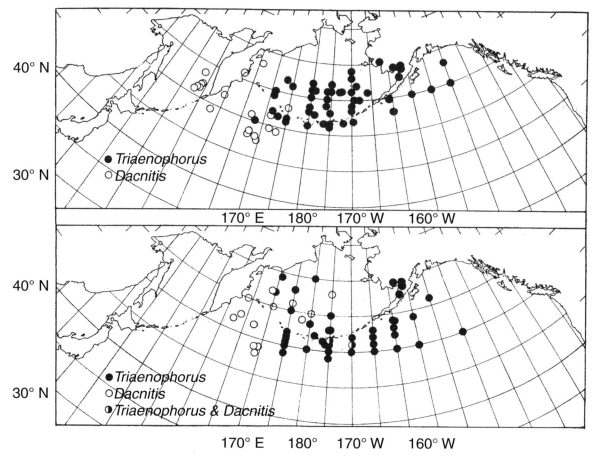

Fig. 14.6 Distribution of sockeye salmon *Oncorhynchus nerka*, infected with *Triaenophorus crassus* and *Dacnitis truttae* in the North Pacific. Top: maturing fish. Bottom: immature fish. (Modified from Margolis, 1965, with permission, *Journal of the Fisheries Research Board of Canada*, 22, 1387–1395.)

2000 km from their rivers of origin, overlapping in their distributions (Fig. 14.6). In contrast, stocks of salmon that originated from different streams in Alaska remained as distinctive schools in the sea. Margolis (1965) was able to recognize these different stocks based on contrasting patterns of infection of *Triaenophorus* that were characteristic of different streams in western Alaska. The geopolitical implications of these parasite-based salmonid stock delineations are still in place today. Baldwin *et al.* (2012) list almost 30 other examples in which trophically transmitted macroparasites were used to distinguish stocks of marine fish and shellfish.

The use of parasites as tags can even extend into the courtroom. Thus, parasite tags may be used to identify poached animals in fisheries and wildlife forensics. One example of 'forensic parasitology' revolved around a court case related to the alleged illegal capture and marketing of sockeye salmon on Vancouver Island on Canada's west coast (Margolis, 1993). Based on prevalence data of the myxozoan *Myxobolus arcticus*, the precise river of origin of the confiscated salmon in question was determined. The salmon had, indeed, been caught illegally and were being offered for sale – two serious offences under the Canadian Fisheries Act. The parasitological evidence presented in court contributed to the conviction of the accused vendor!

These 'success stories' indicate that the use of parasites as tags in fisheries has advantages over conventional tagging methods (MacKenzie, 2002). Indeed, parasite tags tend to be cheaper, less labor intensive, and less time consuming because they do not require the catch–mark–release of the host. Further, parasites used as tags do not interfere with the normal behavior of the host. Current focus tends to be on the use of helminths that have a fairly long life span within the host (larval trematodes, cestodes, and nematodes), that cannot be transferred between hosts at sea (precluding monogeneans and parasitic arthropods), are at least moderately prevalent within their host populations, and are easily identified. Currently, selected macroparasites are being used as markers in fisheries' studies to distinguish host stocks, to indicate migratory routes, to identify feeding and breeding areas, and to monitor recruitment patterns of juveniles.

Molecular methods are increasingly and successfully being used to address these same questions, particularly in the context of fisheries management (Baldwin *et al.*, 2012). In fact, modern molecular tools not only can assign individual smolt salmon to their river of origin, but also to their parents. Given that DNA-based fingerprinting methods have become routine, it may come as a surprise that parasite tags continue to be used. The problem is that assignments based upon molecular data are often inaccurate when the level of genetic differentiation among source populations is low, as is often the case for species of marine fish. Criscione *et al.* (2006) showed that parasite genotypes of a trematode (*Plagioporus shawi*) more accurately assigned an individual steelhead trout (*Oncorhynchus mykiss*) to its natal stream than genotypic data of the fish. In this case, the probability of correctly assigning an individual to its correct source was four times higher with the parasite's genotypes than with the host's genotypes. Results from a study involving parasite communities of redfish (*Sebastes marinus*) collected along the coast of eastern Canada showed that three fish stocks defined by parasite communities confirmed the boundaries previously defined by molecular markers (Marcogliese *et al.*, 2003).

However, the use of parasite communities could subdivide the overall population into four smaller groups, a result that had important management implications.

14.4.2 Parasite phylogeographical distributions as probes for host evolutionary history

Just as parasite distributions can be used as tags to provide insights into host demographic events, they can also be used to reconstruct host evolutionary history. This extension of 'parasites as tags' to evolutionary history is especially effective for parasites that have had a long coevolutionary association with their hosts, where in some cases (e.g., *Plagioporus* in steelhead trout, *Pediculus* in humans), the phylogeographical distributions of parasites can provide a better resolution of evolutionary events than the hosts themselves. Molecular phylogeographical data indicate the evolutionary origin of the human lice, taeniid tapeworms, and *Plasmodium* sp. is in Africa, confirming the patterns indicated by fossils and molecular sequences. In such cases, parasite biogeographical data have provided strong tests of evolutionary hypotheses that are independent of host data. Over the past decade, molecular tools have been used extensively to reconstruct the evolutionary history of various hosts, sometimes leading to striking new interpretations of traditional ideas (reviews in Page, 2003; Nieberding & Olivieri, 2006).

Early examples of the use of parasite distributions to infer host evolutionary history were necessarily observational, typically involving attempts to resolve the marine versus freshwater origins of particular groups of fish. Margolis (1965) observed that parasites of Pacific salmonids, *Oncorhynchus* spp., were dominated by host-specific freshwater parasites, many of which have complex life cycles. Only two marine parasites were specific to salmon, a monogenean and a copepod, both with direct life cycles. These observations, in addition to other evidence, led Margolis (1965) to suggest that Pacific salmonids had a freshwater origin, not a marine one as some biologists had argued. Similarly, Manter (1955) examined the parasites of eels, *Anguilla* spp. from Europe, North

America, Japan, and New Zealand, indicating that eels likely had a marine and, more specifically, a Pacific origin. His arguments were based on the observation that of nine species of parasites found in Pacific eels, eight were highly host specific. Atlantic eels that occur in Europe and North America had a larger number of parasites than those in the Pacific, but these were mostly generalist trematodes common to other fishes.

A landmark example of the use of parasite biogeographical distributions as probes for host evolutionary history involves the parasites of freshwater stingrays in the major river systems of eastern South America (Brooks, 1992). In this case, host and parasite biogeographical data were combined to develop alternative hypotheses regarding the Pacific versus Atlantic origin of this group. Traditional dogma held that stingrays were secondary invaders of freshwater systems, being originally a marine group. Because each drainage system inhabited by potamotrygonids drain into the Atlantic side of South America, the common assumption was that their ancestor was an Atlantic marine or euryhaline stingray that dispersed into freshwater habitats 3 to 5 million years ago. However, phylogenetic analyses of their parasites indicated that the parasites likely originated in the Pacific Ocean and that their closest relatives parasitize marine stingrays along the Pacific coast of South America. The phylogenetic analysis of the parasites, together with the geological history of the region, suggested that the ancestor of these stingrays was a marine stingray present in the Pacific Ocean that was trapped in freshwater by the uplifting of the Andes Mountains, probably beginning in the early Cretaceous. Accordingly, during the origin of the Andes, river drainages were radically altered (i.e., parts of the Amazon flowed west, not east!), a new Continental Divide was established, and the rays became isolated from the ancestral group along the Pacific coast of South America. An independent data set involving nematodes of stingrays (Hoberg et al., 1998) also supports the view that the freshwater stingrays from rivers in South America are likely descendants of stingrays that inhabited the Pacific Ocean. Despite the data in support of a Pacific origin for freshwater

stingrays, it is important to recognize that definitive phylogenetic studies on the hosts have not been done, and alternative scenarios are plausible (e.g., Lovejoy, 1997).

The tools of molecular phylogeography have complemented earlier approaches, particularly in the context of terrestrial host–parasite interactions. Our earlier examples involving *Trichinella* spp., lice (*Pediculus* spp.), and cestodes (*Taenia* spp.) showed that data derived from these specialized parasites can provide important probes for human evolutionary history out of Africa, independent of host-derived data. In each case, the parasites signal a complex history of host migration, contraction, and differentiation, often associated with repeated periods of isolation in glacial refugia. Similar approaches involving non-human hosts, especially those involving helminths of small mammals, show a similar pattern (review in Nieberding & Olivieri, 2006). In one striking example, Galbreath & Hoberg (2012) constructed phylogenies of five lineages of host-specific parasites found in collared and American pikas (*Ochotona collaris* and *O. princeps*). As for many North American mammals, the traditional view of pika origins is one of eastward expansion via the Beringia land bridge, followed by southward colonization. In this case, the collared pika (currently restricted to Alaska, Northern British Columbia, and the Yukon) was thought to have originated from ancestral Beringian populations, while the American pika originated from individuals that expanded southward along the coastal and Rocky Mountains. Results from three of the five parasite phylogenies reversed this traditional view, indicating a southern origin of these hosts, followed by northward expansion into Beringia prior to extensive periods of glaciation during the Holocene. Examples such as this highlight the real-world complexity of both historical and contemporary distributions of species, populations, and entire faunas. Yet perhaps more importantly, these examples highlight the extraordinarily integrative nature of parasites, and what they contribute to our understanding of the biosphere.

References

Araujo, A., Reinhard, K. J., Ferreira, L. & Gardner, S. L. (2008) Parasites as probes for prehistoric human migrations? *Trends in Parasitology*, **24**, 112–115.

Avise, J. C. (2000) *Phylogeography: The History and Formation of Species*. Cambridge: Harvard University Press.

Baldwin, R. E., Banks, M. A. & Jacobson, K. C. (2012). Integrating fish and parasite data as a holistic solution for identifying the elusive stock structure of Pacific sardines (*Sardinops sagax*). *Reviews in Fish Biology and Fisheries*, **22**, 137–156.

Brooker, S., Rowlands, M., Haller, L., *et al.* (2000) Towards an atlas of human helminth infection in sub-Saharan Africa: The use of geographical information systems (GIS). *Parasitology Today*, **16**, 303–307.

Brooks, D. R. (1979) Testing hypothesis of evolutionary relationships among parasites: the digeneans of crocodilians. *American Zoologist*, **19**, 1225–1238.

Brooks, D. R. (1992) Origins, diversification, and historical structure of the helminth fauna inhabiting neotropical freshwater stingrays (Potamotrygonidae). *Journal of Parasitology*, **78**, 588–595.

Brooks, D. R. & McLennan, D. A. (1993) *Parascript: Parasites and the Language of Evolution*. Washington, D.C.: Smithsonian Institution Press.

Brooks, D. R. & McLennan, D. A. (2002) *The Nature of Diversity: An Evolutionary Voyage of Discovery*. Chicago: University of Chicago Press.

Brown, J. H. & Lomolino, M. V. (1998) *Biogeography*. Sunderland: Sinauer Associates.

Cameron, T. W. (1929) The species of *Enterobius* Leach in Primates. *Journal of Helminthology*, **7**, 161–182.

Criscione, C. D., Cooper, B. & Blouin, M. S. (2006) Parasite genotypes identify source populations of migratory fish more accurately than fish genotypes. *Ecology*, **87**, 823–828.

Dittmar, K. (2010) Paleogeography of parasites. In *The Biogeography of Host–Parasite Interactions*, ed. S. Morand & B. R. Krasnov, pp. 21–29. Oxford: Oxford University Press.

Edwards, D. D. & Bush, A. O. (1989) Helminth communities in avocets: importance of the compound community. *Journal of Parasitology*, **75**, 225–238.

Fox, N. J., White, P. C. L., McClean, C. J., *et al.* (2011) Predicting impacts of climate change on *Fasciola hepatica* risk. *PLoS One*, **6**: e16126.

Galbreath, K. E. & Hoberg, E. P. (2012) Return to Beringia: parasites reveal cryptic biogeographic history of North American pikas. *Proceedings of the Royal Society, Series B, Biological Sciences*, **279**, 371–378.

Goater, C. P., Baldwin, R. E. & Scrimgeour, G. (2005) Physico-chemical determinants of helminth community structure in whitefish (*Coregonus clupeoformis*) from adjacent lakes in northern Alberta, Canada. *Parasitology*, **131**, 713–722.

Hay, S. I., Tucker, C. J., Rogers, D. J. & Packer, M. J. (1996) Remotely sensed surrogates of meteorological data for the study of the distribution and abundance of arthropod vectors of disease. *Annals of Tropical Medicine and Parasitology*, **90**, 1–19.

Hay, S. I., Tatem, A. J., Graham, A. J., *et al.* (2006) Global environmental data for mapping infectious disease distribution. *Advances in Parasitology*, **62**, 37–77.

Hoberg, E. P. (1992) Congruent and synchronic patterns in biogeography and speciation among seabirds, pinnipeds and cestodes. *Journal of Parasitology*, **78**, 601–615.

Hoberg, E. P. (1995) Historical biogeography and modes of speciation across high-latitude seas of the Holarctic: concepts for host–parasite coevolution among the Phocini (Phocidae) and Tetrabothriidae (Eucestoda). *Canadian Journal of Zoology*, **73**, 45–57.

Hoberg, E. P. & Brooks, D. R. (2010) Beyond vicariance: integrating taxon pulses, ecological fitting, and oscillation in evolution and historical biogeography. In *The Biogeography of Host–Parasite Interactions*, ed. S. Morand & B. R. Krasnov, pp. 7–20. Oxford: Oxford University Press.

Hoberg, E. P. & Klassen, G. J. (2002) Revealing the faunal tapestry: co-evolution and historical biogeography of hosts and parasites in marine systems. *Parasitology*, **124**, 3–22.

Hoberg, E. P., Brooks, D. R., Molina-Ureña, H. & Erbe, E. (1998) *Echinocephalus janzeni* n. sp. (Nematoda: Gnathostomatidae) in *Himantura pacifica* (Chondrichthyes: Myliobatiformes) from the Pacific Coast of Costa Rica and Mexico, with historical biogeographic analysis of the genus. *Journal of Parasitology*, **84**, 571–581.

Hoberg, E. P., Alkire, N. L., de Queiroz, A. & Jones, A. (2001) Out of Africa: origins of the *Taenia* tapeworms in humans. *Proceedings of the Royal Society, Series B, Biological Sciences*, **268**, 781–787.

Hoberg, E. P., Galbreath, K. E., Cook, J. A., *et al.* (2012) Northern host–parasite assemblages: history and biogeography on the borderlands of episodic climate and environmental transition. *Advances in Parasitology*, **79**, 1–97.

Hugot, J. P. (1999) Primates and their pinworm parasites: the Cameron Hypothesis revisited. *Systematic Biology*, **48**, 523–546.

Klassen, G. J. (1992) Coevolution: a history of the macro-evolutionary approach to studying host–parasite associations. *Journal of Parasitology*, **78**, 573–587.

Lovejoy, N. R. (1997) Stingrays, parasites, and neotropical biogeography: a closer look at Brooks *et al.*'s hypothesis concerning the origins of neotropical freshwater rays (Potamotrygonidae). *Systematic Biology*, **46**, 218–230.

MacKenzie, K. (2002) Parasites as biological tags in population studies of marine organisms: an update. *Parasitology*, **124**, S153–S163.

Malone, J. B. (2005) Biology-based mapping of vector-borne parasites by Geographic Information Systems and Remote Sensing. *Parassitologia*, **47**, 27–50.

Manter, H. W. (1934) Some digenetic trematodes from the deep-water fish at Tortugas, Florida. *Carnegie Institute Publication no. 435, Papers from Tortugas Laboratory*, **27**, 257–345.

Manter, H. W. (1955) The zoogeography of trematodes of marine fishes. *Experimental Parasitology*, **4**, 62–86.

Manter, H. W. (1963) The zoogeographical affinities of trematodes of South American freshwater fishes. *Systematic Zoology*, **12**, 45–70.

Marcogliese D. J., Albert, E., Gagnon, P. & Sevigny, J-M. (2003) Use of parasites in stock identification of the deep-water redfish (*Sebastes mentella*) in the Northwest Atlantic. *Fisheries Bulletin*, **101**, 183–188.

Margolis, L. (1963) Parasites as indicators of the geographical origin of sockeye salmon, *Oncorhynchus nerka* (Walbaum) occurring in the North Pacific Ocean and adjacent seas. *Bulletin of the International North Pacific Fish Community*, **11**, 101–156.

Margolis, L. (1965) Parasites as an auxiliary source of information about the biology of Pacific salmon (genus *Oncorhynchus*). *Journal of the Fisheries Research Board of Canada*, **22**, 1387–1395.

Margolis, L. (1993) A case of forensic parasitology. *Journal of Parasitology*, **79**, 461–462.

Martinson, D. G., Pisias, N. G., Hays, J. D., *et al.* (1987) Age, dating, and orbital theory of the Ice Ages: development of a high resolution 0–300,000 year chronostratigraphy. *Quaternary Research*, **27**, 1–29.

Metcalf, M. (1940) Further studies on the opalinid ciliate infusorians and their hosts. *Proceedings of the United States Museum*, **87**, 465–634.

Morand, S. & Krasnov, B. R. (2010) *The Biogeography of Host–Parasite Interactions*. Oxford: Oxford University Press.

Nieberding, C. M. & Olivieri, I. (2006) Parasites: proxies for host genealogy and ecology? *Trends in Ecology and Evolution*, **22**, 156–165.

Ollerenshaw, C. B. & Rowlands, W. T. (1959) A method of forecasting the incidence of fascioliasis in Anglesey. *Veterinary Record*, **71**, 591–598.

Page, R. D. M. (2003) *Tangled Trees: Phylogeny, Cospeciation, and Coevolution*. Chicago: University of Chicago Press.

Pozio, E., La Rosa, G., Hoberg, E. P. & Zarlenga, D. S. (2009) Molecular taxonomy, phylogeny, and biogeography of nematodes belonging to the *Trichinella* genus. *Infection, Genetics and Evolution*, **9**, 606–616.

Pullan, R. L. & Brooker, S. J. (2012) The global limits and population at risk of soil-transmitted helminth infections in 2010. *Parasites and Vectors*, **5**, 81.

Reed, D. L., Smith, V. S., Hammond, S. L., *et al.* (2004) Genetic analysis of lice supports direct contact between modern and archaic humans. *PLoS Biology*, **2**, 1972–1983.

Reed, D. L., Light, J. E., Allen, J. M. & Kirchman, J. J. (2007) Pair of lice lost or parasites regained: the evolutionary history of anthropoid primate lice. *BMC Biology*, **5**, 7.

Sehgal, R. N. M., Buermann, W., Harrigan, R. J., *et al.* (2010) Spatially explicit predictions of blood parasites in a widely distributed African rainforest bird. *Proceedings of the Royal Society, Series B, Biological Sciences*, **278**, 1025–1033.

Simoonga, C., Utzinger, J., Brooker, S., *et al.* (2009) Remote sensing, geographical information system and spatial analysis for schistosomiasis epidemiology and ecology in Africa. *Parasitology*, **136**, 1683–1693.

Spear, R. C., Seta, E., Liang, S., *et al.* (2004) Factors influencing the transmission of *Schistosoma japonicum* in the mountains of Sichuan Province of China. *American Journal of Tropical Medicine and Hygiene*, **70**, 48–56.

Waltari, E., Hoberg, E. P., Lessa, E. P. & Cook, J. A. (2007) Eastward ho: phylogeographic perspectives on colonization of hosts and parasites across the Beringian nexus. *Journal of Biogeography*, **34**, 561–574.

Zarlenga, D. S., Rosenthal, B. M., La Rosa, G., *et al.* (2006) Post Miocene expansion, colonization, and host switching drove speciation among extant nematodes of the archaic genus *Trichinella*. *Proceedings of the National Academy of Sciences USA*, **103**, 7354–7359.

15 Effects of parasites on their hosts: from individuals to ecosystems

15.1 General considerations

Parasitologists have long been interested in the extent to which parasites affect their hosts. While the historical focus has been at the scale of host individuals and host populations, enquiries that span the range from genome-level effects to ecosystem-level effects are increasingly common. The early focus on host individuals and populations likely arose from the need for clinical parasitologists to characterize and control important human and veterinary pathogens. Indeed, a key direction of modern studies, aided by the explosion in molecular methodologies, is to determine the mechanisms of pathology that occur at the host–parasite interface and to assess their consequences on host individuals and populations. We included this perspective in our coverage of the general biology of human parasites such as *Plasmodium* spp., *Trypanosoma* spp., *Schistosoma* spp., and *Trichinella spiralis* in earlier chapters.

Beginning in the 1970s, interest in the influence of parasites on their hosts was extended to include wildlife and other animals. An important milestone was the theoretical treatments initiated by Crofton (1971) and later by Anderson & May (1978) and May & Anderson (1978) that included *alpha* (the rate of parasite-induced host mortality) as a key epidemiological parameter (see Chapter 12). Although difficult to do, research focused on estimating *alpha* for a range of host–parasite interactions and evaluating how ecological conditions could affect its magnitude. This emphasis had a powerful impact on the integration of parasitology and wildlife disease ecology into mainstream ecology, conservation biology, and wildlife management. But perhaps most importantly, these theoretical treatments shifted the emphasis away from

medical and host-centered views of parasite-induced effects to more parasite-centered views that placed variation in *alpha* into the context of variation in the ways in which parasites exploit their hosts (Poulin, 2007). In so doing, the question of 'effects' has developed into one of the leading questions in our field, requiring integration among parasitologists and researchers from every conceivable subdiscipline in biology, particularly ecology and evolutionary biology.

In this chapter, we take a parasite-centered and parasitologist's perspective of the varied ways in which parasites can influence their hosts. We begin at the scale of individuals, focusing on empirical studies that demonstrate the range of direct and indirect effects that parasites can have on host fitness. We then take a similar approach to explore the extent to which the effects on individual hosts, when present, scale up to the level of host populations, communities, and ecosystems. Although our main aim is to synthesize empirical evidence underlying particular phenomena, we also provide insight into the ways in which leading researchers are tackling the challenging problem of detecting parasite-induced effects on their hosts.

15.2 Effects of parasites on host individuals

The requirement for parasites to cause 'harm' to their hosts is implicit within most definitions of parasitism. Since parasites must exploit host resources to support their development and reproduction, parasites have the potential to reduce the fitness of their hosts. A central challenge for parasitologists and ecologists is to understand the degree to which that potential is met in natural host–parasite interactions. As we have seen throughout the biodiversity chapters, the precise nature of effects on individual hosts can take many forms. At one extreme, the potential for 'harm' is clearly realized. The fecundity of trematode-infected snails, cestode-infected sticklebacks, microparasite-infected water fleas, and barnacle-infected crustaceans can decline to zero under ecological conditions that are common in nature. The obligate development of the gall-like nurse cells in humans infected with larval *Trichinella spiralis* causes permanent damage to individual muscle cells (see Chapter 8). Mortality associated with *Schistosoma mansoni* in humans is associated with the deposition of eggs in the host's liver and other tissues (see Chapter 6). In such cases, the effects on individuals are easily detected and can be attributed to the manner in which parasites exploit host resources. Yet, in many other cases that we covered in earlier chapters, negative effects on host individuals are undetectable. Our view is that even these seemingly benign parasites have the potential to reduce the fitness of their individual hosts but, under ecologically relevant conditions, they do not. Understanding where a particular host–parasite interaction fits along this continuum, and the extent to which it shifts position along it relative to environmental conditions, is an area of intense research activity. This focus should not be surprising, since consistent effects on host individuals will determine effects at higher scales, e.g., host populations and communities, and will determine the strength and direction of parasite-mediated natural selection, evolution, and coevolution (see Chapter 16).

The literature on the effects of parasites on host individuals is enormous. Examples that span the range of potential effects can be found in the first 10 chapters. The complex, but integral, role of the host immune system in contributing to within-host effects was covered in Chapter 2. Further examples that span an even broader taxonomic scope are incorporated into modern parasitology texts (e.g., Roberts & Janovy, 2009). One of the messages to arise from this rich background is the enormous variation that exists between species of parasites in the effects they have on their hosts. *Plasmodium falciparum* is more

pathogenic in humans than the four or five other species of *Plasmodium* that infect us. *Trypanosoma b. brucei* is benign in many species of antelope, but is 100% lethal in domesticated hooved mammals. Introduced *Plasmodium relictum* rapidly kills most species of native honeycreeper in Hawaii, but it is inconsequential to species of introduced birds (see Section 15.4). Thus, at least anecdotally, the same species of parasite can have dramatically different effects in different species of hosts. Similarly, different species of parasite can have different effects on the same species of host. In one of the few experimental demonstrations of this idea, Ebert (2005) compared differences in lifetime reproduction and overall survival between infected and uninfected water fleas exposed to controlled doses of four different species of microparasite (Fig. 15.1). The results confirm the large numbers of anecdotal reports in the literature showing that some species of parasite cause catastrophic reductions in host reproduction, and thus fitness, whereas others do not. Excellent reviews on patterns of interspecific variation in parasite virulence can be found in Ebert (2005) and Poulin (2007).

We begin our coverage of this broad area by emphasizing two points. The first is a cautionary note regarding methodology. For students, your first query into the question of 'harm' may have occurred during an incidental encounter with a carcass riddled with parasites. In such cases, we may intuitively link parasitism as the cause of the host's death. Alternatively, one of your biology laboratories may have involved counts of parasites in wild-caught hosts, followed by analyses of associations between parasite numbers and indicators of host fitness, such as body size. This general approach is akin to a wildlife biologist assessing parasite-induced mortality based upon the post-mortem detection of parasites in host tissues. There are two pitfalls with this approach. The first is that conclusions regarding negative effects on the host must distinguish the mere presence of a parasite with disease. Yet, this is extraordinarily difficult, if not impossible, with survey data alone (Scott, 1988). The second is that without experimental manipulation, it is impossible to distinguish whether infection caused

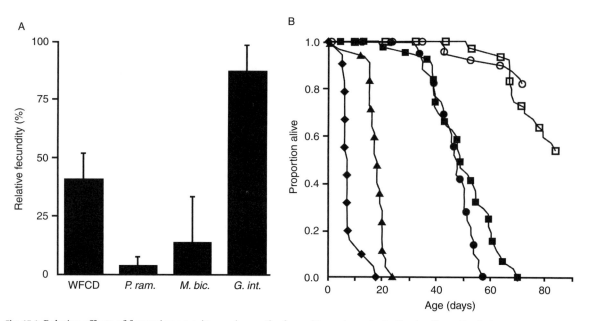

Fig. 15.1 Relative effects of four microparasite species on the fecundity and survival of water fleas, *Daphnia magna*. (A) Effects of White Fat Cell Disease (WFCD), *Pasteuria ramosa* (*P. ram.*), *Metschnikowia bicuspidata* (*M. bic.*), and *Glucoides intestinalis* (*G. int.*) on relative host fecundity. (B) Effects of WFCD (♦), *P. ramosa* (●), *M. bicuspidata* (▲), and *G. intestinalis* (■) on host survival. (□, ○ = controls.) (Figure courtesy of Dieter Ebert; modified from Ebert, 2005, *Ecology, Epidemiology and Evolution of Parasitism in Daphnia* [Internet], National Center for Biotechnological Information, USA.)

any observed effects on host fitness, e.g., reduced size, altered activity, altered behavior, or the effects on host fitness led to increased infection. As we will see below, manipulative experiments are not the only way forward, but in their absence we must be cautious with our interpretations. Our examples through this section are primarily drawn from empirical studies that allow comparisons in performance between infected and uninfected hosts.

We also wish to emphasize the context-dependent nature of parasite-induced effects. For many of the human parasites that we covered earlier, the magnitude of negative effects often result from complex interactions between malnutrition, reduced immunity, and parasite-induced pathology. Similarly, pathogenicity of the nematode *Trichostrongylus tenuis* in red grouse is higher on wet moors than on dry moors, because the survival of infective larvae requires moisture (Hudson *et al.*, 1985). One explanation for the context-dependent nature of these outcomes is that a host's ability to compensate for the effects of

infection will vary between individuals, with the result depending on factors such as body size, general condition, parasite numbers, and immune status (Holmes & Zohar, 1990). Variation in compensatory ability will mean that a given number of parasites, or a parasite with a given virulence, is likely to have less effect on a host in good condition than one in poorer condition.

In a test of this idea, Munger & Holmes (1988) infected Richardson's ground squirrels, *Spermophilus richardsonii*, from the Canadian prairies with the blood protist *Trypanosoma otospermophili*. In most populations, prevalence of trypanosomes within a squirrel colony is 100%. The authors evaluated the context-dependent nature of trypanosome-induced pathology in squirrels with a factorial experiment that crossed exposure to *T. otospermophili*, diet quality (the presence or absence of a particular vitamin), and diet quantity (restricted or *ad libitum* access to food). Under conditions of high quantity and quality food, infected hosts were stunted. When food and vitamin rations were restricted, effects on host growth were undetectable. In an unexpected twist, squirrels that were fed a vitamin-restricted *ad libitum* diet grew significantly larger than uninfected squirrels! In this extreme example, infection led to zero, negative, and even positive effects, depending on the resources available to the host. In another experimental study, Hawlena *et al.* (2008) showed that the main effect of fleas on rodents was to enhance the negative effects of food restriction on the body mass of juveniles. These results emphasize the general point that effects on individual hosts are relevant only within the ecological context within which they are measured.

15.2.1 Parasites and host metabolism

In our definition of parasitism, metabolic dependence of the parasite on the host is obligate and long term relative to the life span of the parasite. Logically, it follows that parasites must utilize host resources to meet their life needs. In so doing, we can expect alterations to energy budgets to arise in infected hosts, either directly by 'stealing' host energy reserves, or indirectly via the reallocation of limited host resources to defense and/or damage protection. The degree to which parasites affect host energy expenditure is, therefore, an important element of host–parasite interactions, with significant implications. From an applied perspective, the types and doses of anti-parasite drugs are often administered to domestic stock to offset production losses due to infection. From an ecological perspective, questions regarding host compensation and context-dependent outcomes are fundamentally linked to potential energy losses imposed by parasites. Moreover, alterations in energy expenditure and allocation in chronically infected hosts may be the underlying cause of other negative consequences of infection, such as those on growth, behavior, and reproduction.

Understanding the effect of parasites on host energy budget is, therefore, fundamental to an understanding of the effects of parasites on individual hosts. One common approach is to use standard physiological assays to compare resting energy expenditure in infected and uninfected hosts. Many such experiments are designed to test the hypothesis that, due to the diverse energetic demands imposed by infection, resting metabolic rate (RMR) is enhanced in infected individuals. Supportive evidence comes from the results of laboratory studies that demonstrate RMR increases up to 90% in systems as diverse as crabs infected with larval acanthocephalans, rats infected with adult nematodes and trematodes, and fish infected with microsporidians (review in Robar *et al.*, 2011). Studies involving mammals and birds infected with their host-specific ectoparasitic arthropods are also illuminating (e.g., Giorgi *et al.*, 2001), with results consistently demonstrating a 10–20% dose-dependent increase in RMR relative to uninfected controls. In many of these examples, enhanced RMR is associated with declines in other aspects of host performance, such as growth rate, grooming behavior, and thermoregulation.

Although laboratory artefacts may confound the interpretation of these results, parasite-induced enhancement of RMR has also been demonstrated in free-ranging hosts. Booth *et al.* (1993) evaluated the energetics of wild rock doves, *Columba livia*, infected with feather-feeding ischnoceran lice (see Chapter 11), a group traditionally considered benign. In one group of wild-caught birds, the researchers reduced louse loads by fumigation. In a second group, they increased louse loads by implanting metal 'bits' that impaired the birds' normal preening ability. When individuals were re-caught 9 months later, high-load birds had 8% higher RMR compared to low-load birds. The former also suffered from decreased thermoregulatory capacity due to enhanced feather damage. These results indicate that chronic infections, even with seemingly benign parasites, can negatively impact host fitness, especially when measurements are made over the long term.

However, our intuitive notion that parasites impose energetic demands that increase host RMR is too simplistic. Equally convincing evidence from studies on an equally wide range of host–parasite interactions has produced non-significant effects of infection on RMR, or parasite-induced decreases. In a synthesis of studies that compared RMR between infected and uninfected hosts, Robar *et al.* (2011) showed that the overall effect of parasites was small and insignificant. Further, the high variation in RMR reported between studies could not be explained by variation in features such as sample size, experimental methods, host taxa, or parasite taxa. The researchers did not interpret their results to indicate that parasites have little effect on host energetics. Rather, they suggest that 'no net effect' resulted from the highly inconsistent magnitude and direction of parasite-induced effects on host metabolism, with the number of studies demonstrating increased host RMR approximately balanced by an equal number of studies signifying the opposite result.

How can such contrasting results be explained? One possibility is that snapshot, laboratory-based assessments of RMR are inaccurate for parasites that have long-term, chronic associations with their hosts (Robar *et al.*, 2011). But this explanation is unsatisfactory for cases where the same parasites demonstrate contrasting results in similar species of hosts. Powell *et al.* (2005) showed that a microsporidian caused a 90% increase in RMR in one species of trout, but a 55% reduction in another. Similar contrasting results were observed for two species of lizards infected with gregarine protists (Oppliger *et al.*, 1996). An alternative explanation lies with the variation in the ability of different hosts to compensate for the effects of infection. Thus, in cases where compensation is possible (perhaps via increased host activity or feeding), we might expect elevated rates of metabolism. In contrast, decreased host metabolism can be expected in cases involving reduced compensation (e.g., reduced host activity). A further alternative lies in the range of mechanisms by which parasites may influence RMR. For the ectoparasites of birds and mammals discussed above, enhanced RMR was due to parasite-induced alterations to host grooming behavior, compensation for losses in thermoregulatory abilities, and to enhanced innate immune responsiveness, depending on the system. Thus, increased understanding of the mechanisms that link infection, host metabolism, and probably immunity, are required to fully interpret the

magnitude and direction of parasite-induced effects on host metabolism. As we will continue to emphasize throughout this section of the text, parasite-induced effects are more complex, and more interesting, than the results of earlier studies initially indicated.

15.2.2 Parasites and host fitness

Assessment of host fitness in infected versus uninfected hosts provides another starting point for understanding parasite-induced effects on individual hosts. Manipulative experiments such as those involving the microparasites of water fleas that assess differences in lifetime reproduction between infected and uninfected hosts provide a gold standard (Fig. 15.1), but these are understandably rare. Nonetheless, over the last two decades, ecologists and parasitologists have made great strides in their efforts to determine the extent to which parasites decrease host fitness over a range of ecological conditions.

Tests of parasite-induced reductions in host survival are important since dead hosts pay the ultimate fitness cost. In laboratory mesocosms, a bacterium species reduced overall survival of *Daphnia* by about 60% relative to controls, whereas a microsporidian parasite reduced survival by about 20% (Fig. 15.1). Frogs exposed as tadpoles to cercariae of the trematode *Ribeiroia ondatrae*, suffer similarly high mortality (Johnson *et al.*, 1999; Fig. 15.2). All exposed tadpoles ultimately develop the characteristic limb malformations that we describe in Chapters 6 and 17 (see Color plate Fig. 7.2), whereas those exposed to another species of cercariae found in the same ponds did not. Four weeks later, and even in the absence of predation, 60% of the froglets exposed to *Ribeiroia* cercariae had died. In a follow-up study, Johnson *et al.* (2012) showed that the general pattern in Fig. 15.2 was consistent for eight of 13 other species of amphibian exposed to *R. ondatrae* cercariae. These results imply that infected froglets rarely survive to reproductive maturity, an observation that explains the low prevalence of malformations in adult frogs. In this case, the encystment of metacercariae at intensities that closely

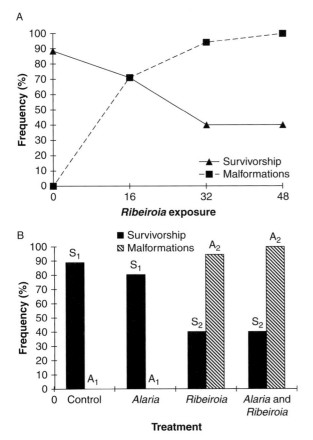

Fig. 15.2 Effects of metacercariae of *Ribeiroia ondatrae* on Pacific treefrog *Pseudacris regilla* survivorship and hindlimb malformation frequency. Survivorship is calculated as number of tadpoles surviving to metamorphosis divided by initial sample size. Malformation frequency is calculated as number of metamorphosing frogs divided by the total number of metamorphosing frogs within a given treatment. (A) *Ribeiroia* density shows a significant, positive relationship to malformation frequency and a significant, negative relationship to survivorship. (B) *Pseudacris regilla* survivorship and malformation frequency across parasite treatments. Survivorship and malformation frequency are calculated as above. Within survivorship, differences in the designations above bars (S_1 or S_2) indicate significant differences between treatments. For malformation frequency, significance groupings are denoted by A_1 or A_2. (Figure courtesy of Pieter Johnson; modified from Johnson *et al.*, 1999, *Science*, **284**, 802–804.)

parallel those found in natural populations had a catastrophic effect on host fitness. Robar *et al.* (2010) compiled results from over 50 experimental tests of parasite-induced effects on host survival, showing that the probability of host mortality more than doubled in infected versus uninfected hosts.

Studies that combine field and laboratory experiments provide another powerful approach for testing parasite-induced effects on survival. Thousands of lakes in North America contain populations of bluegill sunfish, *Lepomis macrochirus*, that are 100% infected with metacercariae of the trematode *Uvulifer ambloplitis*. Metacercariae of this and many other species of digenean trematode cause the characteristic 'blackspot' that arises from a host-induced encapsulation and melanization of developing metacercariae in the muscles. In a field study, Lemly & Esch (1984) monitored the survival of infected and uninfected juvenile bluegill housed in screened cages through their first winter. In a parallel laboratory study, they maintained groups of infected and uninfected juveniles under simulated winter conditions. Regardless of whether the fish were housed in the laboratory or in a natural pond, fish with more than about 50 metacercariae died when water temperature dropped below 10 °C. Further, they showed that blackspot intensity was negatively correlated with the total body lipid concentration of its host. The authors concluded that the energetic costs associated with the development of high numbers of metacercariae reduced the critical energy stores that were required for these young, non-feeding fish to survive their first winter. In this example of context-dependency, interactions among parasite intensity, host metabolism, host age, and water temperature determined the effects of parasitism on host survival. Similarly, years when mass mortalities of juvenile sticklebacks in Alaskan lakes could be attributed to the cestode *Schistocephalus solidus* required a 'perfect storm,' when high rates of exposure coincided with high host densities and extreme winter conditions (Heins *et al.*, 2011).

Comparisons of reproductive performance between infected and uninfected individuals provide powerful tests for fitness-related costs. Such tests can take many forms. At one extreme, parasites may ingest host reproductive tissue, e.g., rediae of some larval trematodes in snails, and/or functionally replace the gonads of the host, e.g., rhizocephalan barnacles of decapod crustaceans, and larval acanthocephalans and nematomorphs in arthropods. In such cases, parasite-induced effects can operate in an all-or-none fashion, such that the fitness of castrated hosts is zero, at least temporarily. At the other extreme, infection is associated with quantitative reductions in the quantity or quality of eggs or offspring, to changes in the timing of host reproduction, or to modifications in elements of the process of sexual selection, e.g., mate choice, parental care. The results from experiments involving hosts, particularly birds, with experimentally enhanced or reduced worm numbers have provided key insights.

Experiments by Peter Hudson and his colleagues involving the nematode *Trichostrongylus tenuis* in populations of red grouse, *Lagopus scoticus*, in northern England are some of the best known (e.g., Hudson *et al.*, 1998). Adult worms develop within the intestinal cecae, releasing eggs that develop into the infective L3 stage on moorland. Infection is via ingestion. Laboratory studies have shown that feeding worms cause hemorrhaging in the walls of the cecae and alter enzyme activity along the brush border, leading to reduced growth rates, reduced muscle mass, and anorexia. In a field test of individual-level effects, Hudson and colleagues compared worm intensities in birds that died naturally with those that were killed by predators (foxes and ravens) or hunters (Hudson *et al.*, 1992). Despite extensive variation in worm counts, the results indicated that birds killed by predators contained the highest intensities. They then performed a field experiment to directly assess the effects of the nematode on host breeding performance. In the spring for 7 years, a sample of birds was caught, tagged, and then treated with an anthelmintic to reduce worm numbers. A control sample of birds was treated with water only. The birds were released and then monitored over the subsequent year for clutch size, hatch success, chick production, and overall survival. Results showed that in each year, the clutches of drug-treated females had

more eggs, higher hatch success, and higher survival compared to sham-treated females. Based upon the strength and consistency of the results from both experimental and observational studies involving parasites of birds (mostly ectoparasites), Møller (1997) concluded that the evidence is unequivocal for a general role for parasites in reducing host fitness. The results of experimental studies involving parasites of insects, snails, and fish provide further support (review in Combes, 2001).

Given that reproduction is size-dependent in many hosts, parasite-induced effects on host growth and development can further reduce host fitness. At least two of the microparasites of *Daphnia* reduce adult growth, such that reproduction of infected hosts occurs later and at smaller sizes (Ebert, 2005). Likewise, parasite-induced stunting occurs in sticklebacks infected with plerocercoids of *S. solidus*, likely contributing towards observed reductions in the egg production of females and breeding performance in males. In such cases, reduced host growth arises from the theft of host nutrients during parasite development, although redistribution of host energy reserves to immunological defenses cannot be ruled out. There is also a marked, dose-dependent reduction in the growth rates of young toads *Bufo bufo* infected with the lungworm *Rhabdias bufonis*, followed by an associated decrease in host survival (Goater & Ward, 1992). In this case, the mechanism of growth reduction was associated with a striking reduction in food intake that occurred even in hosts that had few parasites (Fig. 15.3). Although surviving toads recover towards normal feeding following the initial period of worm development, previously infected hosts remain stunted, even after they have lost their worms (Goater, 1994). Voluntary reduction in feed intake, i.e., parasite-induced anorexia, of 30–60% has been reported in other invertebrate and vertebrate hosts infected with protist, helminth, and arthropod parasites (Kyriazakis *et al.*, 1998). The paradoxical association between infection and reduced feeding may be one of the most important consequences of parasites for individual hosts.

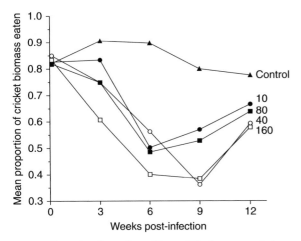

Fig. 15.3 Intensity-dependent effects of the lung nematode *Rhabdias bufonis* on food intake of European toads *Bufo bufo*. Points represent the average proportion of cricket biomass ingested by individual toads at each interval, relative to the total biomass offered. (Modified from Goater & Ward, 1992, with permission, *Oecologia*, **89**, 161–165, Springer Science & Business Media.)

We conclude that the potential for negative effects on host fitness is realized in many natural host–parasite combinations. Often, the mechanisms leading to such effects are precisely what we would expect from organisms that must appropriate host nutrients to meet their own ends. Thus, infection with parasites can often be seen to reduce host growth, reproduction, and survival because of their interference with the host's normal physiological processes. However, as we have seen (and as we will see in the next sections), the range of negative effects on host fitness is much more complex than mere thievery of host nutrients or parasite-induced tissue damage. In other cases, variation in the magnitude of negative effects is closely tied to disparity in the numbers of parasites present. Thus, any discussion of variation in parasite-induced effects among host individuals in nature must consider the aggregated nature of parasites within a host population (see Chapter 12). In other cases, variation in effects is linked to variation in environmental conditions, and by association, to variation in host immunity. Clearly, a host's fitness is determined by factors in addition to infection with parasites, e.g.,

predation, weather, and sexual selection. Perhaps the most biologically relevant effect of most parasites lies in how they interact with these other factors to determine overall variation in host condition.

15.2.3 Parasites and host life history variation

The examples we have discussed up to now show that disparity in the negative effects of infection on individuals often arise from variation in parasite-induced nutrient robbery and/or tissue damage. Such effects are often referred to as side-effects of infection (e.g., Minchella, 1985). However, this view provides a restrictive framework for the full range of effects that parasites have on their individual hosts. For instance, following a period of development of the trematode *Trichobilharzia ocellata* in its snail intermediate host, *Lymnaea stagnalis*, a host-derived molecule called schistosomin is secreted from the cells of the snails' immune system. Because of its cytokine-like properties, the molecule mediates a cascade of events within infected snails that includes inhibition of host growth and reproduction (de Jong-Brink *et al.*, 2001). Likewise, shortly after a female beetle, *Tenebrio molitor*, is exposed to larvae of the cestode *Hymenolepis diminuta*, she produces an anti-gonadotropin that reduces the protein concentration of her eggs (Webb & Hurd, 1999). When this material is injected into uninfected recipient females, they not only exhibit reduced fecundity, they also live longer (Hurd et al., 2001). Here, parasite-induced reduction in fecundity can be considered a host adaptation to increase lifetime reproduction. Taken together, the results of experiments such as these indicate that hosts can alter their schedules of development and reproduction, i.e., their life histories, following exposure to parasites.

Parasite-induced alterations in host life histories are observed most consistently for castrating parasites of crustaceans, insects, and especially snails (reviews in Forbes, 1993; Michalakis, 2009). In previous chapters, we included examples involving enhanced host growth and altered reproduction caused by larval trematodes, rhizocephalan barnacles, and some parasitic isopods. In such cases, radical alterations in many key physiological parameters of the host are the norm, and we might expect concomitant counter-adaptations involving host life histories. Recall that in the case of larval trematodes, the development of sporocysts and/or rediae typically occur at the expense of host reproduction, often leading to outright castration of their mollusc first intermediate hosts (Chapter 6). This cessation in reproductive activity may be accompanied by enhanced host growth in a phenomenon known as **gigantism**. The phenomenon is most common in juvenile, short-lived freshwater snails that are exposed to trematode miracidia (Minchella, 1985), although enhanced host growth also has been reported in other experimental host–parasite interactions (Michalakis, 2009). In the case of snails, the magnitude of growth enhancement, while highly species- and context-dependent, can exceed 30% in infected versus uninfected hosts.

Despite over 100 years of dedicated effort, the adaptive significance of gigantism, to the host and/or parasite, is still poorly understood. One possibility is that enhanced host growth is not adaptive to the host or parasite, and simply arises from a redistribution of host calcium into the host's shell that normally would have gone to support growth in offspring. Alternatively, enhanced shell growth will be host adaptive in cases where the host can outlive the infection and then presumably enjoy the fitness advantages of size-dependent reproduction. Results from a mark–recapture study involving the snail *Helisoma anceps* collected from a small pond in North Carolina showed that several spring-collected snails had lost their infections over winter (Goater *et al.*, 1989). These previously uninfected snails had fully functional reproductive organs, indicating a reversal in castration. Finally, since large snails tend to produce more cercariae than small ones, it is straightforward to interpret gigantism as a parasite adaptation that increases parasite transmission. Thus, empirical studies on diverse trematode–snail interactions, together with field studies involving marked snails, provide support for all three scenarios.

Adjustments to the timing of reproduction have also been observed in infected hosts. Empirical studies

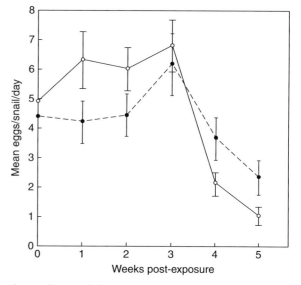

Fig. 15.4 Temporal changes in weekly *per capita* fecundity of the snail *Biomphalaria glabrata* exposed (solid line) and unexposed (dashed line) to miricidia of the trematode *Schistosoma mansoni*. (Modified from Minchella & Loverde, 1981, with permission, *The American Naturalist*, 118, 876–881, The University of Chicago Press.)

involving snails exposed to miracidia of human schistosomes provided initial evidence for parasite-induced alterations in reproductive schedules. Minchella & Loverde (1981) showed that snails, *Biomphalaria glabrata*, exposed to miricidia of *Schistosoma mansoni* reproduced earlier compared to unexposed snails (Fig. 15.4). In a phenomenon known as **fecundity compensation**, infected snails experience a burst of egg laying soon after exposure to miracidia, followed by a decline in fecundity towards zero as the infection proceeds. In an intriguing extension to this general pattern, snails that were exposed to miracidia, but did not develop mature infections (likely due to innate immunity), also exhibited fecundity compensation (Minchella & Loverde, 1981). However, exposed, but uninfected, snails produce fewer eggs over their lifetime, indicating that early egg production to reduce the anticipated losses of future reproduction is a costly strategy. Nonetheless, this alteration in host life history provides an appealing mechanism that enables the host to reduce the future reproduction costs of infection, without incurring the substantial costs of resistance that must be paid even in the absence of exposure (see Chapter 2).

Follow-up studies have shown that the extent and nature of fecundity compensation in snails is closely tied to the age of the host upon exposure to miracidia. One consistent pattern for *Schistosoma* sp.-infected snails is early reproduction (and enhanced shell growth) during the approximate 4-week period of larval development within juvenile snails, followed by growth and reproductive cessation once the host and parasite have reached maturity (Théron & Gerard, 1994). One explanation is that for juvenile snails, infection leads to a re-allocation (or commandeering?) of host resources away from growth and development, whereas in adult hosts, resources normally allocated to reproduction are re-directed. Thus, the occurrence and magnitude of compensation is probably closely tied to the survivorship and reproductive value of snails at the moment they are exposed (Minchella & Loverde, 1981).

It is difficult to evaluate the adaptive significance of gigantism or fecundity compensation in natural populations of hosts. The simple prediction that the prevalence of trematode infection should increase with snail size is supported in most surveys involving freshwater snails in temperate habitats (e.g., Minchella *et al.*, 1986). Further, a handful of field studies have shown that snails mature earlier and smaller in locations where exposure to trematode larvae is high, supporting the predictions of the fecundity compensation hypothesis. Lafferty (1993) compared life history characteristics of marked snails in estuarine populations that varied in their risk of exposure, revealing that after controlling for various confounding factors, snails from high-risk populations matured significantly earlier and smaller than snails from low-risk populations. Similarly, the marine snail *Zeacumantus subcarinatus* develops to maturity faster and at smaller sizes in populations where the risk of trematode exposure is high (Fredensborg & Poulin, 2006).

To conclude, evidence from empirical studies and field surveys is consistent with a diverse role of parasites in determining variation in key host life history traits. Recent results that demonstrate life history modifications in infected freshwater crustaceans infected with microparasites, small-bodied fishes infected with larval cestodes, and larval insects infected with parasitoids indicate that the phenomenon is not restricted to snails castrated by larval trematodes (Michalakis, 2009). However, our understanding of the nature and extent of parasite-induced life history variation remains unclear. For example, we do not understand the factors leading to the high interspecific variability in parasite-induced gigantism or fecundity compensation. While the effects seem to be consistent among several species of schistosomes in their snail hosts, empirical results involving other snail–trematode interactions show no life history alterations, or show effects that are simple consequences of pathology. In addition, care is needed to interpret host life history alterations as active defense strategies by infected hosts. In a theoretical treatment, Hall *et al.* (2007) showed that fecundity compensation arose, at least for castrators, from simple re-allocation of host resources away from reproduction, and did not require active modification by the host. Herein lies another example that highlights the challenges of interpreting single adaptive scenarios regarding the effects of parasites on individual hosts (Poulin, 2007).

15.2.4 Parasites and other host phenotypes

Our coverage in this chapter has emphasized parasites whose negative effects on host growth, reproduction, and survival arise from unavoidable pathology or their requirement for specific types or amounts of host nutrients within specific sites. However, there are many parasites that exploit their hosts in more subtle, and specific, ways. While these may not have detectable effects on host growth, short-term reproduction, and survival, their impact on host and parasite fitness can be profound. Human RBCs that are infected with merozoites of *Plasmodium falciparum* provide an example. In this case, parasite gene products coordinate the development of raised 'knobs' on the surface of host blood cells (see Chapter 2). Why? It turns out that specific receptors located on the knobs bind infected blood cells to the walls of oxygen-rich host capillaries. Here, merozoite development within the blood cells can proceed without risk of filtration by the host's spleen. This example is an illustration of Richard Dawkins' (1982) **extended phenotype**, in which the phenotype of one organism arises, in part, as a consequence of natural selection operating on the genome of another. In a similar fashion, parasite gene products are required for the development of the sophisticated nurse cell that envelops the larvae of *Trichinella spiralis* in host muscle cells, providing a long-term nutrient source and protection from host immunity. Examples of other striking alterations in host phenotype can be found throughout the biodiversity chapters.

The best known of these alterations are those involving specific modifications that facilitate parasite transmission between trophic levels. Most chapters in the biodiversity section of this text include examples of parasite-induced changes in host morphology, color, conspicuousness, and/or behavior that are known to enhance parasite transmission. The altered appearances of *Leucochloridium*-infected terrestrial snails and *Ribeiroia*-infected froglets have been covered earlier, as were the altered behaviors of nematomorph-infected crickets, *Toxoplasma*-infected rodents, and *Dicrocoelium*-infected ants. Some of the most spectacular examples are in Color plates 7 & 8 and some of these are prominent on the text cover. In Box 11.5, we described the bizarre case whereby a parasitoid wasp fools a spider into spinning a special, protective cocoon that enhances its own survival. In Box 7.1, we described the strikingly different anti-predator behaviors and appearances of *Gammarus* infected with larval acanthocephalans. The bizarre case of the 'berry ants' is described in Box 15.1. Here, conspicuous and oddly behaving ants with red abdomens are always infected with gravid nematodes; their drab nest mates are not. As this range of examples shows, parasite-induced alterations of specific host phenotypes are common in nature, they occur within a wide range of host and parasite taxa, and they can be spectacular in their expression.

The phenomenon of parasite manipulation, particularly as it relates to host behavior, has captivated and inspired parasitologists, ecologists, and other biologists for four decades. What began as a series of field-based behavioral observations on two species of larval acanthocephalan in their freshwater hosts (see Box 7.1) has developed into one of our most sophisticated and integrative areas of enquiry. Indeed, the path that enquiry has taken from shrimp in the sloughs of Alberta, to limb development in *Ribeiroia*-infected amphibians, to *Toxoplasma*-infected humans, is one of the triumphs of modern integrative biology. Extensive reviews are available that cover hundreds of examples of the phenomenon, together with key historical, empirical, and theoretical developments (Moore, 2002; Poulin, 2007; Lefèvre *et al.*, 2009). In this section, we restrict our coverage to two aspects that have dominated this area – the adaptive nature of parasite-induced changes in host phenotypes and the mechanisms underlying the phenotypic changes.

One key requirement is to distinguish phenotypic alterations that are true parasite adaptations enhancing parasite reproduction from alterations arising from unavoidable pathology or simple nutrient robbery. As an extreme example, consider the gross malformation of the lower limbs of some humans infected with the filariid nematode *Wuchereria bancrofti* – the causative agent of elephantiasis. While it might be tempting to view the malformation as a parasite adaptation to restrict host mobility (to enhance vector biting rates?), it is more likely a consequence of immunopathology of the lymphoid circulatory system (see Chapter 8). In a similar fashion, we might interpret diarrhea resulting from *Giardia* or *Cryptosporidium* as a clever parasite adaptation to increase the transmission rates of cysts, or the sneeze reflex as a strategy to transmit influenza particles to your brother or sister. Speculative discussions of this sort emphasize the difficulty in determining whether a particular altered phenotype is a true parasite adaptation that increases parasite fitness, a host adaptation, or a coincidental side-effect of the manner in which parasites exploit their hosts. Definitive answers require manipulations within carefully selected model systems.

Compelling evidence that the alterations are parasite adaptations arise from the highly specific and

Box 15.1 │ 'Berry ants': parasite-induced fruit mimicry in neotropical rainforests

In 2005, Stephen Yanoviak and his colleagues observed foraging workers of canopy-dwelling, neotropical ants, *Cephalotes atratus*, with bright red abdomens (Color plate Fig. 7.4). Like earlier investigators, Yanoviak and his team considered that the arboreal ants with red abdomens represented a different taxon. Indeed, in the late nineteenth century, the peculiar ant morphs were identified as *Cepahalotes atratus* var. *rufiventris*, a variety described solely on the basis of their red abdomens. The mystery was solved when these morphs were taken back to the laboratory and dissected. The red abdomens contained gravid females of a new nematode species! In collaboration with George Poinar, the world's expert on insect nematodes, a new genus and species of tetrodonematid nematode, *Myrmeconema neotropicum*, was described (Poinar & Yanoviak, 2008).

Like other parasites of eusocial animals, the life cycle of this nematode is closely linked to the life cycle and age-related duties of the ants making up the colony. Juvenile and adult stages of the nematode were found in the larvae and pupae of *C. atratus*; however, only the older forager caste contained gravid females and only these had red abdomens. Remarkably, the color change in the abdomen from black to red coincided with the period when nematode eggs were most infective and when workers were foraging away from the nest. The color change is caused by localized thinning of the exoskeleton associated with the presence of gravid nematode and not by their selective deposition of color pigment (Verble *et al.*, 2012). The vivid red coloration arises from a combination of the translucent cuticle and the yellow nematode eggs inside the female worms. Such a remarkable change in host appearance inspired Yanoviak to speculate that the ants with bright red abdomens resemble a ripe angiosperm berry to fruit-eating birds, and facilitate transmission. If so, it represents the first case of a parasite causing a host to resemble a plant part, or, as it became known, parasite-induced fruit mimicry (Yanoviak *et al.*, 2008).

It turns out that several behaviors of the 'berry ants' were also significantly altered. The infected foragers were more sluggish and tended to continuously hold their red abdomens in an elevated position (see Color plate Fig. 7.4). Moreover, compared to uninfected ants, the abdomens of berry ants readily detached due to a weakened petiole–abdomen junction. Infected ants were also 40% heavier than uninfected ants due to the parasite load within the abdomen. Ants with maximum 'redness' were also the least aggressive, and produced no detectable alarm pheromones when disturbed (Yanoviak *et al.*, 2008). These diverse effects were proposed to facilitate ingestion by omnivorous and frugivorous birds, including several species of honey-creepers, tanagers, and small flycatchers. In short, the birds are duped to ingest the berry look-alike, thus disseminating the parasite's eggs in their feces and potentially increasing dispersal opportunities. In this case, the birds act as paratenic hosts, since the parasite matures in ants and no development of the parasite occurs in the bird.

It is likely that parasite-induced fruit mimicry was more likely to evolve in this host–parasite system since members of this ant species have large, spherical abdomens and they forage extensively

Box 15.1 | (continued)

on bird feces. Recently emerged larvae within the nests are exposed when bird feces containing nematode eggs are fed to them by the workers. Although natural bird predation on the red abdomens of ants was not directly observed, field studies confirmed that birds were more likely to attack red clay balls resembling the abdomens of infected ants. Furthermore, the eggs of *M. neotropicum* survived intact and were found in the feces when a chicken was fed infected ants. Finally, nematode eggs were found in bird feces carried by *C. atratus* workers (Yanoviak *et al.*, 2008).

Is this merely an unconventional spin on parasite-induced altered behavior and increased conspicuousness, similar to that described across the Animal Kingdom? This interpretation is unlikely. First, *C. atratus* is already one of the most conspicuous arboreal ant species in neotropical rainforests. Second, the foraging workers are typically ignored by almost all insectivorous predators due to their unpalatable pheromones and vicious spines (Stephen Yanoviak, personal communication). Thus, although the infected workers stand out from uninfected ones, this difference is unlikely to increase predation on a common but unpalatable ant that is already easy prey. Lastly, the red-on-black coloration is a warning combination in many insects, typically leading to avoidance by most predators. To overcome this warning signal, infected ants should provide a tasty reward (an assumption of the increased conspicuousness hypothesis), but this is not the case. Nor is there a negative consequence or penalty of abdomen consumption, such as a noxious chemical or a vicious sting. Consequently, the error of eating a 'berry ant' will be approximately cost-neutral, and should persist in the bird's behavioral repertoire (Yanoviak, 2010). Thus, the most parsimonious explanation for enhanced transmission to birds is parasite-induced fruit mimicry, coupled with changes in ant defense behaviors (Yanoviak *et al.*, 2008; Yanoviak, 2010).

seemingly purposeful nature of the most spectacular cases (Poulin, 1995). Consider the berry ants described in Box 15.1. Gravid nematodes chauffeur infected ants into the canopy, where the worms somehow induce the abdomens to mimic the shape, color, and motion of the most common fruit in that habitat. It would be difficult to interpret this highly specific alteration as a coincidental side-effect of infection. Rather, it is precisely what we might envision as a strategy to enhance this nematode's transmission. It would be equally difficult to consider that the circuitous route up the canopy or the conspicuous coloration might enhance the ant's fitness. Similarly, while an ant infected with *Dicrocoelium* metacercariae

spends the night clinging to a flower petal, its uninfected kin are being fed, warmed, and protected in their nest. It is difficult to conceive how the host might benefit from this altered behavior. Furthermore, the odds of a grazer accidentally ingesting a non-clinging infected ant must be extremely small. Thus, clinging tightly to vegetation for an extended period of time is precisely the behavior we might expect to facilitate metacercariae transmission into a grazing animal. Examples such as these that involve complex and purposeful alterations of host phenotype seem unlikely to arise by chance or as side-effects of infection. These examples of adaptive manipulation nicely illustrate Dawkins' (1982) extended phenotype because they likely arise from the expression of parasite genes in another organism.

Despite the compelling arguments for adaptation in these extreme examples, we require explicit evidence that parasites gain a fitness advantage from the alteration. Evidence in support of the parasite manipulation hypothesis would arise from results showing increased rates of predation by an appropriate definitive host on infected (or more heavily infected) prey compared to uninfected prey sampled in the same area. Summary results from three field studies indicate that predators consumed a significantly higher proportion of infected prey than the background prevalence of infection (Fig. 15.5). In each of these cases, the results of earlier studies had indicated that infection caused altered host behaviors. Thus, the prediction of

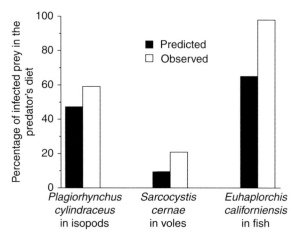

Fig. 15.5 Predicted and observed percentage of parasitized prey in the diets of definitive hosts of three trophically transmitted parasites. Predicted values are the prevalences of parasites in the populations of available prey and observed values are from actual prey captures under natural conditions. The acanthocephalan *Plagiorhynchus cylindraceus* passes from terrestrial isopod intermediate hosts to starlings, the apicomplexan *Sarcocystis cernae* is transmitted from voles to kestrels, and the digenean *Euhaplorchis californiensis* is transmitted from killifish to piscivorous birds. (Data from Moore, 1983; Hoogenboom & Dijkstra, 1987; Lafferty & Morris, 1996; modified from Poulin, R., 2007; *Evolutionary Ecology of Parasites*, 2nd edition, Princeton University Press, Reprinted by permission of Princeton University Press.)

enhanced parasite fitness of the manipulation hypothesis is met in a handful of cases.

While both observational and empirical evidence is consistent with the idea that altered host phenotypes enhance parasite transmission, it remains the case that only a very small number of studies have been designed to assess fitness benefits (Poulin, 2007). While we can be reasonably confident in our conclusions regarding the adaptive nature of some extreme examples, manipulations that involve only slight changes in features such as host activity, responses to predators, and coloration/morphology are more difficult to evaluate. It is these types of minor behavior and color changes that comprise most reports of altered phenotypes in the literature (Moore, 2002; Poulin, 2007). Moreover, as parasite ecologists turn more and more to tests involving experimentally infected hosts, there is evidence that some alterations do not always enhance predation rates by definitive hosts and that some can be better explained as non-adaptive outcomes of persistent and unavoidable pathology (Robb & Reid, 1996; Edelaar *et al.*, 2003; Shirakashi & Goater, 2005). The key message to draw from the mounting empirical evidence is that we should expect enormous variation in the effort that parasites devote to manipulating their hosts' phenotypes. In an important theoretical treatment involving behavior phenotypes, Poulin (2007) emphasized the context-dependent nature of manipulative effort, where variation in effort was associated with variation in factors such as parasite infrapopulation size and structure, parasite genetic structure, and host predation strategy. Another key feature determining manipulative effort will be the intensity and life cycles of the other species of parasites that comprise the regular infracommunity of manipulated hosts. Currently, there is extensive research focus devoted to understanding altered phenotypes of single versus multiple species of potential manipulator (Poulin, 2007).

One way to clarify the adaptive nature of parasite-induced phenotypic alterations is to determine the underlying mechanisms. While our understanding of proximal mechanisms lags behind ecological and evolutionary aspects, the results of some ingenious early efforts have provided key advances (review in

Poulin, 2007). In the case of the dramatic limb mal-formations induced by *Ribeiroia* metacercariae in amphibians, the mechanism seems to be fairly straightforward. Whether tadpoles were exposed to cercariae, or implanted with metacercariae-sized latex beads, the frequency of *Ribeiroia* malformations in froglets was the same (Johnson *et al.*, 1999). These results show that at least a part of the mechanism underlying the altered phenotype is mechanical inter-ference with normal development. The distention to the abdomens of small-bodied fishes infected with diphylobothriidean cestodes and the cataracts that develop within the lens of freshwater fish infected with *Diplostomum* spp. metacercariae can be explained similarly, although additional mechanisms involving biochemical interference cannot be ruled out. In such systems, the challenge is to distinguish those altera-tions that are non-adaptive side-effects of the presence of parasites in key tissues (e.g., Edelaar *et al.*, 2003; Shirakashi & Goater, 2005) from those that enhance parasite transmission.

Understanding the proximate mechanisms of alter-ations in behavioral phenotypes is another key research area (review in Lefèvre *et al.*, 2009). Although site selection within regions of the central nervous system is common for many parasites, especially within their intermediate hosts, there is little evidence that mechanical interference alone causes adaptive alterations to host behavior. Instead, the alterations appear to result from active interference with host neural function. In a pioneering study, Helluy & Holmes (1990) showed that *Gammarus* injected with serotonin, a powerful neuromodulator in invertebrates and vertebrates, demonstrated the same escape behavior (positive phototaxis) as those infected with larvae of the acanthocephalan *Polymorphus para-doxus* (see Box 7.1). Subsequent studies involving *Gammarus* infected with other species of larval acanthocephalan (Tain *et al.*, 2006) or trematode metacercariae (Helluy & Thomas, 2003) used staining methods to show that the brains of infected hosts have increased serotonergic activity, often markedly so. In these cases, infection is associated with a suite of behavior alterations, including phototaxis that leads to increased rates of predation by the definitive host. These results are consistent with the notion that alteration to certain escape behaviors, e.g., phototaxis, arises from increased serotonin release, either by the parasite, or the host, or both. Alterations to the serotonergic system by potential manipulators have also been reported in interactions involving verte-brates as hosts (review in Perrot-Minnot & Cezilly, 2009).

Post-genomic technologies such as proteomics provide a powerful new approach for understanding proximate mechanisms (review in Lefèvre *et al.*, 2009). The results of studies involving two species of nema-tomorph in their cricket and grasshopper hosts show the potential of parasito-proteomics tools to discrim-inate between host and parasite-derived molecules during the process of manipulation (Biron *et al.*, 2005; 2006). The authors characterized the proteomes of parasite and host brain tissue prior to, during, and after the main period of manipulation. Results showed that the proteome derived from host brain tissue remained constant through the manipulation process. This is important because it shows that the manipulation is not due to the secretion of proteins associated with general host stress or tissue damage. However, the proteomes of worms derived from infected hosts collected just prior to their fateful leap into the water (see Chapter 9) differed from those determined before or after. It turns out that both spe-cies of hairworm produced the same WNT-like proteins during the leaping phase. Furthermore, these proteins share more similarity with other insect WNT proteins than they do with other parasites (Biron *et al.*, 2005; 2006). In insects, these proteins are important for circadian rhythms, neurogenesis, and neurotransmis-sion. The implication is that hairworms produce host mimetic WNT proteins that act on the host's central nervous system to produce the dramatic alterations in host behavior. Few other studies have characterized the proteomes of both partners.

Critical tests involving the injection of potential parasite-derived neuromodulators into uninfected hosts are on the near horizon. While there is tremen-dous potential in the work involving serotonin in crustaceans and fish, and the WNT proteins in crickets, the precise mechanisms that parasites use to modulate neurological pathways are unknown. Indeed, it is unlikely that parasites can synthesize and secrete the quantities of serotonin, for example, that the results of injection studies have indicated are required to alter phototaxis in *Gammarus*. In a promising new direction, Adamo (2002) has suggested that biochemical cross-talk between the host's immune system and the nervous system might provide a parsimonious mechanism for some altered behaviors, especially in arthropods. In the case of *Dicrocoelium* in ants, the single unencysted metacercariae located in the head must be countering the insect defense system. Perhaps if we understood that mechanism, and its possible linkages to the ant's nervous system, we might be closer to understanding their altered behavior. In an initial test of this idea, Helluy & Thomas (2010) showed that neuroimmune responses were elevated at the host–parasite interface in metacercariae-infected *Gammarus* that were exhibiting altered behavior.

The results of these pioneering experiments indicate that the proximate mechanisms underlying parasite manipulation are diverse and complex. We should expect similar complexity to trigger the manipulations of *Dicrocoelium*-infected ants, *Ribeiroia*-infected froglets, and so on. In the context of behavioral phenotypes, the early evidence supports the notion that manipulation arises from widespread alterations to general neural pathways, e.g., the serotonergic system in crustaceans and the WNT system in insects, rather than selective 'hits' on specific neurons (Lefèvre *et al.*, 2009). For some of the more spectacular alterations, e.g., the berry ants in Box 15.1, we predict even more complexity. In the case of hairworm-infected crickets and grasshoppers, the manipulation via WNT proteins is a one-way phenomenon that results in the host's death. Yet for *Dicrocoelium*-infected ants and other infected 'zombies,' we need to understand the mechanisms that turn on the temperature-dependent altered behavior, i.e. climbing and clinging, and then just as dramatically turn them off – and then repeat it day after day. Currently, we can only speculate on the nature of those mechanisms.

15.3 Effects of parasites on host populations

Theory predicts that variation in the effects of infection on the survival and fecundity of individual hosts should translate to variation in effects at the population level. In a test of this idea, Ebert *et al.* (2000) showed that variation in the risk of extinction of *Daphnia* populations within small mesocosms was associated with variation in the extent to which the microparasites in Fig. 15.1 reduced the survival and fecundity of individual hosts. These results are consistent with the intuitive expectation that so long as parasites affect the survival and fecundity of individual hosts, they have the potential to reduce the size of their host populations. In this section, we consider whether that potential is realized in natural populations of hosts. Understanding the precise role that parasites play in their host populations has been a central question for parasitologists and ecologists for at least 50 years (reviews in Grenfell & Dobson, 1995; Hudson *et al.*, 2001). The answer to the question drives current enquiry into the role of parasites as biological control agents, into the role of introduced parasites within new host populations, into the conservation of animal biodiversity in the face of disease and parasite threats, and into the role of parasites in the management of our domestic livestock, fish, and wildlife. Thus, the question has never been more important than at present.

On a superficial level, the answer to the question of parasite-induced effects on host populations must be 'Yes'. Ecologists, epidemiologists, parasitologists, and anthropologists have long known that parasites and disease can decrease the size of host populations. Jared Diamond (1997) in his influential book entitled *Guns, Germs and Steel: The Fates of Human Societies* provides a fascinating history of humankind's interaction with parasites and pathogens, and warns of the perils of those yet to come. Based on our coverage in earlier sections of this text, it is equally clear that some parasites can devastate animal populations as well, e.g., the mite *Varroa destructor* on honey bees (see Box 11.3), the myxozoan *Myxobolus cerebralis* in salmonid fish (see Box 5.1), the microsporidian *Nosema ceranae* in honey bees (see Box 4.1), and the nematode *Parelaphostrongylus tenuis* in North American moose (see Chapter 8), to list just a few. But these are spectacular cases, often leading to catastrophic epizootics within their host populations. Perhaps these are highly detectable exceptions because they are in 'new' hosts that are located within highly altered, anthropogenic habitats. In the two subsections that follow, we consider two of our most vexing problems: the extent to which typical parasites influence their host populations and the extent to which parasites regulate host population sizes.

15.3.1 Parasites and host demography

Direct assessments of the consequences of parasites on natural host populations are rare. Consider the practical difficulties associated with adding or removing *Myxobolus cerebralis* from a population of rainbow trout, or *Dermacentor albipictus* from a population of moose, or *Ribeiroia ondatrae* from a population of frogs. You would need to collect a sample of individuals from the population, add or remove targeted parasites, mark the treated hosts, and then monitor their survival over time. You would need to perform the same procedure on a sample of sham-treated hosts. Meeting this 'gold standard' presents enormous challenges for parasitologists and ecologists. For some host–parasite interactions, e.g., most microparasites, the challenges seem insurmountable. For others, we must often relax certain elements of the gold standard, a strategy that can result in broiling controversy among seasoned scientists and even politicians. In the case of wild salmon on Canada's Pacific coast, assessment of the true population-level consequences of a potentially devastating parasite must be done in the context of a declining resource, the conservation of unique biodiversity, the preservation of a pristine habitat, and economic realities – a 'perfect storm' of parasites and politics (Box 15.2).

The sea lice–salmon controversy emphasizes the logistical constraints associated with manipulating parasite burdens in free-ranging host populations. Even in an interaction that is amenable to

---

Box 15.2 | **The politics of parasitism: sea lice, aquaculture, and the decline of salmon populations in coastal British Columbia**

Canada's western coastline is a hotspot of temperate marine and rainforest biodiversity. Among the key components of that biodiversity are the five species of native salmon that undergo their famous migrations between freshwater and marine habitats. It is populations of these salmon that are the cornerstone of the world-renowned recreational and commercial fishing economy in British Columbia. Likewise, the salmon's spectacular habitat supports a thriving ecotourism economy. As so often happens around the world, these two conflicting priorities lead to inevitable regional tensions among politicians, conservationists, and the public. Yet in the Broughton Archipelago in coastal British Columbia, two additional elements have further inflamed these tensions. First, it is within this network of pristine islands where sea-cage aquaculture of introduced Atlantic salmon has dramatically expanded over the past 20 years. Second, it is here where pathogenic ectoparasitic arthropods that are thought to originate from Atlantic salmon farms are suspected to cause catastrophic losses of pink salmon. It is the contentious role that these parasites play on the demography of wild salmon that is behind one of the most scientifically and politically tumultuous periods in Canada's coastal history.

Box 15.2 | (continued)

Fig. 15.6 Parasitic sea lice *Lepeophtheirus salmonis* on a juvenile pink salmon *Oncorhynchus gorbuscha* from the Broughton Archipelago, British Columbia, Canada. (Photograph courtesy of Simon Jones.)

The parasite of greatest concern is the 'salmon louse,' *Lepeophtheirus salmonis* (Fig. 15.6; Color plate Fig. 4.1). We introduced the general biology of this important marine copepod (remember, it is not a 'louse'!) in Chapter 11. It has a direct life cycle involving at least 10 stages, four to five of which are parasitic, and a free-swimming copepodite that is the infective stage. It is common on both wild and farmed adult Atlantic salmon where feeding activities associated with development on the hosts' epidermis can lead to physiological dysfunctions, tissue damage, and secondary viral or bacterial infections (reviews in Pike & Wadsworth, 1999; Jones & Beamish, 2011). The economic impact of *L. salmonis* on the aquaculture industry is significant, especially in northern Europe (on Atlantic salmon and brown trout), where annual losses of $5 billion US have been estimated due to lost productivity and control measures (Costello, 2009). Not surprisingly, there has been a long history of field and laboratory-based research on *L. salmonis*, leading to a solid understanding of key aspects of host specificity, transmission dynamics, and host immune responses (Jones & Beamish, 2011).

Pink salmon, *Oncorhynchus gorbuscha*, is the most abundant native salmon species in the Broughton Archipelago. Their life history differs from other sympatric salmonids in a manner that has an important impact on their interaction with *L. salmonis*. Pink salmon enter the marine environment just weeks after hatching (0.2 g), then return to their natal streams to breed when they are 2 years old. Thus, in contrast to other species of salmon, juvenile pinks are present along the coast during the period when *L. salmonis* copepodites, potentially originating from salmon farms, are most abundant in the water column. Pink salmon returns into rivers of the Archipelago reached record numbers in 2001, the same year they were reported for the first time to be heavily infected with lice. But the peak returns in that year were followed by a 97% decline in the next. Thus, a catastrophic decline in returns of pinks coincided with the peak expansion of salmon farming in the region, and with the first reports of lice infections on this host. This observation has led to the key question that has embroiled scientists and politicians for the past decade: did sea lice from farms kill juvenile pink salmon and did this cause the run collapse of 2002?

Martin Krkošek and his colleagues initiated a series of empirical and modeling studies in 2002 to answer the question of farm-to-wild salmon transmission and the question of lice-induced

Box 15.2 | (continued)

salmon mortality. Results from their assessments of thousands of juvenile salmon collected along three different migration paths were clear. Fish sampled after they passed regions of intensive salmon farming had much higher abundances of *L. salmonis* than those collected as they approached the farms. Further, the mortality of naturally infected juveniles raised within screened enclosures ranged between 9% and 95% over a 2-month period, whereas mortality of uninfected fish was near zero (Krkošek *et al.*, 2006). In their modifications of the classical Anderson & May macroparasite models (see Chapter 12), Krkošek coupled a model of lice transmission dynamics with a model of lice-induced effects on salmon survival to estimate the overall mortality of wild salmon caused by farm-origin lice. The outcome of their combined model was dire, culminating in their bold prediction for a 99% collapse in Broughton pink salmon populations within 8 years (Krkošek *et al.*, 2007).

Yet, the mortality results reported by Krkošek and colleagues contrast sharply with the results from controlled exposures of naïve pink salmon to *L. salmonis* copepodids. There is little evidence from laboratory experiments for lice-induced mortality of pink salmon (even for food-deprived juveniles) and little evidence for a reduction in other aspects of host performance compared to controls (Jones & Beamish, 2011). Further, juvenile Pacific salmon, especially pinks, can mount an effective immune defense that rejects colonizing and feeding lice from the epidermis. These contrasting results imply that either benign laboratory conditions involving single, brief exposures underestimate true salmon mortality, or mortality estimates based upon naturally exposed salmon in screened enclosures are unrealistic. When Krkošek *et al.* (2009) repeated their approach with a different cohort of young pink salmon and slightly larger pens, mortality of infected salmon over a 2-month period was similar to controls. These results suggest that great care is needed to distinguish true additive mortality due to infection from compensatory mortality due to factors such as starvation or other stressors.

Detailed discussions of the sea lice–salmon controversy on Canada's west coast are provided in Brooks & Jones (2008), Marty *et al.* (2010), and Krkošek *et al.* (2011). Currently, consensus among parasitologists, fisheries biologists, mathematicians, and ecologists seems distant. Additional field and laboratory experiments combined with field surveys and modeling are required to clarify the role that *L. salmonis* plays in juvenile pink salmon.

experimental manipulation and involves a host species that drives regional economies, assigning a role for additive parasite-induced host mortality is not straightforward. Similarly, for those many cases where manipulative experiments indicate significant additive mortality in laboratory or captive populations (review in Robar *et al.*, 2010), it is rarely possible to extrapolate the results to natural populations of hosts. Despite these difficulties, Hudson *et al.* (1992) combined data on grouse population sizes on moors with data from their manipulation experiments on the additive effect of *T. tenuis* on grouse survival (see previous section), to determine that nematode parasitism significantly reduced overall bird densities. Studies that combine laboratory tests on parasite-induced pathology, field surveys that link worm numbers with host mortality, and field-based manipulations of parasite numbers provide our most convincing cases for the demographic consequences of infection in free-ranging systems.

In the absence of manipulation experiments, field surveys that document changes in host and parasite populations over time are illuminating. Although these are observational studies that draw inferences based upon correlation data, they are compelling if they are supported by supplementary laboratory or field experiments. Thus, population declines in North American moose populations are correlated with high densities of the winter tick *Dermacentor albipictus* (see Chapter 11). In this case, the declines that follow high-tick years are associated with severe nutrient restriction due to the preoccupation of infected moose with grooming (Delgiudice *et al.*, 1997). Similarly, the ongoing decline in moose population sizes in northern Minnesota is attributed to high intensities with the liver fluke *Fascioloides magna* (Murray *et al.*, 2006), and the declines in population sizes of lesser scaup ducks throughout their migratory range in central North America are attributed to two species of intestinal trematode (Herrmann & Sorensen, 2011). In each of these cases, as for sea lice in salmon (see Box 15.2), complex wildlife management decisions must be made relative to the role of parasites in their host populations.

As we discussed in the previous section, plerocercoids of diphyllobothriidean cestodes reduce the growth, reproduction, and survival of their fish hosts, leading to periodic, context-dependent die-offs. In a series of studies conducted within small, pristine lakes in Alaska, David Heins monitored annual changes in the population sizes of threespine sticklebacks, *Gasterosteus aculeatus*, infected with *Schistocephalus solidus* (Heins *et al.*, 2010). Not surprisingly, estimates of fish density varied over his 8-year study. In one of the lakes, stickleback density peaked in 1997, only to crash the following year when just a few 2-year-olds were collected (these sticklebacks breed at 2 years of age, then die) (Fig. 15.7). Analysis of length–frequency distributions shows that the decline of this year class was due to poor overwinter survival of 1-year-olds. It turns out that in 1997, almost all (90%) sticklebacks entered their first winter infected with *S. solidus*, sometimes heavily so. Few of these fish survived to breed the following year. In this particular lake, the epizootic lasted for a further 2–3 years, until more and more uninfected 2-year-old fish were caught. Presumably, these mature fish contributed to the successful 1-year-old cohort the subsequent year. Heins *et al.* (2010) suggested that the mechanism leading to the epizootic was the high physiological cost of development of these large plerocercoids during a stickleback's first winter. Results such as those depicted in Fig. 15.7 are consistent with those from a parallel study completed in an adjacent lake (Heins *et al.*, 2010), and with those involving roach, *Rutilus rutilus*, infected with *L. intestinalis* in Slapton Ley, a small lake in southern England (Kennedy *et al.*, 2001). In the latter case, monitoring of fish and parasite populations over 31 years showed three separate and independent epizootic events.

The interpretations of plerocercoid-induced epizootics in these small fish seem straightforward, but they are not! Heins and his colleagues simultaneously monitored fish and parasite populations in three adjacent lakes in Alaska. Each is small and pristine, and each lacks piscivorous fish. Despite their superficial similarities, the epizootics in the lakes were not concordant in time, the duration of emergent and decline phases of the epizootics were inconsistent within and between lakes, and the magnitudes in the decline of fish populations differed strongly between lakes. Heins *et al.* (2011) concluded that the epizootics in these lakes were context-dependent, independent, and inherently unstable events. These authors refer to the parasite-induced epizootics as a "conundrum of deceptive simplicity concealing complexity." Kennedy *et al.* (2001) also emphasized the stochastic and unpredictable nature of each of the three plerocercoid-induced epizootics that occurred in their study system.

The application of computer models provides an additional tool to assess parasite-induced effects on host populations. This indirect approach is particularly

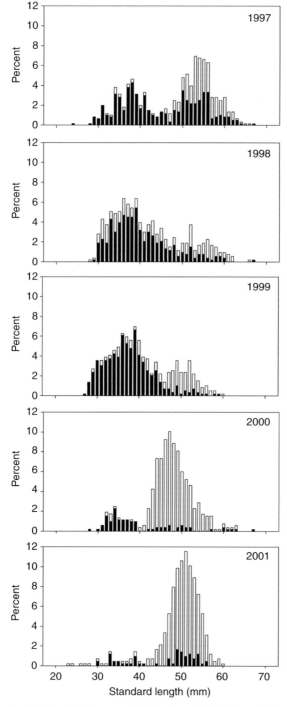

Fig. 15.7 Length–frequency distributions for *Schistocephalus solidus*-infected threespine sticklebacks from Scout Lake, Alaska, between 1997 and 2001. Black portions of stacked

appropriate for systems that are not amenable to experimentation. The complex and diverse communities of larval trematodes in freshwater snails are a good example. As we discussed in Chapter 6, infection tends to reduce the survival of infected snails and host fecundity usually drops to zero due to castration. Thus, castrated hosts are 'reproductively dead' and we might expect strong population-level effects, particularly in situations where a high proportion of snails are infected. Negovetich & Esch (2008a) tested this idea by applying a matrix model to census data for the snail *Helisoma anceps* collected from a small pond in North Carolina. The natural history of this model system is described in Chapter 6. To parameterize their model, the authors used data from a prior field study that combined mark-recapture data on infected and uninfected snails, with data on snail fecundity, survival, and infection (Negovetich & Esch, 2008b). Their results showed that castration of individual snails (mostly by the dominant trematode *Halipegus occidualis*) caused additive host mortality, leading to a 40% decrease in the growth rate of the snail population. These results provide indirect evidence that trematode-induced castration has strong population-level consequences, confirming a pattern that field parasitologists had long suspected. More studies like this one are needed to confirm the generality of these results, particularly in systems where prevalence of larval trematodes is both higher and lower than those normally seen in *H. anceps* in this pond.

Our discussion to this point has been restricted to simple interactions involving single species of parasite in single species of host. Moreover, our examples have considered the demographic consequences of parasites in populations free of predation, competition, resource limitation, and exploitation by humans. Both of these shortcomings reflect the intractability of most host–

Caption for Fig. 15.7 (cont.)
bars represent infected fish; open portions represent uninfected fish. (Figure courtesy of David Heins; modified from Heins, 2011, with permission, *Journal of Parasitology*, 97, 371–376.)

parasite interactions under natural conditions. Several bold attempts to assess the role of parasites in combination with other key factors have provided important results. A large-scale field experiment involving penned Canadian hares, *Lepus americanus*, infected with the nematode *Obeliscoides cuniculi* provides a good example (Murray *et al.*, 1997). The experiment involved over 600 individual hares that were radio-tracked over 6 months, one-half of which had been treated with ivermectin, the other half with water. In this system, about 95% of natural mortality is due to predation by lynx, owls, raptors, and squirrels. Thus, it would be statistically impossible to detect an additive effect of parasites, especially as *O. cuniculi* causes little pathology to individual hosts. Nonetheless, results of the experiment showed that untreated hares (those with normal nematode burdens) doubled their risk of predation compared to treated hares. These results point to a cryptic, but important, role of subclinical parasitism in free-ranging animals. Indirect evidence from field studies involving population crashes of bighorn sheep, *Ovis canadensis*, in Alberta, Canada and moose in Quebec also indicate the role of complex interactions between parasites (*Protostrongylus* spp. lungworms and hydatid cysts of larval *Echinococcus granuolosus*, respectively), predators, and nutrition in determining (and perhaps regulating) population sizes of large mammals (review in Holmes, 1995).

Another example involves an isolated population of Soay sheep that occurs on the island of St. Kilda off the northern coast of Scotland. The sheep occur in the absence of predators or competitors, and have done so for at least 2000 years since their introduction to the island. Their population sizes wax and wane dramatically, with crashes every 3–4 years caused by density-dependent resource limitation. It turns out that the sheep are infected with a guild of direct life cycle trichostrongyle nematodes, with *Ostertagia circumcincta* (see Chapter 8) dominating. Gulland (1992) conducted a field experiment that crossed a parasite reduction treatment (through administration of ivermectin) with a food restriction treatment. Her results showed that mortality was highest in untreated, undernourished sheep, whereas those that were

untreated, but well nourished, survived as well as controls. These results suggest that nematodes alone are not sufficient to cause the periodic crashes in sheep populations. Rather, it is the combined effect of poor nutrition due to overgrazing and high nematode intensities that act together to cause declines. The underlying mechanism is likely a relaxation of immunity-mediated control of nematode intensities in undernourished sheep (Gulland, 1992).

We summarize this section by emphasizing three features. First, by using all tools at their disposal, from manipulation experiments, to field surveys of host and parasite populations, to computer modeling, researchers have boldly advanced our understanding of parasite-induced effects on host populations. The effects range from obvious and strong, e.g., fish epizootics, to cryptic and subtle, e.g., nematodes in hares. We currently cannot predict the characteristics that place particular interactions along this broad continuum. However, available evidence indicates that interactions at the 'strong' end of the continuum tend to be those in anthropogenically altered habitats (nematodes in red grouse and Soay sheep, ticks on moose) or those that have especially stark effects on host energy budgets or host morphology (e.g., larval cestodes in fish). Second, results from factorial experiments and from careful field surveys have emphasized the context-dependent nature of effects on host populations, just as they did for effects on individual hosts. Third, much of the evidence we have discussed above shows that the negative effects of parasites appear to be strongest in young, non-reproductive hosts. Thus, a key demographic consequence of infection lies in the restriction of the number of individuals within a population that reach breeding status.

15.3.2 Parasites and host population regulation

Population regulation refers to the tendency of a population to decrease in size when it is above a certain level, and to increase in size when it is below that level. The central feature behind any process leading to population regulation is **density**

dependence (see Chapter 12). Thus, processes such as the rates of birth, death, immigration, and emigration can regulate a population, so long as they are affected by density.

The disciplines of parasite ecology and wildlife epidemiology were transformed and catalyzed by two ground-breaking theoretical papers, again by Roy Anderson and Robert May (Anderson & May, 1978; May & Anderson, 1978). Following on the earlier theoretical work of Crofton (1971), they showed that parasites alone had the potential to regulate their host populations so long as they reduced host survival or fecundity in a density-dependent manner. This meant that host mortality due to parasites could act alone, or in addition to other agents of mortality, to regulate host populations. For the case of macroparasites (Anderson & May, 1978), they developed a simple mathematical model with two differential equations that described changes over time in the host population:

$$dH/dt = (a - b)H - \alpha P \qquad (15.1)$$

and in the parasite population:

$$dP/dt = \beta wN - (b + u + \alpha)P - \alpha(k + 1)P^2/kN \qquad (15.2)$$

where H and P are the host and parasite populations. In equation 15.1, a and b represent the birth and death rates of uninfected hosts, respectively. The term α is one of the critical parameters representing the rate that parasites induce mortality on their hosts (see also Chapter 12). In effect, this equation calculates the typical rate of growth of an uninfected population (based on the typical logistic growth where populations grow up to a carrying capacity set by resources) and subtracts deaths due to parasites. The second equation looks cumbersome, but it too is a simple abstraction of how we might anticipate parasite numbers changing over time within a host population. The first term represents a gain in parasite numbers, where β is the transmission rate of infective stages into hosts and w is the number of infective stages in the

environment. The second is a loss term that combines parasite death rates due to natural mortality (u), natural host deaths (b), and deaths of infected hosts (α). The final term represents the loss of worms according to their statistical distribution within a host population, where small values of k represent high clumping of parasites in few hosts and larger values represent random distribution of parasites among hosts. Chapter 12 presents a discussion of the biological significance of k in the negative binomial distribution.

Anderson and May then combined the two equations to develop their Basic Model to determine the conditions whereby parasites could regulate the growth of their host population. To do so, they used computer simulations to vary individual rate parameters, e.g., α, β etc., while keeping the others constant. Details regarding the outcomes of these simulations are provided in their original publications and are further elaborated upon subsequently (e.g., Roberts et al., 1995; Tompkins et al., 2001a). Their results provided novel insights, some of which were counter-intuitive. On the surface, we might expect that the most pathogenic parasites (high α) would be most likely to regulate their host populations. This is incorrect. Under their scenario, infection will impact host demography (especially if k is large and most hosts are infected), but long-term stability is unlikely when infected hosts die and take their parasites with them. Thus, low to intermediate values of k tend to lead to regulation. Similarly, highly pathogenic parasites that cause outright host mortality tend also to destabilize host–parasite dynamics. Instead, parasites with moderate α that have a greater effect on host reproduction than on host survival tend to be the types that are more likely to lead to the long-term co-regulation (Fig. 15.8). In a nutshell, the theoretical framework developed by Anderson and May directs us to the conclusion that host population regulation by parasites can be viewed as an interaction between parasite pathogenicity (α) on the one hand, and the statistical distribution of parasites within their host populations (k), on the other. This key prediction is testable.

The Anderson and May models are obviously abstractions of reality. Estimating the values for any of

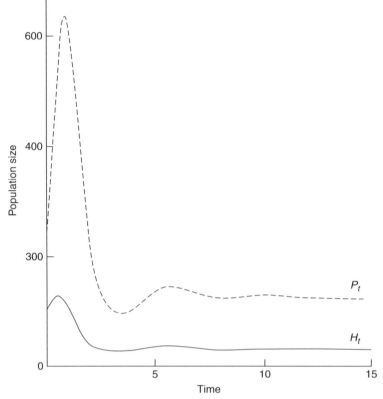

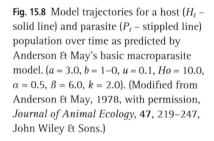

Fig. 15.8 Model trajectories for a host (H_t – solid line) and parasite (P_t – stippled line) population over time as predicted by Anderson & May's basic macroparasite model. ($a = 3.0$, $b = 1–0$, $u = 0.1$, $Ho = 10.0$, $\alpha = 0.5$, $\beta = 6.0$, $k = 2.0$). (Modified from Anderson & May, 1978, with permission, *Journal of Animal Ecology*, **47**, 219–247, John Wiley & Sons.)

the parameters in equations 15.1 and 15.2 is a challenge, especially for natural host–parasite interactions (for a critical perspective, see Barlow, 1995). Moreover, direct evidence for the existence of co-regulation as depicted in Figure 15.8 requires that a host population be perturbed away from its regulated (or equilibrial) state. Regardless of whether parasite population size or host population size is manipulated, it is critical to resolve whether the population subsequently returns to its regulated state. Herein lies the challenge in determining whether the hypothetical scenario indicated in Figure 15.8 is applicable to real host–parasite interactions. Full discussions of the role of parasites as regulators of animal populations are provided in Scott & Dobson (1989), Grenfell & Dobson (1995), and Tompkins et al. (2001a).

Keymer (1981) was among the first to quantify parameters of the Anderson and May macroparasite model. She monitored changes in laboratory populations of the common grain beetle, *Tribolium confusum*, exposed to eggs of the cestode *Hymenolepis diminuta*. We introduced this host–parasite interaction as a laboratory model for studies in parasite ecology in Chapter 6. Experimental populations of size- and age-matched beetles were periodically exposed to suspensions of infected eggs originating from rats and allowed to breed under standard conditions. Results showed that the peak densities of infected beetle populations were reduced up to 50% compared to unexposed populations and that final host population sizes were always lower in the exposed groups (Fig. 15.9). These results showed that the parasite has the potential to reduce the size of its host population and that the parasite was regulatory under constrained laboratory conditions.

Scott (1987) extended this laboratory paradigm by following changes in host population size over time within arenas of mice exposed to the nematode

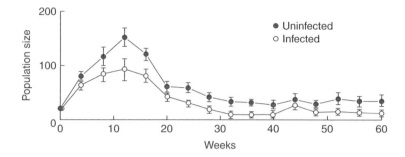

Fig. 15.9 Change in population size of laboratory populations of *Tribolium confusum* infected with larval *Hymenolepis diminuta* (Cestoda). The points represent observed values of total population size; the vertical bars indicate 95% confidence limits of the means. (Modified from Keymer, 1981, with permission, *Journal of Animal Ecology*, 50, 941–950, John Wiley & Sons.)

Heligmosomoides polygyrus. This species has a direct life cycle in which eggs shed in the feces hatch and free-living larvae become infective within the soil in approximately 7 days (see Chapter 8). The larvae are accidentally ingested by mice where they migrate into the intestinal wall, stay in the area of the serosa for about a week, and then migrate back into the lumen of the intestine before maturing sexually. In this experiment, the large indoor arenas were constructed so that individual mice could be repeatedly exposed to infective larvae as they carried out their normal activities. In the uninfected colony, reproduction and mortality stabilized at a density determined by density-dependent mortality among young mice, some cannibalism, and a decline in *per capita* female reproduction (Fig. 15.10).

Following introduction of the nematode into immunologically naïve mouse populations, the impact was dramatic (Fig. 15.10). In the three arenas housing the exposed colonies, mouse densities were reduced by 50% approximately 7 weeks after initial exposure. By 9 weeks, 90% of the mice in the exposed arenas had died. Mean parasite intensities in necropsied mice from these three arenas ranged between 200 and 1500 nematodes. Mice in one of the three arenas were then transferred into a habitat in which high transmission rates were prevented. Mortality of mice continued at a high rate. However, when these mice were treated with an anthelmintic, the population growth rate increased significantly (Fig. 15.10). These results show that *H. polygyrus* is capable of regulation in the sense that it reduced host population equilibria to levels lower than those that existed without the parasite.

We return to the red grouse-*T. tenuis* interaction studies by Peter Hudson and his colleagues for the best evidence of host regulation in a free-ranging system (Hudson *et al.*, 1998). We covered the life cycle and general background of this system in Chapter 8 and in the previous section. Recall that the owners of private 'grouse estates' in northern Britain keep accurate counts of birds on their moorlands. These bag records indicate that most grouse populations, especially those on wet moorlands where *T. tenuis* transmission conditions are ideal, undergo regular cycles of abundance that peak every 4–8 years. In a clever twist of the classical paradigm of population regulation, they tested the hypothesis that experimental reduction of *T. tenuis* would prevent population cycling. Their hypothesis was based upon two predictions of the Anderson and May macroparasite model. First, their mark–recapture studies on de-wormed birds determined that the relative impact of *T. tenuis* was higher on breeding performance than on host mortality. Second, extensive field surveys of natural patterns of infection demonstrated that the distribution of worms in grouse populations can be described by the negative binomial frequency distribution, with an approximately intermediate value of *k* relative to patterns of infection in other species of bird.

Hudson *et al.* (1998) studied six populations of grouse on adjacent, independent moors for their 10-year study. At two designated control sites, grouse numbers showed their characteristic cycle of abundance, with precipitous drops occurring in 1989 and 1993 (Fig. 15.11). At the other sites, between 15 and 50% of all birds (about 3000 in total) were captured

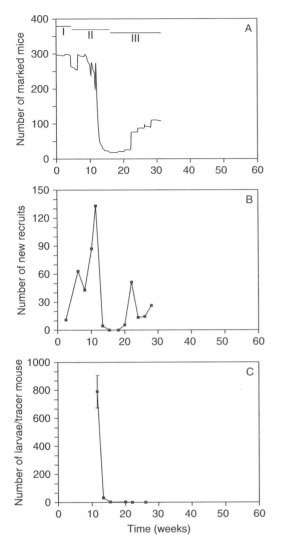

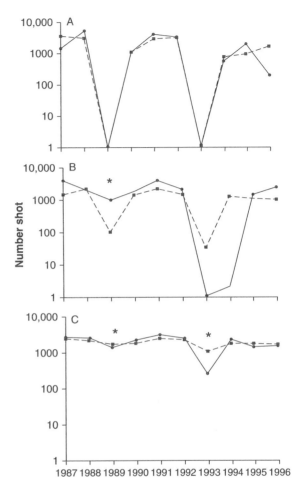

Fig. 15.10 Effect of the introduction of the nematode *Heligmosomoides polygyrus* into an immunologically naïve mouse population. (A) Change over time in the number of marked mice (2-week-old) during three phases of the experiment. Phase I represents the mouse population prior to parasite transmission. Phase II represents the mouse population during the initial period of *H. polygyrus* transmission. Phase III represents the mouse population during continued low-level transmission of *H. polygyrus*. (B) Numbers of newly recruited, 2-week-old mice at each census. (C) Mean (±S.E.) numbers of *H. polygyrus* larvae in tracer mice exposed to the contaminated arena for 24 hours following each census. (Modified from Scott, 1987, with permission, *Parasitology*, **95**, 111–124, Cambridge University Press.)

Fig. 15.11 Changes in the population sizes of red grouse *Lagopus lagopus scoticus* in response to anthelmintic removal of the nematode *Trichostrongylus tenuis* as represented through annual bag records in (A) two control sites; (B) two populations with a single anthelmintic treatment each; and (C) two populations with two anthelmintic treatments each. The solid and dashed lines represent population responses at the each of the two sites, respectively. Asterisks represent the years of treatment when burdens of *T. tenuis* in adult grouse were reduced. (Modified from Hudson *et al.*, 1998, with permission, *Science*, **282**, 2256–2258, The American Association for the Advancement of Science.)

and orally administered an anthelmintic using the procedures described above. At two sites, the manipulations were performed in 1989 only, whereas at another two sites the procedure was repeated in 1993. Thus, the researchers attempted to reverse the 1989 crash at two sites, and both crashes at the other two sites. The results of the post-treatment population censuses were clear, revealing that the experimental reduction in nematode burdens led to a cancellation of the 1989 crash on three moors, and a cancellation of both declines at the site where the birds were treated twice (Fig. 15.11). The results of these manipulations supported observations from field studies on other moors. Thus, *T. tenuis* tends to be absent, or less common, on moors where grouse show regular cycles of abundance, and at one site bird population numbers tended to crash in years when mean intensity approached 10 000 worms/host. Taken together, results from the field manipulations and from parallel field observations supported their hypothesis that these nematodes were solely responsible for causing population cycles of grouse. The likely mechanism responsible for the observed demographic effect lies in density-dependent reduction in grouse fecundity.

Results from these manipulations of *T. tenuis* in red grouse are important because they distinguish the role of additive parasite-induced mortality from possible compensatory mortality that might occur in birds weakened by factors such as competition for food. However, it is important to recognize that these results provide the only direct evidence for a regulatory role of macroparasites in free-ranging host populations. One shortcoming is that red grouse populations in Britain are managed for high density to support local hunting economies; predators are controlled and supplementary food is often provided. Such artificially high host densities likely contribute to the high *T. tenuis* burdens on these moorlands. Thus, it is inappropriate to extend the results from the grouse–*T. tenuis* system to unmanaged populations of hosts. Furthermore, field experiments involving other vertebrate–nematode interactions have provided inconsistent results. Using a combination of parasite reduction experiments and dynamic modeling, Townsend *et al.* (2009) concluded that nematodes did not directly contribute to the regulation of mountain hare, *Lepus timidus*, populations in Scotland. This study is significant because previous work on this system indicated that aggregation of nematodes in the hare population was also

intermediate and parasite-induced effects on fecundity were higher than on host survival. In this example, results contradict the predictions arising from the Anderson & May macroparasite model for the regulation of host population size.

Overall, the evidence in support of parasites as regulators of natural animal populations is limited to a handful of studies involving laboratory host/parasite interactions and semi-natural systems that are highly managed. The lack of successful instances of parasites as biocontrol agents, despite extensive attempts by researchers to find them, is further evidence that parasites alone are unlikely to regulate host populations. Of course, this is not to say that parasites do not play any role in regulation. Evidence from Soay sheep infected with gastrointestinal nematodes indicates that infection, in combination with nutrient limitation, may regulate population sizes (Gulland, 1992). Interactions between nutrient limitation and winter ticks may contribute to regulation of moose populations in North America, especially in the absence of wolves. Holmes (1995) cautions that single causes of population regulation are probably exceedingly rare in natural ecosystems.

15.4 Effects of parasites on host communities and ecosystems

Given the negative effects of parasites on host populations, coupled with their roles in trophic interactions and food webs, we might expect that they also exert impacts on host communities and ecosystems. The effects of parasitism can be particularly pronounced when the hosts are keystone species with vital ecological roles in an ecosystem (reviews in Combes, 1996; Marcogliese, 2004; Preston & Johnson, 2010). Recall, for example, the cascading ecological impacts of the microsporidian *Nosema ceranae* (see Box 4.1) and the *Varroa* mite (see Box 11.3), both introduced and highly pathogenic parasites of honey bees, and the invasive bopyrid isopod *Orthione griffenis* of estuarine mud shrimp (Color plate Fig. 3.3). These and other parasites can be termed 'keystone parasites' or 'ecosystem engineers' due to their effects on biodiversity, and their influence on community and ecosystem structure. Introduced parasites infecting naïve hosts, in particular, can play prominent ecological roles at community and ecosystem scales. The devastating impacts of introduced

avian malaria on several endemic species of Hawaiian birds provide an excellent example (Box 15.3).

Parasites may also influence biodiversity and the structure of host communities by altering the competitive interactions between free-living animals, a phenomenon termed parasite-mediated competition,

---

Box 15.3 | **Avian malaria determines the distributions, biodiversity, and community structure of Hawaii's native birds**

The introduction of *Plasmodium relictum*, the causative agent of mosquito-borne avian malaria, to the Hawaiian archipelago provides a case study for the effect of invasive parasites on naïve natural populations and communities (Woodworth *et al.*, 2005). The introduction of *P. relictum* and its primary vector, the southern house mosquito, *Culex quinquefasciatus*, to Hawaii early in the nineteenth century contributed to the decimation of populations of endemic bird species, including the famous Hawaiian honeycreepers (van Riper *et al.*, 1986). These native birds evolved without natural exposure to avian malaria. Early experimental studies showed that most endemic birds lacked genetically based defenses and were susceptible to *P. relictum*, suffering high parasitemia, significant pathologies, and mortality rates of up to 90% (Atkinson *et al.*, 1995). Consequently, avian malaria currently limits the geographical distributions, abundances, and biodiversity of native Hawaiian birds. Moreover, the disease impacts recovery efforts of threatened and endangered species of Hawaii's native birds.

Based upon a series of field and laboratory experiments, van Riper *et al.* (1986) proposed a model to explain the distributions of native birds along an elevation gradient (Fig. 15.12) that has guided conservation efforts (Woodworth *et al.*, 2005; Foster *et al.*, 2007). van Riper found that endemic bird species are restricted to high-elevation forests where mosquito vector densities are low, and where few introduced bird species are present to serve as reservoirs. The mosquito vectors are most common in wet, low-elevation habitats with standing water that provides suitable microhabitat for larval development. However, transmission of malaria declines as elevation increases because there are fewer mosquitoes and because cooler temperatures restrict *P. relictum* development. Furthermore, malaria transmission becomes increasingly seasonal at higher elevations. Thus, malaria transmission is highest during the warmest time of the year at mid-elevations (September–December) when mosquito population sizes reach their peak. This period immediately follows the nesting season for most native birds. The abundance of recently fledged, susceptible juvenile birds, coupled with increased mosquito populations, can lead to epizootic outbreaks. All of these factors help explain the absence of most native honeycreeper species at low elevations, and why communities of susceptible endemic forest birds at high elevation are more diverse (Fig. 15.12).

Box 15.3 | (continued)

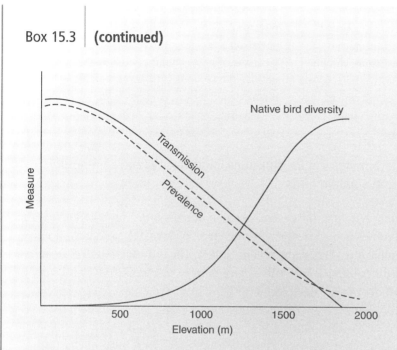

Fig. 15.12 Schematic representation of avian malaria (*Plasmodium relictum*) prevalence, vector transmission, and native bird diversity along an elevation gradient on Mauna Loa and Kilauea Volcanoes, Hawaii. Malaria prevalence and transmission decline as elevation increases. Communities of native birds at high elevation are more diverse, and include species that have little or no resistance to malaria. (Figure courtesy of Carter Atkinson, U.S. Geological Survey; modified from van Riper *et al.*, 1986, *Ecological Monographs*, 56, 327–344.)

It turns out that some native honeycreepers have rapidly evolved adaptations to reduce the severity of malaria. Some species have evolved novel roosting behaviors at night and forage in the lowlands only during the day; both behaviors serve to protect the birds from the nocturnally active mosquito vectors. There is also evidence for the rapid evolution of genetic resistance, or more likely tolerance, in some species of honeycreepers. The amakihi, once greatly reduced at elevations below 1000 m, now tolerates high parasitemia and has made a remarkable recovery in lowland habitats, despite high rates of local malaria transmission. Thus, the recovery of amakihi is particularly promising from a conservation perspective. Woodworth *et al.* (2005) state: "Understanding the biocomplexity of genetic, immunological, epidemiological, demographic, ecological, and landscape factors in the persistence of low-elevation amakihi populations may hold the key to the preservation of the remaining endemic Hawaiian avifauna."

or **apparent competition** (reviews in Price *et al.*, 1986; Hudson & Greenman, 1998). In short, theory predicts that if two competing host species share a parasite, and this parasite is pathogenic in one species and not in the other, then the parasite may shift competitive interactions, and thus alter community structure. In this scenario, the parasite can be a particularly powerful 'biological weapon' if it has greater pathological effects on the superior competitor. Competition mediated through a shared parasite has considerable relevance given the rise in emerging diseases and the introduction of parasites into new regions (Hudson & Greenman, 1998).

Given the complexity of host–parasite interactions and the complex nature of free-living ecological communities, conclusive evidence of parasite-mediated competition occurring in nature is difficult to generate. One often-cited example is the so-called meningeal or brain worm, *Parelaphostrongylus tenuis*, a nematode we first introduced in Chapter 8. Recall that the parasite infects the meninges of the brain of white-tailed deer and uses terrestrial gastropods as intermediate hosts in eastern and central North America. It causes little to no pathology in white-tailed deer, but causes severe neurological disease (parelaphostrongylosis) in ungulates such as caribou and moose. Following the clearing of land in northern and western North America, white-tailed deer (and their parasites) have expanded their ranges into regions of sympatry with moose. It seems that when ecological conditions favor consistently high population densities of white-tailed deer and also favor larval transmission via infected gastropods, parelaphostrongylosis can cause moose densities to decline, sometimes to local extinction (review in Lankester, 2010). Similarly, high intensities of giant liver fluke, another parasite originating in white-tailed deer, were the leading cause of mortality in a moose population in northern Minnesota (Murray *et al.*, 2006). These results suggest that in areas of sympatry, white-tailed deer have a competitive advantage over moose because of their parasites.

Parasite-mediated competition may impact host biodiversity by allowing a competitively inferior species to co-occur with a dominant species. Schall (1992) studied the effects of malaria on two lizard congeners on a Caribbean island. Both lizard species, *Anolis gingivinus* and *Anolis wattsi*, are host to *Plasmodium azurophilum*, and the former lizard is the superior competitor. However, the two lizards coexist only where *A. gingivinus* is heavily parasitized, suggesting that malaria reduces the competitive ability of the dominant lizard, allowing the inferior lizard to coexist (Schall, 1992).

Direct tests for parasite-mediated competition are difficult, if not impossible (Hudson & Greenman, 1998). A possible exception involves a cecal nematode, *Heterakis gallinarum*, in its game-bird hosts in the United Kingdom. As for the red grouse that we discussed earlier, ring-necked pheasants are reared and then released on private sporting estates. It turns out that as the numbers of pheasants released increased,

the numbers of sympatric gray partridge declined. Evidence supports the idea that *H. gallinarum* may be contributing to population declines of gray partridge via parasite-mediated apparent competition. Evidence from field collections and from experimental infections confirm that the nematode causes reduced weight gain, reduced food consumption, and impaired cecal function in partridges, but not in pheasants (Tompkins *et al.*, 1999, 2001b). In a follow-up field study, parasite-naïve partridges were maintained in pens and then released onto game bird estates (Tompkins *et al.*, 2000). Notably, nematode intensities in partridge were correlated with nematode intensities recorded in the previous year from pheasants on the estates. Moreover, infections originating from pheasants negatively influenced partridge body condition, and probably fitness (Tompkins *et al.*, 2000). Thus, parasite-mediated apparent competition with the ring-necked pheasant may be a factor contributing to the decline of wild partridges in the UK.

Parasites may also function as ecosystem engineers by directly or indirectly modifying the environment of other organisms (reviews in Thomas *et al.*, 1999; Lefèvre *et al.*, 2009). In essence, parasite-mediated alteration of a particular phenotype may create new habitat or available ecological niches for other organisms, or they may have strong effects on ecosystem processes by influencing energy flow, for example. Thomas *et al.* (1999) illustrate an example involving rhizocephalan barnacles. Recall from Chapter 11 that rhizocephalans inhibit molting of their crab hosts. It turns out that rhizocephalan-infected crabs, e.g., *Sacculina carcini* and its host, *Carcinus maenas*, offer a more permanent substrate for several sessile invertebrates, e.g., barnacles and polychaete worms, than that of uninfected crabs. Thus, a rich epibiont community can form on the surface of rhizocephalan-infected crabs. By altering a specific trait, e.g., inhibiting molting in this case, parasites can favor the emergence of engineering functions (Thomas *et al.*, 1999). Lefèvre *et al.* (2009) review further examples of how manipulative parasites can create habitats and act as ecosystem engineers.

Sato *et al.* (2011a; 2011b) provide the first empirical evidence to show that a manipulative parasite can alter the overall flow of energy in ecosystems. Recall from Chapter 9 that nematomorphs manipulate their terrestrial orthopteran hosts to seek out, and then jump into, water. Terrestrial insects can be important components of trout diets in streams. This leads to the possibility that hairworms might indirectly provide a subsidy for trout by manipulating orthopterans to water (Sato *et al.*, 2008). Remarkably, Sato *et al.* (2011a) found that Japanese trout were more likely to eat *Gordionus chinensis*-infected, energy-rich crickets and grasshoppers that fall onto the surface of streams. They showed that such predation accounted for an incredible 60% of the annual caloric intake of the trout population (Fig. 15.13). Furthermore, fish growth rates were highest in the fall months when the nematomorphs manipulated orthopterans into the stream. By

moving crickets from the forest to the stream, the parasite diverts substantial energy across habitats (Fig. 15.13; Sato *et al.*, 2011a; 2011b). Other cascading ecological effects were observed. When infected orthopterans were available, trout were satiated and ate fewer benthic invertebrates, potentially altering other stream–forest ecosystem processes, such as attached algal production in streams and predator–prey interactions within streams and the surrounding riparian habitat.

Marcogliese & Cone (1997) were among the first to draw attention to the dynamic and integral roles parasites can play within food webs. As we have discussed, many helminth parasites depend on trophic interactions for transmission. Thus, parasites within a single host population reflect many different trophic links within a food web. These associations are central to many ecological concepts, including stability,

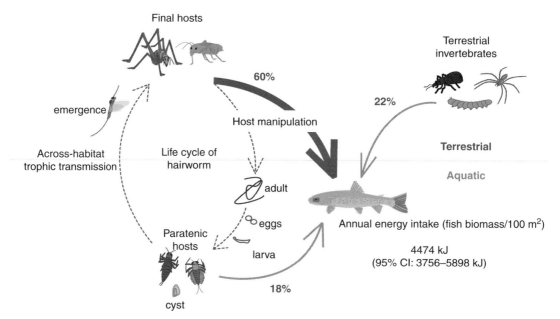

Fig. 15.13 Schematic representation of the life cycle of the hairworm *Gordionus chinensis* and the nematomorph-induced energy flow from forest to stream. Orthopteran final hosts of *G. chinensis* include camel crickets *Tachycines* spp., which contributed 95% of trout energy intake of orthopterans. Across-habitat trophic transmission of nematomorphs may be facilitated by orthopteran scavenging upon stonefly and mayfly paratenic hosts. Due to predation on these nematomorph-manipulated insects, endangered Japanese trout (*Salvelinus leucomaenis japonicus*) are provided with a key diet subsidy. Percentages of the energy obtained annually by the trout population from each prey category (infected orthopterans, terrestrial invertebrates, and stream benthic invertebrates) are shown. (Figure courtesy of Takuya Sato; modified from Sato *et al.*, 2011a, with permission, *Ecology*, **92**, 201–207, Ecological Society of America.)

diversity, and complexity of ecosystems. Given the diversity of parasites in food webs and the prominent roles they play, it is important to understand the influence they may have on the structure, function, and dynamics of food webs (review in Lafferty *et al.*, 2008). A case in point is provided by the Carpinteria Salt Marsh food web in California (Lafferty *et al.*, 2006). In a ground-breaking study, the authors demonstrated that the inclusion of parasites into this food web revealed remarkable changes in its structure, including increased species richness, doubling the overall connections between members of the food web, and quadrupling the overall number of linkages. Parasites were involved in approximately 75% of the salt marsh food web links. Additionally, incorporation of parasites into food webs revealed that mid-trophic levels are most vulnerable to natural enemies. This is the group most at risk from both predators and trophically transmitted parasites (Lafferty *et al.*, 2006).

Biomass is a measure of productivity and energetics in an ecosystem. One of the reasons parasites have been largely ignored in food web ecology is based on the assumption that they contribute negligible biomass to ecosystems (Preston & Johnson, 2010). Kuris *et al.* (2008), in a comprehensive study of three estuarine systems in California, have forever changed this notion. Their study included an astounding 199 species of free-living animals, 15 species of vascular plants, and 138 parasitic species! They found that the biomass of parasites is comparable to that of several groups of free-living animals and greater than the top predators in these estuaries. Parasitic castrators, e.g., epicaridean isopods, and trophically transmitted parasite stages dominated the parasite biomass. The biomass of trematodes in snail intermediate hosts, especially *Cerithidea californica* which is host to 18 trematode species, was incredibly high, being comparable to, or exceeding, that of abundant birds and fishes, as well as certain free-living benthic invertebrates. The functional group that included parasitic castrators, e.g., larval trematodes in *C. californica*, comprised most of the parasite biomass. We have already learned of the cascading ecological roles of such parasites in determining community and ecosystem structure and function. Kuris *et al.* (2008) conclude: "The substantial biomass and productivity attributed to parasites in these estuaries calls for the full integration of parasite ecology into the general body of ecological theory."

References

Adamo, S. A. (2002) Modulating the modulators: parasites, neuromodulators and host behavioral change. *Brain, Behaviour, and Evolution*, 60, 370–377.

Anderson, R. M. & May, R. M. (1978) Regulation and stability of host–parasite interactions. I. Regulatory processes. *Journal of Animal Ecology*, 47, 219–247.

Atkinson, C. T., Woods, K. L., Dusek, R. J., *et al.* (1995) Wildlife disease and conservation in Hawaii: Pathogenicity of avian malaria (*Plasmodium relictum*) in experimentally infected Iiwi (*Vestiaria coccinea*). *Parasitology*, 111, S59–S69.

Barlow, N. D. (1995) Critical evaluation of wildlife disease models. In *Ecology of Infectious Diseases in Natural Populations*, ed. B. T. Grenfell & A. P. Dobson, pp. 230–259. Cambridge: Cambridge University Press.

Biron, D. G., Marché, L., Ponton., F., *et al.* (2005) Behavioural manipulation in a grasshopper harbouring hairworm: a proteomics approach. *Proceedings of the Royal Society, Series B, Biological Sciences*, 272, 2117–2126.

Biron, E., Poncet, J., Brown, S. P., *et al.* (2006). 'Suicide' of crickets harbouring hairworms: a proteomics investigation. *Insect Molecular Biology*, 15, 731–742.

Booth, D. T., Clayton, D. H. & Block, B. A. (1993) Experimental demonstration of the energetic cost of parasitism in free-ranging hosts. *Proceedings of the Royal Society, Series B, Biological Sciences*, 253, 125–129.

Brooks, K. M. & Jones, S. R. M. (2008) Perspectives on pink salmon and sea lice: scientific evidence fails to support the extinction hypothesis. *Reviews in Fisheries Sciences*, 16, 403–412.

Combes, C. (1996) Parasites, biodiversity and ecosystem stability. *Biodiversity and Conservation*, 5, 953–962.

Combes, C. (2001) *Parasitism: The Ecology and Evolution of Intimate Interactions*. Chicago: University of Chicago Press.

Costello, M. J. (2009) How sea lice from salmon farms may cause wild salmonid declines in Europe and North America and be a threat to fishes elsewhere. *Proceedings of the Royal Society, Series B, Biological Sciences*, 276, 3385–3394.

Crofton, H. D. (1971) A quantitative approach to parasitism. *Parasitology*, 62, 179–193.

Dawkins, R. (1982) *The Extended Phenotype*. Oxford: Oxford University Press.

De Jong-Brink M., Bergamin-Sassen M. & Solis-Soto, M. (2001) Multiple strategies of schistosomes to meet their requirements in the intermediate snail host. *Parasitology* 123, S129–S141.

Delgiudice, G. D., Peterson, R. O. & Samuel, W. M. (1997) Trends of winter nutritional restriction, ticks, and numbers of moose on Isle Royale. *Journal of Wildlife Management*, 61, 895–903.

Diamond, J. (1997) *Guns, Germs and Steel: The Fates of Human Societies*. New York: W. W. Norton.

Ebert, D. (2005) Ecology, epidemiology, and evolution of parasitism in *Daphnia* (Internet). Bethesda: National Center for Biotechnology Information.

Ebert, D., Lipsitch, M. & Mangin, K. L. (2000) The effect of parasites on host population density and extinction: experimental epidemiology with *Daphnia* and six microparasites. *The American Naturalist*, 156, 459–477.

Edelaar, P., Drent, J. & de Goeij, P. (2003) A double test of the parasite manipulation hypothesis in a burrowing bivalve. *Oecologia*, 134, 66–71.

Forbes, M. R. I. (1993) Parasitism and host reproductive effort. *Oikos*, 67, 444–450.

Foster, J. T., Woodworth, B. L., Eggert, L. E., *et al.* (2007) Genetic structure and evolved malaria resistance in Hawaiian honeycreepers. *Molecular Ecology*, 16, 4738–4746.

Fredensborg, B. L. & Poulin, R. (2006) Parasitism shaping host life-history evolution: adaptive responses in a marine gastropod to infection by trematodes. *Journal of Animal Ecology*, 75, 44–53.

Giorgi, M. S., Arlettaz, R., Christe, P. & Vogel, P. (2001) The energetic grooming costs imposed by a parasitic mite (*Spinturnix myoti*) upon its bat host (*Myotis myotis*). *Proceedings of the Royal Society, Series B, Biological Sciences*, 268, 2071–2075.

Goater, C. P. (1994) Growth and survival of post-metamorphic toads: interactions among larval history, density, and parasitism. *Ecology*, 75, 2264–2274.

Goater, C. P. & Ward, P. I. (1992) Negative effects of *Rhabdias bufonis* (Nematoda) on the growth and survival of toads (*Bufo bufo*). *Oecologia*, 89, 161–165.

Goater, T. M., Shostak, A. W., Williams, J. A., *et al.*, (1989) A mark recapture study of trematode parasitism in over-wintered *Helisoma anceps* (Pulmonata), with special reference to *Halipegus occidualis* (Hemiuridae). *Journal of Parasitology*, 75, 553–560.

Grenfell, B. T. & Dobson, A. P. (1995) *Ecology of Infectious Diseases in Natural Populations*. Cambridge: Cambridge University Press.

Gulland, F. M. D. (1992) The role of nematode parasites in Soay sheep (*Ovis aries* L.): mortality during a population crash. *Parasitology*, 105, 493–503.

Hall, S. R., Becker, C. & Caceras, C. E. (2007) Parasitic castration: a perspective from a model of dynamic energy budgets. *Integrative and Comparative Biology*, 47, 295–309.

Hawlena H., Krasnov, B. R., Abramsky, Z., *et al.* (2008) Effects of food abundance, age, and flea infestation on the body condition and immunological variables of a rodent host, and their consequences for flea survival. *Comparative Biochemistry and Physiology*, 150, 66–74.

Heins, D. C., Birden, E. L. & Baker, J. A. (2010) Host mortality and variability in epizootics of *Schistocephalus solidus* infecting the threespine stickleback, *Gasterosteus aculeatus*. *Parasitology*, 137, 1681–1686.

Heins, D. C., Baker. J. A., & Green, D. M. (2011) Processes influencing the duration and decline of epizootics in *Schistocephalus solidus*. *Journal of Parasitology*, 97, 371–376.

Helluy, S. & Holmes, J. C. (1990) Serotonin, octopamine, and the clinging behavior induced by the parasite *Polymorphus paradoxus* (Acanthocephala) in *Gammarus lacustris* (Crustacea). *Canadian Journal of Zoology*, 68, 1214–1220.

Helluy, S. & Thomas, F. (2003) Effects of *Microphallus papillorobustus* (Platyhelminthes: Trematoda) on serotonergic immunoreactivity and neuronal architecture in the brain of *Gammarus insensibilis* (Crustacea: Amphipoda). *Proceedings of the Royal Society, Series B, Biological Sciences*, 270, 563–568.

Helluy, S. & Thomas, F. (2010) Parasitic manipulation and neuroinflammation: evidence from the system *Microphallus papillorobustus* (Trematoda)–*Gammarus* (Crustacea). *Parasites and Vectors*, 3, 38–49.

Herrmann, K. K. & Sorensen, R. E. (2011) Differences in natural infections of two mortality-related trematodes in lesser scaup and American coot. *Journal of Parasitology*, 97, 555–558.

Holmes, J. C. (1995) Population regulation: a dynamic complex of interactions. *Wildlife Research*, 22, 11–19.

Holmes, J. C. & Zohar, S. (1990) Pathology and host behaviour. In *Parasitism and Host Behaviour*, ed. C. J. Barnard & J. M. Behnke, pp. 34–63. London: Taylor & Francis.

Hoogenboom, I. & Dijkstra, C. (1987) *Sarcocystis cernae*: a parasite increasing the risk of predation of its intermediate host, *Microtis arvalis*. *Oecologia*, **74**, 86–92.

Hudson, P. J. & Greenman, J. V. (1998) Competition mediated by parasites: biological and theoretical progress. *Trends in Ecology and Evolution*, **13**, 387–390.

Hudson, P. J., Dobson, A. P. & Newborn, D. (1985) Cyclic and non-cyclic populations of red grouse: a role for parasitism? In *Ecology and Genetics of Host-Parasite Interactions*, ed. D. Rollinson & R. M. Anderson, pp. 77–89. London: Academic Press.

Hudson, P. J., Dobson, A. P. & Newborn, D. (1992) Do parasites makes prey vulnerable to predation: red grouse and parasites. *Journal of Animal Ecology*, **61**, 681–692.

Hudson, P. J., Dobson, A. P. & Newborn, D. (1998) Prevention of population cycles by parasite removal. *Science*, **282**, 2256–2258.

Hudson, P. J., Rizzoli A., Grenfell, B. T., *et al.* (2001) *The Ecology of Wildlife Diseases*. Oxford: Oxford University Press.

Hurd, H., Warr, E. & Polwart, A. (2001) A parasite that increases host lifespan. *Proceedings of the Royal Society, Series B, Biological Sciences*, **268**, 1749–1753.

Johnson, P. T. J., Lunde, K. B., Ritchie, E. G. & Launer, A. E. (1999) The effect of trematode infection on amphibian limb development and survivorship. *Science*, **284**, 802–804.

Johnson, P. T. J., Rohr, J. R., Hoverman, J. T., *et al.* (2012) Living fast and dying of infection: host life history drives interspecific variation in infection and disease. *Ecology Letters*, **15**, 235–242.

Jones, S. R. M. & Beamish, R. (2011) *Salmon Lice: An Integrated Approach to Understanding Parasite Abundance and Distribution*. Chichester: Wiley-Blackwell.

Kennedy, C. R., Shears, P. C. & Shears, J. A. (2001) Long-term dynamics of *Ligula intestinalis* and roach *Rutilus rutilus*: a study of three epizootic cycles over 31 years. *Parasitology*, **123**, 257–269.

Keymer, A. E. (1981) Population dynamics of *Hymenolopis diminuta* in the intermediate host. *Journal of Animal Ecology*, **50**, 941–950.

Krkošek M., Lewis, M. A., Morton A., *et al.* (2006) Epizootics of wild fish induced by farm fish. *Proceedings of the National Academy of Sciences USA*, **103**, 15506–15510.

Krkošek, M., Ford, J. S., Morton, A., *et al.* (2007) Declining wild salmon populations in relation to parasites from farm salmon. *Science*, **318**, 1772–1775.

Krkošek, M., Morton, A., Volpe, J. P. & Lewis, M. A. (2009) Sea lice and salmon population dynamics: effects of exposure time for migratory fish. *Proceedings of the Royal Society, Series B, Biological Sciences*, **276**, 2819–2828.

Krkošek, M., Conners, B. M., Morton, A., *et al.* (2011) Effects of parasites from salmon farms on productivity of wild salmon. *Proceedings of the National Academy of Sciences USA*, **108**, 14700–14704.

Kuris, A. M., Hechinger, R. F., Shaw, J. C., *et al.* (2008) Ecosystem energetic implications of parasite and free-living biomass in three estuaries. *Nature*, **454**, 515–518.

Kyriazakis, I, Tolkamp, B. J. & Hutchings, M. R. (1998) Towards a functional explanation for the occurrence of anorexia during parasitic infections. *Animal Behaviour*, **56**, 265–274.

Lafferty, K. D. (1993) The marine snail, *Cerithidea californica*, matures at smaller sizes where parasitism is high. *Oikos*, **68**, 3–11.

Lafferty, K. D. & Morris, K. (1996) Altered behavior of parasitized killifish increases susceptibility to predation by bird and final hosts. *Ecology*, **77**, 1390–1397.

Lafferty, K. D., Dobson, A. P. & Kuris, A. M. (2006) Parasites dominate food web links. *Proceedings of the National Academy of Sciences USA*, **103**, 11211–11216.

Lafferty, K. D., Allesina, S., Arim, M., *et al.* (2008) Parasites in food webs: the ultimate missing links. *Ecology Letters*, **11**, 533–546.

Lankester, M. W. (2010) Understanding the impact of meningeal worm, *Parelaphostrongylus tenuis*, on moose populations. *Alces*, **46**, 53–70.

Lefèvre, T., Adamo, S. A., Biron, D. G., *et al.* (2009) Invasion of the body snatchers: the diversity and evolution of manipulative strategies in host–parasite interactions. *Advances in Parasitology*, **68**, 45–83.

Lemly, A. D. & Esch, G. W. (1984) Effects of the trematode *Uvulifer ambloplitis* on juvenile bluegill sunfish, *Lepomis macrochirus*: ecological implications. *Journal of Parasitology*, **70**, 475–492.

Marcogliese, D. J. (2004) Parasites: small players with crucial roles in the ecological theatre. *EcoHealth*, 1, 151–164.

Marcogliese, D. J. & Cone, D. K. (1997) Food webs: a plea for parasites. *Trends in Ecology and Evolution*, **12**, 320–325.

Marty, G. D., Saksida, S. M. & Quinn, T. J. (2010) Relationship of farm salmon, sea lice, and wild salmon populations. *Proceedings of the National Academy of Sciences USA*, www.pnas.org/cgi/doi/10.1073/pnas.1009573108

May, R. M. & Anderson, R. M. (1978) Regulation and stability of host–parasite interactions. II. Destabilizing processes. *Journal of Animal Ecology*, **47**, 249–267.

Michalakis, Y. (2009) Parasitism and the evolution of life-history traits. In *Ecology and Evolution of Parasitism*, ed.

F. Thomas, J-F Guegan & F. Renaud, pp. 19–30. Oxford: Oxford University Press.

Minchella, D. J. (1985) Host life-history variation in response to parasitism. *Parasitology*, 90, 205–261.

Minchella, D. J. & Loverde, P. T. (1981) A cost of increased early reproductive effort in the snail *Biomphalaria glabrata. The American Naturalist*, 118, 876–881.

Minchella, D. J., Leathers, B. K., Brown, K. M., *et al.* (1986). Host and parasite counteradaptations: an example from a freshwater snail. *The American Naturalist*, 126, 843–854.

Møller, A. P. (1997) Parasitism and the evolution of host life history. In *Host–Parasite Evolution: General Principles and Avian Models*, ed. D. H. Clayton & J. Moore, pp. 105–127. Oxford: Oxford University Press.

Moore, J. (1983) Responses of an avian predator and its isopod prey to an acanthocephalan parasite. *Ecology*, 64, 1000–1015.

Moore, J. (2002) *Parasites and the Behavior of Animals.* Oxford: Oxford University Press.

Munger, J. C. & Holmes, J. C. (1988) Benefits of parasitic infection: a test using a ground squirrel/trypanosome system. *Canadian Journal of Zoology*, 66, 222–227.

Murray, D.L., Cary, J.R. & Keith, L.B. (1997) Interactive effects of sublethal nematodes and nutritional status on snowshoe hare vulnerability to predation. *Journal of Animal Ecology* 66, 250–264.

Murray, D. L., Cox, E. W., Ballard, W. B., *et al.* (2006) Pathogens, nutritional deficiency, and climate influences on a declining moose population. *Wildlife Monographs*, 166, 1–30.

Negovetich, N. J. & Esch, G. W. (2008a) Quantitative estimation of the cost of parasitic castration in a *Helisoma anceps* population using a matrix population model. *Journal of Parasitology*, 94, 1022–1030.

Negovetich, N. J. & Esch, G. W. (2008b) Life-history cost of trematode infection in *Helisoma anceps* using mark-recapture in Charlie's Pond. *Journal of Parasitology*, 94, 314–325.

Oppliger, A., Celerier, M. L., & Clobert, J. 1996. Physiological and behaviour changes in common lizards parasitized by haemogregarines. *Parasitology*, 113, 433–438.

Perrot-Minnot, M. & Cezilly, F. (2009) Parasites and behavior. In *Ecology and Evolution of Parasitism*, ed. F. Thomas, J-F Guegan & F. Renaud, pp. 49–68. Oxford: Oxford University Press.

Pike, A. W. & Wadsworth, S. (1999) Sealice on salmonids, their biology and control. *Advances in Parasitology*, 44, 233–337.

Poinar, Jr., G. O. & Yanoviak, S. P. (2008) *Myrmeconema neotropicum* n. g., n. sp., a new tetradonematid nematode parasitising South American populations of *Cephalotes*

atratus (Hymenoptera: Formicidae), with the discovery of an apparent parasite-induced host morph. *Systematic Parasitology*, 69, 145–153.

Poulin, R. (1995) "Adaptive" changes in the behaviour of parasitized animals: a critical review. *International Journal for Parasitology*, 25, 1371–1383.

Poulin, R. (2007) *Evolutionary Ecology of Parasites*, 2nd edition. Princeton: Princeton University Press.

Powell, M. D., Speare, D. J., Daley, J., & Lovy, J. (2005) Differences in metabolic response to *Loma salmonae* infection in juvenile rainbow trout *Oncorhynchus mykiss* and brook trout *Salvelinus fontinalis. Diseases of Aquatic Organisms*, 67, 233–237.

Preston, D. L. & Johnson, P. T. J. (2010) Ecological consequences of parasitism. *Nature Education Knowledge*, 1, 39–42.

Price, P. W., Westoby, M., Rice, B., *et al.* (1986) Parasite mediation in ecological interactions. *Annual Reviews in Ecology and Systematics*, 17, 487–505.

Robar, N., Burness, G. & Murray, D. L. & (2010) Tropics, trophics, and taxonomy: the determinants of parasite-associated host mortality. *Oikos*, 119, 1273–1280.

Robar, N., Murray, D. L. & Burness, G. (2011) Effects of parasites on host energy expenditure: the resting metabolic rate stalemate. *Canadian Journal of Zoology*, 89, 1146–1155.

Robb, T. & Reid, M. L. (1996) Parasite-induced changes in the behaviour of cestode-infected beetles: adaptation or simple pathology? *Canadian Journal of Zoology*, 74, 1268–1274.

Roberts, L. S. & Janovy, J. (2009) *Foundations of Parasitology*, 8th edition. New York: McGraw-Hill.

Roberts, M. G., Smith, G., & Grenfell, B. (1995) Mathematical models for macroparasites of wildlife. In *Ecology of Infectious Diseases in Natural Populations*, ed. B. T. Grenfell & A. P. Dobson, pp. 177–208. Cambridge: Cambridge University Press.

Sato, T., Arizono, M., Sone, R. & Harada, Y. (2008) Parasite-mediated allochthonous input: do hairworms enhance subsidized predation of stream salmonids on crickets? *Canadian Journal of Zoology*, 86, 231–235.

Sato, T., Watanabe, K., Kanaiwa, M., *et al.* (2011a) Nematomorph parasites drive energy flow through a riparian ecosystem. *Ecology*, 92, 201–207.

Sato, T., Watanabe, K., Tokuchi, N., *et al.* (2011b) A nematomorph parasite explains variation in terrestrial subsidies to trout streams in Japan. *Oikos*, 120, 1595–1599.

Schall, J. J. (1992) Parasite-mediated competition in *Anolis* lizards. *Oecologia*, 92, 58–64.

Scott, M. E. (1987) Regulation of mouse colony abundance by *Heligmosomoides polygyrus. Parasitology*, 95, 111–124.

Scott, M. E. (1988) The impact of infection and disease on animal populations: implications for conservation biology. *Conservation Biology*, 2, 40–55.

Scott, M. E. & Dobson, A. (1989) The role of parasites in regulating host abundance. *Parasitology Today*, 5, 176–183.

Shirakashi, S. & Goater, C. P. (2005) Chronology of parasite-induced alteration of minnow behaviour: effects of parasite maturation and host experience. *Parasitology*, 130, 177–183.

Tain, L., Perrot-Minnot, M. & Thomas, F. (2006) Altered host behaviour and brain activity caused by acanthocephalans: evidence for specificity. *Proceedings of the Royal Society, Series B, Biological Sciences*, 273, 3039–3045.

Théron, A. & Gerard, C. (1994) Development of accessory sexual organs in *Biomphalaria glabrata* (Planorbidae) in relation to timing of infection with *Schistosoma mansoni*: consequences for energy utilization patterns by the parasite. *Journal of Molluscan Studies*, 60, 25–31.

Thomas, F., Poulin, R. & de Meeüs, T., *et al.* (1999) Parasites and ecosystem engineering: what roles could they play? *Oikos*, 84, 167–171.

Thomas, F., Ulitsky, P., Augier, R., *et al.*, (2003) Biochemical and histological changes in the brain of the cricket *Nemobius sylvestris* infected by the manipulative parasite *Paragordius tricuspidatus* (Nematomorpha). *International Journal for Parasitology*, 33, 435–443.

Tompkins, D. M., Dickson, G. & Hudson, P. J. (1999) Parasite-mediated competition between pheasant and grey partridge: a preliminary investigation. *Oecologia*, 119, 378–382.

Tompkins, D. M., Draycott, R. A. H. & Hudson, P. J. (2000) Field evidence for apparent competition mediated via the shared parasites of two gamebird species. *Ecology Letters*, 3, 10–14.

Tompkins, D. M., Dobson, A. P., Arneberg, P., *et al.* (2001a) Parasites and host population dynamics. In *The Ecology of Wildlife Diseases*, ed. P. J. Hudson, A. Rizzoli, B. T. Grenfell, *et al.*, pp. 45–62. Oxford: Oxford University Press.

Tompkins, D. M., Greenman, J. V. & Hudson, P. J. (2001b) Differential impact of a shared nematode parasite on two gamebird hosts: implications for apparent competition. *Parasitology*, 122, 187–193.

Townsend, S. E., Newey, S., Thirgood, S. J., *et al.* (2009) Can parasites drive population cycles in mountain hares? *Proceedings of the Royal Society, Series B, Biological Sciences*, 269, 1611–1617.

van Riper, C., III, van Riper, S. G., Goff, M. L. & Laird, M. (1986) The epizootiology and ecological significance of malaria in Hawaiian land birds. *Ecological Monographs*, 56, 327–344.

Verble, R. M., Meyer, A. D., Kleve, M. G., *et al.* (2012) Exoskeletal thinning in *Cephalotes atratus* ants (Hymenoptera: Formicidae) parasitized by *Myrmeconema neotropicum* (Nematoda: Tetradonematidae). *Journal of Parasitology*, 98, 226–228.

Webb, T. J. & Hurd, H. (1999) Direct manipulation of insect reproduction by agents of parasite origin. *Proceedings of the Royal Society, Series B, Biological Sciences*, 266, 1537–1341.

Woodworth, B. L., Atkinson, C. T., LaPointe, D. A., *et al.* (2005) Host population persistence in the face of introduced vector-borne diseases: Hawaii amakihi and avian malaria. *Proceedings of the National Academy of Sciences USA*, 102, 1531–1536.

Yanoviak, S. P. (2010) 'Berry' ants: an eye-popping symbiosis from the tropical rainforest canopy. In: *Ant Ecology*, ed. L. Lach, C. L. Parr & K. L. Abbott, pp. 98–99. Oxford: Oxford University Press.

Yanoviak, S. P., Kaspari, M., Dudley, B. & Poinar, Jr., G. O. (2008) Parasite-induced fruit mimicry in a tropical canopy ant. *The American Naturalist*, 171, 536–544.

16 Evolution of host–parasite interactions

16.1 General considerations

This chapter synthesizes the evolutionary significance of parasites on their hosts. As such, it marks the logical transition from the previous chapter by asking whether parasite-induced effects on host individuals and populations translate to evolutionary-level effects. Our focus is on microevolutionary phenomena. In the first section, we evaluate the empirical evidence that parasites mediate natural selection on specific host traits, and whether this process leads to an evolutionary response. In the next section, we synthesize studies on the population genetic structure of parasites. Here, our focus is on the manner in which genetic variation is distributed within and among parasite populations, and how patterns of genetic structuring can provide insight into the roles that microevolutionary phenomena such as **genetic drift** and **gene flow** play in natural host–parasite systems. In the final section, we introduce the phenomenon of host–parasite coevolution. Here, we evaluate the evidence for reciprocal evolution between hosts and parasites within natural host–parasite combinations. We conclude this section by considering how the microevolutionary interplay between hosts and parasites over many generations can help us understand how parasites evolved, and potentially cospeciated with their hosts, over longer time scales.

16.2 Parasite-mediated natural selection and evolution

In an address to the Royal Society, the great evolutionary biologist John Haldane described how he and his father were immune to the bites of bed bugs. Perhaps anecdotal observations such as this provided the impetus for his essay entitled 'Disease and evolution' (Haldane, 1949). This paper heralded the idea that parasites (specifically diseases of humans) could act as significant selective agents on their hosts, with important consequences. Thus, parasites could produce and maintain rare host genotypes, they could drive host speciation, and human malaria could maintain polymorphism of the hemoglobin molecule. Sixty years later, Haldane's original ideas have become a central focus for biologists interested in the evolutionary significance of parasites on their hosts (Little, 2002; Bernatchez & Landry, 2003; Piertney & Oliver, 2006).

Haldane's (1949) ideas are based on the central assumption that parasites exert selective pressure on host phenotypes, and that this phenotypic selection can lead to an evolutionary response in a host population. In our view, the most convincing examples of parasite-mediated natural selection include both of these processes. Thus, to be of evolutionary significance, phenotypic selection against parasitized individuals should lead to gene frequencies for a trait that differ from those expected for that trait in the absence of parasites (review in Goater & Holmes, 1997). It is important to understand to what degree this process occurs in natural systems. From a theoretical perspective, parasite-mediated selection is a central assumption of models of parasite–host coevolution, parasites and the evolution of sex, parasites and sexual selection, and parasite-mediated host speciation. From an applied perspective, the manner in which hosts respond to selective pressures imposed by parasites can determine the severity of parasites and disease, and can determine the outcome of interactions when 'new' parasites are introduced into naïve host populations.

The classic example of parasite-mediated natural selection is the maintenance of the gene for sickle hemoglobin, which provides partial protection against *Plasmodium falciparum* malaria (Box 16.1). The general biology, life cycle, and pathology of this devastating human parasite were described in Chapter 3. In this case, the parasite provides a selective advantage to heterozygotes, which maintains the gene at far higher proportions than expected where the parasite is absent. Falciparum-mediated selection for the maintenance of the sickle-cell gene appears to be a worldwide phenomenon, such that the local frequency of the gene is

---

Box 16.1 | Parasite-mediated natural selection on the human HbS gene

Evidence from molecular studies indicate that all extant *P. falciparum* populations originated from a single host transfer of *P. reichenowi* from chimpanzees to humans (Rich *et al.*, 2009; see Chapter 3). This transfer may have occurred as early as 3 000 000 years ago, or as recently as 10 000 years ago. Regardless of the precise timing, the parasite transfer to humans led to an enormous opportunity for *P. falciparum*, and the parasite has had, and continues to have, a strong impact on human evolution.

The source of variation that allows falciparum-mediated natural selection to occur is a homozygous mutational change in the hemoglobin molecule within red blood cells (RBCs). People who are heterozygous for the trait are normal, but they are resistant to *P. falciparum*. Normally, glutamic acid occurs at position 6 on the ß-chains of the hemoglobin molecule. In persons with the mutant gene, valine substitutes for glutamic acid. While an apparent minor switch, the molecular structure of hemoglobin is altered to the extent that it leads to a devastating human blood disease known as sickle-cell anemia. When RBCs occur in regions of the body with low-oxygen tensions, the mutant hemoglobin aggregates into chains, creating long and stiff fibrils that cause the RBCs to become sticky and sickled in shape. The distorted cells are then trapped in narrow capillaries, blocking the flow of blood, and interfering with the oxygen-carrying capacity of the RBCs. Blockage will produce debilitating symptoms, eventually leading to tissue damage, excruciating pain, and severe anemia. The mutant form of hemoglobin has the designation 'HbS' ('S' for sickle). Patients that carry HbS have 100% mutant hemoglobin and inherit both defective copies of the HbS allele – one from their mother and one from their father. Tragically, most children born with sickle-cell anemia die before reaching 5 years of age. However, heterozygote carriers have up to 50% abnormal hemoglobin, but suffer no deleterious effects.

Given the lethal effect of HbS, natural selection should have eliminated it. Indeed, under most circumstances, its frequency is extremely low. Yet, the prevalence of the allele in some native African populations reaches as high as 40%; frequencies of 15–20% are common. To maintain such high frequencies in a population there must be an extremely strong selective advantage of being a heterozygote or a carrier of sickle-cell anemia. It turns out that the selective advantage is among the highest ever measured for any trait in any species (Carroll, 2009). It was a young physician, Anthony Allison, working in Africa in the 1950s, who resolved the paradox of the sickle-cell trait's commonness in certain African populations. Allison first observed that the prevalence of HbS coincided with the geographical distribution of *P. falciparum* (Allison, 1954).

Box 16.1 | (continued)

Thus, wherever falciparum malaria was common in Africa, the allele was present in significantly higher proportions. Using a combination of modern geospatial tools, e.g., remote sensing, and molecular diagnoses, the linkage between the frequency of HbS and risk of infection with *P. falciparum* has been confirmed on a global scale (Fig. 16.1; Hay *et al.*, 2004; Piel *et al.*, 2010).

Thus, carriers of sickle cell anemia are genetically protected from *P. falciparum*. It turns out that in humans infected with falciparum malaria (but not other species of *Plasmodium*), infected blood cells bind to capillary walls within internal tissues (see Chapter 3). Within these low-oxygen sites, about 40% of heterozygous cells sickle, killing the parasite. Thus, *P. falciparum* provides a selective advantage to heterozygotes. In evolutionary terminology, this is an example of a **balanced polymorphism** – the maintenance of two alleles in a population arising from a fitness advantage to the heterozygote. It is also a case of **heterozygote advantage** – a situation in which a heterozygote has increased fitness relative to either homozygote. In its coevolutionary interaction with humans, *P. falciparum* has acted, and continues to act, as an agent of natural selection.

closely matched with the local risk of exposure to *P. falciparum* (Fig. 16.1; Piel *et al.*, 2010). Yet, despite the clarity of this classic example, it is an anthropogenic system involving a single gene and a single parasite. The extent to which outcomes such as this one apply to more natural parasite–host interactions involving multiple species of parasite and polygenic host responses is an active area of enquiry.

Here, we examine the evidence for parasite-mediated natural selection within naturally occurring host–parasite interactions. First, we examine whether the general requirements for parasite-mediated selection are met in wild systems and then we evaluate the evidence for evolutionary responses to selection. We conclude by describing characteristics of host–parasite interactions that may constrain strong evolutionary responses.

16.2.1 Requirements for parasite-mediated natural selection

Phenotypic selection on a particular trait requires variation in the trait, at least some of which must be genetically determined. Further, variation in the trait must be associated with variation in fitness. Parasite-mediated selection differs only in that selection should be directed against parasitized hosts (or, more specifically, parasitized genotypes), which indirectly selects for some other host trait (e.g., immunity, avoidance). For this type of selection, attention should be focused first on the parasites and their effects on the host (e.g., the effects of *P. falciparum* in Box 16.1), and secondarily on the host traits undergoing indirect selection (e.g., altered hemoglobin in Box 16.1). This double focus is the crux of parasite-mediated selection.

One key conclusion from the preceding chapter holds that the effect of micro- and macroparasites on their hosts is tightly linked to the numbers of parasites present. Aspects of parasite quality such as biomass, reproductive status, and genetic structure can also influence the nature of effects on host fitness, but these are rarely assessed. Given our focus on parasite numbers, one of the requirements for parasite-mediated selection is that the numbers of parasites must vary among individuals within a host population. As we have seen throughout the text, and especially in Chapter 12, this requirement is virtually always met in natural host–parasite interactions. Most micro- and

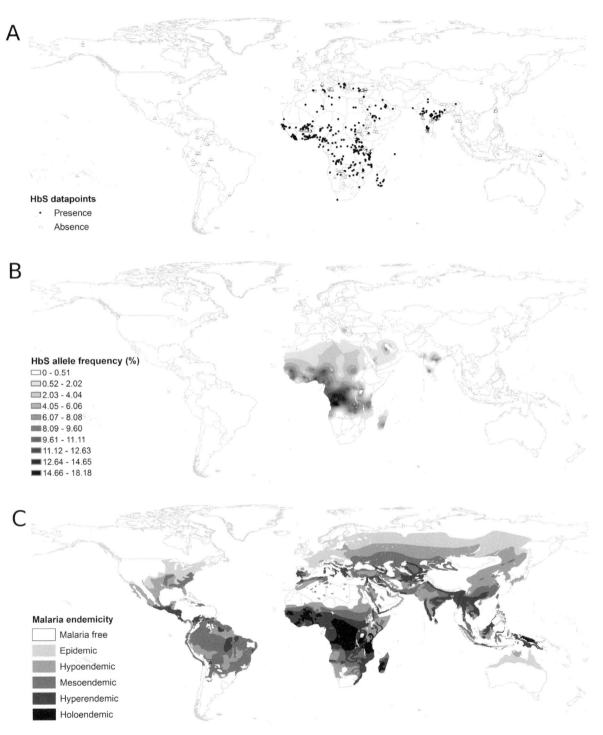

Fig. 16.1 Global distribution of the sickle-cell gene and geographical confirmation of the malaria hypothesis. (A) Distribution of data points for presence (●) and absence (Δ) of the HbS gene. (B) Map of HbS allele frequencies generated by a Bayesian model-based geostatistical framework. HbS allele frequencies of > 5% were present throughout much of Africa, the Middle East,

macroparasites show an aggregated distribution of numbers within their host populations, whereby the variance in parasite numbers are almost always much higher than the mean (see Chapter 12). Even when exposure and other conditions are held constant in the laboratory (e.g., within congenic strains of mice) considerable variation among individual hosts is still present (e.g., Wassom & Kelly, 1990). Thus, the requirement for variation in parasite numbers between individual hosts in a population is almost always met in natural host–parasite interactions.

Parasite-mediated selection also requires an association between parasite numbers and parameters associated with host fitness, such as mortality or fecundity. Such an association is assumed in the theoretical literature. Evidence for the role of parasite quantity and quality in determining a wide range of fitness-related parameters was presented in Chapter 15. Further, the magnitude of their effects on energy budget, life history characteristics, growth, survival, reproduction, and the expression of altered phenotypes, is often linked to their numbers within a host. Thus, parasites do not always have demonstrable effects on host fitness, but when they do, variation in the number of parasites often leads to variation in those effects. Tests regarding the strength of parasite-mediated natural selection must demonstrate that variation in parasite numbers leads to variation in host fitness.

Finally, for parasite-mediated selection to occur on any trait, the trait must be linked to the numbers of parasites, or with their effects. Because of the central role of the immune system in defense against parasites (see Chapter 2), components of that system provide an obvious suite of traits on which selection might work.

Genes and DNA sequences associated with the MHC (see Chapter 2) have received the most attention, particularly the easily manipulated models involving mice and rats as hosts. Compelling evidence supports the idea that specific MHC genes, or alleles, are correlated with resistance against specific parasites. In humans, specific HLA haplotypes are associated with malaria (Hill, 1998). Further, allele *Bw53* (an HLA variant) is nearly absent from populations without a history of endemic malaria. In chickens, the B21 allele is tightly associated with resistance to Marek's disease (a type of leukemia), and to the coccidian, *Eimeria maxima* (Bumstead *et al.*, 1991). The results of studies involving laboratory mice have demonstrated the key role of MHC genes in determining susceptibility and resistance to a wide range of micro- and macroparasites (review in Apanius *et al.*, 1997). Studies on non-model systems, e.g., helminths and mites of mole rats, nematodes of Soay sheep and wild mice, and several parasites of sticklebacks, have also detected close linkages between MHC genes and parasite intensities (review in Piertney & Oliver, 2006). Overall, these results highlight the role of genes with a known function in host immunity as potent targets for parasite-mediated natural selection.

Traits associated with host tolerance have also received attention as targets of parasite-mediated selection (review in Råberg *et al.*, 2009). Here, focus is on traits that enhance a host's ability to compensate for the damage done by parasites. Humans with a hemoglobin disorder known as α^+-thalassemia are protected from anemia caused by *P. falciparum* (Wambua *et al.*, 2006). Individuals that are heterozygous for the mutation are not protected from high

---

Caption for Fig. 16.1 (cont.)

and India. A maximum of 18.18% was found in northern Angola, Africa. (C) Historical map (*circa* 1900) of malaria transmission intensity (endemicity) prior to malaria control interventions. The six endemicity classes are defined by parasite rates (the proportion of 2- to 10-year-olds with malaria in their peripheral blood): malaria-free PR=0; epidemic, PR≈0; hypoendemic, PR ≤ 0.10; mesoendemic, PR ≥ 0.10 and PR < 0.50; hyperendemic, PR ≥ 0.50 and 0.75; holoendemic, PR ≥ 0.75. HbS allele frequencies were higher in hyperendemic and holendemic areas, providing evidence for a geographical link between the global distribution of HbS and malaria endemicity; the relationship was especially strong in Africa. Detailed methods are presented in Hay *et al.* (2004) and Piel *et al.* (2010). (Figure courtesy of Fred Piel; modified from Piel *et al.*, 2010, *Nature Communications*, 1, 104 DOI: 10.1038/ncomms1104.)

falciparum parasitemias, but the extent of anemia is reduced to a level that enhances their survival during infection. Thus, the mutation affects tolerance to infection, but not resistance. In reviewing breeding for resistance of sheep to the nematode *Haemonchus contortus*, Albers *et al.* (1987) concluded that 'resilience,' i.e., tolerance, or the ability to retain a relatively undepressed production level during infection, was distinct from resistance. One potential mechanism is an increased ability to produce RBCs in tolerant hosts. The same mechanism enhanced the survival and growth rates of resistant genotypes of mice when infected with a pathogenic strain of malaria (Råberg *et al.*, 2007), and is likely to apply to other anemia-producing parasites, e.g., *Leucocytozoon* and other protists, and *Trichostrongylus tenuis*, in birds.

Although much less studied, parasite-mediated selection may occur on other non-immunological traits, such as host avoidance behaviors that reduce risk of exposure. Such features are difficult to study and their underlying genetic basis is rarely known. Following the introduction of avian malaria to the Hawaian islands in the 1940s, some native species of honeycreepers learned to avoid nocturnal mosquito vectors by roosting at night at higher elevations and feeding in the lowlands during the day (see Box 15.3). These birds may have also adopted a new roosting posture, with the head tucked under the wings, to reduce exposure to mosquitoes. In other birds, behaviors such as preening and 'anting' can reduce ecto-parasite intensities. Grazing mammals avoid fecal pats containing nematode eggs or larvae (Hart, 1997) and minnows and tadpoles reduce their overall activity in the presence of trematode cercariae (Wisenden *et al.*, 2009). Although the expression of traits such as these undoubtedly influence the duration of exposure, their role as targets of parasite-mediated selection is unknown.

To summarize, the conditions necessary for parasite-mediated selection are met in natural host–parasite combinations, probably frequently. Variation in parasite numbers is virtually universal. Further, high numbers of parasites frequently, but not always, reduce host fitness. Testing for a reduction in fitness should be a requirement in any study of parasite-mediated selection. Finally, there are many genetically determined traits that could influence parasite numbers that are available for selection, or be influenced by those numbers, particularly those associated with host immunity and host tolerance.

16.2.2 Parasite-mediated evolution

Parasites can be important agents of selection. A more fundamental question is whether the selective effect of parasites results in evolution of their hosts. A continued evolutionary response to selection requires not only that a trait has a genetic basis, i.e., is heritable, but that the genetic component be directly responsible for some of the current phenotypic variation in the population.

The clearest demonstration of an evolutionary response is to document changes over time in one, or more, host traits that result in an effective response to a specific parasite. One of the best opportunities to observe such responses is to monitor changes following the introduction of a new parasite into a naïve host population. A classic example is the rapid evolution of resistance in European rabbits following the introduction of the myxoma virus into Australia, England, and France (review in Kerr, 2012). The virus is benign in the two species of rabbit it infects in South America, but it caused up to 99% mortality when it was introduced into Australia in 1950 to control exploding populations of European rabbits. Similar mortality was observed when the virus was later introduced into populations of rabbits in France and England. In the best-studied case involving a population of rabbits in New South Wales, Australia, the mortality rate of wild rabbits exposed to a strain of intermediate virulence decreased from 90% to 26% within 7 years (Kerr, 2012). The mechanism underlying the consistent, rapid decrease in mortality following introduction is poorly known. 'Resistant' rabbits develop a rapid and highly effective antibody (IgG and IgM) response that persists for at least 2 years after initial exposure. Thus, extremely strong directional selection on traits associated with host immunity provides the most likely mechanism to explain this classic case of parasite-mediated evolution.

Another example of parasite-mediated evolution involves the native bird–*Plasmodium relictum* interaction on Hawaii (see Box 15.3). As we discussed earlier, avian malaria has caused extinctions and declines of several species of low-elevation Hawaiian birds over the past 60 years, especially the honeycreepers. Experimental studies have shown that these birds are highly susceptible to introduced malaria, rapidly developing parasitemias that are orders of magnitude higher than for species of birds on the mainland or for introduced birds in Hawaii. Mortality up to 90% is not uncommon. However, one species of honeycreeper, the amakihi, has made a remarkable recovery, particularly within lowland sites (Woodworth *et al.*, 2005). Although transmission rates are high in these regions (see Box 15.3), amakihi have apparently evolved mechanisms to reduce the severity of malaria, rather than (or in addition to) increase immunity. This is another case where parasite-mediated selection has favored the rapid evolution of tolerance.

Similar observations on the results of evolution have been seen when an introduced host species is exposed to selection by endemic species of parasites. One example involves the evolution of resistance in striped bass on the Pacific coast of California (Sakanari & Moser, 1990). Striped bass from the Atlantic coast were introduced over 100 years ago. In these bass, but not in the other endemic hosts, larvae of the shark cestode, *Lacistorhynchus dollfusi*, stimulate a severe tissue response, causing considerable mortality. Sakanari and Moser showed that the west coast stock of striped bass responded less, and suffered less mortality, to the cestode than the east coast stock from which it was derived. In this case, severe immunopathology developed as a response against the parasite, and what has evolved is not increased resistance, but increased tolerance in the form of reduced response to infection.

Very few phenotypic traits are wholly dependent on genetic control. Even though variant hemoglobins and MHC molecules are probably exceptions, the expression of most traits will result from a mixture of environmental and genetic factors. Such traits can evolve only if they show additive genetic variation, that is, if variation in the character is heritable. Evolutionary biologists have focused on measuring heritabilities of countless traits, because their magnitude can give insight into both the direction and strength of the evolutionary response to natural selection. The heritability of factors associated with host defense against parasites has been estimated in a number of ways (review in Little, 2002). One is through a detailed analysis of controlled breeding studies designed to partition the total variance in a phenotypic trait by the degree of relatedness. Most applications of this approach to host–parasite interactions have involved domesticated hosts. One extension of this technique that has been used in more natural systems is to look for patterns in numbers of parasites among family groups or in groups with contrasting genotypes identified with molecular tools. Significant differences in parasite intensities among families or genotypes imply genetic differences, if environmental influences associated with exposure can be ruled out. Møller (1990), Chan *et al.* (1994), and Grosholz (1994) used this approach to demonstrate a genetic component to the determination of ectoparasite numbers in swallows, whipworms in humans, and metacercariae in molluscs, respectively. The results of these studies provide evidence for a significant genetic component upon which parasite-mediated selection can act.

Another method used to evaluate heritability exerts artificial selection of known intensity and calculates heritability from the change in response. Although artificial selection has been used for many years to develop strains of animals resistant to parasites (reviews in Bishop *et al.*, 2011), only rarely have heritabilities been calculated. When individual mice were infected with known doses of the nematode *Heligmosomoides polygyrus*, heritability for resistance (calculated from artificial selection of hosts that produced high and low numbers of parasite eggs) was about 50% (Sitepu & Dobson, 1982). Heritability of this magnitude explains how artificial selection for 'high' and 'low' responding mice can be achieved within one or two generations. Similarly high estimates of heritability for resistance occurred in the myxoma–rabbit interaction in Australia (Kerr, 2012) and for snails infected with *S. mansoni* (Webster & Woolhouse, 1998). These results indicate that when selection pressure imposed by parasites is high, directional responses can be extremely rapid.

The previous examples have demonstrated, or inferred, directional selection for increased responsiveness. In addition, there is convincing evidence for strong balancing selection for heterogeneity of the immune response. One focus has been to determine whether balancing selection imposed by parasites is responsible for maintaining the high diversity characteristic of MHC genes. Recall from our coverage in Chapter 2 that the central feature of the MHC family of genes is the high genetic diversity of the antigen recognition sites located on their surface. It is variation in these sites that is responsible for binding an infinite range of foreign antigens, presenting them to macrophages, and initiating and regulating the complex adaptive immune response that is specific to that antigen (see Chapter 2). For the human MHC, hundreds of alleles have been described for at least three HLA loci, and nucleotide diversity can be up to two orders of magnitude higher than the genome-wide average (Garrigan & Hedrick, 2003). Similarly high levels of diversity are characteristic of MHC loci in non-model hosts (review in Piertney & Oliver, 2006). Other characteristics of the MHC are significantly lower levels of homozygosity than expected by neutral models, non-random association between alleles at different loci, and, in some isolated host populations, an excess of heterozygotes over those expected by Hardy–Weinberg proportions. Each of these patterns points to balancing selection for diversity of response. "The logical conclusion is that the selective agent is the bearer of the antigen against which the immune response is directed – the parasites" (Klein, 1991).

The notion that natural selection maintains high diversity at MHC loci is intuitively obvious. Individual hosts possessing high sequence variation at MHC loci should recognize and bind more antigens than those with lower sequence variation, thereby supporting immune competency to a wider range of parasites.

While the results from model species of host tend to support this prediction, the picture emerging from studies on non-model species is more complex (Little, 2002; Piertney & Oliver, 2006). Two contrasting examples emphasize this complexity. In sockeye salmon, between-population variation was much higher at the MHC-*B1* locus than it was for neutral genetic markers and there was an excess of *B1* heterozygotes in about half of the populations sampled (Miller *et al.*, 2001). These results likely reflect spatial variation in the risk of exposure of fish to a range of pathogens and parasites. In contrast, in populations of bighorn sheep, patterns of genetic variation at neutral microsatellite loci were higher than for MHC loci, and neither loci demonstrated deviations from Hardy-Weinberg expectations (Boyce *et al.*, 1997). The authors concluded that despite the occurrence of severe lungworm-induced mortality in some of these bighorn populations, a signature of parasite-mediated selection on MHC loci was absent. This range of outcomes is common for studies involving non-model hosts (Piertney & Oliver, 2006), reflecting the context-dependent nature of parasite-mediated selection and variation in the timing of selection episodes imposed on MHC loci.

The wide variation in outcome observed in natural systems likely also reflects the relative importance of the mechanisms that determine MHC diversity within particular systems. As we have discussed, some form of balancing selection such as overdominant selection (heterozygote advantage) may be responsible for MHC polymorphism due to the advantage that diversity provides for an individual to bind diverse antigens. In contrast to overdominant selection, frequency-dependent selection can occur when rare or new alleles are favored during periods when hosts are exposed to new species or strains of parasite (Slade & McCallum, 1992). If the new allele is less likely to be infected than a common allele, it will increase in frequency. As it becomes more frequent, selection will favor new parasite alleles that are not recognized by the new allele, and so on. In this way, MHC diversity is enhanced when yet another allele is favored that can recognize the new parasite allele. In contrast to these two mechanisms of parasite-mediated selection, MHC polymorphism can also be enhanced if individuals choose their mates on the basis of the diversity of their MHC genes (Penn & Potts, 1999). In mice, mating preferences are dependent on the genetic makeup of the MHC of the mating pair, favoring matings that produce progeny with heterozygous MHC. Indeed, products of the MHC have been identified in urine, conferring specific odors to potential mates. Mating is based on the recognition of olfactory signals that provide specific information about the MHC genotypes, favoring matings that increase polymorphism and avoid genome-wide inbreeding. In practice, distinguishing among these alternative mechanisms is difficult, particularly in natural systems where it is possible that all three play a role in determining overall patterns of MHC diversity.

16.2.3 Constraints

Environmental complexity and variation impose a significant constraint on the rate of evolution. Parasite-mediated natural selection exerted by avian malaria on the native birds of Hawaii differs markedly between islands, between high versus low elevations, and in habitats with contrasting moisture regimes (see Box 15.3). The pressure exerted on red grouse by *Trichostrongylus tenuis* is highly variable in space and time and is also affected by the availability of moisture (see Chapter 15). As we saw in Chapter 15, episodic crashes in population sizes of small-bodied freshwater fishes infected with larval *Ligula intestinalis* and *Schistocephalus solidus* are highly unpredictable, even among adjacent lakes and consecutive years. In Chapter 14, GIS-based risk maps show that individual sunbirds vary by orders of magnitude in their risk of exposure to *Plasmodium* spp. or *Trypanosoma* spp. in west-central Africa, depending on humidity and precipitation (see Color plate 6). The numbers and kinds of helminths in American avocets differ between lakes, depending on whether they are sympatric with lesser scaup and their helminths (see Chapter 13). Most well-studied natural systems will show similar spatial and temporal variation, likely leading to highly localized and temporally restricted patterns of parasite-mediated selection (Thompson, 1994).

Even in years and locations when parasite transmission rates are high, the strength of parasite-mediated selection will be constrained when rates of exposure are determined primarily by stochastic and unpredictable factors. The results of studies involving plant–insect interactions show little evidence for directional selection in cases where many of the factors determining the development of epidemics are stochastic (Burdon *et al.*, 1989). For humans exposed to geohelminths such as *Trichuris*, *Ascaris*, and *Necator*, factors such as gender, family, and body size often play minor roles in determining worm numbers (Chan *et al.*, 1994). Instead, high worm counts are attributed to chance encounters with infective stages, often mediated by human hygiene behavior (see Chapter 12). Similarly, Scott (2006) showed that when two strains of mice are exposed to the nematode *H. polygyrus* in arenas, the characteristic differences between the two strains in their genetically determined immunological defenses disappear under most exposure conditions. In these systems, the genetically based differences in resistance that could support parasite-mediated selection were masked by environmental conditions that determined exposure. In systems where chance exposure plays a primary role in determining parasite intensity, especially when it overwhelms genetic factors, there will be little opportunity for sustained selection on a particular trait or group of traits.

Strong directional selection for immunological resistance or other defense traits will be constrained if the expression of the traits comes at a cost to host fitness. We discussed the nature and extent of these costs in Chapter 2. Strong evidence comes from the results of artificial selection experiments, whereby reproductive rates of resistant mice, snails, and insects demonstrate precipitous declines in reproduction following one or two generations of selection. In addition, the energetic costs of mounting and maintaining functional immunological defenses are substantial, and typically must be traded off with the expression of other life history characteristics such as productivity, growth, and life span. It is certainly conceivable that in any one host–parasite interaction, directional selection for resistance will be constrained by the combined costs of resistance, immunopathology, and energy expenditure acting in concert (see Chapter 2).

A final cost of resistance may be that the specific mechanism that confers resistance to one parasite may increase susceptibility to another parasite. The elevated production of new RBCs in mice, which confers resistance to one isolate of malaria, but susceptibility to another (Sayles & Wassom, 1988), was covered earlier. Infections of snails with larval echinostomes are known to influence their susceptibility to schistosomes. Helminth infections that disrupt the balance of Th1 and Th2 cell types (see Chapter 2) that, in turn, impact susceptibility to other infections, provide another example.

In addition to the trade-offs mentioned above, there may be a trade-off between the evolution of resistance and evolution for tolerance to parasites (review in Råberg et al., 2009). This would be analogous to the possible trade-offs that exist between the evolution for defense against herbivory and that for tolerance to herbivory in plant–insect interactions. Thus, an adaptation that allowed a host to tolerate the effects of a parasite would reduce selection for resistance to that parasite, particularly if resistance is costly. Råberg et al. (2007) tested this idea using several strains of mice infected with different strains of rodent malaria. Their results showed that genetic variation in resistance negatively correlated with genetic variation in tolerance. These results provide initial evidence that parasite-mediated selection for resistance may come at a cost to selection for tolerance, and vice versa. More generally, the results demonstrating genetic variation for both resistance and tolerance means that parasite-mediated selection can take a range of evolutionary paths.

These constraints may account for the absence of strong evidence for sustained evolution of resistance in natural host–parasite interactions. This is despite the evidence that indicates sufficient heritability in a wide range of traits exists to allow such directional evolution. In contrast, certain newly established interactions in which parasites can impose an initially strong selection pressure provide evidence for a directional evolutionary response, often leading to increased resistance. It is likely that in such systems, increased resistance will reduce the selection pressure exerted by the parasite. When selection by the parasite is roughly equivalent to other selection pressures imposed on the host, we might expect a shift from the evolution of resistance to the evolution of diversity of responses, in addition to evolution of tolerance to the negative consequences of infection. Thus, uncritical acceptance of the assumption that such pressures regularly lead to increasing resistance is unwarranted. The role of parasites in the evolution of their hosts is far more complex.

16.3 Genetic structure of parasite populations

Starting in the early 1990s, parasitologists began applying the tools of molecular biology to study the genetic structure of parasite infra- and component populations (reviews in Criscione et al., 2005; Gorton et al., 2012). This early effort coincided with a period when evolutionary biologists, geneticists, and ecologists were applying the theory and analyses of population genetics data to provide insights on microevolutionary and macroevolutionary phenomena. In Chapter 14, we showed how population genetics data could provide information regarding phylogeographical patterns of parasite distribution at various spatial and temporal scales. Such data can also provide population-level information regarding parasite mating systems, dispersal, and transmission dynamics. Indeed, for many host–parasite interactions, this information is often unavailable by any other means. From a microevolutionary perspective, information on population genetic structure is needed to estimate the evolutionary potential of populations of parasites and thus to understand the evolutionary dynamics of speciation, local adaptation, and evolution of resistance to immunity and to anti-parasite drugs. From a conservation perspective, population genetics analyses are central to understanding ecological and evolutionary responses of threatened host populations to parasites, and they are one of the only tools available to determine invasion pathways of introduced parasites.

The genetic structure of a population can be defined as the distribution of genetic variation among individuals sampled over different spatial scales (Anderson et al., 1998). Thus, it is the nature of this structure that provides the basis for the microevolutionary phenomena that we covered in the previous section. At the microevolutionary level, genetic drift and gene flow are the key processes driving the genetic variability of populations. Genetic drift is particularly important for evolutionary dynamics within small populations. The extent of genetic drift in a deme is determined only by the number of individuals contributing genes to the next generation (the effective population size, N_e) and not by the total number of individuals in the deme. With a small effective population size, the heterozygosity of the population will be reduced due to inbreeding, and allelic frequencies will change randomly due to genetic drift. Gene flow, the movement of genes within and between local populations, occurs via migration of individuals among these populations. The amount of gene flow among local populations, together with other population parameters, affects the rate of evolutionary change of these populations. Low gene flow among local populations increases their chance of evolving independently. High levels of gene flow, on the other hand, tend to homogenize local populations into one evolutionary unit. Price (1980) emphasized the former perspective by arguing that, because parasites are adapted to exploit small and discontinuous environments, i.e., hosts, that occur within a matrix of inhospitable habitats, they should have small and homogeneous populations with little gene flow between them. If this is the case, parasite infra- and component populations will depart significantly from **panmixia**, whereby all members of the population have an equal chance of mating with any other individual. Understanding the extent to which this prediction holds within various host–

parasite combinations continues to be a key focus for parasite population geneticists (review in Gorton et al., 2012).

In this section, we synthesize studies that illustrate the variation that exists in population genetic structures of selected parasites, and the underlying causative factors. Molecular approaches using PCR and direct sequencing of DNA polymorphisms with techniques such as DNA fingerprinting of microsatellite regions and randomly amplified polymorphic DNA (RAPD), together with whole genome approaches, e.g., amplified fragment length polymorphisms (AFLP), are contributing enormously to the study of genetic variation at the DNA level. Methodological details regarding the application of these tools to host–parasite interactions are reviewed in Lefèvre et al. (2007). Genetic variation in populations is commonly quantified using the proportion of polymorphic loci in a population (P) and heterozygosity (H), the average frequency of heterozygous individuals per locus. Polymorphism and heterozygosity can also be estimated at the nucleotide level as the proportion of polymorphic nucleotide sites, and as the number of nucleotide substitutions per site. It is the nature of these polymorphisms that provides key information on features such as genetic relatedness, reproductive mode, and gene flow.

16.3.1 Genetic variability of parasites

Most parasite populations have levels of genetic variation (measured as P and H) that are similar to species of free-living invertebrates. Thus, not surprisingly, the parasitic life style is not associated with contrasting patterns of genetic variation relative to free-living animals. However, patterns of genetic variation span an enormous range, even within parasite taxa. Heterozygosity estimates for 22 species of ascaridoid nematode ranged from the very high end of observed values for other parasites and invertebrates (H = about 0.3) to virtually undetectable (Fig. 16.2). For these nematodes, very little of this variation could be explained by features such as life cycle variation (direct versus indirect, vectored

Table 16.1 **Examples of ecological and natural history factors that may influence the population genetic structure of animal parasites**

Factors increasing genetic structure	Factors reducing genetic structure
Sedentary definitive host or extreme morbidity of all infected hosts	Highly vagile hosts (definitive, intermediate, paratenic) or vectors
Life cycle includes a large number of specific obligate hosts	Persistent (long-lived) life cycle stages in environment or definitive host
Suitable parasite niches patchily distributed in space or time	Low definitive host specificity/many reservoir hosts
Small effective size for parasite population	Uniform distribution of parasites among hosts
Parasite predominantly self-fertilizing	Life history with frequent metapopulation extinction followed by reestablishment
Physical contact between definitive hosts required for transmission	

(Modified from Nadler, 1995, with permission, *Journal of Parasitology*, **81**, 395–403.)

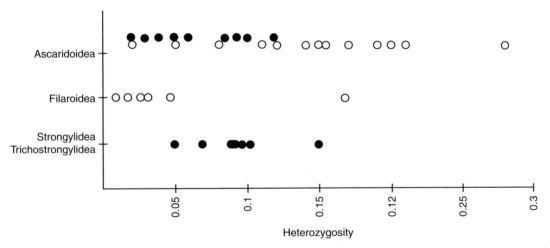

Fig. 16.2 Heterozygosity of nematode species with direct (●) and indirect (○) life cycles. Each point represents one nematode species. In those species in which heterozygosity has been measured several times, the point represents the average value. (Modified from Anderson *et al.*, 1998, with permission, *Advances in Parasitology*, **41**, 219–283, Elsevier.)

versus non-vectored), body size, or host species (Anderson *et al.*, 1998). Instead, a large number of ecological and natural history factors likely interact to determine observed pattern of population genetic structure of animal parasites (Table 16.1) – some acting to decrease variation, others acting to increase it (Nadler, 1995).

Only a few of the factors hypothesized to determine variation in parasite genetic structure (Table 16.1) have been directly assessed. The role of parasite life cycle variation and reproductive strategies has received some attention. Anderson *et al.* (1998) compared heterozygosity in nematodes with direct and indirect life cycles in four families of nematodes (Fig. 16.2). The direct life cycle strongylids and trichostrongylids showed levels of variation overlapping both direct and indirect life cycle ascarids. Within the ascaridoids, species with indirect life cycles tend to have higher heterozygosity levels than species with direct life cycles, indicating a possible linkage between parasite heterozygosity and environmental heterogeneity in terms of life-cycle complexity.

Parasites that incorporate an obligate asexual stage into their life cycle may demonstrate contrasting patterns of genetic structure (e.g., Prugnolle *et al.*, 2005). Studies on digenean trematodes and some taeniid cestodes, e.g., *Echinococcus* spp., support this idea. Results of population genetic studies on the giant liver fluke, *Fascioloides magna*, in populations of white-tailed deer, *Odocoileus virginianus*, showed low differentiation within local populations, but high differentiation between isolated populations, consistent with an isolation-by-distance effect (Mulvey *et al.*, 1991). Further, there was a deficiency of heterozygous individuals within localities. These results suggest extensive selfing of worms in deer, or the ingestion of single genotypes of metacercariae arising from asexual

reproduction in snails. Similar mechanisms may explain the relatively low levels of genetic variation found in *E. granulosus* in mainland Australia and Tasmania (Lymbery *et al.*, 1997). Most genetic variation in *E. granulosus* occurs within local populations, with little evidence of genetic variation between populations. Although self-fertilization is ubiquitous in this genus, it appears that genetic differentiation between populations is prevented by the dispersal of the parasite over wide geographical areas by multiple species of intermediate and definitive hosts, and probably by occasional outcrossing of the worms.

Population genetics theory indicates that effective population size (N_e) can have a strong effect on levels of genetic diversity in populations and on the fate of alleles. Tests of this hypothesis are difficult for parasites, particularly those that have complex life cycles with dormant stages that can persist in intermediate hosts or as larvae/eggs. In an initial test, Criscione *et al.* (2005) showed that mtDNA diversity of several species of direct life cycle, parasitic nematodes was positively correlated with mean intensities in their mammalian host populations (Fig. 16.3). This is an important result for this particular group of parasites because it identifies those species, e.g., *Ostertagia ostertagi*, most likely to evolve drug resistance within local populations of domestic animals. Additional comparative

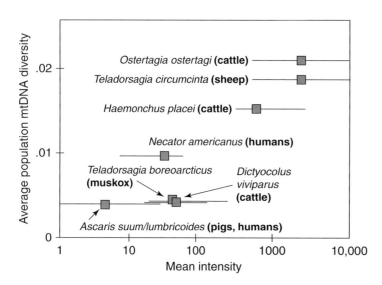

Fig. 16.3 Average population mtDNA diversity (average number of substitutions per site) vs. mean intensities for 8 species of nematodes. Host species are in brackets. Horizontal bars give the range of typical intensities for that species. mtDNA diversities are based on either whole-molecule RFLP, or on *Cox1* or *Nad4* gene sequences. (Figure courtesy of Charles Criscione; modified from Criscione *et al.*, 2005, with permission, *Molecular Ecology*, 14, 2247–2257, John Wiley & Sons.)

tests that incorporate a range of parasite taxa and a range of ecological conditions are needed to establish general patterns.

16.3.2 Genetic structure of parasite infra- and component populations

The results of a large number of studies are consistent with the idea that parasite population structure is closely tied to the nature of parasite transmission. Criscione & Blouin (2004) hypothesized that panmixia will be promoted in habitats where parasite offspring mix extensively during the transmission process. Extensive mixing is likely to result from factors such as host behavior, e.g., migration and territoriality, parasite life cycle patterns, e.g., the use of vectors or highly mobile intermediate hosts, and the dispersal ability of parasites (motile larvae versus eggs) and their hosts. The nature of the transmission environment is another key feature. Thus, many aquatic parasites demonstrate panmixia, probably because the aquatic environment favors the mixing of free-living stages and infected hosts. Complete panmixia was observed for adults of *Plagioporus shawi*, a trematode whose intermediate and definitive hosts are confined to localized aquatic habitats (Criscione & Blouin, 2004). Similarly, genetic structure was absent among infrapopulations of cercariae of *Schistosoma mansoni* sampled from five populations of snails across an 1800-km² region of Lake Victoria in Africa (Steinauer *et al.*, 2009). Here, despite the distances involved, the continuity of the habitat, together with the dispersal abilities of cercariae and snails, place constraints on the development of local transmission foci that could promote genetic structuring. Panmixia was also observed between populations of ascaridoid nematodes *Contracaecum* and *Pseudoterranova* (see Chapter 8) sampled from seals across their extensive range in the Atlantic Ocean (e.g., Nascetti *et al.*, 1993). The authors hypothesized that the migration of seal definitive hosts, in combination with the transport of larvae in intermediate and paratenic hosts, facilitates gene flow over extensive geographical areas.

At the other extreme, parasites that have direct life cycles confined to a single or a few hosts will have restricted opportunities for dispersal and gene flow. Studies of chewing lice, *Geomydoecus actuosi*, from pocket gophers in the USA, and lice, *Heterodoxus octoseriatus*, from rock-wallabies in Australia, indicate that lice populations are highly structured with high differentiation among populations, a pattern consistent with Price's (1980) earlier ideas (Nadler *et al.*, 1990; Barker *et al.*, 1991). Both groups of lice showed low polymorphism and heterozygosity probably as a consequence of population bottlenecks, **founder effects** and inbreeding. In both studies, the social organization of the host and the fragmented habitat they inhabit reduces contact and limits transmission of the parasite that can only occur by direct contact between the hosts. These examples of strong genetic structuring involving lice are probably exceptions reflecting the life histories of these lice on their social hosts.

Reports of patterns of genetic structuring between complete panmixia and high differentiation are common in the literature. Studies of genetic variability in the nematode *Ascaris suum* provided one of the first insights into the nature of variation in allele frequencies on a large geographical scale. Nadler *et al.* (1995) used both protein electrophoresis and RAPD markers to assess the population genetic structure of this nematode sampled from pigs at three spatial scales – individual hosts, individual farms, and geographical regions (Michigan, Illinois, Indiana, USA). They showed that 91% of the total genetic diversity was found within infrapopulations and 9% was partitioned among infrapopulations within a farm. On a larger geographical scale, genetic diversity of all pooled infrapopulations within a region accounted for 92% of the diversity and only 8% was partitioned among the geographical areas. The data showed excess homozygosity for infrapopulations of *A. suum*, as indicated by high inbreeding coefficients, and significant genetic isolation among farms within geographical regions. On a geographical scale, the results can be explained by low gene flow between close populations and by genetic isolation among the infrapopulations within geographical regions. It is likely that the genetic differentiation among infrapopulations was promoted by genetic drift as a consequence of their small effective population size.

Studies of variation in mitochondrial DNA sequences in populations from different geographical regions of the trichostrongylid nematodes *Ostertagia ostertagi* and *Haemonchus placei* (from cattle), *H. contortus* and *Teladorsagia circumcincta* (from sheep), and *Mazamastrongylus odocoilei* (from deer), revealed a different kind of genetic structure (Blouin *et al.*, 1995). The parasites of sheep and cattle had high within-population diversities, with a pattern of high gene flow and low genetic differentiation among populations. These results are likely determined by anthropogenic factors, such as the long-distance transport of domestic animals and/or the spread of contaminated manure. The parasites of deer also had high within-population diversity, but genetic differentiation among populations was high, with populations more subdivided than those of the domestic animals. The pattern observed in the deer parasite was one of isolation-by-distance among the populations, meaning that gene flow in the parasites is a function of host movement. The genetic structure of infrapopulations of ticks *Ixodes uriae* on seabirds is also strongly dependent on the movement patterns of their hosts. McCoy *et al.* (2003) showed that the population genetic structure of the tick was panmictic on kittiwakes, but not on puffins. Moreover, the genetic similarity between pairs of tick component populations on kittiwakes, but not puffins, increased as the distance between them increased. It turns out that puffins make more frequent and longer range visits to widely scattered colonies, whereas kittiwakes are more faithful to the cliffs where they were raised. These results support the findings on the nematodes of domestic animals, indicating that host vagility plays a key role in determining gene flow, and thus population differentiation, between parasite populations.

The relationship between high parasite dispersal or host vagility and low genetic structure of parasite populations is not restricted to helminths. Complete panmixia was observed for infrapopulations of fleas sampled from different colonies of black-tailed prairie dog (Jones & Britten, 2010). In this case, gene flow between colonies likely occurred via the other mammals that these fleas can infect. Similarly, different populations of ticks such as *Amblyomma americanum*, *Ornithodoros erraticus*, and *O. senrai* in the USA revealed a pattern with low levels of genetic variability within, as well as between, different geographical populations (Hilburn & Sattler, 1986a; 1986b). In the case of *A. americanum*, a small degree of genetic structuring was observed between nine geographical populations. Since ticks are relatively sedentary when they are not on a host, the rate of gene exchange will depend on host vagility. The low host specificity and the number of hosts in the life cycle also increase the amount of gene flow between geographical populations. For *A. americanum*, each tick attaches and feeds on three hosts during its lifetime but, each time the ticks molt or lay eggs, they leave the host and remain on the ground. This behavior likely enhances its chances of infecting a different individual or host species each time it reattaches, thereby increasing its chances of dispersal and gene flow.

Our coverage in this section emphasizes the variation and complexity of parasite population genetics. Although the identification of general patterns remains elusive, early results are consistent relative to the roles of parasite (and/or host) dispersal, mating systems, and isolation-by-distance in determining the rates of gene flow and drift that underlie genetic structure. Indeed, given the variation we typically observe in features such as parasite dispersal, life cycles, and life histories, we should not be surprised that Price's (1980) early predictions only describe one end on a broad continuum of parasite population genetic structure. More comparative studies, especially those involving genes with a known function (e.g., MHC, sickle-cell anemia) within varied host–parasite systems, will help further unravel the many ecological determinants and consequences of parasite population genetic structure.

16.4　Introduction to host–parasite coevolution

Our earlier coverage in this chapter demonstrated that parasites can mediate natural selection on various genetically determined host traits, and that such selection can lead to evolutionary responses. It is equally clear that host traits such as immunity, tolerance, and avoidance have a strong impact on the evolution of various parasite traits. Thus, parasites and hosts can have an evolutionary impact on each other. This process of reciprocal, adaptive genetic change in two or more species is known as **coevolution** (Thompson, 1994). It occurs at many biological levels, from molecular interactions between amino acids in a protein, to interactions between males and females, to covarying traits between species in a population, e.g. pollinators and flowers, and some predators and prey. In the special case involving hosts and their parasites, **antagonistic coevolution** refers to reciprocal evolution between potentially matching traits, such as those associated with host defense on the one hand, and parasite infectivity on the other (Thompson, 1994). The process has been implicated to play a central role in the evolution of virulence, the maintenance of host sexual reproduction, and the determination of between-population differentiation of hosts and parasites (reviews in Thompson, 1994; Woolhouse *et al.*, 2002; Schmid-Hempel, 2011). The process is also important as a mechanism maintaining genetic polymorphisms of key host and parasite loci.

Modern approaches to the study of coevolution in host–parasite interactions trace back to theoretical treatments by Roy Anderson and Robert May in the early 1980s (Anderson & May, 1982). In an important series of papers, they dispelled the long-held notion held especially by medical and veterinary workers that 'well-adapted' parasites inevitably evolve towards commensalism with their hosts. Indeed, parasitologists were among the last to 'let go' of this obligate, unidirectional view of antagonistic coevolution (Thompson, 1994). This change in viewpoint stemmed from Anderson and May's clear demonstration of the complex linkage between parasite fitness and parasite virulence. Since the two features are often positively associated in natural host–parasite interactions, evolution towards commensalism can be expected to be rare. Anderson & May (1982) concluded that the complex interplay between virulence and reproduction (specifically, transmission) "leaves room for many

coevolutionary pathways to be followed, with many endpoints." Variation in outcome is a hallmark of modern studies of host–parasite coevolution.

Interactions involving agricultural crops and their pathogens, and bacteria and their pathogens, have played a prominent role in the development of the theory of host–parasite coevolution. Central to this theory are empirical tests of contrasting models of coevolution, such as 'gene-for-gene,' 'matching allele,' and 'matching genotype' (review in Thompson, 1994). Unifying and consistent results of such tests indicate that the outcomes of antagonistic coevolution vary extensively, that they depend strongly on the combination of host and parasite genotypes involved, and that genetic changes in one antagonist can cause genetic changes in the other. Thus, in these tractable and simple host–parasite interactions, the key requirements for antagonistic coevolution are met. The extent to which such interactions occur in more complex systems involving parasites of animals (especially vertebrates) is one of the most challenging and fascinating disciplines in modern evolutionary parasitology (Woolhouse *et al.*, 2002; Schmid-Hempel, 2011).

The arms race and Red Queen paradigms of antagonistic coevolution provide the traditional framework for ecologists and evolutionary biologists to assess the complex coevolutionary interplay between hosts and parasites. Theoretically, arms race dynamics occur when new alleles arise by mutation in a host population, conferring an advantage relative to defense against exposure or parasite-induced damage. This allele then sweeps through the population. A new allele then arises by mutation in the parasite population, conferring an advantage in terms of infectivity or rate of host exploitation. An arms race occurs when this process repeats, leading to the accumulation of 'improvements' in both host and parasite populations (Woolhouse *et al.*, 2002). In contrast, 'Red Queen' dynamics occur when natural selection acts upon existing genetic variation in host and parasite populations. In this case, coevolution is non-directional. Here, host genotypes, e.g., for resistance, are at an advantage when they are rare in a host population, but become increasingly 'visible' by parasite genotypes, e.g., for infectivity, as they become more common. So long as parasite genotypes can track host genotypes, Red Queen dynamics ("running as fast as you can to stay in the same place") can occur indefinitely.

An extension to the Red Queen paradigm predicts a source of new host genotypes that can potentially combat exploitation by diverse parasites. The **Red Queen hypothesis** predicts that in a dynamic, coevolutionary arms race with parasites (especially highly virulent ones), sex and recombination are favored in hosts because they produce rare genotypes, which are expected to have a greater chance of escaping coevolving parasites. The parasites can then track these rare genotypes as they become common, initiating a negative frequency-dependent selection process (Hamilton *et al.*, 1990). Thus, sexually produced offspring should produce new genotypes that confer resistance to coevolving parasites. Offspring produced asexually, however, will have the same resistance genes as their parents and may suffer increased exploitation by parasites. Specifically, the Red Queen hypothesis predicts that parasites will prevent clones with a reproductive advantage from eliminating a diverse sexual population (Jokela *et al.*, 2009). The hypothesis also predicts continued reciprocal evolutionary change of both host and parasite, i.e., antagonistic coevolution, under conditions of constancy in the physical environment.

16.4.1 Antagonistic coevolution in the wild

Direct evidence for antagonistic coevolution comes from studies that monitor reciprocal changes in selected trait combinations over time. Field-based assessments of this sort are understandably rare, especially in host–macroparasite interactions. Even rarer are the parallel laboratory tests involving model strains of parasite and host that can 'freeze' one of the antagonists so that the responses of the other can be isolated. One example involves the European rabbit–myxoma system that we introduced earlier in this chapter. Recall that rabbits evolved an immunological response to infection with a lethal strain of the virus one or two generations following its introduction to

eastern Australia (Kerr, 2012). Although the precise nature of the host responses is unknown, they led to a decrease in host mortality from 99% to less than 30% within 7 years. The results of experiments involving the inoculation of laboratory rabbits with a standardized strain of virus showed that the drop in mortality was due to a host response, and not to changes in the virus.

The results of experiments involving rabbits infected with different strains of virus confirmed the rapid evolution of decreased virulence. These laboratory results supported observations made in the field (Kerr, 2012). During initial attempts to rid Australia of their 'gray carpet' in the summer and fall of 1951 and 1952, the introduced strain killed an estimated 99.8% of rabbits. Just a few months later (following the first winter), mortality had dropped to 90% and almost half of over 800 sampled rabbits were seropositive. Thus, many exposed rabbits had recovered. This result is best explained as due to strong selection imposed by the initial overwintering period for a form of virus with reduced virulence. Results from laboratory trials showed that the introduced strain could still kill over 99% of exposed hosts. Yet, each time the lethal strain was introduced into rabbit populations, it disappeared within a few years, to be replaced by a strain of intermediate virulence that killed a maximum of about 70% of hosts. Natural selection likely favored the evolution of less virulent forms because mosquito-to-rabbit transmission was higher in hosts that lived longer. Taken together, temporal changes in host and parasite responses of this nature are consistent with antagonistic coevolution (review in Kerr, 2012), although the extent to which the host and parasite responses were mutually induced remains a key knowledge gap.

Evidence in support of Red Queen dynamics provides additional support for antagonistic coevolution. Curtis Lively and his students have reasoned that if sexual reproduction is an adaptation against coevolving parasites, then it should be most common where parasites are most common (review in Lively, 2010). Conversely, asexual forms of reproduction such as parthenogenesis should be common where parasites are rare. These predictions can be tested by careful selection of appropriate model systems. In New Zealand, populations of the dioecious freshwater snail *Potamopyrgus antipodarum* contain both sexual and parthenogenetic individuals. The snail is also host to several species of larval trematodes, most of which are castrators. This system has become a model for empirical and theoretical tests of key predictions of the Red Queen hypothesis (Box 16.2). First, results from a series of field surveys in New Zealand lakes and

---

Box 16.2 | **Parasites, sex, and the Red Queen: castrating trematodes maintain sexual reproduction in New Zealand snails**

The Red Queen hypothesis predicts that sex persists because natural selection favors genetically diverse hosts that are engaged in arms races with coevolving parasites. One host–parasite system that has played a pivotal role in addressing several key predictions of the hypothesis involves a tiny species of freshwater snail, *Potamopyrgus antipodarum*, in New Zealand and its suite of larval trematodes (Lively, 1987; 1992; review in Lively, 2010). Most populations of this dioecious snail contain two kinds of females, i.e., obligate sexual types that produce a mixture of male and female offspring, and parthenogenetic types that produce daughters that are clones of themselves. It turns out that the proportion of sexual versus asexual females varies from population to population, as does the prevalence of larval trematodes. This scenario provides a powerful setting for empirical and field-based tests of the Red Queen hypothesis.

The snail is the first intermediate host for at least 12 species of larval trematodes, most of which are obligate castrators. The most prevalent species within lake populations is a species of *Microphallus*. As we have seen elsewhere in this text, parasite-mediated selection by castrating larval trematodes can be exceptionally strong on snail defense and life history traits. Thus, in the *Potamopyrgus*–trematode system, support for the Red Queen hypothesis would arise from results showing that sexual snails are more common in populations experiencing higher infection rates with trematodes, especially by the most common local parasite, *Microphallus* sp. In an initial test, Lively (1992) determined the sex and infection status of snails sampled from over 60 lakes. His results were consistent with the hypothesis, showing that males were more frequent in snail populations where the risk of infection with trematodes was highest. Thus, since males can only be produced by sexual females, the results pointed to an association between host sexual reproduction and risk of infection. The association between the frequency of sexual reproduction and trematodes has been shown in several studies since, including snails collected from widely scattered stream populations (Fig. 16.4). Indeed, over the geographical range of *Potamopyrgus* in New Zealand, the best predictor of the frequency of sexual reproduction within snail populations is the most common species (usually *Microphallus* sp.) present in a particular habitat (King *et al.*, 2011).

The Red Queen hypothesis also predicts that parasite-induced selection is frequency dependent. Thus, for Red Queen dynamics to operate as depicted in Fig. 16.4, parasites, especially common ones, must track host genotypes as they wax and wane over time. Supportive evidence in the

Box 16.2 | (continued)

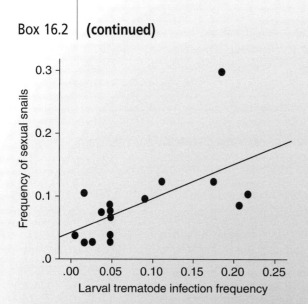

Fig. 16.4 Scatterplot showing significant linear relationship between proportions of snails infected with larval trematodes and the frequency of sexual snails *Potamopyrgus antipodarum* for 17 stream populations in New Zealand. The frequency of sexual snails increased with the prevalence of larval trematodes. (Figure courtesy of Kayla King; Modified from King *et al.*, 2011, with permission, *Evolution*, 65, 1474–1481, John Wiley & Sons.)

Potamopyrgus–trematode interaction would arise if *Microphallus* sp., for example, had higher rates of infection within snail clones that were rare in populations (and vice versa), then increased their rate of infection as clones increased in frequency over time. Supportive evidence for this prediction came from a study by Dybdahl & Lively (1998), in which genetic markers were used to identify clonal lineages of *Potamopyrgus* within one lake. Because the snails in the lake only reproduced asexually, the genetic markers remain unchanged within each snail clone. The authors identified rare and common clones of snails and tracked their frequencies, and their parasites, over a 5-year period.

It turned out that most snail clones oscillated over time in a manner consistent with frequency-dependent selection, with cycling of host clones and time-lagged correlated responses by the parasites (Dybdahl & Lively, 1998). One of the infected common clones was driven down and replaced in the lake by what was, initially, a rare clone. This clone was also driven down over time, and replaced. These clones were then experimentally exposed to the common parasite. In the laboratory, the host clones that had been rare during the previous 5 years were significantly less infected by the parasite when compared to the common clones. In a follow-up study, Jokela *et al.*, (2009) showed that the most common snail clones were almost completely replaced within 7–10 years by initially rare clones, whereas sexual snails persisted throughout the study. The common clones also became more susceptible to infection by sympatric trematodes (*Microphallus* sp.) over the course of the study. Further, *Microphallus* sp. appears to be highly locally adapted to snail genotypes within shallow-water regions of a lake where the snails maintained sexual reproduction (King *et al.*, 2009). These shallow-water habitats are probably coevolutionary 'hotspots' since transmission is highest in shallow water where the snails are most likely to be ingested by waterfowl definitive hosts. Thus, parasite-mediated selection appears to act against common host clones, preventing asexual clones from expanding within mixed snail populations, as predicted by the Red Queen hypothesis.

streams consistently showed that sexual reproduction was rare when castrating trematodes were absent, but widespread when they were common. Second, results from molecular fingerprinting studies showed that the most common larval trematodes in a population tracked host genotypes as they became more common. The latter result is especially important in the context of coevolution because it demonstrates that parasite and host genotypes can track each other, and respond to each other over ecological timescales. It is this reciprocal evolution between matching traits that is the crux of antagonistic coevolution (Thompson, 1994).

Additional support for Red Queen dynamics comes from recent studies on carefully selected model systems involving microparasites (review in Schmid-Hempel, 2011). Direct evidence for the 'parasites and sex' prediction comes from a study by Morran *et al.* (2011), who manipulated the mating system of the nematode *Caenorhabditis elegans* to produce populations that were either sexual, asexual (via self-fertilization), or both. When the contrasting populations were exposed to pathogenic bacteria, the asexual populations were driven to extinction, whereas the sexual populations kept pace with the coevolving bacteria. Additional evidence for the 'frequency-dependent selection' prediction comes from a clever experiment involving *Daphnia* and one of its pathogenic bacteria. Decaestecker *et al.* (2007) revived resting *Daphnia* eggs from different depths within pond sediment, together with the spores of one of their bacterial parasites. They estimated that 2 cm of sediment corresponded to about 4 years of coevolutionary time. This scenario permitted a series of reciprocal exposures between parasites and hosts from different time periods. The results showed that the bacteria adapted to their hosts within only a few years. Thus, infectivity of current parasites in current hosts was significantly higher than for past parasites in current hosts and for novel parasites in current hosts. Based upon the results involving several host–parasite isolates, and a theoretical coevolutionary model, the authors concluded that negative frequency-dependent selection consistent with Red Queen dynamics was common within these ponds. These results are similar to those described for the *Potamopyrgus–Microphallus* system and a handful of others involving pathogenic microparasites of animals (review in Schmid-Hempel, 2011).

Results from these experiments provide tantalizing support for the role of antagonistic coevolution in determining the outcomes of host–parasite interactions. However, the lack of comparable manipulative approaches on other model systems makes generalizations regarding the role of long-term coevolutionary dynamics within natural systems very difficult. One key difficulty lies in the detection of *reciprocal* variation in host and parasite phenotypes (and ideally, genotypes) over time. Demonstration of this key requirement is possible with laboratory models, e.g. bacteria/phage, where both arms race and Red Queen dynamics have been consistently demonstrated. However, it remains exceedingly difficult within natural systems, especially for those involving macroparasites of vertebrates, including humans.

The difficulty in detection is compounded further by the highly localized nature of parasite transmission, meaning that coevolutionary dynamics may only be detectable at restricted spatial and temporal scales. Earlier examples in this chapter include variation between adjacent watersheds in the transmission of avian malaria in Hawaii, variation between adjacent regions in myxoma transmission into Australian rabbits, and variation in miracidia transmission into *Potamopyrgus* at different sites within a New Zealand lake. The localized nature of transmission was further emphasized in Chapters 12, 14, and 15. John Thompson recognizes the highly localized nature of intimate interactions in his **geographical mosaic of coevolution hypothesis** that predicts a network of coevolutionary interactions across a landscape (Thompson, 1999). Supportive evidence comes from studies that demonstrate the presence of coevolutionary 'hotspots,' imbedded within a matrix of 'coldspots' where coevolutionary dynamics are infrequent or completely absent (e.g., King *et al.*, 2009).

In a similar fashion, we should expect extensive temporal variation in coevolutionary dynamics. In an experimental study, Hall *et al.* (2011) monitored the expression of host resistance and parasite infectivity phenotypes over multiple generations in bacteria infected with one of their parasitic phages. Their results were consistent with arms race dynamics between these matching traits during the first few generations of the interaction, but not thereafter. In this case, highly resistant host phenotypes and highly infective parasite phenotypes evolved, but they rapidly died out. Thus, arms race dynamics between the two antagonists subsided over time. The authors argue that escalating arms race dynamics cannot continue indefinitely due to the inevitable costs of evolving enhanced resistance or infectivity. These results further emphasize the dynamic and complex nature of antagonistic coevolution, leading to recognition of a continuum of coevolutionary dynamics, with escalation of defense and counter-defense at one extreme, and no change in mean defense and counter-defense at the other. An ongoing challenge is to determine the ecological conditions that lead to the placement of particular interactions along the continuum (Gandon *et al.*, 2008).

16.4.2 Macroevolutionary perspectives

If the microevolutionary processes discussed in the previous section occur over many generations, they should leave a distinctive signature at the macro-evolutionary scale. Parasitologists have a long tradition of using phylogenetic tools to detect patterns of host–parasite associations in shallow and deep evolutionary time. In one of the oldest ideas in evolutionary parasitology, Fahrenholz (1913) reasoned that evolutionary relationships among hosts could be inferred from evolutionary relationships among their parasites (reviews in Klassen, 1992; Page, 2003). Using this scenario, primitive hosts within a taxon have primitive parasites, derived hosts have derived parasites, and host extinction results in parasite extinction. Thus, according to **Fahrenholtz's rule**, speciation in parasites occurs in response to, and in concert with, speciation of their hosts. Such a pattern of strict host–parasite **cospeciation** is recognized as one of several

possible macroevolutionary outcomes arising from long-term coevolutionary dynamics between parasites and their hosts.

Daniel Brooks and Deborah McLennan have played prominent roles in developing ideas and methods for the assessment of how microevolutionary processes of coevolution can determine macroevolutionary patterns (Brooks & McLennan, 2002). In the simplest case, their model of allopatric cospeciation assumes that parasites and hosts share space and energy. When gene flow between two host populations is interrupted by a vicariant event (see Chapter 14), so too is the gene flow of the parasites, resulting in allopatric speciation of both the host and parasite populations. A reconstruction of the host and parasite phylogenies can reveal congruent branching patterns (Fig. 16.5). Here, cospeciation can be viewed as a neutral process involving parasite speciation in or on isolated and diverging hosts. The timing of speciation for the host and the parasite, however, is not necessarily the same, and cospeciation can be synchronous or delayed (Hafner & Nadler, 1990). In synchronous cospeciation, the host and parasite speciate simultaneously and their lineages

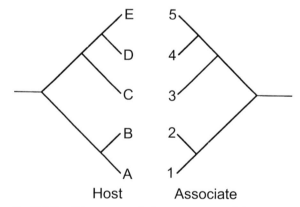

Host   Associate

Fig. 16.5 Possible macroevolutionary outcome of the allopatric model of coevolution. Complete congruence between the phylogeny for the hosts (taxa represented by letters) and the phylogeny for their parasites (taxa represented by numbers) is due to simultaneous cospeciation. (Modified from Brooks & McLennan, 1991, *Phylogeny, Ecology and Behavior: A Research Program in Comparative Biology*, with permission, University of Chicago Press.)

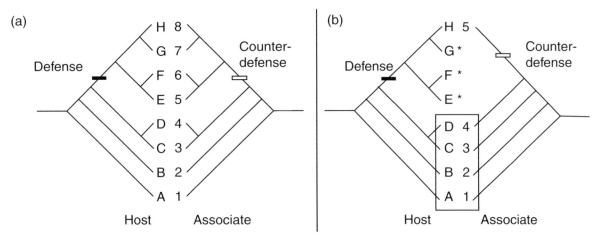

Fig. 16.6 Possible macroevolutionary outcomes of the evolutionary arms race model of coevolution. (A) Host and parasite phylogenies are congruent because traits for defense and counter-defense appear at the same point in the common phylogenies. (B) Host and parasite phylogenies are congruent (boxed area) up to the point at which the defense trait appears; if the origination of a counter-defense lags behind the origination of a defense while the host continues to speciate, the parasites without the counter-defense will not be able to parasitize the new host species (asterisks) until the counter-defense is acquired. Once the counter-defense appears, the host and parasite phylogenies rejoin (host H – parasite 5). (Modified from Brooks & McLennan, 1991, *Phylogeny, Ecology and Behavior: A Research Program in Comparative Biology*, with permission, University of Chicago Press.)

show similar degrees of evolutionary divergence. In delayed cospeciation, one of the members of the association lags behind the other.

The evolutionary arms race model of antagonistic coevolution that we discussed in the previous section can produce congruent or incongruent phylogenies (Fig. 16.6) (Brooks & McLennan, 2002). For example, in systems where the defense and counter-defense mechanisms arise in a short evolutionary time, these traits will probably appear at the same point in the phylogeny of host and parasite, producing roughly congruent phylogenies (Fig. 16.6A). However, if the time frame in which the traits for defense and counter-defense originate is longer than the time between the host speciation events, the macroevolutionary pattern will show the parasite group missing from most members of the host clade that possessed the defense trait (Fig. 16.6B). Partial incongruence between host and parasite phylogenies also can be found when some parasites with the counter-defense colonize relatively more ancestral members of the host clade (Brooks &

McLennan, 2002). Thus, we should expect the macroevolutionary outcomes of antagonistic coevolution between hosts and parasites to be highly variable, with congruent phylogenies indicative of cospeciation at one extreme, to completely incongruent phylogenies indicative of extensive host switching and extinction events occurring at the other.

Empirical evidence confirms this range of outcomes. One of the best known examples involves species of pocket gophers infected with their host-specific species of chewing lice. In one of a series of studies, Hafner et al. (1994) constructed gopher (n=15 species) and lice (n=17 species) phylogenetic trees from sequences of the CO1 mitochondrial gene. Further analyses of these data confirmed that although the host and parasite phylogenetic trees were not perfectly congruent (Fig. 16.7), the degree of similarity was statistically significant and the evolutionary rates of hosts and parasites were strongly correlated (Huelsenbeck et al., 1997). When a restricted section of the phylogenies was analyzed separately (the top five

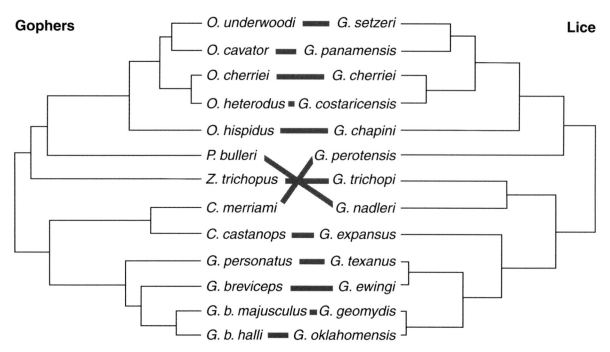

Fig. 16.7 Phylogenetic relationships between 13 species of pocket gopher and their parasitic lice using DNA sequence data from Hafner *et al.* (1994). Lines connecting extant species indicate host–parasite associations. The top five gopher and lice species indicate true cospeciation with synchronous speciation for hosts and parasites. The differences in the topology of the remainder of the tree are probably due to host switching by the lice or persistence of multiple ancestral louse lineages, or both. The pocket gopher species are in the genera *Orthogeomys*, *Pappogeomys*, *Zygogeomys*, *Cratogeomys*, and *Geomys*; *G.b.* stands for *Geomys bursarius*; the species of lice belong to the genus *Geomydoecus*. (Modified from Huelsenbeck & Rannala, 1997, with permission, *Science*, **276**, 227–232, American Association for the Advancement of Science.)

gopher and lice species in Fig. 16.7), the results indicated that these species had cospeciated and that speciation times for hosts and parasites were identical. The differences in the topology of the remainder of the tree were probably due to host switching by the lice. Studies involving species of lice on seabirds (Paterson & Gray, 1997) and on doves and pigeons (Clayton *et al.*, 2003) show similar macroevolutionary patterns consistent with adaptive cospeciation.

Lice provide an ideal model for tests of cospeciation. These permanent ectoparasites tend to be highly host specific, relying on host body temperature and humidity for survival and reproduction (see Chapter 11). Their life cycle is usually completed on one individual host and transmission often occurs from parents to offspring and between mates,

minimizing the chances for interspecific transmission and probably minimizing host switching. Yet, not all host–lice interactions demonstrate strict cospeciation. Weckstein (2004) found incongruent phylogenies of chewing lice on subspecies of toucans collected in South America, such that some closely related species of lice were found on distantly related, but sympatric, species of host. These results were attributed to enhanced opportunities for host switching. First, these social birds nest in tree cavities, where high humidity and temperature likely enhances louse survival during potential periods of inter-host transfer. Second, this taxa of lice is known to have phoretic associations with dipteran insects that fly between their bird hosts, providing a route for lice to disperse between potential host species.

Results of the toucan–lice study parallel the results from an increasing number of studies involving a wide range of host–parasite combinations. Highly incongruent host–parasite phylogenies have been reported for the specialist cestodes of seabirds, the parasites of arctic mammals, *Trichinella* spp. and filariid nematodes in mammals, monogeneans in amphibians and fish, and *Plasmodium* in passerine birds, to list just a few (review in Hoberg & Brooks, 2008). As more host and parasite cophylogenies have become available, the evidence supports the notion that adaptive cospeciation as indicated in Fig. 16.7 is probably the exception, and not the rule. Thus, host switching and parasite colonization are key determinants of the complex macroevolutionary patterns of host–parasite assemblages observed in modern biotas.

This conclusion parallels our summary in Chapter 14 regarding the complex processes that determine phylogeographical patterns of hosts and their parasites. Thus, for many host–parasite associations, the evidence supports the idea that episodic periods of radical environmental change, e.g., creation of land bridges and glaciation, were key drivers of the diversification of parasite assemblages. These periods of geographical expansion enhanced opportunities for colonization via host switching among temporarily sympatric hosts. For parasite assemblages that developed in locations where such environmental upheavals were common, e.g., arctic vertebrates, we might expect little evidence of tight host–parasite cospeciation (see also Chapter 14). In contrast, intervening periods of geographical isolation and stability likely allowed for periods during which coevolution *and* cospeciation could occur. As emphasized by Hoberg & Brooks (2008), it is these cyclical and episodic periods of geological upheaval that provided opportunities for the dual processes of coevolution *and* colonization that determine complex macroevolutionary patterns of hosts and their parasites.

References

Albers, G. A. A., Gray, G. D., Piper, L. R., *et al.* (1987) The genetics of resistance and resilience to *Haemonchus contortus* infection in young Merino sheep. *International Journal for Parasitology*, 17, 1355–1363.

Allison, A. C. (1954) Protection afforded by sickle-cell trait against subtertian malarial infection. *British Medical Journal*, 1, 290–294.

Anderson, R. M. & May, R. M. (1982) Coevolution of hosts and parasites. *Parasitology*, 85, 411–426.

Anderson, T. J. C., Blouin, M. S. & Beech, R. N. (1998) Population biology of parasitic nematodes: applications of genetic markers. *Advances in Parasitology*, 41, 219–283.

Apanius, V., Penn, D., Slev, P. R., *et al.* (1997) The nature of selection on the major histocompatibility complex. *Critical Reviews in Immunology*, 17, 179–224.

Barker, S. C., Briscoe, D. A., Close, R. L. & Dallas, P. (1991) Genetic variation in the *Heterodoxus octoseriatus* group (Phthiraptera): a test of Price's model of parasite evolution. *International Journal for Parasitology*, 21, 555–563.

Bernatchez, L. & Landry, C. (2003) MHC studies in nonmodel vertebrates: what have we learned about natural selection in 15 years? *Journal of Evolutionary Biology*, 16, 363–377.

Bishop, S. C., Axford, R. F. E., Nicholas, F. W. & Owen, J. B. (2011) *Breeding for Disease Resistance in Farm Animals*. 3rd edition. Wallingford: CAB International.

Blouin, M. S., Yowell, C. A., Courtney, C. H. & Dame, J. B. (1995) Host movement and the genetic structure of populations of parasitic nematodes. *Genetics*, 141, 1007–1014.

Boyce, W. M., Hedrick, P. W., Muggli-Cockett, N. E., *et al.* (1997) Genetic variation of major histocompatibility complex and microsatellite loci: a comparison in bighorn sheep. *Genetics*, 145, 421–433.

Brooks, D. R. & McLennan, D. A. (2002) *The Nature of Diversity: An Evolutionary Voyage of Discovery*. Chicago: University of Chicago Press.

Bumstead, N., Millard, R. M., Barrow, P., *et al.* (1991) Genetic basis of disease resistance in chickens. In *Breeding for Disease Resistance in Farm Animals*, ed. J. B. Owen &

R. F. E. Exelrod, pp. 10–13. Wallingford: CAB International.

Burdon, J. J., Jarosz, A. M. & Kirby, G. C. (1989) Pattern and patchiness in plant–pathogen interactions: causes and consequences. *Annual Review of Ecology and Systematics*, 20, 119–136.

Carroll, S. B. (2009) *Into the Jungle: Great Adventures in the Search for Evolution*. San Francisco: Pearson Benjamin Cummings.

Chan, L., Bundy. D. A. P. & Kao, S. P. (1994) Genetic relatedness as a determinant of predisposition to *Ascaris lumbricoides* and *Trichuris trichura* infection. *Parasitology*, 108, 77–80.

Clayton, D. H., Bush, S. E., Goates, B. M. & Johnson, K. P. (2003) Host defense reinforces host–parasite cospeciation. *Proceedings of the National Academy of Sciences USA*, 100, 15694–15699.

Criscione, C. D. & Blouin, M. S. (2004) Life cycles shape parasite evolution: comparative population genetics of salmon trematodes. *Evolution*, 58, 198–202.

Criscione, C. D., Poulin, R. & Blouin, M. S. (2005) Molecular ecology of parasites: elucidating ecological and microevolutionary processes. *Molecular Ecology*, 14, 2247–2257.

Decaestecker, E., Gaba, S., Raeymaekers, J. A. M., *et al.* (2007) Host–parasite 'Red Queen' dynamics archived in pond sediment. *Nature*, 450, 870–873.

Dybdahl, M. F. & Lively, C. M. (1998) Host–parasite coevolution: evidence for rare advantage and time lagged selection in a natural population. *Evolution*, 52, 1057–1066.

Fahrenholz, H. (1913) Ectoparasiten unde abstammungslehre. *Zoologishe Anzieger*, 41, 371–374.

Gandon, S., Buckling, A., Decaestecker, E. & Day, T. (2008) Host–parasite coevolution and patterns of adaptation across time and space. *Journal of Evolutionary Biology*, 21,1861–1866.

Garrigan, D. & Hedrick, P. W. (2003) Perspective: detecting adaptive molecular polymorphism: lessons learned from the MHC. *Evolution*, 57, 1707–1722.

Goater, C. P. & Holmes, J. C. (1997) Parasite-mediated natural selection. In *Host–Parasite Evolution: General Principles and Avian Models*, ed. D. Clayton & J. Moore, pp. 9–29, Oxford: Oxford University Press.

Gorton, M. J., Kasl, E. L., Detwiler, J. T. & Criscione, C. D. (2012) Testing local-scale panmixia provides insights into the cryptic ecology, evolution, and epidemiology of metazoan animal parasites. *Parasitology*, 139, 981–997.

Grosholz, E. D. (1994). The effects of host genotype and spatial distribution on trematode parasitism in a bivalve population. *Evolution*, 48, 1514–1524.

Hafner, M. S. & Nadler, S. A. (1990) Cospeciation in host–parasite assemblages: comparative analysis of rates of evolution and timing of cospeciation events. *Systematic Zoology*, 39, 192–204.

Hafner, M. S., Sudman, P. D., Villablanca, F. X., *et al.* (1994) Disparate rates of molecular evolution in cospeciating hosts and parasites. *Science*, 265, 1087–1090.

Haldane, J. B. S. (1949) Disease and evolution. *La Ricerca Scientifici* (Supplement), 19, 68–76.

Hall, A. R., Scanlan, P. D., Morgan, A. D. & Buckling, A. (2011) Host–parasite coevolutionary arms races give way to fluctuating selection. *Ecology Letters*, 14, 635–642.

Hamilton, W. D., Axelrod, R. & Tanese, R. (1990) Sexual reproduction as an adaptation to resist parasites (a review). *Proceedings of the National Academy of Sciences USA*, 87, 3566–3573.

Hart, B. L. (1997) Behavioural defence. In *Host–Parasite Evolution: General Principles and Avian Models*, ed. D. Clayton & J. Moore, pp. 59–77, Oxford: Oxford University Press.

Hay, S. I., Guerra, C. A., Tatem, A. J., *et al.* (2004) The global distribution and population at risk of malaria: past, present, and future. *Lancet Infectious Disease*, 4, 327–336.

Hilburn, L. R. & Sattler, P. W. (1986a) Electrophoretically detectable protein variation in natural populations of the lone star tick, *Amblyomma americanum* (Acari: Ixodidae). *Heredity*, 57, 67–74.

Hilburn, L. R. & Sattler, P. W. (1986b) Are tick populations really less variable and should they be? *Heredity*, 57, 113–117.

Hill, A. V. S. (1998) The immunogenetics of human infectious diseases. *Annual Review of Immunology*, 16, 593–617.

Hoberg, E. P. & Brooks, D. R. (2008) A macroevolutionary mosaic: episodic host switching, geographical colonization and diversification in complex host–parasite systems. *Journal of Biogeography*, 35, 1533–1550.

Huelsenbeck, J. P., Rannala, B. & Yang, Z. (1997) Statistical tests of host–parasite cospeciation. *Evolution*, 51, 410–419.

Jokela, J., Dybdahl, M. F. & Lively, C. M. (2009) The maintenance of sex, clonal dynamics, and host–parasite coevolution in a mixed population of sexual and asexual snails. *The American Naturalist*, 174, S43–S53.

Jones, P. H. & Britten, H. B. (2010) The absence of concordant population genetic structure in the black-tailed prairie dog and the flea, *Oropsylla hirsuta*, with implications for the spread of *Yersinia pestis*. *Molecular Ecology*, 19, 2038–2049.

Kerr, P. J. (2012) Myxomatosis in Australia and Europe: a model for emerging infectious diseases. *Antiviral Research*, 93, 387–415.

King, K. C., Delph, L. F., Jokela, J. & Lively, C. M. (2009) The geographic mosaic of sex and the Red Queen. *Current Biology*, **19**, 1–4.

King, K. C., Jokela, J. & Lively, C. M. (2011) Parasites, sex, and clonal diversity in natural snail populations. *Evolution*, **65**, 1474–1481.

Klassen, G. J. (1992) Coevolution: a history of the macro-evolutionary approach to studying host–parasite associations. *Journal of Parasitology*, **78**, 573–587.

Klein, J. (1991) Of HLA, tryps, and selection: an essay on coevolution of MHC and parasites. *Human Immunology*, **30**, 247–258.

Lefèvre, T., Ris, N. & Mitta, G. (2007) Methods. In *Ecology and Evolution of Parasitism*, ed. F. Thomas, J-F. Guégan & F. Renaud, pp. 165–181. Oxford: Oxford University Press.

Little, T. (2002) The evolutionary significance of parasitism: do parasite-driven genetic dynamics occur *ex silico*? *Journal of Evolutionary Biology*, **15**, 1–9.

Lively, C. M. (1987) Evidence from a New Zealand snail for the maintenance of sex by parasitism. *Nature*, **328**, 519–521.

Lively, C. M. (1992) Parthenogenesis in a freshwater snail: reproductive assurance versus parasitic release. *Evolution*, **46**, 907–913.

Lively, C. M. (2010) A review of Red Queen models for the persistence of obligate sexual reproduction. *Journal of Heredity*, **101**, S13–S20.

Lymbery, A. J., Constantine, C. C. & Thompson, R. C. A. (1997) Self-fertilization without genomic or population structuring in a parasitic tapeworm. *Evolution*, **51**, 289–294.

McCoy, K. D., Boulinier, T., Tirard, C. & Michalakis, Y. (2003) Host-dependent genetic structure of parasite populations: differential dispersal of seabird tick host races. *Evolution*, **57**, 288–296.

Miller, K. M., Kaukinen, K. H., Beacham, T. D. & Withler, R. E. (2001) Geographic heterogeneity in natural selection on an MHC locus in sockeye salmon. *Genetica*, **111**, 237–257.

Møller, A. P. (1990) Effects of parasitism by a haematophagous mite on the barn swallow (*Hirundo rustica*): a test of the Hamilton–Zuk hypothesis. *Evolution*, **44**, 771–784.

Morran, L. T., Schmidt, O. G., Gelardern, I. A., *et al.* (2011) Running with the Red Queen: host–parasite coevolution selects for biparental sex. *Science*, **333**, 216–218.

Mulvey, M., Aho, J. M., Lydeard, C., *et al.* (1991) Comparative population genetic structure of a parasite (*Fascioloides magna*) and its definitive host. *Evolution*, **45**, 1628–1640.

Nadler, S. A. (1995) Microevolution and the genetic structure of parasite populations. *Journal of Parasitology*, **81**, 395–403.

Nadler, S. A., Hafner, M. S., Hafner, J. C. & Hafner, D. J. (1990) Genetic differentiation among chewing louse populations (Mallophaga: Trichodectidae) in a pocket gopher contact zone (Rodentia: Geomyidae). *Evolution*, **44**, 942–951.

Nadler, S. A., Lindquist, R. L. & Near, T. J. (1995) Genetic structure of midwestern *Ascaris suum* populations: a comparison of isoenzyme and RAPD markers. *Journal of Parasitology*, **81**, 385–394.

Nascetti, G., Cianchi, R., Mattiucci, S., *et al.* (1993) Three sibling species within *Contracaecum osculatum* (Nematoda, Ascaridida, Ascaridoidea) from the Atlantic Arctic-Boreal region: reproductive isolation and host preferences. *International Journal for Parasitology*, **23**, 105–120.

Page, R. D. M. (2003) *Tangled Trees: Phylogeny, Cospeciation, and Coevolution*. Chicago: University of Chicago Press.

Paterson, A. M. & Gray, R. D. (1997) Host–parasite cospeciation, host switching, and missing the boat. In *Host–Parasite Evolution: General Principles and Avian Models*, ed. D. H. Clayton & J. Moore, pp. 236–250. Oxford: Oxford University Press.

Penn, D. & Potts, W. K. (1999) The evolution of mating preferences and major histocompatibility complex genes. *The American Naturalist*, **153**, 145–164.

Piel, F. B., Patil, A. P., Howes, R. E., *et al.* (2010) Global distribution of the sickle cell gene and geographical confirmation of the malaria hypothesis. *Nature Communications*, DOI: 10.1038/ncomms1104.

Piertney, S. B. & Oliver, M. K. (2006) The evolutionary ecology of the major histocompatibility complex. *Heredity*, **96**, 7–21.

Price, P. W. (1980) *Evolutionary Biology of Parasites*. Princeton: Princeton University Press.

Prugnolle, F., Liu, H., de Meeûs, T. & Balloux, F. (2005) Population genetics of complex life-cycle parasites: an illustration with trematodes. *International Journal for Parasitology*, **35**, 255–263.

Råberg, L., Sim, D. & Read, A. F. (2007) Disentangling genetic variation for resistance and tolerance to infectious disease in animals. *Science*, **318**, 812–814.

Råberg, L., Graham, A. L. & Read, A. F. (2009) Decomposing health: tolerance and resistance to parasites in animals. *Philosophical Transactions of the Royal Society, Series B, Biological Sciences*, **364**, 37–49.

Rich, S. M., Leendertz, F. H., Xu, G., *et al.* (2009) The origin of malignant malaria. *Proceedings of the National Academy of Sciences USA*, **106**, 14902–14907.

Sakanari, J. A. & Moser, M. (1990) Adaptation of an introduced host to an indigenous parasite. *Journal of Parasitology*, **76**, 420–423.

Sayles, P. C. & Wassom, D. L. (1988) Immunoregulation in murine malaria: susceptibility of inbred mice to infection with *Plasmodium yoelii* depends on the dynamic interplay of host and parasite genes. *Journal of Immunology*, 141, 241–248.

Schmid-Hempel, P. (2011) *Evolutionary Parasitology: The Integrated Study of Infections, Immunology, Ecology, and Genetics*. Oxford: Oxford University Press.

Scott, M. E. (2006) High transmission rates restore expression of genetically determined susceptibility of mice to nematode infections. *Parasitology*, 132, 669–679.

Sitepu, P. & Dobson, C. (1982) Genetic control of resistance to infection with *Nematospiroides dubius* in mice: selection of high and low immune responder populations of mice. *Parasitology*, 85, 73–84.

Slade, R. W. & McCallum, H. I. (1992) Overdominant vs. frequency-dependent selection at MHC loci. *Genetics*, 132, 861–862.

Steinauer, M. L., Hanelt, B., Agola, L. E., *et al.* (2009) Genetic structure of *Schistosoma mansoni* in western Kenya: the effects of geography and host sharing. *International Journal for Parasitology*, 39, 1353–1362.

Thompson, J. N. (1994) *The Coevolutionary Process*. Chicago: University of Chicago Press.

Thompson, J. N. (1999) Specific hypotheses on the geographical mosaic of coevolution. *The American Naturalist*, 153, 1–14.

Wambua, S., Mwangi, T. W., Kortok, M., *et al.* (2006) The effect of alpha (+)-thalassaemia on the incidence of malaria and other diseases in children on the coast of Kenya. *PLoS Medicine*, 3, 643–651.

Wassom, D. L. & Kelly, E. A. (1990) The role of the major histocompatibility complex in resistance to parasite infections. *Critical Reviews in Immunology*, 10, 31–52.

Webster, J. P. & Woolhouse, M. E. J. (1998) Selection and strain specificity of compatability between snail intermediate hosts and their parasitic schistosomes. *Evolution*, 52, 1627–1634.

Weckstein, J. D. (2004) Biogeography explains cophylogenetic patterns in toucan chewing lice. *Systematic Parasitology*, 53, 154–164.

Wisenden, B. D., Goater, C. P. & James, C. T. (2009) Behavioral defenses against parasites and pathogens. In: *Fish Defenses: Pathogens, Parasites and Predators*, ed. G. Zaccone, A. Perrière, A. Mathis & B. G. Kapoor, pp. 151–168. New York: Science Publishers.

Woodworth, B. L., Atkinson, C. T., LaPointe, D. A., *et al.* (2005) Host population persistence in the face of introduced vector-borne diseases: Hawaii amakihi and avian malaria. *Proceedings of the National Academy of Sciences USA*, 102, 1531–1536.

Woolhouse, M. E. J., Webster, J. P., Domingo, E., *et al.* (2002) Biological and biomedical implications of the co-evolution of pathogens and their hosts. *Nature Genetics*, 32, 569–577.

17 Environmental parasitology: parasites as bioindicators of ecosystem health

17.1 General considerations

Parasites comprise the most significant component of our planet's biodiversity, and are integral components of all ecosystems, often playing pivotal ecological and evolutionary roles. We hope you have been convinced by previous chapters that, for example, parasites may influence the biology of their hosts in a myriad of ways. Many parasites manipulate the phenotypes of their hosts dramatically. Some parasites have been shown to also regulate host populations. Others can impact the evolution of their hosts and act as powerful agents of natural selection. Parasites can mediate the competitive interactions between free-living animals and act as 'cryptic determinants of animal community structure' (Minchella & Scott, 1991). It is not surprising that concepts such as 'keystone parasite' and 'ecosystem engineer' have been applied to parasitic animals, alluding to their significant ecological roles in nature (e.g., Thomas *et al.*, 1999; see Chapter 15). Indeed, there is increasing evidence that, paradoxically, the 'healthiest' ecosystems are those which are rich in parasites, due to their influence on a range of ecosystem functions, their important roles in food web structure and function, and as 'drivers' of biodiversity (reviews in Marcogliese, 2005; Hudson *et al.*, 2006).

In addition to these diverse ecological and evolutionary roles, recall that parasites are also widely studied from an applied perspective, e.g., as biological tags in fisheries stock management (see Chapter 14), and as indicators of complex food web interactions (e.g., Marcogliese & Cone, 1997a). This chapter reviews yet another contribution of parasites, as posed by Lafferty (1997), "What can parasites tell us about human impacts on the environment?"

Parasites are increasingly being used as bioindicators of environmental quality due to the variety of ways in which they may respond to anthropogenic changes on the environment (Khan & Thulin, 1991; Poulin, 1992; MacKenzie *et al.*, 1995; Overstreet, 1997; Sures *et al.*, 1998; MacKenzie, 1999; Marcogliese, 2005). In particular, the use of parasites as sensitive biological indicators and early warning tags of pollutant contamination in aquatic environments has received substantial attention. Freshwater fish endoparasites have dominated this literature (review in Sures, 2001), although other host-parasite systems have also been explored in this context, especially marine host-parasite systems (e.g., Williams & MacKenzie, 2003; Huspeni & Lafferty, 2004) and amphibian parasite systems (e.g., King *et al.*, 2007; Marcogliese *et al.*, 2009; King *et al.*, 2010; review in Koprivnikar *et al.*, 2012). In addition, ectoparasitic protists, monogeneans, and copepods may also be useful bioindicators. These parasites are in direct contact with the aquatic environment throughout their life cycle, and are thus continuously exposed to potential anthropogenic changes (Williams & MacKenzie, 2003; Khan, 2004; Sanchez-Ramirez et al., 2007).

'Environmental parasitology' is a relatively new discipline that integrates aspects of ecosystem health with the intricate complexities of parasite biology (reviews in Lafferty, 1997; Sures, 2001; 2004). It is a complex, interdisciplinary field that requires extensive knowledge of the parasite(s), (e.g., life cycle and transmission dynamics), and biology of the host(s) (e.g., immune status and trophic position). Furthermore, detailed knowledge of anthropogenic changes to an ecosystem, host-parasite interactions, and environmental toxicology is required. It is no wonder that the research in this field is truly multidisciplinary, requiring that the parasitologist also be an ecologist, ecotoxicologist, and immunologist!

Ecologists and environmental toxicologists recognize that there are two broad categories of pollution-indicator organisms. 'Effect indicators' and 'accumulation indicators' (sentinels) can both be used to indicate quality of the environment (Sures, 2004). In this chapter we review both, as they apply to parasites as bioindicators of pollutant stress. Throughout, we also discuss the possible interactive effects of pollution and parasite-induced stressors, and that parasites can also be utilized as sensitive biomonitors of the physiological/immunological state of their hosts (Morley et al., 2006; Marcogliese et al., 2009, 2010; review in Marcogliese & Pietrock, 2011).

17.2 Parasites as effect indicators of pollutant stress

Because of their life cycles, many parasites (especially those that are trophically transmitted) provide important ecological information concerning host diet, migration, and ontogenetic niche shifts, as well as food web structure and function, and the nature of predator-prey interactions (Goater, 1990; Marcogliese, 2004). Consequently, a trophically transmitted parasite's presence or absence likely indicates a great deal to an environmental parasitologist. For example, the presence of a digenean metacercaria in a freshwater fish host indicates that specific snail hosts are present in the ecosystem and that certain piscivorous birds or mammals must visit the aquatic habitat, at least periodically. Moreover, the critical abiotic factors, e.g., temperature, oxygen, and pH levels, which may impact a fluke's free-living miracidia and cercariae stages, must be conducive to transmission for the parasite to be present. The premise that parasites may be used as effect indicators, or natural tags of ecosystem quality, is based on the possibility that anthropogenic change may alter the presence and/or ecology of critical intermediate or definitive hosts and impact trophic interactions. If so, such alterations may be detected by studying parasites at their population and/or

community levels. Additionally (or alternatively), human impacts may affect parasite transmission, and therefore prevalence and intensity, by influencing the free-living stages of endoparasites and/or the larvae/adults of ectoparasites. If these stages are indeed susceptible to pollution stressors it may help explain patterns observed at parasite population and community levels in hosts within environmentally degraded ecosystems. In short, effect indicators may serve as an 'early warning system' of environmental deterioration (MacKenzie, 1999).

17.2.1 Susceptibility of free-living parasitic stages to pollutants

Most toxicological studies looking at the susceptibility of parasite larval stages to pollutants have used digenean miracidia and cercariae. In such studies, the survival, viability, and/or infectivity of the life history stages can be analyzed to determine if there are toxic effects (review in Pietrock & Marcogliese, 2003). These authors stress that parasite infectivity is the most sensitive parameter to measure toxic effects, as it is affected earlier then viability in the presence of pollutants. For example, Pietrock & Goater (2005) demonstrated that cadmium is toxic to cercariae of two trematode species, *Posthodiplostomum minimum* and *Ornithodiplostomum ptychocheilus*; short-term exposure to low concentrations of this heavy metal significantly reduced infectivity of both digeneans (Fig. 17.1). The reduced activity of enzymes necessary for the penetration and migration of cercariae into the fathead minnow second intermediate host may be a viable explanation (Pietrock & Goater, 2005). Cadmium is a potent enzyme inhibitor. Binding of heavy metal toxins to the active sites of proteolytic penetration enzymes, leading to their impairment, may be the mechanism responsible for the observed reduced infectivity. Results of the experiment demonstrate that short-term exposure to environmentally realistic cadmium concentrations (such as those aquatic environments heavily affected by industrial activities) significantly interferes with the process of cercariae transmission in these two trematodes.

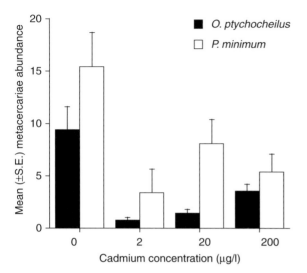

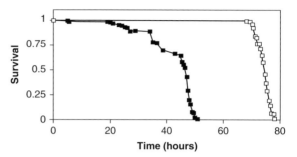

Fig. 17.2 Survival of *Cryptocotyle lingua* cercariae released by periwinkle snails *Littorina littorea* from 'clean' (□) and 'polluted' (■) sites. The polluted site contained elevated levels of metals such as copper, zinc, lead, iron, and manganese relative to the unpolluted site. (Modified from Cross *et al.*, 2001, with permission, *Parasitology*, **123**, 499–507, Cambridge University Press.)

Fig. 17.1 Differences in the mean abundance of *Posthodiplostomum minimum* and *Ornithodiplostomum ptychocheilus* metacercariae recovered from fathead minnows *Pimephales promelas* exposed to cercariae treated with various concentrations of cadmium. (Modified from Pietrock & Goater, 2005, with permission, *Journal of Parasitology*, **91**, 854–856.)

Another example of the impact of heavy metal toxicity on cercariae transmission comes from the marine environment. Cross *et al.* (2001) conducted experimental studies examining the swimming behavior, survival, and longevity of cercariae of the heterophyid fluke *Cryptocotyle lingua* in response to various heavy metals. The snail first intermediate host, the periwinkle, *Littorina littorea*, is known to accumulate heavy metals, especially zinc, copper, and iron; consequently, the snails have been used for monitoring marine heavy metal pollution. Cercariae that develop within *L. littorea* are subjected to the heavy metals that have accumulated in their snail hosts (Cross *et al.*, 2001). As well, released cercariae are potentially exposed to heavy metal pollutants in the water. Both mechanisms may impact cercariae transmission to fish second intermediate hosts. The fluke's cercariae are strong swimmers and positively phototactic, swimming directly towards a horizontal light source. They constructed a special swimming chamber and observed the behavior of cercariae over time.

Cercariae horizontal swimming rates were then compared for cercariae released from snails collected from heavy metal-polluted and unpolluted sites. The horizontal swimming rate of cercariae from the 'polluted' and 'clean' snails differed significantly. They also compared survival and longevity of the cercariae from 'clean' and 'polluted' snails, and they were also significantly different. Cercariae from 'clean' hosts lived longer with low mortality for over 60 hours, while those from polluted hosts suffered increased mortality after 20 hours (Fig. 17.2). Furthermore, longevity was significantly reduced at the higher concentration of each metal tested, especially manganese and zinc. Absorption and effect of the metals occurred very rapidly, within 1 minute. Overall, these results demonstrate that cercariae are vulnerable to heavy metals, and that metals negatively impact swimming speed and survival of the cercariae while they are within heavy metal-exposed snails. Cercariae develop within rediae within the snails. Presumably, having accumulated heavy metal pollutants from the snail, the rediae pass them on to the developing cercariae within. Cercariae then absorb the metals across their tegument. The efficient absorption of heavy metals from their host and/or from the environment in which they swim makes these cercariae "excellent rapid indicators for heavy metal pollution and reliable models for studying the effects of aquatic pollutants on ecologically important behavior" (Cross *et al.*, 2001).

Miracidia of digeneans have also been shown to be impacted negatively by heavy metals. For example, hatching of *Schistosoma mansoni* miracidia from eggs was inhibited by short-term exposure to high concentrations of cadmium and zinc. Moreover, miracidia survival was reduced by high cadmium and zinc concentrations, as well as mixtures of both metals (Fig. 17.3). Finally, miracidia demonstrated rapid avoidance behavior when briefly exposed to heavy metals (Morley *et al.*, 2001). Obviously, all three responses may influence transmission to snail hosts, and thus influence the transmission dynamics of *S. mansoni* in metal-contaminated aquatic habitats. This has potential implications for disease transmission from snails to humans.

Thus far, we have shown that the survival, behavior, and longevity of free-living parasitic stages are clearly impacted by pollutants, such as heavy metals. Whether or not miracidia and cercariae can be carefully monitored and used as reliable indicators of environmental health is a decidedly complex issue. For example, cercariae are short-lived, making monitoring programs very difficult. In addition, potential combined effects must be recognized, since the toxicity of chemicals depends on numerous environmental variables such as pH, water hardness, salinity, and microbial activity, to name a few (Pietrock & Marcogliese, 2003). Contrasting results for combined effects of metals and environmental variables have been observed for larval trematodes, complicating interpretations immensely. In other words, the influence of pollutants on parasite transmission is highly complex, with many of the effects observed under laboratory conditions often masked by interactions with other factors in the field (Morley *et al.*, 2003). Nonetheless, the fact that free-living parasitic stages are 'at the mercy of environmental conditions' (Pietrock & Marcogliese, 2003) and thus have been demonstrated to be vulnerable to pollutants is significant. It means free-living stages can be used to evaluate the effects of pollutant stress on parasite transmission and help explain observed

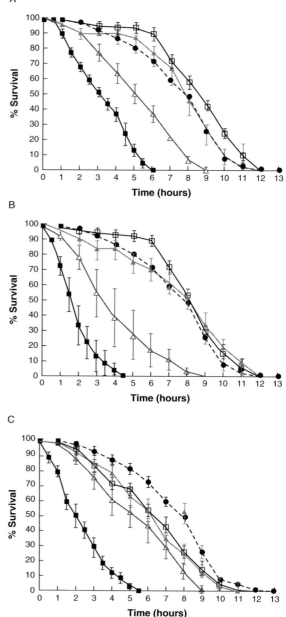

Fig. 17.3 Survival of miracidia of *Schistosoma mansoni* in various concentrations of cadmium (A), zinc (B), and a cadmium/zinc mixture (C). (●) control; (□)10 μg/l; (▲) 100 μg/l; (Δ) 1000 μg/l; (■)10 000 μg/l. Error bars are standard errors. (Modified from Morley *et al.*, 2001, with permission, *Parasitology*, 122, 81–85, Cambridge University Press.)

patterns in the subsequent composition of parasite populations and communities in intermediate and definitive hosts.

17.2.2 Pollution-induced changes in parasite populations

Changes in parasite populations within host populations as a result of numerous anthropogenic environmental perturbations have also been the focus of considerable attention. The potential impact of environmental stressors such as pesticides, heavy metals, sewage and pulp mill effluent, crude oil pollution, and eutrophication have all been examined at the host and parasite population levels. Differences in the prevalence and intensity of parasite infrapopulations in perturbed habitats relative to reference habitats implies that parasites may be used as biomarkers of environmental quality. In essence, parasites can be used to discriminate among host populations inhabiting sites with different levels of pollutant stress. This 'environmental warning tag' approach is conceptually the same as the biological tag approach used to discriminate commercially important shellfish and fish stocks (see Chapter 14; Marcogliese, 2005). As we have stressed, both applied areas of parasitology require extensive knowledge of both the host(s) and the parasite, and the biological and abiotic nature of the environments in which they live.

Ectoparasites of fish possess a variety of criteria that make them especially suitable as bioindicators of environmental degradation. Ectoparasites have direct life cycles with rapid reproductive rates; in addition, adults and larvae, as well as the fish hosts are exposed continuously to the aquatic environment. The ectoparasites may respond to changes in the host's physiology or immunity status as a result of an environmental stressor, or they may be impacted directly. Generally, several studies indicate that ectoparasite intensities increase after exposure to sublethal levels of chemical pollutants, but decrease at high levels (e.g., Khan & Thulin, 1991). Increases in abundances of ectoparasitic ciliates due to chronic exposure to pollution-contaminated sediments were observed

on gills of winter flounder and shorthorn sculpin (Khan, 2004). The higher protist abundance was attributed to the interactive effects of excess secretion of mucus, which acted to enhance ciliate reproduction, as well as suppress the immune system.

The gill monogenean *Cichlidogyrus sclerosus* and its Nile tilapia host have been shown to be a good model system for examining aquatic environmental quality in tropical ecosystems. After 15 days of exposure to low to medium waste water-contaminated sediments polluted with various hydrocarbons, biphenyls, and heavy metals, parasite abundance increased significantly compared to control fishes, but declined at high pollutant concentrations (Sanchez-Ramirez *et al.*, 2007). They attribute this result largely to a pollution-induced weakening of the tilapia's overall general health at low concentrations, but at high concentrations, the parasites themselves were directly affected. As an indicator of fish health, they measured possible pollution-induced damage to the gills using histological methods. They found that fish exposed to highly concentrated polluted sediments showed significantly higher frequencies of gill hyperplasia and hypertrophy, relative to controls. By comparing blood and spleen lymphocyte numbers they found that the immune response of tilapia was negatively impacted at the highest polluted sediment concentrations. Lymphocyte numbers were lower at these concentrations. The immunosuppression observed in the fish may provide the conditions for the persistence of the monogeneans at high pollutant concentrations (Sanchez-Ramirez *et al.*, 2007). This study and several others (e.g., Barker *et al.*, 1994; Gendron *et al.*, 2003; Jacobson *et al.*, 2003; Morley *et al.*, 2006; Marcogliese *et al.*, 2005; 2010) stress the importance of considering aspects of host immunology and physiology in order to fully understand the interactive effects of environmental stressors on parasite infections (review in Marcogliese & Pietrock, 2011).

The effects of increased nutrient load and eutrophication of aquatic habitats on parasite populations has received considerable attention. For example, a long-term ecological study of the allocreadid trematode, *Crepidostomum cooperi*, in its second intermediate host, the burrowing mayfly, in Gull Lake, Michigan provides a particularly fine example of how eutrophication, and then its reversal, can significantly alter parasite population dynamics (Esch *et al.*, 1986; Marcogliese *et al.*, 1990). Recall from Chapter 6 that *Crepidostomum cooperi* uses sphaeriid clams as first intermediate hosts and centrarchid fishes such as bass as their definitive hosts. For the first 18 years (1969–1984) of their study, Esch *et al.* (1986) found high prevalence (80–90%) and intensities of the fluke in the mayfly hosts. Heavy eutrophication of the lake caused by extensive nutrient loading due to shoreline development resulted in the deeper water becoming anoxic. The external gills of mayflies are extremely sensitive to reduced oxygen levels. This physiological requirement in turn altered the distribution of the mayflies such that they moved to shallow-water sediments and were thus in greater spatial overlap with sphaeriid clams, resulting in increased transmission of *C. cooperi* (Fig. 17.4). Construction of a sewage system in 1984 led to a reversal in eutrophication; subsequently, mayflies returned to deeper depths, reducing the spatial transmission window available for transmission of cercariae from sphaeriid clams. This ecological interpretation was used to explain the dramatic declines in prevalence and intensities of the metacercariae of the fluke in mayflies from 1984 to 1989 (Marcogliese *et al.*, 1990; Marcogliese, 2005).

Increases in prevalence and intensities of the nematode *Eustrongylides ignotus* in fish second intermediate and avian definitive hosts as a result of high nutrient loads and eutrophication have also been observed (Coyner *et al.*, 2003). The parasite was implicated as the cause of up to 80% mortality in a colony of herons and egrets. The likely cause was eutrophication from intense human activities in Florida, which caused cascading biomagnifying effects up the food web. Eutrophication led to increased population densities of oligochaete first intermediate hosts, which then led to higher infection levels in small fishes like mosquitofish. Many benthic oligochaete species are often used as pollution-tolerant indicators as they thrive in sediments with high nutrient loads and low oxygen levels.

Mosquitofish infected with *E. ignotus* are more susceptible to predation (Coyner *et al.*, 2001). Thus, *E. ignotus* has impacts at three different trophic levels; significantly, these can be dramatically amplified by nutrient enrichment of aquatic habitats. This example stresses the necessity of understanding food web dynamics and ecological interactions to make meaningful interpretations of the role of anthropogenic environmental perturbations on parasites and their hosts.

Perhaps the most famous parasite from environmental and ecological perspectives is the trematode *Ribeiroia ondatrae* (see Chapters 6 and 15). Recall that this is the parasite that has been shown to be responsible for the severe limb deformities observed in several populations of North American frogs and toads. If the malformations are caused by a naturally occurring parasite, how can we account for the increasingly high frequencies of amphibian deformities observed in many populations? Extensive research on the causes of amphibian malformations has revealed that the deformity issue is extraordinarily complex (see Johnson *et al.*, 2010), and may be linked to cultural eutrophication of aquatic habitats (Blaustein & Johnson, 2003; Johnson & Chase, 2004). This problem provides, once again, a beautiful example of the multidisciplinary nature of environmental parasitology, for it is an issue that integrates concepts in aquatic ecology, amphibian biology, immunology, and developmental biology (Box 17.1).

17.2.3 Pollution-induced effects in parasite communities

If populations of parasites are affected by pollution stress, it is reasonable to predict that such effects would also be detectable at the infracommunity and component community levels. If so, comparing parasite communities of a dominant host species in an ecosystem may provide a valuable bioindicator for monitoring pollution stressors within that ecosystem (Marcogliese & Cone, 1997b). Recall that parasite communities reflect the presence of many different types of organisms due to the diversity of life cycles

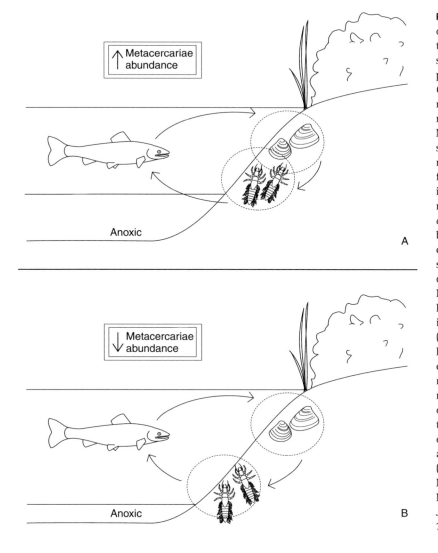

Fig. 17.4 Schematic representation of the sequence of events depicting the result of eutrophication and its subsequent reversal on the population dynamics of *Crepidostomum cooperi* metacercariae in burrowing mayflies in Gull Lake, Michigan. The life cycle involves sphaeriid clams, burrowing mayflies, and centrarchid fishes as first intermediate, second intermediate, and definitive hosts, respectively. (A) Abundance of C. *cooperi* metacercariae was high between 1969 and 1984 because eutrophication, due to extensive shoreline development, caused deeper waters to be anoxic. Mayflies were forced to inhabit less-preferred shallower depths, increasing overlap with clams. (B) Construction of a sewage system led to a reversal in eutrophication. Burrowing mayflies returned to their preferred depths, reducing the temporal and spatial overlap with clams. This resulted in the reduction of transmission opportunities and metacercariae abundance declined after 1984. (Figure courtesy of Danielle Morrison; modified from Marcogliese, 2005, *International Journal for Parasitology*, 35, 705–716.)

displayed by different parasites. An example is provided by Cone *et al.* (1993) and Marcogliese & Cone (1996) who studied the effects of acid rain on the helminth parasites of American eels, *Anguilla rostrata*, in Nova Scotia, Canada. Cone *et al.* (1993) took advantage of an experimental deacidification program that took place in the late 1980s. One river was artificially buffered through application of lime, while others were left alone. As such, they were able to sample eels from rivers that differed in their levels of acidity (high acidity, mixed and limed). They tested the

hypothesis that parasite communities would be altered in much the same way as in macroinvertebrate aquatic communities. Their results showed that eels sampled from the limed river harbored more parasites than did those eels collected from the acidified one. Moreover, it was rare to find eels infected with more than one parasite species in acidic rivers. Multiple infections were limited to eels sampled from the limed river. The reduction in parasite richness was largely attributable to the lack of digeneans at the most acidic sites. This result, in turn, reflects the sensitivity of the molluscan

Box 17.1 | *Ribeiroia* and the complexity of amphibian deformities: a multidisciplinary approach to understanding the possible link to eutrophication

One of the most significant environmental issues in recent years involves the alarming declines and extinctions observed for many amphibian populations, occurring worldwide (reviews in Alford & Richards, 1999; Blaustein & Kiesecker, 2002). As global sentinels of environmental health, a tremendous amount of research has been conducted in attempts to understand the causes of amphibian population declines. Of the many factors that are likely to play a significant role is infectious disease, such as that caused by the pathogenic chytrid fungus. *Batrachochytrium dendrobatidis*, and parasites, especially the trematode *Ribeiroia ondatrae*. As we have seen in prior chapters, this parasite is the cause of limb deformities observed in several amphibian populations and species throughout North America.

The history of the amphibian deformity problem is intriguing. While reports of deformed frogs have been known for centuries, the increase in the frequency and extent of the deformities was what sparked a phenomenal amount of media and scientific attention. Was some sort of pollutant causing mutations and the observed grotesque deformities? While no single cause can explain all of the various amphibian malformations observed today, it does appear that *R. ondatrae* is certainly a major factor in several instances.

Sessions & Ruth (1990) were the first to show experimentally that the supernumerary limb malformations could be explained by mechanical disruption of the developing limb buds of frog tadpoles and salamander larvae by trematode metacercariae. When a group of school children sampled a pond in Minnesota, USA in 1995 and found over 50% of the metamorphosing leopard frogs exhibited severe abnormalities, the issue came to the forefront. Pieter Johnson and collaborators published an important paper showing experimentally that *R. ondatrae* metacercariae were responsible for the limb abnormalities observed in metamorphs of Pacific treefrogs. Moreover, they showed conclusively that limb abnormality frequency increased and tadpole survivorship declined with an increasing density of metacercariae. They also suggested that most abnormal metamorphosing frogs die before reaching sexual maturity (Johnson *et al.*, 1999). Similarly, experimental exposure of western toad tadpoles to cercariae of *R. ondatrae* resulted in high frequencies of severe limb deformities, as well as parasite-induced mortality (Johnson *et al.*, 2001). The types of limb abnormalities that were induced experimentally were identical to those observed in amphibian populations across North America. In addition, subsequent field studies have corroborated the laboratory studies; the more heavily infected the amphibian population was with metacercariae of *R. ondatrae*, the greater the frequency of deformities.

The cercariae target the developing limb buds of amphibians. The metacercariae either mechanically disturb the arrangement of the growing limb bud cells and/or the parasite produces a compound that interferes with a retinoid-sensitive signaling pathway. The window of opportunity to induce limb deformities is narrow; cercariae must penetrate and encyst as metacercariae during the precise time when limbs are developing.

Box 17.1 (continued)

To date, over 60 species of frogs, toads, and salamanders throughout North America have been reported with deformities, mostly affecting the hind limbs and causing extra, missing, and/or often severely malformed limbs (Fig. 17.5; see Color plate Fig. 7.2). It is clear that *R. ondatrae* is a common cause of these amphibian limb malformations (reviews in Johnson *et al.*, 2002; Johnson & Sutherland, 2003; Johnson & Chase, 2004). But, how do we explain the increase in amphibian deformities in recent years? At some field sites, over 90% of the individuals in a population have severe limb deformities. Moreover, the problem is far more widespread, and the nature of the deformities much more severe, than historical museum records and natural history observations indicate (Johnson *et al.*, 2003). It is clearly an emerging parasitic disease of amphibians in some locations.

There is increasing evidence that cultural eutrophication due to human agricultural activities has altered critical wetland habitats. It is hypothesized that by doing so, such anthropogenic change has caused cascading ecological effects that, in turn, impacted the distribution and densities of the specific snail first intermediate hosts (Fig. 17.6). As a result, there is increased transmission of *R. ondatrae* to amphibians, which may help explain the higher occurrence of malformed amphibians. The significant clue was provided when ponds that could be termed 'deformity hotspots' were carefully examined. These tended to be phosphorous-rich farm ponds that were created to irrigate crops or to provide watering holes for cattle. Algal production rates were high in these ponds relative to others, as were population densities, growth rates, and reproduction rates of the snail intermediate host. In a survey of several small ponds throughout Midwestern and Western United States, higher densities of snails were associated with increased

Fig. 17.5 Severe hind limb malformations of metamorphs of northern leopard frogs *Lithobates pipiens* induced by the trematode *Ribeiroia ondatrae*. (Photographs courtesy of Pieter Johnson.)

Box 17.1 | **(continued)**

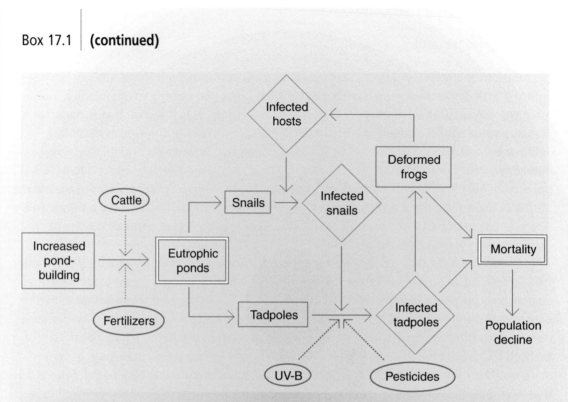

Fig. 17.6 Schematic diagram illustrating the cascading ecological effects of eutrophication and other possible anthropogenic impacts on amphibian limb deformities and population declines. (Figure courtesy of Danielle Morrison; modified from Blaustein & Johnson, 2003, *Frontiers in Ecology and the Environment*, **1**, 87–94.)

R. ondatrae intensities in amphibians (Johnson & Chase, 2004). Johnson *et al.* (2007) also showed experimentally that eutrophication caused an increase in per-snail production of cercariae, providing another mechanism for the observed increase in deformities.

Additional stressors are also likely to play a role in explaining the increasing frequency of amphibian deformities. For example, the interactive effects of ultraviolet radiation and heavy pesticide use may also act in a synergistic manner to increase the frequency of deformed amphibians (Fig. 17.6). Thus, Kiesecker (2002) demonstrated that amphibian larvae exposed to both *R. ondatrae* cercariae and low levels of pesticides exhibited increased levels of infection and a suppressed immune response compared to those exposed only to the parasite. In addition, the frequency of limb malformations increased in long-toed salamanders exposed to both predation and *R. ondatrae* compared to either factor alone, demonstrating synergistic biotic effects (Johnson *et al.*, 2006).

As stressed throughout this discussion and further emphasized by Johnson *et al.* (2004), multidisciplinary, collaborative efforts, as well as long-term monitoring studies will continue to be required to decipher the amphibian deformity mystery. The linkages between environmental nutrient enrichment and emergence of diseases will remain a hot topic for many years.

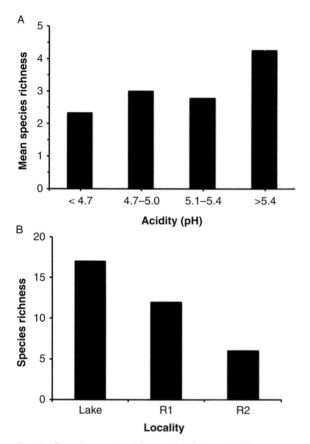

A

B

Fig. 17.7 Parasite species richness in habitats of different acidity. (A) Mean species richness of helminth parasites of American eels *Anguilla rostrata* from rivers of different acidity in Nova Scotia, Canada. (Data from Marcogliese & Cone, 1996). (B) Species richness of protozoan and metazoan parasites in perch, *Perca fluviatilis*, from a lake and reservoirs of different pH in Finland. Lake pH = 6.4; R1 = 5.9; R2 = 5.3. (Data from Halmetojoa *et al.*, 2000.) (Figure courtesy of David Marcogliese; modified from Marcogliese, 2005, with permission, *International Journal for Parasitology*, 35, 705–716, Elsevier.)

intermediate hosts (snails and sphaeriid clams) to acid stress. Decline or elimination of these intermediate hosts in acidic rivers may account for the observed local extinction of digeneans in parasite component communities of eels from acidic sites (Cone *et al.*, 1993; Lafferty, 1997). In a subsequent study, Marcogliese & Cone (1996) looked at eels from several rivers spanning an acidity gradient from highly

acidified to naturally buffered. Their results from the experimental watershed study were confirmed at the regional scale; parasite species richness and diversity was reduced in the eels from the most acidic rivers (Fig. 17.7A). Similarly, Halmetoja *et al.* (2000) found that species richness decreased in the parasite component community of perch from acidified reservoirs in Finland (Fig 17.7B).

Cone *et al.* (1993) and Marcogliese & Cone (1996; 1997b) conclude that helminth parasite communities of fish may be useful as environmental indicators of pollution stress, and also provide information on the dynamics of pollution-altered food webs. These parasitologists also studied the diverse myxozoan faunas of spottail shiners in the Great Lakes and the St. Lawrence River (Marcogliese & Cone, 2001; Cone *et al.*, 2004). The most myxozoan species-rich localities were those that were also subject to nutrient loading and eutrophication, likely stemming from intensive urbanization and sewage effluent near, or downstream, of major cities. Thus, transmission of myxozoan triactinomyxon stages to fish is enhanced in localized regions of high organic enrichment, probably because oligochaete annelid alternate host abundances are increased under eutrophic conditions (Marcogliese & Cone, 2001).

Although freshwater fish and their parasite communities have dominated the literature, other host–parasite systems have generated considerable interest as well. Given the importance of amphibians as global sentinels of environmental degradation (see Box 17.1), it perhaps comes as no surprise that the impact of pollutant stress on amphibian parasite communities has become a recent research focus (King *et al.*, 2007; Koprivnikar *et al.*, 2006a; McKenzie, 2007; Marcogliese *et al.*, 2009; King *et al.*, 2010). The comprehensive study by King *et al.* (2007) of the impacts of agricultural land use practices on parasite communities of the northern leopard frog is a case in point (see also King *et al.*, 2008; 2010). Wetlands provide crucial habitat for amphibians and these vital habitats are increasingly becoming more fragmented, as well as being subjected to agricultural runoff and possible pesticide accumulation. They compared parasite communities of leopard frogs from wetlands exposed to varying degrees of agricultural activity and also with varying amounts of forest cover. The lowest parasite infracommunity and component community species richness and diversity were found in frogs sampled from the two most agriculturally modified wetlands, one of which contained high concentrations of the herbicide, atrazine. A major reason for the reduced diversity of parasites from these two sites was the rarity or absence of parasites that use leopard frogs as intermediate hosts and birds and mammals as definitive hosts. Thus, avian trematode parasites such as *Clinostomum* sp. and *Diplostomum* sp. and mammalian trematodes, including *Alaria* sp. and *Fibricola* sp., use frogs as second intermediate host; metacercariae of these trematodes were rarely encountered or absent from frogs sampled from the two wetlands with the most landscape modification.

King *et al.* (2007) acknowledge that extensive landscape modification may cause reductions in abundance of intermediate hosts for these parasites, and other trematode species (e.g., *Haematoloechus* spp., *Gorgoderina attenuata*) maturing in leopard frogs, which would then result in the observed reductions in species diversity. Additionally, direct toxicity of atrazine on snail and arthropod intermediate hosts, as well as this herbicide's toxicity on free-living miracidia and cercariae stages may also explain the reduced parasite community diversity in frogs from the most polluted wetland. Atrazine is known to reduce the survival and infectivity of trematode cercariae (Koprivnikar *et al.*, 2006b). One of the most important factors accounting for their results is related to the ecology and behavior of the definitive avian and mammalian hosts for the amphibian trematode parasites (King *et al.*, 2007). Recall that the presence of larval helminths in amphibians reflects the nature of predator–prey interactions in the ecosystem. For parasites to be present, the natural definitive hosts must also be present and prey on these frogs, at least periodically, to complete their complex life cycles (Goater, 1990). The two wetlands surrounded by large agricultural landscapes and limited in forest cover leads to fewer predatory birds and mammals using these wetland habitats. Reduction in transmission of the trematodes to frogs would result, leading to the observed patterns of decreased parasite community diversity in frogs from the most agriculturally perturbed landscapes. For this reason, forested buffer zones adjacent to wetlands may be essential for the natural transmission of many amphibian parasites, since they provide habitat and foraging areas for the definitive hosts that prey on amphibians.

Parasite communities can also be used as indicators of recovery from pollutant stresses. In a superb example, Huspeni & Lafferty (2004) examined the diverse digenean communities of the California horn snail, *Cerithidea californica*, to determine the consequences of salt marsh restoration. They found that prevalences and species richness increased at restored sites. This was attributed to increased bird presence in these habitats, since birds were the definitive hosts for several of the digenean species. The larval digeneans of this dominant estuarine snail provided an excellent indication of changes in food web structure and function associated with restoration activities in a coastal ecosystem.

17.3 Parasites as environmental sentinels

Certain 'sentinel' organisms can provide valuable information concerning the chemical and ecological states of their environment. For example, invertebrates such as freshwater bivalves have been widely used as accumulation indicators to assess or to monitor the pollutant levels in aquatic ecosystems. Similarly, certain parasites have been studied as potential environmental sentinels to determine the anthropogenic stressors acting on an ecosystem, including organic and heavy metal contamination. Sures (2004) reviews several criteria for the 'ideal sentinel organism.' Foremost among these is that the organism must efficiently accumulate the pollutant without toxic effects. While most parasites might not be 'ideal' sentinels, there are several that have been successfully utilized as indicators of heavy metal pollution.

Heavy metals, such as lead and cadmium, are often important environmental contaminants in freshwater, marine, and terrestrial ecosystems. Many free-living invertebrate species are known to accumulate these metals in significantly higher concentrations than those measured in the environment (Dallinger, 1994). Likewise, a number of intestinal helminth endoparasite species, mainly acanthocephalans and cestodes, have been shown to accumulate heavy metals in higher concentrations than those measured both in their environment and in their host. These observations are consistent with traditional studies of bioaccumulation within food webs. Since the parasite feeds at a high trophic level and lives in intimate contact with their host tissues, they not only act as sinks for host nutrients, but also, potentially, heavy metals from the environment.

Bernd Sures and his colleagues have pioneered the use of certain helminth endoparasites as sensitive bioaccumulation indicators of heavy metal contamination in freshwater fish hosts (review in Sures, 2003). Acanthocephalans of fish have been shown to be particularly promising in this regard. Remarkably, Sures (2001) showed that lead concentrations were 2700× higher in *Pomphorhynchus laevis* than those measured in the muscle tissues of its chub host and 11 000× higher than in the water (Fig. 17.8). Similar results were found for lead and cadmium in an acanthocephalan in perch. Further studies showed that this acanthocephalan accumulated heavy metal toxins, especially copper, cadmium and lead, from the aquatic environment at concentrations much higher than those accumulated by established free-living sentinel species, like zebra mussels (Fig. 17.9).

Clearly, there is now overwhelming evidence that fish acanthocephalans can serve as sensitive bioindicators of the state of heavy metal pollution in

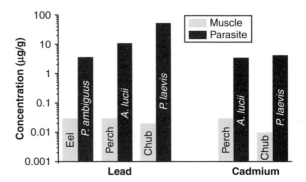

Fig. 17.8 Comparison of heavy metal (lead and cadmium) concentrations in muscle tissue of three fish host species with three acanthocephalan species: *Paratenuisentis ambiguus* in eels *Anguilla anguilla*, *Acanthocephalus lucii* in perch *Perca fluviatilis*, and *Pomphorhynchus laevis* in chub *Leuciscus cephalus*. (Figure courtesy of Bernd Sures; modified from Sures, 2001, with permission, *Aquatic Ecology*, 35, 245–255, Springer Science & Business Media.)

freshwater environments. As Sures (2004) states, "by using acanthocephalans in environmental impact studies, very low concentrations of metals can be detected in the environment owing to the enormous accumulation capacity of these worms." Such efficient sentinels are of particular relevance when looking for heavy metal pollution in remote and/or supposedly pristine regions such as the Arctic and Antarctic. For example, significantly higher concentrations of several heavy metal elements, especially lead, silver and cadmium, were demonstrated in the acanthocephalan *Aspersentis megarhynchus* from an Antarctic fish species, even though some of the metals were virtually undetectable in fish tissues (Sures, 2004). Similar results documenting heavy metal accumulation in parasites has been shown for a diversity of host–parasite systems, including lung nematodes in marine mammals (Szefer *et al.*, 1998), intestinal cestodes in birds (Baruš *et al.*, 2000) and sharks (Malek *et al.*, 2007), and liver flukes in terrestrial mammals (Sures *et al.*, 1998). With respect to this latter group, Sures (2004) stresses the urgent need of environmental sentinels in terrestrial ecosystems, especially in urban settings. Experimental studies with the common rat parasites, the acanthocephalan *Moniliformis moniliformis* and the cestode

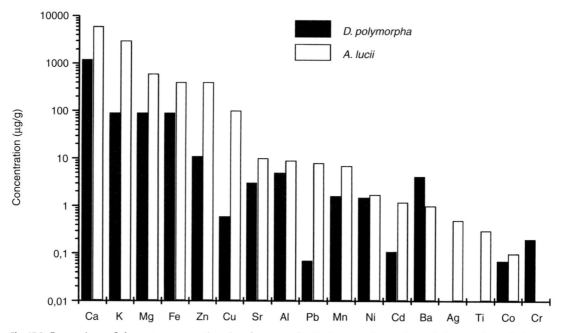

Fig. 17.9 Comparison of element concentrations in zebra mussels, *Dreissena polymorpha*, and the acanthocephalan *Acanthocephalus lucii* from perch, *Perca fluviatilis*. (Figure courtesy of Bernd Sures; modified from Sures, 2001, with permission, *Aquatic Ecology*, 35, 245–255, Springer Science & Business Media.)

Hymenolepis diminuta, have demonstrated that the parasites accumulate higher levels of lead than do host organs such as the kidney and liver (Sures, 2002). These studies indicate that rats and their parasites may be particularly promising biomonitors of heavy metal pollution, especially in heavily populated urban environments.

The phenomenon of heavy metal accumulation by parasites shown from field studies has also been studied experimentally in certain fish acanthocephalans. A lead exposure study of *Pomphorhynchus laevis* in chub, for example, demonstrated that lead accumulation in the parasite occurs much more rapidly and reaches a steady state than in host muscle, liver, or intestinal tissues (Fig. 17.10). Thus, it appears that lead concentrations in this parasite species are very likely to respond rapidly to changes in environmental exposure (Sures, 2001). Furthermore, since it is clear that the acanthocephalans serve as lead sinks, there is evidence that lead uptake by *P. laevis* may reduce lead

accumulation in the organs and tissues of chub. Sures & Siddall (1999) have shown that the presence of *P. laevis* in the intestine of chub hosts reduces the lead concentrations in the intestinal wall; the intestinal walls of infected fish had less than half of the lead levels than did uninfected fish. In short, hosts and parasites compete for available heavy metals.

How exactly do acanthocephalans (and other parasites) act as sinks for heavy metals? Sures & Siddall (1999) propose that lead ions first enter the fish through the gills and pass into the bloodstream and bind to RBCs, where they are then transported to the liver via the circulatory system. Much of the lead is removed from the blood and excreted into the intestine via the bile, as bile contains steroids with which the lead ions form organometallic complexes. In the small intestine, these complexes can either be reabsorbed across the intestinal wall or excreted via the feces of the fish. Acanthocephalans are known to be efficient at

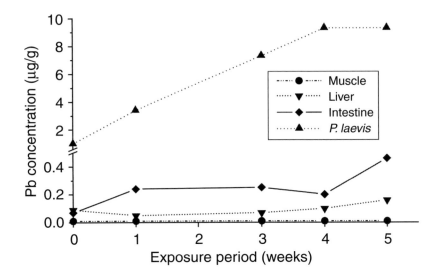

Fig. 17.10 Uptake of lead by chub *Leuciscus cephalus* experimentally infected with the acanthocephalan *Pomphorhynchus laevis*, showing the rapid accumulation relative to host tissues. (Figure courtesy of Bernd Sures; modified from Sures, 2001, with permission, *Aquatic Ecology*, 35, 245–255, Springer Science & Business Media.)

taking up bile salts as well as bile-bound lead from the intestinal lumen of infected fish. In short, bile plays an essential role in the uptake of lead. The observed reduction in lead within the intestinal wall with which the parasites are competing supports the hypothesis. Given their efficiency at accumulating heavy metals in their tissues, acanthocephalan–host systems offer several advantages over free-living bioindicators in numerous environmental situations.

References

Alford, R. A. & Richards, S. J. (1999) Global amphibian declines: a problem in applied ecology. *Annual Review of Ecology and Systematics*, 30, 133–165.

Barker, D. E., Khan, R. A. & Hooper, R. (1994) Bioindicators of stress in winter flounder, *Pleuronectes americanus*, captured adjacent to a pulp and paper mill in St. George's Bay, Newfoundland. *Canadian Journal of Fisheries and Aquatic Sciences*, 51, 2203–2209.

Baruš, V., Tenora, F. & Kráčmar, S. (2000) Heavy metal (Pb, Cd) concentrations in adult tapeworms (Cestoda) parasitizing birds (Aves). *Helminthologia*, 37, 131–136.

Blaustein, A. R. & Johnson, P. T. J. (2003) The complexity of deformed amphibians. *Frontiers in Ecology and the Environment*, 1, 87–94.

Blaustein, A. R. & Kiesecker, J. M. (2002) Complexity in conservation: lessons from the global decline in amphibian populations. *Ecology Letters*, 5, 597–608.

Cone, D. K., Marcogliese, D. J. & Watt, W. D. (1993) Metazoan parasite communities of yellow eels (*Anguilla rostrata*) in acidic and limed rivers of Nova Scotia. *Canadian Journal of Zoology*, 71, 177–184.

Cone, D. K., Marcogliese, D. J. & Russell, R. (2004) The myxozoan fauna of spottail shiner in the Great Lakes basin: membership, richness, and geographic distribution. *Journal of Parasitology*, 90, 921–932.

Coyner, D. F., Schaack, S. R., Spalding, M. G. & Forrester, D. J. (2001) Altered predation susceptibility of mosquitofish infected with *Eustrongylides ignotus*. *Journal of Wildlife Diseases*, 3, 556–560.

Coyner, D. F., Spalding, M. G. & Forrester, D. J. (2003) Influence of treated sewage on infection of *Eustrongylides ignotus* (Nematoda: Dioctophymatoidea) in eastern mosquitofish (*Gambusia holbrooki*) in an urban watershed. *Comparative Parasitology*, 70, 205–210.

Cross, M. A., Irwin, S. W. B. & Fitzpatrick, S. M. (2001) Effects of heavy metal pollution on swimming and longevity in cercariae of *Cryptocotyle lingua* (Digenea: Heterophyidae). *Parasitology*, **123**, 499–507.

Dallinger, R. (1994) Invertebrate organisms as biological indicators of heavy metal pollution. *Applied Biochemistry and Biotechnology*, **48**, 27–31.

Esch, G. W., Marcogliese, D. J., Goater, T. M. & Crews, A. E. (1986) Long-term study on the population biology of *Crepidostomum cooperi* (Trematoda: Allocreadidae) in the burrowing mayfly, *Hexagenia limbata* (Ephemeroptera). *American Midland Naturalist*, **116**, 304–316.

Goater, T. M. (1990) Helminth parasites indicate predator–prey relationships in desmognathine salamanders. *Herpetological Review*, 21, 32–33.

Gendron, A. D., Marcogliese, D. J., Barbeau, S., *et al.* (2003) Exposure of leopard frogs to a pesticide mixture affects life history characteristics of the lungworm *Rhabdias ranae*. *Oecologia*, **135**, 469–476.

Halmetoja, A., Valtonen, E. T. & Koskenniemi, E. (2000) Perch (*Perca fluviatilis* L.) parasites reflect ecosystem conditions: a comparison of a natural lake and two acidic reservoirs in Finland. *International Journal for Parasitology*, **30**, 1437–1444.

Hudson, P. J., Dobson, A. P. & Lafferty, K. D. (2006) Is a healthy ecosystem one that is rich in parasites? *Trends in Ecology and Evolution*, 21, 381–385.

Huspeni, T. C. & Lafferty, K. D. (2004) Using larval trematodes that parasitize snails to evaluate a saltmarsh restoration project. *Ecological Applications*, **14**, 795–804.

Jacobson, K. C., Arkoosh, M. R., Kagley, A. N., *et al.* (2003) Cumulative effects of natural and anthropogenic stress on immune function and disease resistance in juvenile Chinook salmon. *Journal of Aquatic Animal Health*, **15**, 1–12.

Johnson, P. T. J. & Chase, J. M. (2004) Parasites in the food web: linking amphibian malformations and aquatic eutrophication. *Ecology Letters*, **7**, 521–526.

Johnson, P. T. J. & Sutherland, D. R. (2003) Amphibian deformities and *Ribeiroia* infection: an emerging helminthiasis. *Trends in Parasitology*, **19**, 332–335.

Johnson, P. T. J., Lunde, K. B., Ritchie, E. G. & Launer, A. E. (1999) The effect of trematode infection on amphibian limb development and survivorship. *Science*, **284**, 802–804.

Johnson, P. T. J., Lunde, K. B., Haight, R. W., *et al.* (2001) *Ribeiroia ondatrae* (Trematoda: Digenea) infection induces severe limb malformations in western toads (*Bufo boreas*). *Canadian Journal of Zoology*, **79**, 370–379.

Johnson, P. T. J., Lunde, K. B., Thurman, E. M., *et al.* (2002) Parasite (*Ribeiroia ondatrae*) infection linked to amphibian malformations in the western United States. *Ecological Monographs*, **72**, 151–168.

Johnson, P. T. J., Lunde, K. B., Zelmer, D. A. & Werner, J. K. (2003) Limb deformities as an emerging parasitic disease in amphibians: evidence from museum specimens and resurvey data. *Conservation Biology*, **17**, 1724–1737.

Johnson, P. T. J., Sutherland, D. R., Kinsella, J. M. & Lunde, K. B. (2004) Review of the trematode genus *Ribeiroia* (Psilostomidae): ecology, life history and pathogenesis with special emphasis on the amphibian malformation problem. *Advances in Parasitology*, **57**, 191–253.

Johnson, P. T. J., Preu, E. R., Sutherland, D. R., *et al.* (2006) Adding infection to injury: synergistic effects of predation and parasitism on amphibian malformations. *Ecology*, **87**, 2227–2235.

Johnson, P. T. J., Chase, J. M., Dosch, K. L., *et al.* (2007) Aquatic eutrophication promotes pathogenic infection in amphibians. *Proceedings of the National Academy of Sciences USA*, **104**, 15781–15786.

Johnson, P. T. J., Townsend, A. R., Cleveland, C. C., *et al.* (2010) Linking environmental nutrient enrichment and disease emergence in humans and wildlife. *Ecological Applications*, **20**, 16–29.

Khan, R. A. (2004) Parasites of fish as biomarkers of environmental degradation: a field study. *Bulletin of Environmental Contamination and Toxicology*, **72**, 394–400.

Khan, R. A. & Thulin, J. (1991) Influence of pollution on parasites of aquatic animals. *Advances in Parasitology*, **30**, 201–238.

Kiesecker, J. M. (2002) Synergism between trematode infection and pesticide exposure: a link to amphibian deformities in nature? *Proceedings of the National Academy of Sciences USA*, **99**, 9900–9904.

King, K. C., McLaughlin, J. D., Gendron, A. D., *et al.* (2007) Impacts of agriculture on the parasite communities of northern leopard frogs (*Rana pipiens*) in southern Quebec, Canada. *Parasitology*, **134**, 2063–2080.

King, K. C., Gendron, A. D., McLaughlin, J. D., *et al.* (2008) Short-term seasonal changes in parasite community structure in northern leopard froglets (*Rana pipiens*) inhabiting agricultural wetlands. *Journal of Parasitology*, **94**, 13–22.

King, K. C., McLaughlin, J. D., Boily, M., & Marcogliese, D. J. (2010). Effects of agricultural landscape and pesticides on parasitism in native bullfrogs. *Biological Conservation*, **143**, 302–310.

Koprivnikar, J., Forbes, M. R. & Baker, R. L. (2006a) Environmental factors influencing trematode prevalence in grey tree frog (*Hyla versicolor*) tadpoles in southern Ontario. *Journal of Parasitology*, **92**, 997–1001.

Koprivnikar, J., Forbes, M. R. & Baker, R. L. (2006b) Effects of atrazine on cercarial infectivity, activity, and infectivity. *Journal of Parasitology*, **92**, 306–311.

Koprivnikar, J., Marcogliese, D. J. & Rohr, J. R. (2012) Macroparasite infections of amphibians: what can they tell us? *EcoHealth*, **9**, 342–360.

Lafferty, K. D. (1997) Environmental parasitology: what can parasites tell us about human impacts on the environment? *Parasitology Today*, **13**, 251–255.

MacKenzie, K. (1999) Parasites as pollution indicators in marine ecosystems: a proposed early warning system. *Marine Pollution Bulletin*, **38**, 955–959.

MacKenzie, K., Williams, H. H., Williams, B., *et al.* (1995) Parasites as indicators of water quality and the potential use of helminth transmission in marine pollution studies. *Advances in Parasitology*, **35**, 85–144.

Malek, M., Haseli, M., Mobedi, I., *et al.* (2007) Parasites as heavy metal bioindicators in the shark *Carcharhinus dussumieri* from the Persian Gulf. *Parasitology*, **134**, 1053–1056.

Marcogliese, D. J. (2004) Parasites: small players with crucial roles in the ecological theater. *EcoHealth*, **1**, 151–164.

Marcogliese, D. J. (2005) Parasites of the superorganism: are they indicators of ecosystem health? *International Journal for Parasitology*, **35**, 705–716.

Marcogliese, D. J. & Cone, D. K. (1996) On the distribution and abundance of eel parasites in Nova Scotia: influence of pH. *Journal of Parasitology*, **82**, 389–399.

Marcogliese, D. J. & Cone, D. K. (1997a) Food webs: a plea for parasites. *Trends in Ecology and Evolution*, **12**, 320–325.

Marcogliese, D. J. & Cone, D. K. (1997b) Parasite communities as indicators of ecosystem stress. *Parassitologia*, **39**, 227–232.

Marcogliese, D. J. & Cone, D. K. (2001) Myxozoan communities parasitizing *Notropis hudsonius* (Cyprinidae) at selected localities on the St. Laurence River, Quebec: possible effects of urban effluents. *Journal of Parasitology*, **87**, 951–956.

Marcogliese, D. J. & Pietrock, M. (2011) Combined effects of parasites and contaminants on animal health: parasites do matter. *Trends in Parasitology*, **27**, 123–130.

Marcogliese, D. J., Goater, T. M. & Esch, G. W. (1990) *Crepidostomum cooperi* (Allocreadidae) in the burrowing mayfly, *Hexagenia limbata* (Ephemeroptera) related to trophic status of a lake. *American Midland Naturalist*, **124**, 309–317.

Marcogliese, D. J., Gagnon Brambilla, L., Gagne, F. & Gendron, A. D. (2005) Joint effects of parasitism and pollution on oxidative stress biomarkers in yellow perch *Perca flavescens*. *Diseases of Aquatic Organisms*, **63**, 77–84.

Marcogliese, D. J., King, K. C., Salo, H. M., *et al.* (2009) Combined effects of agricultural activity and parasites on biomarkers in the bullfrog, *Rana catesbeiana*. *Aquatic Toxicology*, **91**, 126–134.

Marcogliese, D. J., Dautremepuits, C., Gendron, A. D., & Fournier, M. (2010) Interactions between parasites and pollutants in yellow perch (*Perca flavescens*) in the St. Lawrence River, Canada: implications for resistance and tolerance to parasites. *Canadian Journal of Zoology*, **88**, 247–258.

McKenzie, V. J. (2007) Human land use and patterns of parasitism in tropical amphibian hosts. *Biological Conservation*, **137**, 102–116.

Minchella, D. J. & Scott, M. E. (1991) Parasitism: a cryptic determinant of animal community structure. *Trends in Ecology and Evolution*, **6**, 250–254.

Morley, N. J., Crane, M. & Lewis, J. W. (2001) Toxicity of cadmium and zinc to miracidia of *Schistosoma mansoni*. *Parasitology*, **122**, 81–85.

Morley, N. J., Irwin, S. W. B. & Lewis, J. W. (2003) Pollution toxicity to the transmission of larval digeneans through their molluscan hosts. *Parasitology*, **126**, S5–S26.

Morley, N. J., Lewis, J. W. & Hoole, D. (2006) Pollutant-induced effects on immunological and physiological interactions in aquatic host-trematode systems: implications for parasite transmission. *Journal of Helminthology*, **80**, 137–149.

Overstreet, R. M. (1997) Parasitological data as monitors of environmental health. *Parassitologia*, **39**, 169–175.

Pietrock, M. & Goater, C. P. (2005) Infectivity of *Ornithodiplostomum ptychocheilus* and *Posthodiplostomum minimum* (Trematoda: Diplostomidae) cercariae following exposure to cadmium. *Journal of Parasitology*, **91**, 854–856.

Pietrock, M. & Marcogliese, D. J. (2003) Free-living endo-helminth stages: at the mercy of environmental conditions. *Trends in Parasitology*, **19**, 293–299.

Poulin, R. (1992) Toxic pollution and parasitism in freshwater fish. *Parasitology Today*, **8**, 58–61.

Sanchez-Ramirez, C., Vidal-Martinez, V. M., Aguirre-Macedo, M. L., *et al.* (2007) *Cichlidogyrus sclerosus* (Monogenea: Ancyrocephalinae) and its host, the Nile tilapia (*Oreochromis niloticus*), as bioindicators of chemical pollution. *Journal of Parasitology*, **93**, 1097–1106.

Sessions, S. K. & Ruth, S. B. (1990) Explanation for naturally occurring supernumerary limbs in amphibians. *Journal of Experimental Zoology*, **254**, 38–47.

Sures, B. (2001) The use of fish parasites as bioindicators of heavy metals in aquatic ecosystems: a review. *Aquatic Biology*, **35**, 245–255.

Sures, B. (2002) Experimental studies on the lead accumulation in the cestode *Hymenolepis diminuta* and its final rat host *Rattus norvegicus*. *Ecotoxicology*, **11**, 365–368.

Sures, B. (2003) Accumulation of heavy metals by intestinal helminths in fish: an overview and perspective. *Parasitology*, **126**, S53–S60.

Sures, B. (2004) Environmental parasitology: relevancy of parasites in monitoring environmental pollution. *Trends in Parasitology*, **20**, 170–177.

Sures, B. & Siddall, R. (1999) *Pomphorhynchus laevis*: the intestinal acanthocephalan as a lead sink for its fish host, chub (*Leuciscus cephalus*). *Experimental Parasitology*, **93**, 66–72.

Sures, B., Jurges, G. & Taraschewski, H. (1998) Relative concentrations of heavy metals in the parasites *Ascaris suum* (Nematoda) and *Fasciola hepatica* (Digenea) and their respective porcine and bovine definitive hosts. *International Journal for Parasitology*, **28**, 1173–1178.

Szefer, P., Rokicki, J., Frelek, K., *et al.* (1998) Bioaccumulation of selected trace elements in lung nematodes, *Pseudoalius inflexus*, of harbor porpoise (*Phocoena phocoena*) in a Polish zone of the Baltic Sea. *Science of the Total Environment*, **220**, 19–24.

Thomas, F., Poulin, R., de Meelüs, T., *et al.* (1999) Parasites and ecosystem engineering: what roles could they play? *Oikos*, **84**, 167–171.

Williams, H. & Mackenzie, K. (2003) Marine parasites as pollution indicators: an update. *Parasitology*, **126**, S27–S41.

GLOSSARY

acanthella: The larval stage of acanthocephalans; it develops between the acanthor and cystacanth stages.

acanthor: The acanthocephalan larva that hatches from an egg.

acetabula: The sucker-like attachment structures on the scolex of some cestodes.

acetabulum: Another name for the ventral holdfast sucker on most digeneans.

actinospore: The life cycle stage of myxozoans which develops within invertebrate hosts.

aedeagus: The penis or male copulatory organ of insects and acarines.

aegaltroid: The larval stage that succeeds the manca stage in the life cycle of parasitic cymothoid isopods.

aesthetases: Chemoreceptive hairs usually found on the antennae of crustaceans that function in food, mate, and probably host recognition.

alae: The lateral wing-like expansions found on several groups of organisms, e.g., the cuticular structures found on the mouth region of some nematodes.

alveoli: A series of sacs found under the cell surface membranes of members of the protozoan clade Alveolata which includes the dinoflagellates, ciliates, and apicomplexans.

amastigote: The intracellular form of some trypano-somatids lacking a flagella, e.g., *Trypanosoma cruzi* and *Leishmania* spp.

ametabolous: Development in insects in which no metamorphosis occurs.

amphid: The sensory organ on each side of the anterior region of nematodes.

annuli: Superficial, false segmentation of leeches and pentastomes.

antagonistic coevolution: Reciprocal evolution-ary change between two antagonists through selective pressures imposed by one species (e.g., a host) on another (e.g., a parasite).

antennae: A pair of chemosensory appendages on the head in crustaceans and insects; second pair often modified as prehensile holdfast in parasitic copepods.

antennules: The most anterior pair of appendages in Crustacea.

antigenic variation: A protozoan adaptation to evade the vertebrate immune system; in the African trypanosomes it involves continual changes of their cell surfaces (variant surface glyoproteins).

apical complex: A complex of anteriorly located organelles in some life cycle stages of apicomplexan parasites.

apicoplast: A membrane-bound organelle in the Apicomplexa.

apparent competition: Ecological interaction in which the presence of one host species adversely affects another via a shared parasite/pathogen, e.g., a shared parasite may mediate the competitive interactions between sympatric host species.

areole: The rounded protrusion on the cuticle of nematomorphs.

axostyle: A prominent, rod-like, supporting organelle in some flagellated protists.

axoneme: The microtubules in a cilium or flagellum.

balanced polymorphism: The maintenance of two alleles in a population by a fitness advantage for the heterozygote.

basal body: The same as kinetosome or centriole from which an axoneme arises.

Basic Reproductive Rate (Ro): An epidemiological term denoting the number of infected hosts that arise, on average, from an infection in a single host over the course of its infectious period. For macroparasites, the term is extended to describe the number of reproductively mature adults that an individual parasite produces over its lifetime. An infection can spread through a population when Ro > 1, but dies out when Ro < 1.

binary fission: The simple division of one cell into two daughter cells. Before division, the organelles of the original cell are duplicated.

blastocyst: The portion of a larval tapeworm, into which the body may be withdrawn.

bothria: The grooved attachment structures on the scolex of certain cestodes, e.g., diphyllobothriideans.

bothridia: The leaf-like structures associated with the scolices of some cestodes.

bradyzoites: The slowly dividing merozoites in the life cycle of certain coccidians, e.g., *Toxoplasma gondii*.

brood parasite: Birds which lay their eggs in nests of other birds, depending on the host bird to raise the young, e.g., cowbirds.

buccal capsule: The portion of the mouth of many nematodes that is lined by cuticle.

buccal cone: A section of the mouthparts of mites and ticks formed by the hypostome and labrum.

bulla: The non-living structure secreted by the head and maxillary glands of lernaeopodid copepods used to securely anchor the copepod to tissues of the host.

bursa: In some male nematodes and all-male acanthocephalans the posterior part of the body is drawn out into a fleshy appendage that is used to grasp the female's body during copulation.

cadre: The sclerotized mouth lining of pentastomids.

calotte: The light-colored area at the anterior end of nematomorphs.

capitulum: The anterior region of mites and ticks, carrying the mouthparts and feeding appendages.

carapace: The structure formed by the extension of the dorsal sclerites of the head in many Crustacea, usually covering and/or fusing with one or more thoracic somites.

cement glands: The glands in a male acanthocephalan that produce secretions to seal the female reproductive tract after copulation.

cephalization: The evolutionary process by which specialization of sensory organs and appendages occurs; became localized in the head region of animals.

cephalothorax: In arachnids and crustaceans, the portion of the body formed by the fusion of the head and thoracic segments; can be highly modified in parasitic copepods.

cercaria: An immature digenean, usually free-swimming, produced by asexual reproduction in a sporocyst or a redia within mollusc intermediate hosts.

cercomer: An embryonic structure on certain cestode procercoids on which occur the oncosphere's hooks.

chagoma: The nodule that forms at the site where *Trypanosoma cruzi* enters human skin.

chalimus: Specialized stage of parasitic copepods that succeeds the copepodid stage; in siphonostomatoid copepods the chalimus attaches to hosts via an anterior frontal filament.

chelicerae: The anterior-most pair of feeding appendages in chelicerate arthropods.

cilia: The short, cylindrical organelles consisting of nine pairs of outer and two central microtubules arising from an internal base, the kinetosome.

cirri: The thoracic suspension-feeding appendages of barnacles.

cirrus: The protrusible penis or copulatory organ of a male platyhelminth.

clade: In phylogenetic systematics, a taxon consisting of a particular ancestral lineage and all of its descendants; forms a distinct branch on a phylogenetic tree.

cladistics: A system of arranging taxa by analysis of primitive and derived features so that the arrangement reflects phylogenetic relationships.

cladogram: An evolutionary tree reflecting the results of a cladistic analysis; shows the pattern of sharing of evolutionarily derived characters (synapomorphies) among taxa.

coelomocyte: In nematodes and other invertebrates, cells of unknown function that are found in the pseudocoelom.

coenurus: A cysticercus-type cestode larval stage in possession of many scolices that remain attached to the bladder wall; always found in a mammalian intermediate host.

coevolution: Reciprocal evolutionary change in interacting species; one partner acts as an agent of natural selection on the other, and vice versa.

commensalism: A symbiotic relationship in which the commensal lives in or on a host, and obtaining nutrients from the host without causing harm.

component community: All of the parasite species in a population of hosts.

component population: All of the individuals of a specified life history phase at a particular place and time; all of the infrapopulations in a single host species in an ecosystem.

concomitant immunity: Resistance to re-infection of a host by a specific parasite when the host is currently infected with that parasite.

conjugation: Reciprocal fertilization; a form of sexual reproduction (exchange of micronuclei) that occurs between members of different mating types in ciliated protists.

conoid: The cone of fibrils within the polar rings of some apicomplexan protists.

contractile vacuole: A fluid-filled cytoplasmic organelle involved in osmoregulation in many protozoans.

copepodid: The juvenile stage that follows the nauplius stage in copepods.

copulatory cap: The cement-like cap applied to the genital opening of the female by a male acanthocephalan following copulation.

coracidium: The free-swimming, ciliated larva of certain cestodes, e.g., diphyllobothriideans.

cordons: The cuticular ridges at the anterior end of some nematodes.

cospeciation: Speciation by one species in response to the speciation of another.

coxal glands: The excretory organs of chelicerates. A sac, a tubule, and an opening on the coxa form the gland.

cryptogonochorism: The condition in which the sexes are separate, but hidden, giving the appearance of hermaphroditism, e.g., rhizocephalan barnacles.

cryptoniscus: The free-swimming larval stage of epicaridean isopods which succeeds the microniscus stage; leaves the copepod intermediate host and attaches to decapod crustacean definitive hosts.

ctenidia: A series of stiff, peg-like holdfast spines on the head and body of fleas.

cyprid (=cypris): The larval stage that follows the nauplius stage of barnacles.

cyst: An inactive, dormant stage in the life cycle of many animals, usually with a secreted wall. It is more environmentally resistant than other stages of the life cycle and is often the stage important in transmission and dispersal for parasites, e.g., cyst-forming protists.

cystacanth: The juvenile stage of acanthocephalans that is infective to the definitive host.

cysticercoid: A metacestode larval stage in possession of a non-invaginated scolex and a 'tail' to which may be attached the embryonic hooks; almost always found in invertebrate intermediate hosts.

cysticercus: A metacestode larval stage possessing an invaginated scolex and a bladder; always found in a mammalian intermediate host.

cyton: The body of the syncytial tegument of parasitic flatworms containing the nuclei and other organelles.

dauer juvenile: A juvenile nematode in which development is arrested during unfavorable conditions.

definitive host: That host in a parasite's life cycle in which the parasite reaches sexual maturity.

deirid: A sensory papilla on either side near the anterior end of some nematodes.

density–dependent: Attributes such as growth, fecundity, etc. that are affected by population density.

density–independent: Attributes such as growth, fecundity, etc. that are unaffected by population density.

deutonymph: The nymphal stage of mesostigmatid ticks succeeding the protonymph stage.

developmental arrest: Some invertebrates (e.g., nematodes) have resistant stages that cease to develop, remaining viable and persisting in a form of diapause.

diapause: A brief interruption, genetically programmed, in the developmental progression of an organism.

dilator organ: The male copulatory structure housing the penis in pentastomes.

diplostomulum: The metacercaria of diplostomid digeneans.

dorsal organ: The glandular structure of pentastome larvae. It secretes a mucoid substance that flows through the facette, becoming part of the first protective egg layer surrounding the larva.

ecdysis (=molting): The process of shedding of the cuticle during the life cycle of several groups of invertebrates, e.g., nematodes, arthropods.

ecological biogeography: The branch of biogeography that attempts to understand the distribution patterns of organisms in terms of ecological processes occurring over short temporal and spatial scales.

ecsoma: A retractile 'tail' of some hemiurid digeneans that functions as a feeding organ.

ectoparasite: A parasitic organism that lives on the surface of its host.

endoparasite: A parasitic organism that lives inside its host.

endosome: A nucleolus-like organelle within the nucleus of some protozoans, e.g., *Entamoeba*.

epicaridium: The first larval stage of epicaridean isopods. It attaches to free-living copepod intermediate hosts in bopyrid isopods.

epidemiology: A term used to describe the biology of a disease process affecting humans.

epimastigote: A form in the life cycle of trypanosomes usually found within the gut of the insect vector.

epimerite (=mucron): The attachment organelle of a gregarine.

epizootiology: A term used to describe the biology of a disease process affecting animals other than humans.

eutely: Condition of a body composed of a constant number of cells or nuclei in all adult members of a species, e.g., nematodes.

exflagellation: The rapid transformation of microgametes from a macrogametocyte; in human *Plasmodium* spp. the process occurs within the midgut of specific mosquitoes.

extended phenotype: A general term used to describe the effects of a gene on the external world that can enhance its probability of replication. Such effects can occur on the organism in which the gene resides, or on other organisms. Alterations in behavior, morphology, and color arising from infection from trophically transmitted parasites can be considered extended host phenotypes.

externa: The external reproductive structure of the female rhizocephalan barnacle; located on ventral abdominal surface of decapod crustacean host.

extrusomes: The membrane-bound organelles located beneath the pellicle of ciliates, and which can be released to the cell's exterior upon appropriate stimulation.

facette: The funnel-shaped opening in the inner membrane of the egg through which secretions, produced by the dorsal organ of the larva of pentastomes, flow.

Fahrenholz's rule: The assertion that the phylogeny of parasites closely matches the phylogeny of their hosts.

fecundity compensation: Alteration in host reproductive schedule due to parasitic infection; e.g., the observed early reproduction of snails exposed to castrating larval trematodes.

flagellum: A long, thread-like locomotory organelle of protozoans that has nine pairs of outer, and two central microtubules.

founder effect: A type of genetic drift that occurs as the result of the founding of a population by a small number of individuals carrying a fraction of the total genetic variation of the source population.

frontal filament: The unique larval attachment structure characteristic of the chalimus larvae of parasitic siphonostomatoid copepods. This structure assures attachment to a host when the larva is undergoing metamorphosis to the adult.

fundamental niche: The maximum possible distribution of an organism in an *n*-dimensional hypervolume. Often, abiotic and/or biotic factors limit the distribution to a subset of the fundamental niche.

gametogony: The process by which gametes are formed in apicomplexan protozoans.

gene flow: The movement of alleles into, or out of, a population, typically resulting from the dispersal of individuals.

genetic drift: The random changes in the frequencies of two or more alleles or genotypes within a population; can be especially pronounced in small populations due to sampling error.

geographical mosaic of coevolution hypothesis: John Thompson's hypothesis that predicts divergent coevolutionary selection at different spatial scales. The process leads to genetic differentiation among populations of interacting species.

gigantism: Increase in host body size due to parasitic infection; e.g., the observed temporary increase in body size of snails infected with larval trematodes.

glycocalyx: A secreted layer external to the cell membrane and covering the cell surface; consists of a variety of different macromolecules, e.g., polysaccharides and glycoproteins.

glycosome: An organelle involved in glycolysis in kinetoplastid trypanosomes.

gnathopods: The first two pairs of pereopods modified as clawed holdfasts in ectoparasitic cyamid amphipods.

gnathosoma: The anterior region of mites and ticks, carrying the mouthparts and feeding appendages.

Gordian knot: The tangled mating mass of nematomorphs.

granuloma: A site of chronic inflammation usually caused by the continued presence of any infectious agent that results in tissue necrosis. The site contains numerous macrophages, T-cells, and fibroblasts. It may also include B-cells, neutrophils, and eosinophils.

gubernaculum: An accessory, cuticularized structure found in some male nematodes having spicules. It guides the spicules out through the cloaca.

guild: A group of organisms that exploit resources in a similar fashion and occupy a similar ecological position within the same habitat. The nature of the exploitation and the resources used will determine the identity of the guild.

gynecophoric canal: The external grooved surface of male schistosomes in which the female resides *in copula*.

haemozoin: A malaria pigment; a by-product of hemoglobin metabolism that accumulates in the host's reticuloendothelial system during a malaria infection.

Haller's organ: A large set of different types of receptive sensilla clumped together on the first tarsi of ticks; it includes mechano-, thermo-, hygro-, and chemoreceptors.

halteres: The reduced pair of hind wings in dipteran flies; used for balance and flight stability.

haptor (=opisthaptor) A large attachment structure located on the ventral surface at the posterior end of monogeneans.

hemimetabolous: Development with simple, or incomplete metamorphosis, in which the newly hatched nymph resembles the adult, but lacks wings, gonads, and external genitalia.

heterozygote advantage: A situation in which the heterozygote has higher fitness than the two homozygotes in a population.

hexacanth (=oncosphere): The six-hooked embryo of the Eucestoda.

historical biogeography: The branch of biogeography that attempts to understand the sequences of origin, dispersal, and extinction of organisms.

holometabolous: The development with complete metamorphosis in which the immature stages of insects (larva and pupa) are morphologically and ecologically markedly different from the adults.

homology: Similarity between species that results from inheritance of traits from a common ancestor.

hydatid cyst: The proliferative metacestode characteristic of the cestode genus *Echinococcus*. Many protoscolices are produced in brood capsules that float freely in the fluid of the parental cyst; forms in the organs of herbivore intermediate hosts as well as humans.

hydrogenosome: A single, membrane-bound cytoplasmic organelle characterized by containing hydrogenases. The hydrogenosome is found in some anaerobic protozoans, such as the trichomonads.

hyperparasite: A parasite that uses another parasite as a host.

hypnozoite: Dormant sporozoite found in the liver cells of certain *Plasmodium* species, e.g., *Plasmodium vivax*.

hypopharynx: The tongue-like structure arising from the floor of the mouth in insects.

hypostome: A portion of the mouthparts of mites and ticks formed by the fusion of the coxae of the pedipalps.

idiosoma: The posterior part of the two basic body parts of mites and ticks, comprising the fused cephalothorax and abdomen. It bears the legs and most internal organs.

imaginal discs: Groups of cells set aside in the insect embryo and maintained through larval development, as sites from which adult structures arise.

imogochrysalis: The quiescent, transitional stage between the deutonymph and adult in the life cycle of water mites.

immunological memory: Phenomenon expressed as a more rapid response upon secondary exposure to an antigen; e.g., mammalian B- and T-cells mediate specific immunological memory and lead to rapid secondary immunological responses.

infraciliature: All of the cilia and their associated organelles in a ciliated protozoan.

infracommunity: All of the parasites of all species in a single host individual.

infrapopulation: All of the parasites of a single species within a single host.

interactive site segregation: Refers to the specialization, or segregation, of niches by two species in which the realized niche of one, or both, is reduced by the presence of the second species.

intermediate host: That host in a parasite's life cycle required by the parasite to complete its life cycle, and in which some morphological change or development occurs.

interna: The internal nutrient-absorbing structure of the female rhizocephalan barnacle; forms a highly invasive, root-like network inside decapod crustacean hosts.

ionocytes (=chloride cells): Specialized, epidermal ion-transporting cells found in pentastomes.

kentrogon: The specialized larval stage of many rhizocephalan barnacles that follows the cypris stage, and injects a mass of cells into crustacean hosts, eventually forming the interna and externa.

kinetodesmata: The fibers comprising the complex infraciliature of ciliated protozoans.

kinetoplast: The portion of the mitochondrion of kinetoplastid trypanosomes containing kDNA.

kinetosome (basal body): A complex, subpellicular ultrastructure consisting of nine sets of three microtubules forming a tubular cylinder plus a core containing microfilaments.

labium: The fused second maxillae, forming the lower lip of the mouth of insects.

labrum: The plate forming an upper lip in the mouth in insects.

lacunar system: The system of fluid-filled canals in the body wall of acanthocephalans. It functions as a circulatory system.

larva: Developmental stage in the life history of many animals that is morphologically and ecologically different from the adult and requires metamorphosis for further development.

Laurer's canal: A non-functional tube connecting the surface of some digeneans to the seminal receptacle; it may be a vestigial vagina.

lemnisci: The paired, sac-like structures found attached to the inner margin of the neck and extending into the body cavity of acanthocephalans.

ligament sac: The envelope surrounding the gonads and accessory organs in acanthocephalans.

lycophore: The ten-hooked larva of a cestodarian (gyrocotylean) tapeworm.

macrogamete: The larger of paired gametes, usually not flagellated, and generally considered the female gamete.

macroparasite: A parasite that is usually visible with the naked eye, e.g., helminths, arthropods.

macrophage: Large, migrating mononuclear phagocytic cells derived from bone marrow precursors. Important in innate immunity and as antigen-presenting cells.

malacospores: The spores of malacosporean myxozoans that develop within bryozoan hosts.

Malpighian tubules: Elongated, blind, excretory tubes that arise near the anterior end of the hindgut and extend into the body cavity in many arthropod groups, e.g., insects.

manca: The earliest, free-swimming larval stage of some cymothoid isopods of fish.

mandibles: The paired mouthpart appendages of arthropods used primarily for feeding.

marsupium: Specialized, ventral chamber of female parasitic isopods and amphipods in which eggs and young are brooded.

maxilla: The first and second maxillae comprise the fourth and fifth feeding appendages of the head of free-living crustaceans. Also, the second pair of jaws derived from the fifth head segment and variously modified in insects. In parasitic branchiurans and copepods the first or second maxillae are often greatly modified into attachment structures.

maxilliped: In crustaceans, one or more pairs of mouthpart appendages posterior to the maxillae and originating from the thorax. In free-living crustaceans they are used usually for feeding functions. In some parasitic copepods they have been modified for prehension.

Mehlis' gland: The specialized cells surrounding the ootype and which play a role in the formation of the egg shell in digeneans.

merogony: The portion of the protozoan life cycle involved in the formation of merozoites from a precursor cell.

merozoite: The final life cycle stage in the process of merogony. It develops into either a new meront following invasion of a new host cell or, at the appropriate time and place, into either a micro- or a macrogamete.

mesocercaria: A life-cycle stage of digeneans between a cercaria and a metacercaria in which there is no developmental change, e.g., some diplostomid flukes.

mesoparasite: Parasites intermediate between ecto- and endoparasites. It generally applies to copepods, where part of the transformed female copepod's body remains outside the host, and other structures, usually a holdfast, extends deep into the body of the host, e.g. Pennellidae.

metacercaria: A developmental stage of digeneans between a cercaria and an adult; usually sequestered within a cyst in a second intermediate host that is a prey item of the definitive host.

metamerism: The division of the body into segments (metameres), and serial repetition of organs and tissues.

metamorphosis: A marked, abrupt change in morphology, physiology, and ecology during animal development.

metapopulation: All of the parasites of a single species within a population of hosts.

microfilaria: The first-stage juveniles found in blood or tissue fluids of filariid nematodes.

microgamete: The smaller of paired gametes in protozoans, often flagellated, and generally considered the male gamete.

microniscus: The feeding larva of bopyrid isopods found on copepod intermediate hosts.

microparasite: Small parasites, e.g., viruses, bacteria, protozoans, which replicate extensively within their hosts.

microtriches: The finger-like projections on the tegument of a tapeworm or digenean; function to increase surface area.

miracidium: The ciliated, free-swimming larval stage of a digenean that emerges from an egg. It either seeks out a mollusc first intermediate host in water or is ingested by the mollusc while within the egg.

mitosome: An organelle of mitochondrial origin found in some protists and microsporidians.

molecular mimicry: An immunoevasion adaptation of some parasites (e.g., human schistosomes) involving acquisition of host molecules onto their surface.

molting (=ecdysis): The process of shedding of the cuticle during the life cycle of several groups of invertebrates, e.g., nematodes, arthropods.

monophyletic: A group of taxa descended from a common ancestor; contains the most recent ancestor of the group and all of its descendants.

multiple fission: The simultaneous division of a cell into many daughter cells.

mutualism: The type of symbiosis in which two different species benefit from their partnership, and usually cannot survive without each other.

mycetome: The specialized organ of some insects that harbors mutualistic microorganisms.

myiasis: An infection caused by fly larvae (maggots).

myocyton: The non-contractile cell body of a muscle cell of a nematode.

myxospore: The life cycle stage of myxozoans possessing valves, sporoplasms, and polar capsules; develop within and are released from vertebrate hosts.

nauplius: The earliest free-swimming, planktonic larval stage of crustaceans. It typically has three pairs of appendages, the antennules, antennae, and mandibles.

neodermis: The 'new skin' comprising the complex, syncitial tegument of parasitic platyhelminths; the synapomorphy defining the clade Neodermata.

neoteny: The retention of larval characteristics past the time of reproductive maturity.

nurse cell: The complex structure formed when a *Trichinella spiralis* 1st stage juvenile invades a mammalian skeletal muscle cell.

nymphs (=naiads): The juvenile instars of insects with hemimetabolous development and incomplete metamorphosis; also, the immature stage of mites and ticks, and of pentastomes.

ommatidia: The optical units of the compound eye of arthropods.

onchocercoma: The subcutaneous nodule containing several adults of the filarid nematode *Onchocerca volvulus*.

oncomiracidium: The ciliated, free-living larval stage of most monogeneans.

oncosphere: The hexacanth embryo of the Eucestoda.

oocyst: An encysted zygote in the life cycle of apicomplexan protists. The zygote undergoes sporogony to form sporozoites.

ookinete: The motile zygote formed by the union of microgamete and macrogamete in some apicomplexan protists; e.g., the ookinete of *Plasmodium* spp. forms in the midgut of mosquitoes.

ootype: A dilation of the female oviduct in parasitic platyhelminths, usually surrounded by the Mehlis' gland.

outgroup: In phylogenetic systematic studies, a species or group of species closely related to, but not included, within the taxon being studied; a taxonomic group that diverged prior to the rest of the taxa in a phylogenetic analysis.

ovijector: A highly muscular uterine structure of some nematodes through which the eggs are ejected forcibly by muscular contraction.

ovipositor: The egg-laying structure on the posterior end of the abdomen of female insects.

panmixia: A population genetics term to describe an interbreeding population in which mating between individuals occurs at random.

pansporoblast: The myxosporidian sporoblast that gives rise to more than one spore.

paraphyletic: A taxonomic group that does not include all of the descendants of a common ancestor.

parasite fauna: The total list of parasites in populations of hosts sampled across their distributional range.

parasitoid: A diversity of primarily hymenopteran and dipteran insects whose females lay their eggs, and larvae develop in other insect hosts, eventually killing the host.

parasitophorous vacuole: A membrane-bound vacuole within a host cell containing parasite stages; amastigotes of *Leishmania* spp. reside within a parasitophorous vacuole (phagolysosome) within macrophages.

paratenic host (=transport host): A host in which development does not occur, but which may serve to bridge an ecological, or trophic, gap in a parasite's life cycle.

paraxial rod (=paraflagellar rod): An electron-dense ribbon or rod consisting of a three-dimensional fibrillar lattice associated with the flagellum, often for most of its length. It is contained within the flagellar membrane, and is present in many parasitic flagellates.

parenchyma: The mass of mesenchymal cells forming a cellular matrix and filling spaces between the organ systems, muscles, or epithelia.

parthenogenesis: A mode of unisexual reproduction involving development of an unfertilized egg into a new individual.

pedipalps: The second pair of appendages in chelicerate arthropods.

pellicle: The thin, secreted envelope and associated infrastructure of many protozoans.

pereopods: The thoracic appendages of crustaceans. Used primarily for locomotion in free-living species and modified into prehensile, sharp claws for attachment to a host in parasitic isopods and amphipods.

periodicity: The phenomenon whereby the microfilariae of filariid nematodes undergo periodic migrations into, and out of, the peripheral blood-stream; coincides with feeding behavior of blood-feeding vectors.

peritrophic membrane: A chitinous, non-cellular membrane that surrounds the food in the midgut of most insects.

phasmid: The sensory pit on either side near the end of the tail of some nematodes.

phoresy: The type of symbiosis in which the smaller of the two partners (the phoront) is mechanically carried by the host; type of commensalism involving no trophic transfer.

phylogenetic systematics (=cladistics): An analytical method used to infer evolutionary histories by arranging taxa based on analysis of primitive and derived characteristics; the arrangement reflects phylogenetic relationships.

phylogeography: The use of evolutionary trees in answering questions regarding the biogeographical distribution of organisms.

pinocytosis: The intake of fluid by a cell through the formation of small, fluid-filled vacuoles derived from the cell membrane.

plasmodia: The multinucleate, syncitial trophic stages of myxozoans; produces infective spores in verte-brate tissues.

plasmolemma: The cell membrane enveloping a cell, composed of lipids, carbohydrates, and proteins.

pleopods: The five pairs of abdominal appendages of the pleon region of crustaceans; function in loco-motion and respiration.

plerocercoid: A second larval stage (after the procer-coid) in the life cycle of certain cestodes; often found in fish intermediate hosts, e.g., diphyllobothriideans.

poecilogyny: In some oxyurid nematodes, this is a process where two different forms of females are produced. One form produces thin-shelled eggs that may play a role in autoinfection, while the other female form produces thick-shelled eggs that play a role in colonizing new hosts.

polar capsule: The compartment bearing the eversible polar filaments in multicellular myxozoan spores.

polar filament: The thread-like organelle in spores which extrudes the sporoplasm in myxozoans and microsporidians.

polaroplast: A series of stacked membranes at one end of the spore of microsporidians.

polydnanviruses: A family of mutualistic viruses found in the female reproductive system of braconid and ichneumonid wasp parasitoids; involved in induc-ing immunosuppression of insect hosts.

polyembryony: The formation of two or more 'embryos' from a single zygote, e.g., asexual mode of devel-opment and replication in digeneans within mol-luscan first intermediate hosts.

polyphyletic: Arising from more than one evolutionary lineage; taxonomic grouping that does not share a common ancestor.

postcloacal crescent: A posterior cuticular structure of male nematomorphs.

praniza: The parasitic, blood-feeding larva of gnathiid isopods of fishes.

preoral stylet: The anterior piercing structure of argulid branchiuran crustaceans.

primary larva: The first larval stage of pentastomes. It remains inside the egg and excysts after the egg is ingested by an intermediate host.

proboscis: In acanthocephalans, the anterior retractile organ bearing hooks and used for attachment to the host.

procercoid: A larval stage of certain cestodes, e.g., diphyllobothriideans, usually parasitic in the hemocoel of microcrustaceans and usually in pos-session of three pairs of embryonic hooks.

progenesis: The advanced development of genitalia without maturation in a larva.

proglottid: The segment in a tapeworm's strobila con-taining a set of reproductive organs.

prohaptor: The anterior end of monogeneans housing feeding and attachment structures.

promastigote: A stage in the life cycle of trypanosomes in which the kinetoplast is anterior to the nucleus.

protandrous hermaphroditism (=protandry): Type of hermaphroditism in which the male reproductive system develops first and the female reproductive system develops later.

protogyny: Development of the female reproductive system and ova first, followed by the development of the male reproductive system in a hermaphrodite.

protonephridium: The unit of structure and function in the osmoregulatory system of the Platyhelminthes.

protonymph: The life-cycle stage succeeding the larva of mesostigmatid mites.

pseudocoelom: The fluid-filled body cavity of several groups of triploblastic invertebrates incompletely lined by mesoderm; embryonically derived from the blastocoel, e.g., nematodes.

pseudocyst: A cyst-like structure, often containing numerous protozoan parasites, that is formed by host cells, not by the parasites themselves.

pseudopodia: The temporary, retractile cytoplasmic protrusions that are characteristic of amoeboid protozoans.

pupa: The stage between the larva and the adult in insects with complete metamorphosis.

pygidium: The sensory organ on the posterior dorsal segment of fleas.

realized niche: The actual distribution of an organism in an *n*-dimensional hypervolume.

Red Queen hypothesis: Predicts continued, reciprocal, evolutionary change in an assemblage of interacting organisms, allowing them to maintain their level of interaction.

redia: An intramolluscan developmental stage produced by a digenean sporocyst.

renette cells: The ventral gland cells unique to the Nematoda that open to the exterior via a mid-ventral pore.

reservoir host: The host in which a parasite can mature and reproduce but is not considered the normal host. Reservoir hosts help maintain a parasite when the normal hosts are not available.

rete network: The highly branched system of tubules associated with the musculature of acanthocephalans.

rostellum: A muscular attachment organ located at the tip of the scolex of some cestodes and which may or may not be armed with spines or hooks.

Saefftigen's pouch: The internal muscular sac near the posterior end of male acanthocephalans; it aids in the functioning of the copulatory bursa.

sarcocyst: The last generation meront of *Sarcocystis* spp., found in the muscle tissue of the intermediate host.

schistosomule: The migratory stage of schistosomes in the vertebrate host that follows penetration of the cercaria and shedding of its tail, and preceding maturation to the adult.

schizogony: A form of asexual reproduction (multiple fission) in which multiple cell divisions result in the formation of many daughter cells.

scolex: The muscular attachment organ located at the anterior end of cestodes.

scutum: The sclerotized dorsal plate of hard ticks.

seminal receptacle: A structure associated with the female reproductive system where sperm from another individual is stored prior to being used for fertilization.

seminal vesicle: A structure associated with the male reproductive system where sperm, produced by the individual, is stored.

shell valves: The external structures comprising the spores of myxozoans; these split to release the polar filaments and sporoplasm when in contact with host tissues.

siphonostome: A term describing the mouthparts of parasitic siphonostomatoid copepods; mouth forms a short, cylindrical tube enclosing a pair of mandibles.

sister taxa: Taxa that are each other's closest relatives; they are derived from the same ancestor.

species diversity: A derived value that relates both the number of species and their relative abundances within a community.

species richness: The actual number of different species in a prescribed area of interest.

spermatophore: A specially formed capsule or package containing sperm that is produced by males of several invertebrate groups and delivered to the female.

spicules: The cuticularized structures found in some male nematodes that, during copulation, are inserted into the female vulva keeping it open against the strong hydrostatic pressure.

spiracles: The external openings of the tracheal system of insects to allow gas exchange with the environment.

sporoblasts: These are found within sporonts and differentiate to produce spores in the microsporidians.

sporocyst: An intramolluscan, asexual developmental stage of digeneans, or a stage present in the life cycle of most coccidians. The latter refers to a cyst that contains sporozoites that is formed within the oocyst.

sporogony: A type of schizogony in which a zygote undergoes a single meiotic division followed by mitotic divisions. This results in the formation of sporocysts and sporozoites during the life cycle of many apicomplexans.

sporont: An undifferentiated cell mass which can divide to form sporoblasts in microsporidians.

sporoplasm: The amoeboid cells produced within myxozoan and microsporidian spores; when released from the polar filament, migrates to and invades host tissues to initiate infection.

sporophorous vesicle: The membrane-bound envelope containing spores in some microsporidians.

sporoplasmosomes: The electron-dense bodies found in sporoplasms of myxozoans.

sporozoite: The motile, infective stage often present within a cyst or shell produced during sporogony.

stichosome: A glandular esophagus with one to three rows of individual cells (stichocytes) characteristic of the trichocephalid nematodes.

strobila: The ribbon-like arrangement of the body of cestodes, consisting of proglottids, and beginning immediately posterior to the scolex.

strobilation: The process of budding leading to the formation of proglottids and the strobila of cestodes.

strobilocercus: A cestode larval stage, possessing a fully developed scolex and a strobila-like neck to which is attached a bladder; always found in a vertebrate intermediate host, usually a rodent.

stylets: The modified mouthparts forming a needle-like structure used by sucking insects and branchiuran crustaceans, or the structures found on the anterior of trematode cercariae and used to aid in penetration of the host.

stylostome: A specialized, tube-like structure used for feeding by chiggers.

subparietal glands: The glands associated with the cuticle of pentastomes.

suprapopulation: A population that includes all of the developmental phases of a parasite species at a particular place and time.

symbioses: The diversity of partnerships in nature in which two different organisms are living together in an intimate relationship; the symbiont always benefits, while the host may benefit (mutualism), be unaffected (commensalism), or be harmed (parasitism).

synapomorphy: In cladistics, a shared derived character state that suggests a monophyletic grouping.

syncytium (=syncytial layer): The term refers to a single cell possessing many nuclei; the tegument of parasitic flatworms is said to be syncytial, with nuclei scattered in cytons extending below the surface of the tegument.

synlophe: The cuticular ridges found in the trichostrongyle nematodes.

systematics: The scientific study of the taxonomy of organisms and reconstruction of their phylogeny; investigates relationships among species to understand the Tree of Life.

syzygy: The side-by-side or end-to-end association of gamonts prior to the formation of gametocysts and gametes; generally used in reference to the gregarines.

tachyzoites: The rapidly developing merozoites that are produced in aggregate during the acute phase of a *Toxoplasma gondii* infection.

tagmata: The distinct region of the body of arthropods formed by the embryonic fusion of segments (metameres).

tagmatization (=tagmosis): The specialization of body segments into functional groups to form different regions of the body, e.g., head, thorax, and abdomen of insects.

taxon pulse model: A model of evolutionary radiations resulting from colonization and within-area diversification.

tegument: The external covering of parasitic platyhelminths.

telson: The posterior part of the last abdominal segment in many crustaceans.

tetracotyle: The metacercaria of some diplostomid digeneans.

tetrathyridium: A metacestode larval stage of the cestode *Mesocestoides*.

tomites: The free-swimming, non-feeding stages in the life cycle of some parasitic ciliates, e.g., *Ichthyophthirius multifiliis*.

tracheae: A system of branching tubules through which gaseous exchange occurs via diffusion in insects.

transmission threshold: An epidemiological term denoting the situation when a parasite's Basic Reproductive Rate (Ro) equals one. A parasite is unable to maintain itself within a host population when the rate of transmission is below this threshold.

tribocytic organ: A secretory structure used in the release of proteolytic enzymes on the external surface of strigeid digeneans.

trichogon: The male larva of a rhizocephalan barnacle that penetrates the female externa and implants itself in a male receptacle.

trichonymph: The third and last nymphal stage of acarine chelicerates.

triungulin: The first-instar larva of parasitic Strepsiptera.

trophonts: The visible white pustules on the skin of fish caused by mature trophozoites of the ectoparasitic ciliate, *Ichthyophthirius multifiliis*.

trophosome: A food-storage organ in mermithid nematodes.

trophozoite: The active, feeding stage in the life cycle of protozoans and myxozoans.

trypomastigote: A form in the life cycle of trypanosomes with the kinetoplast located posterior to the nucleus.

undulating membrane: In parasitic flagellates, the extension of the body's surface membrane that is joined for part, or all, of its length with the membrane that surrounds the flagellar substructure and the axoneme.

uropod: One of the terminal pair of abdominal appendages in malacostracan crustaceans.

uterine bell: The egg-sorting structure found in female acanthocephalans; it allows fully developed eggs to leave the body while retaining immature ones.

vector: A micropredator that transmits a parasite from one host to the next. Parasite development in the vector may, or may not, occur.

vermigon: The motile, worm-shaped larval stage of rhizocephalan barnacles that is injected from the kentrogon larva into crustacean hosts, eventually forming the interna and externa.

vertical transmission: The transmission of a parasite from host parent to its offspring, often during host reproduction.

vicariant biogeography: The branch of historical biogeography that attempts to explain the distribution of organisms based on the assumption that their present-day distribution results from fragmentation and isolation due to the emergence of geographical barriers.

vitellaria: The glandular cells that contribute to the formation of egg shells and yolk.

xenoma: A combination of parasite and hypertrophied cell in microsporidians.

zoitocyst: The tissue cyst containing many bradyzoites of coccidian protozoans, e.g., *Toxoplasma gondii*, *Isopora* spp.

zoonosis: A disease normally cycling through wild animals which, under certain conditions, can be transmitted to humans.

INDEX